Geometric Formulas

Rectangle
   Area: $A = \ell w$
   Perimeter: $P = 2\ell + 2w$

Square
   Area: $A = s^2$
   Perimeter: $P = 4s$

Parallelogram
   Area: $A = bh$

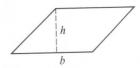

Trapezoid
   Area: $A = \frac{1}{2}h(a + b)$

Triangle
   Area: $A = \frac{1}{2}bh$

   Area: $A = \sqrt{s(s - a)(s - b)(s - c)}$,
   where $s = \frac{1}{2}(a + b + c)$
   Angle sum: $A + B + C = 180°$

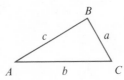

Right Triangle
   Pythagorean theorem: $a^2 + b^2 = c^2$

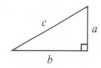

Circle
   Area: $A = \pi r^2$
   Circumference: $C = \pi d = 2\pi r$

Fundamentals of Algebra and Trigonometry

Fundamentals of Algebra and Trigonometry

Dennis T. Christy
Nassau Community College

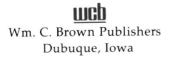

Wm. C. Brown Publishers
Dubuque, Iowa

Book Team

Editor *Earl McPeek*
Developmental Editor *Nova A. Maack*
Designer *K. Wayne Harms*
Production Editor *Eugenia M. Collins*
Visuals Processor *Joseph P. O'Connell*

wcb group

Chairman of the Board *Wm. C. Brown*
President and Chief Executive Officer *Mark C. Falb*

wcb

Wm. C. Brown Publishers, College Division

President *G. Franklin Lewis*
Vice President, Editor-in-Chief *George Wm. Bergquist*
Vice President, Director of Production *Beverly Kolz*
Vice President, National Sales Manager *Bob McLaughlin*
Director of Marketing *Thomas E. Doran*
Marketing Communications Manager *Edward Bartell*
Marketing Information Systems Manager *Craig S. Marty*
Marketing Manager *David F. Horwitz*
Executive Editor *Edward G. Jaffe*
Manager of Visuals and Design *Faye M. Schilling*
Production Editorial Manager *Colleen A. Yonda*
Production Editorial Manager *Julie A. Kennedy*
Publishing Services Manager *Karen J. Slaght*

A study guide for this textbook is available in your college
bookstore. Its title is *Fundamentals of Algebra and Trigonometry,
Student Study Guide and Solutions Manual.* It has been written to
help you review and study the course material. Ask the bookstore
manager to order a copy for you if it is not in stock. For a more
detailed description of the study guide, refer to the preface in this
book.

Library of Congress Catalog Card Number: 88–70346

ISBN 0–697–05323–7

Printed in the United States of America by Wm. C. Brown Publishers
2460 Kerper Boulevard, Dubuque, IA 52001

10 9 8 7 6 5 4 3 2

To Ellen, Michael, and Ilene

Contents

Preface

Audience

This book is intended for students who need a concrete approach to mathematics. The presentation assumes that the student has completed two years of high school algebra or a college course in intermediate algebra. However, a detailed review of all necessary ideas is given early in the text so that students whose basic skills need improvement have a wealth of helpful material.

Approach

Problem-Solving Approach

My experience is that students who take college algebra and trigonometry learn best by "doing." Examples and exercises are crucial since it is usually in these areas that the students' main interactions with the material take place. The problem-solving approach contains brief, precisely formulated paragraphs, followed by many detailed examples. In most cases the discussion proceeds from the specific to the general, and each section has many exercises for both in-class practice and homework. The problem sets are carefully graded and contain an unusual number of routine manipulative problems. (There is nothing more frustrating than being stuck on the beginning exercises!) So that each student may be challenged fairly, difficult problems have been included. There is also a wide variety of questions and discussions that show the student the usefulness of mathematics. Many reviewers felt that the readability of the text and the quality and abundance of the exercises and examples were outstanding features.

Functions Theme

The organization takes a functional approach to algebra and trigonometry in that after developing some basic algebra, the function concept plays the unifying role in the study of polynomial, rational, exponential, logarithmic, and trigonometric functions.

Algebra Review

The first two chapters, which review basic algebra, are particularly detailed to help ensure proficiency in basic skills.

Trigonometry Development

The trigonometric functions are presented in order of abstraction and sophistication, and the text proceeds from the right triangle definitions to the general angle definitions and then to the unit circle definitions.

Calculus Preparation

To prepare students for calculus, there is an emphasis on explaining how to deal with some of the expressions encountered in a calculus course. For instance, Chapter 1 shows how to simplify algebraic expressions that will be found in calculus problems. Chapter 3 expands the discussion of slope to include the difference quotient, Chapter 6 shows problems dealing with inverse trigonometric functions that are important for integration by trigonometric substitution, and Chapter 10 considers the sense in which certain infinite geometric series converge to a sum.

Calculator Use

The text encourages the use of calculators and discusses how they can be used effectively. It is assumed in the discussion that students have scientific calculators that use the algebraic operating system (AOS). Calculator illustrations show primarily the keystrokes required on a Texas Instruments TI-30 SLR+. Tables are included and discussed in the text in case students choose not to use a calculator.

Features

- Problem-solving approach with intuitive concept developments
- Over 4,300 exercises and 500 examples
- Problem sets of graduated difficulty with enough routine problems to give students experience, confidence, and skill with basic procedures
- Extensive and varied applications problems
- Student-oriented and mathematically sound explanations
- Boxes with margin labels for important definitions and rules
- Second color to highlight important ideas
- Avoidance of awkward page and line breaks
- Chapter introductions that include an interest-getting problem that is solved as an example in the text
- Unique chapter overviews that highlight key concepts to review at the end of each chapter
- Abundant chapter review exercises
- Anticipates the needs of students who continue to calculus and other higher level math courses
- Emphasis on functions and graphs
- Right triangle introduction to trigonometry
- Instructions on calculator use
- Important formulas and tables on endsheets
- Complete instructional package

Pedagogy

Chapter Introductions

In the spirit of problem solving, each chapter opens with a problem that should quickly involve students and teachers in a discussion of an important chapter concept. Some of the problems are applications, some are puzzles, and one is a proof. None of the problems requires a lot of sophisticated mathematics and it is hoped that students will try to solve the problem either initially or after covering the relevant section in the text. These problems are later worked out as examples in each chapter.

Chapter Overviews and Review Exercises

At the end of each chapter there is a detailed list of key concepts to review that is organized in a section-by-section format so the material does not seem overwhelming. There is also a collection of review exercises that not only review the basic ideas but also expose students to slightly different question wording and formats, including multiple choice questions. From these exercises it should be easy for instructors to choose questions for review assignments or for sample tests at an appropriate level of difficulty.

Instructional Package

- *Student Study Guide and Solutions Manual*
- *Instructor's Manual* (includes transparency masters)
- *Instructor's Solutions Manual*
- **wcb** *Math TestPak*
- *Test Item File*

Student Study Guide and Solutions Manual

Students for whom the textbook is not enough need more than just a lot of solved problems in a solutions manual. Primarily they need help focusing on the key objectives and concepts in the course. To provide some of this help, an accompanying study guide and solutions manual is available that covers the following key aspects for each section in the text.

1. Specific objectives for the section.
2. A list of important terms.
3. A summary of the key rules and formulas.
4. Detailed solutions to selected even-numbered exercises with at least one example from each exercise group. Exercise numbers for these problems are printed in color in the text for easy identification.
5. Margin exercises matched to the solved problems so students can check their progress.

The *Student Study Guide and Solutions Manual* also contains sample test questions (with answers) for each chapter in the text.

Instructor's Manual   The *Instructor's Manual* contains three tests for each chapter, three final exams, and the answers to the even-numbered problems in the section exercise sets. Transparency masters of the chapter introductory problems and important theorems and definitions are also included in this manual. The problems in the three tests for each chapter and in the three final exams also appear in **wcb** Math TestPak.

Instructor's Solutions Maual   This manual contains solutions to every problem in the text. These solutions are intended for the use of the instructor only and are basic outlines of possible problem solutions.

wcb Math TestPak   **wcb** Math TestPak was developed expressly for **wcb** math texts. It is a free computerized testing service with two convenient options. First, you may use your own Apple IIe, IIc, Macintosh, or IBM PC (both 5¼'' and 3½'' disks) to produce your tests by using items available in the test bank or by editing these items, deleting them, or adding your own. Questions may be chosen by number or at random. Secondly, you may use the call-in service offered by the publisher. Contact your local **wcb** sales representative for details.

Test Item File   The printed *Test Item File* in an 8½'' × 11'' format contains all of the questions on the **wcb** Math TestPak. It will serve as a ready-reference if you use your own computer to generate tests. The *Test Item File* contains the problems in the *Instructor's Manual* along with additional items that have been incorporated into the Math TestPak bank.

Acknowledgments

I wish to thank the many users and reviewers of my texts who have suggested improvements. At this point it is hard to separate my original ideas from the many valuable observations they made, and I am indebted to all of them. For help with this and related projects I am especially grateful to Tom Carnevale; Glen Goodale and John Boyarchuk, Dawson College; Lynda Morton, University of Missouri–Columbia; Ross Rueger, College of the Sequoias; Frank Kocher, Pennsylvania State University; James Magliano, Union College; Earl McPeek, Nova A. Maack, Eugenia M. Collins, and K. Wayne Harms, Wm. C. Brown Publishers; and Louis Gioia, Deborah Levine, James Malone, Abraham Weinstein, and Gene Zirkel, Nassau Community College. My wife, Margaret, once again typed, proofread, and "understood," with the last contribution being irreplaceable. So, to my family, my colleagues at Nassau Community College, the staff at Wm. C. Brown Publishers, and the many users and reviewers, thank you.

Dennis T. Christy

Reviewers

Harry Fainsilber
Dawson College

Pat Gilbert
Diablo Valley College

Marion Glasby
Anne Arundel Community College

Gerald Hahm
College of St. Thomas

Gail Koplin
Ocean County College

Stanley Lukawecki
Clemson University

Jerry McDonald
Loyola University of Chicago

B. K. Michael
University of Pittsburgh

Wing Park
College of Lake County

Marilyn Peacock
Tidewater Community College

Jack Pease
West Valley Community College

Jan Rizzuti
University of Pittsburgh at Bradford

Charles Searcy
New Mexico Highlands

Marsha Self
El Paso Community College

Marvin Shubert
Hagerstown Junior College

Deborah Vrooman
USC–Coastal Carolina College

Richard Watkins
Tidewater Community College

Earl Zwick
Indiana State University

Calculator Use
To the Student

A scientific hand-held calculator is now standard equipment for precalculus mathematics and beyond. These ten dollar wonders provide you with the benefits of electronic computation that is fast, accurate, and easy to learn. Most important, efficient calculator use helps you focus on important mathematical ideas. To understand and apply mathematical concepts is our fundamental aim and calculators are marvelous aids in attaining this goal. Tables are included at the back of the book in case you choose not to use a calculator. But, since calculators are inexpensive, easy to use, and a significant learning aid, we recommend you obtain one.

A scientific calculator (the type you need) contains at least the following special features: algebraic keys x^2, $\sqrt{x}$, $1/x$, y^x or x^y, $\sqrt[x]{y}$; trigonometric keys sin, cos, tan, $\sin^{-1}$, $\cos^{-1}$, $\tan^{-1}$, degree and radian angular modes; logarithmic and exponential keys log, ln, 10^x, e^x; parentheses keys (,); a scientific notation key EE or EXP; and one memory that can store and recall.

In this book we also assume a scientific calculator using the algebraic operating system (AOS). Texas Instruments, Sharp, and Casio produce scientific calculators using this system. With AOS you can key in the problem exactly as it appears, and the calculator is programmed to use the order of operations discussed in Section 1.2. For example, since multiplication is done before addition 2 $\boxed{+}$ 3 $\boxed{\times}$ 4 $\boxed{=}$ 14. If your calculator displays 20 when you key in this sequence, it is operating on left to right logic. You must then be careful to key in the problem so the correct order of operations is followed. Calculator illustrations in this text show primarily the key strokes required on a Texas Instruments TI-30-SLR+. In any case, you should read the owner's manual that comes with your calculator to familiarize yourself with its specific keys and limitations.

One other introductory note—a calculator COMPUTES, that's all. You do the important part—you THINK. You analyze the problem, decide on the significant relationships, and determine if the solution makes sense in the "real world." It's nice not to get bogged down in certain calculations and tables, but that never was the major problem.

Fundamentals of Algebra and Trigonometry

1 *Fundamentals of Algebra*

A race car driver must average at least 150 mi/hour for two laps around a track to qualify for the finals. The driver averages 180 mi/hour on the first lap, but mechanical trouble reduces the average speed on the second lap to only 120 mi/hour. Does the driver qualify for the finals? (See Example 10 of Section 1.4. *Hint:* average speed = total distance/total time.)

A common student lament goes something like, "I understand the new concepts, but the algebra is killing me!" In this chapter we hope to remedy this problem by reviewing *in detail* some basic rules in algebra about real numbers, exponents, factoring, fractions, and radicals. Success here will go a long way toward success in this course and in higher mathematics.

In this text we take a problem-solving approach which emphasizes that one learns mathematics by *doing* mathematics, while *thinking* mathematically. That is, you need to actively work through the problems (with pencil and paper), while *focusing on the definitions, relationships, and procedures* that link together all steps in the solution. In this spirit of problem solving we open each chapter with a problem. Some are applications, some are puzzles, and one is a proof. Taken together, they illustrate the varied nature of problem solving. Since none of them require a lot of sophisticated mathematics, we hope you will take a stab at an answer either initially or after covering the relevant section in the text.

1.1 REAL NUMBERS

Mathematics is a basic tool in analyzing concepts in every field of human endeavor. In fact, the primary reason you have studied this subject for at least a decade is that mathematics is the most powerful instrument available in the search to understand the world and to control it. Mathematics is essential for full comprehension of technological and scientific advances, economic policies and business decisions, and the complexities of social and psychological issues. At the heart of this mathematics is algebra. Calculus, statistics, and computer science are but a few of the areas in which a knowledge of algebraic concepts and manipulations is necessary.

Algebra is a generalization of arithmetic. In arithmetic we work with specific numbers, such as 5. In algebra we study numerical relations in a more general way by using symbols, such as *x*, that may be replaced by a number from some collection of numbers. Since the symbols represent numbers, they behave according to the same rules that numbers must follow. Consequently, instead of studying specific numbers, we study symbolic representations of numbers and try to define the laws that govern them.

We begin our study of algebra by giving specific names to various sets* of numbers. The collection of the counting numbers, zero, and the negatives of the counting numbers is called the **integers.** Thus, the set of integers may be written as

$$\{ \ldots, -3, -2, -1, 0, 1, 2, 3, \ldots \}.$$

The set of fractions with an integer in the top of the fraction (numerator) and a nonzero integer in the bottom of the fraction (denominator) is called the **rational numbers.** Symbolically, a rational number is a number that may be written in the form a/b, where a and b are integers, with b not equal to ($\neq$) zero. The numbers $\sqrt{2}/3$ and $2/\pi$ are fractions but they are not rational numbers because they cannot be written as the quotient of two integers. All integers are rational numbers because we can think of each integer as having a 1 in its denominator. (Example: $4 = 4/1$.)

Our definition for rational numbers specified that the denominator cannot be zero. To see why, you need to know that

$$\frac{8}{2} = 4 \text{ is equivalent to saying that } 8 = 4 \cdot 2 \text{ and}$$

$$\frac{55}{11} = 5 \text{ is equivalent to saying that } 55 = 5 \cdot 11.$$

If $8/0 = a$, where a is some rational number, this would mean that $8 = a \cdot 0$. But $a \cdot 0 = 0$ for any rational number. There is no rational number a such that $a \cdot 0 = 8$. Thus, we say that $8/0$ is *undefined*.

Now consider $0/0 = a$. This is equivalent to $0 = a \cdot 0$. But $a \cdot 0 = 0$ for *any* rational number. Thus, not just one number a will solve the equation—any a will. Since $0/0$ does not name a particular number, it is also **undefined.** Consequently, division by zero is undefined in every case, so the denominator in a rational number cannot be zero.

To define our next set of numbers, we now consider the decimal representation of numbers. We may convert rational numbers to decimals by long division. Consider the following examples of repeating decimals. A bar is placed above the portion of the decimal that repeats.

$$\frac{3}{4} = \begin{matrix} 0.7500 \ldots \\ \text{or} \\ 0.75\overline{0} \end{matrix} \qquad \frac{2}{3} = \begin{matrix} 0.6666 \ldots \\ \text{or} \\ 0.\overline{6} \end{matrix} \qquad \frac{8}{7} = 1.\overline{142857}$$

*A **set** is simply a collection of objects, and we may describe a set by listing the objects or members of the collection within braces.

The decimals repeat because at some point we must perform the same division and start a cycle. For example, when converting $\frac{8}{7}$, the only possible remainders are 0, 1, 2, 3, 4, 5, and 6. In performing the division, as shown in Figure 1.1, we had remainders of 1, 3, 2, 6, 4, and 5. In the next step we must obtain one of these remainders a second time and start a cycle, or obtain 0 as the remainder, which results in repeating zeros. Thus, if a/b is a rational number, it can be written as a repeating decimal.

$$
\begin{array}{r}
1.\overline{142857} \\
7{\overline{\smash{\big)}\,8.000000}} \\
\underline{7} \\
1\,0 \\
\underline{7} \\
30 \\
\underline{28} \\
20 \\
\underline{14} \\
60 \\
\underline{56} \\
40 \\
\underline{35} \\
50 \\
\underline{49} \\
1
\end{array}
$$

Figure 1.1

It is also true that any repeating decimal may be converted to a ratio between two integers, as shown in Example 1.

EXAMPLE 1   Express the repeating decimal $0.\overline{17}$ as the ratio of two integers.

Solution   First, let $x = 0.1717.\ \ldots$ Multiplying both sides of this equation by 100 moves the decimal two places to the right, so we obtain

$$
\begin{aligned}
100x &= 17.1717\ldots \\
x &= 0.1717\ldots
\end{aligned}
$$

now subtracting yields $99x = 17$ or $x = \dfrac{17}{99}$.

Thus, the repeating decimal $0.\overline{17}$ is equivalent to the fraction $\frac{17}{99}$. $\square$

In Example 1 we multiplied by 100 because the decimal repeated after every two digits. If the decimal repeats after one digit, we multiply by 10; if it repeats every three digits, we multiply by 1,000; and so on. In summary, we have illustrated that we may define a rational number either as the quotient of two integers or as a repeating decimal.

There are decimals that do not repeat, and the set of these numbers is called the **irrational numbers.**

EXAMPLE 2   The numbers $\sqrt{2}$, $\sqrt{3}$, $\sqrt{5}$, $\sqrt{6}$, and $\sqrt{7}$ are irrational because they have nonrepeating decimal forms. A proof that $\sqrt{2}$ cannot be written as the quotient of two integers (and equivalently as a repeating decimal) is considered in Exercise 29 of Section 4.5.   □

EXAMPLE 3   The number $\sqrt{4}$ is not irrational because $\sqrt{4} = 2$, which is a rational number. (*Note:* The symbol $\sqrt{}$ denotes the nonnegative square root of a number. Thus, $\sqrt{4} \neq -2$. We discuss this concept in detail in Section 1.5.)   □

EXAMPLE 4   The number π, which represents the ratio between the circumference and the diameter of a circle, is a nonrepeating decimal (irrational number). The fraction $\frac{22}{7}$ is only an approximation for π ($\frac{355}{113}$ is a much better one).   □

Since an irrational number is a nonrepeating decimal and a rational number is a repeating decimal, there is no number that is both rational and irrational. The set of numbers that are either repeating decimals (rational numbers) or nonrepeating decimals (irrational numbers) constitutes the **real numbers.** Real numbers are used extensively in this text, and unless it is stated otherwise, you may assume that the symbols in algebra (such as x) may be replaced by any real number. Consequently, the rules that govern real numbers determine our methods of computation in algebra. A graphical illustration of the relationships among the various sets of numbers is given in Figure 1.2. Note that all of the sets we have discussed are subsets* of the set of real numbers.

We now list the most important properties of the real numbers with respect to addition and multiplication. These properties are the basis for the justification of many algebraic manipulations.

Real numbers	
Rational numbers	Irrational numbers
$\frac{2}{3}$ $0.\overline{17}$	$\sqrt{2}$
	$\sqrt{7}$
Integers -3 0 $1,000$	π

Figure 1.2

*Set A is a **subset** of set B if every element of A is an element of B.

Properties of the Real Numbers

Let a, b, and c be real numbers.

	Addition	**Multiplication**
Closure Properties	$a + b$ is a unique real number.	ab is a unique real number.
Commutative Properties	$a + b = b + a$	$ab = ba$
Associative Properties	$(a + b) + c$ $= a + (b + c)$	$(ab)c = a(bc)$
Identity Properties	There exists a unique real number 0 such that $a + 0 = 0 + a = a$.	There exists a unique real number 1 such that $a \cdot 1 = 1 \cdot a = a$.
Inverse Properties	For every real number a, there is a unique real number, denoted by $-a$, such that $a + (-a)$ $= (-a) + a = 0$.	For every real number a (except zero), there is a unique real number, denoted by $1/a$, such that $a \cdot \dfrac{1}{a} = \dfrac{1}{a} \cdot a = 1$.
Distributive Properties	$a(b + c) = ab + ac$ $(a + b)c = ac + bc$	

EXAMPLE 5   Name the property of real numbers illustrated in each statement.

a. $(2 + 3) + 4 = 2 + (3 + 4)$ **b.** $5 + (-5) = 0$

c. $2(3 + 4) = 2 \cdot 3 + 2 \cdot 4$ **d.** $4 \cdot 7 = 7 \cdot 4$

Solution

a. This statement illustrates the associative property of addition. This property indicates that we obtain the same result if we change the grouping of the numbers in an addition problem.

b. This statement illustrates the addition inverse property. The number $-a$ is called the **negative** or **additive inverse** of a. Note that there is a difference between a negative number and the negative of a number. Although -5 is a negative number, the negative of -5 is 5.

c. This statement illustrates the distributive property (or more technically, the distributive property of multiplication over addition).

d. This statement illustrates the commutative property of multiplication. This property indicates that the order in which we write numbers in a multiplication problem does not affect their product. □

Real Number Line

The real numbers may be interpreted geometrically by considering a straight line. Every point on the line can be made to correspond to a real number, and every real number can be made to correspond to a point. The first point that we designate is zero. It is the dividing point between positive and negative real numbers. Any number to the right of zero is called positive, and any number to the left of zero is called negative. Figure 1.3 is the result of assigning a few positive and negative real numbers to points on the line.

Figure 1.3

Order

The **trichotomy property** states that if a and b are real numbers, then either a is less than b, a is greater than b, or a equals b. By definition, a is **less than** b, written $a < b$, if and only if $b - a$ is positive; while a is **greater than** b, written $a > b$, if and only if $a - b$ is positive. We also define $a \leq b$ (read "a is **less than or equal to** b") to mean $a < b$ or $a = b$. Alternatively, we can write $b \geq a$ and say that b is **greater than or equal to** a. Relations of "less than" and "greater than" can be seen very easily on the number line as shown in Figure 1.4. The point representing the larger number will be to the right of the point representing the smaller number. (*Note:* The definitions use the phrase "if and only if." This phrase is commonly used in mathematical definitions and theorems and the statement "p **if and only if** q" means "if p then q, and conversely, if q then p.")

Figure 1.4

Absolute Value

Positive and negative numbers in mathematics designate direction or indicate whether a result is above or below some reference point. However, sometimes the numerical size of a number is more important than its sign, and in such cases, the concept of absolute value is useful. The **absolute value** of a real number a, denoted $|a|$, is the distance between a and 0 on the number line. For

Figure 1.5

example, $|3| = 3$ and $|-3| = 3$, as shown in Figure 1.5. Note that if $a \geq 0$, then $|a| = a$; but if $a < 0$, then $|a| = -a$ since $-a$ is positive in this case. Thus, algebraically we can define the absolute value of a real number a in tabular form as follows:

$$|a| = \begin{cases} a \text{ if } a \geq 0 \\ -a \text{ if } a < 0. \end{cases}$$

The concept of absolute value is discussed in detail in Section 2.6.

Addition

Figure 1.6

Addition of two real numbers may be interpreted as movement on the number line. The initial starting point is zero. To add a positive number, we move to the right the desired distance. To add a negative number, we move to the left. Consider Figure 1.6, which illustrates the procedure. The addition of $2 + (-4)$ is shown on the real number line by drawing an arrow with a length of 2 units starting at zero and pointing to the right. From the tip of the arrow at 2 we draw an arrow pointing to the left with a length of 4 units. Since the tip of this arrow is at -2, we conclude that $2 + (-4) = -2$.

EXAMPLE 6 $(-2) + (-4) = -6$ (Figure 1.7)

Figure 1.7   □

Subtraction

We define subtraction in terms of addition. If a and b are any real numbers, then $a - b = a + (-b)$. To subtract b from a, we add to a the negative of b.

EXAMPLE 7 $(-5) - 2 = (-5) + (-2) = -7$ □

EXAMPLE 8 $5 - (-\frac{1}{2}) = 5 + \frac{1}{2} = \frac{10}{2} + \frac{1}{2} = \frac{11}{2}$ □

EXAMPLE 9 $[(-7) - 2] - (-1) = [(-7) + (-2)] + 1$
$$= (-9) + 1 = -8$$ □

Multiplication

It seems natural to define the product of two positive numbers as positive. Thus, $4 \cdot 3 = 12$. To determine what sign to use for the product of a positive number and a negative number, we use the distributive property.

$$5[2 + (-2)] = 5 \cdot 2 + 5(-2)$$
$$5 \cdot 0 = 10 + 5(-2)$$
$$0 = 10 + ?$$

We must define $5(-2)$ so that $10 + 5(-2) = 0$. Therefore, $5(-2)$ must equal -10. In every case the product of a positive number and a negative number is negative.

To determine the sign of the product of two negative numbers, consider this problem.

$$-5[2 + (-2)] = (-5)2 + (-5)(-2)$$
$$(-5)0 = -10 + (-5)(-2)$$
$$0 = -10 + ?$$

We must define $(-5)(-2)$ so that $-10 + (-5)(-2) = 0$. Therefore, $(-5)(-2)$ must equal 10. In every case the product of two negative numbers is positive.

EXAMPLE 10 $9(-2)(-\pi) = 18\pi$ □

EXAMPLE 11 $(-5)(-\frac{1}{5})(-2) = -2$ □

EXAMPLE 12 $(-4)(-4)(-4) = -64$ □

If two or more numbers are multiplied together, each number is a **factor** of the product. In Example 12 the factor -4 is used three times. An alternative way of writing $(-4)(-4)(-4)$ is $(-4)^3$. The number $(-4)^3$, or -64, is called the third **power** of -4. In general, by a^n, where n is a positive integer, we mean to use a as a factor n times.

$$a^n = \underbrace{a \cdot a \cdot a \cdots a}_{n \text{ factors}}$$

In the expression a^n, n is called the **exponent**.

EXAMPLE 13   By definition, 2^5 means to use 2 as a factor five times. Thus, $2^5 = 2 \cdot 2 \cdot 2 \cdot 2 \cdot 2 = 32$. The number 2^5, or 32, is the fifth power of 2. The exponent in 2^5 is 5.   □

EXAMPLE 14 $(-3)^4(-1) = (-3)(-3)(-3)(-3)(-1) = -81$ □

Division

We define division in terms of multiplication. If a and b are any real numbers $(b \neq 0)$, then $a \div b = a(1/b)$. To divide a by b, we multiply a by the reciprocal of b. The sign in division is then determined in the same manner as multiplication.

EXAMPLE 15 $(-9) \div (-3) = (-9)(-\frac{1}{3}) = 3$ □

EXAMPLE 16 $6 \div (-\frac{12}{5}) = 6(-\frac{5}{12}) = -\frac{5}{2}$ □

Order of Operations

If an expression involves more than one operation, the following rule has been established to avoid confusion.

1. Perform all operations within parentheses first.
2. Evaluate powers of a number.
3. Multiply or divide working from left to right.
4. Add or subtract working from left to right.

Follow this order of operations carefully. For example, $2 + 3 \cdot 4 = 14$, not 20.

EXAMPLE 17   Evaluate $1 + 7(2 - 5)^2$.

Solution   By following the order of operations, we have

$$1 + 7(2 - 5)^2 = 1 + 7(-3)^2 = 1 + 7(9) = 1 + 63 = 64.$$ □

EXERCISES 1.1

In Exercises 1–8 express each rational number as a repeating decimal.

1. $\frac{4}{5}$ 2. $\frac{5}{4}$ 3. $\frac{5}{11}$ 4. $\frac{2}{9}$

5. $\frac{37}{6}$ 6. $\frac{26}{11}$ 7. $\frac{10}{7}$ 8. $\frac{100}{99}$

In Exercises 9–18 express each repeating decimal as the ratio of two integers.

9. $0.\overline{2}$ 10. $0.\overline{07}$ 11. $0.\overline{321}$

12. $0.\overline{6332}$ 13. $0.3\overline{0}$ 14. $1.7\overline{0}$

15. $5.\overline{9}$ 16. $4.8\overline{1}$ 17. $2.14\overline{3}$

18. $2.1\overline{43}$

In Exercises 19–30 select all correct classifications from the following categories: real number, irrational number, rational number, integer, or none of these.

19. 0 20. $\frac{0}{3}$ 21. $\frac{3}{0}$

22. $\frac{3}{4}$ 23. -9 24. $\sqrt{9}$

25. $-\sqrt{9}$ 26. $\sqrt{-9}$ 27. $\sqrt{7}$

28. π 29. $0.\overline{01}$ 30. $0.101001 \ldots$

In Exercises 31–50 name the property illustrated in the statement.

31. $2 + 7 = 7 + 2$ 32. $11 + 0 = 11$

33. $4(5 \cdot 11) = (4 \cdot 5)11$

34. $6(4 + 3) = 6 \cdot 4 + 6 \cdot 3$

35. $\sqrt{2} \cdot 1 = \sqrt{2}$

36. $-7.3 + 7.3 = 0$

37. $(2.5)3 = 3(2.5)$

38. $17(\frac{1}{17}) = 1$

39. $\pi \cdot 3 + \pi \cdot 8 = \pi(3 + 8)$

40. $(5 + 3) + (2 + 1) = [(5 + 3) + 2] + 1$

41. $z(xy) = (zx)y$

42. $(xy)z = z(xy)$

43. $x + 0 = 0 + x$

44. $(-z) + z = 0$

45. $y(z + x) = (z + x)y$

46. $y \cdot 1 = 1 \cdot y$

47. $x(1/x) = 1$ if $x \neq 0$

48. $ax + ay = a(x + y)$

49. $y + (x + z) = (x + z) + y$

50. $(x + z) + y = x + (z + y)$

In Exercises 51–70 use the proper symbol ($<, >, =$) between the pairs of numbers to indicate their correct order.

51. $6 \underline{\quad} 9$

52. $-1 \underline{\quad} -4$

53. $-7 \underline{\quad} 1$

54. $42.8 \underline{\quad} -91$

55. $|14| \underline{\quad} |-16|$

56. $|1.46| \underline{\quad} |-1.46|$

57. $|-16.2| \underline{\quad} |-18.1|$

58. $\frac{1}{2} \underline{\quad} 0.5$

59. $-0.0001 \underline{\quad} -0.00001$

60. $0.0001 \underline{\quad} 0.00001$

61. $100\% \underline{\quad} 1$

62. $5\% \underline{\quad} 0.5$

63. $\frac{5}{8} \underline{\quad} \frac{2}{3}$

64. $\frac{3}{7} \underline{\quad} \frac{5}{9}$

65. $2 \underline{\quad} \sqrt{3}$

66. $\sqrt{77} \underline{\quad} 8$

67. $\frac{1}{3} \underline{\quad} 0.33$

68. $\frac{2}{3} \underline{\quad} 0.67$

69. $\pi \underline{\quad} 3.14$

70. $\frac{1}{9} \underline{\quad} 0.\overline{1}$

In Exercises 71–114 evaluate each expression by performing the indicated operations.

71. $(-4) + (-3)$

72. $43 + (-21)$

73. $|3 - 8|$

74. $|0 - 6|$

75. $(-6) - (-4)$

76. $(-12) - 5$

77. $(-\frac{1}{5}) + 3$

78. $7 - (-\frac{3}{8})$

79. $(-0.87) + 0.33$

80. $0.17 - (-0.48)$

81. $(-0.1) - 4$

82. $(-3.3) + (-0.67)$

83. $(-11)6$

84. $(-8)(-3)$

85. $(-14) \div (-7)$

86. $89 \div (-89)$

87. $-(\frac{1}{8}) \div 4$

88. $(-\frac{2}{3}) \div (-\frac{4}{15})$

89. $(-5)(-\frac{4}{9})$

90. $(\frac{3}{16})(-2)$

91. $0 \div (-4)$

92. $(-4) \div 0$

93. $0.2(-0.2)$

94. $(-0.64) \div (-0.16)$

95. $(-32.4) \div (-3)$

96. $10.1(-11)$

97. $(-2) \div (6 - 5)$

98. $(5 - 3)(5 + 2)$

99. $-4 + 2(3 - 8)$

100. $(-4 + 2)(3 - 8)$

101. $(2 - 4)(3 - 6)$

102. $2 - 4(3 - 6)$

103. $(-3)^3$

104. $(-2)^4$

105. -2^4

106. $-4 \cdot 5^2$

107. $2(-\frac{1}{2})^5$

108. $1 + 2 \cdot 3^4$

109. $2 - 8(3 - 7)^2$

110. $(-5 + 2)^2 + (-2 - 6)^2$

111. $-5(11 - 6) - [7 - (11 - 19)]$

112. $7 - 3[2(13 - 5) - (5 - 13)]$

113. $3(2)(-1) + (-1)(-3)(2) + (6)(1)(-3)$ $- 6(2)(2) - 3(-3)(-3) - (-1)(1)(-1)$

114. $3[2(-1) - (-3)(-3)] - (-1)[1(-1)$ $- (-3)(2)] + 6[1(-3) - 2(2)]$

115. Find, by two methods, the area of the rectangle and explain how the rectangle can be used to illustrate geometrically the distributive property.

116. If x is a negative number, the negative of x is a $\underline{\quad}$ number.

117. If $xy = x$, what are the possibilities for x and y?

118. If $x^2 + y^2 = 0$, what are the possible values for x and y?

In Exercises 119–128 answer true or false. If false, give a specific counterexample.

119. All rational numbers are integers.

120. All rational numbers are real numbers.

121. All real numbers are irrational numbers.

122. All integers are irrational numbers.

123. The quotient of two integers is always an integer.

124. The quotient of two integers is always a rational number.

125. Every real number is either a rational number or an irrational number.

126. A number that can be written as a repeating decimal is called a rational number.

127. The nonnegative integers are the positive integers.

128. The nonpositive real numbers are the negative real numbers and zero.

1.2 ALGEBRAIC EXPRESSIONS AND INTEGER EXPONENTS

In algebra, two types of symbols are used to represent numbers: variables and constants. A **variable** is a symbol that may be replaced by different numbers in a particular problem. Generally, letters near the end of the alphabet, such as x, y, and z, are variables. A **constant** is a symbol that represents the same number throughout a particular problem. Numbers, such as 2, $-\sqrt{5}$, and π, never change value and are called **absolute constants.** If we do not know the fixed number until we are given specific information about the problem, the symbol is called an **arbitrary constant.** Generally k or letters near the beginning of the alphabet, such as a, b, and c, are arbitrary constants.

EXAMPLE 1   The sales tax T on a purchase is related to the price p of the item by the formula $T = kp$, where k is the sales tax rate. T and p are variables that may take on different values. The sales tax rate is fixed for any particular location so that k is a constant. For example, in New York City k is fixed at $8\frac{1}{4}$ percent. In another location k may be fixed at a different percent. □

An expression that combines constants and variables using the operations of addition, subtraction, multiplication, division, or extracting roots is called an **algebraic expression.** For example,

$$2x^3 - 5x + 3 \quad \text{and} \quad \frac{-b + \sqrt{b^2 - 4ac}}{2a} \quad \text{and} \quad \tfrac{1}{2}gt^2$$

are algebraic expressions.

If we are given numerical values for the symbols, we can evaluate the expression by substituting the given values and performing the indicated operations.

EXAMPLE 2   Find the numerical value of $-x^2 + 5yz^2$ if $x = -2$, $y = 3$, and $z = -4$.

Solution $-x^2 + 5yz^2$, when $x = -2$, $y = 3$, and $z = -4$, equals

$$-(-2)^2 + 5(3)(-4)^2 = -(4) + 5(3)(16)$$
$$= -4 + 240$$
$$= 236.$$

Pay close attention to how the expression $-x^2$ was evaluated in this example. Because powers precede multiplications in the order of operations, $-x^2$ means you must first square x and then take the negative of your result. □

EXAMPLE 3   If

$$m = \frac{y_2 - y_1}{x_2 - x_1},$$

evaluate m when $x_1 = 2$, $x_2 = -1$, $y_1 = -5$, and $y_2 = 3$. (*Note:* In the symbols x_1, x_2, y_1, and y_2 the numbers 1 and 2 are called **subscripts.** In this case x_1 and x_2 are used to denote various x values.)

Solution

$$m = \frac{y_2 - y_1}{x_2 - x_1}$$

so when $x_1 = 2$, $x_2 = -1$, $y_1 = -5$, and $y_2 = 3$, we have

$$m = \frac{(3) - (-5)}{(-1) - (2)} = \frac{8}{-3} = -\frac{8}{3}. \qquad \square$$

Calculator Computation

A scientific calculator is ideal for evaluating expressions, but you need to understand the algebraic rules that have been programmed into your calculator. A calculator using the algebraic operating system (AOS) calculates in the following order (which applies whether or not a calculator is used).

1. Keys that operate on the single number in the display are done immediately. Such keys are called function keys and include square $\boxed{x^2}$; square root $\boxed{\sqrt{}}$; reciprocal $\boxed{1/x}$; sign change $\boxed{+/-}$; trigonometric $\boxed{\sin}$, $\boxed{\sin^{-1}}$, and so on; logarithmic $\boxed{\log}$, $\boxed{\ln}$; and exponential $\boxed{10^x}$, $\boxed{e^x}$.
2. Powers $\boxed{y^x}$ or $\boxed{x^y}$ and roots $\boxed{\sqrt[x]{y}}$ or $\boxed{y^{1/x}}$ follow the functions keys.
3. Multiplication $\boxed{\times}$ and division $\boxed{\div}$ have the next priority.
4. Addition $\boxed{+}$ and subtraction $\boxed{-}$ come last.

The equals key $\boxed{=}$ completes all operations. Parenthesis keys $\boxed{(}$, $\boxed{)}$ can be used to change these built-in priorities when required. Pressing close parenthesis $\boxed{)}$ automatically completes the operations since the previous open parenthesis $\boxed{(}$. Consider carefully the *logic* involved in the following examples, and keep in mind that in many cases other keystroke sequences are possible.

EXAMPLE 4   Calculate $m = \dfrac{(3) - (-5)}{(-1) - (2)}$ from Example 3 using

a. parenthesis keys
b. memory keys
c. the reciprocal key

Solution

a. With parentheses we group the numerator and the denominator.

$\boxed{(}\,3\,\boxed{-}\,5\,\boxed{+/-}\,\boxed{)}\,\boxed{\div}\,\boxed{(}\,1\,\boxed{+/-}\,\boxed{-}\,2\,\boxed{)}\,\boxed{=}\ -2.6666667$

Note that negative numbers are entered with the change sign key $\boxed{+/-}$, not the $\boxed{-}$ key, which is used for the operation of subtraction.

b. Compute the denominator and store it. Then compute the numerator and divide by the stored number.

$1\,\boxed{+/-}\,\boxed{-}\,2\,\boxed{=}\,\boxed{\text{STO}}\,3\,\boxed{-}\,5\,\boxed{+/-}\,\boxed{=}\,\boxed{\div}\,\boxed{\text{RCL}}\,\boxed{=}\ -2.6666667$

On some calculators the store key looks like $\boxed{\text{Min}}$ or $\boxed{x \to \text{M}}$, while the recall key is labeled $\boxed{\text{MR}}$ or $\boxed{\text{RM}}$.

c. Since $a \div b$ can be defined as a times the reciprocal of b, the reciprocal key $\boxed{1/x}$ is often useful in computing fractional expressions. In this case we start with the denominator as follows:

$1\,\boxed{+/-}\,\boxed{-}\,2\,\boxed{=}\,\boxed{1/x}\,\boxed{\times}\,\boxed{(}\,3\,\boxed{-}\,5\,\boxed{+/-}\,\boxed{)}\,\boxed{=}\ -2.6666667.$ □

EXAMPLE 5   Evaluate $2x^3 + 4x^2 - x + 1$ if

a. $x = 5$
b. $x = -5$

Solution

a. The key $\boxed{y^x}$ or $\boxed{x^y}$ computes powers (here 5^3) with the $\boxed{x^2}$ key being the most direct way to square x. If $x = 5$, we can key in the problem exactly as it appears.

$2\,\boxed{\times}\,5\,\boxed{y^x}\,3\,\boxed{+}\,4\,\boxed{\times}\,5\,\boxed{x^2}\,\boxed{-}\,5\,\boxed{+}\,1\,\boxed{=}\ 346$

Instead of keying in a number several times, it can be useful to enter x just once and use the memory keys.

$5\,\boxed{\text{STO}}\,2\,\boxed{\times}\,\boxed{\text{RCL}}\,\boxed{y^x}\,3\,\boxed{+}\,4\,\boxed{\times}\,\boxed{\text{RCL}}\,\boxed{x^2}\,\boxed{-}\,\boxed{\text{RCL}}\,\boxed{+}\,1\,\boxed{=}\ 346$

b. If $x = -5$, we may have a problem computing $(-5)^3$ because some calculators restrict usage of the power key $\boxed{y^x}$ to a positive base. That is because the program for the power key often involves logarithms (see Chapter 5) which apply only to positive numbers. If your calculator works as follows,

$$5 \boxed{+/-} \boxed{y^x} 3 \boxed{=} \text{error}$$

then you cannot use the power key with a negative base. You can evaluate $(-5)^3$ by computing 5^3 and then changing the sign to negative because there is an odd number of negative factors. If the exponent is even, the answer is positive, so there is no need to change the answer from the power key. Thus, we evaluate the given expression as follows:

$$2 \boxed{\times} 5 \boxed{y^x} 3 \boxed{=} \boxed{+/-} \boxed{+} 4 \boxed{\times} 5 \boxed{+/-} \boxed{x^2} \boxed{-} 5 \boxed{+/-} \boxed{+} 1 \boxed{=} \quad -144.$$

If your calculator accepts a negative base in conjunction with the $\boxed{y^x}$ key, then you may key in the numbers directly as in part **a.** □

Combining Algebraic Expressions

Those parts of an algebraic expression separated by plus ($+$) signs are called **terms** of the expression.

EXAMPLE 6

a. $x + \frac{2}{5}x$ is an algebraic expression with two terms, x and $\frac{2}{5}x$.

b. $-4x^3 + x^2 - 3x + 7$ may be written as $-4x^3 + x^2 + (-3x) + 7$ and is an algebraic expression with four terms, $-4x^3$, x^2, $-3x$, and 7.

c. 6 is an algebraic expression with one term. □

If a term is the product of some constants and variables, the constant factor is called the (**numerical**) **coefficient** of the term. For example, the coefficient of the term $2x$ is 2, and the coefficient of ax^2 is a. Every term has a coefficient. If the term is x, the coefficient is 1 since $x = 1 \cdot x$. Similarly, if the term is $-x$, the coefficient is -1 since $-x = -1 \cdot x$. If two terms have identical variable factors (such as $3x^2y$ and $-2x^2y$), they are called **similar terms.** The distributive property indicates that we combine similar terms by combining their coefficients.

EXAMPLE 7

a. $3x + 7x = (3 + 7)x = 10x$

b. $yz^2 - 10yz^2 = (1 - 10)yz^2 = -9yz^2$

c. $2p - 3p + 4p = (2 - 3 + 4)p = 3p$

d. $7x + 2y - 3x = (7 - 3)x + 2y = 4x + 2y$

e. $-3x^2y + xy^2 - 7xy$ There are no similar terms, so we cannot simplify the expression. □

To combine algebraic expressions, it is sometimes necessary to remove the parentheses or brackets that group certain terms together. Parentheses are removed by applying the distributive property; that is, we multiply each term inside the parentheses by the factor in front of the parentheses. If the grouping is preceded by a minus sign, the factor is -1, so the sign of each term inside the parentheses must be changed. If the grouping is preceded by a plus sign, the factor is 1, so the sign of each term inside the parentheses remains the same. If there is more than one symbol of grouping, it is usually better to remove the innermost symbol of grouping first.

EXAMPLE 8   Remove parentheses and combine similar terms.

$$-2(3x - z) - (z - y) + 5(x + 2y)$$

Solution

$$
\begin{array}{ll}
\quad -6x \qquad\quad + 2z & \text{Each term is multiplied by } -2. \\
\qquad\quad + \quad y - z & \text{The sign of each term is changed.} \\
\text{add} \quad \underline{5x + 10y} & \text{Each term is multiplied by } 5. \\
\quad -x + 11y + \; z &
\end{array}
$$

☐

EXAMPLE 9   Remove the symbols of grouping and combine similar terms.

$$7 - 2[4x - (1 - 3x)]$$

Solution

$$
\begin{array}{ll}
7 - 2[4x - (1 - 3x)] & \\
= 7 - 2[4x - 1 + 3x] & \text{The sign of each term in the} \\
& \text{parentheses is changed.} \\
\\
= 7 - 2[7x - 1] & \\
= 7 - 14x + 2 & \text{Each term within the brackets is} \\
& \text{multiplied by } -2. \\
\\
= 9 - 14x &
\end{array}
$$

(*Note:* $9 - 14x \neq -5x$.)

☐

Integer Exponents

Recall that a positive integer exponent is a shortcut way of expressing a repeating factor. That is,

a^n, where n is a positive integer, is shorthand for

$$\underbrace{a \cdot a \cdot a \cdots a.}_{n \text{ factors}}$$

From this definition it is easy to obtain the following laws of exponents that we will illustrate in specific terms in Example 10.

Laws of Exponents   Let a and b be any real numbers and m and n be any positive integers.

1. $a^m \cdot a^n = a^{m+n}$

2. $(a^m)^n = a^{mn}$

3. $(ab)^n = a^n b^n$

4. $\left(\dfrac{a}{b}\right)^n = \dfrac{a^n}{b^n}$ $(b \neq 0)$

5. $\dfrac{a^m}{a^n} = \begin{cases} a^{m-n} & \text{if } m > n \\ \dfrac{1}{a^{n-m}} & \text{if } n > m \end{cases}$ $(a \neq 0)$

EXAMPLE 10   Simplify by the laws of exponents and check your result.

a. $2^2 \cdot 2^3$ **b.** $(2^3)^2$ **c.** $(2 \cdot 3)^2$ **d.** $2^5/2^2$ **e.** $2^2/2^5$

Solution

a. $2^2 \cdot 2^3 = 2^{2+3} = 2^5 = 32$

b. $(2^3)^2 = 2^{3 \cdot 2} = 2^6 = 64$

c. $(2 \cdot 3)^2 = 2^2 \cdot 3^2 = 4 \cdot 9 = 36$

d. $\dfrac{2^5}{2^2} = 2^{5-2} = 2^3 = 8$

e. $\dfrac{2^2}{2^5} = \dfrac{1}{2^{5-2}} = \dfrac{1}{2^3} = \dfrac{1}{8}$

Check

$2^2 \cdot 2^3 = 4 \cdot 8 = 32$

$(2^3)^2 = 8^2 = 64$

$(2 \cdot 3)^2 = 6^2 = 36$

$\dfrac{2^5}{2^2} = \dfrac{32}{4} = 8$

$\dfrac{2^2}{2^5} = \dfrac{4}{32} = \dfrac{1}{8}$ □

EXAMPLE 11   Simplify each expression. Assume variables as exponents denote positive integers.

a. $(6x^5y^3)(-2xy^7)$

b. $(2x^a)(7x^{2a})$

c. $\dfrac{18a^3b^2c^2}{12ab^4c}$

d. $\dfrac{2x^{r+3}}{-4x^r}$

e. $[(a^x)^{3x}]^2$

Solution

a. $(6x^5y^3)(-2xy^7) = 6(-2)x^{5+1}y^{3+7} = -12x^6y^{10}$

b. $(2x^a)(7x^{2a}) = 2 \cdot 7x^{a+2a} = 14x^{3a}$

c. $\dfrac{18a^3b^2c^2}{12ab^4c} = \dfrac{18a^{3-1}c^{2-1}}{12b^{4-2}} = \dfrac{3a^2c}{2b^2}$

d. $\dfrac{2x^{r+3}}{-4x^r} = \dfrac{2x^{(r+3)-r}}{-4} = -\dfrac{1}{2}x^3$

e. $[(a^x)^{3x}]^2 = [a^{3x^2}]^2 = a^{6x^2}$ □

We now wish to extend our definition of exponents to zero and negative integers. Note that it is meaningless to use x as a factor either zero times or a negative number of times, so we must define these exponents in a different manner. However, our guideline in these new definitions is to retain the laws of exponents developed for positive integers.

We start by considering the first law of exponents.

$$a^m \cdot a^n = a^{m+n}$$

If $n = 0$, we have

$$a^m \cdot a^0 = a^{m+0} = a^m.$$

When we multiply a^m by a^0, our result is a^m. Thus, a^0 must equal 1, and we make the following definition.

Zero Exponent   If a is a nonzero real number, then

$$a^0 = 1.$$

EXAMPLE 12

$$3^0 = 1$$
$$(a + b)^0 = 1$$
$$(2x)^0 = 1$$
$$2x^0 = 2(1) = 2$$ □

To obtain a definition for exponents that are negative integers, we will again consider the first law of exponents.

$$a^m \cdot a^n = a^{m+n}$$

If $m = 5$ and $n = -5$, we have

$$a^5 \cdot a^{-5} = a^{5+(-5)} = a^0 = 1.$$

When we multiply a^5 by a^{-5}, the result is 1. Thus, a^{-5} is the reciprocal of a^5 or $a^{-5} = 1/a^5$. In general, our previous laws of exponents may be extended by making the following definition.

Negative Exponent If a is a nonzero real number and n is a positive integer, then

$$a^{-n} = \frac{1}{a^n}.$$

EXAMPLE 13   Simplify $2^{-5} \cdot 2^3$.

Solution

Method 1 $2^{-5} \cdot 2^3 = 2^{(-5)+3} = 2^{-2} = \dfrac{1}{2^2}$ or $\dfrac{1}{4}$

Method 2 $2^{-5} \cdot 2^3 = \dfrac{1}{2^5} \cdot 2^3 = \dfrac{2^3}{2^5} = \dfrac{1}{2^{5-3}} = \dfrac{1}{2^2}$ or $\dfrac{1}{4}$ ☐

EXAMPLE 14   Simplify $(2x)^{-3}$.

Solution

Method 1 $(2x)^{-3} = \dfrac{1}{(2x)^3} = \dfrac{1}{2^3 \cdot x^3} = \dfrac{1}{8x^3}$

Method 2 $(2x)^{-3} = 2^{-3} \cdot x^{-3} = \dfrac{1}{8} \cdot \dfrac{1}{x^3} = \dfrac{1}{8x^3}$ ☐

Note from Examples 13 and 14 that there are different ways of simplifying these expressions. An important principle to keep in mind is that any *factor* of the numerator may be made a factor of the denominator (and vice versa) by changing the sign of the exponent. For example:

$$\frac{3}{5x^{-2}} = \frac{3x^2}{5}, \quad \frac{2^{-3}}{5^{-4}} = \frac{5^4}{2^3}, \quad \frac{3^{-2}x^2}{y+1} = \frac{x^2}{3^2(y+1)},$$

$$\text{and } \left(\frac{2}{3}\right)^{-1} = \frac{2^{-1}}{3^{-1}} = \frac{3}{2}.$$

It is important to remember that this principle applies only to factors.

EXAMPLE 15   Simplify and write the result using only positive exponents.

a. $-2(x + 2)^{-2}(x + 2)$ **b.** $(3x^{-2})^{-3}$ **c.** $b^x \div b^{-x}$

d. $(a^2)^{1-x} \cdot a^{2x-3}$ **e.** $(3^{-1}x^2y^{-3}) \div (2x^{-5}y)$

Solution

a. $-2(x + 2)^{-2}(x + 2) = -2(x + 2)^{-2+1} = -2(x + 2)^{-1}$
$= -2/(x + 2)$

b. $(3x^{-2})^{-3} = 3^{-3} \cdot x^6 = x^6/3^3 = x^6/27$

c. $b^x \div b^{-x} = b^{x-(-x)} = b^{2x}$

d. $(a^2)^{1-x} \cdot a^{2x-3} = a^{2-2x} \cdot a^{2x-3} = a^{(2-2x)+(2x-3)} = a^{-1} = 1/a$

e. $\dfrac{3^{-1}x^2y^{-3}}{2x^{-5}y} = \dfrac{x^{2-(-5)}}{3 \cdot 2y^{1-(-3)}} = \dfrac{x^7}{6y^4}$ □

Integer exponents are generally evaluated on a calculator with the power function key $\boxed{y^x}$ in the usual way. For example, 2^{-3} is computed as

$2\boxed{y^x}3\boxed{+/-}\boxed{=}$ 0.125.

Since negative exponents involve reciprocals, an alternative method is

$2\boxed{y^x}3\boxed{=}\boxed{1/x}$ 0.125.

An exponent of -1 is best computed with the reciprocal key. For instance, 5^{-1} is simply

$5\boxed{1/x}$ 0.2.

As discussed in Example 5, if your calculator does not permit usage of the power key $\boxed{y^x}$ with a negative base, you will need to use a positive base and then insert the correct sign. Thus, $(-2)^{-3}$ is computed as above with you inserting the negative sign in the answer -0.125 because 3 is an odd exponent.

EXERCISES 1.2

In Exercises 1–20 evaluate each expression after setting $x = -2$, $y = 3$, and $z = -4$.

1. $x + y + z$

2. $4x - z$

3. x^3x^2

4. y^3/y^2

5. $(x^2 + y^2)/z^2$

6. $20 - xyz^3$

7. $(-x)^2 + 2y$

8. $-x^2 + 2y$

9. $(x - y)^2 \div (y - x)^2$

10. $2x^3 + 3x^2 - x + 1$

11. $x - 2(3y - 4z)$

12. $(x - 2)(3y - 4z)$

13. $(z - x)(y - 3z)$

14. $z - x(y - 3z)$

15. $(2y - x)x - z$

16. $(2y - x)(x - z)$

17. $(x + y + z)(x + y - z)$

18. $x + y + z(x + y - z)$

19. $(x + 2)(y - 3) + (3 - z)(4 + x)$

20. $(5 - y)(z + 1) - (y - 2)(3 + x)$

In Exercises 21–26 $m = (y_2 - y_1)/(x_2 - x_1)$. Evaluate m for the following values of x_1, y_1, x_2, and y_2.

	x_1	y_1	x_2	y_2
21.	5	4	1	2
22.	1	6	4	8
23.	-3	-1	-6	-1
24.	2	-3	2	-6
25.	2	-1	-6	3
26.	0	3	-2	-4

In Exercises 27–50 simplify each expression.

27. $6a + 4a + 9a$

28. $-3b - b + 7b$

29. $8a - 3b + 6a - 5b$

30. $5k - 6m - 7 + 3k - 4m + 2$

31. $5xy - 4cd - 2xy + 9cd$

32. $4p - 5q + 4p + q + 4p - 2q$

33. $2x^3 + 6y^2 - 5x^3 + 2y^2$

34. $2a^2b + 3ab - 4ab^2$

35. $x - 2(x + y) + 3(x - y)$

36. $k + (m - k) - (m - 2k)$

37. $-2(a - 2b) - 7(2a - b)$

38. $2a - (7a + 3) + (4a - 6)$

39. $-(x + y) + 4(x - y) + 7x - 2y$

40. $a(b - c) - b(a + c)$

41. $(2x^3 + 7x^2y^2 + 9xy^3) + (6x^2y^2 - 3x^3y)$

42. $(3c^2d + 2cd - 5d^3) - (9d^3 - 6c^2d - 2cd)$

43. $(3y^3 - 2y^2 + 7y - 1) + (4y^3 - y^2 - 3y + 7)$

44. $(2n - 4n^3 + 3n^2 + 4) - (6 - 2n^2 + 7n - n^3)$

45. $(b^3 - 2b^2 + 3b + 4) - (b^2 + 2b - 1)$

46. $(4x^3 + 7x^2y^2 - 2y^3) + (-7x^3 + 2x^2y^2 + 2y^3)$

47. $2[a + 5(a + 2)] - 6$

48. $10 - 4[3x - (1 - x)]$

49. $3a - [3(a - b) - 2(a + b)]$

50. $-[x - (x + y) - (x - y) - (y - x)]$

In Exercises 51–70 evaluate each expression and check your result with a calculator.

51. $3^2 \cdot 3^3$ **52.** $(3^2)^3$ **53.** 7^0

54. -7^0 **55.** 3^{-2} **56.** $(-2)^{-3}$

57. $(4/3)^{-1}$ **58.** $(2/5)^{-3}$ **59.** $4^5 \cdot 4^{-2}$

60. $5^{-4} \cdot 5$ **61.** $(-3)^2 \cdot (-3)^3$ **62.** $(-2)^{-1} \cdot (-2)^{-3}$

63. $\dfrac{2}{2^4}$ **64.** $\dfrac{3^6}{3^2}$ **65.** $\dfrac{1}{2^{-3}}$

66. $\dfrac{2}{2^{-2}}$ **67.** $\dfrac{3^{-1}}{3}$ **68.** $\dfrac{5^{-2}}{5^{-4}}$

69. $\dfrac{6^{-1} \cdot 2^{-4}}{3^{-2} \cdot 2^{-3}}$ **70.** $\dfrac{2^{-3} \cdot 7^0}{4^{-1}}$

In Exercises 71–98 use the laws of exponents to perform the operations and write the result in the simplest form that contains only positive exponents.

71. $x \cdot x^4$ **72.** $a \div a^3$

73. $(-2p)^2$ **74.** $(a + b)^2(a + b)^4$

75. $x^5 \cdot x^{-3}$ **76.** $3y \cdot y^{-4}$

77. $(3x^3y^2)^4$ **78.** $(-5y^5z)^3$

79. $(-3x^2)(-3x)^2$ **80.** $(-4s^2c)^3(3sc^2)^5$

81. $(1 - n)^{-1}(1 - n)$ **82.** $(2x^{-3})(-3x^{-2})$

83. $(4x^0)(-2x^{-3})$ **84.** $(2a^{-3})^{-2}$

85. $[3(x + h)^{-1}]^2$ **86.** $(1/x^{-1})^{-1}$

87. $(2a^{-1}/x)^{-2}$ **88.** $(3x^{-2}/y)^{-1}$

89. $2a^{-1}/a$ **90.** $-x^{-2}/x^{-3}$

91. $\dfrac{(-x^3)^2(4yz)}{(-2x^2)(2y^2z^3)^3}$ **92.** $\dfrac{(4x^2yz^3)^4}{(-20x^6yz^2)^2}$

93. $\dfrac{3^{-1}x^2y^{-3}}{9x^{-2}y^{-3}}$ **94.** $\dfrac{2^2x^{-3}y^{-4}}{2^{-1}x^3y^{-2}}$

95. $\dfrac{(xyz)^{-1}}{x^{-2}yz^{-3}}$ **96.** $\dfrac{x^0y^{-2}z^3}{(xy^{-1}z^{-3})^{-1}}$

97. $\dfrac{(2ax^2)^{-2}(a^3x^{-1})^2}{2(ax)^{-1}(ax^5)}$ **98.** $\dfrac{[a^2(a + y)^3]^{-1}}{a(a + y)^2}$

In Exercises 99–124 use the laws of exponents to simplify each expression. Variables as exponents denote integers.

99. $2^x \cdot 2^y$ **100.** $5^a \cdot 5$

101. $2/2^n$ **102.** $5^{2x}/5^x$

103. $(5^x)^{2x}$ **104.** $3^a \cdot 3^{-a}$

105. $y^{2a} \cdot y^{5a}$ **106.** $x^{b+2} \cdot x^{2b}$

107. $(a - b)^x(a - b)^y$ **108.** $[(b - a)^x]^y$

109. $a^x \cdot a^{-x}$

110. $2x^{1-a} \cdot x^{a-1}$

111. $x \div x^{-x}$

112. $a^{-x} \div a^0$

113. $[(y^x)^{2x}]^3$

114. $(2x^a)^c (3x^c)^b$

115. $\dfrac{y^x}{y^{x+1}}$

116. $\dfrac{x^{2n+2}}{x^2}$

117. $(x^a)^{p-1} \cdot x^{a-1}$

118. $(b^x)^{1-a} \div b^x$

119. $\dfrac{x^{2a}y^{a+1}}{x^a y^{a+2}}$

120. $\dfrac{x^a + b y^b}{x^a - b y^{b-1}}$

121. $\dfrac{(1-x)^{2a}(1-x)^2}{(1-x)^a}$

122. $\dfrac{(y+2)^{1+m}}{(y+2)^m (y+2)}$

123. $\dfrac{x^k \cdot 0 - a(kx^{k-1})}{(x^k)^2}$

124. $\dfrac{x^{n-1} \cdot 0 - 2(n-1)x^{n-2}}{(x^{n-1})^2}$

In Exercises 125–134 determine if the given calculator sequence produces the correct answer to the problem. If not, do two things. First, write the arithmetic expression being computed by the given sequence. Then change the sequence so it computes the given problem.

125. $\dfrac{5+4}{3}$; $\boxed{5}\boxed{+}\boxed{4}\boxed{\div}\boxed{3}\boxed{=}$

126. $\dfrac{6}{7} - 3$; $\boxed{6}\boxed{\div}\boxed{7}\boxed{-}\boxed{3}\boxed{=}$

127. $\dfrac{2}{9} \div 15$; $\boxed{2}\boxed{\div}\boxed{9}\boxed{\div}\boxed{15}\boxed{=}$

128. $2 \div \dfrac{15}{9}$; $\boxed{2}\boxed{\div}\boxed{15}\boxed{\div}\boxed{9}\boxed{=}$

129. $\dfrac{2}{9(15)}$; $\boxed{2}\boxed{\div}\boxed{9}\boxed{\times}\boxed{15}\boxed{=}$

130. $2^{3(5)}$; $\boxed{2}\boxed{y^x}\boxed{3}\boxed{\times}\boxed{5}\boxed{=}$

131. $2 \cdot 3^5$; $\boxed{2}\boxed{\times}\boxed{3}\boxed{y^x}\boxed{5}\boxed{=}$

132. $(3+2)^4$; $\boxed{3}\boxed{+}\boxed{2}\boxed{y^x}\boxed{4}\boxed{=}$

133. $\sqrt{5^2 + 6^2}$; $\boxed{5}\boxed{x^2}\boxed{+}\boxed{6}\boxed{x^2}\boxed{\sqrt{\,}}\boxed{=}$

134. $4\sqrt{7+2}$; $\boxed{4}\boxed{\times}\boxed{(}\boxed{7}\boxed{+}\boxed{2}\boxed{=}\boxed{\sqrt{\,}}\boxed{)}$

135. The Greek mathematician Archimedes (250 B.C.) is usually considered one of the greatest mathematicians of all time. Archimedes considered his most important achievement to be the discovery that whenever a sphere is circumscribed by a cylinder, the ratio of their volumes is 3:2. He even asked that this figure and ratio be engraved on his tombstone. If the volume of a cylinder is $\pi r^2 h$ and the volume of a sphere is $\frac{4}{3}\pi r^3$, verify Archimedes' discovery.

136. A rope is wrapped tightly around the equator of the earth. A second rope is suspended in the air 1 ft directly above the first all the way around the earth. Is the difference between the rope lengths about 6 ft, 600 ft, 60,000 ft, or 6 million ft? Justify your answer.

1.3 PRODUCTS OF ALGEBRAIC EXPRESSIONS AND FACTORING

To multiply algebraic expressions, we use the laws of exponents and the distributive properties.

$$a(b + c) = ab + ac \quad \text{and} \quad (a + b)c = ac + bc$$

Example 1 shows the basic procedure.

EXAMPLE 1   Multiply.

a. $3x(x^2 - 2x + 1)$

b. $(2c^2 + t)ct$

Solution

a. $3x(x^2 - 2x + 1) = (3x)(x^2) + (3x)(-2x) + (3x)(1)$
$$= 3x^3 - 6x^2 + 3x$$

b. $(2c^2 + t)ct = (2c^2)(ct) + (t)(ct)$
$$= 2c^3t + ct^2$$ $\square$

Since division is defined in terms of multiplication, a similar procedure is used to simplify division problems in which the divisor is a single term.

EXAMPLE 2   Divide $\dfrac{2xh + h^2 + 2h}{h}$.

Solution $\dfrac{2xh + h^2 + 2h}{h} = \dfrac{2xh}{h} + \dfrac{h^2}{h} + \dfrac{2h}{h}$
$$= 2x + h + 2$$ $\square$

If both factors in the multiplication contain more than one term, the distributive property must be used more than once. For example, no matter what expression is inside the parentheses

$(\blacksquare)(x + 2)$ means $(\blacksquare)x + (\blacksquare)2$.

Thus,

$(x + 3)(x + 2)$ means $(x + 3)x + (x + 3)2$.

Using the distributive property the second time, we get

$(x + 3)x + (x + 3)2 = x^2 + 3x + 2x + 6 = x^2 + 5x + 6$.

Therefore,

$(x + 3)(x + 2) = x^2 + 5x + 6$.

EXAMPLE 3   Multiply.

a. $(2t + 3)(5t - 1)$ b. $(x^n - 2)(x^n - 1)$

Solution

a. $(2t + 3)(5t - 1) = (2t + 3)(5t) + (2t + 3)(-1)$
$$= 10t^2 + 15t - 2t - 3$$
$$= 10t^2 + 13t - 3$$

b. $(x^n - 2)(x^n - 1) = (x^n - 2)(x^n) + (x^n - 2)(-1)$
$$= x^{2n} - 2x^n - x^n + 2$$
$$= x^{2n} - 3x^n + 2$$ $\square$

Notice that this method of multiplication is equivalent to multiplying each term of the first factor by each term of the second factor and then combining similar terms. This observation leads to the arrangement in Example 4, which is good for multiplying longer expressions since similar terms are placed under each other.

EXAMPLE 4   Multiply $(c^2 - 5c + 25)(c + 3)$.

Solution

$$c^2 - 5c + 25$$
$$c + 3$$

$$c^3 - 5c^2 + 25c \qquad \text{This line equals } c(c^2 - 5c + 25).$$
add $\qquad 3c^2 - 15c + 75 \qquad \text{This line equals } 3(c^2 - 5c + 25).$

$$c^3 - 2c^2 + 10c + 75 \qquad\qquad\qquad\qquad\qquad \square$$

When multiplying expressions that contain two terms, a mental shortcut is often used as shown in Example 5. This method is called the FOIL method, and the letters F, O, I, and L denote the products of the first, outer, inner, and last terms, respectively.

EXAMPLE 5   Multiply using the FOIL method: $(x^3 + y^3)^2$.

Solution $(x^3 + y^3)^2 = (x^3 + y^3)(x^3 + y^3)$. Then

$$
\begin{array}{l}
F + \quad O \quad + \quad I \quad + L \\
x^6 + \ x^3y^3 + x^3y^3 + y^6 \\
= x^6 + 2x^3y^3 + \quad y^6
\end{array}
$$

[*Note:* $(x^3 + y^3)^2 \neq x^6 + y^6$, for this result leaves out the middle term, $2x^3y^3$.] $\square$

We now consider the process of factoring which reverses the process of multiplication as shown below.

If we change a product into a sum, we are multiplying.

$$(x + 5)(x + 1) \xrightarrow{\text{multiplying}} x^2 + 6x + 5$$

If we change a sum into a product, we are factoring.

$$(x + 5)(x + 1) \xleftarrow{\text{factoring}} x^2 + 6x + 5$$

In general, the procedures for factoring are not as straightforward as those for multiplication, and we limit our discussion of factoring to algebraic expressions called polynomials. Examples of polynomials are

$$64x^6 - y^6, \quad 9x^2 + 24x + 16, \quad \text{and} \quad 12ct^2 + 9c^2t,$$

and a **polynomial** is defined as an algebraic expression which may be written as a finite sum of terms that contain only nonnegative integer exponents on the variables. Thus, algebraic expressions like

$$4x^{-2}, \quad \sqrt{x} - 1, \quad \text{and} \quad 5x^{1/3} + 5$$

are not polynomials. A polynomial with just one term is called a **monomial;** one that contains exactly two terms is a **binomial;** and one with exactly three terms is a **trinomial.** The **degree of a monomial** is the sum of the exponents on the variables in the term. The **degree of a polynomial** is the same as the degree of its highest monomial term. Thus, the degree of $5x^4y^2$ is 6, the degree of $2xy^2$ is 3, and the degree of $5x^4y^2 + 2xy^2$ is 6, since $5x^4y^2$ is the highest-degree term.

There are several techniques that enable us to factor certain polynomials, and you need to consider carefully each of the factoring methods that follow.

Common Factors

The first method that should always be employed in factoring a polynomial is to attempt to find a factor that is common to each of the terms. For example:

Polynomial	Common Factor
$9s + 6t$	3
$15x^2 + (-5x)$ (or $15x^2 - 5x$)	$5x$
$2a^2b + 8ab^2$	$2ab$

In each case we attempt to pick the greatest common factor that will divide into each term of the polynomial. Therefore, although 5 is a common factor of $15x^2 - 5x,$ a preferable common factor is $5x$ since $5x$ is the largest factor that divides both terms. After we determine the greatest common factor, we divide this factor into the polynomial and obtain a second factor of our original expression. The product of these two factors represents the polynomial in factored form. You will notice that the directions in factoring problems use the phrase "factor completely." This expression directs us to continue factoring until the polynomial contains no factors of two or more terms that can be factored again. The restrictions we place on the form of the factors will determine if a polynomial is factorable, so it is important to note that, unless otherwise specified, we are interested only in polynomial factors with integer coefficients.

EXAMPLE 6   Factor completely.

a. $12ct^2 + 9c^2t$

b. $s(c + 2) - t(c + 2)$

Solution

a. Greatest common factor: $3ct$

$$\frac{12ct^2 + 9c^2t}{3ct} = 4t + 3c$$

Therefore, $12ct^2 + 9c^2t = 3ct(4t + 3c)$.

b. Greatest common factor: $c + 2$

$$\frac{s(c + 2) - t(c + 2)}{c + 2} = s - t$$

Therefore, $s(c + 2) - t(c + 2) = (c + 2)(s - t)$. □

Difference of Squares

A second method of factoring comes from reversing the following product.

$$(x + y)(x - y) \quad \overset{\text{multiplying}}{\underset{\text{factoring}}{\longleftrightarrow}} \quad x^2 - y^2$$

Note that the factors on the left differ only in the operation between x and y, while the result of the multiplication is two perfect squares with a minus sign between them. (Remember that x^2, x^4, x^6, x^{even}, and 1, 4, 9, 16, 25, and so on, are perfect squares.) Thus, in factoring two perfect squares with a minus sign between them, we obtain two factors that consist of the sum and the difference of the square roots of each of the squared terms.

EXAMPLE 7   Factor completely.

a. $x^2 - 81$ **b.** $y^4 - x^4$

Solution

a. $x^2 = x \cdot x$
    $81 = 9 \cdot 9$
    Therefore, $x^2 - 81 = (x + 9)(x - 9)$.

b. $y^4 = y^2 \cdot y^2$
    $x^4 = x^2 \cdot x^2$
    Therefore, $y^4 - x^4 = (y^2 + x^2)(y^2 - x^2)$. But

$$y^2 - x^2 = (y + x)(y - x).$$

Therefore, $y^4 - x^4 = (y^2 + x^2)(y + x)(y - x)$. □

Factoring Trinomials (Leading Coefficient 1)

In some cases trinomials of the form $ax^2 + bx + c$ can be factored (with integer coefficients) by reversing the FOIL multiplication process shown earlier. Example 8 discusses this factoring procedure in the simplest case when the leading coefficient a equals 1.

EXAMPLE 8   Factor completely $x^2 - 5x - 6$.

Solution
The first term, x^2, is the result of multiplying the first terms in the FOIL method. Thus,

$$x^2 - 5x - 6 = (x + ?)(x + ?).$$

The last term, -6, is the result of multiplying the last terms in the FOIL method. Since -6 equals $(-6)(1)$, $(6)(-1)$, $(-3)(2)$, and $(3)(-2)$, we have four possibilities. We want the pair whose sum is -5, which is the coefficient of the middle term. The combination of -6 and 1 satisfies this condition and produces the middle term of $-5x$.

$$(x - 6)(x + 1)$$

$$x + (-6x) = -5x$$

Thus, $x^2 - 5x - 6 = (x - 6)(x + 1)$. We summarize our method in this example with the factoring formula

$$x^2 + (a + b)x + ab = (x + a)(x + b)$$

which you can use as a model for factoring such expressions.   □

Factoring Trinomials (General Case)

When the coefficient of the squared term is not 1, the trinomial is harder to factor. The reason for this difficulty can be seen in the following product.

$$(ax + b)(cx + d) = (ac)x^2 + (ad + bc)x + bd$$

Note that the factors for the last term, b and d, add up to the middle coefficient only when a and c are 1. In factoring these more complicated trinomials, if you keep in mind that the middle term is the sum of the inside product and the outside product, then a little trial and error will eventually produce the answer. Although there are often many combinations to consider, experience and practice will help you *mentally* eliminate many of the possibilities.

EXAMPLE 9   Factor completely $9y^2 + 12y + 4$.

Solution
The first term, $9y^2$, is the result of multiplying the first terms in the FOIL method.

$$9y^2 + 12y + 4 \overset{?}{=} \begin{array}{c} (9y + ?)(y + ?) \\ \text{or} \\ (3y + ?)(3y + ?) \end{array}$$

The last term, 4, is the result of multiplying the last terms in the FOIL method. Since 4 equals (4)(1) and (2)(2), we have two possibilities. We eliminate negative factors of 4 since the middle term is positive.

$$9y^2 + 12y + 4 \overset{?}{=} \begin{array}{c} (9y + 1)(y + 4) \\ (9y + 4)(y + 1) \\ (9y + 2)(y + 2) \\ (3y + 4)(3y + 1) \\ (3y + 2)(3y + 2) \end{array}$$

The middle term, $12y$, is the sum of the inner and the outer terms.

$$(3y + 2)(3y + 2)$$
$$6y + 6y = 12y$$

Thus, $9y^2 + 12y + 4 = (3y + 2)(3y + 2)$ or $(3y + 2)^2$.

Note: Very often when the first term and the last term of the trinomial to be factored are perfect squares, the factored form is also a perfect square. Try this possibility first. The factoring model in such cases (called perfect square trinomials) is

$$x^2 + 2xy + y^2 = (x + y)^2. \qquad \square$$

It should be noted that many trinomials cannot be factored with integer coefficients, and if you suspect that you are trying to find a factorization that might not exist, then apply the following test.

Factoring Test for Trinomials	A trinomial of the form $ax^2 + bx + c$, where a, b, and c are integers, is factorable into binomial factors with integer coefficients if and only if $b^2 - 4ac$ is a perfect square.

The expression $b^2 - 4ac$ is called the **discriminant,** and when we derive the discriminant in Section 2.3, the rationale for this test will be easy to understand.

EXAMPLE 10   Calculate $b^2 - 4ac$ and determine if $x^2 - 10x - 30$ can be factored with integer coefficients.

Solution   In this trinomial $a = 1$, $b = -10$, and $c = -30$. Therefore,

$$b^2 - 4ac = (-10)^2 - 4(1)(-30)$$
$$= 220.$$

Since 220 is not a perfect square, $x^2 - 10x - 30$ cannot be factored into binomial factors with integer coefficients.   □

A polynomial is said to be **irreducible** over the set of integers if it cannot be written as the product of two polynomials of positive degree with integer coefficients. Thus, we say $x^2 - 10x - 30$ is irreducible over the set of *integers*. Note, however, that $x^2 - 10x - 30$ is not irreducible over the set of *real numbers* since we can show that

$$x^2 - 10x - 30 = (x - (5 + \sqrt{55}))(x - (5 - \sqrt{55}))$$

by using techniques that we will develop. Similarly,

$$x^2 - 7 = (x + \sqrt{7})(x - \sqrt{7})$$

if we allow real number coefficients in the factors. Thus, whether a polynomial is irreducible depends on the number system from which the coefficients may be selected.

Sums or Differences of Cubes

To factor sums or differences of cubes, we rely heavily on these factoring models.

$$a^3 + b^3 = (a + b)(a^2 - ab + b^2)$$
$$a^3 - b^3 = (a - b)(a^2 + ab + b^2)$$

The idea is to identify appropriate replacements for a and b in the expression to be factored, and then substitute in these formulas.

EXAMPLE 11   Factor completely.

a. $8x^3 + 27$ **b.** $64y^6 - x^6$

Solution

a. Here we use the formula for the sum of two cubes. Since $8x^3 = (2x)^3$ and $27 = (3)^3$, we replace a with $2x$ and b with 3. The result is

$$8x^3 + 27 = (2x)^3 + 3^3$$
$$= (2x + 3)[(2x)^2 - (2x)(3) + (3)^2]$$
$$= (2x + 3)(4x^2 - 6x + 9).$$

b. First, factor $64y^6 - x^6$ as the difference of squares.

$$64y^6 - x^6 = (8y^3)^2 - (x^3)^2$$
$$= (8y^3 + x^3)(8y^3 - x^3)$$

Now replace a by $2y$ and b with x in the formulas for both the sum and difference of cubes.

$$64y^6 - x^6 = (2y + x)[(2y)^2 - 2yx + x^2](2y - x)$$
$$[(2y)^2 + 2yx + x^2]$$
$$= (2y + x)(2y - x)(4y^2 - 2yx + x^2)$$
$$(4y^2 + 2yx + x^2) \qquad \square$$

Factoring by Grouping

Factoring by grouping is not really a different way of factoring for it relies on the methods we have already discussed. However, before those methods can be applied it is sometimes necessary to rewrite or group together terms in the expression to be factored. For example,

$$3xy + 2y + 3xz + 2z$$

can be factored by rewriting the expression as

$$(3xy + 2y) + (3xz + 2z).$$

If we factor out the common factor in each group, we have

$$y(3x + 2) + z(3x + 2).$$

Now $3x + 2$ is a common factor and the final result is

$$(3x + 2)(y + z).$$

You might check that in this case different groupings of the original expression achieve the same result.

EXAMPLE 12   Factor completely $x^2 + y^2 - z^2 - 2xy$.

Solution   We need to recognize that if we group together $x^2 - 2xy + y^2$, it factors into the perfect square $(x - y)^2$. We then have $(x - y)^2 - z^2$, which factors as the difference of two squares. Thus,

$$x^2 + y^2 - z^2 - 2xy = (x^2 - 2xy + y^2) - z^2$$
$$= (x - y)^2 - z^2$$
$$= (x - y + z)(x - y - z). \qquad \square$$

Summary

The basic formulas we discussed in this section are listed in the box for reference purposes. Some people find it helpful to stick very closely to these forms, while others have a feel for the basic patterns involved and find that the formulas only complicate the issue. You decide which group you belong to. As a final note, remember you can always check your factoring by multiplying out your answer.

Factoring and Product Models

1. Common factor:

   $$ax + ay = a(x + y)$$

2. Difference of squares:

   $$x^2 - y^2 = (x + y)(x - y)$$

3. Trinomial (leading coefficient 1):

   $$x^2 + (a + b)x + ab = (x + a)(x + b)$$

4. Perfect square trinomial:

   $$x^2 + 2xy + y^2 = (x + y)^2$$

5. General trinomial:

   $$(ac)x^2 + (ad + bc)x + bd = (ax + b)(cx + d)$$

6. Sum of cubes:

   $$a^3 + b^3 = (a + b)(a^2 - ab + b^2)$$

7. Difference of cubes:

   $$a^3 - b^3 = (a - b)(a^2 + ab + b^2)$$

EXERCISES 1.3

In Exercises 1–40 perform the multiplication or division and combine similar terms.

1. $2(x - y)$

2. $-6(z + y)$

3. $-5x(x^3 - x^2 - x)$

4. $y(3y^2 + 5y - 6)$

5. $-2xyz(4x - y + 7z)$

6. $4x^2yz^3(3x^3 - 2yz + 5xz^4)$

7. $(p^2 + q^2)p^2q^2$

8. $(a^2 - b^2)ab$

9. $(24n^2x + 18nx^2) \div 6nx$

10. $(6dt^2 - 12d^2t) \div 2dt$

11. $\dfrac{2hx + h^2}{h}$

12. $\dfrac{6xh + 3h^2}{h}$

13. $2x^4\left(\dfrac{3}{x} + \dfrac{4}{x^2} - \dfrac{5}{x^3}\right)$

14. $-12x^2\left(3x^2 - \dfrac{1}{3} + \dfrac{3}{4x} - \dfrac{1}{x^2}\right)$

15. $(a + 3)(a + 4)$

16. $(z - 2)(z - 6)$

17. $(3x - 4)(2x - 1)$

18. $(5y - 4x)(3y + x)$

19. $(k - 2)^2$

20. $(2x + 3y)^2$

21. $(y - 4)(y^2 + 5y - 1)$

22. $(2a - 1)(a^3 + a + 1)$

23. $(x - y)(x^2 + xy + y^2)$

24. $(2y + 3z)(y^2 - 2yz + z^2)$

25. $(2y - 1)(3y + 2)(1 - y)$

26. $(x + 1)^3$

27. $(x^{-1} - 1)(x^{-1} + 1)$

28. $(y^{-1} + 2)(y^{-1} + 3)$

29. $(z^a + 3)^2$

30. $(3x^n - 2)(x^n - 1)$

31. $(x^a + y^b)^2$

32. $(x^a - y^b)(x^a + y^b)$

33. $(x^3 - y^3)(x^3 + y^3)$

34. $(3x^3 - 1)(x - x^2)$

35. $[(x - y) - 1]^2$

36. $[3 + (4a - b)]^2$

37. $(x^2 - 2xy + y^2)(x^2 + 2xy + y^2)$

38. $(x^2 - x - 1)(x^2 + x + 1)$

39. $(x^n + y^n)(x^{2n} - x^ny^n + y^{2n})$

40. $(x^n - y^k)(x^{2n} + x^ny^k + y^{2k})$

In Exercises 41–94 factor each expression completely.

41. $yx + yz$

42. $ab - ac$

43. $9a^2x^2 + 3ax$

44. $22x^3y^3 - 2x^2y^2$

45. $2x^2z^2 + 4xy - 5x^2y^2$

46. $3a^{10}x^6 - 9a^7x^7 + 6a^9x^4$

47. $b(a - c) + d(a - c)$

48. $2x(y + z) - 5z(y + z)$

49. $36x^2 - 1$

50. $100 - 81y^2$

51. $36r^6 - k^4$

52. $25x^{12} - 36a^{10}$

53. $a^2 + 5a + 4$

54. $y^2 + 7y + 10$

55. $x^2 + x - 12$

56. $p^2 - 3p - 18$

57. $6 + x - x^2$

58. $9 - 8y - y^2$

59. $7z^2 - 13z + 6$

60. $5r^2 - 4r - 12$

61. $9b^2 - 25b - 6$

62. $12x^2 + 19x - 18$

63. $x^2 + 5xy + 4y^2$

64. $2a^2 - 7ad + 6d^2$

65. $16y^2 - 24y + 9$

66. $9x^2 + 24x + 16$

67. $9a^2 - 12ab + 4b^2$

68. $16x^2y^2 + 8xy + 1$

69. $x - xy^2$

70. $n^3 - n$

71. $3k^2 - 6k - 24$

72. $5y^2 - 15y - 50$

73. $3x^3 - 12x^2 + 9x$

74. $2x^3y - 18x^2y + 40xy$

75. $1 - x^3$

76. $27 + y^3$

77. $a^3b^3 + c^3$

78. $8x^3 - 27y^3$

79. $2y^3 - 16x^6$

80. $(x - 2)^3 + 1$

81. $x^6 - 1$

82. $1 - x^{12}$

83. $x^6 - y^6$

84. $x^8 - y^8$

85. $x^{2a} - y^{2b}$

86. $x^{3a} + y^{3b}$

87. $3xy + 6x + y + 2$

88. $2a + 2b + ka + kb$

89. $2xy - 3y + 2xz - 3z$

90. $4xy - 6xz - 6y + 9z$

91. $x^2 - y^2 - z^2 - 2yz$

92. $4x^2 + y^2 + 4xy - 1$

93. $x^5 + x^3 + x^2 + 1$

94. $x^3 - 3x^2 - x + 3$

95. Verify through multiplication the factoring model
$a^3 + b^3 = (a + b)(a^2 - ab + b^2)$.

96. **a.** Multiply $(x - 1)(x + 1)$.

   b. Multiply $(x - 1)(x^2 + x + 1)$.

   c. Multiply $(x - 1)(x^3 + x^2 + x + 1)$.

   d. Look at your results from parts a–c. Without multiplying, simplify the product $(x - 1)(x^7 + x^6 + x^5 + x^4 + x^3 + x^2 + x + 1)$.

   e. What is a factoring model for $x^n - 1$?

In Exercises 97–100 determine whether each trinomial can be factored (with integer coefficients) by calculating $b^2 - 4ac$.

97. $6x^2 + 7x + 2$ **98.** $12s^2 - s + 1$

99. $12y^2 - 11y - 18$ **100.** $18x^2 + 19x - 12$

101. Show that $\dfrac{(x + h)^2 + 1 - (x^2 + 1)}{h}$
simplifies to $2x + h$.

102. Show that $\dfrac{(x + h)^3 - x^3}{h}$
simplifies to $3x^2 + 3xh + h^2$.

103. Simplify $\dfrac{[1 - (-3 + h)^2] - [1 - (-3)^2]}{h}$.

104. Simplify $\dfrac{(2 + h)^3 - 2^3}{h}$.

105. Simplify $\dfrac{[(x + h) + 1]^2 - (x + 1)^2}{h}$.

106. Simplify
$$\dfrac{[128(t + h) - 16(t + h)^2] - [128t - 16t^2]}{h}.$$

107. Consider the square with side length $a + b$.

a. Algebraically, calculate the area $(a + b)^2$.
b. Find the area by adding the areas of regions 1, 2, 3, and 4.
c. Does $(a + b)^2 = a^2 + b^2$?

108. The Pythagorean relation is one of the most famous ideas in mathematics. Several hundred different proofs of this theorem have been recorded, and the National Council of Teachers of Mathematics published a book, *The Pythagorean Proposition* (1968), that presents 370 demonstrations of this statement. Early proofs, which were geometric in nature, gradually gave way to analytical proofs that used algebra to verify geometric ideas. Two well-known proofs of this variety follow.

a. Consider the square with side length $a + b$. Calculate the area of the square in two ways and establish the Pythagorean relation $c^2 = a^2 + b^2$. (*Note:* It must be shown that $\theta = 90°$.)

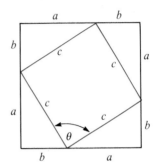

b. The trapezoid in the diagram was used by James Garfield (the twentieth president of the United States) to prove the Pythagorean theorem. Once again the idea is to find the area in two ways and establish $c^2 = a^2 + b^2$. Try it.

109. An odd number is a positive integer that can be written in the form $2k + 1$, where k is an integer. Show that the product of two odd numbers is odd.

1.4 ALGEBRAIC FRACTIONS

Algebraic fractions are the quotients of algebraic expressions and the same principles that govern a fraction in arithmetic also apply when the numerator and the denominator contain algebraic expressions. Two fraction principles of particular importance follow.

Fraction Principles

Let a, b, c, d, and k be real numbers with b, d, and $k \neq 0$.

Equality of fractions $\dfrac{a}{b} = \dfrac{c}{d}$ if and only if $ad = bc$

Fundamental principle $\dfrac{ak}{bk} = \dfrac{a}{b}$

The fundamental principle can be established from the criterion for the equality of two fractions since $(ak)b$ is equal to $(bk)a$. As its name suggests, the fundamental principle is applied often, and two common uses of this principle are shown below.

Simplifying fractions $\dfrac{6}{8} = \dfrac{3 \cdot 2}{4 \cdot 2} = \dfrac{3}{4}$

fundamental principle

Adding fractions $\dfrac{1}{8} + \dfrac{3}{4} = \dfrac{1}{8} + \dfrac{3 \cdot 2}{4 \cdot 2} = \dfrac{1}{8} + \dfrac{6}{8} = \dfrac{7}{8}$

When simplifying fractions by the fundamental principle, it is important to recognize that we may divide out only nonzero factors of the numerator and the denominator. Thus, we have a general procedure for expressing a fraction in simplest form.

To Simplify Fractions

1. Factor completely the numerator and the denominator of the fraction.
2. Divide out nonzero factors that are common to the numerator and the denominator according to the fundamental principle.

Example 1 illustrates this procedure.

EXAMPLE 1   Express in simplest form.

a. $\dfrac{x^2 - 3x - 4}{x^2 - 2x - 8}$

b. $\dfrac{(x^2 + 2) - (a^2 + 2)}{x - a}$

Solution   Factor and use the fundamental principle.

a. $\dfrac{x^2 - 3x - 4}{x^2 - 2x - 8} = \dfrac{(x + 1)(x - 4)}{(x + 2)(x - 4)} = \dfrac{x + 1}{x + 2}, \quad x \neq 4$

b. $\dfrac{(x^2 + 2) - (a^2 + 2)}{x - a} = \dfrac{x^2 - a^2}{x - a} = \dfrac{(x + a)(x - a)}{x - a} = x + a, \ x \neq a$

□

In Example 1 the restrictions $x \neq 4$ and $x \neq a$ are necessary because we may divide out only nonzero factors since division by zero is undefined. With the understanding that such restrictions always apply, we will not continue to list them from this point on.

When working with fractions, it is important to keep in mind what happens in addition and subtraction if we change the order of the terms. In addition $a + b = b + a$. However, in subtraction $a - b = -(b - a)$. Thus,

$$\frac{a + b}{b + a} = 1 \text{ while } \frac{a - b}{b - a} = \frac{-(b - a)}{b - a} = -1.$$

Although $a - b$ does not equal $b - a$, the expressions differ only by a factor of -1 and can be simplified accordingly.

EXAMPLE 2   Express in simplest form $\dfrac{x^2 - y^2}{y - x}$.

Solution   Since $y - x = -(x - y)$ we proceed as follows:

$$\frac{x^2 - y^2}{y - x} = \frac{(x + y)(x - y)}{(-1)(x - y)}$$

$$= \frac{x + y}{-1} \text{ or } -x - y.$$

□

Multiplication

In arithmetic we know that the product of two or more fractions is the product of their numerators divided by the product of their denominators. In symbols, this principle is

$$\frac{a}{b} \cdot \frac{c}{d} = \frac{ac}{bd} \qquad b, d \neq 0.$$

Similarly, we will use this procedure for multiplying algebraic fractions. To express products in simplest form, factoring and the fundamental principle are usually required.

EXAMPLE 3   Simplify $\dfrac{5x^2 + 15x}{4 - x^2} \cdot \dfrac{12 - 6x}{(3 + x)^2}$.

Solution   We factor, multiply, and then simplify as shown.

$$\frac{5x^2 + 15x}{4 - x^2} \cdot \frac{12 - 6x}{(3 + x)^2} = \frac{5x(x + 3)}{(2 + x)(2 - x)} \cdot \frac{6(2 - x)}{(3 + x)^2}$$

$$= \frac{5 \cdot 6x(x + 3)(2 - x)}{(2 + x)(2 - x)(3 + x)(3 + x)}$$

$$= \frac{30x}{(2 + x)(3 + x)} \qquad \square$$

Division

To divide two fractions, we must recall that division is defined in terms of multiplication. That is, to divide a by b, we multiply a by the reciprocal of b.

$$\frac{a}{b} \div \frac{c}{d} = \frac{a}{b} \cdot \frac{d}{c} \qquad b, d, c/d \neq 0$$

Note the reciprocal of a fraction can be found by inverting the fraction. For example, the reciprocal of $\frac{2}{7}$ is $\frac{7}{2}$. Thus, to divide two fractions, we invert the fraction by which we are dividing to find its reciprocal and then we multiply.

EXAMPLE 4   Simplify $\dfrac{n + 1}{2^n + 1} \div \dfrac{n}{2^n}$.

Solution   We convert to multiplication and simplify.

$$\frac{n + 1}{2^{n + 1}} \div \frac{n}{2^n} = \frac{n + 1}{2^{n + 1}} \cdot \frac{2^n}{n} = \frac{1}{2^{(n + 1) - n}} \cdot \frac{n + 1}{n} = \frac{n + 1}{2n} \qquad \square$$

Addition and Subtraction

In arithmetic we know that the sum (or difference) of two or more fractions that have the same denominator is given by the sum (or difference) of the numerators divided by the common denominator.

$$\frac{a}{b} \pm \frac{c}{b} = \frac{a \pm c}{b} \qquad b \neq 0.$$

We add or subtract algebraic fractions in a similar way as shown in Example 5.

EXAMPLE 5   Simplify $\dfrac{2x + 1}{x - 3} - \dfrac{x - 2}{x - 3}$.

Solution   The fractions have the same denominator, so

$$\frac{2x + 1}{x - 3} - \frac{x - 2}{x - 3} = \frac{(2x + 1) - (x - 2)}{x - 3}$$

$$= \frac{2x + 1 - x + 2}{x - 3}$$

$$= \frac{x + 3}{x - 3}.$$

Note: Be careful in this and other subtraction problems to remove correctly the parentheses in an expression like $-(x - 2)$. □

When fractions have different denominators, we can change them into equivalent fractions with the same denominator and then add them. In such cases computation is simpler if we use the smallest possible common denominator, called the **least common denominator** or **LCD.** One way of finding the LCD follows.

To Find the LCD	1. Factor *completely* each denominator.
	2. The LCD is the product of all the different factors with each factor raised to the highest power to which it appears in any one factorization.

The procedure is illustrated in Example 6.

EXAMPLE 6   Find the LCD of $\dfrac{2}{x^2 - 1}$ and $\dfrac{1}{x^2 + 2x + 1}$.

Solution   First, factor each denominator completely.

$$x^2 - 1 = (x + 1)(x - 1)$$
$$x^2 + 2x + 1 = (x + 1)(x + 1) = (x + 1)^2$$

The LCD will contain the factors $x + 1$ and $x - 1$. The highest power of $x + 1$ is 2, and the highest power of $x - 1$ is 1. Thus,

$$\text{LCD} = (x + 1)^2(x - 1). \qquad \square$$

We can now combine several principles of fractions and formulate a general procedure for adding or subtracting two or more fractions.

To Add or Subtract Fractions

1. Completely factor each denominator and find the LCD.
2. For each fraction, obtain an equivalent fraction by applying the fundamental principle and multiplying the numerator and the denominator of the fraction by the factors of the LCD that are not contained in the denominator of that fraction.
3. Add or subtract the numerators and divide this result by the common denominator.

EXAMPLE 7   Simplify $\dfrac{2a + 3}{a^2 + a} - \dfrac{a}{a^2 - 1}$.

Solution   The denominators $a^2 + a$ and $a^2 - 1$ factor as $a(a + 1)$ and $(a + 1)(a - 1)$, respectively. Since the highest power of each different factor is 1, the LCD is $a(a + 1)(a - 1)$. Then,

$$
\begin{aligned}
\frac{2a + 3}{a^2 + a} - \frac{a}{a^2 - 1} &= \frac{2a + 3}{a(a + 1)} - \frac{a}{(a + 1)(a - 1)} \\
&= \frac{(2a + 3)(a - 1)}{a(a + 1)(a - 1)} - \frac{a(a)}{(a + 1)(a - 1)(a)} \\
&= \frac{2a^2 + a - 3 - a^2}{a(a + 1)(a - 1)} \\
&= \frac{a^2 + a - 3}{a(a + 1)(a - 1)}. \qquad \square
\end{aligned}
$$

We often need to add or subtract two fractions for which the LCD is simply the product of the denominators. This special situation is best handled by the following formula.

$$\frac{a}{b} \pm \frac{c}{d} = \frac{ad \pm bc}{bd} \qquad b, d \neq 0$$

We can verify this property from our previous methods since

$$\frac{a}{b} \pm \frac{c}{d} = \frac{a \cdot d}{b \cdot d} \pm \frac{c \cdot b}{d \cdot b} = \frac{ad \pm bc}{bd}.$$

EXAMPLE 8   Simplify $\dfrac{x + h + 1}{x + h} - \dfrac{x + 1}{x}$.

Solution   We apply the above formula and then simplify.

$$\frac{x + h + 1}{x + h} - \frac{x + 1}{x} = \frac{(x + h + 1)x - (x + h)(x + 1)}{(x + h)x}$$

$$= \frac{x^2 + hx + x - x^2 - x - hx - h}{(x + h)x}$$

$$= \frac{-h}{(x + h)x} \qquad\qquad \square$$

Complex Fractions

A **complex fraction** is a fraction in which the numerator or the denominator, or both, involve fractions. A procedure for simplifying complex fractions follows.

To Simplify Complex Fractions

1. Find the LCD of all the fractions that appear in the numerator and the denominator of the complex fraction.
2. Multiply the numerator and the denominator of the complex fraction by this LCD and simplify your results.

EXAMPLE 9   Simplify.

a. $\dfrac{x^{-1} + y^{-1}}{(x/y) - (y/x)}$

b. $\left(\dfrac{1}{x + h} - \dfrac{1}{x}\right) \div h$

Solution

a. $x^{-1} = 1/x$, $y^{-1} = 1/y$, and the LCD is xy.

$$\frac{\dfrac{1}{x} + \dfrac{1}{y}}{\dfrac{x}{y} - \dfrac{y}{x}} = \frac{xy\left(\dfrac{1}{x} + \dfrac{1}{y}\right)}{xy\left(\dfrac{x}{y} - \dfrac{y}{x}\right)}$$

$$= \frac{y + x}{x^2 - y^2}$$

$$= \frac{y + x}{(x + y)(x - y)}$$

$$= \frac{1}{x - y}$$

b. The LCD is $x(x + h)$.

$$\frac{\dfrac{1}{x + h} - \dfrac{1}{x}}{h} = \frac{x(x + h)\left(\dfrac{1}{x + h} - \dfrac{1}{x}\right)}{x(x + h) \cdot h}$$

$$= \frac{x - (x + h)}{x(x + h)h}$$

$$= \frac{-h}{x(x + h)h}$$

$$= \frac{-1}{x(x + h)} \qquad \qquad \square$$

EXAMPLE 10   Solve the problem in the chapter introduction on page 2.

Solution   Let d represent the distance around the track. Since $d = rt$, the times required for the first and second laps are denoted by $d/180$ and $d/120$, respectively. Then,

$$\text{average speed} = \frac{\text{total distance}}{\text{total time}} = \frac{2d}{\dfrac{d}{180} + \dfrac{d}{120}}.$$

Multiplying each term in the numerator and the denominator by 360 gives

$$\frac{360(2d)}{360\left(\dfrac{d}{180} + \dfrac{d}{120}\right)} = \frac{720d}{2d + 3d} = \frac{720d}{5d} = 144.$$

Thus, the average speed is 144 mi/hour and the driver does not qualify for the finals. Note that the average speed is closer to 120 mi/hour because more *time* was needed for the second lap. An interesting follow up to this question is given in the introduction to Chapter 2.   □

EXERCISES 1.4

In Exercises 1–28 express each fraction in simplest form.

1. $\dfrac{4x - 4}{11x - 11}$ **2.** $\dfrac{y^2 + 2y}{y^2 + 3y}$ **3.** $\dfrac{4b}{12b^2 + 4b}$

4. $\dfrac{20z^2 - 5z}{10z}$ **5.** $\dfrac{y - x}{x - y}$ **6.** $\dfrac{(x - 4)^2}{4 - x}$

7. $\dfrac{x - a}{ax(a - x)}$ **8.** $\dfrac{-x + a}{x - a}$ **9.** $\dfrac{1 - x}{x^2 - 1}$

10. $\dfrac{i - I}{-(I^2 - i^2)}$ **11.** $\dfrac{a^2 - 3a - 4}{a^2 - 8a + 16}$

12. $\dfrac{(x^2 + 1) - (a^2 + 1)}{x - a}$ **13.** $\dfrac{49 - 14y + y^2}{49 - y^2}$

14. $\dfrac{2m^2 - 5m + 2}{4m - 2}$ **15.** $\dfrac{6x^3 - 3x^2 - 30x}{2x^2 - 4x - 16}$

16. $\dfrac{z^3 - z^2 - 6z}{z^3 + z^2 - 12z}$ **17.** $\dfrac{x + y - z}{y^2 - x^2 + xz - yz}$

18. $\dfrac{a^3 - 8}{a^3 - 2a^2 + 4a - 8}$ **19.** $\dfrac{2(2 + h) - 3(2 + h)}{2h(2 + h)}$

20. $\dfrac{(2 + h)^2 - 4}{(2 + h) - 2}$ **21.** $\dfrac{(x^2 - x) - (a^2 - a)}{x - a}$

22. $\dfrac{(x^2 + 2x) - (a^2 + 2a)}{x - a}$ **23.** $\dfrac{x^3 - a^3}{x - a}$

24. $\dfrac{x^4 - a^4}{x - a}$ **25.** $\dfrac{x^3 - (x - 1) \cdot 3x^2}{x^6}$

26. $\dfrac{x^4(-2) - (3 - 2x)4x^3}{x^8}$

27. $\dfrac{(x + 1)^2 - x \cdot 2(x + 1)}{(x + 1)^4}$

28. $\dfrac{(x^2 - 1)^2(-2) - (-2x)(x^2 - 1)4x}{(x^2 - 1)^4}$

In Exercises 29–56 express each product or quotient in simplest form.

29. $\dfrac{2x}{5y} \cdot \dfrac{5y^2}{8x^3}$ **30.** $\dfrac{14z}{10y} \cdot \dfrac{15y}{35z}$

31. $\dfrac{5a}{a^2 - 9} \cdot \dfrac{6a - 18}{15a^2}$ **32.** $\dfrac{n^2 - n - 12}{n^2} \cdot \dfrac{n}{n - 4}$

33. $\dfrac{(y - 1)^2}{(y + 2)^2} \cdot \dfrac{y^2 - 4}{y^2 - 1}$ **34.** $(x + 4)^2 \cdot \dfrac{x}{16 - x^2}$

35. $\dfrac{x^2 + 3x - 4}{x^2 - 3x - 4} \cdot \dfrac{x^2 - 5x + 4}{x^2 + 5x + 4}$

36. $\dfrac{2a^2 - 7a + 6}{a^2 - a - 6} \cdot \dfrac{5a - 15}{a^2 - 4}$

37. $\dfrac{xy}{z} \div \dfrac{x^2y}{z}$ **38.** $\dfrac{zy^2}{2x} \div \dfrac{yz^2}{x}$

39. $\dfrac{n^2 + n - 2}{n - 3} \div (n + 2)$ **40.** $\dfrac{4x^2 - 9}{4x - 9} \div (2x - 3)$

41. $\dfrac{x^4 - 1}{7x + 7} \div \dfrac{x^3 + x}{x^2 - 1}$

42. $\dfrac{3b^2 + 6b - 24}{b^2 - 7b + 10} \div \dfrac{3b^2 + 4b}{b^3 - 5b^2}$

43. $\dfrac{1 - a}{x^2 + x} \div \dfrac{a^2 - 1}{x^2 - 1}$

44. $\dfrac{x^2 - 7x + 10}{5x - 25} \div \dfrac{4 - 2x}{25 - x^2}$

45. $\dfrac{x^3 - y^3}{(x - y)^3} \div \dfrac{x^2 + xy + y^2}{x^2 - 2xy + y^2}$

46. $\dfrac{2x^2 - 2ax + 2a^2}{(ax)^3} \div \dfrac{a^3 + x^3}{ax^3}$

47. $\dfrac{n+1}{3^{n+1}} \div \dfrac{n}{3^n}$

48. $\left(\dfrac{n+2}{n+1} \cdot \dfrac{1}{4^n}\right) \div \left(\dfrac{n+1}{n} \cdot \dfrac{1}{4^{n-1}}\right)$

49. $\dfrac{x-y}{4a} \div \left(\dfrac{y-x}{a} \div \dfrac{x-y}{a^2}\right)$

50. $\left(\dfrac{x-y}{4a} \div \dfrac{y-x}{a}\right) \div \dfrac{x-y}{a^2}$

51. $(x-y)\left(2 - \dfrac{2x-y}{x-y}\right)$

52. $\left(\dfrac{x^2}{1-x^2} + \dfrac{2x}{1+x}\right) \div \dfrac{3}{x^2-1}$

53. $-(4-x^2)^{-1}(8-4x)$

54. $2(x-y)^{-3} \div 3(2y-2x)^{-2}$

55. $\dfrac{5x-5}{2}\left(\dfrac{x-1}{4}\right)^{-1}$

56. $\dfrac{3x+6}{x^2+4x} \div \left(\dfrac{x^2+3x-4}{x^2-4}\right)^{-1}$

In Exercises 57–96 combine each expression into a single fraction in simplest form.

57. $\dfrac{4a}{xy} + \dfrac{2a}{xy} - \dfrac{3a}{xy}$

58. $\dfrac{1}{a^2x^2} - \dfrac{5}{a^2x^2} + \dfrac{11}{a^2x^2}$

59. $\dfrac{3x+4}{x+2} - \dfrac{2x+5}{x+2}$

60. $\dfrac{4y+1}{2y-6} - \dfrac{2y+7}{2y-6}$

61. $\dfrac{2n-3}{n-6} - \dfrac{n+1}{6-n}$

62. $\dfrac{2x-5}{4-3x} + \dfrac{2-x}{3x-4}$

63. $\dfrac{4a}{5} + a$

64. $z + \dfrac{1}{z}$

65. $\dfrac{1}{3x} + \dfrac{1}{4x}$

66. $\dfrac{5}{8y} - \dfrac{2}{12y}$

67. $\dfrac{2}{k^3} - \dfrac{3}{k^2} + \dfrac{5}{k}$

68. $\dfrac{16}{a^2b} + \dfrac{1}{ab} - \dfrac{6}{ab^2}$

69. $\dfrac{2x+3}{5x} - \dfrac{2x-1}{10x} + \dfrac{4}{x}$

70. $\dfrac{7}{n^2} - \dfrac{5n-2}{n} + 6$

71. $\dfrac{2-y}{9y+6} + \dfrac{y-2}{6y+4}$

72. $\dfrac{2x+3}{2x^3-4x^2} - \dfrac{1}{x-2}$

73. $\dfrac{3a}{a-3} - \dfrac{3}{a}$

74. $\dfrac{d}{b+d} - \dfrac{b}{b-d}$

75. $\dfrac{3}{2x-3} - \dfrac{2}{9-4x^2}$

76. $3k+1 + \dfrac{5}{k+2}$

77. $\dfrac{n}{n-1} - \dfrac{1}{n} + 1$

78. $\dfrac{4x}{x^2+2x-3} - \dfrac{5x}{x^2+5x+6}$

79. $\dfrac{5}{y+1} - \dfrac{1}{y^2-1} + \dfrac{2}{y-1}$

80. $\dfrac{2a}{a^2-1} - \dfrac{a+1}{a-1} + 7$

81. $\dfrac{y}{y-x} - \dfrac{x}{x+y} + \dfrac{x^2+y^2}{x^2-y^2}$

82. $\dfrac{a}{a^2+ab} - \dfrac{b}{a^2-b^2} + \dfrac{a+b}{ab-b^2}$

83. $\dfrac{1}{x} - \dfrac{1}{a^x-1}$

84. $\dfrac{2}{x^a} + \dfrac{3}{y^b}$

85. $\dfrac{1}{x^n} + \dfrac{2}{x^{n+1}}$

86. $\dfrac{a}{x^{n-1}} + \dfrac{b}{x^n}$

87. $\left(\dfrac{1}{x} + \dfrac{1}{y}\right) \div \left(\dfrac{x}{y} - \dfrac{y}{x}\right)$

88. $\left(4 + \dfrac{1}{a-1}\right) \div \left(\dfrac{2}{a-1} + 3\right)$

89. $\dfrac{1}{h}\left(\dfrac{1}{x+h} - \dfrac{1}{x}\right)$

90. $\dfrac{1}{h}\left(\dfrac{2}{(x+h)^2} - \dfrac{2}{x^2}\right)$

91. $\left(\dfrac{a^x+a^{-x}}{2}\right)^2 - \left(\dfrac{a^x-a^{-x}}{2}\right)^2$

92. $1 - \left(\dfrac{a^x-a^{-x}}{a^x+a^{-x}}\right)$

93. $2 + 3y^{-1} - 4y^{-2}$

94. $(x-4)^{-2} - 5(x-4)^{-1}$

95. $4(3x+1)^{-1} - (4x-1)(3x+1)^{-2}$

96. $(2x^{-3} - 3x^{-2} + x^{-1})(x^{-2} - x^0)$

In Exercises 97–118 change the complex fraction to a fraction in simplest form.

97. $\dfrac{3 + \frac{1}{4}}{4 - \frac{2}{5}}$

98. $\dfrac{\frac{5}{18} + \frac{11}{12}}{\frac{7}{6} - \frac{2}{9}}$

99. $\dfrac{1 + (1/x)}{1 - (1/x)}$

100. $\dfrac{x + (1/a)}{(2/a) - x}$

101. $\dfrac{n+1 + (2/n)}{n-1 - (2/n)}$

102. $\dfrac{z+4 - (5/z)}{z+1 - (2/z)}$

103. $\left(\dfrac{y^2}{x^2} - 1\right) \div \left(\dfrac{y}{x} - 1\right)$

104. $\left(k - \dfrac{1}{x^2}\right) \div \left(2 - \dfrac{k}{x}\right)$

105. $\dfrac{3P - \dfrac{(3 + \pi)3P}{6 + \pi}}{6}$

106. $\dfrac{4P - \dfrac{2P}{2 + \pi} - \dfrac{4P}{2 + \pi}}{2}$

107. $\dfrac{1 + \dfrac{3}{x + 1}}{\dfrac{4}{x^2 - 1}}$

108. $\dfrac{\dfrac{b}{b - a} - \dfrac{a}{b + a}}{\dfrac{b^2 + a^2}{b^2 - a^2}}$

109. $\left[\dfrac{x + h}{x + h + 1} - \dfrac{x}{x + 1}\right] \div h$

110. $\left[\dfrac{3}{(x + h)^2} - \dfrac{3}{x^2}\right] \div h$

111. $\dfrac{\dfrac{(x + h)^2 + 1}{x + h} - \dfrac{x^2 + 1}{x}}{h}$

112. $\dfrac{\dfrac{x + h - 4}{x + h + 1} - \dfrac{x - 4}{x + 1}}{h}$

113. $\dfrac{x^{-1} + y^{-1}}{(xy)^{-1}}$

114. $\dfrac{xy^{-1} + x^{-1}y}{x^{-1}y - xy^{-1}}$

115. $(x^{-1} + y^{-1})^{-1}$

116. $1 + (x + x^{-1})^{-1}$

117. A person drives from her home to work at an average speed of 30 mi/hour. On the way home she takes the same route and averages 20 mi/hour. What is her average speed for the round trip?

118. We can generalize the idea in Exercise 117. If we let s_g represent the speed going and s_h represent the speed coming back, then the following formula can be used to find the average speed for the round trip.

$$s_{\text{avg}} = \dfrac{2}{\dfrac{1}{s_g} + \dfrac{1}{s_h}}$$

Change this complex fraction to a fraction in simplest form.

1.5 RATIONAL EXPONENTS

In Section 1.2 we extended our definition of exponents to include zero and negative integers by defining them so they satisfied the laws of exponents that we developed for positive integers. We will now follow the same procedure to develop a consistent definition for rational exponents.

If the laws of exponents are to hold for rational numbers, consider the result of the following product:

$$9^{1/2} \cdot 9^{1/2} = 9^{1/2 + 1/2} = 9^1.$$

Here we have two equal factors whose product is 9. Since the square root of a number is one of its two equal factors, $9^{1/2}$ must mean the square root of 9. That is,

$$9^{1/2} = \sqrt{9}.$$

Similarly,

$$8^{1/3} \cdot 8^{1/3} \cdot 8^{1/3} = 8^{1/3 + 1/3 + 1/3} = 8.$$

The factor $8^{1/3}$ is one of three equal factors whose product is 8. Since the cube root of a number is one of its three equal factors, $8^{1/3}$ must mean the cube root of 8.

$$8^{1/3} = \sqrt[3]{8}$$

In general, our previous laws of exponents may be extended by making the following definitions.

Definitions of $a^{1/n}$ and $\sqrt[n]{a}$ For any positive integer n,

$$a^{1/n} = \sqrt[n]{a} \begin{cases} \text{for } a \geq 0, \text{ if } n \text{ is even} \\ \text{for any real number } a, \text{ if } n \text{ is odd} \end{cases}$$

where

$$\sqrt[n]{a} = b \text{ if and only if } b^n = a \begin{cases} \text{for } a \geq 0, b \geq 0, \text{ if } n \text{ is even} \\ \text{for any real number } a \text{ if } n \text{ is odd.} \end{cases}$$

It is important to note the following ideas concerning the root of a number.

1. In the expression $\sqrt[n]{a}$, which is called a **radical**, we say $\sqrt{}$ is the **radical sign**, a is the **radicand**, and n is the **index** of the radical. The index is usually omitted from the square root radical.

2. There are two square roots for a positive number such as 4 since $(2)(2) = 4$ and $(-2)(-2) = 4$. To avoid ambiguity, we define the **principal square root** of a positive number to be its *positive* square root. Thus, the principal square root of 4 is 2 and not -2. The radical sign, $\sqrt{}$, is used to symbolize the principal square root, so $\sqrt{4} = 2$. To symbolize the negative square root of 4 (which is -2), we use $-\sqrt{4}$. In general, when n is even, $\sqrt[n]{a}$ means the nonnegative nth root of a.

3. No real number is the square root of a negative number such as -4 since the product of two equal real numbers is never negative. In general, when n is even, the nth root of a negative number does not exist in the set of real numbers.

4. The square root of any positive number that is not a perfect square is an irrational number. Consequently, if we wish to express the square root as a decimal, we can only *approximate* the number to some desired number of significant digits. In most cases we will leave these numbers in radical form.

EXAMPLE 1   Simplify.

a. $9^{1/2}$ **b.** $8^{1/3}$ **c.** $(-8)^{1/3}$ **d.** $-8^{1/3}$ **e.** $16^{-1/4}$

Solution

a. $9^{1/2} = \sqrt{9} = 3$ **b.** $8^{1/3} = \sqrt[3]{8} = 2$

c. $(-8)^{1/3} = \sqrt[3]{-8} = -2$ **d.** $-8^{1/3} = -\sqrt[3]{8} = -2$

e. $16^{-1/4} = \dfrac{1}{16^{1/4}} = \dfrac{1}{\sqrt[4]{16}} = \dfrac{1}{2}$ □

When the numerator in the rational exponent is not 1, we can evaluate the expression by considering the following law of exponents:

$(a^m)^n = a^{mn}.$

For example, consider

$9^{3/2}.$

Rewrite the expression in either of these forms and simplify.

$(9^{1/2})^3$ or $(9^3)^{1/2}$
$=(\sqrt{9})^3$ or $\sqrt{9^3}$
$= (3)^3$ or $\sqrt{729}$
$= 27$

Thus, $9^{3/2} = 27$. (*Note:* Although either form may be used, you should notice that if the root results in a rational value, it is easier to find the root first.) This example leads us to the following definition.

Rational Exponent   If m and n are integers with $n > 0$, and if m/n represents a reduced fraction such that $a^{1/n}$ represents a real number, then

$a^{m/n} = (a^{1/n})^m = (a^m)^{1/n}$

or equivalently

$a^{m/n} = (\sqrt[n]{a})^m = \sqrt[n]{a^m}.$

Notice that in this definition we avoid the situation where the base is negative and the exponent is a reduced fraction in which n is an even number. This is done because such a condition deals with numbers that are not real numbers and, in such a case, the laws of exponents do not necessarily hold.

EXAMPLE 2   Simplify.

a. $8^{2/3}$ **b.** $49^{3/2}$ **c.** $(-32)^{4/5}$ **d.** $125^{-2/3}$

Solution

a. $8^{2/3} = (8^{1/3})^2 = (\sqrt[3]{8})^2 = (2)^2 = 4$

b. $49^{3/2} = (49^{1/2})^3 = (\sqrt{49})^3 = (7)^3 = 343$

c. $(-32)^{4/5} = [(-32)^{1/5}]^4 = (\sqrt[5]{-32})^4 = (-2)^4 = 16$

d. $125^{-2/3} = \dfrac{1}{125^{2/3}} = \dfrac{1}{(125^{1/3})^2} = \dfrac{1}{(\sqrt[3]{125})^2} = \dfrac{1}{(5)^2} = \dfrac{1}{25}$ □

EXAMPLE 3   Use the laws of exponents to simplify.

a. $6^{7/5} \cdot 6^{8/5}$ **b.** $(a^8)^{-3/16}$ **c.** $\sqrt{2}/\sqrt[3]{2}$ **d.** $(x^{1/a})^{a/3}$

Solution

a. $6^{7/5} \cdot 6^{8/5} = 6^{7/5 + 8/5} = 6^{15/5} = 6^3 = 216$

b. $(a^8)^{-3/16} = a^{8(-3/16)} = a^{-3/2} = 1/a^{3/2}$ or $1/\sqrt{a^3}$

c. $\sqrt{2}/\sqrt[3]{2} = 2^{1/2}/2^{1/3} = 2^{1/2 - 1/3} = 2^{1/6}$ or $\sqrt[6]{2}$

d. $(x^{1/a})^{a/3} = x^{(1/a)(a/3)} = x^{1/3}$ or $\sqrt[3]{x}$ □

EXAMPLE 4   Simplify by performing the indicated operations and writing your results with only positive exponents. Also, express the result in radical form.

a. $\dfrac{2^{-1/2} \cdot 3x^{1/3}}{2 \cdot 3^{1/4}x}$ **b.** $\left(\dfrac{4x^{-1}y^{1/5}}{32x^5y^0}\right)^{1/3}$

Solution

a. $\dfrac{2^{-1/2} \cdot 3x^{1/3}}{2 \cdot 3^{1/4}x} = \dfrac{3^{1 - 1/4}}{2^{1 - (-1/2)}x^{1 - (1/3)}}$

$= \dfrac{3^{3/4}}{2^{3/2}x^{2/3}}$ or $\dfrac{\sqrt[4]{3^3}}{\sqrt{2^3}\sqrt[3]{x^2}}$

b. $\left(\dfrac{4x^{-1}y^{1/5}}{32x^5y^0}\right)^{1/3} = \left(\dfrac{y^{1/5}}{8x^{5 - (-1)} \cdot 1}\right)^{1/3}$

$= \left(\dfrac{y^{1/5}}{8x^6}\right)^{1/3}$

$= \dfrac{y^{(1/5)(1/3)}}{8^{1(1/3)}x^{6(1/3)}}$

$= \dfrac{y^{1/15}}{2x^2}$ or $\dfrac{\sqrt[15]{y}}{2x^2}$ □

EXAMPLE 5   Simplify by performing the indicated operations and writing your results with only positive exponents. Also, express the result in radical form.

a. $\dfrac{x^{1/2} \cdot 2 - (2x - 1) \cdot \frac{1}{2}x^{-1/2}}{x}$ **b.** $\frac{1}{2}x^{-1/2} - x^{-3/2}$

Solution

a. The main problem is how to take care of the expression $\frac{1}{2}x^{-1/2}$ or $1/(2x^{1/2})$. One way to simplify this expression is to multiply it by $2x^{1/2}$, since the resulting product is 1. Thus, in this case we will multiply both the numerator and the denominator of the given fraction by $2x^{1/2}$.

$$\left(\frac{2x^{1/2}}{2x^{1/2}}\right)\left(\frac{x^{1/2} \cdot 2 - (2x - 1) \cdot \frac{1}{2}x^{-1/2}}{x}\right) = \frac{4x - (2x - 1)(1)}{2x^{3/2}}$$

$$= \frac{2x + 1}{2x^{3/2}} \text{ or } \frac{2x + 1}{2\sqrt{x^3}}$$

b. $\frac{1}{2}x^{-1/2} - x^{-3/2} = \dfrac{1}{2x^{1/2}} - \dfrac{1}{x^{3/2}}$

$$= \frac{1}{2x^{1/2}}\left(\frac{x}{x}\right) - \frac{1}{x^{3/2}}\left(\frac{2}{2}\right)$$

$$= \frac{x}{2x^{3/2}} - \frac{2}{2x^{3/2}}$$

$$= \frac{x - 2}{2x^{3/2}} \text{ or } \frac{x - 2}{2\sqrt{x^3}} \qquad \qquad \square$$

Alternative Solution

Expressions with negative exponents are often simplified by factoring out the smallest power of the variable from each term. In this case $x^{-3/2}$ is the smallest power of x, so

$$\frac{1}{2}x^{-1/2} - x^{-3/2} = x^{-3/2}(\tfrac{1}{2}x - 1) \quad (\textit{Note: } x^{-3/2} \cdot \tfrac{1}{2}x^1 = \tfrac{1}{2}x^{-1/2}.)$$

$$= \frac{\frac{1}{2}x - 1}{x^{3/2}}$$

$$= \frac{x - 2}{2x^{3/2}} \quad \text{or} \quad \frac{x - 2}{2\sqrt{x^3}}. \qquad \qquad \square$$

You can calculate rational powers of numbers on a scientific calculator in several ways. The most straightforward general method is to use the power key $\boxed{y^x}$ with parentheses around the exponent. In this way $8^{2/3}$ is computed as

$$8\ \boxed{y^x}\ \boxed{(}\ \boxed{2}\ \boxed{\div}\ \boxed{3}\ \boxed{)}\ \boxed{=}\ 4.$$

The parentheses are important. Without them, the power takes precedence over division and we compute $8^2 \div 3$.

If the problem involves only the taking of a root, it is easier to use the $\boxed{\sqrt{}}$ key for a 1/2 exponent and $\boxed{INV}\ \boxed{y^x}$ or $\boxed{y^{1/x}}$ for the xth root of y. Once again, negative exponents may be computed through use of the reciprocal key $\boxed{1/x}$, and you may have to adjust the sign for problems involving the power key $\boxed{y^x}$ and a negative base.

EXERCISES 1.5

In Exercises 1–50 simplify the expression.

1. $25^{1/2}$
2. $81^{1/2}$
3. $4^{-1/2}$
4. $49^{-1/2}$
5. $1^{-1/3}$
6. $27^{-1/3}$
7. $8^{1/3}$
8. $64^{1/3}$
9. $(-27)^{1/3}$
10. $(-64)^{1/3}$
11. $16^{1/4}$
12. $81^{1/4}$
13. $(\frac{1}{4})^{1/2}$
14. $(\frac{1}{16})^{1/2}$
15. $(\frac{9}{25})^{-1/2}$
16. $(\frac{49}{4})^{-1/2}$
17. $(-8)^{-1/3}$
18. $(-32)^{-1/5}$
19. $(0.04)^{-1/2}$
20. $(0.027)^{-1/3}$
21. $-4^{1/2}$
22. $-16^{1/4}$
23. $9^{3/2}$
24. $4^{5/2}$
25. $8^{5/3}$
26. $16^{3/4}$
27. $(-8)^{5/3}$
28. $(-32)^{3/5}$
29. $(-27)^{2/3}$
30. $(-27)^{-2/3}$
31. $(-32)^{-4/5}$
32. $(-1)^{-7/3}$
33. $(\frac{4}{9})^{-3/2}$
34. $(\frac{27}{8})^{-2/3}$
35. $9^{-3/2} + 7^0$
36. $4(4)^0 + 2(4)^{-1/2}$
37. $8^{2/3} - 2^{-1} + 4^0$
38. $(\frac{1}{9})^{-2} + 2(3)^0 - 27^{2/3}$
39. $4(16)^{-1/2} + (\frac{1}{64})^{-2/3} - 16^0$
40. $(3^2 + 4^2)^{-1/2}$
41. $2^{1/2} \cdot 2^{3/2}$
42. $5^{1/2} \cdot 5^{1/2}$
43. $(-3)^{5/3} \cdot (-3)^{4/3}$
44. $(-4)^{7/5} \cdot (-4)^{8/5}$
45. $(4^2)^{1/2}$
46. $[(-4)^{10}]^{2/5}$
47. $[(-8)^6]^{-1/3}$
48. $(9^6)^{-2/3}$
49. $\dfrac{5^{1/2}}{5^{3/2}}$
50. $\dfrac{6^{1/5} \cdot 6^{3/5}}{6^{-1/5}}$

In Exercises 51–100 simplify each expression by performing the indicated operations and writing your result with only positive exponents. Also, show your result in radical form. (*Note:* When the denominator in a fractional exponent is even, assume that the base is a positive number.)

51. $x^{4/3} \cdot x^{-1}$
52. $x^{1/2} \cdot x^{2/3}$
53. $\dfrac{x^{-1/3}}{x^{5/6}}$
54. $\dfrac{x}{x^{1/2}}$
55. $(8x^3y^6)^{1/3}$
56. $(3^{1/2}x^{3/4})^{-2}$
57. $(4a^{-2}b^6)^{-1/2}$
58. $(2x^{1/3}y^{5/6})^6$
59. $\dfrac{x^{1/2}yz^{1/3}}{x^0y^{-1}z}$
60. $\dfrac{2x^{4/5}y^{-2/3}}{6^0x^{-1}y}$
61. $(x + y)^{5/3}(x + y)^{-2/3}$
62. $(a - b)^{1/2} \div (a - b)^{3/4}$
63. $(x^{a/2}y^{a/5})^{1/a}$
64. $(a^{3/n})^{n/6}(a^{n/2})^{4/n}$
65. $k^{1/x}k^{-y/x}$
66. $(x^{1/q})^{1-q}$
67. $(x^{1/q})^{p-1}x^{(1/q)-1}$
68. $(a^{x^2 + 2x} \cdot a)^{1/(x + 1)}$
69. $3u^{1/2}(2u^{1/2} + u^4)$
70. $-a^{2/3}(a^3 - a^{1/2})$
71. $\dfrac{(4x^2)^{1/2} - x^{1/2}}{(16x)^{1/2}}$
72. $\left(\dfrac{b^{1/3}}{b^{4/3} + b^{7/3}}\right)^{-1}$
73. $\left(\dfrac{x^{2/3}y^{-3/2}}{x^{-1/3}y}\right)^{-1}$
74. $\left(\dfrac{27x^{-3}}{64y^{-3}}\right)^{-1/3}$
75. $\left(\dfrac{49x^{-4}y^6}{25x^{-2}y^{-10}}\right)^{-3/2}$
76. $\left(\dfrac{x^{1/4}y^{-2/3}z^0}{x^{3/4}y^{-2/3}z^{1/2}}\right)^{-3/2}$
77. $\dfrac{3x^{1/2}y^{-2/3}}{5y^{-1}} \cdot \dfrac{20y^{1/4}}{27x^{-2}}$

78. $\dfrac{7^{-1}x^{1/4}}{3^{-2}y^{-1}} \div \dfrac{2(3x^{-1/2})^2}{(49y)^0}$

79. $\sqrt[3]{2} \cdot \sqrt[4]{2}$

80. $\sqrt[3]{3} \cdot \sqrt{3}$

81. $\dfrac{16}{\sqrt[3]{16}}$

82. $\dfrac{\sqrt[4]{2}}{\sqrt[8]{16}}$

83. $\sqrt{3\sqrt{3}}$

84. $\sqrt[4]{\sqrt[3]{x}}$

85. $\sqrt[m]{\sqrt[n]{a}}$

86. $\sqrt[m]{x} \cdot \sqrt[n]{x}$

87. $\sqrt[3]{2} \cdot \sqrt{3}$

88. $\sqrt{2} \div \sqrt[4]{5}$

89. $(x^{1/2} - x^{-1/2})^2$

90. $(x^{2/3} + x^{-2/3})^2$

91. $(x^{1/2} + y^{1/2})^2$

92. $(a^{1/2} - b^{3/2})^2$

93. $x^{-1/2} - \frac{1}{2}x^{-3/2}$

94. $\frac{2}{3}x^{-1/3} + \frac{1}{3}x^{-4/3}$

95. $\dfrac{x^{1/3} + x^{-2/3}}{x^{1/3} - x^{-2/3}}$

96. $\dfrac{x^{1/2}}{2x^{-3/2} + 3x^{-1/2}}$

97. $(x + 2)(-\frac{1}{2}x^{-3/2}) + x^{-1/2} \cdot 1$

98. $x^{1/2} \cdot [-(x + 2)^{-2}] + (x + 2)^{-1} \cdot \frac{1}{2}x^{-1/2}$

99. $\dfrac{x^{1/2} \cdot 1 - (x + 2) \cdot \frac{1}{2}x^{-1/2}}{x}$

100. $\dfrac{1 + \frac{1}{2}(1 + x^2)^{-1/2}(2x)}{x + (1 + x^2)^{1/2}}$

In Exercises 101–108 simplify each expression by factoring out the smallest power of the variable from each term and writing your result with only positive exponents.

101. $x^{-5} + x^{-2}$

102. $5x^{-2} - 10x^{-1}$

103. $3x^{1/3} - 15x^{-2/3}$

104. $4x^{-1/3} + 12x^{-4/3}$

105. $2x + \frac{1}{2}x^{-1/2}$

106. $x^2 - \frac{1}{3}x^{-2/3}$

107. $x^2(x - 5)^{1/2} - x(x - 5)^{-1/2}$

108. $-x^4(1 - x^2)^{-1/2} + 3x^2(1 - x^2)^{1/2}$

1.6 OPERATIONS WITH RADICALS

Although it is possible to use fractional exponents for any operation required with radicals, computations are often performed by use of radical symbolism. For this reason we now state three essential properties of radicals.

Properties of Radicals   (a, b, $\sqrt[n]{a}$, and $\sqrt[n]{b}$ denote real numbers.)

1. $(\sqrt[n]{a})^n = a$ (by definition of $\sqrt[n]{a}$)
2. $\sqrt[n]{a} \cdot \sqrt[n]{b} = a^{1/n} \cdot b^{1/n} = (a \cdot b)^{1/n} = \sqrt[n]{a \cdot b}$
3. $\dfrac{\sqrt[n]{a}}{\sqrt[n]{b}} = \dfrac{a^{1/n}}{b^{1/n}} = \left(\dfrac{a}{b}\right)^{1/n} = \sqrt[n]{\dfrac{a}{b}}$ $(b \neq 0)$

With the aid of these properties we can now use either radicals or fractional exponents, whichever is more convenient.

Simplifying Radicals

When working with radicals, we must be able to change the radical to simplest form. The first procedure for simplifying a radical is to remove any factor of the radicand whose indicated root can be taken exactly.

EXAMPLE 1   Simplify $3\sqrt{98}$.

Solution   Rewrite 98 as the product of a perfect square and another factor and simplify.

$$3\sqrt{98} = 3\sqrt{49 \cdot 2} = 3\sqrt{49}\sqrt{2} = 3 \cdot 7\sqrt{2} = 21\sqrt{2}$$ □

EXAMPLE 2   Simplify $\sqrt{8x^9y^6}$ (assume that x and y denote positive numbers).

Solution   Rewrite $8x^9y^6$ as the product of a perfect square and another factor. Remember that any expression with an even exponent is a perfect square.

$$\sqrt{8x^9y^6} = \sqrt{(4x^8y^6)(2x)} = \sqrt{4x^8y^6}\sqrt{2x} = 2x^4y^3\sqrt{2x}$$ □

The second consideration for simplifying a radical is to eliminate any fractions in the radicand. This procedure is called rationalizing the denominator. The rationale for this name is given later in this section.

EXAMPLE 3   Simplify $\sqrt{\frac{3}{2}}$.

Solution   Rewrite $\frac{3}{2}$ as an equivalent fraction whose denominator is a perfect square and simplify.

$$\sqrt{\frac{3}{2}} = \sqrt{\frac{3 \cdot 2}{2 \cdot 2}} = \sqrt{\frac{6}{4}} = \frac{\sqrt{6}}{\sqrt{4}} = \frac{\sqrt{6}}{2} \text{ or } \frac{1}{2}\sqrt{6}$$ □

EXAMPLE 4   Simplify $\sqrt[3]{\frac{7}{9}}$.

Solution   Rewrite $\frac{7}{9}$ as an equivalent fraction whose denominator is a perfect cube and simplify.

$$\sqrt[3]{\frac{7}{9}} = \sqrt[3]{\frac{7 \cdot 3}{9 \cdot 3}} = \sqrt[3]{\frac{21}{27}} = \frac{\sqrt[3]{21}}{\sqrt[3]{27}} = \frac{\sqrt[3]{21}}{3}$$ □

EXAMPLE 5   Simplify $\sqrt{2x/y}$ (assume that x and y denote positive numbers).

Solution   Rewrite $2x/y$ as an equivalent fraction whose denominator is a perfect square and simplify.

$$\sqrt{\frac{2x}{y}} = \sqrt{\frac{2x \cdot y}{y \cdot y}} = \sqrt{\frac{2xy}{y^2}} = \frac{\sqrt{2xy}}{\sqrt{y^2}} = \frac{\sqrt{2xy}}{y}$$ □

In Example 5 we assumed that y was a positive number, which meant that $\sqrt{y^2}$ simplified to y. Without this assumption it is necessary to use $\sqrt{y^2} = |y|$. For example, $\sqrt{(-2)^2} \neq -2$; instead $\sqrt{(-2)^2} = |-2| = 2$. Keep in mind that while $(\sqrt[n]{a})^n = a$ for any positive integer n, $\sqrt[n]{a^n} = |a|$ when n is even and $\sqrt[n]{a^n} = a$ when n is odd.

Addition and Subtraction of Radicals

We can simplify an algebraic expression such as $7x + 2x - x$ by combining similar terms. That is,

$$7x + 2x - x = (7 + 2 - 1)x = 8x.$$

Similarly, if x is replaced by $\sqrt{2}$, we have

$$7\sqrt{2} + 2\sqrt{2} - \sqrt{2} = (7 + 2 - 1)\sqrt{2} = 8\sqrt{2},$$

or if x is replaced by $\sqrt[3]{5x}$, we have

$$7\sqrt[3]{5x} + 2\sqrt[3]{5x} - \sqrt[3]{5x} = (7 + 2 - 1)\sqrt[3]{5x} = 8\sqrt[3]{5x}.$$

Thus, we can combine similar radicals by adding or subtracting their coefficients. By **similar radicals** we mean radicals that have the same radicand and the same index. In general, to combine radicals, we first express each radical in simplest form and then we combine those that are similar.

EXAMPLE 6   Combine $2\sqrt{32} - 4\sqrt{\frac{1}{2}}$.

Solution

$$2\sqrt{32} = 2\sqrt{16 \cdot 2} = 2\sqrt{16}\sqrt{2} = 2 \cdot 4\sqrt{2} = 8\sqrt{2}$$

$$4\sqrt{\frac{1}{2}} = 4\sqrt{\frac{1 \cdot 2}{2 \cdot 2}} = 4\sqrt{\frac{2}{4}} = \frac{4\sqrt{2}}{\sqrt{4}} = \frac{4\sqrt{2}}{2} = 2\sqrt{2}$$

Thus, $2\sqrt{32} - 4\sqrt{\frac{1}{2}} = 8\sqrt{2} - 2\sqrt{2} = (8 - 2)\sqrt{2} = 6\sqrt{2}$.   □

EXAMPLE 7   Combine $\sqrt[3]{-16} + \sqrt[3]{54}$.

Solution

$$\sqrt[3]{-16} = \sqrt[3]{-8 \cdot 2} = \sqrt[3]{-8}\sqrt[3]{2} = -2\sqrt[3]{2}$$

$$\sqrt[3]{54} = \sqrt[3]{27 \cdot 2} = \sqrt[3]{27}\sqrt[3]{2} = 3\sqrt[3]{2}$$

Thus, $\sqrt[3]{-16} + \sqrt[3]{54} = -2\sqrt[3]{2} + 3\sqrt[3]{2} = (-2 + 3)\sqrt[3]{2} = 1\sqrt[3]{2}$.   □

EXAMPLE 8 Combine $\sqrt{50x^3y} - \sqrt{8xy^3}$ (assume that x and y denote positive numbers).

Solution

$$\sqrt{50x^3y} = \sqrt{(25x^2)(2xy)} = \sqrt{25x^2}\sqrt{2xy} = 5x\sqrt{2xy}$$
$$\sqrt{8xy^3} = \sqrt{(4y^2)(2xy)} = \sqrt{4y^2}\sqrt{2xy} = 2y\sqrt{2xy}$$

Thus, $\sqrt{50x^3y} - \sqrt{8xy^3} = 5x\sqrt{2xy} - 2y\sqrt{2xy} = (5x - 2y)\sqrt{2xy}$. □

Multiplication of Radicals

When multiplying expressions containing radicals, we use the second property of radicals that we developed in this section. That is,

$$\sqrt[n]{a} \cdot \sqrt[n]{b} = \sqrt[n]{ab}.$$

This property, together with the normal rules of algebraic multiplication, enables us to multiply radicals. It should be noted that both radicals must have the same index if the second property of radicals is to be used.

EXAMPLE 9 Multiply and express the result in simplest form.

a. $\sqrt{7} \cdot \sqrt{2}$ **b.** $\sqrt[3]{6x} \cdot \sqrt[3]{3x}$ **c.** $4\sqrt[3]{2} \cdot 3\sqrt[3]{4}$

Solution

a. $\sqrt{7} \cdot \sqrt{2} = \sqrt{7 \cdot 2} = \sqrt{14}$

b. $\sqrt[3]{6x} \cdot \sqrt[3]{3x} = \sqrt[3]{6x \cdot 3x} = \sqrt[3]{18x^2}$

c. $4\sqrt[3]{2} \cdot 3\sqrt[3]{4} = 4 \cdot 3\sqrt[3]{2 \cdot 4} = 12\sqrt[3]{8} = 12 \cdot 2 = 24$

EXAMPLE 10 Multiply and express the result in simplest form:
$(3\sqrt{2} + 2\sqrt{3})(3\sqrt{2} - 2\sqrt{3})$.

Solution

$$
\begin{array}{r}
3\sqrt{2} + 2\sqrt{3} \\
3\sqrt{2} - 2\sqrt{3} \\
\hline
9 \cdot 2 + 6\sqrt{6} \\
-6\sqrt{6} - 4 \cdot 3 \\
\hline
18 \quad + 0 \quad - \quad 12
\end{array}
$$

Thus, $(3\sqrt{2} + 2\sqrt{3})(3\sqrt{2} - 2\sqrt{3}) = 6$. □

Division of Radicals

When dividing expressions containing radicals, we use the third property of radicals that we developed in this section. That is,

$$\frac{\sqrt[n]{a}}{\sqrt[n]{b}} = \sqrt[n]{\frac{a}{b}} \qquad (b \neq 0).$$

It should be noted that both radicals must have the same index if the third property of radicals is to be used.

EXAMPLE 11   Divide and express the result in simplest form.

a. $\dfrac{\sqrt{7}}{\sqrt{2}}$ b. $\dfrac{50\sqrt{32}}{5\sqrt{2}}$ c. $\dfrac{\sqrt[3]{48}}{\sqrt[3]{2}}$

Solution

a. $\dfrac{\sqrt{7}}{\sqrt{2}} = \sqrt{\dfrac{7}{2}} = \sqrt{\dfrac{7 \cdot 2}{2 \cdot 2}} = \dfrac{\sqrt{14}}{\sqrt{4}} = \dfrac{\sqrt{14}}{2}$

b. $\dfrac{50\sqrt{32}}{5\sqrt{2}} = \dfrac{50}{5}\sqrt{\dfrac{32}{2}} = 10\sqrt{16} = 10 \cdot 4 = 40$

c. $\dfrac{\sqrt[3]{48}}{\sqrt[3]{2}} = \sqrt[3]{\dfrac{48}{2}} = \sqrt[3]{24} = \sqrt[3]{8} \cdot \sqrt[3]{3} = 2\sqrt[3]{3}$ □

We have discussed how to simplify a radical so the radicand contains no fractions. A similar procedure may be used to ensure that no radicals appear in a denominator. For example, $1/\sqrt{2}$ can be converted to $\sqrt{2}/2$ as follows.

$$\frac{1}{\sqrt{2}} = \frac{1}{\sqrt{2}} \cdot \frac{\sqrt{2}}{\sqrt{2}} = \frac{\sqrt{2}}{2}$$

One instance when the latter form is more useful occurs with the addition of radicals as shown below.

$$3\sqrt{2} + \frac{1}{\sqrt{2}} = 3\sqrt{2} + \frac{\sqrt{2}}{2} = \frac{7}{2}\sqrt{2}$$

Thus, we adopt the condition that a simplified radical cannot have radicals in the denominator. When we change the denominator from a radical (irrational number) to a rational number, the procedure is called rationalizing the denominator.

EXAMPLE 12   Rationalize the denominator.

a. $\dfrac{2}{\sqrt{5}}$ **b.** $\dfrac{7}{4\sqrt{3}}$ **c.** $\dfrac{9}{2\sqrt{18}}$ **d.** $\dfrac{9}{\sqrt{3x}}$

Solution

a. $\dfrac{2}{\sqrt{5}} = \dfrac{2\sqrt{5}}{\sqrt{5}\sqrt{5}} = \dfrac{2\sqrt{5}}{5}$

b. $\dfrac{7}{4\sqrt{3}} = \dfrac{7\sqrt{3}}{4\sqrt{3}\sqrt{3}} = \dfrac{7\sqrt{3}}{4\cdot 3} = \dfrac{7\sqrt{3}}{12}$

c. $\dfrac{9}{2\sqrt{18}} = \dfrac{9\sqrt{2}}{2\sqrt{18}\sqrt{2}} = \dfrac{9\sqrt{2}}{2\sqrt{36}} = \dfrac{9\sqrt{2}}{2\cdot 6} = \dfrac{3\sqrt{2}}{4}$

d. $\dfrac{9}{\sqrt{3x}} = \dfrac{9\sqrt{3x}}{\sqrt{3x}\sqrt{3x}} = \dfrac{9\sqrt{3x}}{3x} = \dfrac{3\sqrt{3x}}{x}$ □

If the denominator contains square roots and is the sum of two terms, we rationalize the denominator by multiplying the numerator and the denominator of the fraction by the difference of the same two terms, and vice versa. This procedure is effective since if a and b are nonnegative rational numbers, then

$$(\sqrt{a} + \sqrt{b})(\sqrt{a} - \sqrt{b}) = a - \sqrt{ab} + \sqrt{ab} - b = a - b,$$

which is a rational number. The expressions $\sqrt{a} + \sqrt{b}$ and $\sqrt{a} - \sqrt{b}$ are called **conjugates** of each other.

EXAMPLE 13   Rationalize the denominator: $\dfrac{2 + \sqrt{3}}{4\sqrt{3} - \sqrt{2}}$.

Solution   The conjugate of $4\sqrt{3} - \sqrt{2}$ is $4\sqrt{3} + \sqrt{2}$. Then,

$$\frac{2 + \sqrt{3}}{4\sqrt{3} - \sqrt{2}} = \frac{(2 + \sqrt{3})(4\sqrt{3} + \sqrt{2})}{(4\sqrt{3} - \sqrt{2})(4\sqrt{3} + \sqrt{2})}$$

$$= \frac{8\sqrt{3} + 2\sqrt{2} + 4\cdot 3 + \sqrt{6}}{16\cdot 3 + 4\sqrt{6} + (-4\sqrt{6}) + (-2)}$$

$$= \frac{8\sqrt{3} + 2\sqrt{2} + 12 + \sqrt{6}}{46}.$$ □

In some cases, instead of rationalizing the denominator, it is useful to **rationalize the numerator**. The idea is the same and the following example illustrates the procedure.

EXAMPLE 14   Rationalize the numerator: $\dfrac{\sqrt{x} - \sqrt{a}}{x - a}$.

Solution   The conjugate of $\sqrt{x} - \sqrt{a}$ is $\sqrt{x} + \sqrt{a}$. Then,

$$\frac{\sqrt{x} - \sqrt{a}}{x - a} = \frac{(\sqrt{x} - \sqrt{a})(\sqrt{x} + \sqrt{a})}{(x - a)(\sqrt{x} + \sqrt{a})}$$

$$= \frac{x - a}{(x - a)(\sqrt{x} + \sqrt{a})} = \frac{1}{\sqrt{x} + \sqrt{a}}. \qquad \square$$

Now that we have shown how to perform the basic operations with radicals, we consider a problem that requires several of the ideas discussed. It is essential that you understand this type of problem if you intend to continue to higher mathematics.

EXAMPLE 15   Write as a single fraction in simplest form

$$x \cdot \frac{2x}{2\sqrt{x^2 + 5}} + \sqrt{x^2 + 5}.$$

Solution

$$x \cdot \frac{2x}{2\sqrt{x^2 + 5}} + \sqrt{x^2 + 5} = \frac{x^2}{\sqrt{x^2 + 5}} + \sqrt{x^2 + 5}\left(\frac{\sqrt{x^2 + 5}}{\sqrt{x^2 + 5}}\right)$$

$$= \frac{x^2 + (x^2 + 5)}{\sqrt{x^2 + 5}}$$

$$= \frac{2x^2 + 5}{\sqrt{x^2 + 5}} \text{ or } \frac{(2x^2 + 5)\sqrt{x^2 + 5}}{x^2 + 5} \qquad \square$$

EXERCISES 1.6

In Exercises 1–20 express each radical in simplest form (assume that x and y represent positive real numbers).

1. $\sqrt{108}$ **2.** $-\sqrt{80}$ **3.** $5\sqrt{8}$

4. $\dfrac{1}{3}\sqrt{50}$ **5.** $\sqrt[3]{-32}$ **6.** $\sqrt[5]{64}$

7. $\sqrt{12x^7y^4}$ **8.** $\sqrt{54x^{11}y^5}$ **9.** $\sqrt[4]{9x^6y^7}$

10. $\sqrt[5]{16x^3y^5}$ **11.** $9\sqrt{\frac{4}{3}}$ **12.** $-12\sqrt{\frac{20}{27}}$

13. $\sqrt[4]{\frac{5}{8}}$ **14.** $\sqrt[5]{\frac{2}{9}}$ **15.** $\sqrt{9/x}$

16. $\sqrt{5/(3x^2y)}$ **17.** $\sqrt[3]{x/y}$ **18.** $\sqrt[4]{2x/y^5}$

19. $\sqrt{x^2 - (2x^2/4)}$ **20.** $\sqrt{(1/x^2) + (1/y^2)}$

In Exercises 21–50 express each sum or difference in simplest form (assume that x and y represent positive real numbers).

21. $4\sqrt{2} + 7\sqrt{2}$ **22.** $-2\sqrt{7} - \sqrt{7}$

23. $\sqrt{27} + \sqrt{12}$ **24.** $\sqrt{8} - \sqrt{32}$

25. $3\sqrt{72} - 5\sqrt{50}$ **26.** $3\sqrt{28} - 10\sqrt{63}$

27. $-2\sqrt{176} - 3\sqrt{44}$ **28.** $3\sqrt{48} - \sqrt{20}$

29. $4\sqrt{12} - \sqrt{27} + 2\sqrt{48}$

30. $-4\sqrt{45} + 2\sqrt{125} - 6\sqrt{20}$

31. $2\sqrt{50} - 6\sqrt{\frac{1}{2}}$ **32.** $4\sqrt{\frac{1}{5}} + 2\sqrt{125}$

33. $6\sqrt{\frac{1}{6}} - \sqrt{\frac{2}{3}}$ **34.** $5\sqrt{\frac{1}{3}} - 3\sqrt{\frac{1}{12}}$

35. $3\sqrt{\frac{3}{2}} - \sqrt{\frac{2}{3}} + 5\sqrt{6}$

36. $\frac{1}{3}\sqrt{\frac{5}{3}} - \frac{1}{2}\sqrt{60} + \frac{1}{6}\sqrt{15}$

37. $\sqrt[3]{24} + 7\sqrt[3]{3}$ **38.** $2\sqrt[3]{-54} - \sqrt[3]{128}$

39. $\sqrt[3]{\frac{1}{9}} + 2\sqrt[3]{-375}$ **40.** $3\sqrt[4]{32} - 5\sqrt[4]{2}$

41. $\sqrt{8x} + \sqrt{72x}$ **42.** $9\sqrt{3y} - 4\sqrt{12y}$

43. $\sqrt{20xy^2} - 2\sqrt{45x^3}$ **44.** $\sqrt{x^3y/2} + \frac{1}{2}\sqrt{32xy^7}$

45. $\sqrt{50x^3y} + x\sqrt{xy/2} - 2\sqrt{9x^3y/2}$

46. $2\sqrt{x/y} - \sqrt{y/x} + 3\sqrt{1/(xy)}$

47. $2\sqrt{2x/y} - 4\sqrt{y/(2x^3)} + 5\sqrt{\frac{1}{8}x^3y}$

48. $\sqrt[3]{54x^4y} - \sqrt[3]{xy^4/4}$ **49.** $\sqrt[4]{x/y} - \sqrt[4]{xy^3}$

50. $\sqrt{4 + 4x} + \sqrt{16 + 16x}$

In Exercises 51–70 express each product or quotient in simplest form (assume that x and y represent positive real numbers).

51. $\sqrt{6} \cdot \sqrt{2}$ **52.** $(-3\sqrt{5})^2$

53. $\sqrt{24} \div \sqrt{3}$ **54.** $\sqrt{14} \div \sqrt{5}$

55. $\sqrt[3]{-4} \cdot \sqrt[3]{16}$ **56.** $\sqrt[4]{2} \div 2\sqrt[4]{2}$

57. $(\sqrt{15} + \sqrt{60}) \div \sqrt{3}$ **58.** $-3\sqrt{3}(4 - 2\sqrt{27})$

59. $(1 + \sqrt{2})(1 - \sqrt{2})$ **60.** $(1 + \sqrt{2})^2$

61. $(\sqrt{5} - 2\sqrt{6})^2$

62. $(5\sqrt{7} - 3\sqrt{8})(2\sqrt{7} + 4\sqrt{2})$

63. $\sqrt{15x} \cdot \sqrt{6x}$ **64.** $\sqrt{5x^3y} \cdot \sqrt{10y}$

65. $\sqrt[3]{4x^2y} \cdot \sqrt[3]{6x^2y^3}$ **66.** $4\sqrt[4]{27x^3} \cdot 5\sqrt[4]{3x^5}$

67. $\sqrt{18} \div \sqrt{3x}$ **68.** $-2\sqrt{54x^3y} \div 4\sqrt{3xy}$

69. $\sqrt[4]{7x^2} \div \sqrt[4]{2x}$ **70.** $\sqrt[3]{-3y} \div \sqrt[3]{48y}$

In Exercises 71–80 rationalize the denominator.

71. $\dfrac{2}{\sqrt{3}}$ **72.** $\dfrac{7}{2\sqrt{18}}$ **73.** $\dfrac{1 - \sqrt{3}}{\sqrt{3}}$

74. $\dfrac{\sqrt{3}}{1 - \sqrt{3}}$ **75.** $\dfrac{2 - \sqrt{2}}{2 + \sqrt{2}}$ **76.** $\dfrac{\sqrt{3} + \sqrt{5}}{2\sqrt{3} - 7\sqrt{5}}$

77. $\dfrac{\sqrt{x}}{\sqrt{x} + \sqrt{y}}$ **78.** $\dfrac{2\sqrt{y}}{x - 2\sqrt{y}}$ **79.** $\dfrac{x - 1}{\sqrt{x^2 - 1}}$

80. $\dfrac{x}{\sqrt{2x^2 + x}}$

In Exercises 81–86 rationalize the numerator.

81. $\dfrac{\sqrt{x} - 2}{x - 4}$ **82.** $\dfrac{\sqrt{x} - \sqrt{9}}{x - 9}$

83. $\dfrac{\sqrt{x + h} - \sqrt{x}}{h}$ **84.** $\dfrac{\sqrt{1 + h} - 1}{h}$

85. $\dfrac{\sqrt{2x + 2h + 1} - \sqrt{2x + 1}}{h}$

86. $\dfrac{\sqrt{x + h - 1} - \sqrt{x - 1}}{h}$

In Exercises 87–98 write each expression as a single fraction in simplest form (assume that all variables represent positive real numbers).

87. $600\sqrt{5} + \dfrac{360,000}{120\sqrt{5}}$ **88.** $900\sqrt{2} + \dfrac{540,000}{300\sqrt{2}}$

89. $\dfrac{\sqrt{3/x}(1 + 3/x)}{\sqrt{3}/(2x^{3/2})}$ **90.** $-\dfrac{1}{2}\sqrt{\dfrac{x}{y}} \cdot \dfrac{x\sqrt{y/x} - y}{x^2}$

91. $1 - \dfrac{1}{2\sqrt{x + 1}}$ **92.** $\dfrac{u}{2\sqrt{u + 1}} + \sqrt{u + 1}$

93. $a \cdot \dfrac{a}{\sqrt{a^2 + b^2}} + b \cdot \dfrac{b}{\sqrt{a^2 + b^2}}$

94. $\dfrac{2x}{2\sqrt{x^2 - 1}} - \dfrac{1}{x\sqrt{x^2 - 1}}$

95. $\pi\left(h \cdot \dfrac{-2h}{2\sqrt{4R^2 - h^2}} + \sqrt{4R^2 - h^2}\right)$

96. $\sqrt{R^2 - x^2} + (R + x)\dfrac{-x}{\sqrt{R^2 - x^2}}$

97. $\dfrac{x \cdot \dfrac{2x}{2\sqrt{x^2 + 1}} - \sqrt{x^2 + 1}}{x^2}$

98. $\dfrac{\sqrt{1 - x^2} - x \cdot \dfrac{-x}{\sqrt{1 - x^2}}}{1 - x^2}$

99. A useful definition for proving statements about absolute value is $|a| = \sqrt{a^2}$. Use this definition to show that $|ab| = |a| \cdot |b|$.

100. When drawing a rectangle, what should be the ratio of the length to the width to achieve the most satisfying visual effect? The answer is obviously subjective, but there is a leading candidate for the title of most pleasing. This rectangle is called the **golden rectangle,** and the ratio of the length to the width in this figure is called the **golden ratio.** Consider the buildings sketched within rectangles whose side ratios are indicated. Which figure appeals the most to you?

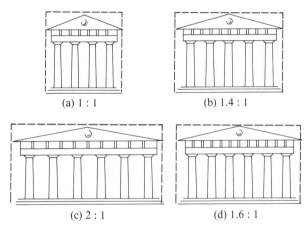

(a) 1 : 1   (b) 1.4 : 1

(c) 2 : 1   (d) 1.6 : 1

If you selected figure (d) you are in good company, for this figure has the same side ratios as the Parthenon in Athens. This fact and other examples of the golden rectangle in art and architecture are nicely illustrated in the Time-Life book by David Bergamini entitled *Mathematics* (1963, 1970). The ratio 1.6:1 (or simply 1.6) is actually an approximation of the golden ratio. The exact number can be determined by considering the simple geometric method used to construct a golden rectangle, which is shown below. We start with square *ABCD* and extend this square to form golden rectangle *ABEF* by drawing arc *CF*, which is centered at the midpoint *M* of *AD*.

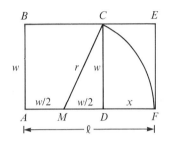

a. Show in rectangle *ABEF* that $\ell/w = (1 + \sqrt{5})/2$, which is the exact value of the golden ratio. [*Hint:* Note that $\ell = (w/2) + r$.]

b. Show that $w/x = (1 + \sqrt{5})/2$, which means that rectangle *DCEF* is also a golden rectangle.

c. Subtracting 1 from the golden ratio yields its reciprocal. Verify this.

d. In some literature the golden ratio is defined as w/ℓ instead of ℓ/w. Using this definition, what is the exact value of the golden ratio in simplest form?

CHAPTER OVERVIEW

Section	Key Concepts to Review
1.1	• Definitions of integers, rational numbers, irrational numbers, real numbers, $a < b$, $a > b$, $a \le b$, absolute value, subtraction, factor, power, exponent, and division
	• Relationships among the various sets of numbers
	• Statements of basic properties of real numbers
	• Methods to add, subtract, multiply, and divide real numbers
	• Order of operations

Section	Key Concepts to Review
1.2	• Definitions of variable, constant, algebraic expression, terms, (numerical) coefficient, and similar terms
	• Methods to evaluate an algebraic expression
	• Methods to add or subtract algebraic expressions
	• Guidelines for using a scientific calculator
	• Laws of exponents (m and n denote integers)

$$\textbf{1. } a^m \cdot a^n = a^{m+n} \qquad \textbf{2. } (a^m)^n = a^{mn} \qquad \textbf{3. } (ab)^n = a^n b^n \qquad \textbf{4. } \left(\frac{a}{b}\right)^n = \frac{a^n}{b^n} \quad (b \neq 0)$$

$$\textbf{5. } \frac{a^m}{a^n} = a^{m-n} = \frac{1}{a^{n-m}} \quad (a \neq 0) \qquad \textbf{6. } a^0 = 1 \quad (a \neq 0) \qquad \textbf{7. } a^{-n} = \frac{1}{a^n} \quad (a \neq 0)$$

Section	Key Concepts to Review
1.3	• Definitions of polynomial, monomial, binomial, trinomial, degree of a monomial, and degree of a polynomial
	• Methods to multiply various types of algebraic expressions
	• FOIL multiplication method
	• Methods to factor an expression that contains a factor common to each term, that is the difference of squares, that is a trinomial, or that is the sum or difference of cubes
	• Method to determine if $ax^2 + bx + c$ can be factored with integer coefficients
	• Factoring and product models
1.4	• Definitions of LCD and complex fractions
	• Methods to simplify, multiply, divide, add, and subtract fractions; to find the LCD; and to simplify a complex fraction
	• Equality principle: $\dfrac{a}{b} = \dfrac{c}{d}$ if and only if $ad = bc$ $(b, d \neq 0)$
	• Fundamental principle: $\dfrac{ak}{bk} = \dfrac{a}{b}$ $(b, k \neq 0)$
	• Operations principles $(b, d, c/d \neq 0)$

$$\frac{a}{b} \cdot \frac{c}{d} = \frac{ac}{bd} \qquad\qquad\qquad \frac{a}{b} + \frac{c}{b} = \frac{a \pm c}{b}$$

$$\frac{a}{b} \div \frac{c}{d} = \frac{a}{b} \cdot \frac{d}{c} \qquad\qquad \frac{a}{b} \pm \frac{c}{d} = \frac{ad \pm bc}{bd}$$

Section	Key Concepts to Review
1.5	• Definitions of $a^{1/n}$, index and radicand of a radical, principal square root, and nth root of a number
	• If m/n represents a reduced fraction such that $a^{1/n}$ represents a real number, then $a^{m/n} = (a^{1/n})^m = (a^m)^{1/n}$, or equivalently, $a^{m/n} = (\sqrt[n]{a})^m = \sqrt[n]{a^m}$

Section	Key Concepts to Review
1.6	• Definitions of similar radicals and conjugates
	• Methods to simplify, add, subtract, multiply, and divide radicals
	• Methods to rationalize the denominator or the numerator
	• Properties of radicals (a, b, $\sqrt[n]{a}$, $\sqrt[n]{b}$ denote real numbers)

1. $(\sqrt[n]{a})^n = a$ 2. $\sqrt[n]{a} \cdot \sqrt[n]{b} = \sqrt[n]{ab}$

3. $\dfrac{\sqrt[n]{a}}{\sqrt[n]{b}} = \sqrt[n]{\dfrac{a}{b}}$ $(b \neq 0)$ 4. $\sqrt[n]{a^n} = \begin{cases} a, & \text{if } n \text{ is odd} \\ |a|, & \text{if } n \text{ is even} \end{cases}$

CHAPTER REVIEW EXERCISES

In Exercises 1–50 perform the indicated operations and simplify.

1. $1 - 3(2 - 4)$

2. $ct^2(3c - 2t + 4)$

3. $\dfrac{(1 - x)^{a + b}}{(1 - x)^{a - b}}$

4. $\dfrac{2k + 1}{k - 1} - \dfrac{3k - 4}{k - 1}$

5. $\dfrac{10ax^2}{21y^4} \cdot \dfrac{7y}{5ax}$

6. $\dfrac{2}{x} + \dfrac{3}{y}$

7. $\dfrac{x^2 + 1}{4x + 4} \cdot \dfrac{(x + 1)^2}{x^2 + x}$

8. $(x^n - 3)(x^n - 1)$

9. $\dfrac{(3x)^2}{y}\left(\dfrac{2zy}{x^2}\right)^3$

10. $\dfrac{(x + h)^4 - x^4}{h}$

11. $(2x + 4)(4 + x - x^2)$

12. $2^{3x}/2^x$

13. $5\sqrt{12} - 2\sqrt{27}$

14. $2^0 - (4/9)^{-3/2}$

15. $\sqrt{(-\pi\sqrt{3})^2 + (3\pi)^2}$

16. $\sqrt{4/3} + \sqrt{3/4}$

17. $-6(-\sqrt{3}/2) - 4(\sqrt{3}/2)(-1)$

18. $(-2)^{-1} \cdot (-2)^{-3}$

19. $\dfrac{6R^2}{\sqrt{3R^2}}, R > 0$

20. $\dfrac{a^{-2/3}}{\sqrt[3]{a}}$

21. $\sqrt{\dfrac{1 + (4/5)}{2}}$

22. $\dfrac{1}{\sqrt{3}} + \dfrac{2\pi}{3}$

23. $\dfrac{(x^{-2/3}x^{1/6})^{-1}}{x^{1/3}}$

24. $2\pi h\sqrt{\dfrac{100}{\pi h}}, \quad h > 0$

25. $\dfrac{\sqrt{3} + \sqrt{2}}{\sqrt{3}}$

26. $\dfrac{2^{-3} - 7^0}{4^{-1/2} + 5^0}$

27. $\dfrac{2\sqrt{2}h\pi}{\sqrt{2\pi h^2}}$

28. $\left[\dfrac{\sqrt{3/2}}{\sqrt{1 - \frac{3}{4}}} - \dfrac{\pi}{3}\right] \div \dfrac{3}{4}$

29. $\dfrac{1 - (1/s)}{1 + (1/s)}$

30. $\left(\dfrac{x}{1 + x} - \dfrac{1 - x}{x}\right) \div \left(\dfrac{x - 1}{x} - \dfrac{x}{x + 1}\right)$

31. $\left[\dfrac{2}{x + h + 1} - \dfrac{2}{x + 1}\right] \div h$

32. $\dfrac{2t}{t^2 + 5t + 6} + \dfrac{5t}{t^2 + 2t - 3}$

33. $\dfrac{(c - 3)(c + 3)(c - 3)}{(3 - c)(-3 + c)(3 + c)}$

34. $\dfrac{(1 + h)^3 - 1}{(1 + h) - 1}$

35. $\dfrac{1}{x^n} + \dfrac{1}{x^{n + 1}}$

36. $x^2 + \dfrac{kx^2}{2x - k}$

37. $\dfrac{(x + y)^2 - 4xy}{x^2 - y^2}$

38. $\dfrac{(2x^2 - 1) - (2a^2 - 1)}{x - a}$

39. $\dfrac{(a - 1)x - (1 - a)y}{(a - 1)xy}$

40. $\left(\dfrac{2^{n + 1}}{n \cdot 3^n}\right) \div \left(\dfrac{2^n}{(n - 1)3^{n - 1}}\right)$

41. $\dfrac{(x^2 + x) - (a^2 + a)}{x - a}$

42. $(x^{1/2} + x^{-1/2})^2$

43. $x^{1/3} + \frac{1}{2}x^{-1/3}$

44. $\dfrac{3}{x^{-3}} - \dfrac{2x^2}{x^{-1}}$

45. $(x - 3)(\frac{1}{2}x^{-1/2}) + x^{1/2} \cdot 1$

46. $[(x + h)^{1/3} - x^{1/3}] \cdot$
$[(x + h)^{2/3} + x^{1/3}(x + h)^{1/3} + x^{2/3}]$

47. $\dfrac{1}{\sqrt{1-x^2}} - x \cdot \dfrac{-x}{\sqrt{1-x^2}} - \sqrt{1-x^2}$

48. $\sqrt{R^2-y^2}(-1) + (R-y) \cdot \dfrac{-2y}{2\sqrt{R^2-y^2}}$

49. $\dfrac{u \cdot \dfrac{2u}{2\sqrt{u^2+4}} - \sqrt{u^2+4}}{u^2}$

50. $\dfrac{\dfrac{1}{\sqrt{x+h}} - \dfrac{1}{\sqrt{x}}}{h}$

51. Rationalize the numerator in the answer to Exercise 50.

52. If $m = (y_2 - y_1)/(x_2 - x_1)$, evaluate m for $x_1 = 2$, $x_2 = -5$, $y_1 = -1$, and $y_2 = 4$.

53. Find the value of the expression $-x^2 + 2xy^3$ when $x = -1$ and $y = 3$.

54. If $x = -2$ and $y = 5$, then evaluate $|2x| + |y|$.

55. Simplify $(1 - \tfrac{1}{2})(1 - \tfrac{1}{3})(1 - \tfrac{1}{4})(1 - \tfrac{1}{5}) \cdots$
$(1 - \tfrac{1}{9})(1 - \tfrac{1}{10})$.

56. Express $4.\overline{71}$ as the ratio of two integers.

57. Name the property of real numbers illustrated in each statement.

  a. $(x + y) + z = z + (x + y)$

  b. $\tfrac{1}{9} \cdot 8 + \tfrac{1}{9} \cdot 2 = \tfrac{1}{9}(8 + 2)$

58. Use the correct symbol ($<, >, =$) between the pairs of expressions.

  a. -0.1 _____ -0.01

  b. 2% _____ 0.2

  c. $|-3 + 7|$ _____ $|-3| + |7|$

  d. $|-2a|$ _____ $|-2| \cdot |a|$

59. The number 3.14 is a member of which of the following collections of numbers: reals, rationals, irrationals, integers?

60. Calculate $b^2 - 4ac$ for $12x^2 - 23x - 18$. Can the expression be factored into binomial factors with integer coefficients?

In Exercises 61–64 factor each of the expressions completely.

61. $t^3 - t^3x^2$

62. $x^2 - 8x + 12$

63. $x^3 + x^2 + x + 1$

64. $2x(h - k) + (h^2 - k^2) - (h - k)$

In Exercises 65 and 66 determine the least common denominator.

65. $\dfrac{5}{18x^2yz^3}, \dfrac{11}{24xy^2z}$

66. $\dfrac{1}{t^2 - 4}, \dfrac{t + 4}{t^2 - 4t + 4}$

In Exercises 67–70 answer true or false. If false, give a specific counterexample.

67. All real numbers are rational numbers.

68. All integers are rational numbers.

69. There is no number equal to $0/4$.

70. If $ab = ac$, then $b = c$.

71. The side of a square is represented by $4x^3$. Write in simplest form the perimeter and area of this square in terms of x.

72. Rationalize the denominator: $(x - 2)/\sqrt{x^2 - 4}$.

73. Rationalize the numerator: $\sqrt{x - 2}/(x^2 - 4)$.

74. Evaluate $x^{3/2} + 2x^0$ when $x = 4$.

75. Show that $\dfrac{(2x + 1)(2x - 1) - (x^2 - x) \cdot 2}{(2x + 1)^2}$

  simplifies to $\dfrac{2x^2 + 2x - 1}{(2x + 1)^2}$.

76. Show that $\dfrac{[(x + h)^2 - (x + h)] - [x^2 - x]}{h}$

  simplifies to $2x + h - 1$.

77. Show that $\left(\dfrac{1}{1 + (1/n)}\right)^p$ is equivalent to $\dfrac{n^p}{(n + 1)^p}$.

78. Show that $\dfrac{(x - 1)^3 - (x + 1) \cdot 3(x - 1)^2}{(x - 1)^6}$

  simplifies to $\dfrac{-2(x + 2)}{(x - 1)^4}$.

79. Simplify $n\left(\dfrac{2}{n}\right) + \tfrac{1}{2}n^2\left(\dfrac{2}{n}\right)^2$
  $+ \tfrac{1}{4}(4n^3 - n)\left(\dfrac{2}{n}\right)^3 + \tfrac{1}{8}(2n^4 - n^2)\left(\dfrac{2}{n}\right)^4$.

80. Simplify $x - [-x - (x - 1)]$.

81. Simplify $[(1/x) - (1/a)] \div (x - a)$.

82. Simplify $\dfrac{(1 + h)^2 + 2(1 + h) - 3}{h}$.

83. Simplify

$$\dfrac{(x - 2y)\left(\dfrac{y}{x - 2y}\right) - (-y)\left(x + \dfrac{2y}{x - 2y}\right)}{(x - 2y)^2}.$$

84. If $x = -2$, find (a) the absolute value of x, (b) the negative of x, and (c) the reciprocal of x.

85. If $x_1 = \dfrac{-b + \sqrt{b^2 - 4ac}}{2a}$ and

$x_2 = \dfrac{-b - \sqrt{b^2 - 4ac}}{2a}$, find $x_1 + x_2$ and

$x_1 \cdot x_2$.

86. The **equal-monthly-payment formula** is used by banks when they finance many common loans. The problem is to find the equal monthly installment, E, that will pay off a loan of P dollars with interest in n months. The formula used by the bank is

$$E = \dfrac{Pr}{1 - (1 + r)^{-n}}.$$

Suppose that you obtain a $10,000 car loan for 24 months with 12 percent annual interest charged on the unpaid balance. Use a calculator to determine the amount of the equal monthly installment that will pay off this debt. [*Note:* If the annual interest rate is 12 percent, the monthly interest rate (r) is 12 percent/12, or 1 percent.]

In Exercises 87–100 select the choice that answers the question or completes the statement.

87. Which statement illustrates the commutative property of multiplication?

  a. $a(b + c) = (b + c)a$

  b. $a(b + c) = ab + ac$

  c. $(ab)c = a(bc)$

  d. $a \cdot 1 = a$

88. If $a < 0$, then $|a|$ equals

  a. a **b.** a^2 **c.** $-a$ **d.** $1/a$

89. The fraction $(x - y)/4$ is *not* equal to

  a. $\dfrac{y - x}{-4}$ **b.** $-\dfrac{x - y}{-4}$

  c. $\dfrac{-x + y}{-4}$ **d.** $\dfrac{y - x}{4}$

90. If $2^a = x$, then 2^{a+3} equals

  a. $x + 3$ **b.** x^3 **c.** $6x$ **d.** $8x$

91. $\left(1 + \dfrac{1}{n}\right)\left(\dfrac{1}{n + 1} - 1\right)$ simplifies to

  a. -1 **b.** 1 **c.** n **d.** $n + 1$

92. The reciprocal of $3 + (1/n)$ is

  a. $(3n + 1)/3$ **b.** $n/(3n + 1)$

  c. $(3n + 1)/n$ **d.** $3 + n$

93. $\dfrac{(R + r)^2 E^2 - RE^2 \cdot 2(R + r)}{(R + r)^4}$ simplifies to

  a. $\dfrac{rE^2(r - 2R)}{(R + r)^4}$ **b.** $\dfrac{E^2(r - 2R)}{(R + r)^3}$

  c. $\dfrac{E^2(r - R)}{(R + r)^3}$ **d.** $\dfrac{E^2(r - R)^2}{(R + r)^4}$

94. Which statement is true for all values of the unknowns (for which it is defined)?

  a. $\dfrac{1}{n} - \dfrac{1}{n + 1} = \dfrac{1}{n(n + 1)}$

  b. $\dfrac{2x + y}{x + y} = 2$

  c. $\dfrac{1/a}{1/b} = \dfrac{1}{ab}$

  d. $\dfrac{1}{p} + \dfrac{1}{q} = \dfrac{2}{p + q}$

95. Which statement is true for all values of x (for which it is defined)?

  a. $(\sqrt{x} + \sqrt{2})^2 = x + 2$ **b.** $\sqrt{x} + \sqrt{x} = \sqrt{2x}$

  c. $(\sqrt{x + 2})^2 = x + 2$ **d.** $\sqrt{x^2 + 1} = x + 1$

96. Which number is an irrational number?

  a. $-\sqrt{4}$ **b.** $\sqrt{-4}$ **c.** $\sqrt[3]{8}$ **d.** $\sqrt{8}$

97. The expression $\dfrac{\sqrt{2} + 1}{\sqrt{2} - 1}$ is equivalent to

  a. 3 **b.** -1 **c.** $5 + \sqrt{2}$ **d.** $3 + 2\sqrt{2}$

98. The expression $x^{-1/3}$ is equivalent to

  a. $\sqrt[3]{x}$ **b.** $-\sqrt[3]{x}$ **c.** $1/\sqrt[3]{x}$ **d.** $-1/\sqrt[3]{x}$

99. The expression $(3^{a+b})(3^{a-b})$ is equivalent to

  a. 9^a **b.** 9^{2a} **c.** $3^{a^2 - b^2}$ **d.** $9^{a^2 - b^2}$

100. The expression $\dfrac{a^{-1} + b^{-1}}{a^{-1}}$ is equivalent to

  a. $1/b$ **b.** $(a + b)/b$

  c. $b/(a + b)$ **d.** $a/(a + b)$

$\mathcal{2}$ *Equations and Inequalities*

A race car driver must average at least 100 mi/hour for two laps around a track to qualify for the finals. Because of mechanical trouble the driver is only able to average 50 mi/hour for the first lap. What speed must the driver average for the second lap to qualify for the finals? (See Example 12 of Section 2.1.)

Setting up and solving equations and inequalities is often an essential part of the mathematical analysis of a problem, and in this chapter we introduce some basic techniques and applications that involve linear, quadratic, radical, fractional, and absolute value equations and inequalities. But this topic does not end here. In subsequent chapters, whenever we introduce a new idea (such as a logarithm), you must always ask yourself if you can solve equations involving this new concept.

2.1 EQUATIONS AND WORD PROBLEMS

Word problems give us our first experience with one of the most challenging parts of mathematics: to obtain a mathematical description (model) of a practical situation. For as soon as we are able to describe the data mathematically, a solution can usually be obtained through algebraic methods. The first step in the process of translating the problem into mathematical language is to assign some variable to represent an unknown. If there is more than one unknown in the problem, try to represent these unknowns in terms of the original variable. Always use as few symbols as possible. Finally, use the information given in the problem to obtain a relationship between two algebraic expressions. In this section we discuss only the relationship of equality. A statement of equality between two expressions is called an **equation.** Here are two examples that illustrate the basic idea. We will give more hints for solving word problems later in the section.

EXAMPLE 1   The sum of two consecutive integers is 67. Translate this statement into an equation.

Solution   If x represents the first integer, $x + 1$ represents the next integer, then

the sum of two consecutive integers is 67

$$x + (x + 1) \qquad\qquad = 67.$$ □

EXAMPLE 2   The width of a rectangular lot is one-half the length and the perimeter of the lot is 5,000 ft. Translate this statement into an equation.

Solution   Let P represent the perimeter, ℓ the length, and w the width; then

$$P = 2(\ell + w) \text{ and } w = \tfrac{1}{2}\ell.$$

Substituting $\tfrac{1}{2}\ell$ for w and 5,000 for P, we have

$$5,000 = 2(\ell + \tfrac{1}{2}\ell).$$ □

Once we represent the problem in mathematical symbols, we can attempt a solution. To solve an equation means to find the set of *all* values for the variable that make the equation a true statement. This set is called the **solution set** of the equation. Equations may be always true, always false, or their truth may depend on the value substituted for the variable. If the equation is a true statement for all admissible values of the variable, as is $2x + x = 3x$, then it is called an **identity**. A **conditional equation** is an equation in which some replacements for the variable make the statement true, while others make it false. For example, the conditional equation $x + 5 = 21$ is true if $x = 16$ and false otherwise. We call 16 the **solution** or **root** of the equation. The concept of solving equations is usually associated with conditional equations.

One of the simplest equations to solve is a linear equation. By definition, a **first-degree** or **linear equation** is an equation that can be written in the form

$$ax + b = 0,$$

where a and b are real numbers with $a \neq 0$. We solve a linear equation by isolating the variable on one side of the equation as described below.

<table>
<tr><td>

Solution of Linear Equations

</td><td>

Two equations are **equivalent** when they have the same solution set. We may change a linear equation to an equivalent equation of the form x = number by performing any combination of the following steps:

1. Adding or subtracting the same expression to (from) both sides of the equation
2. Multiplying or dividing both sides of the equation by the same (non-zero) number

</td></tr>
</table>

EXAMPLE 3   Solve the equation $6(x + 1) = 14x + 2$.

Solution

$$6(x + 1) = 14x + 2$$
$$6x + 6 = 14x + 2$$
$$6x + 6 - 6 = 14x + 2 - 6 \qquad \text{Subtract 6 from both sides.}$$
$$6x = 14x - 4$$
$$6x - 14x = 14x - 14x - 4 \qquad \text{Subtract } 14x \text{ from both sides.}$$
$$-8x = -4$$
$$\frac{-8x}{-8} = \frac{-4}{-8} \qquad \text{Divide both sides by } -8.$$
$$x = \frac{1}{2}$$

Thus, $\frac{1}{2}$ is the solution of the equation and the solution set is $\{\frac{1}{2}\}$. We can check the solution by replacing x with $\frac{1}{2}$ in the original equation.

$$6(\tfrac{1}{2} + 1) \overset{?}{=} 14(\tfrac{1}{2}) + 2$$
$$6(\tfrac{3}{2}) \overset{?}{=} 7 + 2$$
$$9 \overset{\checkmark}{=} 9 \qquad\qquad \square$$

EXAMPLE 4   Solve and check $\dfrac{3x}{8} - \dfrac{1}{4} = \dfrac{x + 5}{2}$.

Solution   The usual procedure for solving equations with fractions is to remove the fractions by multiplying both sides of the equation by the LCD. In this equation the LCD is 8, so

$$8\left(\frac{3x}{8} - \frac{1}{4}\right) = 8\left(\frac{x+5}{2}\right)$$

$$3x - 2 = 4x + 20$$

$$-22 = x.$$

Check

$$\frac{3(-22)}{8} - \frac{1}{4} \overset{?}{=} \frac{-22+5}{2}$$

$$-\frac{17}{2} \overset{\checkmark}{=} -\frac{17}{2}$$

Thus, the solution set is $\{-22\}$. ☐

Note that in Examples 3 and 4 each linear equation had exactly one solution. We can anticipate this result since the general form of a linear equation, $ax + b = 0$ with $a \neq 0$, may be solved as follows:

$$ax + b = 0$$
$$ax = -b \qquad \text{Add } -b \text{ to both sides.}$$
$$x = \frac{-b}{a}. \qquad \text{Divide both sides by } a \neq 0.$$

Thus, the linear equation $ax + b = 0$, $a \neq 0$, has exactly one solution, $-b/a$. The requirement that $a \neq 0$ is important as shown in Example 5.

EXAMPLE 5   Solve $3(x + 3) = 3x$.

Solution   If we proceed in the usual way, we have

$$3(x + 3) = 3x$$
$$3x + 9 = 3x$$
$$9 = 0.$$

When we subtract $3x$ from both sides of $3x + 9 = 3x$, the result is $9 = 0$. This statement is never true and indicates the original equation has no solution. When the solution set contains no elements, we say the solution set is $\emptyset$, called the **empty set**. Note that when an equation may be written in the form $ax + b = 0$, with $a = 0$, then either the equation has no solution (if $b \neq 0$) or the equation is an identity (if $b = 0$). ☐

The procedures for solving a linear equation are fundamental to the solutions of a wide variety of equations. For example, fractional equations involving unknown denominators often lead to linear equations as shown in Examples 6 and 7.

EXAMPLE 6   Solve and check $\dfrac{2}{3a} = \dfrac{3}{2a - 1}$.

Solution   As in Example 4, we first remove the fractions by multiplying both sides of the equation by the LCD, which is $3a(2a - 1)$ in this equation. Note that this step requires the restrictions that $a \neq 0$ and $a \neq \frac{1}{2}$ to ensure that we are multiplying both sides of the equation by a nonzero number.

$$3a(2a - 1)\frac{2}{3a} = 3a(2a - 1)\frac{3}{2a - 1}$$
$$2(2a - 1) = 9a$$
$$4a - 2 = 9a$$
$$-2 = 5a$$
$$\frac{-2}{5} = a$$

Check

$$\frac{2}{3(-2/5)} \overset{?}{=} \frac{3}{2(-2/5) - 1}$$
$$\frac{2}{(-6/5)} \overset{?}{=} \frac{3}{(-9/5)}$$
$$\frac{5}{-3} \overset{\checkmark}{=} \frac{5}{-3}$$

The LCD is nonzero when $a = -\frac{2}{5}$, and the check of this solution is shown above. Thus, the solution set is $\{-\frac{2}{5}\}$.   □

EXAMPLE 7   Solve and check $\dfrac{2}{x + 1} + \dfrac{3}{x - 1} = \dfrac{-4}{x^2 - 1}$.

Solution   The LCD is $(x + 1)(x - 1)$ and if $x \neq -1$ and $x \neq 1$, then

$$(x + 1)(x - 1)\left(\frac{2}{x + 1} + \frac{3}{x - 1}\right) = (x + 1)(x - 1)\frac{-4}{x^2 - 1}$$
$$2(x - 1) + 3(x + 1) = -4$$
$$2x - 2 + 3x + 3 = -4$$
$$5x + 1 = -4$$
$$5x = -5$$
$$x = -1.$$

The restriction $x \neq -1$ eliminates -1 as a solution, so no number satisfies the original equation and the solution set is $\emptyset$. We verify that -1 is not a solution in the following check.

Check

$$\frac{2}{(-1)+1} + \frac{3}{(-1)-1} \overset{?}{=} \frac{-4}{(-1)^2-1}$$

$$\frac{2}{0} + \frac{3}{-2} \neq \frac{-4}{0} \qquad \qquad \Box$$

Now that we can solve some equations, we can determine the final solutions for some word problems. The most difficult part is usually to describe the situation mathematically; but this is a realistic obstacle. Problems you wish to solve from your own areas of interest will probably not start out as math problems but will be stated in words. It's up to you to analyze the situation and apply the appropriate formulas. Although there's no guaranteed procedure for setting up word problems, the guidelines given at the beginning of this section along with the following general approaches should be useful.

1. Be sure you understand the question. What are you being asked to find? What are you given? Try putting it into your own words.
2. If you're having trouble "getting the feel" of the problem, force yourself to get involved by making up a specific example using the given relationships. For example, if it's a problem involving numerical values, make up some illustrations with simple numbers. If it's a problem in which a picture might help, try drawing several simple sketches to illustrate the relationships. See if these simple illustrations suggest a pattern you can test. A little trial and error can go a long way toward clarifying the problem.
3. Check to see if the problem is similar to another problem you already know how to solve. This is using analogy. Since the two problems are alike in some ways, an approach that was useful in the problem you understand might be helpful in the other. Here's where experience really pays off. It's one of the reasons why word problems become gradually easier if you stick with them.
4. When you think you have a solution, be sure to check that your answer satisfies the conditions of the original problem. It is not unusual to find that you have accidently solved a slightly different problem from the one stated and so must adjust your solution accordingly. You might also discover that you're way off the track, so you'll have to examine your attempted solution more carefully.

We all have our preferred ways of solving problems, so don't hesitate to try other methods. Word problems are a common sore point, and you will not have difficulty finding reference material that offers other suggestions on how to get started. Just keep in mind that you should always understand and be able to explain what you are doing.

EXAMPLE 8   Percentage Problem: The total cost (including tax) of a new car is $14,256. If the sales tax rate is 8 percent, how much do you pay in taxes?

Solution   Let x represent the price of the car; then

$$\text{Total cost} = \text{price of car} + \text{tax}$$
$$14{,}256 = x \qquad\qquad + 0.08x$$
$$14{,}256 = 1.08x$$
$$\frac{14{,}256}{1.08} = x$$
$$13{,}200 = x.$$
$$\text{Tax} = 8 \text{ percent of } 13{,}200 = 0.08(13{,}200) = \$1{,}056 \qquad\qquad \square$$

EXAMPLE 9   Proportion Problem: A cylindrical tank holds 480 gal when it is filled to its full height of 8 ft. When the gauge shows that it contains water at a height of 3 ft 1 in., how many gallons are in the tank?

Solution   A **ratio** is a comparison of two quantities by division; a **proportion** is a statement that two ratios are equal. If we let x represent the amount of water in the tank and set up two equal ratios that compare like measurements, we have the proportion

$$\frac{x}{480 \text{ gal}} = \frac{37 \text{ in.}}{96 \text{ in.}}. \qquad (Note: \text{ 3 ft 1 in.} = 37 \text{ in.})$$

To find x multiply both sides of this equation by 480 and simplify.

$$x = \left(\frac{37}{96}\right) \cdot 480$$
$$x = 185$$

Thus, there are 185 gal of water in the tank when the water is at a height of 3 ft 1 in. $\qquad\qquad \square$

EXAMPLE 10   Mixture Problem: A chemist has two acid solutions, one 30 percent acid and the other 70 percent acid. How much of each solution must be used to obtain 100 ml of a solution that is 41 percent acid?

Solution   Let x be the amount of the first solution in the mixture.

Solution	Percent Acid	·	Amount of Solution (ml)	=	Quantity of Acid (ml)
First solution	30		x		$0.3x$
Second solution	70		$100 - x$		$0.7(100 - x)$
New solution	41		100		$0.41(100)$

$$\begin{array}{l} \text{Quantity of acid} \\ \text{in first solution} \end{array} + \begin{array}{l} \text{Quantity of acid} \\ \text{in second solution} \end{array} = \begin{array}{l} \text{Quantity of acid} \\ \text{in new solution} \end{array}$$

$$\begin{aligned} 0.3x + 0.7(100 - x) &= 0.41(100) \\ 0.3x + 70 - 0.7x &= 41 \\ -0.4x &= -29 \\ x &= \frac{-29}{-0.4} = 72.5 \end{aligned}$$

Thus, the chemist must mix 72.5 ml of the first solution with 27.5 ml of the second solution to obtain the desired mixture. $\square$

EXAMPLE 11   Work Problem: One pipe can fill a swimming pool in 4 hours; a second pipe can fill the pool in 12 hours. How long will it take to fill the pool if both pipes operate together?

Solution   Let x represent the number of hours required to fill the pool with both pipes operating. If the first pipe fills the pool in 4 hours, the pipe fills $\frac{1}{4}$ of the tank for each hour it operates. Thus, in 1 hour it fills $\frac{1}{4}$ of the tank; in 2 hours it fills $(\frac{1}{4})2$, or $\frac{2}{4}$, of the tank; and in x hours it fills $\frac{1}{4}x$ of the tank.

Similarly, if the second pipe fills the pool in 12 hours, the pipe fills $\frac{1}{12}$ of the tank for each hour it operates. Thus, in x hours it fills $\frac{1}{12}x$ of the tank. Note in general that if a job requires t units of time to complete, then $1/t$ of the job is completed in 1 unit of time.

To complete the job, the sum of the two fractions must be 1. For example, if the first pipe did $\frac{3}{5}$ of the job and the second pipe did $\frac{2}{5}$ of the job, the job would be completed (for $\frac{3}{5} + \frac{2}{5} = \frac{5}{5}$ of the job, or the whole job). Therefore,

$$\begin{aligned} \tfrac{1}{4}x + \tfrac{1}{12}x &= 1 \quad \text{LCD: 12} \\ 12(\tfrac{1}{4}x + \tfrac{1}{12}x) &= 12(1) \\ 3x + x &= 12 \\ 4x &= 12 \\ x &= 3. \end{aligned}$$

It takes the two pipes 3 hours to fill the pool when they are in operation simultaneously. $\square$

EXAMPLE 12 Solve the problem in the chapter introduction on page 62.

Solution First, review Example 10 of Section 1.4. Now if we let x represent the minimum speed that the driver must average for the second lap to qualify, then

$$100 = \frac{2d}{\dfrac{d}{50} + \dfrac{d}{x}}$$

$$100\left(\frac{d}{50} + \frac{d}{x}\right) = 2d$$

$$2d + \frac{100d}{x} = 2d$$

$$\frac{100d}{x} = 0.$$

Since $100d$ cannot equal 0, the equation has no solution. This means that it is not possible to qualify, no matter how fast the driver goes on the second lap! To illustrate why this answer makes sense, note that the distance around the track is not important in the above equations, so let us suppose d equals 100 miles. Then the driver is required to travel 200 miles in 2 hours or less to qualify. However, the first lap took 2 hours to complete because the average speed was only 50 mi/hour, so the driver cannot qualify. ☐

Formulas

The translation of a relationship into mathematical symbols often results in a general equation or formula. For example, the formula

$$C = \tfrac{5}{9}(F - 32)$$

is often used to convert from degrees Fahrenheit to degrees Celsius. Although equations with more than one variable are very different from the conditional equations we have studied, the equation-solving techniques we learned may be used to rearrange formulas so they may be solved for any variable in the equation. For instance, since the formula $C = \tfrac{5}{9}(F - 32)$ is not very efficient for converting degrees Celsius to degrees Fahrenheit, we may solve for F as follows:

$$C = \tfrac{5}{9}(F - 32)$$
$$\tfrac{9}{5}C = F - 32 \qquad \text{Multiply both sides by } \tfrac{9}{5}.$$
$$\tfrac{9}{5}C + 32 = F. \qquad \text{Add 32 to both sides.}$$

Now we may select the more efficient version of the formula, depending on the given information.

EXAMPLE 13   Solve for y': $\dfrac{1}{2\sqrt{x}} + \dfrac{y'}{2\sqrt{y}} = 0$.

Solution

$$\frac{1}{2\sqrt{x}} + \frac{y'}{2\sqrt{y}} = 0$$

$$\frac{y'}{2\sqrt{y}} = \frac{-1}{2\sqrt{x}} \qquad \text{Subtract } \frac{1}{2\sqrt{x}} \text{ from both sides.}$$

$$y' = \frac{-2\sqrt{y}}{2\sqrt{x}} = \frac{-\sqrt{xy}}{x} \qquad \begin{array}{l}\text{Multiply both sides by } 2\sqrt{y} \\ \text{and simplify.} \qquad \square\end{array}$$

EXAMPLE 14   Solve $n = \dfrac{A + D}{A}$ for A.

Solution   To solve for A we rewrite the equation so A appears only once. One such sequence is as follows:

$$n = \frac{A + D}{A}$$

$$nA = A + D \qquad \text{Multiply both sides by } A.$$

$$nA - A = D \qquad \text{Subtract } A \text{ from both sides.}$$

$$A(n - 1) = D \qquad \text{Factor out the common factor } A.$$

$$A = \frac{D}{n - 1}. \qquad \text{Divide both sides by } n - 1. \qquad \square$$

EXERCISES 2.1

In Exercises 1–42 solve each equation and check your answer.

1. $6x = 18$

2. $-3x + 3 = 0$

3. $3x + 2 = 2x + 8$

4. $7y - 9 = 6y - 10$

5. $2z + 12 = 5z + 15$

6. $-6x + 7 = 5x - 4$

7. $3y - 14 = -4y + 7$

8. $3z - 8 = 13z - 9$

9. $8x - 10 = 3x$

10. $15 - 6y = 0$

11. $9 - 2y = 27 + y$

12. $15x + 5 = 2x + 14$

13. $4x - x = 3x$

14. $x - 1 = x + 3$

15. $2(x + 4) = 2(x - 1) + 3$

16. $2(x + 3) - 1 = 2x + 5$

17. $7x = 2x$

18. $3y - 7 = 2y - 7$

19. $4 - 3y = 3(4 - y)$

20. $7x - 2(x - 7) = 14 - 5x$

21. $5(x - 6) = -2(15 - 2x)$

22. $18(z + 1) = 9z + 10$

23. $3[2x - (x - 2)] = -3(3 - 2x)$

24. $2y - (3y - 4) = 4(y + \frac{3}{4})$

25. $\dfrac{x - 2}{7} = -1$

26. $\dfrac{2y - 7}{9} = 5$

27. $\dfrac{x - 9}{3} = \dfrac{3x}{2} - \dfrac{11}{6}$

28. $\dfrac{2x - 5}{3} - \dfrac{x}{4} = \dfrac{1}{2}$

29. $\dfrac{3}{7} - \dfrac{n - 5}{14} = \dfrac{11n}{14}$

30. $\dfrac{5}{3a} - \dfrac{7}{4a} = \dfrac{5}{6}$

31. $\dfrac{2 + x}{6x} - 2 = \dfrac{3}{5x}$

32. $\dfrac{9y - 5}{7} = \dfrac{2y - 4}{3}$

33. $\dfrac{k}{k + 2} = \dfrac{2}{3}$

34. $\dfrac{x + 4}{x + 1} = \dfrac{x + 2}{x + 3}$

35. $\dfrac{z-1}{z-2} = \dfrac{z+1}{z-3}$ **36.** $\dfrac{1}{n-3} + \dfrac{2}{3-n} = \dfrac{1}{2}$

37. $\dfrac{2}{2-x} + \dfrac{x}{x-2} = 1$

38. $\dfrac{4}{x+1} - \dfrac{3}{x-1} = \dfrac{-6}{x^2-1}$

39. $\dfrac{2}{a^2+a} = \dfrac{1}{a} - \dfrac{2}{a+1}$ **40.** $\dfrac{3}{k+2} - 2 = \dfrac{-5}{2k+4}$

41. $\dfrac{5x}{x-2} - \dfrac{4x}{2x-7} = 3$ **42.** $3 - \dfrac{t-4}{t-3} = \dfrac{4t-1}{2t+3}$

In Exercises 43–63 solve each problem by first setting up an appropriate equation.

43. The sum of two integers is 35. One of the integers is four times the other. Find the integers.

44. During a test run, machine A produced x cans; machine B turned out three times as many; machine C produced 10 more cans than A. The total production during the run was 5,510. How many cans did each machine produce?

45. The sum of the three angles in a triangle is 180°. The second angle is 20° more than the first and the third angle is twice the first. What is the measure of each of the angles in the triangle?

46. The length of a rectangle is three times its width. If the perimeter is 160 in., determine the area of the rectangle.

47. A magazine and a newsletter cost $1.10. If the magazine cost $1.00 more than the newsletter, how much did each cost?

48. If the total cost (including tax) of a new car is $13,054, how much do you pay in taxes if the sales tax rate is 7 percent?

49. A man left one-half of his property to his wife, one-third to his son, and the remainder, which was $10,000, to his daughter. How much money did the man leave?

50. One card sorter can process a deck of punched cards in 1 hour; another can sort the deck in 2 hours. How long would it take the two sorters together to process the cards?

51. An oil storage tank is being filled through two pipes. One pipe working alone can fill the tank in 6 hours; the other pipe can fill the tank in 9 hours. How long will it take to fill the tank when both pipes are used?

52. A sociologist wishes to do a survey in a city where the census report indicates that the ratio of Protestants to Catholics is 5:2. He also determines that a suitable sample size is 420 people. How many of each religion should he interview if his sample is to have the same religious distribution as the city's population?

53. A cylindrical tank holds 370 gal when it is filled to its full height of 9 ft. When the gauge shows that it contains oil at a height of 6 ft 7 in., how many gallons (to the nearest gallon) are in the tank?

54. If 72 gal of water flows through a feeder pipe in 40 minutes, how many gallons of water will flow through the pipe in 3 hours?

55. A map has a scale of 1.5 in. = 7 mi. To the nearest tenth of a mile, what distance is represented by 20 in. on the map?

56. A chemist has 5 liters of a 25 percent sulfuric acid solution. He wishes to obtain a solution that is 35 percent acid by adding a solution of 75 percent acid to his original solution. How much of the more concentrated acid must be added to achieve the desired concentration?

57. One bar of tin alloy is 35 percent pure tin and another bar is 10 percent pure tin. How many pounds of each must be used to make 95 lb of a new alloy that is 20 percent pure tin?

58. An alloy of copper and silver weighs 40 lb and is 20 percent silver. How much silver must be added to produce a metal that is 50 percent silver?

59. A 20 percent antifreeze solution is a solution that consists of 20 percent antifreeze and 80 percent water. If we have 20 qt of such a solution, how much water should be added to obtain a solution that contains 10 percent antifreeze?

60. A race car driver must average 100 mi/hour for two laps around the track if he is to qualify for the finals. Because of mechanical trouble he is only able to average 60 mi/hour for the first lap. What speed must he average for the second lap if he is to qualify for the finals?

61. One of the classic problems from the history of mathematics is the ingenious method used by the Greek mathematician Eratosthenes to estimate the circumference of the earth in about 200 B.C. Eratosthenes obtained his estimate by noting the following information, which is illustrated here. Alexandria (where Eratosthenes was librarian) was 500 mi due north of the city of Syene. At noon on June 21 he knew from records that the sun cast no shadow at Syene, which meant that the sun was directly overhead. At the same time in Alexandria he measured from a shadow that the sun was 7.5° south from the vertical. By assuming that the sun was sufficiently far away for the light rays to be parallel to the earth, he then determined that angle *AOB* was 7.5°. Why? Complete the line of reasoning and obtain Eratosthenes' estimate for the circumference. You will find the result remarkably close to the modern estimate of 24,900 mi, which is obtained with the same basic procedure but with more accurate measurements.

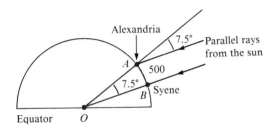

62. Consider this illustration of a piece of a broken chariot wheel found by a group of archaeologists. What is the radius of the wheel?

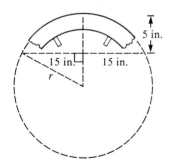

63. Puzzles of all types are popular in magazines and newspapers and have a devout and fanatical following. Mathematical puzzles are no exception. One of America's greatest puzzlists was Sam Loyd (1841–1911) whose work originally appeared in the *Brooklyn Daily Eagle* and the *Woman's Home Companion*. His puzzles, usually richly illustrated, drew national attention and spanned a wide variety of mathematical concepts. Test your skills with the given puzzle (Puzzle number 50, vol. 2, p. 35) from Gardner's collection of Loyd's most famous puzzles. (*Note:* More puzzles may be found in *Mathematical Puzzles of Sam Loyd,* a two-volume series edited by Martin Gardner, published by Dover Publications, © 1960.)

How many children are on the carousel?

WHILE ENJOYING a giddy ride on the carousel, Sammy propounded this problem: "One third of the number of kids riding ahead of me, added to three quarters of those riding behind me gives the correct number of children on this merry-go-round."

How many children were riding the carousel?

64. The following question is a favorite of problem solvers because it makes an important point. A car leaves Washington and heads for New York at 55 mi/hour. Two hours later a bus leaves New York and heads for Washington at 60 mi/hour. When the car and the bus meet, which is closer to New York? What is the point made by this problem?

In Exercises 65–92 solve each equation for the variable indicated.

65. $x + 2y = 4$ for y **66.** $7x - y = 2$ for x

67. $\dfrac{4 - 3x}{c} = 1$ for x **68.** $A = \dfrac{a + b + c}{3}$ for b

69. $T = m \cdot g - m \cdot f$ for f

70. $S = \frac{1}{2}n(a + L)$ for L **71.** $A = \frac{1}{2}h(b + c)$ for b

72. $a = p(1 + rt)$ for t **73.** $a = p + prt$ for p

74. $A = \pi(R^2 - r^2)$ for R^2

75. $R = \dfrac{kL}{d^2}$ for d^2 **76.** $S = \dfrac{a}{1 - r}$ for r

77. $C = \dfrac{E}{R + S}$ for R **78.** $I = \dfrac{2E}{R + 2r}$ for r

79. $\dfrac{I_1}{I_2} = \dfrac{r_1^2}{r_2^2}$ for I_2 **80.** $\dfrac{W_1}{W_2} = \dfrac{L_2}{L_1}$ for L_1

81. $S = \dfrac{a - rL}{1 - r}$ for r **82.** $I = \dfrac{nE}{R + nr}$ for n

83. $Z = \dfrac{Z_1 Z_2}{Z_1 + Z_2}$ for Z_2 **84.** $\dfrac{E}{e} = \dfrac{R + r}{r}$ for r

85. $\dfrac{1}{f} = \dfrac{1}{a} + \dfrac{1}{b}$ for a **86.** $\dfrac{1}{R} = \dfrac{1}{R_1} + \dfrac{1}{R_2}$ for R_2

87. $y = \dfrac{x + 3}{x}$ for x **88.** $x = \dfrac{y - 1}{y - 2}$ for y

89. $2x + xy' + y - 2yy' = 0$ for y'

90. $8x + 3xy' + 3y - 2yy' = 0$ for y'

91. $3x^2 + 3[x^2y' + y(2x)] + 3y^2y' = 0$ for y'

92. $\dfrac{2}{3\sqrt[3]{x}} + \dfrac{2y'}{3\sqrt[3]{y}} = 0$ for y'

93. In the equation $ax = b$, a and b are constants. What combination of values for a and b results in an equation with exactly one solution? No solution? Infinitely many solutions?

94. Consider the following statement: "If two fractions are equal and have equal numerators, they also have equal denominators." Now consider the following equation, which we wish to solve for x.

$$\frac{x - 1}{x + 1} + 2 = \frac{3x + 1}{x + 5}$$

$$\frac{x - 1}{x + 1} + \frac{2(x + 1)}{x + 1} = \frac{3x + 1}{x + 5}$$

$$\frac{x - 1 + 2(x + 1)}{x + 1} = \frac{3x + 1}{x + 5}$$

$$\frac{3x + 1}{x + 1} = \frac{3x + 1}{x + 5}$$

By the above statement we conclude that $x + 1 = x + 5$ or, upon subtracting x from both sides, that $1 = 5$. What is wrong? What value of x satisfies the equation?

2.2 COMPLEX NUMBERS

The real number system is the basis for most of the work both in this course and in higher mathematics. However, to solve equations like $x^2 + 1 = 0$, we need to extend the number system a step beyond the real numbers by considering numbers such as $\sqrt{-1}$. Such numbers are not real numbers, for the product of two equal real numbers never results in a negative number. We thus introduce a different type of number, called an **imaginary number.** The basic unit in imaginary numbers is $\sqrt{-1}$ and it is designated by i. Thus, by definition, we have

$$i = \sqrt{-1} \quad \text{and} \quad i^2 = -1.$$

When working with the square roots of negative numbers, the first step is to express these imaginary numbers as the product of a real number and i. For example:

$$\sqrt{-4} = i\sqrt{4} \quad\;\; = 2i$$
$$\sqrt{-12} = i\sqrt{4}\sqrt{3} = 2i\sqrt{3}.$$

Using real numbers and imaginary numbers, we may extend the number system to include a number like $3 + 2i$, which is needed to solve the equation $(x - 3)^2 + 4 = 0$ (see Example 4). We define this new type of number as follows:

Definition of Complex Number	A number of the form $a + bi$, where a and b are real numbers and $i = \sqrt{-1}$, is called a complex number.

The number a is called the **real part** of $a + bi$, and b is called the **imaginary part** of $a + bi$. Note that both the real part and the imaginary part of a complex number are real numbers. Two complex numbers are **equal** if and only if their real parts are equal and their imaginary parts are equal. That is,

$$a + bi = c + di \text{ if and only if } a = c \text{ and } b = d.$$

Since a and/or b may be zero, the complex numbers include the real numbers and the imaginary numbers. Figure 2.1 illustrates the relationship among the various sets of numbers.

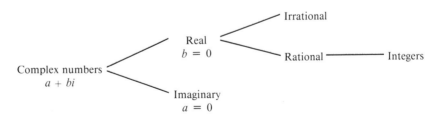

Figure 2.1

In many ways computations with complex numbers are similar to computations with real numbers that contain square roots. The important difference is that we must always express the imaginary part in terms of i, remembering that $i^2 = -1$. The following examples illustrate how to perform the various operations with complex numbers.

EXAMPLE 1   Express in terms of i and combine.

a. $\sqrt{-100} + \sqrt{-64}$ **b.** $3\sqrt{-3} - 2\sqrt{-48}$

Solution

a. $\sqrt{-100} + \sqrt{-64} = i \cdot \sqrt{100} + i \cdot \sqrt{64}$
$$= 10i + 8i$$
$$= 18i$$

b. $3\sqrt{-3} - 2\sqrt{-48} = 3i \cdot \sqrt{3} - 2i \cdot \sqrt{16} \cdot \sqrt{3}$
$$= 3i\sqrt{3} - 8i\sqrt{3}$$
$$= -5i\sqrt{3} \qquad \square$$

EXAMPLE 2   Perform the indicated operation and simplify $\sqrt{-2} \cdot \sqrt{-32}$.

Solution

$$\sqrt{-2} \cdot \sqrt{-32} = (i \cdot \sqrt{2})(i \cdot \sqrt{16} \cdot \sqrt{2})$$
$$= (i\sqrt{2})(4i\sqrt{2}) = 4 \cdot i^2 \cdot 2$$
$$= 4(-1)2 \quad (\text{since } i^2 = -1)$$
$$= -8$$

Note that the property of radicals $\sqrt{x}\sqrt{y} = \sqrt{xy}$ does not hold in this case. That is, it is *not* true that

$$\sqrt{-2}\sqrt{-32} = \sqrt{(-2)(-32)} = \sqrt{64} = 8.$$

The correct result is -8. This is why we must be careful when we work with complex numbers. Always remember to express the complex numbers in terms of i before performing computations. $\qquad \square$

EXAMPLE 3   Perform the indicated operation and simplify $(2 - i\sqrt{3})^2$.

Solution

$$(2 - i\sqrt{3})^2 = (2 - i\sqrt{3})(2 - i\sqrt{3})$$

$$\begin{array}{r} 2 - i\sqrt{3} \\ 2 - i\sqrt{3} \\ \hline 4 - 2i\sqrt{3} \\ -2i\sqrt{3} + i^2 \cdot 3 \qquad (\text{But } i^2 = -1.) \\ \hline 4 - 4i\sqrt{3} + (-1)3 = 1 - 4i\sqrt{3} \end{array}$$

$\qquad \square$

EXAMPLE 4 Verify that the complex number $3 + 2i$ is a solution of the equation $(x - 3)^2 + 4 = 0$.

Solution We replace x by $3 + 2i$ and verify.

$$[(3 + 2i) - 3]^2 + 4 \stackrel{?}{=} 0$$
$$(2i)^2 + 4 \stackrel{?}{=} 0$$
$$4i^2 + 4 \stackrel{?}{=} 0$$
$$4(-1) + 4 \stackrel{?}{=} 0$$
$$0 \stackrel{\checkmark}{=} 0$$

Note in the check above that $4i^2 = -4$, since $i^2 = -1$. Thus, $3 + 2i$ is a solution of $(x - 3)^2 + 4 = 0$. □

EXAMPLE 5 Is $4 - 3i$ a solution of the equation $x^2 - 8x + 25 = 0$?

Solution To check we replace x by $4 - 3i$ in the given equation.

$$(4 - 3i)^2 - 8(4 - 3i) + 25 \stackrel{?}{=} 0$$
$$16 - 24i + 9i^2 - 32 + 24i + 25 \stackrel{?}{=} 0$$
$$9i^2 + 9 \stackrel{?}{=} 0$$
$$0 \stackrel{\checkmark}{=} 0$$

Therefore, $4 - 3i$ is a solution of the equation $x^2 - 8x + 25 = 0$. □

Conjugates and Division

To understand the procedure for dividing complex numbers, we first must define the conjugate of a complex number and then consider some basic properties of conjugates. The complex conjugate, or simply **conjugate,** of $a + bi$ is $a - bi$. It is standard notation to denote the conjugate of a number by placing a bar above the number. For example,

$$\overline{3 + 2i} = 3 - 2i \quad \text{and} \quad \overline{-4 - 0i} = -4 + 0i.$$

We now state some of the basic properties of conjugates that will be needed either here or in Section 4.4.

Properties of Conjugates	If z and w are complex numbers, then

1. $z \cdot \bar{z}$ is a real number.

2. $\bar{z} = z$ if and only if z is a real number.

3. $\overline{z + w} = \bar{z} + \bar{w}$.

4. $\overline{z \cdot w} = \bar{z} \cdot \bar{w}$.

5. $\overline{z^n} = (\bar{z})^n$ for any positive integer n.

Illustrations of these properties with specific complex numbers are given in the exercises, which also outline proofs for properties 2, 3, and 4. (The proof of property 5 requires the use of mathematical induction, which is discussed in the last chapter of the text.) Property 1 is our main interest at this time because it provides the rationale for the procedure for dividing two complex numbers. In words, property 1 states that the product of a complex number and its conjugate is always a real number. To prove this property, we first let $z = a + bi$. Then

$$
\begin{aligned}
z \cdot \bar{z} &= (a + bi)(a - bi) \\
&= a^2 - abi + abi - b^2 i^2 \\
&= a^2 + b^2.
\end{aligned}
$$

Since a and b are real numbers, $a^2 + b^2$ is a real number, which proves the property. Because $z \cdot \bar{z}$ is a real number, the quotient w/z can be written in the form $a + bi$ by multiplying both the numerator and the denominator by $\bar{z}$. Example 6 shows this procedure.

EXAMPLE 6   Write the quotient $\dfrac{4 + 3i}{1 - 2i}$ in the form $a + bi$.

Solution   As discussed, we divide two complex numbers by multiplying the numerator and the denominator by the conjugate of the denominator. Thus,

$$
\begin{aligned}
\frac{4 + 3i}{1 - 2i} &= \frac{4 + 3i}{1 - 2i} \cdot \frac{1 + 2i}{1 + 2i} \\
&= \frac{4 + 8i + 3i + 6i^2}{1 + 2i - 2i - 4i^2} \\
&= \frac{-2 + 11i}{5} = -\frac{2}{5} + \frac{11}{5}i.
\end{aligned}
$$ $\square$

EXAMPLE 7   The reciprocal, or multiplicative inverse, of a complex number z is $1/z$. Write the reciprocal of $4i$ in the form $a + bi$.

Solution   The reciprocal of $4i$ is $1/4i$. Since the conjugate of $0 + 4i$ is $0 - 4i$, we have

$$\frac{1}{4i} = \frac{1}{4i} \cdot \frac{-4i}{-4i} = \frac{-4i}{-16i^2} = \frac{-4i}{16} = -\frac{1}{4}i.$$

Thus, the reciprocal of $4i$ is $-\frac{1}{4}i$, or $0 - \frac{1}{4}i$ in standard form. We can verify our result as follows:

$$(4i)(-\tfrac{1}{4}i) = -1 \cdot i^2 = -1(-1) = 1.$$ □

In summary, note that we have been operating with complex numbers according to the following definitions.

Operations on Complex Numbers

1. Two complex numbers are added by adding separately their real parts and their imaginary parts.

   $$(a + bi) + (c + di) = (a + c) + (b + d)i$$

2. Two complex numbers are subtracted by subtracting separately their real parts and their imaginary parts.

   $$(a + bi) - (c + di) = (a - c) + (b - d)i$$

3. Two complex numbers are multiplied in the usual algebraic method with i^2 being replaced by -1.

   $$(a + bi)(c + di) = ac + adi + bci + bdi^2$$
   $$= (ac - bd) + (ad + bc)i$$

4. Two complex numbers are divided by multiplying the numerator and the denominator by the conjugate of the denominator.

   $$\frac{a + bi}{c + di} = \frac{(a + bi)(c - di)}{(c + di)(c - di)} = \frac{(ac + bd) + (bc - ad)i}{c^2 + d^2}$$

We discuss complex numbers again in Section 7.4, giving both geometric and trigonometric representations of such numbers. Note that complex numbers are made up of two components and can, therefore, convey more information about a quantity. For example, a complex number is often used in physics to describe a force, since such a numerical representation indicates not only the magnitude but also the direction of the force. In electronics complex numbers are used extensively to designate voltage, current, and other electrical quantities, for this representation indicates both the strength and time (or phase) relationships of the quantities.

EXERCISES 2.2

In Exercises 1–20 simplify each expression in terms of i.

1. $\sqrt{-25}$

2. $\sqrt{-49}$

3. $2\sqrt{-16}$

4. $5\sqrt{-121}$

5. $\sqrt{-22}$

6. $-3\sqrt{-7}$

7. $\frac{1}{2}\sqrt{-12}$

8. $-3\sqrt{-98}$

9. $2 - \sqrt{-128}$

10. $-3 + \frac{1}{2}\sqrt{-162}$

11. $\sqrt{(-4)^2 - 4(1)(5)}$

12. $\sqrt{(12)^2 - 4(9)(8)}$

13. $\sqrt{(2)^2 - 4(-1)(-3)}$

14. $\sqrt{(-5)^2 - 4(2)(7)}$

15. $\dfrac{-(-1) - \sqrt{(-1)^2 - 4(2)(3)}}{2(2)}$

16. $\dfrac{-(0) + \sqrt{(0)^2 - 4(1)(2)}}{2(1)}$

17. $\dfrac{-(4) + \sqrt{(4)^2 - 4(1)(5)}}{2(1)}$

18. $\dfrac{-(-2) - \sqrt{(-2)^2 - 4(1)(5)}}{2(1)}$

19. $\dfrac{-(2) - \sqrt{(2)^2 - 4(5)(3)}}{2(5)}$

20. $\dfrac{-(-2) + \sqrt{(-2)^2 - 4(-5)(-3)}}{2(-5)}$

In Exercises 21–40 perform the indicated operations and write each result in the form $a + bi$.

21. $2\sqrt{-9} - 3\sqrt{-9} + 4\sqrt{-9}$

22. $4\sqrt{-1} - 2\sqrt{-4} + 5\sqrt{-16}$

23. $\sqrt{-8} + \sqrt{-18} - \sqrt{-50}$

24. $4\sqrt{-27} - 2\sqrt{-48} + 3\sqrt{-75}$

25. $(-1 - 2i) + (-2 + i)$

26. $(11 + 2i) - (-5 - 7i)$

27. $(3 - i) - (-2 + 5i)$

28. $(6 + 2i) + (2 - 4i)$

29. $\sqrt{-3} \cdot \sqrt{-12}$

30. $\sqrt{-5} \cdot \sqrt{-18}$

31. $(2i)^2$

32. $(-i)^2$

33. $(4 + 2i)(1 - 3i)$

34. $(2 - 5i)(3 - i)$

35. $(2 + 3i)(2 - 3i)$

36. $(6 - 5i)(6 + 5i)$

37. $(1 - 2i)^2$

38. $(5 + i)^2$

39. $(-3 + \frac{1}{3}i)^2$

40. $(-1 - \frac{1}{2}i)^2$

In Exercises 41–50 verify that the given number is a solution of the equation.

41. $x^2 + 25 = 0$; $5i$

42. $x^2 + 9 = 0$; $-3i$

43. $x^2 + 8 = 0$; $-2i\sqrt{2}$

44. $x^2 + 48 = 0$; $4i\sqrt{3}$

45. $x^2 - 2x + 2 = 0$; $1 - i$

46. $x^2 - 2x + 2 = 0$; $1 + i$

47. $2x^2 - x + 3 = 0$; $\dfrac{1}{4} + i\dfrac{\sqrt{23}}{4}$

48. $2x^2 - x + 3 = 0$; $\dfrac{1}{4} - i\dfrac{\sqrt{23}}{4}$

49. $x^4 - 16 = 0$; $2i$

50. $x^4 + 13x^2 - 48 = 0$; $-4i$

In Exercises 51–60 determine by checking if the given number is a solution of the equation.

51. $x^2 + 4 = 0$; $-2i$

52. $x^2 - 4 = 0$; $2i$

53. $x^2 - 4x + 5 = 0$; $2 + i$

54. $x^2 - 4x + 5 = 0$; $2 - i$

55. $2x^2 + x + 1 = 0$; $-1 + 2i$

56. $2x^2 + x + 1 = 0$; $-1 - 2i$

57. $3x^2 - 2x + 1 = 0$; $\dfrac{1}{3} - i\dfrac{\sqrt{2}}{3}$

58. $3x^2 - 2x + 1 = 0$; $\dfrac{1}{3} + i\dfrac{\sqrt{2}}{3}$

59. $x^3 - 27 = 0$; $3i$

60. $x^4 - 81 = 0$; $-3i$

In Exercises 61–66 find the conjugate of the given complex number.

61. $3 + 4i$ **62.** $5 - 7i$ **63.** i

64. $-4i$ **65.** -7 **66.** 7

In Exercises 67–76 write each quotient in the form $a + bi$.

67. $\dfrac{1}{1 + 2i}$ **68.** $\dfrac{-3}{2 - 3i}$ **69.** $\dfrac{-2}{5i}$

70. $\dfrac{1}{i}$ **71.** $\dfrac{i}{-1 - 4i}$ **72.** $\dfrac{-2i}{6 - 2i}$

73. $\dfrac{2 + i}{3 + i}$ **74.** $\dfrac{2 + 5i}{4 + 2i}$ **75.** $\dfrac{\sqrt{2} - i}{\sqrt{2} + i}$

76. $\dfrac{\frac{1}{2} + i}{\frac{1}{2} - i}$

In Exercises 77–80 find the reciprocal and write the answer in the form $a + bi$.

77. $-i$ **78.** $3i$ **79.** $1 + i$ **80.** $-2 - 2i$

In Exercises 81 and 82 verify that **(a)** $\overline{z + w} = \bar{z} + \bar{w}$, **(b)** $\overline{z \cdot w} = \bar{z} \cdot \bar{w}$, and **(c)** $\overline{z^2} = (\bar{z})^2$ for the given values of z and w.

81. $z = 1 + i, w = 2 - 3i$

82. $z = 4 - i, w = 5 + 2i$

In Exercises 83–86 let $z = a + bi$ and $w = c + di$, and use the definition of conjugate to prove each of the following properties.

83. $\overline{z + w} = \bar{z} + \bar{w}$ **84.** $\overline{z \cdot w} = \bar{z} \cdot \bar{w}$

85. If z is a real number, then $\bar{z} = z$.

86. If $\bar{z} = z$, then z is a real number.

87. Use the fact that $\overline{z^2} = \overline{z \cdot z}$ to prove that $\overline{z^2} = (\bar{z})^2$ for any complex number z.

88. Write each expression in a simpler form.

   a. i^3 **b.** i^4 **c.** i^5 **d.** i^6

   e. i^7 **f.** i^9 **g.** i^{93} **h.** i^{100}

89. Complex numbers are used extensively in electrical engineering to designate voltage (E), current (I), and resistance (R). However, since i usually symbolizes current in this discipline, complex numbers are written in the form $a + bj$ with $j = \sqrt{-1}$. In the following exercises use Ohm's law, $E = IR$, to determine the value of the missing variable.

   a. $I = 1 - 2j$ amps, $R = 2 + 3j$ ohms

   b. $E = 2$ volts, $I = 1 + 4j$ amps

   c. $R = 7 - 5j$ ohms, $E = 1 - j$ volts

90. Consider the product

$$\sqrt{-1} \cdot \sqrt{-1} = \sqrt{(-1)(-1)} = \sqrt{1} = 1$$
$$\text{but } \sqrt{-1} \cdot \sqrt{-1} = (\sqrt{-1})^2 = -1.$$

Therefore, $1 = -1$! What is wrong?

2.3 QUADRATIC EQUATIONS

An equation that can be put in the form $ax^2 + bx + c = 0$, where a, b, and c are real numbers with $a \neq 0$, is called a **second-degree** or **quadratic equation.** For example,

$$2x^2 - 7x + 6 = 0$$
$$4t^2 = 9t - 2$$
$$(x - 1)^2 = 5$$

are quadratic equations. Depending on the particular equation, different procedures are used for solving quadratic equations, and you need to consider carefully each of the methods that follow.

The Factoring Method

A technique that can often be used to solve a second-degree equation relies on factoring. For example, let us try to find the numbers that make the equation $x^2 + 5x + 4 = 0$ a true statement.

First, factor the expression on the left-hand side of the equation.

$$(x + 4)(x + 1) = 0$$

We now have the situation where the product of two factors is zero. In multiplication zero is a special number as outlined in the following principle.

Zero Product Principle   For any numbers a and b,

$$ab = 0 \text{ if and only if } a = 0 \text{ or } b = 0.$$

By applying this principle, we know

$$(x + 4)(x + 1) = 0 \text{ if and only if } x + 4 = 0 \text{ or } x + 1 = 0.$$

Since $x + 4 = 0$ when $x = -4$ and $x + 1 = 0$ when $x = -1$, the solutions are -4 and -1, and the solution set is $\{-4, -1\}$. To catch any mistakes, it is recommended that solutions be checked in the original equation, so

$$x^2 + 5x + 4 = 0 \qquad\qquad x^2 + 5x + 4 = 0$$
$$(-4)^2 + 5(-4) + 4 \overset{?}{=} 0 \qquad\qquad (-1)^2 + 5(-1) + 4 \overset{?}{=} 0$$
$$0 \overset{\checkmark}{=} 0 \qquad\qquad 0 \overset{\checkmark}{=} 0.$$

To summarize, quadratic equations that are factorable (with integer coefficients) are usually solved as follows:

Factoring Method for Solving Quadratic Equations

1. If necessary, change the form of the equation so that one side is 0.
2. Factor the nonzero side of the equation.
3. Set each factor equal to zero and obtain the solution(s) by solving the resulting equations.
4. Check each solution by substituting it in the original equation.

EXAMPLE 1   Solve $4x^2 = x$.

Solution

$$
\begin{aligned}
4x^2 &= x \\
4x^2 - x &= 0 \\
x(4x - 1) &= 0
\end{aligned}
$$

$$x = 0 \qquad 4x - 1 = 0$$
$$x = \tfrac{1}{4}$$

Check $4x^2 = x$ $\qquad$ $4x^2 = x$

$$4(0)^2 \overset{?}{=} 0 \qquad 4(\tfrac{1}{4})^2 \overset{?}{=} \tfrac{1}{4}$$

$$4(0) \overset{?}{=} 0 \qquad 4(\tfrac{1}{16}) \overset{?}{=} \tfrac{1}{4}$$

$$0 \overset{\checkmark}{=} 0 \qquad \tfrac{1}{4} \overset{\checkmark}{=} \tfrac{1}{4}$$

Thus, 0 and $\tfrac{1}{4}$ are the solutions of the equation, and the solution set is $\{0, \tfrac{1}{4}\}$.   □

EXAMPLE 2   The height (y) of a ball thrown vertically up from the roof of a building 64 ft high with an initial velocity of 48 ft/second is given by the formula $y = 64 + 48t - 16t^2$. When will the ball strike the ground?

Solution   When the ball strikes the ground, the height (y) of the ball is 0. Therefore, we want to find the values of t for which $64 + 48t - 16t^2$ equals 0.

$$
\begin{aligned}
0 &= 64 + 48t - 16t^2 \\
&= 16(4 + 3t - t^2) \\
&= 16(4 - t)(1 + t)
\end{aligned}
$$

Since the first factor 16 cannot be 0, there are two ways that $0 = 16(4 - t)(1 + t)$ can be true.

$$4 - t = 0 \qquad 1 + t = 0$$
$$\text{or}$$
$$4 = t \qquad t = -1$$

The ball hits the ground 4 seconds later. We reject -1 as a solution since for this problem a negative time has no physical significance.   □

The Quadratic Equation $ax^2 + c = 0$

The easiest quadratic equation to solve is one in which there is no x term and the equation has the form $ax^2 + c = 0$ with $a \neq 0$. For example, to solve $2x^2 - 10 = 0$, we merely transform the equation so that x^2 is on one side of the equation by itself giving $x^2 = 5$. We then conclude the solutions are $\sqrt{5}$ and $-\sqrt{5}$ by applying the following property.

Square Root Property	If n is any real number, then
	$$x^2 = n \text{ implies } x = \sqrt{n} \text{ or } x = -\sqrt{n}.$$

This property is a key step in the general solution to quadratic equations, and we can prove this property as follows:

$$x^2 = n$$
$$x^2 - n = 0$$
$$(x - \sqrt{n})(x + \sqrt{n}) = 0 \qquad \text{Factoring over the set of complex numbers}$$
$$x - \sqrt{n} = 0 \quad \text{or} \quad x + \sqrt{n} = 0 \qquad \text{Zero product principle}$$
$$x = \sqrt{n} \quad \text{or} \qquad x = -\sqrt{n}.$$

We usually abbreviate these solutions as $\pm\sqrt{n}$. Note that there are two real number solutions if $n > 0$, two (conjugate) complex number solutions if $n < 0$, and one solution (namely, 0) if $n = 0$.

EXAMPLE 3   Solve each equation by using the square root property.

a. $x^2 = 27$ **b.** $5(x - 3)^2 + 20 = 0$

Solution

a. $x^2 = 27$ implies $x = \pm\sqrt{27}$. Since $\sqrt{27} = \sqrt{9}\sqrt{3} = 3\sqrt{3}$ the solution set is $\{\pm 3\sqrt{3}\}$.

b. First, transform the equation so $(x - 3)^2$ is on one side of the equation by itself.

$$5(x - 3)^2 + 20 = 0$$
$$5(x - 3)^2 = -20$$
$$(x - 3)^2 = -4$$

Now apply the square root property and simplify.

$$x - 3 = \pm\sqrt{-4}$$
$$x = 3 \pm 2i \qquad \text{Since } \sqrt{-4} = 2i$$

Thus, the solution set is $\{3 \pm 2i\}$.   □

Completing the Square

In Example 3b we used the square root property to solve a quadratic equation of the form

$$(x + \text{constant})^2 = \text{constant}.$$

Actually, any quadratic equation may be placed in this form by a technique called completing the square. Consider an expression like

$$x^2 + 8x.$$

What constant needs to be added to make this expression a perfect square? Since

$$(x + k)^2 = x^2 + 2kx + k^2$$

we set $2k = 8$, so $k = 4$ and $k^2 = 16$. Thus, adding 16 gives

$$x^2 + 8x + 16 = (x + 4)^2.$$

In general, we complete the square for $x^2 + bx$ by adding $(b/2)^2$, which is the square of one half of the coefficient of x. Example 4 shows how completing the square can solve a quadratic equation.

EXAMPLE 4   Solve the equation $x^2 - 4x - 3 = 0$ by completing the square.

Solution   First, rearrange the equation with the x terms to the left of the equals and the constant to the right.

$$x^2 - 4x = 3$$

Now complete the square on the left. Half of -4 is -2 and $(-2)^2 = 4$. Add 4 to both sides of the equation and proceed as follows:

$$x^2 - 4x + 4 = 3 + 4$$
$$(x - 2)^2 = 7$$
$$x - 2 = \pm\sqrt{7}$$
$$x = 2 \pm \sqrt{7}.$$

The solutions are $2 + \sqrt{7}$ and $2 - \sqrt{7}$, and the solution set is abbreviated $\{2 \pm \sqrt{7}\}$.   □

Quadratic Formula

Completing the square is a useful technique in many situations because it often converts an expression to a standard form that is easy to analyze. In this case, the completing the square method works so well that if we apply it to the general quadratic equation $ax^2 + bx + c = 0$, we obtain a powerful formula (the quadratic formula) that solves all quadratic equations. We now derive this formula with the particular equation $2x^2 - 5x + 1 = 0$ being solved on the left to illustrate in specific terms what is happening on the right.

Particular Equation	General Equation
$2x^2 - 5x + 1 = 0$	$ax^2 + bx + c = 0$ (with $a \neq 0$)

We shall attempt to make the left side of the equation a perfect square so that we can use the square root property. To begin, if the coefficient of x^2 is not 1, we must divide by this coefficient before proceeding as before.

$$x^2 - \frac{5}{2}x + \frac{1}{2} = 0 \qquad x^2 + \frac{b}{a}x + \frac{c}{a} = 0$$

Subtract the constant term from both sides of the equation.

$$x^2 - \frac{5}{2}x = \frac{-1}{2} \qquad x^2 + \frac{b}{a}x = \frac{-c}{a}$$

We now add a number to both sides of the equation to make the left side a perfect square. The desired number can always be found by taking half the coefficient of the x term and squaring the result.

Half of $\dfrac{-5}{2}$ is $\dfrac{-5}{4}$ and   Half of $\dfrac{b}{a}$ is $\dfrac{b}{2a}$ and

$\left(\dfrac{-5}{4}\right)^2$ is $\dfrac{25}{16}$ $\left(\dfrac{b}{2a}\right)^2$ is $\dfrac{b^2}{4a^2}$

$$x^2 - \frac{5}{2}x + \frac{25}{16} \qquad\qquad x^2 + \frac{b}{a}x + \frac{b^2}{4a^2}$$

$$= \frac{-1}{2} + \frac{25}{16}. \qquad\qquad = \frac{-c}{a} + \frac{b^2}{4a^2}.$$

Write the left side as a perfect square and simplify the right side.

$$\left(x - \frac{5}{4}\right)^2 = \frac{17}{16} \qquad\qquad \left(x + \frac{b}{2a}\right)^2 = \frac{b^2 - 4ac}{4a^2}$$

Use the square root property.

$$x - \frac{5}{4} = \pm\sqrt{\frac{17}{16}} \qquad\qquad x + \frac{b}{2a} = \pm\sqrt{\frac{b^2 - 4ac}{4a^2}}$$

Simplify the result.

$$x - \frac{5}{4} = \frac{\pm\sqrt{17}}{4} \qquad\qquad x + \frac{b}{2a} = \frac{\pm\sqrt{b^2 - 4ac}}{2a}$$

$$x = \frac{5}{4} + \frac{\pm\sqrt{17}}{4} \qquad\qquad x = \frac{-b}{2a} + \frac{\pm\sqrt{b^2 - 4ac}}{2a}$$

$$= \frac{5 \pm \sqrt{17}}{4} \qquad\qquad = \frac{-b \pm \sqrt{b^2 - 4ac}}{2a}$$

Thus, our two solutions x_1 and x_2 are

$$x_1 = \frac{5 + \sqrt{17}}{4} \qquad\qquad x_1 = \frac{-b + \sqrt{b^2 - 4ac}}{2a}$$

$$x_2 = \frac{5 - \sqrt{17}}{4} \qquad\qquad x_2 = \frac{-b - \sqrt{b^2 - 4ac}}{2a}$$

and we have derived the following formula.

Quadratic Formula   If $ax^2 + bx + c = 0$ and $a \neq 0$, then

$$x = \frac{-b \pm \sqrt{b^2 - 4ac}}{2a}.$$

Note that this formula results in two solutions.

$$x_1 = \frac{-b + \sqrt{b^2 - 4ac}}{2a} \quad \text{and} \quad x_2 = \frac{-b - \sqrt{b^2 - 4ac}}{2a}$$

We can solve *any* quadratic equation with this formula by identifying the values of a, b, and c and substituting these numbers into the formula.

EXAMPLE 5   Solve the equation $x^2 - 4x - 3 = 0$ by using the quadratic formula.

Solution   In this equation $a = 1$, $b = -4$, and $c = -3$. Therefore,

$$x = \frac{-b \pm \sqrt{b^2 - 4ac}}{2a} = \frac{-(-4) \pm \sqrt{(-4)^2 - 4(1)(-3)}}{2(1)}$$

$$= \frac{4 \pm \sqrt{16 + 12}}{2}$$

$$= \frac{4 \pm \sqrt{28}}{2}$$

$$= \frac{4 \pm 2\sqrt{7}}{2}$$

$$= 2 \pm \sqrt{7}.$$

Thus, $x_1 = 2 + \sqrt{7}$, $x_2 = 2 - \sqrt{7}$, and (as in Example 4) the solution set is $\{2 \pm \sqrt{7}\}$.   □

EXAMPLE 6   Solve the equation $3x^2 + 5x + 2 = 0$ by using the quadratic formula.

Solution   In this equation $a = 3$, $b = 5$, and $c = 2$. Therefore,

$$x = \frac{-b \pm \sqrt{b^2 - 4ac}}{2a} = \frac{-(5) \pm \sqrt{(5)^2 - 4(3)(2)}}{2(3)}$$

$$= \frac{-5 \pm \sqrt{25 - 24}}{6}$$

$$= \frac{-5 \pm \sqrt{1}}{6}$$

$$= \frac{-5 \pm 1}{6}.$$

$$x_1 = \frac{-5 + 1}{6} = \frac{-4}{6} = \frac{-2}{3} \qquad x_2 = \frac{-5 - 1}{6} = \frac{-6}{6} = -1$$

Thus, the solution set is $\{-\frac{2}{3}, -1\}$.   □

EXAMPLE 7   Solve the equation $4x^2 = 20x - 25$ by using the quadratic formula.

SECTION 2.3   QUADRATIC EQUATIONS

Solution   First, express the equation in the form $ax^2 + bx + c = 0$.

$$4x^2 - 20x + 25 = 0$$

In this equation $a = 4$, $b = -20$, and $c = 25$. Therefore,

$$x = \frac{-b \pm \sqrt{b^2 - 4ac}}{2a} = \frac{-(-20) \pm \sqrt{(-20)^2 - 4(4)(25)}}{2(4)}$$

$$= \frac{20 \pm \sqrt{400 - 400}}{8}$$

$$= \frac{20 \pm \sqrt{0}}{8}.$$

$$x_1 = \frac{20 + 0}{8} = \frac{20}{8} = \frac{5}{2} \qquad x_2 = \frac{20 - 0}{8} = \frac{20}{8} = \frac{5}{2}$$

The roots x_1 and x_2 both equal $\frac{5}{2}$, so the solution set is $\{\frac{5}{2}\}$.   □

EXAMPLE 8   Solve the equation $2x^2 - x + 3 = 0$ by using the quadratic formula.

Solution   In this equation $a = 2$, $b = -1$, and $c = 3$. Therefore,

$$x = \frac{-b \pm \sqrt{b^2 - 4ac}}{2a} = \frac{-(-1) \pm \sqrt{(-1)^2 - 4(2)(3)}}{2(2)}$$

$$= \frac{1 \pm \sqrt{1 - 24}}{4}$$

$$= \frac{1 \pm \sqrt{-23}}{4}$$

$$= \frac{1 \pm i\sqrt{23}}{4}.$$

$$x_1 = \frac{1 + i\sqrt{23}}{4} \qquad x_2 = \frac{1 - i\sqrt{23}}{4}$$

The solutions of this equation are (conjugate) complex numbers, and the solution set is $\left\{\dfrac{1 \pm i\sqrt{23}}{4}\right\}$.   □

The Discriminant

Notice from the examples that the nature of the solutions to a quadratic equation depends on the value of $b^2 - 4ac$ (called the **discriminant**), which appears under the radical in the quadratic formula. That is,

When a, b, c Are Rational and	The Solutions Are
$b^2 - 4ac < 0$	conjugate complex numbers
$b^2 - 4ac = 0$	real, rational, equal numbers
$b^2 - 4ac > 0$ and a perfect square	real, rational, unequal numbers
$b^2 - 4ac > 0$, not a perfect square	real, irrational, unequal numbers

For example, in the equation $3x^2 + 5x + 2 = 0$ (Example 6), $b^2 - 4ac = (5)^2 - 4(3)(2) = 1$. Since 1 is greater than 0 and a perfect square, we know that there are two different solutions that are rational numbers. As another example, in the equation $2x^2 - x + 3 = 0$ (Example 8), $b^2 - 4ac = (-1)^2 - 4(2)(3) = -23$. Since -23 is less than 0, our solutions are conjugate complex numbers. Thus, without solving the equation, we can determine the nature of the solutions by finding the value of $b^2 - 4ac$.

When using the quadratic formula on a scientific calculator, it is best to start with the radical, store its value, and then compute both roots by recalling the stored number at the appropriate spot in the formula. For example, to solve the equation in Example 5

$$x^2 - 4x - 3 = 0$$

first, compute the radical, $\sqrt{b^2 - 4ac}$, and store it.

4 [+/−] [x^2] [−] 4 [×] 1 [×] 3 [+/−] [=] [√] [STO]   5.2915026

Now, one solution adds the stored number.

4 [+/−] [+/−] [+] [RCL] [=] [÷] 2 [÷] 1 [=]   4.6457513

The other solution subtracts the stored number.

4 [+/−] [+/−] [−] [RCL] [=] [÷] 2 [÷] 1 [=]   −0.6457513

Note that it is unnecessary to multiply or divide by 1, press the change sign key [+/−] twice in succession, and change sign when an entry is to be squared. Thus, the keystrokes shown in color may be omitted. If the roots are not real numbers, an error message appears in the display when computing the radical in the first step.

EXERCISES 2.3

In Exercises 1–16 solve each equation by using the factoring method.

1. $x(x + 5) = 0$ **2.** $x(x - 1) = 0$

3. $x^2 - 5x = 0$ **4.** $y^2 = 2y$

5. $a^2 - 4 = 0$ **6.** $5x^2 - 5 = 0$

7. $y^2 - 2y - 8 = 0$ **8.** $z^2 - 3z - 18 = 0$

9. $k^2 - 3k = 4$ **10.** $b^2 + 3b = 4$

11. $3x^2 - 16x + 5 = 0$ **12.** $5y^2 - 11y + 2 = 0$

13. $3r^2 = 5r - 2$ **14.** $2x^2 = 7x + 4$

15. $2(x^2 - 1) = 3x$ **16.** $3(x^2 + 1) = 10x$

In Exercises 17–28 solve each equation by using the square root property.

17. $x^2 = 9$ **18.** $x^2 = 36$

19. $x^2 = -25$ **20.** $x^2 = -9$

21. $3x^2 + 24 = 0$ **22.** $2y^2 + 48 = 0$

23. $(b - 5)^2 = 4$ **24.** $(x + 3)^2 = 100$

25. $(x + 1)^2 = -1$ **26.** $(a - 1)^2 = -16$

27. $5(k - 7)^2 + 35 = 0$ **28.** $3(z + 4)^2 + 42 = 0$

In Exercises 29–32 determine the number that should be added to make the expression a perfect square.

29. $x^2 + 6x$ **30.** $x^2 - x$

31. $x^2 - \frac{3}{2}x$ **32.** $x^2 + \frac{7}{3}x$

In Exercises 33–40 solve each equation by completing the square.

33. $x^2 - 2x + 2 = 0$ **34.** $x^2 + 4x + 5 = 0$

35. $x^2 + x - 3 = 0$ **36.** $x^2 + 4x + 4 = 0$

37. $3x^2 - 4x - 2 = 0$ **38.** $5x^2 - 3x - 4 = 0$

39. $-5x^2 = 2x + 3$ **40.** $-2x^2 = 6x + 5$

In Exercises 41–52 solve each equation by using the quadratic formula.

41. $x^2 + 5x + 4 = 0$ **42.** $x^2 - 3x + 2 = 0$

43. $x^2 - 6x + 9 = 0$ **44.** $x^2 + 2x - 1 = 0$

45. $x^2 - 4 = 0$ **46.** $x^2 + 2 = 0$

47. $3x^2 - 2x = 0$ **48.** $2x^2 + x = 0$

49. $3x^2 = 2x - 1$ **50.** $4x^2 = 12x - 9$

51. $4x^2 + x = 3$ **52.** $5x^2 - 28x = 12$

In Exercises 53–62 use the discriminant to determine the nature of the solutions of the equation.

53. $2x^2 - x + 2 = 0$ **54.** $-3x^2 + x + 1 = 0$

55. $x^2 + 2x + 1 = 0$ **56.** $5x^2 - 7x + 2 = 0$

57. $x^2 - 10x - 9 = 0$ **58.** $-2x^2 + 4x + 9 = 0$

59. $4x^2 = 2x + 5$ **60.** $3x^2 = -4x - 1$

61. $x^2 - 5x = -7$ **62.** $-2x^2 - x = -2$

63. The sum of a positive number and its square is 56. What is the number?

64. The perimeter of a rectangle is 100 ft and the area is 400 ft². Find the dimensions of the rectangle.

65. The length of a rectangle is 2 ft more than the width and the area is 24 ft². Find the dimensions of the rectangle.

66. The area of a rectangular plate is to be 28 in.² and the sides of the plate are to differ in length by 3 in. Find the dimensions of the plate.

67. A rectangular plate is to have an area of 30 in.² and the sides of the plate are to differ in length by 4 in. To the nearest tenth of an inch, find the dimensions of the plate.

68. When a uniform border is added to a rug that is 6 ft by 4 ft, the area increases by 11 ft². Find the width of the border.

69. The height (y) of a projectile that is shot directly up from the ground with an initial velocity of 96 ft/second is given by the formula $y = 96t - 16t^2$.
   a. When will the projectile strike the ground?
   b. At what time after firing is the projectile 80 ft off the ground?
   c. Physically, why are there two solutions to part **b**?

70. The height (y) above water of a diver t seconds after she steps off a platform 10 ft high is given by the formula $y = 10 - 16t^2$. When will the diver hit the water?

71. The height (y) of a ball that is thrown down from a roof 200 ft high with an initial velocity of 50 ft/second is given by the formula $y = 200 - 50t - 16t^2$. To the nearest tenth of a second, when will the ball hit the ground?

72. The height (y) of a projectile that is shot directly up from the ground with an initial velocity of 100 ft/second is given by the formula $y = 100t - 16t^2$. To the nearest hundredth of a second, when will the projectile initially attain a height of 50 ft?

73. The total sales revenue for a company is estimated by the formula $S = 8x - x^2$, where x is the unit selling price (in dollars) and S is the total sales revenue (1 unit equals \$10,000). What should be the unit selling price if the company wishes the total revenue from sales to be \$150,000 (that is, $S = 15$)? Why are there two solutions to the problem?

74. A manufacturer sells his product for \$10 per unit. The cost of manufacturing x units is estimated by the formula $C = 200 + x + 0.01x^2$. How many units must he manufacture and sell to earn a profit of \$1,200? (*Hint:* profit = income − cost.)

75. From a square sheet of metal an open box is made by cutting 2-in. squares from the four corners and folding up the ends. To the nearest tenth of an inch, how large a piece of metal should be used if the box is to have a volume of 100 in.³?

76. When two resistors R_1 and R_2 are connected in series, their combined resistance is given by $R = R_1 + R_2$. If the resistors are connected in parallel, their combined resistance is given by $1/R = (1/R_1) + (1/R_2)$. Find the value of resistors R_1 and R_2 if their combined resistance is 32 ohms when connected in series and 6 ohms when connected in parallel.

77. Show that the sum of the solutions of the equation $ax^2 + bx + c = 0$ (with $a \neq 0$) is $-b/a$ and that the product of the solutions is c/a.

78. Consider the following expression, which continues indefinitely:

$$x = 1 + \cfrac{1}{1 + \cfrac{1}{1 + \text{etc.}}}$$

This expression is called a **continued fraction.** Continued fractions have been studied by many of the great mathematicians since the seventeenth century and are particularly useful in the analysis of numbers. For example, the continued fraction representation of π shows the following ratios to be succeedingly better approximations for this number: 22/7, 333/106, 355/113, and 103,993/33,102. The list could be continued, but for most purposes 355/113 is adequate since this ratio is accurate to seven significant digits. Determine the real number x in the continued fraction above. The solution requires a clever substitution.

79. If $R > 0$, solve this equation for x.

$$(4x) \left(\frac{-2x}{2\sqrt{R^2 - x^2}} \right) + 4\sqrt{R^2 - x^2} = 0$$

80. If $x^2 - 2xy + y^2 - x = 0$, use the quadratic formula to show that $y = x \pm \sqrt{x}$.

2.4 OTHER TYPES OF EQUATIONS

An equation in which the unknown appears under a radical is called a **radical equation.** Use these steps to solve such equations.

1. Isolate a radical on one side of the equation.
2. Raise both sides of the equation to the power that matches the index of the radical.
3. Solve the resulting equation and check all solutions in the *original* equation.

The check in Step 3 is essential because this procedure is based on the following principle.

Principle of Powers   If P and Q are algebraic expressions, then the solution set of the equation $P = Q$ is a subset of the solution set of $P^n = Q^n$ for any positive integer n.

Thus, every solution of $P = Q$ is a solution of $P^n = Q^n$; but solutions of $P^n = Q^n$ may or may not be solutions of $P = Q$, so checking *is* necessary. Solutions of $P^n = Q^n$ that do not satisfy the original equation are called **extraneous solutions,** and Example 1 illustrates this possibility.

EXAMPLE 1   Solve $x - \sqrt{x + 1} = 1$.

Solution   We must first isolate the radical on one side of the equation.

$$x - \sqrt{x + 1} = 1$$
$$x - 1 = \sqrt{x + 1}$$

Now square both sides of the equation and solve for x.

$$x^2 - 2x + 1 = x + 1$$
$$x^2 - 3x = 0$$
$$x(x - 3) = 0$$
$$x = 0 \text{ or } x = 3$$

Check

$$0 - \sqrt{0 + 1} \overset{?}{=} 1 \qquad\qquad 3 - \sqrt{3 + 1} \overset{?}{=} 1$$
$$0 - 1 \overset{?}{=} 1 \qquad\qquad 3 - 2 \overset{?}{=} 1$$
$$-1 \neq 1 \quad \text{Extraneous solution} \qquad\qquad 1 \overset{\checkmark}{=} 1$$

Only 3 is a solution of the original equation, so the solution set is $\{3\}$.   □

EXAMPLE 2   Solve the equation $\sqrt[3]{x^2 + 15} = 4$.

Solution   The index of the radical is 3, so we raise both sides of the equation to the third power and then solve for x.

$$\sqrt[3]{x^2 + 15} = 4$$
$$(\sqrt[3]{x^2 + 15})^3 = 4^3$$
$$x^2 + 15 = 64$$
$$x^2 = 49$$
$$x = \pm 7$$

Check

$$\sqrt[3]{(7)^2 + 15} \overset{?}{=} 4 \qquad\qquad \sqrt[3]{(-7)^2 + 15} \overset{?}{=} 4$$
$$\sqrt[3]{64} \overset{?}{=} 4 \qquad\qquad \sqrt[3]{64} \overset{?}{=} 4$$
$$4 \overset{\checkmark}{=} 4 \qquad\qquad 4 \overset{\checkmark}{=} 4$$

Since both solutions check, the solution set is $\{7, -7\}$.   □

EXAMPLE 3   Solve the equation $\sqrt{x} + \sqrt{x + 5} = 5$.

Solution   To solve the equation, we first isolate one of the radicals.

$$\sqrt{x} + \sqrt{x + 5} = 5$$
$$\sqrt{x + 5} = 5 - \sqrt{x}$$
$$(\sqrt{x + 5})^2 = (5 - \sqrt{x})^2 \qquad \text{Square both sides of the equation.}$$
$$x + 5 = 25 - 10\sqrt{x} + x$$
$$-20 = -10\sqrt{x}$$
$$2 = \sqrt{x}$$
$$(2)^2 = (\sqrt{x})^2 \qquad \text{Square both sides of the equation.}$$
$$4 = x$$

Check

$$\sqrt{4} + \sqrt{4 + 5} \overset{?}{=} 5$$
$$2 + 3 \overset{?}{=} 5$$
$$5 \overset{\checkmark}{=} 5$$

Thus, the solution set is {4}. Note that it is sometimes necessary to raise both sides of the equation to some power more than once.   □

EXAMPLE 4   Solve the equation $x + \sqrt{x} - 2 = 0$.

Solution   To solve the equation, we must first isolate the radical.

$$x + \sqrt{x} - 2 = 0$$
$$x - 2 = -\sqrt{x}$$
$$(x - 2)^2 = (-\sqrt{x})^2 \qquad \text{Square both sides of the equation.}$$
$$x^2 - 4x + 4 = x$$
$$x^2 - 5x + 4 = 0$$
$$(x - 4)(x - 1) = 0$$
$$x = 4 \quad \text{or} \quad x = 1$$

Check

$$4 + \sqrt{4} - 2 \overset{?}{=} 0 \qquad\qquad\qquad 1 + \sqrt{1} - 2 \overset{?}{=} 0$$
$$4 + 2 - 2 \overset{?}{=} 0 \qquad\qquad\qquad 1 + 1 - 2 \overset{?}{=} 0$$
$$4 \neq 0 \quad \text{Extraneous solution} \qquad\qquad 0 \overset{\checkmark}{=} 0$$

Therefore, 1 is the solution of the equation, and the solution set is {1}.   □

In Example 4 the solution led to a quadratic equation. Although we often encounter quadratic equations in the process of solving other equations, $x + \sqrt{x} - 2 = 0$ is a special case. Equations that are not themselves quadratic but which are equivalent to equations having the form

$$at^2 + bt + c = 0 \qquad (a \neq 0)$$

are called **equations with quadratic form.** In Example 4 if we let $t = \sqrt{x}$, then $x + \sqrt{x} - 2 = 0$ becomes

$$t^2 + t - 2 = 0$$

with solution

$$(t + 2)(t - 1) = 0$$
$$t = -2 \quad \text{or} \quad t = 1.$$

Now we can resubstitute $\sqrt{x}$ for t and solve for x.

$$\sqrt{x} = -2 \quad \text{or} \quad \sqrt{x} = 1$$
$$x = 4 \quad \text{or} \quad x = 1$$

As in our solution above, only 1 is a root of the equation. The extraneous root $x = 4$ is easy to pick out here because $\sqrt{x}$ cannot be -2. To spot equations with quadratic form, look for the exponent in one term to be double the exponent in another term.

EXAMPLE 5   Solve $x^{-2} - 4x^{-1} - 3 = 0$.

Solution   First, note that

$$x^{-2} - 4x^{-1} - 3 = (x^{-1})^2 - 4(x^{-1}) - 3 = 0,$$

so if we let $t = x^{-1}$, we have

$$t^2 - 4t - 3 = 0.$$

By the quadratic formula (see Section 2.3, Example 5),

$$t = 2 \pm \sqrt{7}.$$

Now we resubstitute x^{-1} for t and solve for x.

$$x^{-1} = 2 \pm \sqrt{7}$$
$$x = \frac{1}{2 \pm \sqrt{7}} = \frac{2 \pm \sqrt{7}}{-3}$$

Since x^{-1} may be any real number except 0, both roots check, and the solution set is $\left\{ \dfrac{2 \pm \sqrt{7}}{-3} \right\}$. □

Higher Degree Polynomial Equations

We have considered how to solve linear and quadratic equations which are first- and second-degree polynomial equations. Higher degree polynomial equations are usually much harder to solve, and we will consider this topic in detail in Chapter 4. For now, however, it is sometimes possible to factor higher degree polynomials into linear and/or quadratic factors with integer coefficients so that the equations can be solved by using the zero product principle. The idea is the same as with second-degree equations. First, rearrange the equation (if necessary) so that zero is on one side by itself, then factor. Now set each factor equal to zero and obtain solutions by solving the resulting equations.

EXAMPLE 6   Solve $x^5 = 10x^3 - 9x$.

Solution   First, rewrite the equation so that one side is 0.

$$x^5 = 10x^3 - 9x$$
$$x^5 - 10x^3 + 9x = 0$$

Now factor completely.

$$x(x^4 - 10x^2 + 9) = 0$$
$$x(x^2 - 9)(x^2 - 1) = 0$$
$$x(x + 3)(x - 3)(x + 1)(x - 1) = 0$$

Setting each factor equal to zero gives

$$x = 0 \quad\bigg|\quad \begin{matrix} x + 3 = 0 \\ x = -3 \end{matrix} \quad\bigg|\quad \begin{matrix} x - 3 = 0 \\ x = 3 \end{matrix} \quad\bigg|\quad \begin{matrix} x + 1 = 0 \\ x = -1 \end{matrix} \quad\bigg|\quad \begin{matrix} x - 1 = 0 \\ x = 1. \end{matrix}$$

Thus, the solution set is $\{0, -3, 3, -1, 1\}$.   □

EXERCISES 2.4

In Exercises 1–50 solve each equation and check the solution.

1. $\sqrt{x + 5} = 2$
2. $\sqrt{2x + 1} = 7$
3. $\sqrt{x - 5} = -3$
4. $\sqrt{x - 2} + 3 = 0$
5. $2\sqrt{3x + 3} = 13$
6. $3\sqrt{2x} - 4 = 5$
7. $\sqrt[4]{2x + 1} = \sqrt[4]{3x - 5}$
8. $\sqrt[4]{x^2 + x + 6} = \sqrt[4]{x^2 + 3x + 2}$
9. $\sqrt[3]{x + 1} = -2$
10. $\sqrt[3]{x^2 - 17} = 4$
11. $\sqrt{x^2 + 9} = x + 1$
12. $\sqrt{x^2 - x + 1} = x - 1$

13. $\sqrt{8 - x^2} + x = 0$
14. $\sqrt{x + 1} + 1 = x$
15. $\sqrt{x - 2} + x = 4$
16. $\sqrt{x + 4} - 4 = x$
17. $2\sqrt{x + 1} = 2 - x$
18. $2\sqrt{x - 1} = x - 1$
19. $\sqrt{2x^2 - 2} + x - 1 = 0$
20. $\sqrt{x - 4} = 5 - \sqrt{x + 1}$
21. $\sqrt{x - 8} + \sqrt{x} = 2$
22. $\sqrt{x + 5} - 9 = \sqrt{x - 4}$
23. $\sqrt{x} - \sqrt{2x + 1} = -1$
24. $\sqrt{4x + 2} + \sqrt{2x} - \sqrt{2} = 0$

25. $\sqrt{2x} + \sqrt{2x + 3} = 3$

26. $\dfrac{10}{\sqrt{x - 5}} = \sqrt{x - 5} - 3$

27. $\sqrt{x^2} = x$

28. $\sqrt{(2x - 1)^2} = 2x - 1$

29. $\dfrac{x}{\sqrt{x^2 + 16}} - 1 = 0$

30. $\dfrac{1}{4} \cdot \dfrac{2x}{2\sqrt{x^2 + 16}} - \dfrac{1}{5} = 0$

31. $\dfrac{1}{\sqrt{2}} = \dfrac{6 - a}{\sqrt{5(9 + a^2)}}$

32. $\dfrac{2a + 3}{\sqrt{5a^2 + 45}} = 1$

33. $x + 2\sqrt{x} - 3 = 0$

34. $x - 10\sqrt{x} + 9 = 0$

35. $x + 23 = 10\sqrt{x}$

36. $3x^{-2} + 1 = 4x^{-1}$

37. $x^{-2} + x^{-1} = 0$

38. $x^4 - 3 = 4x^2$

39. $x^{1/2} + 2 = 3x^{1/4}$

40. $x^{1/3} - 1 = 2x^{-1/3}$

41. $y^3 + 9y^2 + 14y = 0$

42. $4a^3 - 13a^2 = -3a$

43. $3a^3 = 3a$

44. $16y^3 = 9y$

45. $x^2(2x - 1)(3x + 1) = 0$

46. $x^3(1 - 2x)(4x + 5) = 0$

47. $x^4 + 4 = 5x^2$

48. $x^6 = 10x^4 - 9x^2$

49. $3x^3 - 11x^2 = 3x - 11$

50. $2x^3 - x^2 - 8x + 4 = 0$

51. If $c > 0$, solve $\sqrt{x + c} = \sqrt{x} + \sqrt{c}$ for x.

52. Solve for d: $r = \sqrt[3]{3a/(4\pi d)}$.

53. Solve for y: $x = \sqrt[3]{y} + 2$.

54. Solve for g: $t = \pi\sqrt{L/g}$.

55. If the perimeter in this right triangle is 14 cm, find x.

2.5 INEQUALITIES

The solution to a problem often requires us to use the inequality signs ($<$, $\leq$, $>$, $\geq$) when expressing relationships. For instance, in Example 2 we will determine the temperatures in degrees Fahrenheit at which water turns to steam by solving the inequality

$$\tfrac{5}{9}(F - 32) \geq 100.$$

As with equations, a **solution** of an inequality is a value for the variable that makes the inequality a true statement, and the set of all such solutions is called the **solution set.** To solve an inequality like $\tfrac{5}{9}(F - 32) \geq 100$, we use basically the same method that we developed to solve linear equations, but an important exception is discussed in Example 1.

EXAMPLE 1 $4 < 8$. If we multiply both sides of the inequality by -3, the result, $-12 < -24$, is a false statement. We can make this statement true by changing the "less than" to "greater than" so that the result is $-12 > -24$. In general, if we multiply an inequality by a negative number, we must change the sense of the inequality if the statement is to be true. The sense of the inequality means the direction in which the inequality symbol is pointing.   □

With this example in mind, we now state the properties of inequalities that are the basis for solving inequalities.

Properties of Inequalities   Let a, b, and c be real numbers.

	Comment
1. If $a < b$, then $a + c < b + c$.	The sense of the inequality is preserved when the same number is added to (or subtracted from) both sides of an inequality.
2. If $a < b$ and $c > 0$, then $ac < bc$.	The sense of the inequality is preserved when both sides of an inequality are multiplied (or divided) by the same positive number.
3. If $a < b$ and $c < 0$, then $ac > bc$.	The sense of the inequality is reversed when both sides of an inequality are multiplied (or divided) by the same negative number.
4. If $a < b$ and $b < c$, then $a < c$.	This property is called the **transitive property.**

Similar properties may be stated for the other inequality signs. We can prove any of these properties by using the appropriate definition from Section 1.1 and the fact that the sum and product of any two positive numbers is positive. For example, to prove that

if $a < b$ and $c < 0$, then $ac > bc$,

we recall that $a < b$ if and only if $b - a$ is positive. Similarly, $c < 0$ implies $0 - c$, or $-c$, is positive. Then

$$\underbrace{(b - a)}_{\text{pos}}\underbrace{(-c)}_{\text{pos}} = \underbrace{-bc + ac}_{\text{pos}}.$$

Thus, $ac - bc$ is positive, so $bc < ac$, or alternatively $ac > bc$, which proves the property.

EXAMPLE 2   Solve the inequality $2x + 1 < 5x - 8$.

Solution   We isolate x on one side of the inequality in the following sequence of equivalent (same solution set) inequalities.

$$2x + 1 < 5x - 8$$
$$2x < 5x - 9 \qquad \text{Subtract 1 from both sides.}$$
$$-3x < -9 \qquad \text{Subtract } 5x \text{ from both sides.}$$
$$\frac{-3x}{-3} > \frac{-9}{-3} \qquad \begin{array}{l}\text{Divide both sides by } -3 \text{ and change}\\ \text{the sense of the inequality.}\end{array}$$
$$x > 3$$

Thus, all real numbers greater than 3 make the original inequality a true statement, and the solution set, written in set-builder notation,* is $\{x: x > 3\}$. The graph of this solution set is the set of all points on the real number line representing numbers greater than 3 (see Figure 2.2). Note that we put a parenthesis at the point representing the number 3. This denotes that 3 is not a solution to the problem. When the endpoint is a member of the solution set, a bracket is used (as shown in Example 3). □

Figure 2.2

EXAMPLE 3   Water turns to steam when the temperature is at least 100 degrees Celsius. At what temperatures in degrees Fahrenheit will water turn to steam if the temperature scales are related by the formula $C = \frac{5}{9}(F - 32)$?

Solution   If water turns to steam when $C \geq 100$, then by the given formula this event occurs when the Fahrenheit temperature satisfies

$$\tfrac{5}{9}(F - 32) \geq 100.$$

The easiest way to solve this inequality is to multiply both sides by 9/5 to clear fractions. Doing this, we have

$$\tfrac{9}{5} \cdot \tfrac{5}{9}(F - 32) \geq \tfrac{9}{5} \cdot 100$$
$$F - 32 \geq 180$$
$$F \geq 212.$$

Therefore, water turns to steam when the Fahrenheit temperature is at least 212 degrees. We express this result in a graph as shown in Figure 2.3 □

Figure 2.3   Fahrenheit temperatures for steam

*__Set-builder notation__ writes sets in the form $\{x: x$ has property $P\}$, which is read "the set of all elements x such that x has property P." The colon : is read "such that."

EXAMPLE 4   Solve the inequality $x > x + 1$.

Solution   If we proceed in the usual way, we have

$$x > x + 1$$
$$x - x > x - x + 1 \qquad \text{Subtract } x \text{ from both sides.}$$
$$0 > 1.$$

When we subtract x from both sides, the resulting inequality is false. Since the original inequality is *equivalent* to a false statement, no replacements of x will make it true, and the solution set is $\emptyset$.   □

Inequalities can be used to describe an interval between two numbers. For example, suppose the average (a) of your grades must be between 80 and 90 if you are to receive a grade of B in the course. We may describe this interval by using a pair of inequalities.

$$80 < a \quad \text{and} \quad a < 90$$

The inequality $80 < a$ specifies that the average must be above 80, while $a < 90$ specifies that the average must be below 90. The two inequalities are both true only for averages between 80 and 90 (see Figure 2.4). The statement "$80 < a$ and $a < 90$" is usually expressed in compact form and written $80 < a < 90$.

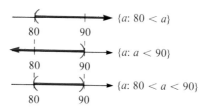

Figure 2.4

Intervals such as $\{a: 80 < a < 90\}$ may be written more simply by using **interval notation.** With this notation, we merely write (80,90) to specify the set of numbers between 80 and 90. More generally, if a and b are real numbers with $a < b$, then by definition,

$$(a,b) = \{x: a < x < b\}.$$

The set (a,b) is called the **open interval** from a to b, and the numbers a and b are called **endpoints** of the interval. As with graphs, interval notation uses a bracket if the endpoint is included in the set and a parenthesis if it is not.

When an interval extends indefinitely, the symbols ∞, read "infinity," and −∞ are used as shown in the following chart, which summarizes our methods for specifying intervals. ∞ and −∞ are not real numbers but convenient symbols to help us designate intervals that are unbounded in the positive or negative direction.

Type of Interval	Interval Notation	Set Notation	Graph
Open interval	(a,b)	$\{x: a < x < b\}$	
Closed interval	$[a,b]$	$\{x: a \leq x \leq b\}$	
Half-open interval	$[a,b)$	$\{x: a \leq x < b\}$	
	$(a,b]$	$\{x: a < x \leq b\}$	
Infinite interval	$(-\infty,a)$	$\{x: x < a\}$	
	$(-\infty,a]$	$\{x: x \leq a\}$	
	(a,∞)	$\{x: x > a\}$	
	$[a,\infty)$	$\{x: x \geq a\}$	
	$(-\infty,\infty)$	$\{x: x \text{ is a real number}\}$	

This chart shows that **intervals** are sets of real numbers that may be represented graphically as line segments, half lines, or the entire number line.

EXAMPLE 5 Solve $-4 < 5 - 3x < 4$ and write the solution in interval notation.

Solution Although the given inequality is actually an abbreviated form of a pair of inequalities, we can remain in compact form provided we use the properties of inequalities carefully.

$$-4 < 5 - 3x < 4$$
$$-9 < -3x < -1 \qquad \text{Subtract 5 from each member.}$$
$$3 > x > \frac{1}{3} \qquad \begin{array}{l}\text{Divide each member by } -3 \text{ and change the} \\ \text{sense of the inequalities.}\end{array}$$

Thus, the solution set is the interval $(\frac{1}{3},3)$, as graphed in Figure 2.5. Note that we do not write $(3,\frac{1}{3})$ since a must be less than b when writing (a,b).

Figure 2.5

Figure 2.6

EXAMPLE 6   Designate the set of numbers graphed in Figure 2.6 by using (a) set-builder notation and (b) interval notation.

Solution

a. We need to specify the set of numbers that are either less than -3 or greater than 1, so in set-builder notation we write

$$\{x: x < -3 \text{ or } x > 1\}.$$

We cannot write $1 < x < -3$ since this statement means $x < -3$ and $x > 1$ *simultaneously,* and no number satisfies both inequalities at the same time.

b. To use interval notation, we first note that the **union** of two sets A and B is the set of all elements belonging to A or B and is denoted by $A \cup B$. Then in interval notation the set of numbers less than -3 is written $(-\infty, -3)$, while the set of numbers greater than 1 is written $(1, \infty)$. Thus, the set of numbers belonging to either of these intervals is

$$(-\infty, -3) \cup (1, \infty). \qquad \square$$

Solving Inequalities by Factoring

Factoring can be the key to the solution of a wide variety of inequalities. In any product or quotient the number of negative factors determines whether the expression is positive (greater than zero) or negative (less than zero) according to the following rule.

Sign Rule	A product or quotient of nonzero factors is positive provided the number of negative factors is even. The product or quotient is negative if the number of negative factors is odd.

For example, $(-2)(-3)(4)(-5)$ has three negative factors and the product, -120, is less than zero; $(-2)(-3)(4)$ contains two negative factors and the product, 24, is greater than zero. When all the factors are positive, the product or quotient is obviously positive. This situation is covered in the sign rule since the number of negative factors in this case is zero—an even number. Examples 7–11 show how this rule is applied.

Sign of $x + 3$

$- \ - \ 0 + + + +$

-3

Sign of $x - 1$

$- \ - \ - \ - \ 0 + +$

1

Figure 2.7

Sign of $(x + 3)(x - 1)$

$+ \ + \ 0 - \ - \ 0 + +$

$-3 \quad\quad 1$

Figure 2.8

Sign of $1 + x$

$- \ - \ 0 + + + + +$

-1

Sign of $1 - x$

$+ \ + \ + \ + \ + \ 0 - \ -$

1

Sign of $(1+x)(1-x)$

$- \ - \ 0 + + \ 0 - \ -$

$-1 \quad\quad 1$

Figure 2.9

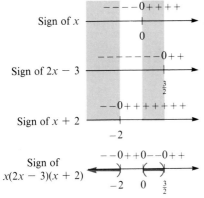

Sign of x

$- - - - 0 + + + +$

0

Sign of $2x - 3$

$- - - - - - 0 + +$

$\frac{3}{2}$

Sign of $x + 2$

$- - 0 + + + + + +$

-2

Sign of $x(2x - 3)(x + 2)$

$- - 0 + + 0 - - 0 + +$

$-2 \quad 0 \quad \frac{3}{2}$

Figure 2.10

EXAMPLE 7   Solve $x^2 + 2x - 3 > 0$.

Solution   As with second-degree equations, we want zero by itself on one side of the inequality. In this case, the right side is already zero, so factor the left side.

$$x^2 + 2x - 3 > 0$$
$$(x + 3)(x - 1) > 0$$

We now make up the chart in Figure 2.7, which shows the signs of both factors for the various values of x. Note the importance of the numbers -3 and 1 in the figure. An x value that makes any factor zero is called a **zero point,** and the solution hinges on these points. The sign of $(x + 3)(x - 1)$ is now easy to determine. There is an even number of negative factors if x is to the left of -3 or to the right of 1. Thus, $(x + 3)(x - 1) > 0$ if $x < -3$ or $x > 1$, as shown in Figure 2.8. We designate the solution set in interval notation by writing

$$(-\infty, -3) \cup (1, \infty)$$

as discussed in Example 6.   □

EXAMPLE 8   For what values of x is the inequality $1 \geq x^2$ a true statement?

Solution   First, rewrite the inequality so that one side is zero. Then factor.

$$1 \geq x^2$$
$$1 - x^2 \geq 0$$
$$(1 + x)(1 - x) \geq 0$$

Now make up the chart in Figure 2.9. Zero points are -1 and 1. The expression is positive or zero for $-1 \leq x \leq 1$, and the solution set is the interval $[-1, 1]$.   □

EXAMPLE 9   Solve $x(2x - 3)(x + 2) < 0$.

Solution   The inequality is already in the right form; that is, it is factored and one side is zero. Therefore, we go straight to the chart in Figure 2.10. Zero points are 0, $\frac{3}{2}$, and -2. Since the question asks for x values that make the expression negative, the inequality is satisfied if $x < -2$ or $0 < x < \frac{3}{2}$, and the solution set is $(-\infty, -2) \cup (0, \frac{3}{2})$.   □

EXAMPLE 10   For what values of x will $\sqrt{\dfrac{3x-1}{x+1}}$ be a real number?

Solution   For the square root to be a real number, the expression under the radical must be greater than or equal to 0. Thus, we need to solve

$$\frac{3x-1}{x+1} \geq 0.$$

Sign of $3x - 1$

Sign of $x + 1$

Sign of $\dfrac{3x-1}{x+1}$

Figure 2.11

We analyze quotients the same as products, since they both follow the same sign rule. The zero points are $\frac{1}{3}$ and -1, and as shown in Figure 2.11, the expression is nonnegative if $x < -1$ or $x \geq \frac{1}{3}$. In interval notation the solution set is $(-\infty, -1) \cup [\frac{1}{3}, \infty)$.   □

EXAMPLE 11   Solve $\dfrac{5x+1}{x+1} \geq 2$.

Solution   First, rewrite the inequality so the right side is zero. Then simplify the resulting expression on the left side into a single fraction.

$$\frac{5x+1}{x+1} \geq 2$$

$$\frac{5x+1}{x+1} - 2 \geq 0$$

$$\frac{5x+1-2(x+1)}{x+1} \geq 0$$

$$\frac{3x-1}{x+1} \geq 0$$

This inequality is solved in Example 10, and in interval notation the solution set is $(-\infty, -1) \cup [\frac{1}{3}, \infty)$.   □

EXERCISES 2.5

In Exercises 1–24 solve the inequality and graph each solution.

1. $2x + 9 < 0$

2. $3 - 4x > 0$

3. $-4x - 5 \geq 6x + 5$

4. $4 - 6x \geq -7 - 9x$

5. $13 - 2x < 7 - 3x$

6. $16x - 7 \leq 17x - 4$

7. $-2x + 5 > 5x - 7$

8. $1 - 4x \geq 2x + 3$

9. $7 - x > 5 - 2x$

10. $214 + 6x \leq -4x + 6$

11. $2(x - 3) \geq 6(x + 1)$

12. $4(3 + x) > 8(x - 4)$

13. $\dfrac{x}{2} + 4 < 8 + x$

14. $24 - 13x \geq -12.5x + 26$

15. $x > x - 1$

16. $x < x - 1$

17. $3x < 4x$

18. $-x \geq 2x$

19. $2(x + 3) \geq 7 + 2x$

20. $7 - 4y \leq 4(7 - y)$

21. $1 - 3y \leq 1 - 4y$

22. $5x + 4 > x + 4$

23. $2(2 - x) + 1 > 5 - 2x$

24. $2(2 - x) + 1 \geq 5 - 2x$

In Exercises 25–34 complete the chart and provide the alternative designations for the given set of real numbers.

	Interval Notation	Set Notation	Graph
25.	$[1,4)$		
26.	$(-\infty,1)$		
27.		$\{x: x \le 3\}$	
28.		$\{x: -1 < x \le 1\}$	
29.			
30.			
31.	$[0,\infty)$		
32.		$\{x: 0 < x\}$	
33.			
34.			

In Exercises 35–80 solve each inequality and write the solution set in interval notation.

35. $-2 < x - 4 < 2$
36. $0 < x + 7 < 10$
37. $-5 \le 2 - x \le 7$
38. $-0.1 \le 1 - x \le 0.1$
39. $-1 \le \dfrac{5 - 2x}{3} < \dfrac{1}{2}$
40. $-\dfrac{9}{4} < \dfrac{3x + 2}{-2} \le -\dfrac{1}{4}$
41. $(x + 1)(x - 2) > 0$
42. $(x - 3)(x + 2) < 0$
43. $a(2a - 1)^2 < 0$
44. $y^2(1 - 3y) > 0$
45. $\dfrac{t - 3}{t + 2} \le 0$
46. $\dfrac{x + 2}{x - 3} > 0$
47. $x(x - 1)(x - 2) \ge 0$
48. $b(4b + 3)(3b - 4) > 0$
49. $(1 - y)(2 - y)(3 - y)^2 < 0$
50. $k(4k + 3)^2(3k - 4) \ge 0$
51. $\dfrac{(x - 1)(x + 1)}{x} \le 0$
52. $\dfrac{a(4 - a)}{2a + 3} \ge 0$
53. $a^2 \ge 0$
54. $(3x - 1)^2 \ge 0$
55. $x^2 + x - 6 < 0$
56. $z^2 - 9 < 0$
57. $k^2 > 5k$
58. $2x \le x^2$
59. $b^2 + b \le 2$
60. $10 - c^2 > -3c$

61. $2x^2 \ge 7x + 4$
62. $3y^2 + 2 < 5y$
63. $t^3 > 4t$
64. $x^4 \le 4x^2$
65. $x^4 - 5x^2 \le -4$
66. $x^3 + 8x^2 + 12x < 0$
67. $c^5 + 9c \ge 10c^3$
68. $a^6 + 9a^2 > 10a^4$
69. $\dfrac{x^2 - 4}{3 - x} \ge 0$
70. $\dfrac{t^2 + 3t - 4}{3t + 2} \le 0$
71. $2x^3 - 5x^2 < 2x - 5$
72. $3n^3 - 2n^2 - 6n + 4 > 0$
73. $\dfrac{1}{a} < 2$
74. $\dfrac{2}{x} > 5$
75. $\dfrac{1}{4} \ge \dfrac{7}{7 - x}$
76. $\dfrac{1}{y + 2} \le \dfrac{1}{3}$
77. $\dfrac{2m}{m - 4} \le 3$
78. $\dfrac{-x}{2x + 1} \ge 1$
79. $\dfrac{3}{x - 2} \le \dfrac{3}{x + 3}$
80. $\dfrac{2}{x - 1} < \dfrac{1}{x + 1}$

Intervals appear in a natural way whenever we analyze data based on measurements. Measurements of one kind or another are essential to both scientific and nontechnical work. Weight, distance, time, volume, and temperature are but a few of the quantities for which measurements are required. However, any reading is an approximation that can only be as accurate as the measuring instruments allow. Thus, if we record a man's weight as 164 lb, we guarantee only that his exact weight is somewhere in the interval [163.5 lb, 164.5 lb). Use this concept to answer Exercises 81 and 82.

81. The circumference of the earth through the North and South Poles is about 25,000 mi. Use interval notation to write an interval that contains the earth's exact polar circumference. If we use a more accurate measuring system and record the polar circumference as 24,860 mi, which interval now contains the exact distance?

82. To the nearest inch, the height of a woman is 69 in. Use interval notation to write an interval that contains the woman's exact height. If we use a more accurate measuring device and record her height as 69.3 in., which interval now contains the woman's exact height?

83. In Lobachevskian geometry the sum of the three angles in a triangle is always less than 180°. If the second angle is 20° more than the first, and the third is twice the first, how many degrees might there be in the first angle?

84. If we categorize any temperature in adults above 98.6 degrees Fahrenheit as a "fever," then what is the temperature range of a fever in degrees Celsius? Use $F = \frac{9}{5}C + 32$.

85. Suppose a student's average grade on four tests must be in the interval [74.5,79.5) to receive a grade of C^+. If the student has grades of 81, 70, and 76 so far, find all possible grades for the last test that result in a grade of C^+.

86. What set of real numbers has the following property: The square of the number is smaller than the number itself?

87. What set of real numbers satisfies the following condition: When the number and its square are added together, the sum is between 6 and 42?

88. The length of a rectangle is 2 ft more than the width, and the area is less than 63 ft². What are the possibilities for the length in such a case?

89. The height (y) of a projectile that is shot directly up from the ground with an initial velocity of 96 ft/second is given by the formula $y = 96t - 16t^2$. During what period is the projectile more than 80 ft off the ground?

90. For what values of x will $\sqrt{4 - x^2}$ be a real number?

91. For what values of x will $\sqrt{x^2 - 4}$ be a real number?

92. For what values of x will $\sqrt{\dfrac{x + 1}{x - 2}}$ be a real number?

93. The geometric mean of a and b is given by $\sqrt{ab}$. Show that the geometric mean is always less than or equal to the arithmetic mean. That is, show $\sqrt{ab} \le (a + b)/2$ for all $a \ge 0$, $b \ge 0$. (*Hint:* Start with $x^2 \ge 0$, which is true for all values of x, and replace x by $\sqrt{b} - \sqrt{a}$. Then work your way to the desired result.)

94. Take two identical sheets of notebook paper. Form two cylinders from these rectangular ($y > x$) sheets: one with the length as the base [diagram (a)], the other with the width as the base [diagram (b)]. Show that the cylinder formed in (a) will always have the greater volume. (*Note:* $V = \pi r^2 h$.)

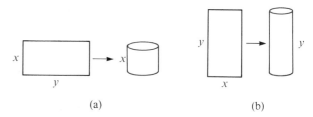

(a)   (b)

In Exercises 95–98 let a, b, c, and d be real numbers and prove each statement by using the definition of less than as demonstrated in this section.

95. If $a < b$, then $a + c < b + c$.

96. If $a < b$ and $c > 0$, then $ac < bc$.

97. If $a < b$ and $b < c$, then $a < c$.

98. If $a < b$ and $c < d$, then $a + c < b + d$.

2.6 ABSOLUTE VALUE EQUATIONS AND INEQUALITIES

We now return to the concept of absolute value that was introduced in Section 1.1. Recall the absolute value of the number a is denoted by $|a|$ and is interpreted geometrically as the distance on the number line from a to zero. For example, if you look at Figure 2.12(a) you can see that $|-4| = 4$, while Figure 2.12(b) shows the general geometric interpretation of $|a|$.

Figure 2.12

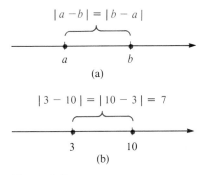

Figure 2.13

We can extend the idea that an absolute value represents a distance on the number line. Given any two points a and b, the distance between them on the number line is $|a - b|$ [see Figure 2.13(a)]. For example, the distance between 3 and 10 is $|3 - 10| = |-7| = 7$ [see Figure 2.13(b)]. Note that $|a - b| = |b - a|$, so it does not matter whether the smaller or larger number is labeled a.

The concept of absolute value is very important in mathematics. Thus, in addition to our geometric interpretation, an algebraic definition of the absolute value concept is essential. In formulating the algebraic rule, we must consider two cases. For positive numbers and zero the absolute value is simply that number. However, we obtain the absolute value of a negative number by taking the negative of that number. Thus, algebraically we can define the absolute value of a real number a as follows:

$$|a| = \begin{cases} a \text{ if } a \geq 0 \\ -a \text{ if } a < 0. \end{cases}$$

For example, we know $|-5| = 5$. We arrive at this answer using the algebraic rule as follows: Since $-5 < 0$ we have $a < 0$, so we use the rule $|a| = -a$; substituting, we get $|-5| = -(-5) = 5$. This rule might seem too complicated for such a simple concept and you probably are tempted to define absolute value as being merely the number without the sign. On an arithmetic level this definition works fine, but does $|-a| = a$? For any negative value of a the answer is no. For instance, $|-(-5)| \neq -5$. Consider the following example, which shows how the above definition is used to solve equations containing absolute value.

EXAMPLE 1   Solve $|x - 2| = 5$.

Solution   By definition, $|a|$ equals either a or $-a$. Thus, $|x - 2|$ equals either $x - 2$ or $-(x - 2)$. Setting these expressions equal to 5, we have

$$x - 2 = 5 \qquad -(x - 2) = 5$$
$$\text{or}$$
$$x = 7 \qquad\qquad x = -3.$$

Thus, 7 and -3 are the solutions of the equation, and the solution set is $\{7, -3\}$. We can also arrive at the answers by geometric considerations. $|x - 2|$ can be interpreted as meaning the distance between 2 and some number x on the number line. When solving $|x - 2| = 5$, we are looking for the two numbers that are 5 units from 2 on the number line. As shown in Figure 2.14 the desired numbers are 7 and -3. □

Figure 2.14

EXAMPLE 2   Solve $|2x + 1| = |3 - x|$.

Solution   By definition, $|2x + 1| = \pm(2x + 1)$ and $|3 - x|$ $= \pm(3 - x)$. Setting these expressions equal to each other produces two distinct cases:

$$2x + 1 = 3 - x \qquad \text{or} \qquad 2x + 1 = -(3 - x).$$

Finally, solving each of these equations separately gives

$$2x + 1 = 3 - x \qquad \text{or} \qquad 2x + 1 = -(3 - x)$$
$$3x = 2 \qquad\qquad\qquad 2x + 1 = -3 + x$$
$$x = \tfrac{2}{3} \qquad\qquad\qquad\quad x = -4.$$

Thus, the solution set is $\{\tfrac{2}{3}, -4\}$. (*Note:* The given equation is also satisfied when $-(2x + 1) = -(3 - x)$ or when $-(2x + 1) = 3 - x$. However, these equations are equivalent to the two we solved and need not be considered.)   □

Intervals on the number line are frequently specified in higher mathematics by inequalities that contain absolute value expressions. The most important such expression is of the form

$$|a| < b,$$

where b is some positive number. Remember that $|a|$ gives the distance between 0 and a on the number line. Geometrically then, $|a| < b$ says a is contained in an open interval that begins and ends b units from 0. As shown in Figure 2.15 this condition means that a must be between $-b$ and b. Thus, we have the following important result, which indicates how we algebraically handle these expressions.

> If $b > 0$, then $|a| < b$ is equivalent to $-b < a < b$.

The following examples show the procedure for applying this rule.

Figure 2.15 $|a| < b$ means a is between $-b$ and b.

EXAMPLE 3   Solve $|x - 4| < 3$.

Solution   Applying the above rule with $a = x - 4$ and $b = 3$, we rewrite

$$|x - 4| < 3$$

as

$$-3 < x - 4 < 3.$$

Now we proceed by obtaining equivalent inequalities through any of the steps given in Section 2.5.

$$-3 < x - 4 < 3$$
$$1 < x \quad\ < 7 \qquad \text{Add 4 to each member.}$$

Figure 2.16 $|x - 4| < 3$ **means x is between 1 and 7.**

Thus, any number between 1 and 7 satisfies the given inequality, and the solution set is $(1,7)$. We can also arrive at the answer by geometric considerations. $|x - 4| < 3$ can be interpreted as meaning that the distance between 4 and some number x on the number line must be less than 3 units. Figure 2.16 shows the interval defined by being within 3 units of the number 4 starts at 1 and ends at 7.   □

EXAMPLE 4   Write the interval $\{x: -3 < x < 7\}$ using inequalities and absolute value.

Figure 2.17

Solution   The interval between -3 and 7 is 10 units long and centered at 2. As shown in Figure 2.17 we are describing numbers that are within 5 units of 2. This condition means that $-3 < x < 7$ is equivalent to $|x - 2| < 5$.   □

EXAMPLE 5   Solve $|3x + 5| < 11$.

Solution   We rewrite $|3x + 5| < 11$ as $-11 < 3x + 5 < 11$ and proceed as follows:

$$-11 < 3x + 5 < 11$$
$$-16 < 3x < \ \ 6 \qquad \text{Subtract 5 from each member.}$$
$$\frac{-16}{3} < \ \ x < \ \ 2. \qquad \text{Divide each member by 3.}$$

Thus, any number between $-\frac{16}{3}$ and 2 satisfies the inequality, and the solution set is $(-\frac{16}{3}, 2)$.   □

Figure 2.18

Another common expression for specifying intervals on the number line is of the form

$$|a| > b,$$

where b is some positive number. Once again remember that $|a|$ gives the distance between 0 and a on the number line. Geometrically then, $|a| > b$ says a is located more than b units from zero. Figure 2.18 shows this condition means that a is either to the right of b or to the left of $-b$. Thus, we have the following algebraic rule for working with these expressions.

> If $b > 0$, then $|a| > b$ is equivalent to $a > b$ or $a < -b$.

Example 6 shows how this rule can be applied.

EXAMPLE 6   Solve $|x - 4| > 3$.

Solution   Applying the above rule with $a = x - 4$ and $b = 3$, we rewrite

$$|x - 4| > 3$$

as

$$x - 4 > 3 \quad \text{or} \quad x - 4 < -3$$

which simplifies to

$$x > 7 \quad \text{or} \quad x < 1$$

Figure 2.19 $|x - 4| > 3$ means $x > 7$ or $x < 1$.

so the solution set is $(-\infty,1) \cup (7,\infty)$. This answer checks with our geometric interpretation since the numbers located more than 3 units from 4 on the number line are either to the left of 1 or the right of 7 (see Figure 2.19). □

Finally there are rules that are important in higher mathematics concerning the absolute values of sums, differences, products, and quotients. We list these rules for reference purposes. Specific examples illustrating each of the properties are contained in the exercises.

Properties of Absolute Value	**1.** $\|a + b\| \leq \|a\| + \|b\|$
	2. $\|a - b\| \geq \|a\| - \|b\|$
	3. $\|ab\| = \|a\|\ \|b\|$
	4. $\left\|\dfrac{a}{b}\right\| = \dfrac{\|a\|}{\|b\|}$ $(b \neq 0)$

EXERCISES 2.6

In Exercises 1–20 use the proper symbol $(<, >, =)$ between the given expressions to indicate their correct order.

1. $\|14\|$ ____ $\|-16\|$ **2.** $\|0\|$ ____ $\|-\sqrt{3}\|$

3. $\|1.46\|$ ____ $\|-1.46\|$ **4.** $\|-0.1\|$ ____ $\|-0.01\|$

5. $\|2 - 10\|$ ____ $\|10 - 2\|$ **6.** $\|2 - 10\|$ ____ $\|2\| - \|10\|$

7. $\|-3/2\|$ ____ $\|-3\|/\|2\|$ **8.** $\|2\|\|-3\|$ ____ $\|2(-3)\|$

9. $\|-1\| + \|5\|$ ____ $\|-1 + 5\|$

10. $\|-1 + (-5)\|$ ____ $\|-1\| + \|-5\|$

11. $\|5 - 3\|$ ____ $\|5\| - \|3\|$ **12.** $\|3 - 5\|$ ____ $\|3\| - \|5\|$

13. $\|a\|$ ____ $\|-a\|$

14. $\|4 - b\|$ ____ $\|b - 4\|$

15. $-\|a\|$ ____ $\|a\|$ $(a \neq 0)$

16. $\|4(x - 1)\|$ ____ $\|4\|\ \|x - 1\|$

17. If $a > b$, then $\|a - b\|$ ____ $b - a$.

18. If $a > b$, then $\|a - b\|$ ____ $a - b$.

19. If $a < b$, then $\|a - b\|$ ____ $b - a$.

20. If $a < b$, then $\|a - b\|$ ____ $a - b$.

In Exercises 21–40 solve each equation.

21. $\|x - 1\| = 2$ **22.** $\|x - 3\| = 7$

23. $\|\frac{1}{2}x - 9\| = 8$ **24.** $\|3x + 7\| = 0$

25. $2\|x\| + 7 = 10$ **26.** $3\|x - 1\| + 2 = 11$

27. $\|x - 3\| = -2$ **28.** $\|x + 2\| = -5$

29. $\|x\| = x$ **30.** $\|x\| = -x$

31. $\|-x\| = x$ **32.** $\|-x\| = -x$

33. $\|x - 5\| = 5 - x$ **34.** $\|x - 10\| = x - 10$

35. $\|2x - 1\| = x$ **36.** $\|-x + 2\| = x + 1$

37. $\|x - 3\| = \|x + 1\|$ **38.** $\|1 - x\| = \|3x - 1\|$

39. $\|2x + 1\| = \|3 + x\|$ **40.** $\|6x - 4\| = \|2x - 7\|$

In Exercises 41–60 solve the inequality. In Exercises 41–50 use both an algebraic and a geometric interpretation.

41. $\|x\| < 4$ **42.** $\|x\| \leq \frac{1}{2}$

43. $\|x - 5\| \leq 3$ **44.** $\|x - 3\| < 0.01$

45. $\|x - a\| < 1$ **46.** $\|x - a\| < d, (d > 0)$

47. $\|x\| > 2$ **48.** $\|x\| > 0$

49. $\|x - 5\| \geq 3$ **50.** $\|x\| < -2$

51. $\|3x + 5\| < 8$ **52.** $\|6x - 7\| \leq 10$

53. $\|1 - 2x\| \leq 13$ **54.** $\|3 - 4x\| < 5$

55. $\|2x - 1\| > 3$ **56.** $\|1 - x\| + 2 \geq 4$

57. $\|x\| < x$ **58.** $x < \|x\|$

59. $\|3x - 7\| < x$ **60.** $\|3x - 7\| > x$

In Exercises 61–80 write the given interval using inequalities and absolute value.

61. $(-2,2)$ **62.** $[-7,7]$

63. $[-1,11]$ **64.** $(0,5)$

65. $\{x: 1.1 < x < 1.2\}$ **66.** $\{x: -0.6 < x < -0.5\}$

67. $(-\infty,-2) \cup (2,\infty)$ **68.** $(-\infty,-0.1) \cup (0.1,\infty)$

69. $\{x: x < 0 \text{ or } x > 6\}$ **70.** $\{x: x \leq -4 \text{ or } x \geq 0\}$

71. $(-\infty,6] \cup [7,\infty)$ **72.** $(-\infty,-7) \cup (-6,\infty)$

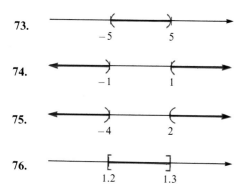

73.

74.

75.

76.

77. x is between 2 and 4.

78. x is within one-half unit of 3.

79. x is within d units of a.

80. x is within d units of a but not equal to a.

81. A projectile fired vertically up from the ground with an initial velocity of 128 ft/second will hit the ground 8 seconds later, and the speed of the projectile in terms of the elapsed time t equals $|128 - 32t|$. For what values of t is the speed of the projectile less than 80 ft/second?

82. Use $|x| = \sqrt{x^2}$ to show that $|a + b| \leq |a| + |b|$. (*Hint:* $a^2 + 2ab + b^2 \leq |a|^2 + 2|a|\,|b| + |b|^2$)

CHAPTER OVERVIEW

Section	Key Concepts to Review
2.1	• Definitions of equation, conditional equation, identity, equivalent equation, solution set, linear equation, and proportion • Methods to obtain an equivalent equation or formula • Methods to solve an equation containing fractions • Guidelines for setting up and solving word problems
2.2	• Definitions of imaginary number, complex number, equality for complex numbers, and the conjugate of a complex number • Methods to add, subtract, multiply, and divide complex numbers • Properties of conjugates • Relationships among the various sets of numbers • $i = \sqrt{-1}$ and $i^2 = -1$
2.3	• Definition of a second-degree or quadratic equation • Zero product principle: $a \cdot b = 0$ if and only if $a = 0$ or $b = 0$ • Square root property: $x^2 = n$ implies $x = \sqrt{n}$ or $x = -\sqrt{n}$ • Methods to solve a quadratic equation by the factoring method, by the square root property, by completing the square, and by using the quadratic formula • Quadratic formula: If $ax^2 + bx + c = 0$ and $a \neq 0$, then $$x = \frac{-b \pm \sqrt{b^2 - 4ac}}{2a}.$$ • Method to determine the nature of the solutions to a quadratic equation from the discriminant ($b^2 - 4ac$)

Section	Key Concepts to Review
2.4	• Definitions of radical equation and equations with quadratic form • Principle of powers • Methods to solve radical equations, equations with quadratic form, and higher degree polynomial equations that factor into linear and/or quadratic factors
2.5	• Definitions of intervals and zero point • Properties of inequalities • Methods to write intervals of numbers using inequalities or interval notation • Sign rule • Method to solve certain types of inequalities by factoring and using the sign rule

2.6	• Definition of absolute value
	• Methods to solve absolute value equations and inequalities
	• If $b \geq 0$, then $|a| = b$ is equivalent to $a = b$ or $a = -b$.
	• If $b > 0$, then $|a| < b$ is equivalent to $-b < a < b$.
	• If $b > 0$, then $|a| > b$ is equivalent to $a > b$ or $a < -b$.
	• Properties of absolute value
	  1. $|a + b| \leq |a| + |b|$ **2.** $|a - b| \geq |a| - |b|$
	  3. $|ab| = |a||b|$ **4.** $\left|\dfrac{a}{b}\right| = \dfrac{|a|}{|b|}$ $(b \neq 0)$

CHAPTER REVIEW EXERCISES

In Exercises 1–33 solve each equation or inequality.

1. $3(x - 6) = -2(12 - 3x)$
2. $1 - x = 1 + x$
3. $1 - x \leq 1 + x$
4. $-x + 1 > 3$
5. $\dfrac{20 - x}{3} = \dfrac{x}{5}$
6. $|x| - 1 = \dfrac{|x|}{2}$
7. $|x| < 5$
8. $|x| > 3$
9. $5 - 2(a - 2) = 2 - 3a$
10. $|5 - x| = 2$
11. $|2x - 3| \geq 1$
12. $85 = \frac{1}{2}(17)(-7 + L)$
13. $3 - 120{,}000x^{-2} = 0$
14. $2\sqrt{x + 1} = x - 2$
15. $(1 - \sqrt{3})x + 1 = \sqrt{3}$
16. $(x + 5)^2 + 9 = 0$
17. $\sqrt{2x + 1} - 3 = 0$
18. $x^{-1} + a^{-1} = b$
19. $\dfrac{x}{2\sqrt{25 + x^2}} - \dfrac{1}{4} = 0$
20. $\dfrac{\sqrt{11} - \sqrt{7}}{x} = \dfrac{x}{\sqrt{11} + \sqrt{7}}$
21. $x(x + 2) = 0$
22. $x^2 + 3x \geq 4$
23. $b^3 < 9b$
24. $(x - 3)^2 - 3(x - 3) - 4 = 0$
25. $2x^2 + 5x = 12$
26. $(x - 2)^2 \geq 0$
27. $a^3 - 5a^2 = a - 5$
28. $x + \sqrt{x} - 12 = 0$
29. $\dfrac{x + 1}{x - 1} < 1$
30. $\dfrac{x}{2x - 5} < 0$

31. $\dfrac{x}{3} - \dfrac{5}{3} = 2x^{-1}$ **32.** $1 - \dfrac{12}{c^2 - 4} = \dfrac{3}{c + 2}$

33. $\dfrac{2}{x + 3} \geq \dfrac{1}{x - 1}$

In Exercises 34–39 answer true or false. If false, give a specific counterexample.

34. If $a < b$, then $a + c < b + c$.

35. If $a < b$, then $ac < bc$.

36. $|-a| = a$ **37.** $|a + b| = |a| + |b|$

38. $|ab| = |a| \cdot |b|$ **39.** $|a - b| = |b - a|$

40. Solve for y: $x = \dfrac{2y + 1}{1 - y}$.

41. Solve for y: $P = (x - y - z)/10$.

42. Solve for L: $t = \pi\sqrt{L/g}$.

43. Perform the indicated operations and simplify.

    a. $(2 + 3i^2) - (3 - 2i)$

    b. $1 \div (3 + i)$

    c. $(4 - i)(4 + i)$

    d. $\sqrt{(-8)^2 - 4(1)(41)}$

44. If $2x^2 - 5x = k$, then what value(s) for k lead to each of the following: (a) the roots are equal; (b) one root is 0?

45. What value of k makes the following statement an identity?

$$(3x - k)(2x + k) = 6x^2 + 7x - k^2$$

46. For what values of x will $\sqrt{x^2 + 2x - 3}$ be a real number?

47. Evaluate $x^2 + 4x - 5$ for $x = 2 - i$.

48. Write the interval $\{x : 1 < x < 10\}$ using inequalities and absolute value.

49. Use inequalities and interval notation to describe the sets of numbers illustrated.

    a.

            0   6

    b.

           -4   -1

50. Is -3 a solution of the equation $\dfrac{2}{2 - c} + \dfrac{c}{c - 2} = 1$?

51. Find the value of t in the formula $a = p(1 + rt)$ if $a = 3{,}000$; $p = 1{,}500$; and $r = 0.05$.

52. Solve for y': $2x + 2y' = 0$.

53. Solve for y': $12x - 5xy' - 5y + 3 - y' = 0$.

54. Solve for y': $-\frac{1}{3}x^{-4/3} - \frac{1}{3}y^{-4/3}y' = 0$.

55. If $s = x - x^2$, solve for x in terms of s.

56. Solve for x: $\dfrac{bx + a}{ab} + \dfrac{1 - x^2}{ax + b} = 0$.

In Exercises 57–64 select the choice that answers the question or completes the statement.

57. If $x^{-2} = -36$, then x equals

    a. ± 6 **b.** $\pm 1/6$ **c.** $\pm 6i$ **d.** $\pm i/6$

58. If $\sqrt{x^2 + 3} = 2x$, then x equals

    a. both 1 and -1 **b.** only 1

    c. only -1 **d.** no solution

59. Pick the statement that describes the set of numbers illustrated.

          -1 3

    a. $|x - 2| < 1$ **b.** $|x + 2| > 1$

    c. $|x - 1| > 2$ **d.** $|x + 1| < 2$

60. If $ax^n - bx = 0$, then b equals

    a. ax^{n-1} **b.** ax^{2n} **c.** a^n **d.** na

61. The solution set to $x^2 - 6 < x$ is

    a. $(-2,3)$ **b.** $(-\infty,-2) \cup (3,\infty)$

    c. $(-\infty,-3) \cup (2,\infty)$ **d.** $(-3,2)$

62. If $\dfrac{1}{a} = \dfrac{1}{x} - \dfrac{1}{b}$, then x equals

    a. $\dfrac{ab}{a - b}$ **b.** $\dfrac{a + b}{ab}$ **c.** $a + b$ **d.** $\dfrac{ab}{a + b}$

63. Pick the statement that corresponds to the diagram.

          -2 2

    a. $|x| < 2$ **b.** $|x| > 2$ **c.** $|x| \geq 2$ **d.** $|x| \leq 2$

64. If $at = \left(\dfrac{a}{k} - 1\right)x$, then x equals

    a. $\dfrac{a(kt - 1)}{a - 1}$ **b.** $\dfrac{akt}{a - k}$ **c.** $\dfrac{akt}{a - 1}$ **d.** $\dfrac{at}{a - k}$

65. The total cost (including tax) of a new car is $9,222. If the sales tax is 6 percent, how much do you pay in taxes?

66. To the nearest inch, the height of a woman is 64 in. Use interval notation to write an interval that contains the woman's exact height. If we use a more accurate measuring device and record her height as 63.9 in., which interval now contains the woman's exact height?

67. Ohm's law states that the current in a circuit (I) is equal to the voltage (E) divided by the resistance (R). Write a formula for the resistance in terms of the current and the voltage.

68. The sum of a number and its reciprocal is $\frac{25}{12}$. What are the numbers?

69. One side of a right triangle is 3 ft longer than the other, and the hypotenuse is 15 ft. What is the length of the shortest side in the triangle?

70. A student needs an average of 90 or better to get an A in a course. Her first four test grades were 94, 91, 88, and 97. What possible grades on her fifth test will result in an A grade?

71. A real estate broker's commission for selling a house is 6 percent of the selling price. If you want to sell your house and receive $80,000, then what must be the selling price of the house?

72. Triangle ABC is similar to triangle CDE. If $\overline{BD} = 25$ ft, $\overline{AB} = 9$ ft, and $\overline{DE} = 11$ ft, then find $\overline{BC}$ to the nearest foot.

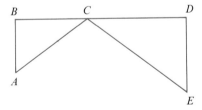

73. A diver jumps from a cliff that is 64 ft above water. Her height above water (y) after t seconds is given by the formula $y = 64 - 16t^2$. When will the diver hit the water? During what period is the diver at least 28 ft above water?

74. a. Because of a company's financial crisis, all employees' weekly salaries are cut a fixed percentage, p. What percent raise, x, is now needed to return the employees' weekly salaries to their original level? (*Hint:* Solve the equation $S(1 - p)(1 + x) = S$ for x.)

b. Use the answer to part **a** and determine the raise needed when the pay cut is 20 percent.

75. Explain the apparent paradox which is obtained in solving the equation $x - 2 = 1$ in the following roundabout way:

$$x - 2 = 1$$
$$x^2 - 8x + 12 = x - 6 \quad \text{Multiply both sides of the equation by } x - 6.$$
$$x^2 - 8x + 15 = x - 3 \quad \text{Add 3 to both sides of the equation.}$$
$$\frac{(x - 5)(x - 3)}{x - 3} = \frac{x - 3}{x - 3} \quad \text{Divide both sides of the equation by } x - 3.$$
$$x - 5 = 1$$
$$x = 6 \quad \text{Add 5 to both sides of the equation.}$$

But 6 is not the correct solution since it does not satisfy the original equation $x - 2 = 1$ (that is, $6 - 2 \neq 1$).

76. The age at which Diophantus (the Greek father of algebra) died is preserved in the following riddle: "Diophantus's youth lasted $\frac{1}{6}$ of his life. He grew a beard after $\frac{1}{12}$ more. After $\frac{1}{7}$ more of his life Diophantus married; five years later he had a son. The son lived exactly $\frac{1}{2}$ as long as his father, and Diophantus died just four years after his son." At what age did Diophantus die?

77. In Exercise 100 in the exercises to Section 1.6 we discussed the golden ratio. One way to determine the exact value for this number is to answer the following question from Euclid's *Elements:* "Divide a line segment such that the ratio of the large part to the whole is equal to the ratio of the small part to the large."

This condition means that for the segment shown below we have

$$\text{golden ratio} = \frac{\ell}{w} = \frac{\ell + w}{\ell}.$$

Set $w = 1$ and determine the exact value of the golden ratio.

3 Functions and Graphs

A rectangular insert of clear glass is placed in a colored semi-circular glass window of radius 1 m to improve the lighting near a stairway. To find the dimensions of the rectangular insert of greatest area that most improves the lighting, it is necessary to first find a formula for the area of the insert in terms of side dimension x as shown. What is this formula? (See Example 5 of Section 3.1.)

We now introduce the idea of a function and its graph. This concept will then become the central theme for the next few chapters as we investigate the behavior of the "elementary functions" studied in higher mathematics. These important functions are called polynomial, rational, exponential, logarithmic, and trigonometric functions. This chapter starts us on the road to analyzing relationships with formulas, tables, and graphs, and when done, it will be easy to see why the function concept is so important.

3.1 FUNCTIONS

One of the most important considerations in mathematics is determining the relationship between two variables. For example:

The postage required to mail a package is related to the weight of the package.
The bill from an electric company is related to the number of kilowatt-hours of electricity that are purchased.
The current in a circuit with a fixed voltage is related to the resistance in the circuit.
The demand for a product is related to the price charged for the product.
The perimeter of a square is related to the length of the side of the square.

In each of these examples we determine the relationship between the two variables by finding a rule that establishes a correspondence between values

Figure 3.1

Figure 3.2   Function

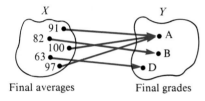

Figure 3.3   Not a function

of each variable. For example, if we know the length of the side of a square, we can determine the perimeter by the formula $P = 4s$. In this case the rule is a formula or equation. Sometimes the rule is given in tabular form. For example, consider the formula table below. This rule assigns to each final average (a) a final grade for the course.

$$\text{Final grade} = \begin{cases} A & \text{if} & 90 \le a \le 100 \\ B & \text{if} & 80 \le a < 90 \\ C & \text{if} & 70 \le a < 80 \\ D & \text{if} & 60 \le a < 70 \\ F & \text{if} & 0 \le a < 60 \end{cases}$$

Since formulas and tables are not always applicable, it is sometimes best to give the rule verbally or to make a list that shows the correspondence. The relationship between students and their social security numbers is such a case. Finally, you are undoubtedly familiar with the type of graph shown in Figure 3.1, which is commonly used to specify relationships in a quick and vivid way.

Whether the rule is given by formula or table, graphically, or by a list, the rule is most useful if we obtain *exactly one* answer whenever we use it. For example, the rule in the formula table above assigns to each final average exactly one final grade. Once we compute the final average, the rule tells us exactly what grade to assign for the course. Some typical assignments are shown in Figure 3.2. Assignments are represented as arrows from the points that represent final averages to the points that represent final grades. However some correspondences do not always give us exactly one answer. For example, if we reverse the assignments in Figure 3.2, then we cannot determine a unique final average when the final grade is A, as shown in Figure 3.3.

We wish to define a function so the correspondence in Figure 3.2 is a function, while the correspondence in Figure 3.3 is not a function. We do this as follows:

Definition of a Function	A **function** is a rule that assigns to each element x in a set X exactly one element y in a set Y. In this definition, set X is called the **domain** of the function, and the set of all elements of Y that correspond to elements in the domain is called the **range** of the function.

We clarify these definitions in Example 1.

EXAMPLE 1   The correspondence in Figure 3.2 assigns to *each* final average *exactly one* final grade, so it is a function with domain {63,82,91,97,100} and range {A,B,D}. Note that the range may or may not be set Y since the definition of a function does not require that all elements of Y be used. Meanwhile, the correspondence in Figure 3.3 is not a function for two reasons. First, an A grade corresponds to three elements in set Y, and second, the final grades of C and F in set X are not associated with any element in set Y.   □

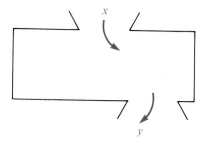

Figure 3.4

The analogy between a function and a computing machine may help to clarify the concept. Consider Figure 3.4, which shows a machine processing domain elements (x values) into range elements (y values). In goes an x value, out comes exactly one y value. With this in mind, you should have a clear image of the three features of a function: (1) the domain (the input), (2) the range (the output), and (3) the rule (the machine). When y is a function of x as just described, the value of y depends on the choice for x, so we call x the **independent variable** and y the **dependent variable.**

It is common practice to refer to a function by stating only the rule; no domain or range is specified. Thus we say, "Consider the circumference function $C = \pi d$." The domain is then assumed to be the collection of values for d that are interpretable in the problem. Since the length of the diameter of a circle must be positive in this case, the domain is the set of positive real numbers. For functions defined by algebraic equations like $y = 1/\sqrt{x - 1}$, the domain is the set of all real numbers for which a real number exists in the range. Thus we exclude from the domain values for the independent variable (x) that result in an even root of a negative number or in division by zero. For $y = 1/\sqrt{x - 1}$ we want $x - 1 > 0$ so that the domain is the interval {$x: x > 1$}, or $(1,\infty)$ in interval notation. We illustrate these ideas in the following examples.

EXAMPLE 2   Indicate whether or not the given equation determines y as a function of x.

a. $y = x^2$ **b.** $x^2 + y^2 = 1$ **c.** $x = y^3$

Solution

a. The equation $y = x^2$ determines y as a function of x, since to each real number input for x there corresponds exactly one y value.

b. Since $x^2 + y^2 = 1$ implies $y = \pm\sqrt{1 - x^2}$, it is possible for an x value to correspond to two different y values. For example, if $x = 0$, then $y = 1$ or $y = -1$. Thus, $x^2 + y^2 = 1$ does not determine y as a function of x. (However, $x^2 + y^2 = 1$ is an important equation, as shown in Section 3.2.)

c. $x = y^3$ implies $y = \sqrt[3]{x}$. Thus, the assignment of a real number to x results in exactly one output for y, so the equation determines y as a function of x.   □

EXAMPLE 3   Determine the domain and range of each function.

a. $y = x^2 + 2$ b. $y = 4$ c. $y = \sqrt{1 - x^2}$

Solution

a. For $y = x^2 + 2$ the domain is the set of all real numbers, since the assignment of any real number to x results in a real number output for y. For all x, $x^2 \geq 0$, so y, or $x^2 + 2$, is greater than or equal to 2. Thus, the range is the interval $[2,\infty)$.

b. In the equation $y = 4$ (or $y = 4 + 0 \cdot x$) the y value is 4 regardless of the x chosen. This type of function in which the y value does not change is called a **constant function.** The domain is the set of all real numbers and the range is $\{4\}$.

c. For the output of $y = \sqrt{1 - x^2}$ to be a real number, x must satisfy $1 - x^2 \geq 0$. As shown in Example 8 of Section 2.5, $1 - x^2 \geq 0$ is true in the interval $[-1,1]$, which is the domain. The corresponding y values vary from 0 (when $x = \pm 1$) to 1 (when $x = 0$), so the range is the interval $[0,1]$.

(*Note:* The range of a function is usually more difficult to determine than the domain, but this topic will become easier in the next section when we show how to read the range from the graph of a function.)   □

EXAMPLE 4   A piece of wire 46 in. long is bent into a rectangle of sides x and y units. Express y as a function of x. Express the area (A) of the rectangle as a function of x. Find the domain of the function.

Solution   First, express y as a function of x by using the perimeter formula

$$P = 2x + 2y$$
$$46 = 2x + 2y$$
$$23 - x = y.$$

To express the area (A) as a function of x we now replace y by $23 - x$ in the area formula

$$A = x \cdot y$$
$$= x(23 - x).$$

Since x and $23 - x$ must both be positive, the domain is the interval $(0,23)$.   □

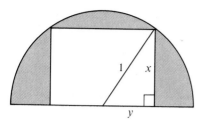

Figure 3.5

EXAMPLE 5   Solve the problem in the chapter introduction on page 116.

Solution   First, consider the sketch of the situation in Figure 3.5 and note that we strategically placed the radius of the semicircle so as to form a right triangle inside the rectangle. We may then use the Pythagorean theorem to write y in terms of x as follows:

$$x^2 + y^2 = 1^2$$
$$y^2 = 1 - x^2$$
$$y = \sqrt{1 - x^2}.$$

Then the area formula in terms of x is

$$A = 2xy$$
$$= 2x\sqrt{1 - x^2}.$$

In the context of this problem, the meaningful replacements for x are the real numbers between 0 and 1, so the domain is the interval $(0,1)$.   □

EXAMPLE 6   Each month a salesperson earns $500 plus 7 percent commission on sales above $2,000. Find a rule that expresses the monthly earnings (e) of the salesperson in terms of the amount (a) of merchandise sold during the month.

Solution   If the salesperson sells less than or equal to $2,000 worth of merchandise for the month, he earns $500. Thus,

$$e = \$500 \quad \text{if} \quad \$0 \leq a \leq \$2,000.$$

If the salesperson sells above $2,000, he earns $500 plus 7 percent of the amount above $2,000. Thus,

$$e = \$500 + 0.07(a - \$2,000) \quad \text{if} \quad a > \$2,000.$$

The following rule may then be used to determine the monthly earnings of the salesperson when we know the amount of merchandise sold:

$$e = \begin{cases} \$500 \quad \text{if} \quad \$0 \leq a \leq \$2,000 \\ \$500 + 0.07(a - \$2,000) \text{ if } a > \$2,000. \end{cases}$$

The domain of the function is $\{a: a \geq \$0\}$, and the range is $\{e: e \geq \$500\}$.   □

Functions as Ordered Pairs

In mathematical notation we use **ordered pairs** to show the correspondence in a function. For example, consider the equation $y = 2x + 1$.

If x Equals	Then $y = 2x + 1$	Thus, the Ordered Pairs
2	$2(2) + 1 = 5$	$(2,5)$
1	$2(1) + 1 = 3$	$(1,3)$
0	$2(0) + 1 = 1$	$(0,1)$
-1	$2(-1) + 1 = -1$	$(-1,-1)$
-2	$2(-2) + 1 = -3$	$(-2,-3)$

In the pairs that represent the correspondence, we list the values of the independent variable first and the values of the dependent variable second. Thus, the order of the numbers in the pair is significant. The pairing $(2,5)$ indicates that when $x = 2$, $y = 5$; $(5,2)$ means when $x = 5$, $y = 2$. In the equation $y = 2x + 1$, $(2,5)$ is an ordered pair that makes the equation a true statement; so $(2,5)$ is said to be a solution of the equation; $(5,2)$ is not a solution of this equation.

EXAMPLE 7   Which of the following ordered pairs are solutions of the equation $y = 3x - 2$?

a. $(-1,-5)$ **b.** $(-5,-1)$

Solution

a. $(-1,-5)$ means that when

$$x = -1, y = -5$$
$$y = 3x - 2$$
$$(-5) \overset{?}{=} 3(-1) - 2$$
$$-5 \overset{\checkmark}{=} -5.$$

Thus, $(-1,-5)$ is a solution.

b. $(-5,-1)$ means that when

$$x = -5, y = -1$$
$$y = 3x - 2$$
$$(-1) \overset{?}{=} 3(-5) - 2$$
$$-1 \neq -17.$$

Thus, $(-5,-1)$ is not a solution.   □

The representation of a correspondence as a set of ordered pairs gives us a different perspective of the function concept. In higher mathematics the following definition of a function is very useful.

Alternate Definition
of a Function

A **function** is a set of ordered pairs in which no two different ordered pairs have the same first component. The set of all first components (of the ordered pairs) is called the **domain** of the function. The set of all second components is called the **range** of the function.

In this section we have defined a function in terms of (1) a rule and (2) ordered pairs. Since the function concept is so important, you should consider both definitions and satisfy yourself that these definitions are equivalent.

EXAMPLE 8   State whether the set of ordered pairs is a function.

a. $\{(-1,0), (0,0), (1,0)\}$ **b.** $\{(0,-1), (0,0), (0,1)\}$

Solution

a. This set of ordered pairs is a function because the first component in the ordered pairs is always different. The domain is $\{-1,0,1\}$ and the range is $\{0\}$.

b. This set of ordered pairs is not a function because the number 0 is the first component in more than one ordered pair.   □

Functional Notation

A useful notation commonly used with functions allows us to represent more conveniently the value of the dependent variable for a particular value of the independent variable. In this notation we write an equation like $y = 2x$ as $f(x) = 2x$, where we replace the dependent variable y with the symbol $f(x)$. The symbol representing the independent variable appears in the parentheses. The term $f(x)$ is read "f of x" or "f at x" and means the value of the function (the y value) corresponding to the value of x. Similarly, $f(3)$ is read "f of 3" or "f at 3" and means the value of the function when $x = 3$. To find $f(3)$, we substitute 3 for x in the equation $f(x) = 2x$.

$$f(3) = 2 \cdot 3 = 6 \quad \text{therefore} \quad f(3) = 6$$

EXAMPLE 9   If $y = f(x) = (x + 2)/x$, find $f(1)$, $f(2)$, and $f(0)$.

Solution

$$y_{\text{when } x = 1} = f(1) = \frac{1 + 2}{1} = \frac{3}{1} = 3$$

$$y_{\text{when } x = 2} = f(2) = \frac{2 + 2}{2} = \frac{4}{2} = 2$$

$$y_{\text{when } x = 0} = f(0) = \frac{0 + 2}{0} = \frac{2}{0} \quad \text{undefined} \qquad □$$

EXAMPLE 10   If $f(x) = x^2$, show that $f(a + b)$ does not equal $f(a) + f(b)$.

Solution   To determine $f(a + b)$, $f(a)$, and $f(b)$, replace x in the function $f(x) = x^2$ by $a + b$, a, and b, respectively.

$$f(a + b) = (a + b)^2 = a^2 + 2ab + b^2$$
$$f(a) = a^2$$
$$f(b) = b^2$$

Since $a^2 + 2ab + b^2 \neq a^2 + b^2$, $f(a + b) \neq f(a) + f(b)$. $\square$

EXAMPLE 11   If $f(x) = \begin{cases} 2 & \text{if } x < 0 \\ x + 1 & \text{if } 0 \leq x < 3, \end{cases}$ find

a. $f(2)$ **b.** $f(-2)$ **c.** $f(3)$

Solution
a. Since 2 is in the interval $0 \leq x < 3$, use $f(x) = x + 1$, so
   $f(2) = 2 + 1 = 3$.
b. Since -2 is less than 0, use $f(x) = 2$, so $f(-2) = 2$.
c. Since 3 is not in the domain of the function, $f(3)$ is undefined. $\square$

EXAMPLE 12   The difference quotient of a function $y = f(x)$ is defined as

$$\frac{f(x + h) - f(x)}{h}, \quad h \neq 0.$$

Computing this ratio is an important consideration when analyzing the rate of change of a function (as discussed in Section 3.6). Find the difference quotient for $f(x) = x^2 + 2x$.

Solution   If $f(x) = x^2 + 2x$, we have

$$f(x + h) = (x + h)^2 + 2(x + h) = x^2 + 2xh + h^2 + 2x + 2h.$$

Then

$$\frac{f(x + h) - f(x)}{h} = \frac{(x^2 + 2xh + h^2 + 2x + 2h) - (x^2 + 2x)}{h}$$

$$= \frac{x^2 + 2xh + h^2 + 2x + 2h - x^2 - 2x}{h}$$

$$= \frac{2xh + h^2 + 2h}{h}$$

$$= \frac{2xh}{h} + \frac{h^2}{h} + \frac{2h}{h}$$

$$= 2x + h + 2.$$ $\square$

In conclusion, here are some important ideas to keep in mind about functional notation.

Functional Notation Concepts

1. If $y = f(x)$ and a is in the domain of f, then $f(a)$ means the value of y when $x = a$. Thus, evaluating $f(a)$ often requires nothing more than a *substitution* of the value a for x.
2. $f(a)$ is a y value; a is an x value. Hence, ordered pairs for the function defined by $y = f(x)$ all have the form $(a, f(a))$.
3. In functional notation we use the symbols f and x more out of custom than necessity, and other symbols work just as well. The notations $f(x) = 2x$, $f(t) = 2t$, $g(y) = 2y$, and $h(z) = 2z$ all define exactly the same function if x, t, y, and z may be replaced by the same numbers.

EXERCISES 3.1

In Exercises 1–4 determine whether or not the given correspondence is a function.

1.

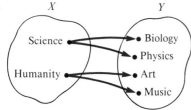

2. X (unit price) Y (revenue)

3.

4.

In Exercises 5–12 indicate whether or not the given equation determines y as a function of x.

5. $x + y = 3$ 6. $x = y^4$

7. $|y| = x$ 8. $y = x^4$

9. $y = -\sqrt{1 - x^2}$ 10. $x^2 + y^2 = 4$

11. $y = \pm\sqrt{x}$ 12. $x^2 y = 1$

In Exercises 13–30 find the domain and range of the given function.

13. $C = 2\pi r$ (circumference of a circle)

14. $C = 100n$ (cost of n radial tires)

15. $y = x^2 + 3$ 16. $f(x) = x^2 + 3,\ x < 0$

17. $g(x) = (x + 3)^2$ 18. $y = 3 - x^2$

19. $y = 1 - |x|$ 20. $y = |1 - x|$

21. $f(x) = \dfrac{1}{x + 1}$ 22. $h(x) = \dfrac{1}{x^2 + 1}$

23. $f(x) = 0$ **24.** $y = x^{1/3}$

25. $y = \sqrt{x - 1}$ **26.** $y = 1/\sqrt{x + 5}$

27. $y = -\sqrt{4 - x^2}$ **28.** $y = \sqrt{x^2 - 9}$

29. $y = \begin{cases} 1 \text{ if } x \le 0 \\ x \text{ if } 0 < x \le 4 \end{cases}$

30. $f(x) = \begin{cases} x + 1 \text{ if } -2 \le x < 1 \\ x - 1 \text{ if } 1 \le x < 5 \end{cases}$

In Exercises 31–34 determine which of the ordered pairs are solutions of the equation.

Equation	Ordered Pairs
31. $y = 3x + 1$	$(4,1), (-2,-5), (1,4), (-5,-2)$
32. $y = \lvert x \rvert$	$(-1,-1), (1,1), (-1,1), (1,-1)$
33. $y = x^2$	$(-1,-1), (1,1), (-1,1), (1,-1)$
34. $y = 1/x$	$(2,0.5), (0.5,2), (-1,-1), (0,0)$

35. Fill in the missing component in each of the following ordered pairs so they are solutions of the equation $y = -3x + 7$: (0,), (,0), (-5,), (,5).

36. Fill in the missing component in each of the following ordered pairs so they are solutions of the equation $y = \sqrt{x}$: (,4), (4,), (0,), (,$\sqrt{5}$).

37. If $(a,1)$ is a solution of the equation $y = 2^x$, then find the value of a.

38. If $(0,b)$ is a solution of the equation $y = 7x + 5$, then find the value of b.

39. If $y = \lvert x - 1 \rvert$, then determine the range element that corresponds to a domain element of -1.

40. If $y = -x - 1$, then determine the domain element that corresponds to a range element of 6.

In Exercises 41–46 state which sets of ordered pairs are functions.

41. $\{(1,-3), (2,0), (3,1)\}$

42. $\{(-3,1), (0,2), (1,3)\}$

43. $\{(2,-1), (3,0), (4,-1)\}$

44. $\{(-1,2), (0,3), (-1,4)\}$

45. $\{(2,1), (2,2), (2,3)\}$

46. $\{(1,2), (2,2), (3,2)\}$

In Exercises 47 and 48 state the domain and range of the function.

47. $\{(-2,4), (-1,1), (1,1), (2,4)\}$

48. $\{(0,5), (1,6), (2,7), (3,8)\}$

49. If $f(x) = -2x + 7$, find $f(-2), f(1)$, and $f(5)$.

50. If $f(t) = t^2 + 1$, find $f(-1), f(0)$, and $f(1)$.

51. If $g(x) = 2x^2 - x + 4$, find $g(3), g(0)$, and $g(-1)$.

52. If $h(t) = -t^2$, find $h(5), h(1)$, and $h(-5)$.

53. If $f(x) = (x + 1)/(x - 2)$, find $f(-3), f(1)$, and $f(2)$.

54. If $h(y) = y/(y + 1)$, find $h(1), h(0)$, and $h(-1)$.

55. If $f(x) = x - 1$, for what value of x will
 a. $f(x) = 2$?   **b.** $f(x) = 0$?   **c.** $f(x) = -2$?

56. If $g(x) = 2x + 7$, for what value of x will
 a. $g(x) = 5$?   **b.** $g(x) = 0$?   **c.** $g(x) = -3$?

57. If $f(x) = x + 1$ and $g(x) = x^2$, find
 a. $f(1) + g(0)$ **b.** $f(-2) - g(-3)$
 c. $4g(0) + 5f(1)$ **d.** $2f(3) - 3g(2)$
 e. $f(2) \cdot g(1)$ **f.** $\dfrac{f(1)}{g(2)}$ **g.** $[f(1)]^2$
 h. $g[f(1)]$ **i.** $f[g(1)]$ **j.** $f[g(x)]$

58. If $f(x) = x + 1$, find
 a. $f(2)$ **b.** $f(3)$
 c. $f(2 + 3)$ **d.** Does $f(2 + 3) = f(2) + f(3)$?
 e. $f(a)$ **f.** $f(b)$
 g. $f(a + b)$ **h.** Does $f(a + b) = f(a) + f(b)$?

59. If $f(x) = x + 1$, find
 a. $f(2)$ **b.** $f(-2)$ **c.** Does $f(-2) = -f(2)$?

60. If $g(x) = x^2$, find
 a. $g(2)$ **b.** $g(-2)$
 c. Does $g(x) = g(-x)$ for all values of x?

61. If $f(x) = \begin{cases} x - 1 & \text{if } x < 0 \\ -1 & \text{if } x > 0 \end{cases}$, find
 a. $f(-1)$ **b.** $f(1)$ **c.** $f(0)$

62. If $f(x) = \begin{cases} -1 & \text{if } x > 0 \\ 0 & \text{if } x = 0 \\ 1 & \text{if } x < 0 \end{cases}$, find
 a. $f(1)$ **b.** $f(0)$ **c.** $f(-1)$ **d.** $f(0.0001)$

63. If $g(x) = \begin{cases} x & \text{if } x \geq 0 \\ -x & \text{if } x < 0 \end{cases}$, find

    a. $g(3)$ **b.** $g(0)$ **c.** $g(-3)$

64. If $h(x) = \begin{cases} x & \text{if } 0 \leq x \leq 1 \\ -1 & \text{if } x > 1 \end{cases}$, find

    a. $h(3)$ **b.** $h(1)$ **c.** $h(\frac{1}{2})$ **d.** $h(0)$ **e.** $h(-3)$

65. If $f(x) = 2x^2 - 1$, find

    a. $f(x + h)$ **b.** $f(x + h) - f(x)$

    c. $\dfrac{f(x + h) - f(x)}{h}$

66. If $f(x) = 1 - x$, find

    a. $f(1 + h)$ **b.** $f(1 + h) - f(1)$

    c. $\dfrac{f(1 + h) - f(1)}{h}$

In Exercises 67–70 write the difference quotient for each function in simplest form (see Example 12).

67. $f(x) = 7x - 5$ 68. $f(x) = 7$

69. $f(x) = x^2 - 2x + 3$ 70. $f(x) = 1 - x^2$

In Exercises 71–95 find a formula or table that defines the functional relationship between the two variables; in each case indicate the domain of the function.

71. Express the area (A) of a square in terms of the length of its side (s).

72. Express the area (A) of a rectangle of width 5 as a function of its length (ℓ).

73. Express the length (ℓ) of a rectangle of width 5 in terms of its area A.

74. Express the area (A) of a circle as a function of its radius (r).

75. Express the length of the side (s) of a square in terms of its perimeter (P).

76. Express the area (A) of a square as a function of the perimeter (P) of the square.

77. Express the distance (d) that a car will travel in t hours if the car is always traveling at 40 mi/hour.

78. Express the cost (c) of y gal of gasoline if the gas cost $1.17/gal.

79. Express the earnings (e) of an electrician in terms of the number (n) of hours worked if she makes $28 per hour.

80. Express the interest (i) earned on $1,000 at 6 percent simple interest as a function of the time (t).

81. Express the earnings (e) of a real estate agent who receives a 6 percent commission as a function of the sale price (p) of a house.

82. Express the monthly cost (c) for a checking account as a function of the number (n) of checks serviced that month if the bank charges 10 cents per check plus a 75 cents maintenance charge.

83. The total cost of producing a certain product consists of paying $400 per month rent plus $5 per unit for material. Express the company's monthly total costs (c) as a function of the number of units (x) they produce that month.

84. A company plans to purchase a machine to package its new product. If the company estimates that the machine will cost $5,000 per year plus $1 to package each unit, express the yearly cost (c) of using the machine as a function of the number of units (x) packaged during the year.

85. A reservoir contains 10,000 gal of water. If water is being pumped from the reservoir at a rate of 50 gal/minute, write a formula expressing the amount (a) of water remaining in the reservoir as a function of the number (n) of minutes the water is being pumped.

86. Express the monthly cost of renting a computer as a function of the number (n) of hours the computer is used if the company charges $200 plus $100 for every hour the computer is used during the month.

87. Express the federal income tax (t) for a single person as a function of his taxable income (i) if the person's taxable income is between $2,000 and $4,000 inclusive and the tax rate is $310 plus 19 percent of the excess over $2,000.

88. Express the monthly earnings (e) of a salesperson in terms of the cash amount (a) of merchandise sold if the salesperson earns $600 per month plus 8 percent commission on sales above $10,000.

89. Express the monthly cost (c) of an electric bill in terms of the number (n) of electrical units purchased (the unit of measure is the kilowatt-hour) if the person used no more than 48 units and the electric company has the following rate schedule:

Amount	Charge
First 12 units or less	$1.75
Next 36 units at	3.82 cents/unit

90. Express the monthly cost (c) of a gas bill in terms of the number (n) of units purchased (1 unit equals 100 ft³ of gas) if the customer used no more than 14 units and the gas company has the following rate schedule:

Amount	Charge
First 2 units or less	$2.06
Next 6 units at	40 cents/unit
Next 6 units at	26.5 cents/unit

91. A long strip of galvanized sheet metal 12 in. wide is to be formed into an open gutter by bending up the edges to form a gutter with rectangular cross section. Write the cross-sectional area of the gutter as a function of the depth (x).

92. A farmer encloses an area by connecting 2,000 ft of fencing in the shape of a rectangle of sides x and y units. Express y as a function of x. Find a formula for the enclosed area (A) as a function of x.

93. A parking lot is adjacent to a building and is to have fencing on three sides, the side on the building requiring no fencing. If 100 yd of fencing is used to form a rectangular parking lot, find a formula expressing the area (A) of the lot as a function of the length (x) of the side of the fencing that is perpendicular to the building.

94. A rectangular sheet of tin 8 in. by 15 in. is to be used to make an open top box by cutting a small square of tin from each corner and bending up the sides. Find an equation for the volume (v) of the box in terms of the length (x) of the side of the square that is to be cut from each corner (see the diagram).

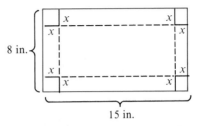

95. A rectangular billboard is to display 100 ft² of advertising material with borders that are 3 ft on the top and bottom and 5 ft on the sides. Express the total area (A) of the billboard as a function of the width (w) of the advertised material (see the diagram).

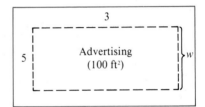

3.2 CARTESIAN COORDINATES AND GRAPHS

One of the sources of information and insight about a relationship is a picture that describes the particular situation. Pictures or graphs are often used in business reports, laboratory reports, and newspapers to present data quickly and vividly. Similarly, it is useful to have a graph that describes the behavior of a particular function, for this picture helps us see the essential characteristics of the relationship.

We can pictorially represent a function by using the **Cartesian coordinate system.** This system was devised by the French mathematician and philosopher René Descartes and is formed from the intersection of two real number lines at right angles. The values for the independent variable (usually x) are represented on a horizontal number line and values for the dependent variable (usually y) on a vertical number line. These two lines are called **axes,** and they intersect at their common zero point, which is called the **origin** [see Figure 3.6(a)].

This coordinate system divides the plane into four regions called **quadrants.** The quadrant in which both x and y are positive is designated the first quadrant. The remaining quadrants are labeled in a counterclockwise direction. Figure 3.6(b) shows the name of each quadrant as well as the sign of x and y in that quadrant.

(a)   (b)

Figure 3.6

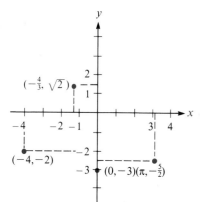

Figure 3.7

Any ordered pair can be represented as a point in this coordinate system. The first component indicates the distance of the point to the right or left of the vertical axis. The second component indicates the distance of the point above or below the horizontal axis. These components are called the **coordinates** of the point.

EXAMPLE 1   Represent the ordered pairs $(0,-3)$, $(-4,-2)$, $(\pi,-\frac{5}{2})$, and $(-\frac{4}{3},\sqrt{2})$ as points in the Cartesian coordinate system.

Solution   See Figure 3.7.   □

When working with the Cartesian coordinate system, it is often necessary to determine the distance between two points in the system. It is easiest to calculate the distance between two points on a horizontal or vertical line. If

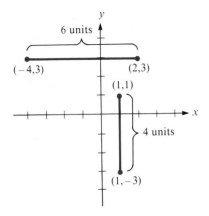

Figure 3.8

the points lie on the same horizontal line, we find the distance between them by taking the absolute value of the difference between their x-coordinates. If two points lie on the same vertical line, we calculate the distance between them by taking the absolute value of the difference between their y-coordinates.

EXAMPLE 2   Find the distance between the given points.

a. $(-4,3)$ and $(2,3)$ **b.** $(1,1)$ and $(1,-3)$

Solution   See Figure 3.8.

a. The points $(-4,3)$ and $(2,3)$ have the same y-coordinate, so

$$d = |x_2 - x_1| = |2 - (-4)| = |6| = 6.$$

b. The points $(1,1)$ and $(1,-3)$ have the same x-coordinate, so

$$d = |y_2 - y_1| = |-3 - 1| = |-4| = 4.$$ ☐

If two points do not lie on the same horizontal or vertical line, we can find the distance between them by drawing a horizontal line through one point and a vertical line through the other, as illustrated in Figure 3.9. The x-coordinate at the point where the two lines intersect will be x_2. The y-coordinate will be y_1. The distance between (x_1,y_1) and (x_2,y_1) will be $|x_2 - x_1|$ because they lie on the same horizontal line. The distance between (x_2,y_2) and (x_2,y_1) will be $|y_2 - y_1|$ because they lie on the same vertical line. The vertical and horizontal lines meet at a right angle $(90°)$ and thus the resulting triangle is a right triangle.

The distance d is the length of the hypotenuse in the right triangle, and by the Pythagorean theorem we get

$$d^2 = |x_2 - x_1|^2 + |y_2 - y_1|^2$$
$$d = \sqrt{(x_2 - x_1)^2 + (y_2 - y_1)^2}.$$

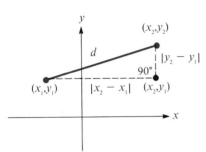

Figure 3.9

Note that we no longer need the absolute value symbols since the square of any real number is never negative. Thus, we have the following formula.

Distance Formula	The distance d between the points $P_1(x_1,y_1)$ and $P_2(x_2,y_2)$ is given by

$$d = \overline{P_1 P_2} = \sqrt{(x_2 - x_1)^2 + (y_2 - y_1)^2}.$$

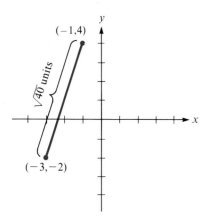

Figure 3.10

EXAMPLE 3   Find the distance between $(-3,-2)$ and $(-1,4)$.

Solution

$$\begin{aligned}
d &= \sqrt{(x_2 - x_1)^2 + (y_2 - y_1)^2} \\
&= \sqrt{[-1 - (-3)]^2 + [4 - (-2)]^2} \\
&= \sqrt{(2)^2 + (6)^2} \\
&= \sqrt{4 + 36} \\
&= \sqrt{40} \quad \text{or} \quad 2\sqrt{10} \quad \text{(See Figure 3.10.)}
\end{aligned}$$

□

EXAMPLE 4   Show that the points $(-4,-2)$, $(-3,2)$, and $(1,1)$ are vertices of an isosceles triangle. Is this triangle a right triangle?

Solution   Label $(-4,-2)$ as point A, $(-3,2)$ as B, and $(1,1)$ as C. Then

$$\begin{aligned}
\overline{AB} &= \sqrt{[-4 - (-3)]^2 + (-2 - 2)^2} = \sqrt{17} \\
\overline{BC} &= \sqrt{(-3 - 1)^2 + (2 - 1)^2} = \sqrt{17} \\
\overline{AC} &= \sqrt{(-4 - 1)^2 + (-2 - 1)^2} = \sqrt{34}.
\end{aligned}$$

Since the lengths of two sides are equal, the triangle is isosceles. Since $(\sqrt{34})^2 = (\sqrt{17})^2 + (\sqrt{17})^2$ the Pythagorean theorem holds, and ABC is a right triangle with angle B as the right angle (see Figure 3.11). □

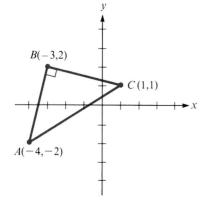

Figure 3.11

We now consider one of the most useful ideas in mathematics. With the Cartesian coordinate system we can represent any ordered pair of real numbers by a particular point in the system. This enables us to draw a geometric picture (graph) of a function.

Graph	The graph of a function is the set of all points in a coordinate system that correspond to ordered pairs in the function.

There are many techniques associated with determining the graph of a function. One technique is simply to assign values to the independent variable and obtain a list of ordered pair solutions. By plotting enough of these solutions, we can establish a trend and then complete the graph by following the established pattern. However, we cannot possibly list all the solutions of most equations because they are an infinite set of ordered pairs. Determining how many and which points to plot is a difficult decision. Therefore, as we proceed in this section and succeeding chapters, we develop the more efficient method of determining the essential characteristics of the graph from the form of the equation. For example, in Section 4.1 we will prove that the graph of a function of the form $y = f(x) = mx + b$, where m and b are real number constants, is a straight line. For now let us use this fact and illustrate many of the important ideas of graphing by considering functions that require only straight lines for their graphs.

EXAMPLE 5   Graph the function $f(x) = -2x + 4$.

Solution

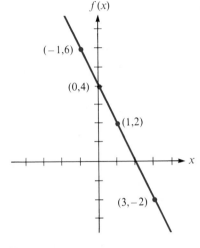

Figure 3.12

If $x =$	Then $f(x) = -2x + 4$	Thus, the Ordered-Pair Solutions
3	$-2(3) + 4 = -2$	$(3,-2)$
1	$-2(1) + 4 = 2$	$(1,2)$
0	$-2(0) + 4 = 4$	$(0,4)$
-1	$-2(-1) + 4 = 6$	$(-1,6)$

The horizontal axis is used to represent the values of the independent variable and thus is called the x-axis. The vertical axis is named for the dependent variable, so it is labeled the $f(x)$- or y-axis. The function is of the form $y = f(x) = mx + b$, with $m = -2$ and $b = 4$. Thus, the graph of $f(x) = -2x + 4$ is a straight line (see Figure 3.12) that passes through the points in the above table. Note that only two points are needed to determine the correct line.   □

EXAMPLE 6   Graph the function $y = 4$ (or $y = 4 + 0x$).

Solution

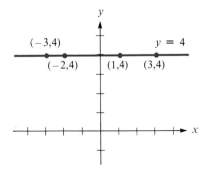

Figure 3.13

If $x =$	Then $y = 4 + 0 \cdot x$	Thus, the Ordered-Pair Solutions
3	$4 + 0(3) = 4$	$(3,4)$
1	$4 + 0(1) = 4$	$(1,4)$
-2	$4 + 0(-2) = 4$	$(-2,4)$
-3	$4 + 0(-3) = 4$	$(-3,4)$

Note that no matter what number we substitute for x, the resulting value for y is 4 (see Figure 3.13). Since the value of the function does not change, $y = 4$ is called a **constant function.** A physical example of a constant function is the acceleration of a falling body as a function of time ($a = 32$) since the acceleration is always 32 ft/second², regardless of the time. In every case the graph of a constant function is a horizontal line.   □

EXAMPLE 7   Graph the function defined as follows:

$$f(x) = \begin{cases} x & \text{if } x \geq 0 \\ -x & \text{if } x < 0. \end{cases}$$

Solution

If $x =$	Then $f(x) = \begin{cases} x & \text{if } x \geq 0 \\ -x & \text{if } x < 0 \end{cases}$	Thus, the Ordered-Pair Solutions
3	since $3 \geq 0$, $f(3) = 3$	(3,3)
2	since $2 \geq 0$, $f(2) = 2$	(2,2)
1	since $1 \geq 0$, $f(1) = 1$	(1,1)
0	since $0 \geq 0$, $f(0) = 0$	(0,0)
-1	since $-1 < 0$, $f(-1) = -(-1) = 1$	(-1,1)
-2	since $-2 < 0$, $f(-2) = -(-2) = 2$	(-2,2)
-3	since $-3 < 0$, $f(-3) = -(-3) = 3$	(-3,3)

See Figure 3.14. (*Note:* This rule defines algebraically the **absolute value function** $f(x) = |x|$.)   □

Figure 3.14

EXAMPLE 8   Graph the function defined as follows:

$$f(x) = \begin{cases} 1 - x & \text{if } x < 0 \\ -1 & \text{if } x \geq 0. \end{cases}$$

Solution   If $x \geq 0$, $f(x) = -1$, which is a constant function whose graph is a horizontal line with such ordered pairs as $(0,-1)$, $(1,-1)$, $(2,-1)$, and so on. If $x < 0$, $f(x) = 1 - x$, which graphs as a line with ordered pairs $(-1,2)$, $(-2,3)$, $(-3,4)$, and so on. The graph is given in Figure 3.15.

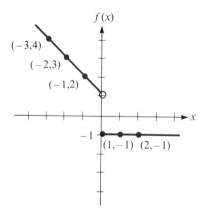

Figure 3.15   □

EXAMPLE 9   Graph the function $y = \dfrac{|x - 10|}{x - 10}$.

Solution   From Section 2.6 we know

$$|x - 10| = \begin{cases} x - 10 & \text{if } x - 10 \geq 0 \\ -(x - 10) & \text{if } x - 10 < 0. \end{cases}$$

Since division by 0 is undefined, $x \neq 10$. If $x > 10$, we have

$$\frac{|x - 10|}{x - 10} = \frac{x - 10}{x - 10} = 1.$$

Similarly, if $x < 10$, we have

$$\frac{|x - 10|}{x - 10} = \frac{-(x - 10)}{x - 10} = -1.$$

Thus, $y = \dfrac{|x - 10|}{x - 10}$ is equivalent to $y = \begin{cases} 1 & \text{if } x > 10 \\ -1 & \text{if } x < 10. \end{cases}$

The function is graphed in Figure 3.16. Note the open circles above and below $x = 10$, which indicate the function is not defined at this point. □

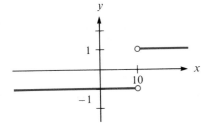

Figure 3.16

We have said that a graph helps us see the essential characteristics of a function. However, the benefits derived from such a picture are directly related to your ability to read the graph. Essential features like the domain and range of a function stand out, but you have to know what you are looking for. For example, consider Figures 3.17 and 3.18, which are the graphs from Examples 7 and 9. In each case the domain and range are specified in the figure. The domain is given by the variation of the graph in a horizontal direction, while the range is given by the variation in the vertical direction.

Figure 3.17   **Figure 3.18**

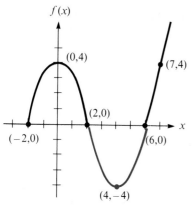

Figure 3.19

Here is another example illustrating how you can read important information from a graph.

EXAMPLE 10   Consider the graph of $y = f(x)$ in Figure 3.19.
a. What is the domain of the function?
b. What is the range of the function?
c. Determine $f(0)$.
d. For what values of x does $f(x) = 0$?
e. For what values of x is $f(x) < 0$?

Solution
a. The x values in the figure start at -2 and keep going to the right. Thus, the domain is the interval $[-2, \infty)$.
b. The minimum y value is -4 and the graph extends indefinitely in the positive y direction so there is no maximum value. Thus, the range is $[-4, \infty)$.
c. The notation $f(0)$ means we want the y value when $x = 0$. The ordered pair $(0,4)$ tells us $y = 4$ when $x = 0$. Thus, $f(0) = 4$.
d. Here $f(x)$ or y is 0 and we look at the ordered pairs $(-2,0)$, $(2,0)$, and $(6,0)$. Thus, $f(x) = 0$ when $x = -2$, 2, or 6.
e. The y values are less than 0 when the figure is below the x-axis. As indicated in color, $f(x) < 0$ for $2 < x < 6$.   □

Examples 5 and 9 showed functions whose graphs require only lines. Graphs are by no means limited to lines, however, and the next two examples consider the parabola—a figure that is associated with the squaring function.

EXAMPLE 11   Graph the squaring function $y = x^2$. Indicate on the graph the domain and range of the function.

Solution

If $x =$	Then $y = x \cdot x = x^2$	Thus, the Ordered-Pair Solutions
3	$y = 3(3) = 9$	(3,9)
2	$y = 2(2) = 4$	(2,4)
1	$y = 1(1) = 1$	(1,1)
0	$y = 0(0) = 0$	(0,0)
-1	$y = -1(-1) = 1$	$(-1,1)$
-2	$y = -2(-2) = 4$	$(-2,4)$
-3	$y = -3(-3) = 9$	$(-3,9)$

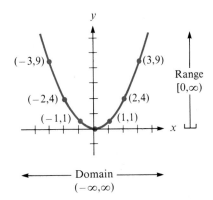

Figure 3.20

The cuplike curve in Figure 3.20 is called a **parabola.** In every case the graph of a function of the form $y = ax^2 + bx + c$, where a, b, and c are real numbers ($a \neq 0$), is a parabola. We discuss these functions in detail in Section 4.2. As shown in the graph, the domain is the set of all real numbers and the range is the interval $[0,\infty)$.

EXAMPLE 12   Graph the function $f(x) = 4 - x^2$. Indicate on the graph the domain and range of the function.

Solution

If $x =$	Then $f(x) = 4 - x^2$	Thus, the Ordered-Pair Solutions
3	$f(3) = 4 - (3)^2 = -5$	$(3,-5)$
2	$f(2) = 4 - (2)^2 = 0$	$(2,0)$
1	$f(1) = 4 - (1)^2 = 3$	$(1,3)$
0	$f(0) = 4 - (0)^2 = 4$	$(0,4)$
-1	$f(-1) = 4 - (-1)^2 = 3$	$(-1,3)$
-2	$f(-2) = 4 - (-2)^2 = 0$	$(-2,0)$
-3	$f(-3) = 4 - (-3)^2 = -5$	$(-3,-5)$

The graph of $f(x) = 4 - x^2$, shown in Figure 3.21, is a parabola. The domain is the set of all real numbers and the range is the interval $(-\infty,4]$.

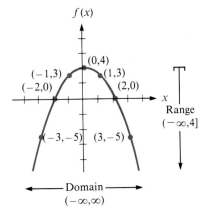

Figure 3.21

In a function no two ordered pairs may have the same first component. Graphically, this means that a function cannot contain two points that lie on the same vertical line. This observation leads to the following simple test for determining if a graph represents a function.

Vertical Line Test   Imagine a vertical line sweeping across the graph. If the vertical line at any position intersects the graph in more than one point, the graph is not the graph of a function.

EXAMPLE 13   Use the vertical line test to determine which graphs in Figure 3.22 represent the graph of a function.

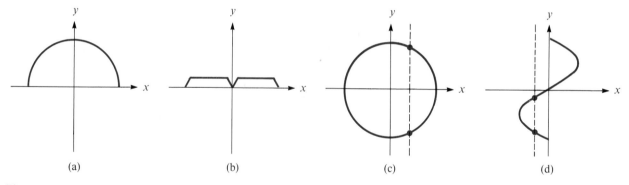

(a)   (b)   (c)   (d)

Figure 3.22

Solution   By the vertical line test graphs (a) and (b) represent functions, whereas (c) and (d) do not.   ☐

In (c) of Figure 3.22 we saw that the graph of a circle is not the graph of a function. However, circles are important graphs, and in particular, we can use the distance formula to show that the equation of a circle with center at the origin and radius r is

$$x^2 + y^2 = r^2.$$

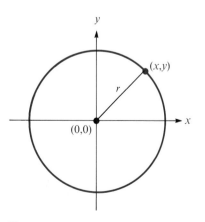

Figure 3.23

To prove this, consider Figure 3.23 and note that (x,y) is a point on the circle if and only if the distance from (x,y) to the origin is equal to the radius r. Thus, by the distance formula, the equation for the specified circle is

$$\sqrt{(x - 0)^2 + (y - 0)^2} = r$$

or, after squaring both sides of the equation,

$$x^2 + y^2 = r^2.$$

This result allows us to graph without difficulty the type of equations given in the next example.

EXAMPLE 14   Graph each equation and indicate the domain and range of the functions in parts **b** and **c**.

a. $x^2 + y^2 = 4$ **b.** $y = \sqrt{4 - x^2}$ **c.** $y = -\sqrt{4 - x^2}$

Solution

a. The equation $x^2 + y^2 = 4$ fits the form $x^2 + y^2 = r^2$, so the graph is a circle centered at the origin. Also, $r^2 = 4$, so the radius is $\sqrt{4}$, or 2. By drawing a circle that satisfies these conditions, we obtain the graph in Figure 3.24(a).

b. We start with $y = \sqrt{4 - x^2}$ and square both sides of the equation to obtain $y^2 = 4 - x^2$, or $x^2 + y^2 = 4$, which is the equation from **a.** However, in the original equation y is restricted to positive numbers or zero, so the graph is a semicircle as shown in Figure 3.24(b). From the graph we read that the domain is $[-2,2]$ and the range is $[0,2]$.

c. As in **b,** the graph of $y = -\sqrt{4 - x^2}$ is a semicircle. But y is restricted to negative numbers or zero in this case, so we draw the semicircle in Figure 3.24(c), which shows that the domain is $[-2,2]$ and the range is $[-2,0]$.

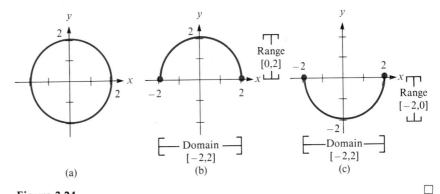

Figure 3.24

In conclusion, keep in mind the following ideas that are associated with graphing:

1. Coordinate (or analytic) geometry bridges the gap between algebra and geometry. The fundamental principle in this scheme is that every ordered pair that satisfies an equation corresponds to a point in its graph, and every point in the graph corresponds to an ordered pair that satisfies the equation. Thus, a picture may be drawn for each

equation relating x and y, which helps us see the essential characteristics of the relationship. On the other hand, a graph can often be represented as an equation, so that the powerful methods of algebra can be used to analyze these curves. This simple, yet brilliant, idea is one of the most useful in mathematics.

2. All necessary information must be shown in the graph so you can see the relationship without consulting any supplementary material. In particular, be sure to identify the appropriate units on both axes. A graph must be able to "stand" by itself!

3. It is not necessary to scale the two axes in the same manner. In fact, different scales are frequently desirable. It is important that you be aware that different scales can have a dramatic effect on the apparent behavior of a relationship.

4. In this section we discussed the Cartesian (or rectangular) coordinate system. There are other important systems that can be used to obtain a picture of a relationship. If you continue in mathematics, you are sure to encounter some of them (for example, see Section 7.5 and Appendix A.4). No one system is ever more correct than another, just more advantageous for the particular situation being analyzed.

EXERCISES 3.2

1. Graph the following ordered pairs.

    a. $(3,1)$ **b.** $(-3,4)$ **c.** $(0,2)$

    d. $(-3,0)$ **e.** $(-1,-2)$ **f.** $(2,-3)$

    g. $(-2,-3)$ **h.** $(-1,2)$ **i.** $(\sqrt{2},3)$

    j. $(1,-\pi)$

2. Approximate (use integers) the ordered pairs corresponding to the points shown in the graph.

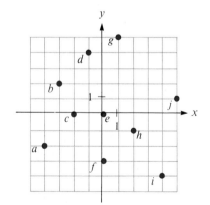

3. Represent the ordered pairs as points in the Cartesian coordinate system and find the distance between them.

    a. $(-1,3), (-1,1)$ **b.** $(2,-4), (2,-1)$

    c. $(4,1), (0,1)$ **d.** $(-3,-2), (2,-2)$

    e. $(-12,5), (0,0)$ **f.** $(0,0), (-4,1)$

    g. $(4,-1), (-2,-3)$ **h.** $(0,2), (4,-3)$

    i. $(-1,-1), (-2,-2)$ **j.** $(5,2), (1,3)$

    k. $(-2,5), (4,-3)$ **l.** $(6,-2), (-6,3)$

4. Three vertices of a square are $(3,3)$, $(3,-1)$, and $(-1,3)$. Find the ordered pair corresponding to the fourth vertex. What is the perimeter of the square?

5. Three vertices of a rectangle are $(4,2)$, $(-5,2)$, and $(-5,-4)$. Find the ordered pair corresponding to the fourth vertex. What is the perimeter of the rectangle?

6. By finding the lengths of *AB, BC,* and *AC,* show that the points $A(-7,-7)$, $B(-1,1)$, and $C(2,5)$ lie on a straight line.

7. Show that the points $(3,5)$, $(-3,7)$, and $(-6,-2)$ are vertices of a right triangle.

8. Show that the points $(-8,-2)$, $(-11,3)$, and $(3,8)$ are vertices of an isosceles triangle. Is this triangle a right triangle?

9. Find the radius of a circle with center at (0,0) and which passes through the point (2,3).
10. Find the area of a circle whose diameter extends from $(-1,3)$ to $(3,1)$.

In Exercises 11–34 graph the function and indicate the range of the function on the graph.

11. $y = x$
12. $y = -3x$
13. $f(x) = x + 3$
14. $g(x) = x - 1$
15. $y = -x + 2$
16. $y = 5x - 5$
17. $f(x) = 1$
18. $h(x) = \pi$
19. $y = x^2 + 1$
20. $f(x) = x^2 - 4$
21. $f(x) = 1 - x^2$
22. $y = x^2 - 2x + 1$
23. $f(x) = x^3$
24. $y = x^3 - 2$
25. $y = |x - 1|$
26. $g(x) = |x + 3|$
27. $y = \dfrac{|x|}{x}$
28. $f(x) = \dfrac{|x - 2|}{x - 2}$
29. $f(x) = \sqrt[3]{x}$
30. $y = \sqrt{x}$
31. $h(x) = 1 - \sqrt{x}$
32. $y = \sqrt{1 - x}$
33. $y = \sqrt{1 - x^2}$
34. $f(x) = -\sqrt{9 - x^2}$

In Exercises 35 and 36 graph each equation and indicate the domain and range of the function in **b**.

35. **a.** $x^2 + y^2 = 1$ **b.** $y = -\sqrt{1 - x^2}$
36. **a.** $x^2 + y^2 - 25 = 0$ **b.** $y = \sqrt{25 - x^2}$

37. The bracket notation $[x]$ means the greatest integer less than or equal to x. For example, $[4.3] = 4$, $[\pi] = 3$, $[-\frac{1}{2}] = -1$, and $[2] = 2$. If $f(x) = [x]$, which is called the **greatest integer function,** find
   a. $f(\frac{1}{2})$ **b.** $f(\sqrt{2})$ **c.** $f(-1.4)$ **d.** $f(-\pi)$
   e. the graph of this function for $-2 \le x < 3$

38. A projectile fired vertically up from the ground with an initial velocity of 128 ft/second will hit the ground 8 seconds later, and the speed of the projectile in terms of the elapsed time t equals $|128 - 32t|$. Graph the function $s = |128 - 32t|$.

39. The height (y) above water of a diver t seconds after she steps off a platform 100 ft high is given by the formula $y = 100 - 16t^2$. Graph this function.

40. The charge for the first 3 minutes of a Monday station-to-station call from New York to Los Angeles depends on the time of day that the call is made. The following formula table indicates the charge in terms of 24-hour time (that is, 1700 hours is equivalent to 5 P.M.):

$$\text{Charge} = \begin{cases} \$0.85 & \text{if} \quad 0 \text{ hours} \le t < 0800 \text{ hours} \\ \$1.45 & \text{if} \quad 0800 \text{ hours} \le t \le 1700 \text{ hours} \\ \$0.85 & \text{if} \quad 1700 \text{ hours} < t \le 2400 \text{ hours.} \end{cases}$$

Graph this function.

41. Because of the need for energy conservation, the local power company advertises to "save a watt." To this end, they calculate the efficiency of air conditioners by finding the ratio between the cooling ability of the machine (Btu) and the amount of watts of electricity that it requires (that is, efficiency = Btu/watt). Their rating in terms of the efficiency (e) of the air conditioner is as follows:

$$\text{Rating} = \begin{cases} 0 \text{ (flunk)} & \text{if} \quad 0 \le e < 6 \\ 1 \text{ (pass)} & \text{if} \quad 6 \le e < 8 \\ 2 \text{ (good)} & \text{if} \quad 8 \le e < 10 \\ 3 \text{ (very good)} & \text{if} \quad 10 \le e. \end{cases}$$

Graph this function.

In Exercises 42–46 graph the function and indicate the domain and range of the function on the graph.

42. $f(x) = \begin{cases} 1 & \text{if} \quad x > 2 \\ 0 & \text{if} \quad x \le 2 \end{cases}$

43. $g(x) = \begin{cases} 1 & \text{if} \quad x \ge 0 \\ -1 & \text{if} \quad x < 0 \end{cases}$

44. $h(x) = \begin{cases} x & \text{if} \quad x \ge 1 \\ 1 & \text{if} \quad x < -1 \end{cases}$

45. $f(x) = \begin{cases} x & \text{if} \quad -2 \le x < 0 \\ 0 & \text{if} \quad x = 0 \\ -x & \text{if} \quad x > 0 \end{cases}$

46. $y = \begin{cases} -x & \text{if} \quad 0 < x < 3 \\ 2 & \text{if} \quad x = 0 \\ x + 1 & \text{if} \quad x < 0 \end{cases}$

47. Graph $y = |x|$ and $y = -|x|$ on the same coordinate system. How are the graphs related? In general, what is the relationship between the graphs of $y = f(x)$ and $y = -f(x)$?

48. a. Graph $y = x^2 + 2$ and $y = x^2 - 2$ on the same coordinate system. How do the graphs relate to $y = x^2$? In general, what is the effect of adding a constant *after* applying the squaring function?
  b. Graph $y = (x + 2)^2$ and $y = (x - 2)^2$ on the same coordinate system. How do these graphs relate to $y = x^2$? In general, what is the effect of adding a constant *before* applying the squaring function?

49. Consider the graph of $y = f(x)$ given here.
  a. What is the domain of the function?
  b. What is the range of the function?
  c. Determine $f(0)$ and $f(5)$.
  d. For what value(s) of x does $f(x) = 4$?
  e. Solve $f(x) = 0$.
  f. For what value(s) of x is $f(x) > 0$?
  g. For what value(s) of x is $f(x) < 0$?
  h. For what value(s) of x is $|f(x)| < 5$?

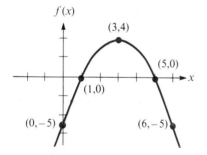

50. Consider the graph of $y = f(x)$ given here.
  a. What is the domain of the function?
  b. What is the range of the function?
  c. Determine $f(a)$ and $f(c)$.
  d. Solve $f(x) = a$.
  e. For what value(s) of x does $f(x) = 0$?
  f. For what value(s) of x is $f(x) < 0$?
  g. For what value(s) of x is $f(x) > 0$?
  h. For what value(s) of x is $|f(x)| < a$?

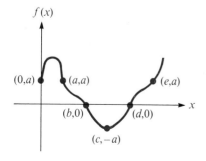

In Exercises 51–60 determine which of the graphs represent the graph of a function?

59.   **60.**

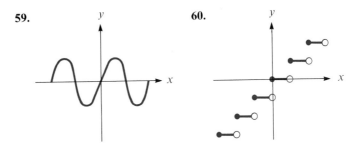

3.3 GRAPHING TECHNIQUES

There are many techniques associated with graphing functions. In the last section we concentrated primarily on plotting points in conjunction with graphs that required only lines or parabolas. We now mention some useful general approaches that revolve around two central themes: (1) how we can use the idea of symmetry to "cut in half" the job of graphing unfamiliar functions and (2) how we can graph variations of familiar functions by somehow adjusting a known curve. Let us begin with the idea of symmetry.

Symmetry

Consider the graph of $y = x^2$ in Figure 3.25. Notice that the y-axis divides the parabola into two segments such that if the curve were folded over on this line, its left half would coincide with its right half. In such a case we say $y = x^2$ has **symmetry about the y-axis.** Such symmetry is a nice feature for this or any other curve. It means we have to plot only the curve for $x \geq 0$ and then reflect our result over to the other side of the y-axis to complete the graph.

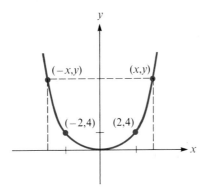

Figure 3.25

EXAMPLE 1   Complete the graph in Figure 3.26 for $x < 0$ if f is symmetric about the y-axis.

Figure 3.26

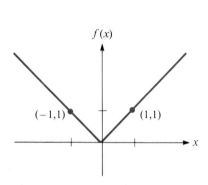

Figure 3.27

Solution   Just reflect the given segment over to quadrant 2 to obtain the graph for $x < 0$. The completed graph is shown in Figure 3.27. □

Now the question is how can we anticipate y-axis symmetry by looking at an equation. Consider Figure 3.27 again and note that y-axis symmetry results when x values that are opposite in sign share the same y value. For example, in the case of $y = x^2$, values of both 2 and -2 for x lead to a y value of 4; and in general, both (x,y) and $(-x,y)$ are on the graph. This observation leads us to the following rule for recognizing y-axis symmetry.

Even Function   A function f is called an even function when $f(-x) = f(x)$ for all x in the domain of f. The graph of an even function is symmetric about the y-axis.

EXAMPLE 2   Show that the absolute value function $f(x) = |x|$ is an even function.

Solution

$$f(x) = |x| \text{ and } f(-x) = |-x| = |x| = f(x)$$

Since $f(-x) = f(x)$ for any x, the absolute value function is an even function. In case you did not recognize it, Example 1 is the graph of $y = |x|$, which has y-axis symmetry. □

Another important type of symmetry is shown in the graph of $y = x^3$ in Figure 3.28. In this graph every point (x,y) in the first quadrant has a matching point $(-x,-y)$ in the third quadrant equidistant from the origin. In such a case we say $y = x^3$ has **symmetry about the origin.** Origin symmetry is a little more subtle than y-axis symmetry in that we must reflect the given segment twice in our shortcut graphing approach. As shown in Figure 3.29, we reflect the curve for $x \geq 0$ about both the x- and y-axes (in either order) to complete the graph.

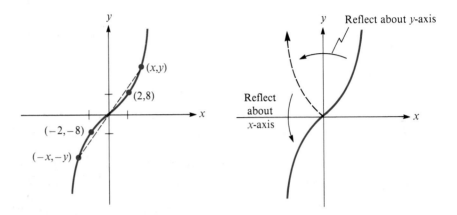

Figure 3.28   **Figure 3.29**

EXAMPLE 3   Complete the graph in Figure 3.30 for $x < 0$ if f is symmetric about the origin.

Figure 3.30

Figure 3.31

Solution   Reflect the given segment over to quadrant 2 and then reflect the result about the x-axis and down to quadrant 3. The completed graph is shown in Figure 3.31.   □

We can anticipate origin symmetry because it results when x values that are opposite in sign produce y values that are opposite in sign. For example, in the case of $y = x^3$, values of 2 and -2 for x produce y values of 8 and -8, respectively; and in general, both (x,y) and $(-x,-y)$ are on the graph. Thus, we have the following rule for recognizing origin symmetry.

Odd Function   A function f is called an odd function when $f(-x) = -f(x)$ for all x in the domain of f. The graph of an odd function is symmetric about the origin.

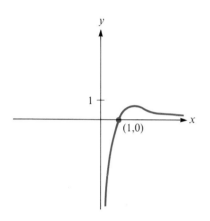

Figure 3.32

EXAMPLE 4   Show that the identity function $f(x) = x$ is an odd function.

Solution

$$f(x) = x \quad \text{and} \quad f(-x) = -x = -f(x)$$

Thus, $f(-x) = -f(x)$ for any x and the identity function is an odd function. Example 3 is the graph of the identity function that has origin symmetry.   □

EXAMPLE 5   Is the function $y = (x^2 - 1)/x^3$ an even function, an odd function, or neither? Figure 3.32 is the graph of this function for $x \geq 0$. Complete the graph.

Solution

$$f(x) = \frac{x^2 - 1}{x^3}$$

and

$$f(-x) = \frac{(-x)^2 - 1}{(-x)^3} = -\frac{x^2 - 1}{x^3} = -f(x)$$

Since $f(-x) = -f(x)$, f is an odd function that is symmetric about the origin. Reflecting the given curve about both the x- and y-axes gives the completed graph in Figure 3.33.   □

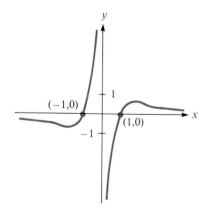

Figure 3.33

Translations

We now come to a second major consideration in graphing: How can we graph variations of a familiar function by manipulating a known curve? Consider Figure 3.34, which shows four variations of the absolute value function $f(x) = |x|$. Do you see a pattern?

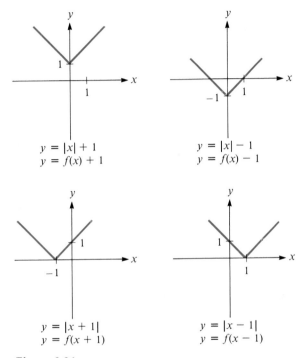

$$y = |x| + 1$$
$$y = f(x) + 1$$

$$y = |x| - 1$$
$$y = f(x) - 1$$

$$y = |x + 1|$$
$$y = f(x + 1)$$

$$y = |x - 1|$$
$$y = f(x - 1)$$

Figure 3.34

All four graphs have the basic $\vee$ shape that characterizes the absolute value function. Our job is merely to translate the $\vee$ to the right spot. Note that when we add or subtract the 1 *after* applying the absolute value rule, the effect is to move the $\vee$ up (if adding) or down (if subtracting) a distance of 1 unit. If we add or subtract the 1 *before* applying the absolute value rule, the $\vee$ moves to the left (if adding) or to the right (if subtracting) a distance of 1 unit. Be careful with the left-right shifts—they are deceiving. The graph of $y = |x - 1|$ is 1 unit to the right (not the left) of $y = |x|$ because x must be one larger in $y = |x - 1|$ than in $y = |x|$ to produce the same y value. For example, $y = 0$ when $x = 1$ in $y = |x - 1|$, while $y = 0$ when $x = 0$ in $y = |x|$. These ideas generalize to other functions and constants and provide us with the following guidelines.

Vertical and Horizontal Shifts	Let c be a positive constant.

1. The graph of $y = f(x) + c$ is the graph of f raised c units.
2. The graph of $y = f(x) - c$ is the graph of f lowered c units.
3. The graph of $y = f(x + c)$ is the graph of f shifted c units to the left.
4. The graph of $y = f(x - c)$ is the graph of f shifted c units to the right.

EXAMPLE 6   Use the graph of $y = x^2$ to graph each of the following functions.

a. $y = x^2 + 3$ **b.** $y = (x - 2)^2$ **c.** $y = (x + 1)^2 - 2$

Solution   The graph of the squaring function $y = x^2$ is a parabola. The rules above tell us to move this basic shape as follows (see Figure 3.35):

a. The constant 3 is added *after* the squaring rule, which means we move the parabola up 3 units.

b. The constant 2 is subtracted *before* the squaring rule, which moves the parabola 2 units to the right.

c. In this case we move the parabola 1 unit to the left and 2 units down.

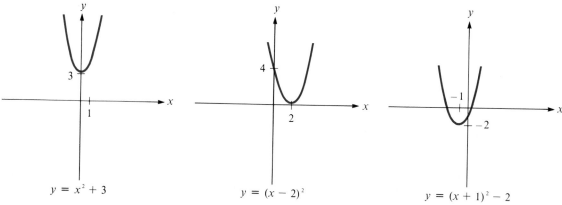

$y = x^2 + 3$ $y = (x - 2)^2$ $y = (x + 1)^2 - 2$

Figure 3.35

Reflecting and Stretching

Consider Figures 3.36 and 3.37, which illustrate the graphs of $y = |x|$, $y = -|x|$, $y = 2|x|$, and $y = \frac{1}{2}|x|$. These graphs show what happens when $|x|$ is multiplied by some constant. The most dramatic effect comes when the sign of the function is changed. As shown in Figure 3.36, switching from $|x|$ to $-|x|$ reflects the graph about the x-axis. Multiplying by a positive constant, such as 2, causes the curve to climb faster, while multiplying by a constant like $\frac{1}{2}$ flattens out the graph (see Figure 3.37). We summarize this pattern in the following rule.

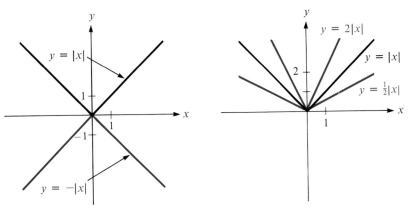

Figure 3.36   **Figure 3.37**

To Graph $y = cf(x)$	**Reflecting** The graph of $y = -f(x)$ is the graph of f reflected about the x-axis. **Stretching** If $c > 1$, the graph of $y = cf(x)$ is the graph of f stretched by a factor of c. **Shrinking** If $0 < c < 1$, the graph of $y = cf(x)$ is the graph of f flattened out by a factor of c.

EXAMPLE 7   Use the graph of $y = x^2$ to graph $y = 2 - x^2$.

Solution   To graph $y = -x^2$, we reflect over the basic $\cup$ shape about the x-axis as in Figure 3.38(a). Now $2 - x^2$ is the same as $-x^2 + 2$, so we raise the graph in Figure 3.38(a) up 2 units to get our answer [see Figure 3.38(b)].

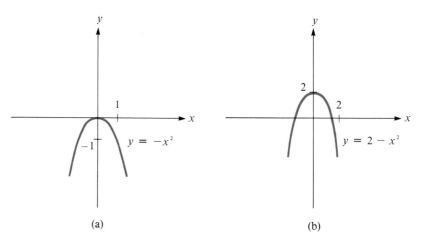

(a)   (b)

Figure 3.38

EXAMPLE 8   Consider the graph of $y = f(x)$ in Figure 3.39. Use this graph to sketch $y = 2f(x)$ and $y = \frac{1}{2}f(x)$.

Figure 3.39

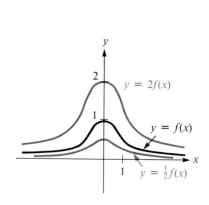

Figure 3.40

Solution   To graph $y = 2f(x)$, we just double each y value in f. Note in particular that the maximum y in f is 1, while the biggest y in $2f$ is 2. Similarly we graph $y = \frac{1}{2}f(x)$ by halving the y values in f. Both graphs are shown in Figure 3.40.

EXERCISES 3.3

In Exercises 1–8 classify each of the functions in the figures shown as even, odd, or neither of these.

1.

2.

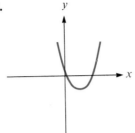

In Exercises 9–12 complete the graphs for $x < 0$ if (a) f is an even function and (b) f is an odd function.

9.

10.

3.

4.

11.

12.

5.

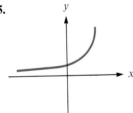

6.

In Exercises 13–22 classify each of the functions as even, odd, or neither of these.

13. $y = -x$

14. $f(x) = x + 1$

15. $f(x) = x^2 - 4$

16. $g(x) = 2$

17. $g(x) = x^2 - x$

18. $y = x^5$

19. $y = x^6$

20. $f(x) = 2 - |x|$

21. $f(x) = 1/x$

22. $y = 1/(x^3 + 1)$

23. Is the function $y = 1/(x^2 + 1)$ an even function, an odd function, or neither? The graph of this function for $x \geq 0$ follows. Complete the graph.

7.

8.

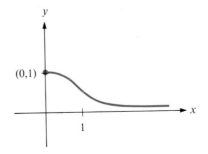

24. Is the function $y = x/(x^2 + 1)$ an even function, an odd function, or neither? The graph of this function for $x \geq 0$ follows. Complete the graph.

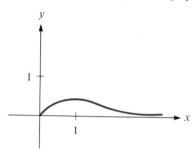

In Exercises 25–30 use the graph of $y = |x|$ to graph the given functions.

25. $y = |x| + 2$

26. $y = |x + 3|$

27. $y = -|x - 1|$

28. $y = 3|x|$

29. $y = |x - 2| - 1$

30. $y = 1 - |x + 1|$

In Exercises 31–36 use the graph of $y = x^2$ to graph the given functions.

31. $y = (x + 2)^2$

32. $y = x^2 - 2$

33. $y = 2x^2$

34. $y = \frac{1}{2}x^2$

35. $y = 1 - x^2$

36. $y = (x + 4)^2 - 3$

In Exercises 37–40 use the graph of $f(x) = \sqrt{x}$ that follows to graph each function.

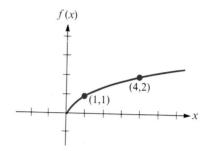

37. $f(x) = \sqrt{x - 1}$

38. $f(x) = \sqrt{x} - 1$

39. $f(x) = 2 - \sqrt{x}$

40. $f(x) = 2 + \sqrt{x + 2}$

In Exercises 41–46 use the graph of $y = f(x)$ that follows to graph each function.

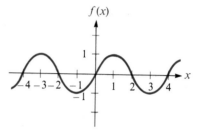

41. $y = 2f(x)$

42. $y = f(x) + 2$

43. $y = f(x - 1)$

44. $y = -f(x)$

45. $y = -\frac{1}{2}f(x + 4)$

46. $y = 3f(x - 1) - 1$

3.4 OPERATIONS WITH FUNCTIONS

We may add, subtract, multiply, and divide functions to form new functions. For example, suppose that an item that costs $5 to manufacture sells for $7. The total cost of x items is given by the function

$$C(x) = 5x.$$

The total sales revenue from x items is given by

$$S(x) = 7x.$$

Since profit is the difference between sales and cost, we form the profit function as the difference of the two other functions.

$$P(x) = S(x) - C(x)$$
$$= 7x - 5x$$
$$= 2x$$

In general, for two functions f and g, the functions $f + g$, $f - g$, $f \cdot g$, f/g are defined as follows:

$$(f + g)(x) = f(x) + g(x)$$
$$(f - g)(x) = f(x) - g(x)$$
$$(f \cdot g)(x) = f(x) \cdot g(x)$$
$$\left(\frac{f}{g}\right)(x) = \frac{f(x)}{g(x)} \qquad g(x) \neq 0$$

The domain of the resulting function is the intersection of the domains of f and g. The domain of the quotient function excludes any x for which $g(x) = 0$. In effect, the definitions say that for any x at which both functions are defined, we combine the y values.

EXAMPLE 1   If $f(x) = x$ and $g(x) = \sqrt{x}$, find $(f + g)(x)$, $(f - g)(x)$, $(f \cdot g)(x)$, and $(f/g)x$. In each case give the domain of the resulting function.

Solution   Using the definitions, we have

$$(f + g)(x) = f(x) + g(x) = x + \sqrt{x}$$
$$(f - g)(x) = f(x) - g(x) = x - \sqrt{x}$$
$$(f \cdot g)(x) = f(x) \cdot g(x) = x\sqrt{x} \quad \text{or} \quad x^{3/2}$$
$$\left(\frac{f}{g}\right)(x) = \frac{f(x)}{g(x)} = \frac{x}{\sqrt{x}} \quad \text{or} \quad \sqrt{x}.$$

The domain of f is the set of all real numbers and the domain of g is $[0,\infty)$. The domain of $f + g$, $f - g$, and $f \cdot g$ is the set of numbers common to both domains, which is $[0,\infty)$. In the quotient function $x \neq 0$. Thus, the domain of f/g is the interval $(0,\infty)$.   □

EXAMPLE 2   If $f = \{(0,1), (1,2), (2,3), (3,4)\}$ and $g = \{(2,0), (3,1), (4,-5), (5,-3)\}$, find $f + g$, $f - g$, $f \cdot g$, and f/g.

Solution   Both functions are defined at $x = 2$ and $x = 3$. Thus,

$$f + g = \{(2, 3 + 0), (3, 4 + 1)\} = \{(2,3), (3,5)\}$$
$$f - g = \{(2, 3 - 0), (3, 4 - 1)\} = \{(2,3), (3,3)\}$$
$$f \cdot g = \{(2, 3 \cdot 0), (3, 4 \cdot 1)\} = \{(2,0), (3,4)\}$$
$$f/g = \{(3, \tfrac{4}{1})\} = \{(3,4)\}; \text{ since } \tfrac{3}{0} \text{ is undefined, we do not list } (2, \tfrac{3}{0}). \quad \square$$

There is another important way of combining functions that has no counterpart in arithmetic. The operation is called **composition.** Basically, composition is a substitution that causes a "chain reaction" in which two functions are applied in succession.

Consider the problem of determining the area of a square whose perimeter is 20 in. You would probably reason as follows: If the perimeter is 20 in., the side is 5 in.; the area is then 25 in.². Given a specific value for the perimeter (P), we first determine the side (s) by using the function

$$s = \frac{P}{4}.$$

The result of this step is then substituted in the function

$$A = s^2$$

to determine the area. Let us put these formulas in functional notation and use the analogy of a function as a machine. Consider Figure 3.41, in which $s = f(P) = P/4$ and $A = g(s) = s^2$. The output of the f machine feeds the input of the g machine. Thus, we first apply the f function to determine s and then apply the g function to determine A. The area is given by the function

$$A = g[f(P)].$$

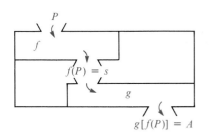

Figure 3.41

This function is said to be a *composition* of f and g. By substitution, we have

$$A = g[f(P)] = g\left(\frac{P}{4}\right) = \left(\frac{P}{4}\right)^2.$$

The symbol "∘" is also used for composition. We could write

$$A = (g \circ f)(P).$$

In general, the composite functions of functions f and g are defined as follows:

$$(g \circ f)(x) = g[f(x)]$$
$$(f \circ g)(x) = f[g(x)].$$

EXAMPLE 3   If $f(x) = x - 1$ and $g(x) = x^2$, find $(g \circ f)(x)$ and $(f \circ g)(x)$.

Solution

$$(g \circ f)(x) = g[f(x)] = g(x - 1)$$
$$= (x - 1)^2 = x^2 - 2x + 1$$
$$(f \circ g)(x) = f[g(x)] = f(x^2)$$
$$= x^2 - 1$$

Note that the order in which the functions are applied is important. In general,

$$(g \circ f)(x) \neq (f \circ g)(x). \qquad \square$$

A number x must satisfy two requirements to belong to the domain of $g \circ f$.

1. Since we first apply the f function, x must be in the domain of f.
2. The output from step 1, $f(x)$, must be in the domain of g.

Thus, the domain of $g \circ f$ is the set of all real numbers x in the domain of f for which $f(x)$ is in the domain of g. This set consists of all the values in the domain of f for which $g \circ f$ is defined.

EXAMPLE 4   If $f(x) = \sqrt{x}$ and $g(x) = x^2$, find $(g \circ f)(x)$ and $(f \circ g)(x)$. Indicate the domain of each function.

Solution

$(g \circ f)(x) = g[f(x)] = g(\sqrt{x}) = (\sqrt{x})^2 = x$
Domain of f: $[0,\infty)$
$(g \circ f)(x) = x$ is defined for all real numbers.

Thus, the domain of $g \circ f$ is $[0,\infty)$.

$(f \circ g)(x) = f[g(x)] = f(x^2) = \sqrt{x^2}$ or $|x|$
Domain of g: $(-\infty,\infty)$
$(f \circ g)(x) = |x|$ is defined for all real numbers.

Thus, the domain of $f \circ g$ is the set of all the real numbers. $\qquad \square$

EXAMPLE 5   If $f = \{(0,1), (1,2), (2,3), (3,4)\}$ and $g = \{(2,0), (3,1),$ $(4,-5), (5,-3)\}$, find $f \circ g$ and $g \circ f$. Indicate the domain of each composite function.

Solution   To determine $f \circ g$, we first apply the g function. As illustrated below, the range elements 0 and 1 are in the domain of f. Thus, $f \circ g$ $= \{(2,1), (3,2)\}$ and the domain of $f \circ g$ is $\{2,3\}$.

$$2 \xrightarrow{g} \quad 0 \xrightarrow{f} 1$$
$$3 \longrightarrow \quad 1 \longrightarrow 2$$
$$4 \longrightarrow -5$$
$$5 \longrightarrow -3$$

To find $g \circ f$, we first apply the f function. As illustrated below, the range elements 2, 3, and 4 are in the domain of g. Thus, $g \circ f = \{(1,0), (2,1), (3,-5)\}$ and the domain of $g \circ f$ is $\{1, 2, 3\}$.

$$0 \xrightarrow{f} 1$$
$$1 \longrightarrow 2 \xrightarrow{g} \quad 0$$
$$2 \longrightarrow 3 \longrightarrow \quad 1$$
$$3 \longrightarrow 4 \longrightarrow -5$$

EXERCISES 3.4

In Exercises 1–12 find $(f + g)(x)$, $(f - g)(x)$, $(f \cdot g)(x)$, $(f/g)(x)$, $(f \circ g)(x)$, and $(g \circ f)(x)$. Indicate the domain of each function.

1. $f(x) = 2x$; $g(x) = x - 1$

2. $f(x) = x - 2$; $g(x) = x + 2$

3. $f(x) = 4x^2 - 5$; $g(x) = -x + 3$

4. $f(x) = -3x + 5$; $g(x) = 1 + 2x - x^2$

5. $f(x) = x^2$; $g(x) = 1$

6. $f(x) = x$; $g(x) = 1/x$

7. $f(x) = 1/x$; $g(x) = x/(1 + x)$

8. $f(x) = \sqrt{x}$; $g(x) = -x^2$

9. $f(x) = x^2 - 1$; $g(x) = \sqrt{x + 1}$

10. $f(x) = \sqrt[3]{x - 5}$; $g(x) = x^3 + 5$

11. $f = \{(0,2), (1,3), (2,4), (3,5)\}$
$g = \{(2,-1), (3,0), (4,1), (5,2)\}$

12. $f = \{(-3,5), (0,1), (2,6), (4,11)\}$
$g = \{(0,0), (2,2), (5,5), (-3,-3)\}$

13. If $f = \{(1,3), (2,-2), (3,5)\}$ and $g = \{(2,-1), (3,4), (4,0)\}$, find
a. $(f + g)(2)$ **b.** $(f \cdot g)(3)$
c. $(g \cdot f)(1)$ **d.** $(g \circ f)(1)$

14. If $f = \{(1,5), (-2,4), (-3,0)\}$ and $g = \{(0,2), (-3,1), (-5,-1)\}$, find
a. $\left(\dfrac{f}{g}\right)(-3)$ **b.** $\left(\dfrac{g}{f}\right)(-3)$
c. $(f \circ g)(-3)$ **d.** $(g \circ f)(-3)$

In calculus it is often necessary to recognize a function as the composition of simpler functions. For example, $h(x) = (2x + 3)^2$ may be thought of as $(f \circ g)(x)$, where $g(x) = 2x + 3$ and $f(x) = x^2$. In Exercises 15–20 find simpler functions $f(x)$ and $g(x)$ so that $h(x) = (f \circ g)(x)$.

15. $h(x) = (4x - 1)^3$ **16.** $h(x) = (x^2 + 1)^2$

17. $h(x) = 2(3 - x)^4$ **18.** $h(x) = \sqrt{1 - x^2}$

19. $h(x) = \sqrt[3]{2x + 1}$ **20.** $h(x) = \dfrac{1}{3x - 5}$

21. If $h(x) = \sqrt{2x^2 + 2}$ and $f(x) = \sqrt{x}$, find $g(x)$ so that $h(x) = (f \circ g)(x)$.

22. If $h(x) = 3x^2 - 1$ and $g(x) = x^2$, find $f(x)$ so that $h(x) = (f \circ g)(x)$.

23. If $f(x) = 2x - 3$ and $g(x) = 5x + b$, find b so that $(f \circ g)(x) = (g \circ f)(x)$.

In Exercises 24–27 find $(f \circ g \circ h)(x)$.

24. $f(x) = 2x; g(x) = x^2; h(x) = x - 2$

25. $f(x) = x - 2; g(x) = 2x; h(x) = x^2$

26. $f(x) = \sqrt{x}; g(x) = 1/x; h(x) = 1 - x$

27. $f(x) = 1/x; g(x) = 1 - x; h(x) = \sqrt{x}$

28. Show that the sum of two odd functions is an odd function.

29. If f and g are odd functions, is $f \circ g$ an odd function or an even function? Verify your answer.

30. If $f(x) = \sqrt{x - 7}$ and $g(x) = \sqrt{2 - x}$, consider the domains of the functions and explain why it does not make sense to form $(f + g)(x)$.

31. If $f(x) = \sqrt{x - 7}$ and $g(x) = 2 - x^2$, explain why it does not make sense to form $(f \circ g)(x)$.

32. **a.** Write a function that converts miles to yards.
  b. Write a function that converts yards to feet.
  c. Use parts **a** and **b** and composition to construct a function that converts miles to feet.

33. **a.** Express the diagonal of a square as a function of the side.
  b. Express the side of a square as a function of the perimeter.
  c. Use parts **a** and **b** and composition to construct a function that expresses the diagonal of a square as a function of the perimeter.

3.5 INVERSE FUNCTIONS

Consider Figure 3.42, which illustrates a few of the assignments in the function $f(x) = x + 2$. Now consider Figure 3.43, which illustrates a few of the assignments in the function $g(x) = x - 2$.

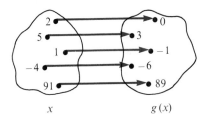

Figure 3.42   **Figure 3.43**

Note that the functions f and g have reverse assignments. For example, $f(0) = 2$ and $g(2) = 0$. Two functions with exactly reverse assignments are called **inverse functions** of each other. Thus, the inverse of $f(x) = x + 2$ is $g(x) = x - 2$, and vice versa. In the notation of ordered pairs, the reverse assignments of inverse functions means that when (a,b) is an element of a function, (b,a) is an element of the inverse function. Since the components in the ordered pairs are reversed, the domain of a function f is the range of its inverse function and the range of f is the domain of its inverse.

EXAMPLE 1   A function f consists of the ordered pairs $(7,-2)$, $(9,1)$, and $(-3,2)$.
a. Find the ordered pairs in the inverse function.
b. Compare the domain and range of the two functions.

Solution
a. Since the inverse function reverses the assignments, the inverse function (say g) consists of the ordered pairs $(-2,7)$, $(1,9)$, and $(2,-3)$.
b. The domain of f is $\{7,9,-3\}$ and the range of f is $\{-2,1,2\}$. The domain of g is $\{-2,1,2\}$ and the range of g is $\{7,9,-3\}$. Thus, the domain of f is the range of g and the range of f is the domain of g.   □

When we reverse the assignments in a function, the result is not always a function. For example, consider the function $y = f(x) = x^2$. Since $f(2) = 4$ and $f(-2) = 4$, an inverse of f would have to assign both 2 and -2 to the number 4. The definition of a function excludes this type of assignment. Thus, there is no function that precisely reverses the assignments of the function $y = x^2$. This example suggests the following guideline for judging if the inverse of a function is a function.

One-to-One Function

A function is **one-to-one** when each x value in the domain is assigned a different y value so that no two ordered pairs have the same second component. If f is one-to-one, then the inverse, denoted f^{-1}, is a function; and if f is not one-to-one, then the inverse is not a function.

EXAMPLE 2   Consider the function $\{(4,-3), (5,-3), (6,-3)\}$. Is the inverse of this function a function?

Solution   Since -3 appears as the second component in more than one ordered pair, the function is not one-to-one. Thus, the inverse of this function is not a function.   □

In a one-to-one function no two ordered pairs have the same second component. Graphically, this means that a one-to-one function cannot contain two points that lie on the same horizontal line. This observation leads to the following simple test for determining from a graph if a function has an inverse function.

Horizontal Line Test   Imagine a horizontal line sweeping down the graph of a function. If the horizontal line at any position intersects the graph in more than one point, the function is not one-to-one and its inverse is not a function.

EXAMPLE 3   Which functions graphed in Figure 3.44 have an inverse function?

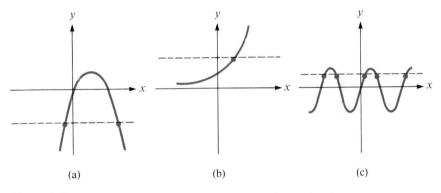

(a)   (b)   (c)

Figure 3.44

Solution
a. This function does not have an inverse function since a horizontal line may intersect the graph at two points.
b. This function has an inverse function because no horizontal line intersects the graph at more than one point.
c. This function does not have an inverse function since a horizontal line may intersect the graph at many points.   ☐

Graphically, there is another useful property of inverse functions. Because the x- and y-coordinates change places, the graphs of inverse functions are related in that each one is the reflection of the other across the line $y = x$. Figure 3.45 shows this relationship between the square root and squaring functions for $x \geq 0$.

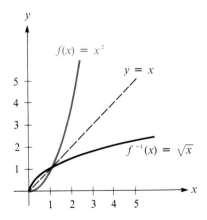

Figure 3.45

EXAMPLE 4   Consider the graph of $y = f(x)$ in Figure 3.46. Use this graph to sketch $y = f^{-1}(x)$.

Figure 3.46

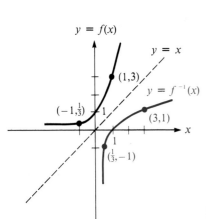

Figure 3.47

Solution   To graph $y = f^{-1}(x)$, we reflect the curve for $y = f(x)$ about the line $y = x$. Both graphs are shown in Figure 3.47.   □

If a function has an inverse function and is defined by an equation, then the fact that f and f^{-1} make reverse assignments means that we can find the rule for the inverse function by reversing the roles of x and y in the defining equation and solving for y. For example, the inverse of the function defined by

$$y = 2x - 5$$

is defined by

$$x = 2y - 5$$

which, when solved for y, becomes

$$y = \tfrac{1}{2}x + \tfrac{5}{2}.$$

Thus, if $f(x) = 2x - 5$, then $f^{-1}(x) = \tfrac{1}{2}x + \tfrac{5}{2}$. A complete statement of the procedure for finding an equation for f^{-1} follows:

To Find an Inverse Function

1. Start with a one-to-one function $y = f(x)$ and interchange x and y in this equation.
2. Solve the resulting equation for y, and then replace y by $f^{-1}(x)$.
3. Define the domain of f^{-1} to be equal to the range of f.

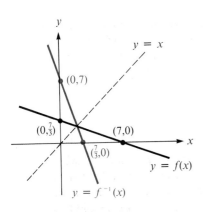

Figure 3.48

EXAMPLE 5   If $f(x) = \dfrac{7 - x}{3}$, find

a. $f^{-1}(x)$ **b.** $f^{-1}(0)$

Solution

a. The function $f(x) = \dfrac{7 - x}{3}$, or $f(x) = -\dfrac{1}{3}x + \dfrac{7}{3}$, graphs as a

straight line (see Figure 3.48) so the function is one-to-one and the inverse is a function. To find an equation for f^{-1}, write the equation for f as

$$y = \frac{7 - x}{3}$$

and then proceed as follows:

$$x = \frac{7 - y}{3} \qquad \text{Interchange } x \text{ and } y.$$

$$\begin{array}{l} 3x = 7 - y \\ y = -3x + 7 \end{array} \Big\} \;\; \text{Solve for } y.$$

$$f^{-1}(x) = -3x + 7. \qquad \text{Replace } y \text{ by } f^{-1}(x).$$

The domain of f^{-1} is equal to the range of f, which is the set of all real numbers.

b. To find $f^{-1}(0)$, replace x by 0 in the equation for f^{-1}.

$$\begin{array}{l} f^{-1}(0) = -3(0) + 7 \\ \phantom{f^{-1}(0)} = 7 \end{array}$$

$\square$

EXAMPLE 6   If $f(x) = \sqrt{x + 1}$, find $f^{-1}(x)$.

Solution   The graph of $f(x) = \sqrt{x + 1}$ in Figure 3.49 shows the function is one-to-one and the range is the interval $[0,\infty)$. We start with $y = f(x)$, so first write the equation as

$$y = \sqrt{x + 1}.$$

Then interchange x and y and solve for y.

$$x = \sqrt{y + 1}$$
$$x^2 = y + 1$$
$$x^2 - 1 = y$$

Now replace y by $f^{-1}(x)$, so the equation for f^{-1} is

$$f^{-1}(x) = x^2 - 1.$$

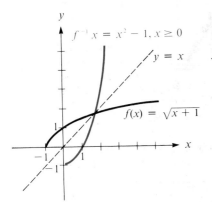

Figure 3.49

Finally, although $y = x^2 - 1$ is defined for any real number, we need to match the domain of f^{-1} to the range of f so that the functions will be inverses of each other. Thus, we restrict x as follows:

$$f^{-1}(x) = x^2 - 1, \quad x \geq 0. \qquad \square$$

When finding an inverse function, you should be aware of two additional concepts. First, it may be easy to show that a function is one-to-one so that an inverse exists, but finding an equation for the inverse may be difficult or impossible because of the algebra involved. (For instance, try finding f^{-1} if $f(x) = x^3 + x$.)

Second, note in Example 5 that $f^{-1}(0) = 7$ and $f(7) = (7 - 7)/3 = 0$. In general, inverse functions offset each other, so after finding $f^{-1}(x)$ you may wish to check your answer and test that applying f and f^{-1}, one after the other, brings you back to your original number. This observation also leads to the following important definition for inverse functions, which may be used to prove that two functions are inverses of each other as shown in Example 7.

Definition of Inverse Functions

Two functions f and g are said to be inverses of each other provided

$$(f \circ g)(x) = f[g(x)] = x \quad \text{for all } x \text{ in the domain of } g$$

and

$$(g \circ f)(x) = g[f(x)] = x \quad \text{for all } x \text{ in the domain of } f.$$

EXAMPLE 7   Verify that $f(x) = 5x - 5$ and $g(x) = \frac{1}{5}x + \frac{4}{5}$ are inverses of each other.

Solution   By the above definition, we first show that $f[g(x)] = x$ for all real numbers (which is the domain of g).

$$\begin{aligned}
f[g(x)] &= f(\tfrac{1}{5}x + \tfrac{4}{5}) \\
&= 5(\tfrac{1}{5}x + \tfrac{4}{5}) - 4 \\
&= x + 4 - 4 \\
&= x
\end{aligned}$$

Next we verify that $g[f(x)] = x$ for all real numbers (which is the domain of f).

$$\begin{aligned}
g[f(x)] &= g(5x - 4) \\
&= \tfrac{1}{5}(5x - 4) + \tfrac{4}{5} \\
&= x - \tfrac{4}{5} + \tfrac{4}{5} \\
&= x
\end{aligned}$$

Thus, f and g are inverses of each other. $\qquad \square$

The fact that f and f^{-1} must satisfy $f[f^{-1}(x)] = x$ provides an alternative method for finding an equation for f^{-1} that does not involve all the symbol switching of our previous method. For example, to redo Example 6 we start with $f(x) = \sqrt{x+1}$ and find $f[f^{-1}(x)]$ by replacing x by $f^{-1}(x)$ to obtain

$$f[f^{-1}(x)] = \sqrt{f^{-1}(x)+1}.$$

Then by definition, $f[f^{-1}(x)]$ equals x, so

$$x = \sqrt{f^{-1}(x)+1}$$
$$x^2 = f^{-1}(x) + 1$$
$$x^2 - 1 = f^{-1}(x),$$

which agrees with our previous result. Of course, the restriction $x \geq 0$ applies as discussed in Example 6. If you feel comfortable working with functional notation, then this method may be helpful.

Finally, scientific calculators use the concept of inverse functions with several important keys, particularly with respect to the trigonometric, logarithmic and exponential functions we are to consider. Pressing the $\boxed{\text{INV}}$ (or $\boxed{\text{2nd}}$) key gives the inverse of the next key pressed or some alternative function programmed for that key. Carefully read the owner's manual to your calculator to learn the second functions associated with some of the keys.

EXERCISES 3.5

In Exercises 1–6 find the inverse of the function. Also, find the domain and range of both functions.

1. $\{(1,7), (2,8), (3,9)\}$
2. $\{(4,-1), (8,-2), (12,-3)\}$
3. $\{(-2,10), (0,0), (5,-1)\}$
4. $\{(-7,1), (-4,11), (1,13)\}$
5. $\{(4,4), (5,5), (6,6)\}$
6. $\{(1,1), (2,\tfrac{1}{2}), (\tfrac{1}{2},2)\}$

In Exercises 7–10 determine if the inverse of the function is a function.

7. $\{(1,4), (2,4), (3,4)\}$
8. $\{(-4,4), (-5,5), (-6,6)\}$
9. $\{(0,0), (5,0.2), (0.2,5)\}$
10. $\{(-1,1), (0,0), (1,1)\}$

In Exercises 11–18 determine if the inverse of the function graphed is a function.

11.

12.

13.

14.

15.

16.

17.

18.

In Exercises 19 and 20 use the graph of $y = f(x)$ to graph $y = f^{-1}(x)$.

19.

20.

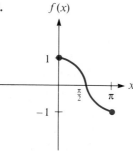

In Exercises 21–38 find the inverse function and graph both the function and its inverse function on the same set of axes. If the given function is not one-to-one so no inverse function exists, state this.

21. $y = x - 10$

22. $g(x) = -x + 5$

23. $f(x) = \dfrac{1}{3}x - \dfrac{7}{3}$

24. $y = \dfrac{4 - x}{11}$

25. $y = x^2$

26. $y = x^3$

27. $f(x) = 2$

28. $f(x) = |x|$

29. $y = \dfrac{1}{x}$

30. $f(x) = \dfrac{1}{x + 3}$

31. $h(x) = \sqrt[3]{x - 4}$

32. $g(x) = \sqrt{x}$

33. $f(x) = x^2 - 4,\ x \geq 0$

34. $g(x) = 1 - x^2,\ x < 0$

35. $y = \sqrt{x + 3}$

36. $f(x) = 2 + \sqrt{x}$

37. $f(x) = \sqrt{1 - x^2}$

38. $h(x) = \sqrt{1 - x^2},\ x \geq 0$

39. If $g(x) = 7x - 3$, find $g^{-1}(-1)$.

40. If $f(x) = \dfrac{1}{x - 1}$, find $f^{-1}(9)$.

41. Let $f(x) = \sqrt{x - 1}$. Find $f^{-1}(x)$. Find the domain and range for f and for f^{-1}.

In Exercises 42–47 verify that the given functions are inverses of each other.

42. $f(x) = x + 5,\ g(x) = x - 5$

43. $f(x) = 3x - 2,\ g(x) = \dfrac{1}{3}x + \dfrac{2}{3}$

44. $f(x) = \sqrt[3]{x},\ g(x) = x^3$

45. $f(x) = \dfrac{1}{x + 4},\ g(x) = \dfrac{1 - 4x}{x}$

46. $f(x) = \dfrac{1}{x + 2},\ g(x) = \dfrac{1}{x} - 2$

47. $f(x) = \sqrt{x + 1},\ x \geq -1$,
$g(x) = x^2 - 1,\ x \geq 0$

48. If $f(x) = \dfrac{1}{x}$, show that f is its own inverse. That is, show $f[f(x)] = x$.

49. If $f(x) = \dfrac{2x + 1}{3x - 2}$, show that f is its own inverse.

50. If $f(x) = x^2$ and $g(x) = \sqrt{x}$, find $f[g(x)]$ and $g[f(x)]$. For what values of x does $g[f(x)] = x$?

51. Use the definition of inverse functions to determine whether $f(x) = 3x - 1$ and $g(x) = \dfrac{1}{3}x + 1$ are inverses of each other.

52. The function $y = \dfrac{5}{9}(x - 32)$ converts degrees Fahrenheit (x) to degrees Celsius (y). Find the inverse function. What formula does the inverse function represent?

53. The function $y = x^2,\ x > 0$ gives the formula for the area (y) of a square in terms of the side length (x). Find the inverse function. What formula does this function represent?

3.6 SLOPE AND RATE OF CHANGE

An understanding of the process of change is vital to the investigation of the laws of nature, for everything is constantly changing in such significant respects as size, position, and temperature. Consequently, an important topic in mathematics is finding the rate at which one variable changes with respect to another. For example:

An investor is interested in the rate of change (interest rate) of his bank deposit so his investment will yield a maximum return.

An economist is interested in knowing the rate at which the demand for a product will change with respect to the price charged for the product.

A motorist is interested in knowing her speed, the rate of change of the distance she is traveling with respect to time.

A businessperson needs to know the rate at which the total cost of a production is changing with respect to each additional unit produced.

Social scientists are groping for a way to analyze the accelerating rate of social processes, which crowds an increasing number of events into an arbitrary interval of time. For example, in *Future Shock* (Random House, 1970, p. 2) Alvin Toffler argues that "unless man quickly learns to control the rate of change in his personal affairs as well as in society at large, we are doomed to a massive adaptational breakdown." This disease of change is labeled "future shock."

A health official helps prevent an epidemic by analyzing the rate at which an infectious disease (such as flu) might spread in a certain community.

These examples illustrate the necessity of analyzing change. The most important tool in this analysis is calculus. In fact, it is this mathematical ability of calculus to analyze movement and change that makes it the principal link between practical science and mathematical thought. To begin, we avoid more complicated functions and study only functions that always change at the same rate so that their graphs are straight lines. To find the rate of change in such cases, pick *any* two points on the given line, calculate the change in the dependent variable, and divide it by the change in the independent variable. This ratio is called the *slope of the line*.

Definition of Slope   If (x_1,y_1) and (x_2,y_2) are any two distinct points on a nonvertical line, as shown in Figure 3.50, then

$$\text{slope} = m = \frac{\Delta y}{\Delta x} = \frac{y_2 - y_1}{x_2 - x_1}.$$

(*Note:* The symbol Δ is the Greek capital letter delta. The symbol is used to indicate a change.)

Figure 3.50

The slope of a line is a fundamental notion in analytic geometry as a measure of the inclination or steepness of a line. In this context we often refer to Δy as the "rise" and Δx as the "run," so

$$\text{slope} = \frac{\text{rise}}{\text{run}}.$$

With this interpretation, it is natural to use the idea of slope to analyze such concrete situations as the steepness of a ramp, the pitch of a roof, or the inclination of a ski slope.

EXAMPLE 1   Find the slope of a line containing the points (2,1) and (5,4).

Solution   If we label (2,1) as point 1 (Figure 3.51),

$$x_1 = 2 \qquad y_1 = 1 \qquad x_2 = 5 \qquad y_2 = 4$$
$$m = \frac{\Delta y}{\Delta x} = \frac{y_2 - y_1}{x_2 - x_1} = \frac{4 - 1}{5 - 2} = \frac{3}{3} = 1.$$

A slope of 1 means that as x increases 1 unit, y increases 1 unit. Notice that the slope is unaffected by the way in which we label the points. If we label (5,4) as point 1,

$$x_1 = 5 \qquad y_1 = 4 \qquad x_2 = 2 \qquad y_2 = 1$$
$$m = \frac{\Delta y}{\Delta x} = \frac{y_2 - y_1}{x_2 - x_1} = \frac{1 - 4}{2 - 5} = \frac{-3}{-3} = 1. \qquad \Box$$

Figure 3.51

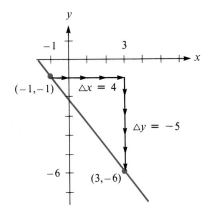

Figure 3.52

EXAMPLE 2   Find the slope of the line containing the points $(-1,-1)$ and $(3,-6)$.

Solution   (See Figure 3.52.)

$$x_1 = -1 \qquad y_1 = -1 \qquad x_2 = 3 \qquad y_2 = -6$$

$$m = \frac{\Delta y}{\Delta x} = \frac{y_2 - y_1}{x_2 - x_1} = \frac{-6 - (-1)}{3 - (-1)} = \frac{-5}{4}$$

A slope of $-5/4$ means that as x increases 4 units, y decreases 5 units. (*Note:* A negative slope indicates that y is decreasing as x increases. If the right end of a line segment is lower than the left end, the slope of the segment is negative. If the right end is higher, the slope of the segment [and of the whole line] is positive.) ☐

EXAMPLE 3   A ball thrown vertically up from a roof with an initial velocity of 128 ft/second will continue to climb for 4 seconds, and its velocity is 96 ft/second when $t = 1$ second and 32 ft/second when $t = 3$ seconds. Find the slope of the line that describes this linear relationship.

Solution   (See Figure 3.53.)

$$t_1 = 1 \qquad v_1 = 96 \qquad t_2 = 3 \qquad v_2 = 32$$

$$m = \frac{\Delta v}{\Delta t} = \frac{v_2 - v_1}{t_2 - t_1} = \frac{32 - 96}{3 - 1} = \frac{-64}{2} = -32$$

A slope of -32 means that for every additional second (up to 4 seconds) the velocity decreases by 32 ft/second. ☐

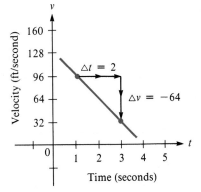

Figure 3.53

EXAMPLE 4   Find the slope of the line containing the points $(-1,4)$ and $(5,4)$.

Solution   (See Figure 3.54.)

$$x_1 = -1 \qquad y_1 = 4 \qquad x_2 = 5 \qquad y_2 = 4$$

$$m = \frac{\Delta y}{\Delta x} = \frac{y_2 - y_1}{x_2 - x_1} = \frac{4 - 4}{5 - (-1)} = \frac{0}{6} = 0$$

The slope of every horizontal line is zero since the numerator of the slope ratio (Δy) is always zero. ☐

Figure 3.54

Figure 3.55

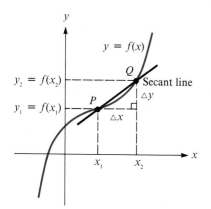

Figure 3.56

EXAMPLE 5   Find the slope of the line containing the points $(3,2)$ and $(3,-4)$.

Solution   (See Figure 3.55.)

$$x_1 = 3 \qquad y_1 = 2 \qquad x_2 = 3 \qquad y_2 = -4$$

$$m = \frac{\Delta y}{\Delta x} = \frac{y_2 - y_1}{x_2 - x_1} = \frac{-4 - 2}{3 - 3} = \frac{-6}{0}$$

m is undefined.

The slope of every vertical line is undefined since the denominator of the slope ratio (Δx) is always zero.   □

     In finding the slope of a line we compute a ratio and determine an average value. What makes linear relationships relatively easy to analyze is that they always change at the same rate. Thus, we obtain the same average no matter which two points we pick on the line. A function need not always change at the same rate to apply the concept of an average rate of change, but we must be careful in our interpretation of this measure. For example, if you drive 165 mi in 3 hours, your average speed is 165/3, or 55 mi *per* hour. This measure does not require or even mean that you go 55 mi *each* hour. It means that over the 3-hour period you average that speed; for another time interval the average speed would probably be different. Since the average varies with the interval chosen, it is important to remember that the average rate of change of a nonlinear relationship has meaning only if we indicate the interval in which we compute the average.

     Geometrically, let us interpret the meaning of the average rate of change of a function as x varies from x_1 to x_2. Consider the graph of $y = f(x)$ in Figure 3.56. Note that from this point on it is helpful to use functional notation and write $f(x_1)$ as well as y_1 to indicate the value of the function at x_1.

     Any line passing through two points on a curve is called a **secant line.** Thus, when we compute

$$m = \frac{\Delta y}{\Delta x} = \frac{y_2 - y_1}{x_2 - x_1} = \frac{f(x_2) - f(x_1)}{x_2 - x_1}$$

to determine the average rate of change of the function from x_1 to x_2, we are finding the slope of the secant line passing through P and Q.

EXAMPLE 6   If $f(x) = 3x^2 + 1$, find the average rate of change of the function as x varies

     a. from $x_1 = 0$ to $x_2 = 2$ **b.** from $x_1 = 1$ to $x_2 = 4$

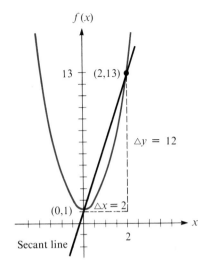

$f(x)$

13 ┤ (2,13)

$\Delta y = 12$

(0,1)

$\Delta x = 2$

2

x

Secant line

Figure 3.57

Solution

a. Since $f(x) = 3x^2 + 1$, we have $f(0) = 3(0)^2 + 1 = 1$ and $f(2) = 3(2)^2 + 1 = 13$. Thus,

$$m = \frac{\Delta y}{\Delta x} = \frac{f(x_2) - f(x_1)}{x_2 - x_1} = \frac{13 - 1}{2 - 0} = 6.$$

The average rate of change of the function $f(x) = 3x^2 + 1$ as x varies from 0 to 2 is 6. The function is graphed in Figure 3.57. Note that $(\Delta y/\Delta x) = 6$ is the slope of the secant line joining the points $(0,1)$ and $(2,13)$ that are located on the graph of the function.

b. Since $f(x) = 3x^2 + 1$, we have $f(1) = 3(1)^2 + 1 = 4$ and $f(4) = 3(4)^2 + 1 = 49$. Thus,

$$m = \frac{\Delta y}{\Delta x} = \frac{f(x_2) - f(x_1)}{x_2 - x_1} = \frac{49 - 4}{4 - 1} = 15.$$

In the interval from 1 to 4, the average rate of change of the function is 15.   ☐

In many applications it is most useful to determine how a function is changing at a particular point, say x_1. For this purpose, we define the average rate of change in terms of the constant x_1 and the variable Δx. Since $\Delta x = x_2 - x_1$, we have $x_2 = x_1 + \Delta x$ and we may write

$$m = \frac{\Delta y}{\Delta x} = \frac{f(x_2) - f(x_1)}{x_2 - x_1} = \frac{f(x_1 + \Delta x) - f(x_1)}{\Delta x}.$$

This concept is illustrated in Figure 3.58. To simplify algebraic computation with this formula, Δx is usually represented by h and the subscript in x_1 is dropped. We therefore have the following important definition.

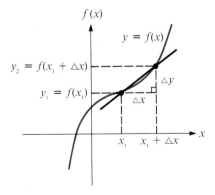

$f(x)$

$y = f(x)$

$y_2 = f(x_1 + \Delta x)$

$y_1 = f(x_1)$

Δy

Δx

x_1 $x_1 + \Delta x$ x

Figure 3.58

Average Rate of Change from x to $x + h$

The average rate of change of a function f as x varies from x to $x + h$ is

$$\frac{\Delta y}{\Delta x} = \frac{f(x + h) - f(x)}{h}, \qquad h \neq 0.$$

Computing this ratio is an essential consideration (and a major stumbling block for students) in calculus. You can probably see on an intuitive level that if we make $|h|$ very small, the above ratio becomes a good approximation for the rate of change of the function at x.

EXAMPLE 7   If $f(x) = 3x^2 + 1$:
a. Find the formula for the average rate of change of this function from x to $x + h$.
b. Evaluate this expression for $x = 1$ and $h = 3$.

Solution
a. If $f(x) = 3x^2 + 1$, we have

$$f(x + h) = 3(x + h)^2 + 1 = 3x^2 + 6xh + 3h^2 + 1.$$

Then

$$\frac{f(x + h) - f(x)}{h} = \frac{(3x^2 + 6xh + 3h^2 + 1) - (3x^2 + 1)}{h}$$

$$= \frac{3x^2 + 6xh + 3h^2 + 1 - 3x^2 - 1}{h}$$

$$= \frac{6xh + 3h^2}{h}$$

$$= 6x + 3h.$$

b. Evaluating this expression for $x = 1$ and $h = 3$, we have

$$\frac{\Delta y}{\Delta x} = 6(1) + 3(3) = 15.$$

In the interval from $x = 1$ to $x + h = 4$, the average rate of change of the function is 15. (See Example 6b.)   □

EXAMPLE 8   If $f(x) = 5$:
a. Find the formula for the average rate of change of the function from x to $x + h$.
b. Evaluate this expression for $x = 2$ and $h = 1$.

Solution
a. The function $f(x) = 5$ is a constant function whose graph is a horizontal line. The function value is therefore 5, regardless of the value for x. Thus, $f(x + h) = 5$ and $f(x) = 5$ so that

$$\frac{\Delta y}{\Delta x} = \frac{f(x + h) - f(x)}{h} = \frac{5 - 5}{h} = \frac{0}{h}.$$

The average rate of change of $f(x) = 5$ is 0. This answer agrees with our result from Example 4.
b. No matter which values we choose for x and h (with $h \neq 0$), the average rate of change of the function is 0.   □

EXERCISES 3.6

In Exercises 1–10 find the slope of the line determined by each pair of points.

1. (1,2) and (3,4)
2. (3,−1) and (5,−7)
3. (5,1) and (−2,−3)
4. (−1,2) and (0,0)
5. (4,−4) and (2,7)
6. (−5,−1) and (−2,−4)
7. (−1,−3) and (2,−3)
8. (−2,0) and (5,0)
9. (1,3) and (1,−1)
10. (−3,2) and (−3,−2)

In Exercises 11–14 find two ordered pairs that satisfy the equation and use them to find the slope of the line.

11. $y = -x + 1$
12. $y = 4x - 5$
13. $y = \frac{1}{2}x + 3$
14. $y = -\frac{5}{4}x - 2$

15. Approximate the slope of each line in the figure.

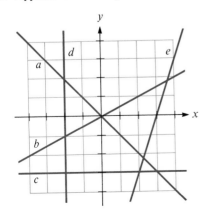

16. Approximate the slope of each line in the figure.

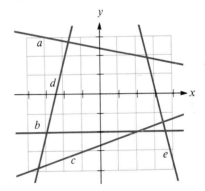

17. On a piece of graph paper draw lines through the point (1,2) with the following slopes:
   a. 2   **b.** −3   **c.** $-\frac{1}{4}$ **d.** $\frac{2}{3}$ **e.** 0

18. On a piece of graph paper draw lines through the point (−2,−1) with the following slopes:
   a. 1   **b.** −4   **c.** $-\frac{1}{3}$ **d.** $\frac{1}{3}$ **e.** $-\frac{3}{2}$

In Exercises 19–26 find the rate of change of each linear relationship by calculating the slope. In each case interpret the meaning of the slope.

19. An object that weighs 24 lb on earth weighs 4 lb on the moon; a different object that weighs 42 lb on earth weighs 7 lb on the moon.

20. The freezing point is 32° Fahrenheit compared to 0° Celsius; the boiling point is 212° Fahrenheit compared to 100° Celsius.

21. A ball thrown vertically up from the ground with an initial velocity of 320 ft/second will continue to climb for 10 seconds; its velocity is 256 ft/second when $t = 2$ seconds and 64 ft/second when $t = 8$ seconds.

22. A projectile fired vertically up from the ground with an initial velocity of 224 ft/second will continue to climb for 7 seconds; its velocity is 96 ft/second when $t = 4$ seconds.

23. The cost of 50 brochures is $11; the cost of 100 brochures is $17.

24. The total cost of manufacturing 100 units of a certain product is $450; the cost of manufacturing 200 units is $600.

25. A moving company charges $50 to move a certain machine 5 mi and $71 to move the same machine 8 mi.

26. The monthly cost of renting a computer is $400 if it is used for 5 hours and $500 if it is used for 7 hours.

In Exercises 27–30 find the average rate of change of each function in the given interval.

27. $f(x) = x^2$ from $x_1 = 1$ to $x_2 = 3$

28. $y = \sqrt{x}$ from $x_1 = 4$ to $x_2 = 9$

29. $g(x) = 1 - x - x^2$ from $x_1 = -2$ to $x_2 = -1$

30. $f(x) = x^3 + 1$ from $x_1 = 0$ to $x_2 = 2$

In Exercises 31–40 compute $\dfrac{f(x + h) - f(x)}{h}$ with $h \neq 0$ to determine the average rate of change of the function from x to $x + h$.

31. $f(x) = x^2$ **32.** $f(x) = 2x^2 + 3$

33. $f(x) = 4 - x^2$ **34.** $f(x) = x^2 + 3x + 5$

35. $f(x) = 2x$ **36.** $f(x) = 4x - 1$

37. $f(x) = 1$

38. $f(x) = \sqrt{x}$ (*Hint:* Rationalize the numerator.)

39. $f(x) = \dfrac{1}{x}$ **40.** $f(x) = x^3$

41. Determine the average rate of change of the area of a circle with respect to its radius as r varies from r to $r + h$.

42. The distance (d) in feet traveled by a freely falling body in t seconds is given by the formula $d = 16t^2$. What is the average velocity of the body in the interval from $t_1 = 1$ to $t_2 = 1 + h$? What is the average velocity if $h = 0.1$? 0.01? 0.001? Intuitively, what do you think we say the velocity of the body is at $t = 1$?

3.7 VARIATION

In many scientific laws the functional relationship between variables is stated in the language of variation. The statement "y varies *directly* as x" means that there is some positive number k such that $y = kx$. The constant k is called the **variation constant.**

EXAMPLE 1   The perimeter of a square varies directly as the side. Algebraically, we write $P = ks$. In this case we know that $k = 4$.   ☐

EXAMPLE 2   The circumference of a circle varies directly as the diameter. Algebraically, we write $C = kd$. In this case we know that $k = \pi$.   ☐

EXAMPLE 3   The sales tax (T) on a purchase varies directly as the price (p) of the item. Algebraically, we write $T = kp$. The value of k depends on the sales tax rate in a given location.   □

EXAMPLE 4   Hooke's law states that the distance (d) a spring is stretched varies directly as the force (F) applied to the spring. Algebraically, we write $d = kF$. The value of k depends on the spring in question (see Figure 3.59).   □

Figure 3.59

After expressing the functional relationship in the language of variation, we may compute the value of k if one pair of values for the variables is known. This value of k may then be used to find other corresponding values of the variables.

EXAMPLE 5   If y varies directly as x and $y = 21$ when $x = 3$, write y as a function of x. Determine the value of y when $x = 10$.

Solution   Since y varies directly as x, we have

$$y = kx.$$

To find k, replace y by 21 and x by 3.

$$21 = k \cdot 3$$
$$7 = k$$

Thus, $y = 7x$. When $x = 10$, $y = 7(10) = 70$.   □

EXAMPLE 6   In a spring to which Hooke's law applies, a force of 20 lb stretches the spring 6 in. How far will the spring be stretched by a force of 13 lb?

Solution   From Example 4 we have

$$d = kF.$$

To find k, replace d by 6 and F by 20.

$$6 = k \cdot 20$$
$$0.3 = k$$

Thus, $d = 0.3F$. When $F = 13$, $d = 0.3(13) = 3.9$ in.   □

When the product of corresponding values of the variables is a constant k, we say we have **inverse variation.** The statement "y varies inversely as x" means that there is some positive number k (the variation constant) such that $xy = k$, or $y = k/x$. Note that with direct variation: as x increases, y increases; with inverse variation: as x increases, y decreases.

EXAMPLE 7   If y varies inversely with x and $y = 27$ when $x = 2$, write y as a function of x. Find y when $x = 9$.

Solution   Since y varies inversely as x, we have

$$y = \frac{k}{x}.$$

To find k, replace y by 27 and x by 2.

$$27 = \frac{k}{2}$$
$$54 = k$$

Thus, $y = 54/x$. When $x = 9$, $y = \frac{54}{9} = 6$.   □

EXAMPLE 8   The speed (S) of a gear varies inversely as the number (n) of teeth. Gear A, which has 12 teeth, makes 400 rpm. How many revolutions per minute are made by a gear with 32 teeth that is connected to gear A?

Solution   Since S varies inversely with n, we have

$$S = \frac{k}{n}.$$

To find k, replace S by 400 and n by 12.

$$400 = \frac{k}{12}$$
$$4,800 = k$$

Thus, $S = 4,800/n$. When $n = 32$, $S = 4,800/32 = 150$ rpm.   □

We may extend the concept of variation to include direct and inverse variation of variables raised to specified powers and relationships that involve more than two variables.

EXAMPLE 9   If y varies directly as x^2 and inversely as z^3 and $y = 18$ when $x = 3$ and $z = 2$, determine the value of y when $x = 1$ and $z = 5$.

Solution   Since y varies directly as x^2 and inversely as z^3, we have

$$y = \frac{kx^2}{z^3}.$$

To find k, replace y by 18, x by 3, and z by 2.

$$18 = \frac{k(3)^2}{(2)^3}$$

$$18 = \frac{9k}{8}$$

$$16 = k$$

Thus, $y = 16x^2/z^3$. When $x = 1$ and $z = 5$,

$$y = \frac{16(1)^2}{(5)^3} = \frac{16}{125}. \qquad \qquad \square$$

EXAMPLE 10   Newton's law of gravitation states that the gravitational attraction between two objects varies directly as the product of their masses and inversely as the square of the distance between their centers of mass. What will be the change in attraction between the two objects if both masses are doubled and the distance between their centers is cut in half?

Solution   If F represents the gravitational attraction, m_1 and m_2 represent the masses of the objects, and d represents the distance between the centers of mass, then algebraically, we write Newton's law as

$$F = \frac{km_1m_2}{d^2}.$$

If the masses are doubled and the distance is cut in half, we have

$$F = \frac{k(2m_1)(2m_2)}{(\frac{1}{2}d)^2}$$

$$= \frac{4km_1m_2}{\frac{1}{4}d^2}$$

$$= \frac{16km_1m_2}{d^2}.$$

Thus, the gravitational attraction becomes 16 times as great. $\qquad \square$

Finally, we point out that any variation problem may also be solved by using a proportion. For example, if y varies directly as x, we know that $y = kx$. Thus, $y/x = k$ and there is a constant ratio between any corresponding values of x and y. This means that

$$\frac{y_1}{x_1} = \frac{y_2}{x_2} \quad \text{or} \quad \frac{y_1}{y_2} = \frac{x_1}{x_2}.$$

The equation $y_1/y_2 = x_1/x_2$ is called a **proportion** and may be used to solve the problem. For this reason, the variation constant k is often called the constant of proportionality, and the expression "varies directly as" is often replaced by "is proportional to." However, the language of variation usually provides a more convenient and informative statement of a relationship.

EXERCISES 3.7

In Exercises 1–30 write an equation relating the variables and solve the problem.

1. If y varies directly as x, and $y = 14$ when $x = 6$, write y as a function of x. Determine the value of y when $x = 10$.
2. If y varies directly as x, and $y = 2$ when $x = 5$, write y as a function of x. Determine the value of y when $x = 11$.
3. If y varies directly as the square of x, and $y = 3$ when $x = 2$, write y as a function of x. Find y when $x = 4$.
4. If y varies directly as x^3, and $y = 5$ when $x = 3$, write y as a function of x. Find y when $x = 2$.
5. If y varies inversely as x, and $y = 9$ when $x = 8$, write y as a function of x. Find y when $x = 24$.
6. If y varies inversely as x, and $y = 3$ when $x = 7$, write y as a function of x. Find y when $x = 21$.
7. If y varies inversely as x^3, and $y = 3$ when $x = 2$, write y as a function of x. Find y when $x = 3$.
8. If y varies inversely as the square of x, and $y = 2$ when $x = 4$, write y as a function of x. Find y when $x = 8$.
9. If y varies directly as x and z, and $y = 105$ when $x = 7$ and $z = 5$, find y when $x = 10$ and $z = 2$.
10. If y varies directly as x and inversely as z, and $y = 10$ when $x = 4$ and $z = 3$, find y when $x = 7$ and $z = 15$.
11. If y varies directly as x and inversely as z^2, and $y = 36$ when $x = 4$ and $z \doteq 7$, find y when $x = 9$ and $z = 9$.

12. If y varies inversely as x^2 and z^3, and $y = 0.5$ when $x = 3$ and $z = 2$, find y when $x = 2$ and $z = 3$.
13. In a spring to which Hooke's law applies (see Examples 4 and 6), a force of 15 lb stretches the spring 10 in. How far will the spring be stretched by a force of 6 lb?
14. The weight of an object on the moon varies directly with the weight of the object on earth. An object that weighs 114 lb on earth weighs 19 lb on the moon. How much will a person who weighs 174 lb on earth weigh on the moon?
15. The amount of garbage produced in a given location varies directly with the number of people living in the area. It is known that 25 tons of garbage is produced by 100 people in 1 year. If there are 8 million people in New York City, how much garbage is produced by New York City in 1 year?
16. Property tax varies directly as assessed valuation. The tax on property assessed at \$12,000 is \$400. What is the tax on property assessed at \$40,000?
17. If the area of a rectangle remains constant, the length varies inversely as the width. The length of a rectangle is 9 in. when the width is 8 in. If the area of the rectangle remains constant, find the width when the length is 24 in.
18. The speed of a gear varies inversely as the number of teeth. Gear A with 48 teeth makes 40 rpm. How many revolutions per minute are made by a gear with 120 teeth that is connected to gear A?

19. The speed of a pulley varies inversely as the diameter of the pulley. The speed of pulley A, which has an 8-in. diameter, is 450 rpm. What is the speed of a 6-in. diameter pulley connected to pulley A?

20. The time required to complete a certain job varies inversely as the number of machines that work on the job (assuming each machine does the same amount of work). It takes 5 machines 55 hours to complete an order. How long will it take 11 machines to complete the same job?

21. The volume of a sphere varies directly as the cube of its radius. The volume is 36π cubic units when the radius is 3 units. What is the volume when the radius is 5 units?

22. The distance an object falls due to gravity varies directly as the square of the time of fall. If an object falls 144 ft in 3 seconds, how far did it fall the first second?

23. The weight of an object varies inversely as the square of the distance from the object to the center of the earth. At sea level (4,000 mi from the center of the earth) a man weighs 200 lb. Find his weight when he is 200 mi above the surface of the earth.

24. The intensity of light on a plane surface varies inversely as the square of the distance from the source of light. If we double the distance from the source to the plane, what happens to the intensity?

25. The exposure time for photographing an object varies inversely as the square of the lens diameter. What will happen to the exposure time if the lens diameter is cut in half?

26. The resistance of a wire to an electrical current varies directly as its length and inversely as the square of its diameter. If a wire 100 ft long with a diameter of 0.01 in. has a resistance of $10\ \Omega$, what is the resistance of a wire of the same length and material but 0.03 in. in diameter?

27. The general gas law states that the pressure of an ideal gas varies directly as the absolute temperature and inversely as the volume. If $P = 4$ atm when $V = 10$ cm^3 and $T = 200°$ Kelvin, find P when $V = 30$ cm^3 and $T = 250°$ Kelvin.

28. The safe load of a beam (the amount it supports without breaking), which is supported at both ends, varies directly as the width and the square of the height and inversely as the distance between supports. If the width and height are doubled and the distance between supports remains the same, what is the effect on the safe load?

29. Coulomb's law states that the magnitude of the force that acts on two charges q_1 and q_2 varies directly as the product of the magnitude of q_1 and q_2 and inversely as the square of the distance between them. If the magnitude of q_1 is doubled, the magnitude of q_2 is tripled, and the distance between the charges is cut in half, what happens to the force?

30. Newton's law of gravitation states that the gravitational attraction between two objects varies directly as the product of their masses and inversely as the square of the distance between their centers of mass. What will be the change in attraction between the two objects if both masses are cut in half and the distance between their centers is doubled?

CHAPTER OVERVIEW

Section	Key Concepts to Review
3.1	• Rule definitions of a function and its domain and range
	• Ordered pair definitions of a function and its domain and range
	• The term $f(x)$ is read "f of x" or "f at x" and means the value of the function (the y value) corresponding to the value of x.
	• If a is in the domain of f, then ordered pairs for the function defined by $y = f(x)$ all have the form $(a, f(a))$. *Note: a is an x value and $f(a)$ is a y value.*

Section	Key Concepts to Review
3.2	• Definition of the graph of a function
	• Distance formula: $d = \sqrt{(x_2 - x_1)^2 + (y_2 - y_1)^2}$ (for any two points)
	• Statement of the fundamental principle in coordinate (or analytic) geometry
	• Vertical line test
	• Basic graphs: Let a, b, c, m, and r be real number constants with $a \neq 0$.
	$y = mx + b$: straight line $y = ax^2 + bx + c$: parabola
	$y = c$: horizontal line $x^2 + y^2 = r^2$: circle (center (0,0) radius r)
	$y = |x|$: $\vee$ shape
3.3	• Definitions of an even function and an odd function
	• The graph of an even function is symmetric about the y-axis.
	• The graph of an odd function is symmetric about the origin.
	• Methods to graph variations of a familiar function by using vertical and horizontal shifts, reflecting, stretching, and shrinking
3.4	• Definitions of $f + g$, $f - g$, $f \cdot g$, and f/g for two functions f and g
	• The symbol "$\circ$" denotes the operation of composition. The composite functions are $(f \circ g)(x) = f[g(x)]$ and $(g \circ f)(x) = g[f(x)]$.
	• Methods to determine the domain of $f + g$, $f - g$, $f \cdot g$, f/g, $f \circ g$, $g \circ f$
3.5	• Definitions of a one-to-one function and inverse functions
	• The special symbol f^{-1} is used to denote the inverse function of f.
	• Methods to determine if the inverse of a function is a function and how to find f^{-1}, if it exists.
	• Horizontal line test
	• f and f^{-1} interchange their domain and range.
	• The graphs of f and f^{-1} are reflections of each other across the line $y = x$.
3.6	• Slope formula: $m = \dfrac{\Delta y}{\Delta x} = \dfrac{y_2 - y_1}{x_2 - x_1}$ $(x_2 \neq x_1)$
	• Average rate of change from x to $x + h$: $\dfrac{\Delta y}{\Delta x} = \dfrac{f(x + h) - f(x)}{h}$, $h \neq 0$
3.7	• The statement "y varies directly as x" means there is some positive number k (variation constant) such that $y = kx$.
	• The statement "y varies inversely as x" means there is some positive number k such that $xy = k$ or $y = k/x$.

CHAPTER REVIEW EXERCISES

1. Fill in the missing component in each of the following ordered pairs so they are solutions of the equation $y = -2x + 4$: (0,), (,−1), (−1,), (,0), (3,).
2. Give three ordered pairs in f if $f(x) = 2$.
3. Is the set of ordered pairs $\{(-1,3), (0,3), (1,3)\}$ a function?
4. Write a formula expressing the circumference of a circle as a function of its radius.
5. Is the following graph the graph of a function? If yes, is the inverse also a function?

6. If $f(x) = x - 1/x$, find $f(0)$.
7. If $g(x) = 2x + 1$, for what values of x does $g(x) = 0$?
8. Three vertices of a rectangle are $(-4,-3)$, $(-4,2)$, and $(3,2)$. Determine the ordered pair corresponding to the fourth vertex.
9. If y varies directly as x, and $y = 5$ when $x = 12$, write y as a function of x. Determine the value of y when $x = 20$.
10. Find the distance between $(-2,1)$ and $(3,-1)$.

In Exercises 11–12 state the domain of the function.

11. $g(x) = 1/[x(x + 1)]$ 12. $y = \sqrt{x + 7}$
13. Write a formula that expresses the area of a square as a function of the diagonal. (*Hint:* Use the drawing and the Pythagorean theorem.)

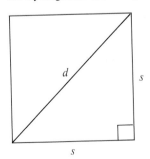

14. The speed of a pulley varies inversely as the diameter of the pulley. The speed of pulley A, which has a 7-in. diameter, is 320 rpm. What is the speed of a 5-in. diameter pulley that is connected to pulley A?
15. Are the following points the vertices of a right triangle: $(1,-3)$, $(8,-7)$, and $(5,-1)$?

In Exercises 16–21 let $f(x) = x - 1$ and $g(x) = 2x$.

16. Evaluate $3f(3) - 4g(-2)$.
17. Does $f[g(1)] = g[f(1)]$?
18. Does $f(a + b) = f(a) + f(b)$?
19. Does $g(a + b) = g(a) + g(b)$?
20. Is f an even function, an odd function, or neither?
21. Is g an even function, an odd function, or neither?

In Exercises 22–25 use the graph of $y = x^3$ to graph each function.

22. $y = (x + 1)^3$ 23. $y = x^3 + 1$
24. $y = -x^3$ 25. $y = (x - 2)^3 - 1$
26. Is the function $y = 3/x^2$ even, odd, or neither of these? The following figure gives the graph of this function for $x \geq 0$. Complete the graph.

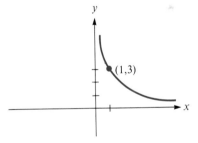

27. The intensity of light on a surface is inversely proportional to the square of the distance from the source of light. You are reading a book 6 ft from an electric light. If you move 2 ft closer, how many times greater is the intensity of light on the book?
28. A circle of radius 5 has its center at $(-1,-1)$. Is the point $(3,2)$ on this circle?
29. Express the area (A) of a circle as a function of its diameter (d).

In Exercises 30–33 consider the function

$$f(x) = \begin{cases} -x & \text{if } 0 \le x \le 3 \\ 3 & \text{if } x > 3. \end{cases}$$

30. Find $f(3)$.
31. Graph the function.
32. What is the domain?
33. What is the range?

In Exercises 34–37 use the following graph of $y = f(x)$ to graph each function.

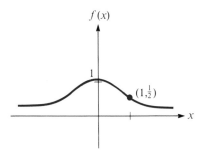

34. $y = 2f(x)$
35. $y = -\frac{1}{2}f(x)$
36. $y = f(x - 2)$
37. $y = f(x) - 2$

In Exercises 38–41 graph the given function.
38. $y = -3$
39. $f(x) = 2x - 4$
40. $g(x) = 9 - x^2$
41. $y = |x + 5|$

In Exercises 42 and 43 classify each function as even, odd, or neither of these.
42. $f(x) = |x|/x$
43. $y = x^4 + 1$

44. If $f = \{(3,-1), (4,-2)\}$, find f^{-1}. Determine the domain and range of each function.

45. If $g(x) = -\frac{1}{2}x + \frac{5}{2}$, find
  a. $g^{-1}(x)$ **b.** $g^{-1}(-1)$

In Exercises 46–53 use the following figure.

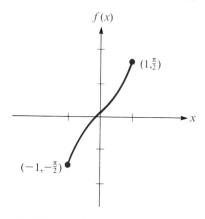

46. What is the domain?
47. What is the range?
48. Is f even, odd, or neither?
49. Find $f(0)$.
50. Solve $f(x) = 0$.
51. Solve $f(x) > 0$.
52. Solve $f(x) \le 0$.
53. Graph $y = f^{-1}(x)$.
54. Write a different function that has the same domain and range as the function $\{(1,4), (2,5)\}$.

In Exercises 55–58 compute in simplest form

$$\frac{\Delta y}{\Delta x} = \frac{f(x + h) - f(x)}{h}, h \ne 0$$

to determine the average rate of change of the function from x to $x + h$.

55. $f(x) = 3x^2 - 1$
56. $f(x) = -3$
57. $f(x) = 3 - x$
58. $f(x) = 1/(x + 1)$

59. Express the area of an equilateral triangle as a function of its side length (x).

60. A rectangle of sides x and y units is inscribed in a circle of radius 5 units. Express the area of the rectangle as a function of x. What is the domain of the function?

61. If $f(x) = x^2$ and $g(x) = 1/x$, find $(f \cdot g)(x)$. What is the domain of $f \cdot g$?

62. A rectangular field is completely enclosed by x feet of fencing. If the length of the rectangular region is twice the width, then express the area of the field as a function of x.

63. Express the monthly earnings (e) of a salesperson in terms of the cash amount (a) of merchandise sold if the salesperson earns $500 per month plus 9 percent commission on sales above $2,000.

64. If $h(x) = 3(1 - x)^2$, find simpler functions $f(x)$ and $g(x)$ so that $h(x) = (f \circ g)(x)$.

In Exercises 65–72 use the following figure.

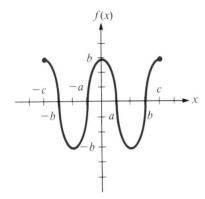

f(x)

65. What is the domain?
66. What is the range?
67. Is f even, odd, or neither?
68. Find $f(b)$.
69. Solve $f(x) = b$.
70. Solve $f(x) < 0$.
71. Graph $y = -\frac{1}{2}f(x)$.
72. Graph $y = f(x - c)$.
73. Find the domain and range of the functions shown.

a.

b.

c.

d.
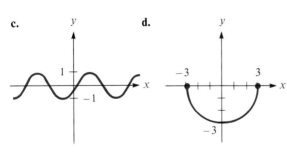

74. Find all real numbers a for which the distance between $(-1,1)$ and $(5,a)$ is 10.

75. Find x, given that the point $(x,0)$ is equidistant from $(-4,-4)$ and $(2,3)$.

In Exercises 76–85 select the choice that answers the question or completes the statement.

76. Which number is not in the domain of $y = \dfrac{x + 2}{x + 3}$?

    a. 2   **b.** -2 **c.** -3 **d.** 3

77. The graph of $y = f(x) - 3$ is the graph of $y = f(x)$ shifted 3 units

    a. up   **b.** down   **c.** left   **d.** right

78. If $f(x) = 5x^2$, then the difference quotient $\dfrac{f(x + h) - f(x)}{h}$ with $h \neq 0$ in simplest form is

    a. $10x + 5h$ **b.** $5h$ **c.** 1   **d.** $10xh + 5h$

79. The area of a circle whose radius extends from $(-2,5)$ to $(5,1)$ is

    a. 11π **b.** 25π **c.** 45π **d.** 65π

80. If y varies inversely with x, then when x is doubled y is

    a. decreased by 2   **b.** increased by 2
    c. multiplied by 2   **d.** divided by 2

81. If $f = \{(0,-3), (-1,-2), (2,-1)\}$ and $g = \{(-1,2), (-2,1), (-3,3)\}$, then $(g \circ f)(-1)$ equals

    a. -1 **b.** 1   **c.** 2   **d.** -2

82. The graph of an odd function is symmetric with respect to the

    a. x-axis   **b.** y-axis   **c.** origin

83. If $f(x) = 2x + b$ and $f(-2) = 3$, then b equals

    a. 7   **b.** -1 **c.** 1   **d.** -8

84. If a square field is completely enclosed by x feet of fencing, then the area of the field as a function of x equals

    a. x^2 **b.** $x^2/4$ **c.** $4x^2$ **d.** $x^2/16$

85. If $f(x) = 2x$, then which one of the following is true?

    a. $f(-a) = -f(a)$ **b.** $f(-a) = f(a)$
    c. $f(ab) = f(a) \cdot f(b)$ **d.** $f(a/b) = f(a)/f(b)$

4 *Polynomial and Rational Functions*

A real-life version of the Monopoly game for small investors is the purchase of a single family home for use as a rental property. An important benefit of this investment is that one can *depreciate* the cost of the building (but not the land) for tax benefits while the home (hopefully) *appreciates* in value. Usually the investment depreciates linearly by the straight line method, which means the cost of the building is deducted equally over a specified number of years as defined in the tax laws. If such an investment depreciates linearly from $200,000 to $0 in 30 years, then how much is the investment worth (for tax purposes) after 7 years? (See Example 7 of Section 4.1.)

I n this chapter we continue with the theme of a function and discuss many ideas associated with a large and important class of functions called polynomial functions. Basically, a polynomial function is characterized by terms of the form

$$(\text{real number})x^{\textbf{nonnegative integer}}$$

so that a polynomial function may contain terms like 4, $2x$, $-3x^2$, $\sqrt{5}x^3$, and so on. More technically, a **polynomial function of degree n** is a function of the form

$$y = P(x) = a_n x^n + a_{n-1} x^{n-1} + \cdots + a_1 x + a_0 \qquad (a_n \neq 0),$$

where n is a nonnegative integer and $a_n, a_{n-1}, \ldots, a_1$, and a_0 are real number constants. For example, the function $P(x) = 5x^3 + 2x^2 - 1$ is a polynomial function of degree 3 in which $a_3 = 5$, $a_2 = 2$, $a_1 = 0$, and $a_0 = -1$. Since n must be a nonnegative integer, note that functions with terms such as $x^{1/2}$ (or $\sqrt{x}$) and x^{-2} (or $1/x^2$) are not polynomial functions.

Although $y = 1/x^2$ is not a polynomial function, it can be expressed as a quotient of two polynomials and is called a **rational function.** This chapter concludes by showing how to graph rational functions. We begin by noting that the constant function $y = P(x) = a_0$ (with $a_0 \neq 0$) is a polynomial function of degree 0. For technical reasons the constant function $P(x) = 0$ is called the zero polynomial function and is not assigned a degree. We now consider polynomial functions of degree 1.

4.1 LINEAR FUNCTIONS

First-degree polynomial functions are called linear functions so the following definition applies.

Definition of Linear Function	A function of the form $$f(x) = ax + b,$$ where a and b are real numbers with $a \neq 0$, is called a **linear function.**

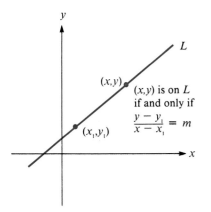

Figure 4.1

(x,y) is on L if and only if $\dfrac{y - y_1}{x - x_1} = m$

The term "linear" is used because such functions graph as straight lines. Two important formulas for an equation of a line, called the *point-slope equation* and the *slope-intercept equation,* are used often when working with linear functions, so we begin by developing these formulas.

Consider any nonvertical line L with slope m that passes through a point (x_1,y_1) as shown in Figure 4.1. (We eliminate vertical lines since their slope is undefined.) Then, any other point (x,y) with $x \neq x_1$ is on L if and only if the slope of the line segment joining (x,y) and (x_1,y_1) is m; that is, if and only if

$$\frac{y - y_1}{x - x_1} = m.$$

If we write this equation in the form

$$y - y_1 = m(x - x_1),$$

then (x_1,y_1) also satisfies the equation, which means that the points on L match the solutions of $y - y_1 = m(x - x_1)$. We summarize this result as follows:

Point-Slope Equation	An equation of the line with slope m passing through (x_1, y_1) is

$$y - y_1 = m(x - x_1).$$

This equation is called the **point-slope** form of the equation of a line.

EXAMPLE 1   Find an equation for the line that contains the point $(1,3)$ and whose slope is 2.

Solution   We are given that $x_1 = 1$, $y_1 = 3$, and $m = 2$. Substituting these numbers in the point-slope equation, we have

$$y - 3 = 2(x - 1)$$
$$y - 3 = 2x - 2$$
$$y = 2x + 1.$$ ☐

EXAMPLE 2   Find an equation for the line that contains the points $(1, -2)$ and $(4,5)$. Write the answer in the form $y = ax + b$.

Solution   First, we find the slope.

$$m = \frac{y_2 - y_1}{x_2 - x_1} = \frac{5 - (-2)}{4 - 1} = \frac{7}{3}$$

Now use the point-slope equation with $m = \frac{7}{3}$ and either $(1, -2)$ or $(4,5)$ for (x_1, y_1). Using $x_1 = 4$ and $y_1 = 5$, we have

$$y - 5 = \tfrac{7}{3}(x - 4),$$

which is converted to the form requested as follows:

$$y - 5 = \tfrac{7}{3}x - \tfrac{28}{3}$$
$$y = \tfrac{7}{3}x - \tfrac{13}{3}.$$ ☐

There are many ways of writing the equation of a line. The point-slope equation is used extensively for finding the equation of a line, but it is not very helpful for graphing lines or interpreting linear relationships when the equation is known. To develop a form that is useful in such cases, consider Figure 4.2 and note that any nonvertical line L must cross the y-axis at some point.

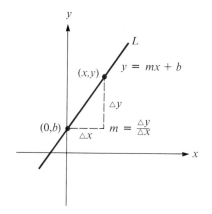

Figure 4.2

The x-coordinate of this point is zero and we will label the y-coordinate b. This point $(0,b)$, where the line crosses the y-axis, is called the **y-intercept.** If we apply the point-slope equation in the case when (x_1, y_1) is the y-intercept, we find that (x,y) is on L if and only if

$$y - b = m(x - 0)$$

or

$$y = mx + b.$$

To emphasize the geometric aspects of our discussion, we state this result as follows:

Slope-Intercept Equation   The graph of the equation

$$y = mx + b$$

is a line with slope m and y-intercept $(0,b)$. The equation $y = mx + b$ is called the **slope-intercept** form of the equation of a line.

Note that a linear function f may be defined by an equation of the form $y = f(x) = mx + b$ with $m \neq 0$, so that it follows from the above result that the graph of a linear function is a line.

EXAMPLE 3   Find the slope and the y-intercept of the line defined by the equation $2x + 3y = 6$.

Solution   First, transform the equation to the form $y = mx + b$.

$$2x + 3y = 6$$
$$3y = -2x + 6$$
$$y = -\tfrac{2}{3}x + 2$$

Matching this equation to the form $y = mx + b$, we conclude that

$$m = -\tfrac{2}{3} \qquad b = 2.$$

Thus, the slope is $-\tfrac{2}{3}$ and the y-intercept is $(0,2)$.   □

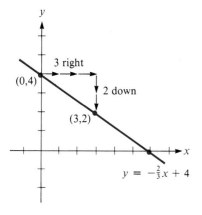

Figure 4.3

EXAMPLE 4   Graph the line whose slope is $-\frac{2}{3}$ and whose y-intercept is (0,4). Also, find an equation for the line.

Solution   We are given the y-intercept so we know one point on the line is (0,4). To find another point, we may interpret a slope of $-\frac{2}{3}$ to mean that when x increases 3 units, y decreases 2 units. By starting at (0,4) and going 3 units to the right and 2 units down, we obtain a second point on the line at (3,2). Drawing a line through these two points produces the graph in Figure 4.3. By substituting $m = -\frac{2}{3}$ and $b = 4$ in the slope-intercept equation $y = mx + b$, we determine that

$$y = -\tfrac{2}{3}x + 4$$

is an equation for the line.   □

EXAMPLE 5   Find the equation that defines the linear function f if $f(1) = 6$ and $f(-1) = 2$.

Solution   If $f(1) = 6$, then when $x = 1$, $y = 6$. Similarly, if $f(-1) = 2$, then when $x = -1$, $y = 2$. Thus, (1,6) and (-1,2) are points on the graph. First, calculate the slope.

$$m = \frac{y_2 - y_1}{x_2 - x_1} = \frac{6 - 2}{1 - (-1)} = 2$$

Now use the point-slope form with one of the points, say (1,6), as follows:

$$y - y_1 = m(x - x_1)$$
$$y - 6 = 2(x - 1)$$
$$y - 6 = 2x - 2$$
$$y = 2x + 4.$$

The equation that defines the function f is

$$f(x) = 2x + 4.$$ □

EXAMPLE 6   Find an equation that defines the linear relationship between Fahrenheit and Celsius temperatures if the freezing point of water is 32° Fahrenheit and 0° Celsius and the boiling point is 212° Fahrenheit and 100° Celsius.

Solution   Arbitrarily, let F be the independent variable (like x), and C be the dependent variable (like y). Then, find the slope.

$$m = \frac{\Delta C}{\Delta F} = \frac{100 - 0}{212 - 32} = \frac{100}{180} = \frac{5}{9}$$

Now use the point-slope equation with C replacing y and F replacing x. By selecting $C_1 = 0$ and $F_1 = 32$, we have

$$C - 0 = \tfrac{5}{9}(F - 32),$$

so an equation that defines the linear relationship is

$$C = \tfrac{5}{9}(F - 32).$$

(*Note:* If the roles of F and C are reversed in step 1, then $m = \Delta F / \Delta C$ and the resulting equation is $F = \tfrac{9}{5}C + 32$.)   □

EXAMPLE 7   Solve the problem in the chapter introduction on page 180.

Solution   If we let y represent the depreciated value of the building, then we are given that $y = 200{,}000$ when $t = 0$, and $y = 0$ when $t = 30$. From these corresponding values of y and t, we calculate the slope.

$$m = \frac{\Delta y}{\Delta t} = \frac{200{,}000 - 0}{0 - 30} = -\frac{20{,}000}{3} \qquad (\approx -\$6{,}667)$$

Since $(0, 200{,}000)$ is the y-intercept and $-\frac{20{,}000}{3}$ is the slope, the equation of the line is

$$y = -\tfrac{20{,}000}{3}t + 200{,}000.$$

When $t = 7$, we have

$$y = -\tfrac{20{,}000}{3}(7) + 200{,}000$$
$$= \$153{,}333.33.$$

Thus, the value of the building (for tax purposes) after 7 years is \$153,333. Note that the depreciation benefit in each year is about \$6,667.   □

Parallel and Perpendicular Lines

We often work with parallel and perpendicular lines in analytic geometry, and the slope concept may be used to establish conditions that parallel and perpendicular lines must satisfy.

Conditions for Parallel and Perpendicular Lines	**1.** Two nonvertical lines are parallel if and only if their slopes are equal. **2.** Two nonvertical lines are perpendicular if and only if the product of their slopes is -1.

Intuitively, it is easy to comprehend that parallel lines have the same slope or inclination, and the proof of statement 1 is outlined in Exercise 70. We can establish statement 2 by using the definition of slope and elementary geometry. Consider Figure 4.4 in which L_1 is perpendicular to L_2 and two right triangles have been formed as shown. From the definition of slope, the length of BC (denoted $\overline{BC}$) is m_1, while $\overline{CD} = -m_2$ (since m_2 is negative). Then since L_1 is perpendicular to L_2, both α_1 and α_2 are complementary to θ_1, so

$$\alpha_1 = \alpha_2 \quad \text{and} \quad \theta_1 = \theta_2$$

and the triangles are similar. In such triangles, corresponding side lengths are proportional, so

$$\frac{\overline{BC}}{\overline{AC}} = \frac{\overline{AC}}{\overline{CD}} \quad \text{or} \quad \frac{m_1}{1} = \frac{1}{-m_2}.$$

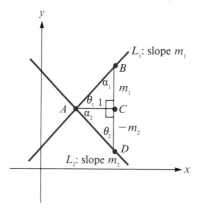

Figure 4.4

It then follows that $m_1 \cdot m_2 = -1$, which establishes that if L_1 and L_2 are nonvertical perpendicular lines, then the product of their slopes is -1. By reversing each of the steps given, we can also show that if $m_1 \cdot m_2 = -1$, then L_1 and L_2 are perpendicular, thereby establishing statement 2.

EXAMPLE 8   Find an equation for the line passing through the point $(-4,1)$ that is

a. parallel to $5x - 2y = 3$ **b.** perpendicular to $5x - 2y = 3$

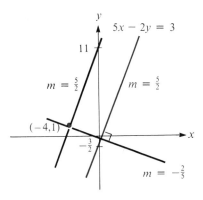

Figure 4.5

Solution   See Figure 4.5.

a. First, find the slope of the given line by changing the equation to the form $y = mx + b$.

$$5x - 2y = 3$$
$$-2y = -5x + 3$$
$$y = \tfrac{5}{2}x - \tfrac{3}{2}$$

By inspection, the slope of this line is $\tfrac{5}{2}$. Since the slopes of parallel lines are equal, we want an equation for the line through $(-4, 1)$ with slope $\tfrac{5}{2}$. Using the point-slope equation, we have

$$y - 1 = \tfrac{5}{2}(x - (-4))$$
$$y = \tfrac{5}{2}x + 11.$$

b. As shown in part **a,** the slope of $5x - 2y = 3$ is $\tfrac{5}{2}$. It follows from statement 2 that the slope m of every line perpendicular to the given line must satisfy $\tfrac{5}{2}m = -1$, so $m = -\tfrac{2}{5}$. Then we find an equation for the line through $(-4, 1)$ with slope $-\tfrac{2}{5}$ as follows:

$$y - y_1 = m(x - x_1)$$
$$y - 1 = -\tfrac{2}{5}(x - (-4))$$
$$y = -\tfrac{2}{5}x - \tfrac{3}{5}.$$

EXAMPLE 9   Show that if two nonvertical lines $A_1x + B_1y + C_1 = 0$ and $A_2x + B_2y + C_2 = 0$ are perpendicular, then $A_1A_2 + B_1B_2 = 0$.

Solution   First, write the given equations in the form $y = mx + b$. Since neither line is vertical, $B_1 \neq 0$ and $B_2 \neq 0$, which means

$$A_1x + B_1y + C_1 = 0 \text{ is equivalent to } y = \frac{-A_1}{B_1}x - \frac{C_1}{B_1}$$

$$\text{and } A_2x + B_2y + C_2 = 0 \text{ is equivalent to } y = \frac{-A_2}{B_2}x - \frac{C_2}{B_2}.$$

By inspection, the slopes of the lines are $-A_1/B_1$ and $-A_2/B_2$. Then, if the lines are perpendicular, the product of their slopes is -1, so

$$\left(\frac{-A_1}{B_1}\right)\left(\frac{-A_2}{B_2}\right) = -1$$
$$A_1A_2 = -B_1B_2$$
$$A_1A_2 + B_1B_2 = 0.$$

This last equation is the result we needed to establish. (*Note:* The equation $Ax + By + C = 0$ with A and B not both zero is often called the **general form** of the equation of a line.)

EXERCISES 4.1

In Exercises 1–10 is the function a polynomial function? If yes, state the degree.

1. $f(x) = x^3 + \sqrt{5}x - 3$ 2. $f(x) = x^3 + 5\sqrt{x} - 3$

3. $f(x) = 4x - 3$ 4. $f(x) = 7 + 4x - x^2$

5. $f(x) = x^2 + x^{1/2} - 1$

6. $f(x) = x^{-3} + 2x^2 - x + 3$

7. $f(x) = \dfrac{1}{x}$ 8. $f(x) = 3^{-1}$

9. $f(x) = \pi$ 10. $f(x) = x^{100} - 1$

In Exercises 11–16 find an equation for the line that has the given slope m and passes through the given point.

11. $m = 5$; $(4,3)$ 12. $m = -4$; $(2,-1)$

13. $m = \frac{1}{2}$; $(-2,0)$ 14. $m = -\frac{3}{4}$; $(-5,2)$

15. $m = -\frac{1}{2}$; $(0,0)$ 16. $m = 0$; $(1,2)$

In Exercises 17–20 find an equation for the line that contains the given points.

17. $(3,2)$ and $(4,1)$ 18. $(-4,3)$ and $(-2,5)$

19. $(-3,-3)$ and $(-4,-2)$ 20. $(2,-3)$ and $(3,-3)$

In Exercises 21–30 find the slope and the y-intercept of the line defined by the following equations.

21. $y = x + 7$ 22. $y = -\frac{2}{3}x - 5$

23. $y = 5x$ 24. $y = -x$

25. $y = -2$ 26. $y = 0$

27. $3y + 2x = -2$ 28. $3x + 5y = 2$

29. $6x - y = -7$ 30. $2x - 7y = -4$

In Exercises 31–36 graph the line whose slope and y-intercept are given and find an equation for the line.

31. $m = \frac{1}{2}$; $(0,3)$ 32. $m = 5$; $(0,1)$

33. $m = -3$; $(0,-1)$ 34. $m = -\frac{3}{4}$; $(0,-2)$

35. $m = 0$; $(0,-5)$ 36. $m = 0$; $(0,0)$

In Exercises 37–42 find the equation that defines the function if f is a linear function or a constant function.

37. $f(3) = 0$ and $f(0) = -2$

38. $f(0) = -1$ and $f(2) = 0$

39. $f(-1) = -3$ and $f(2) = -7$

40. $f(-6) = -1$ and $f(-2) = -5$

41. $f(-3) = 1$ and $f(0) = 1$

42. $f(4) = -6$ and $f(-1) = -6$

In Exercises 43–46 determine the given function value; f is a linear function.

43. $f(3) = 1$ and $f(5) = -3$; find $f(4)$.

44. $f(-2) = 0$ and $f(1) = 6$; find $f(0)$.

45. $f(-4) = -1$ and $f(-1) = -3$; find $f(0)$.

46. $f(-5) = 2$ and $f(1) = 2$; find $f(5)$.

In Exercises 47–58 find an equation for the line passing through the given point that is

  a. parallel to the given line

  b. perpendicular to the given line

47. $y = x - 2$; $(1,2)$ 48. $y = 3x + 1$; $(-2,-1)$

49. $y = -2x + 7$; $(3,0)$ 50. $y = -x$; $(-4,2)$

51. $x + 7y = 2$; $(0,0)$ 52. $3x - 2y = 5$; $(0,-5)$

53. $5x - 3y = 2$; $(1,-3)$ 54. $4x - y = -8$; $(-1,-4)$

55. $3x + 12y = 10$; $(5,2)$ 56. $x + 5y = -3$; $(-2,3)$

57. $2y - 3x + 1 = 0$; $(\frac{1}{2},-\frac{1}{3})$

58. $4x - y - 1 = 0$; $(-\frac{3}{4},-\frac{1}{2})$

59. A moving company charges $46 to move a certain machine 10 mi and $58 to move the same machine 30 mi.
  a. Find an equation that defines this relationship if it is linear.
  b. What will it cost to move the machine 25 mi?
  c. What is the minimum charge for moving the machine?
  d. What is the rate for each mile the machine is moved?

60. The total cost of producing a certain item consists of paying rent for the building and paying a fixed amount per unit for material. The total cost is $250 if 10 units are produced and $330 if 30 units are produced.
  a. Find an equation that defines this relationship if it is linear.
  b. What will it cost to produce 100 units?
  c. How much is paid in rent?
  d. What is the cost of the material for each unit?

61. A ball thrown vertically down from a building has a velocity of -252 ft/second when $t = 1$ second and -316 ft/second when $t = 3$ seconds.
  a. Find an equation that defines this relationship if it is linear.
  b. Find the velocity of the ball when $t = 4$ seconds.

c. What is the initial velocity of the ball?

d. What is the significance of whether the velocity is positive or negative?

62. A spring that is 24 in. long is compressed to 20 in. by a force of 16 lb and to 15 in. by a force of 36 lb.

   a. Find an equation that defines this relationship if it is linear.

   b. What is the length of the spring if a force of 28 lb is applied?

   c. How much force is needed to compress the spring to 10 in.?

63. Show that the points $A(-4,-2)$, $B(-3,2)$, and $C(1,1)$ are the vertices of a right triangle by using the concept of slope.

64. Show that the points $A(-7,-7)$, $B(-1,1)$, and $C(2,5)$ lie on a straight line by using the concept of slope.

65. If the line through $(-2,-3)$ and $(a,4)$ is perpendicular to the graph of $y = 4 - x$, find a.

66. Write an equation for the line with y-intercept $(0,1)$ and parallel to the graph of $y = x$.

67. The following figure shows five lines whose equations are as follows:

   a. $y = ax$

   b. $y = bx + c$

   c. $y = dx + e$

   d. $y = mx + k$

   e. $y = px + n$

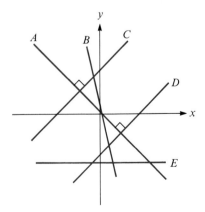

If $d = n = 0$, $b = m$, and $a < k < p$, then match each equation to the corresponding line.

68. Two points A and B are said to be *symmetric about the line L* if line segment AB is perpendicular to L and points A and B are equidistant from L (see illustration (a)). Use this definition to show in illustration (b) that the points (a,b) and (b,a) are symmetric about the line $y = x$. (*Note:* This exercise demonstrates why the graphs of f and f^{-1} are symmetric about the line $y = x$.)

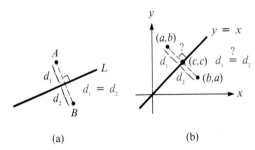

(a)   (b)

69. If two nonvertical lines $A_1x + B_1y + C_1 = 0$ and $A_2x + B_2y + C_2 = 0$ are parallel, then show that $A_1B_2 = A_2B_1$.

70. Use the illustration below and the fact that L_1 is parallel to L_2 if and only if $\theta_1 = \theta_2$ to prove the following statements.

   a. If two nonvertical lines are parallel, then their slopes are equal.

   b. If two nonvertical lines have equal slopes, then the lines are parallel.

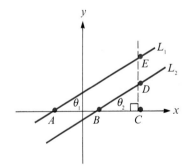

71. For the function $f(x) = mx + b$ compute $\dfrac{f(x + h) - f(x)}{h}$ with $h \neq 0$ and determine the average rate of change of the function from x to $x + h$. Interpret the meaning of your result.

72. Linear functions are often used to approximate the relationship between two variables so that predictions can be made about future events or trends. The history of the record for the 1-mi run is an interesting case in point. Following are a table and graph that show the evolution of the record in this event. Many people are surprised to see that the times have not been leveling off. Instead, they have dropped in a rather predictable linear pattern.

Time (seconds)

Year

Bannister

a. The color line in the figure nicely "fits" the points and can be used to predict with some reliability what will happen in the future in this event. This line passes through the points (1900,258) and (1950,239). Write an equation for this line.

b. Use the answer to part **a** and predict the time for the mile run in the year 2000.

c. Use the answer to part **a** and predict the year in which the record will be 3:42.

Time	Runner	Year
4:36.5	Richard Webster, Britain	1865
4:29	William Chinnery, Britain	1868
4:28.8	W.C. Gibbs, Britain	1868
4:26	Walter Slade, Britain	1874
4:24.5	Walter Slade, Britain	1875
4:23.2	Walter George, Britain	1880
4:21.4	Walter George, Britain	1882
4:19.4	Walter George, Britain	1882
4:18.4	Walter George, Britain	1884
4:18.2	Fred Bacon, Scotland	1894
4:17	Fred Bacon, Scotland	1895
4:15.6	Thomas Conneff, U.S.	1895
4:15.4	John Paul Jones, U.S.	1911
4:14.6	John Paul Jones, U.S.	1913
4:12.6	Norman Taber, U.S.	1915
4:10.4	Paavo Nurmi, Finland	1923
4.09.2	Jules Ladoumegue, France	1931
4:07.6	Jack Lovelock, New Zealand	1933
4:06.8	Glen Cunningham, U.S.	1934
4:06.4	Sydney Wooderson, Britain	1937
4:06.2	Gunder Haegg, Sweden	1942
4:06.2	Arne Andersson, Sweden	1942
4:04.6	Gunder Haegg, Sweden	1942
4:02.6	Arne Andersson, Sweden	1943
4:01.6	Arne Andersson, Sweden	1944
4:01.4	Gunder Haegg, Sweden	1945
3:59.4	Roger Bannister, Britain	1954
3:58	John Landy, Australia	1954
3:57.2	Derek Ibbotson, Britain	1957
3:54.5	Herb Elliott, Australia	1958
3:54.4	Peter Snell, New Zealand	1962
3:54.1	Peter Snell, New Zealand	1964
3:53.6	Michel Jazy, France	1965
3:51.3	Jim Ryun, U.S.	1966
3:51.1	Jim Ryun, U.S.	1967
3:50	Filbert Bayi, Tanzania	1975
3:49.4	John Walker, New Zealand	1975
3:48.9	Sebastian Coe, Britain	1979
3:48.8	Steve Ovett, Britain	1980
3:48.5	Sebastian Coe, Britain	1981
3:48.4	Steve Ovett, Britain	1981
3:48.3	Sebastian Coe, Britain	1981
3:46.3	Steve Cram, Britain	1985

State in each of the following cases whether you would expect the slope of the line that describes the relationship between the given variables to be positive or negative.

d. An individual's height and weight

e. The number of hours that a runner practices and his time for a 1-mi race

f. A student's cumulative average and the number of hours per week that she works at a job

g. The number of automobile accidents and the amount of insurance premiums that the driver has to pay

h. The percentage of nitrogen in a fertilizer and the height to which a treated plant will grow

i. The amount of alcohol consumed by an individual and the length of time in which he responds to a given stimulus

4.2 QUADRATIC FUNCTIONS

We now consider polynomial functions of degree 2, which are called quadratic functions.

Definition of Quadratic Function	A function of the form $$f(x) = ax^2 + bx + c,$$ where a, b, and c are real numbers with $a \neq 0$, is called a **quadratic function.**

To see the essential properties of a quadratic function consider the graphs of $f(x) = x^2 - 4x + 3$ (Figure 4.6) and $f(x) = 4x - 2x^2$ (Figure 4.7). We obtain these graphs by finding some ordered pairs that satisfy the relationships as shown.

Figure 4.6

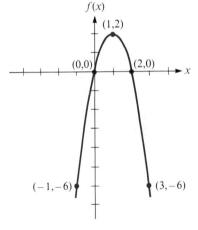

Figure 4.7

x	$x^2 - 4x + 3 = f(x)$	Solutions
-1	$(-1)^2 - 4(-1) + 3 = 8$	$(-1,8)$
0	$(0)^2 - 4(0) + 3 = 3$	$(0,3)$
1	$(1)^2 - 4(1) + 3 = 0$	$(1,0)$
2	$(2)^2 - 4(2) + 3 = -1$	$(2,-1)$
3	$(3)^2 - 4(3) + 3 = 0$	$(3,0)$
4	$(4)^2 - 4(4) + 3 = 3$	$(4,3)$
5	$(5)^2 - 4(5) + 3 = 8$	$(5,8)$

x	$4x - 2x^2 = f(x)$	Solutions
-1	$4(-1) - 2(-1)^2 = -6$	$(-1,-6)$
0	$4(0) - 2(0)^2 = 0$	$(0,0)$
1	$4(1) - 2(1)^2 = 2$	$(1,2)$
2	$4(2) - 2(2)^2 = 0$	$(2,0)$
3	$4(3) - 2(3)^2 = -6$	$(3,-6)$

The graph in both instances is a curve that is called a **parabola.** For $f(x)$ $= x^2 - 4x + 3$, the parabola has a minimum turning point and opens upward like a cup. This occurs because the coefficient of the dominant term (x^2) is a positive number. Similarly, the graph of $f(x) = 4x - 2x^2$ has a turning point at the maximum y value. The parabola opens down since the coefficient of x^2 is a negative number. In general:

1. If a is positive $(a > 0)$, the parabola opens upward and turns at the lowest point on the graph.
2. If a is negative $(a < 0)$, the parabola opens downward and turns at the highest point on the graph.

Another essential property of this graph is that a vertical line that passes through the turning point divides the parabola into two segments such that if the curve were folded over on this line, its left half would coincide with its right half (see Figure 4.8). This line is called the **axis of symmetry** for the parabola. To find the equation of the axis of symmetry, note that this line is halfway between any pair of points on the parabola that have the same y-coordinate. The simplest pair of points to work with is the pair of points on the graph of $y = ax^2 + bx + c$ whose y-coordinate is c. Thus,

$$y = ax^2 + bx + c.$$

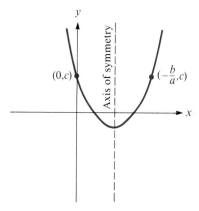

Figure 4.8

Let $y = c$; then

$$c = ax^2 + bx + c$$
$$0 = ax^2 + bx$$
$$0 = (ax + b)x$$
$$ax + b = 0 \qquad x = 0$$
$$ax = -b$$
$$x = \frac{-b}{a}.$$

The x-coordinate halfway between these two x values is

$$x = \frac{0 + (-b/a)}{2}$$

so the equation of the axis of symmetry is

$$x = \frac{-b}{2a}.$$

Since the maximum or minimum point lies on the axis of symmetry, the x-coordinate of this turning point is $-b/2a$. We determine the y-coordinate of this point by finding the y value when $x = -b/2a$. Thus, the coordinates of the turning point of the parabola (which is also called the **vertex**) are

$$\left(\frac{-b}{2a}, f\left(\frac{-b}{2a}\right)\right).$$

The last important property to the graph of a quadratic function that we consider is the coordinates of the points where the curve crosses the axes. The point where the parabola crosses the y-axis (that is, the y-intercept) can be found by substituting 0 for x in the equation $y = ax^2 + bx + c$.

$$y = a(0)^2 + b(0) + c$$
$$= c$$

Thus, the y-intercept is always $(0,c)$.

To find the points where the parabola crosses the x-axis (that is, the x-intercepts), we substitute 0 for y in the equation $y = ax^2 + bx + c$. Thus, the x-coordinates of the x-intercepts are found by solving the equation $ax^2 + bx + c = 0$.

EXAMPLE 1   Graph the function defined by $f(x) = x^2 - 7x + 6$ and indicate
  a. the coordinates of the x- and y-intercepts
  b. the equation of the axis of symmetry
  c. the coordinates of the maximum or minimum point
  d. the range of the function

Solution
  a. To find the x-intercepts, set $f(x) = 0$ and solve the resulting equation.

$$x^2 - 7x + 6 = 0$$
$$(x - 6)(x - 1) = 0$$
$$x - 6 = 0 \qquad x - 1 = 0$$
$$x = 6 \qquad\qquad x = 1$$

x-intercepts: $(1,0)$ and $(6,0)$; y-intercept: $(0,6)$ since $c = 6$

b. Axis of symmetry:

$$x = \frac{-b}{2a} = \frac{-(-7)}{2(1)} = \frac{7}{2}.$$

c. Since $a > 0$, we have a minimum point. The x-coordinate of the minimum point is $\frac{7}{2}$ since the minimum point lies on the axis of symmetry. We find the y-coordinate of the minimum point by finding the y value when $x = \frac{7}{2}$.

$$f(x) = x^2 - 7x + 6$$
$$f(\tfrac{7}{2}) = (\tfrac{7}{2})^2 - 7(\tfrac{7}{2}) + 6$$
$$= \tfrac{49}{4} - \tfrac{49}{2} + 6$$
$$= \tfrac{49}{4} - \tfrac{98}{4} + \tfrac{24}{4}$$
$$= -\tfrac{25}{4}$$

Minimum point: $(\tfrac{7}{2}, -\tfrac{25}{4})$

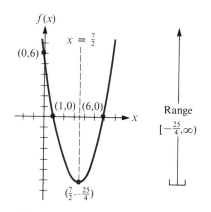

Figure 4.9

d. As illustrated in Figure 4.9, the range is the interval $[-\tfrac{25}{4}, \infty)$.   □

EXAMPLE 2   Graph the function defined by $f(x) = 3 - x - x^2$ and indicate
a. the coordinates of the x- and y-intercepts
b. the equation of the axis of symmetry
c. the coordinates of the maximum or minimum point
d. the range of the function

Solution

a. To find the x-intercepts, set $y = 0$ and solve the resulting equation

$$3 - x - x^2 = 0.$$

By the quadratic formula (with $a = -1$, $b = -1$, and $c = 3$), we have

$$x = \frac{-(-1) \pm \sqrt{(-1)^2 - 4(-1)(3)}}{2(-1)}$$
$$= \frac{1 \pm \sqrt{13}}{-2}.$$

x-intercepts: $\left(\dfrac{1 + \sqrt{13}}{-2}, 0\right)$ and $\left(\dfrac{1 - \sqrt{13}}{-2}, 0\right)$, which are abou..

$(-2.3, 0)$ and $(1.3, 0)$

y-intercept: $(0, 3)$ since $c = 3$

b. Axis of symmetry:

$$x = \frac{-b}{2a} = \frac{-(-1)}{2(-1)} = -\frac{1}{2}.$$

c. Since $a < 0$, we have a maximum point. The x-coordinate of the maximum point is $-\frac{1}{2}$, since the maximum point lies on the axis of symmetry. We find the y-coordinate of the maximum point by finding $f(-\frac{1}{2})$, the value of the function when $x = -\frac{1}{2}$.

$$f(x) = 3 - x - x^2$$
$$f(-\tfrac{1}{2}) = 3 - (-\tfrac{1}{2}) - (-\tfrac{1}{2})^2$$
$$= 3 + \tfrac{1}{2} - \tfrac{1}{4}$$
$$= \tfrac{13}{4}$$

Maximum point: $(-\frac{1}{2}, \frac{13}{4})$

d. As illustrated in Figure 4.10, the range is the interval $(-\infty, \frac{13}{4}]$.

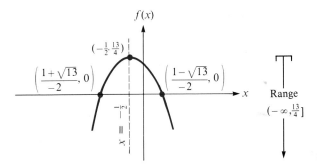

Figure 4.10

The vertex of a parabola can be found more efficiently in some cases by employing the graphing techniques discussed in Section 3.3. For instance, in Example 6c of that section we graphed $y = (x + 1)^2 - 2$ by moving the graph of $y = x^2$, 1 unit to the left and 2 units down, which placed the vertex of the parabola at $(-1,2)$. More generally, any quadratic function may be placed in the form

$$f(x) = a(x - h)^2 + k \qquad (\text{with } a \neq 0)$$

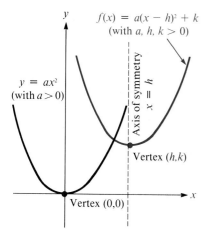

$f(x) = a(x - h)^2 + k$
(with $a, h, k > 0$)

$y = ax^2$
(with $a > 0$)

Axis of symmetry
$x = h$

Vertex (h, k)

Vertex $(0, 0)$

Figure 4.11

by completing the square, and in this form our graphing techniques tell us that the graph of f is the graph of $y = ax^2$ (a parabola) shifted so the vertex is the point (h, k) and the axis of symmetry is the vertical line $x = h$ (see Figure 4.11). Example 3 illustrates this result using the functions analyzed in Examples 1 and 2.

EXAMPLE 3   Determine the vertex and axis of symmetry of the graph of f by matching the function to the form $f(x) = a(x - h)^2 + k$.

a. $f(x) = x^2 - 7x + 6$ **b.** $f(x) = 3 - x - x^2$

Solution

a. First, write the equation as

$$f(x) = (x^2 - 7x \quad) + 6.$$

As discussed in Section 2.3, we complete the square for $x^2 + bx$ by adding $(b/2)^2$, which is the square of one half of the coefficient of x. Half of -7 is $-\frac{7}{2}$ and $(-\frac{7}{2})^2 = \frac{49}{4}$. Now inside the parentheses both add $\frac{49}{4}$ (so we can complete the square) and subtract $\frac{49}{4}$ (so $f(x)$ does not change), and proceed as follows:

$$f(x) = (x^2 - 7x + \tfrac{49}{4} - \tfrac{49}{4}) + 6$$
$$= (x^2 - 7x + \tfrac{49}{4}) - \tfrac{49}{4} + 6$$
$$= (x - \tfrac{7}{2})^2 - \tfrac{25}{4}.$$

Matching this result to the form $f(x) = a(x - h)^2 + k$, we determine $h = \frac{7}{2}$ and $k = -\frac{25}{4}$, so the vertex of the parabola is $(\frac{7}{2}, -\frac{25}{4})$ and the axis of symmetry is $x = \frac{7}{2}$.

b. To complete the square, the coefficient of x^2 must be 1, so first factor out -1 from the x terms and then proceed as in part **a.**

$$f(x) = 3 - x - x^2$$
$$= -1(x^2 + x) + 3$$
$$= -1(x^2 + x + \tfrac{1}{4} - \tfrac{1}{4}) + 3$$
$$= -1(x^2 + x + \tfrac{1}{4}) + (-1)(-\tfrac{1}{4}) + 3$$
$$= -1(x + \tfrac{1}{2})^2 + \tfrac{13}{4}$$

Once again, we match this equation to the form $f(x) = a(x - h)^2 + k$, which gives $h = -\frac{1}{2}$ and $k = \frac{13}{4}$. Thus, the vertex is $(-\frac{1}{2}, \frac{13}{4})$ and the axis of symmetry is $x = -\frac{1}{2}$.   □

The following examples illustrate how quadratic functions can be applied in practical situations. Notice, in particular, the significance of the maximum or minimum value, which is an important topic in applied mathematics.

EXAMPLE 4   The height (y) of a projectile shot vertically up from the ground with an initial velocity of 144 ft/second is given by the formula $y = 144t - 16t^2$. Graph the function defined by this formula and indicate
a. when the projectile will strike the ground
b. when the projectile attains its maximum height
c. the maximum height

Solution
a. When the projectile hits the ground, the height (y) of the projectile is 0. Thus, we want to find the values of t for which $144t - 16t^2$ equals 0.

$$144t - 16t^2 = 0$$
$$16t(9 - t) = 0$$
$$16t = 0 \qquad 9 - t = 0$$
$$t = 0 \qquad\quad t = 9$$

The projectile will hit the ground 9 seconds later.
b. Since $a < 0$, we have a maximum point. Axis of symmetry:

$$t = \frac{-b}{2a} = \frac{-(144)}{2(-16)} = \frac{-144}{-32} = 4.5.$$

The projectile reaches its highest point when $t = 4.5$ seconds.
c. We can find the maximum height by finding the value of y when $t = 4.5$ (or $t = \frac{9}{2}$) seconds.

$$y = f(t) = 144t - 16t^2$$
$$f(\tfrac{9}{2}) = 144(\tfrac{9}{2}) - 16(\tfrac{9}{2})^2$$
$$= 72 \cdot 9 - 16(\tfrac{81}{4})$$
$$= 324$$

Thus, the projectile reaches a maximum height of 324 ft, and Figure 4.12 shows the graph of this function. □

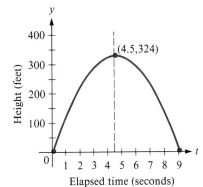

Figure 4.12

EXAMPLE 5   A businessperson wants to fence in a parking lot in which one side of the lot is bounded by a building. If a total of 300 ft of fencing is available, what are the dimensions of the largest rectangular parking lot that can be enclosed with the available fencing?

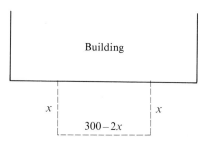

Figure 4.13

Solution   First, draw a sketch of the situation (Figure 4.13). If we let x represent the length of one side of the lot, the opposite side also has length x. The length of the side opposite the building is then $300 - 2x$. We wish to maximize the area of this rectangular region, which is given by the following quadratic function.

$$\text{Area} = x(300 - 2x)$$
$$A = 300x - 2x^2$$

Since $a < 0$, we have a maximum point. Axis of symmetry:

$$x = \frac{-b}{2a} = \frac{-300}{2(-2)} = \frac{-300}{-4} = 75.$$

Length of the side opposite building:

$$300 - 2x = 300 - 2(75) = 300 - 150 = 150.$$

The area is a maximum when the dimensions are 75 ft by 150 ft.   ☐

In Section 2.5 we discussed the meaning and importance of inequalities. We now extend our ability to work with these statements by considering a graphical approach to quadratic inequalities. They are expressed (with $a \neq 0$) in the forms

$$ax^2 + bx + c > 0, \qquad ax^2 + bx + c \geq 0,$$
$$ax^2 + bx + c < 0, \quad \text{or} \quad ax^2 + bx + c \leq 0.$$

For example, let us find the values of x for which $x^2 - x - 6 > 0$ is a true statement.

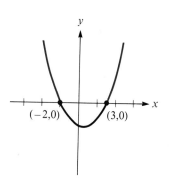

Figure 4.14

1. Sketch the graph of $y = x^2 - x - 6$ with emphasis on the x-intercepts and on whether the curve has a maximum or minimum turning point.
   a. To find the x-intercepts, set $y = 0$ and solve the resulting equation.

   $$0 = x^2 - x - 6$$
   $$0 = (x - 3)(x + 2)$$
   $$x = 3 \qquad x = -2$$

   x-intercepts: $(3,0)$ and $(-2,0)$
   b. Since $a > 0$, the graph has a minimum turning point and opens up (see Figure 4.14).
2. To solve the inequality $x^2 - x - 6 > 0$, we want only those values of x for which $x^2 - x - 6$ (or y) is positive. Observing the graph, we see that values of x which are to the right of 3 (that is, $x > 3$) or values of x that are to the left of -2 (that is, $x < -2$) make $x^2 - x - 6$ (or y) positive. Therefore, the solution set in interval notation is $(-\infty, -2) \cup (3, \infty)$.

EXAMPLE 6   For what values of x is the inequality $4 + 3x - x^2 > 0$ a true statement?

Solution
a. Sketch the graph of $y = 4 + 3x - x^2$.
  (1) To find the x-intercepts, set $y = 0$ and solve the resulting equation, $0 = 4 + 3x - x^2$.
$$4 + 3x - x^2 = 0$$
$$(4 - x)(1 + x) = 0$$
$$4 - x = 0 \qquad 1 + x = 0$$
$$x = 4 \qquad\qquad x = -1$$

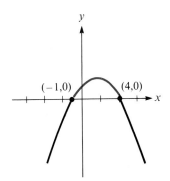

Figure 4.15

  x-intercepts: $(-1,0)$ and $(4,0)$
  (2) Since $a < 0$, the graph has a maximum turning point and opens down (see Figure 4.15).
b. To solve the inequality $4 + 3x - x^2 > 0$, we want only those values of x for which $4 + 3x - x^2$ (or y) is positive. Observing the graph, we see that values of x between -1 and 4 (that is, $-1 < x < 4$) make $4 + 3x - x^2$ (or y) positive. Thus, the solution set is the interval $(-1,4)$. □

EXAMPLE 7   Solve the inequality $2x^2 + 5x \leq 2$.

Solution
a. $2x^2 + 5x \leq 2$ in the form $ax^2 + bx + c \leq 0$ is $2x^2 + 5x - 2 \leq 0$. Therefore, we want to sketch the graph of $y = 2x^2 + 5x - 2$.
  (1) To find the x-intercepts, set $y = 0$ and solve the resulting equation, $0 = 2x^2 + 5x - 2$.

$$x = \frac{-b \pm \sqrt{b^2 - 4ac}}{2a} = \frac{-(5) \pm \sqrt{(5)^2 - 4(2)(-2)}}{2(2)}$$
$$= \frac{-5 \pm \sqrt{25 + 16}}{4}$$
$$= \frac{-5 \pm \sqrt{41}}{4}$$

$$x_1 = \frac{-5 + \sqrt{41}}{4} \qquad x_2 = \frac{-5 - \sqrt{41}}{4}$$

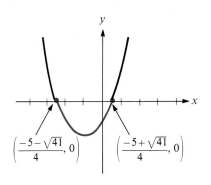

Figure 4.16

x-intercepts: $\left(\dfrac{-5 + \sqrt{41}}{4},0\right)$ and $\left(\dfrac{-5 - \sqrt{41}}{4},0\right)$

  (2) Since $a > 0$, the graph has a minimum point and opens up (see Figure 4.16).

b. To solve the inequality $2x^2 + 5x - 2 \leq 0$, we want only those values of x for which $2x^2 + 5x - 2$ (or y) either equals 0 or is negative. Observing the graph, we see that $2x^2 + 5x - 2$ is less than or equal to zero in the interval

$$\left[\frac{-5 - \sqrt{41}}{4}, \frac{-5 + \sqrt{41}}{4} \right].$$ ☐

EXAMPLE 8   Solve the inequality $x^2 + 2 < 0$.

Solution

a. Sketch the graph of $y = x^2 + 2$.
   (1) To find the x-intercepts, set $y = 0$ and solve the resulting equation.

$$0 = x^2 + 2$$
$$-2 = x^2$$
$$\pm\sqrt{-2} = x$$

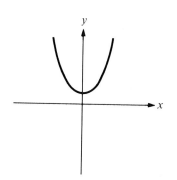

The solutions are not real numbers. Therefore, there are no x-intercepts.
   (2) Since $a > 0$, the graph has a minimum turning point and opens up (see Figure 4.17).

b. To solve the inequality $x^2 + 2 < 0$, we want only those values of x for which $x^2 + 2$ (or y) is negative. Observing the graph, we see $x^2 + 2$ (or y) is never negative. Therefore, there are no values of x for which the inequality will be a true statement, and the solution set is $\emptyset$.   ☐

Figure 4.17

Note from the examples that only a very rough sketch of the graph, which includes the x-intercepts and the direction that the curve opens, is needed. In particular, it is not necessary to find the coordinates of the turning point of the parabola.

EXERCISES 4.2

In Exercises 1–20 graph the quadratic function defined by the equation. On the graph indicate

a. the coordinates of the x- and y-intercepts
b. the equation of the axis of symmetry
c. the coordinates of the maximum or minimum point
d. the range of the function

1. $y = x^2 - 3x + 2$ **2.** $y = x^2 + 6x + 5$

3. $f(x) = x^2 - 4$ **4.** $g(x) = 3x^2 - 3$

5. $y = x^2 - 2x$ **6.** $f(x) = 2x^2 - 3x$

7. $g(x) = 4x - x^2$ **8.** $y = 1 - x^2$

9. $y = x^2 + 2x + 1$ **10.** $y = x^2 - 4x + 4$

11. $f(x) = -x^2 + x + 2$ **12.** $f(x) = -x^2 + 3x - 4$

13. $y = x^2 - 5$ **14.** $y = 2 - x^2$

15. $y = 5 - 4x^2$ **16.** $y = 2x^2 - 3$

17. $f(x) = 2x^2 - x - 1$ **18.** $y = 3x^2 - 5x + 1$

19. $y = 3 + 6x - 2x^2$ **20.** $g(x) = -3x^2 + 5x + 2$

In Exercises 21–30 determine the vertex and the axis of symmetry of the graph of f by matching the function to the form $f(x) = a(x - h)^2 + k$.

21. $f(x) = (x + 2)^2 - 3$ **22.** $f(x) = -2(x - 1)^2 + 5$

23. $f(x) = 3x^2 - 4$ **24.** $f(x) = 3(x - 4)^2$

25. $f(x) = x^2 - 2x + 2$ **26.** $f(x) = x^2 + 6x + 5$

27. $f(x) = 3x^2 - 9x + 1$ **28.** $f(x) = 2x^2 + 4x - 5$

29. $f(x) = 2 + x - x^2$ **30.** $f(x) = 1 - 3x - 4x^2$

31. What is the domain of all the functions in Exercises 1–30?

32. The height of a ball thrown directly up from a roof 144 ft high with an initial velocity of 128 ft/second is given by the formula $y = 144 + 128t - 16t^2$. What is the maximum height attained by the ball? When does the ball hit the ground?

33. The height of a projectile shot vertically upward from the ground with an initial velocity of 96 ft/second is given by the formula $y = 96t - 16t^2$. What is the maximum height attained by the projectile? When does the projectile hit the ground?

34. The total sales revenue for a company is estimated by the formula $S = 8x - x^2$, where x is the unit selling price (in dollars) and S is the total sales revenue (1 unit equals $10,000). What should be the unit selling price if the company wishes to maximize the total revenue?

35. In a 120-volt line having a resistance of 10 ohms, the power W in watts when the current I is flowing is given by the formula $W = 120I - 10I^2$. What is the maximum power that can be delivered in this circuit?

36. Find two positive numbers whose sum is 8 and whose product is a maximum.

37. Find two positive numbers whose sum is 20 and whose product is a maximum.

38. What positive number exceeds its square by the largest amount?

39. Find two positive numbers whose sum is 20 such that the sum of their squares is a minimum.

40. The sum of the base and altitude of a triangle is 12 in. Find each dimension if the area is to be a maximum.

41. A rectangular field is adjacent to a river and is to have fencing on three sides, the side on the river requiring no fencing. If 400 yd of fencing is available, what are the dimensions of the largest rectangular section that can be enclosed with the available fencing?

42. A farmer wants to make a rectangular enclosure along the side of a barn and then divide the enclosure into two pens with a fence constructed at a right angle to the barn. If 300 yd of fencing is available, what are the dimensions of the largest section that can be enclosed with the available fencing?

43. A sheet of metal 12 in. wide and 20 ft long is to be made into a gutter by turning up the same amount of material on each edge at right angles to the base. Determine the amount of material that should be turned up to maximize the volume that the gutter can carry.

44. Find the maximum possible area of a rectangle with a perimeter of 100 ft.

45. A bus tour charges a fare of $10 per person and carries 200 people per day. The manager estimates that she will lose 10 passengers for each $1 increase in fare. Find the most profitable fare for her to charge.

46. An airline offers a charter flight at a fare of $100 per person if 100 passengers sign up. For *each* passenger above 100, the fare for *each* passenger is reduced by 50 cents. What is the maximum total revenue that the airline can obtain if the plane has 200 seats?

In Exercises 47–66 solve the quadratic inequality.

47. $x^2 - 4x - 5 > 0$ **48.** $x^2 + x - 6 < 0$

49. $x^2 < 4x + 5$ **50.** $x^2 + 5 < 6x$

51. $-x^2 + x + 2 \leq 0$ **52.** $-10 \geq 3x - x^2$

53. $x^2 \geq 4$ **54.** $x^2 - 5 \leq 0$

55. $x^2 < 3x$ **56.** $0 > 5x - 10x^2$

57. $2x^2 + x \leq 6$ **58.** $4 + x > x^2$

59. $3x^2 - 2x < 5$ **60.** $5x^2 \geq 4x - 3$

61. $x^2 + 2x > -1$ **62.** $4x^2 \leq 12x - 9$

63. $x^2 - 5x + 7 < 0$ **64.** $x^2 + 1 \geq 0$

65. $1 - x^2 \geq 0$ **66.** $-2x^2 + x > 4$

4.3 SYNTHETIC DIVISION AND THE REMAINDER THEOREM

To discuss polynomial functions of degree greater than 2, we must first consider the division of polynomials. We divide one polynomial by another in a manner similar to the division of two integers. Consider the following arrangement for dividing $3x^3 + x - 4$ by $x + 4$.

$$
\begin{array}{r}
3x^2 - 12x\ + 49 \quad \text{(quotient)} \\
\text{(divisor)} \quad x + 4 \,\overline{)\,3x^3 +\ 0x^2 + x - 4} \quad \text{(dividend)} \\
\underline{3x^3 + 12x^2} \\
-12x^2 + x - 4 \\
\underline{-12x^2 - 48x} \\
49x -\ \ 4 \\
\underline{49x + 196} \\
-\,200 \quad \text{(remainder)}
\end{array}
$$

First, we arrange the terms of the dividend and the divisor in descending powers of x. If a term is missing, write 0 as its coefficient. Then divide the first term of the dividend by the first term of the divisor to obtain the first term of the quotient. Next, we multiply the entire divisor by the first term of the quotient and subtract this product from the dividend. Use the remainder as the new dividend and repeat the above procedure until the remainder is of lower degree than the divisor.

As with the division of numbers,

$$\text{dividend} = \text{(divisor)(quotient)} + \text{remainder}.$$

For example, the division of 7 by 3 may be expressed as

$$7 = 3 \cdot 2 + 1.$$

In the case of the preceding polynomial division, we have

$$3x^3 + x - 4 = (x + 4)(3x^2 - 12x + 49) - 200.$$

The division of any polynomial by a polynomial of the form $x - b$ is of theoretical and practical importance. This division may be performed by a shorthand method called **synthetic division.** Consider the arrangement for dividing $2x^3 + 5x^2 - 1$ by $x - 2$.

$$
\begin{array}{r}
2x^2 + 9x\ + 18 \\
x - 2 \,\overline{)\,2x^3 + 5x^2 +\ 0x - 1} \\
\underline{2x^3 - 4x^2} \\
9x^2 + 0x - 1 \\
\underline{9x^2 - 18x} \\
18x - 1 \\
\underline{18x - 36} \\
35
\end{array}
$$

When the polynomials are written with terms in descending powers of x, there is no need to write all the x's. Only the coefficients are needed. Also, notice that the encircled coefficients entailed needless writing. Using only the necessary coefficients, we may abbreviate this division as follows:

$$
\begin{array}{r}
 2 \quad\ 9 \quad\ 18 \\
-2\,\overline{)\,2 \quad\ 5 \quad\ 0 \quad -1\,} \\
 -4 \ -18 \ -36 \\
\hline
 9 \quad\ 18 \quad\ 35.
\end{array}
$$

If we bring down 2 as the first entry in the bottom row, all the coefficients of the quotient appear. The arrangement may then be shortened to

$$
\begin{array}{rrrrcll}
-2\,\underline{|} & \boxed{2 \quad 5 \quad 0 \quad -1} & \leftarrow & \text{coefficients of dividend} & \text{(row 1)} \\
 & \quad -4 \ -18 \ -36 & & & \text{(row 2)} \\
\text{coefficients of quotient} \rightarrow & \boxed{2 \quad 9 \quad 18} \ \ 35 & \leftarrow & \text{remainder} & \text{(row 3)}
\end{array}
$$

Finally, if we replace -2 by 2, which is the value of b, we may change the sign of each number in row 2 and add at each step instead of subtracting. The final arrangement for synthetic division is then as follows:

$$
\begin{array}{rrrrll}
2\,\underline{|} & 2 \quad\ 5 \quad\ 0 \quad -1 & & \text{(row 1)} \\
 & \quad\ 4 \quad 18 \quad 36 & & \text{(row 2)} \\
\hline
 & 2 \quad\ 9 \quad 18 \quad\ 35 & \rightarrow \ \text{remainder} & \text{(row 3)} \\
 & \downarrow \quad\ \downarrow \quad\ \downarrow & \\
\text{quotient} \rightarrow & 2x^2 + 9x + 18.
\end{array}
$$

Let us summarize the procedure for synthetic division. First, write the coefficients of the terms in the dividend. Be sure the powers are in descending order and enter a 0 as the coefficient of any missing term. Write the value of b (in this case 2) to the left of the dividend. Bring down the first dividend entry (2). Multiply this number by $b(2)$ and place the result (4) under the next number (5) in row 1. Add, then multiply the resulting 9 by $b(2)$. Place the result (18) under the next number (0) in row 1. Add, then multiply the resulting 18 by $b(2)$. Place the result (36) under the next number (-1) in row 1. Add and obtain 35. The last number 35 is the remainder. The other numbers in row 3 are the coefficients of the quotient. The degree of the polynomial in the quotient is always one less than the degree of the polynomial in the dividend.

A summary cannot do justice to the simple procedure in synthetic division. Careful consideration of Examples 1 and 2 should help clarify the process. Remember, synthetic division applies only when we divide by a polynomial of the form $x - b$.

EXAMPLE 1   Use synthetic division to divide $x^5 - 1$ by $x - 1$. Express the result in the form dividend = (divisor)(quotient) + remainder.

Solution   We use 0's as the coefficients of the missing x^4, x^3, x^2, and x terms. We divide by $x - 1$ so that $b = 1$ (not -1). The arrangement is as follows:

$$
\begin{array}{r|rrrrrr}
1\rfloor & 1 & 0 & 0 & 0 & 0 & -1 \\
 & & 1 & 1 & 1 & 1 & 1 \\
\hline
 & 1 & 1 & 1 & 1 & 1 & 0 \quad \leftarrow \text{ remainder}
\end{array}
$$

quotient $\rightarrow$ $x^4 + x^3 + x^2 + x + 1$.

The answer in the form requested is

$$x^5 - 1 = (x - 1)(x^4 + x^3 + x^2 + x + 1) + 0. \qquad \square$$

EXAMPLE 2   Use synthetic division to divide $4x^3 - x^2 + 2$ by $x + 3$. Express the result in the form dividend = (divisor)(quotient) + remainder.

Solution   We use 0 as the coefficient of the missing x term. We divide by $x + 3$ so that $b = -3$ (not 3). The arrangement is as follows:

$$
\begin{array}{r|rrrr}
-3\rfloor & 4 & -1 & 0 & 2 \\
 & & -12 & 39 & -117 \\
\hline
 & 4 & -13 & 39 & -115 \quad \leftarrow \text{ remainder}
\end{array}
$$

quotient $\rightarrow$ $4x^2 - 13x + 39$.

The answer in the form requested is

$$4x^3 - x^2 + 2 = (x + 3)(4x^2 - 13x + 39) - 115. \qquad \square$$

In Example 2 we found that the function

$$P(x) = 4x^3 - x^2 + 2$$

may be written as

$$P(x) = (x + 3)(4x^2 - 13x + 39) - 115.$$

When $x = -3$, the factor $x + 3 = 0$; thus,

$$P(-3) = 0 - 115$$
$$= -115.$$

The value of the function when $x = -3$ is the same as the remainder obtained when $P(x)$ is divided by $x - (-3)$. This discussion suggests the following theorem.

Remainder Theorem   If a polynomial $P(x)$ is divided by $x - b$, the remainder is $P(b)$.

We now prove this theorem. Let $Q(x)$ and r represent the quotient and remainder when $P(x)$ is divided by $x - b$. Then

$$\text{dividend} = (\text{divisor})(\text{quotient}) + \text{remainder}$$
$$P(x) = (x - b)Q(x) + r.$$

This statement is true for all values of x. If $x = b$, then

$$P(b) = (b - b)Q(b) + r$$
$$= 0 \cdot Q(b) + r.$$

Thus,

$$P(b) = r.$$

We know that $P(b)$ may be found by substituting b for x in the function. The remainder theorem provides an alternative method. That is, we find $P(b)$ by determining the remainder when $P(x)$ is divided by $x - b$. Since this remainder may be obtained by synthetic division, this approach is often simpler than direct substitution. In the next section other advantages of the remainder theorem method will be discussed.

EXAMPLE 3   If $P(x) = 3x^5 + 5x^4 + 7x^3 - 4x^2 + x - 24$, find $P(-2)$ by (a) direct substitution and (b) the remainder theorem.

Solution
a. By direct substitution, we have

$$P(-2) = 3(-2)^5 + 5(-2)^4 + 7(-2)^3 - 4(-2)^2 + (-2) - 24$$
$$= -96 + 80 - 56 - 16 - 2 - 24$$
$$= -114.$$

b. By the remainder theorem, we have

$$
\begin{array}{r|rrrrrr}
-2 & 3 & 5 & 7 & -4 & 1 & -24 \\
   &   & -6 & 2 & -18 & 44 & -90 \\
\hline
   & 3 & -1 & 9 & -22 & 45 & -114.
\end{array}
$$

Since the remainder is -114, $P(-2) = -114$.   □

EXAMPLE 4   If $P(x) = 2x^4 - 5x^3 + 11x^2 - 3x - 5$, find $P(-\frac{1}{2})$ by (a) direct substitution and (b) the remainder theorem.

Solution

a. By direct substitution, we have

$$P(-\tfrac{1}{2}) = 2(-\tfrac{1}{2})^4 - 5(-\tfrac{1}{2})^3 + 11(-\tfrac{1}{2})^2 - 3(-\tfrac{1}{2}) - 5$$
$$= \tfrac{1}{8} + \tfrac{5}{8} + \tfrac{11}{4} + \tfrac{3}{2} - 5$$
$$= 0.$$

b. By the remainder theorem, we have

$$
\begin{array}{r|rrrrr}
-\frac{1}{2} & 2 & -5 & 11 & -3 & -5 \\
 & & -1 & 3 & -7 & 5 \\
\hline
 & 2 & -6 & 14 & -10 & 0.
\end{array}
$$

Since the remainder is 0, $P(-\frac{1}{2}) = 0$. □

Notice in Examples 3 and 4 that the solution is obtained more easily by the remainder theorem method.

EXERCISES 4.3

In Exercises 1–6 perform the indicated divisions by using long division. Express the result in the form dividend = (divisor)(quotient) + remainder.

1. $(x^2 + 7x - 2) \div (x + 5)$

2. $(x^2 - 4) \div (x - 1)$

3. $(6x^3 - 3x^2 + 14x - 7) \div (2x - 1)$

4. $(4x^3 + 5x^2 - 10x + 4) \div (4x - 3)$

5. $(3x^4 + x - 2) \div (x^2 - 1)$

6. $(2x^3 - 3x^2 + 10x - 5) \div (x^2 + 5)$

In Exercises 7–12 perform the indicated division by using synthetic division. Express the result in the form dividend = (divisor)(quotient) + remainder.

7. $(x^3 - 5x^2 + 2x - 3) \div (x - 1)$

8. $(4 - x + 3x^2 - x^3) \div (x - 4)$

9. $(2x^3 + x - 5) \div (x + 1)$

10. $(3x^4 - x^2 + 7) \div (x + 3)$

11. $(x^4 - 16) \div (x - 2)$

12. $(x^3 + 27) \div (x - 3)$

In Exercises 13–20 find the given function value by (a) direct substitution and (b) the remainder theorem.

13. $P(x) = x^3 - 4x^2 + x + 6$; find $P(2)$ and $P(3)$.

14. $P(x) = 2x^3 + 5x^2 - x - 7$; find $P(4)$ and $P(-3)$.

15. $P(x) = 2x^4 - 7x^3 - x^2 + 4x + 11$; find $P(2)$ and $P(-2)$.

16. $P(x) = x^4 - 5x^3 - 4x^2 + 17x + 15$; find $P(-1)$ and $P(5)$.

17. $P(x) = 2x^4 + 5x^3 - 20x - 32$; find $P(-2)$ and $P(2)$.

18. $P(x) = 2x^4 - x^3 + 2x - 1$; find $P(\frac{1}{2})$ and $P(-\frac{1}{2})$.

19. $P(x) = 6x^4 + 2x^3 - 5x - 5$; find $P(\frac{1}{3})$ and $P(-\frac{1}{3})$.

20. $P(x) = 3x^4 + x^2 - 7x + 2$; find $P(0.1)$ and $P(-0.1)$.

4.4 THEOREMS ABOUT ZEROS

A **zero** of a function f is a value of x for which $f(x) = 0$. For example, the zeros of the function

$$f(x) = x^2 - 5x + 4$$

are 1 and 4 since

$$f(1) = (1)^2 - 5(1) + 4 = 0$$

and

$$f(4) = (4)^2 - 5(4) + 4 = 0.$$

Zeros is merely a new name applied to a familiar concept. Depending on the frame of reference, consider three closely related names applied to the preceding example.

1. The **solutions** or **roots** of the **equation** $x^2 - 5x + 4 = 0$ are 1 and 4.
2. The **zeros** of the **function** $f(x) = x^2 - 5x + 4$ are 1 and 4.
3. The **x-intercepts** of the **graph** of the function $y = x^2 - 5x + 4$ are (1,0) and (4,0).

Finding zeros for polynomial functions of degree greater than 2 is often difficult. There is no easy formula (like the quadratic formula) that can be used to produce them, and numerical methods that require a computer are frequently used to approximate such zeros. Nevertheless, there are some theorems you should be aware of concerning zeros of polynomial functions. The first one we consider is called the factor theorem.

Factor Theorem   If b is a zero of the polynomial function $y = P(x)$, then $x - b$ is a factor of $P(x)$; and conversely, if $x - b$ is a factor of $P(x)$, then b is a zero of $y = P(x)$.

We now prove the first part of this theorem. As in the proof of the remainder theorem, let $Q(x)$ and r represent the quotient and remainder when $P(x)$ is divided by $x - b$. Then

$$P(x) = (x - b)Q(x) + r.$$

By the remainder theorem, $r = P(b)$, so we have

$$P(x) = (x - b)Q(x) + P(b).$$

If b is a zero of $y = P(x)$, then $P(b) = 0$. Thus,

$$P(x) = (x - b)Q(x) + 0$$

and $x - b$ is a factor of $P(x)$. The proof of the second part of this theorem is left as Exercise 47. Following are two examples that illustrate the usefulness of the factor theorem.

EXAMPLE 1   Write a polynomial function (in factored form) of degree 3 with zeros of 2, -4, and 3.

Solution   If 2, -4, and 3 are zeros of $y = P(x)$, then by the factor theorem, $x - 2$, $x - (-4)$, and $x - 3$ are factors of $P(x)$. Thus, a possible polynomial function is $P(x) = (x - 2)(x + 4)(x - 3)$.   □

EXAMPLE 2   If $P(x) = x(x + 2)(x - 5)^2$, what are the zeros of the function?

Solution   Since x, $x + 2$, and $x - 5$ are factors of $P(x)$, we conclude by the factor theorem that 0, -2, and 5 are zeros of the function. Remember that from a slightly different viewpoint we can also conclude that 0, -2, and 5 are roots or solutions of the equation $x(x + 2)(x - 5)^2 = 0$, and that the graph of $P(x) = x(x + 2)(x - 5)^2$ intercepts the x-axis at 0, -2, and 5.   □

Note in Example 2 that the function $P(x) = x(x + 2)(x - 5)^2$ is shorthand for $P(x) = x(x + 2)(x - 5)(x - 5)$. In this case we have four factors but only three zeros, with the zero of 5 being repeated from two different factors. It is useful to adopt a convention that indicates how many factors of $P(x)$ result in the same zero. Using this convention in this example, we say 5 is a **zero of multiplicity** 2, since there are two factors of $x - 5$. Similarly, for $P(x) = (x - 3)^4$ we say 3 is a zero of multiplicity 4. In general, the multiplicity of zero b is given by the highest power of $x - b$ that is a factor of $P(x)$.

EXAMPLE 3   If $P(x) = x^4(x - 2)(x + 7)^3$, find the zeros of the function as well as the multiplicity of each zero. What is the degree of the polynomial function?

Solution $P(x)$ contains only three different factors: x, $x - 2$, and $x + 7$. Thus, 0, 2, and -7 are the three distinct zeros of the function. However, since x appears as a factor four times, 0 is a zero of multiplicity 4. Similarly, $(x + 7)^3$ means -7 is a zero of multiplicity 3, while $(x - 2)^1$ means 2 is a zero of multiplicity 1 (called a simple zero). The degree of the polynomial function is given by the sum of the multiplicities of the zeros. Thus, the polynomial function is of degree 8.   □

Knowing the multiplicity of a real number zero is useful when graphing polynomial functions. We have already indicated that the real number zeros of $y = P(x)$ are also the x-intercepts in the graph of $y = P(x)$. We now add to this that the multiplicity of each real number zero indicates whether or not the graph crosses the x-axis at such intercepts according to the following theorem.

Graph of $y = P(x)$ near x-intercepts	If b is a real number zero with multiplicity n of $y = P(x)$, then the graph of $y = P(x)$ crosses the x-axis at $x = b$ if n is odd, while the graph turns around and stays on the same side of the x-axis at $x = b$ if n is even.

To use this theorem, we need to express $P(x)$ in factored form. Then the x-intercepts (or real number zeros) can be obtained from the factor theorem, while the behavior of the graph at an x-intercept, say $(b,0)$, can be determined from the multiplicity of zero b (or the highest power of $x - b$ that is a factor of $P(x)$). We illustrate this theorem in the next two examples. The smooth unbroken type of curves shown in the examples are characteristic of the graphs of polynomial functions.

EXAMPLE 4 Graph $y = (x + 1)(x - 2)^2$ based on information obtained from the intercepts.

Solution By setting $x = 0$, we determine that the y-intercept is $(0,4)$. Since $x + 1$ is a factor with an odd exponent, we conclude that $(-1,0)$ is an x-intercept at which the graph crosses the x-axis. Since $(x - 2)^2$ is a factor with an even exponent, we conclude that $(2,0)$ is an x-intercept at which the graph touches the x-axis and then turns around. We also note that as x gets very large, so does y; while as x increases in magnitude in the

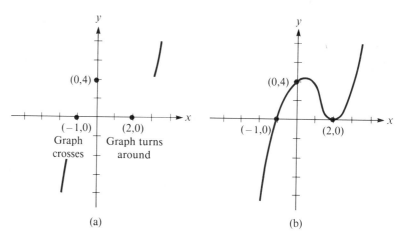

Figure 4.18

negative direction, then y becomes very small. Figure 4.18(a) illustrates our results so far. Finally, since the function is a polynomial function, we draw a smooth unbroken curve that satisfies the conditions in Figure 4.18(a), and we obtain the graph in Figure 4.18(b).   □

EXAMPLE 5   Graph $y = x^5 - 4x^3$ based on information obtained from the intercepts.

Solution   By setting $x = 0$, we determine that $(0,0)$ is a y- (and x-) intercept. To find x-intercepts, we next factor the polynomial as follows:

$$y = x^5 - 4x^3 = x^3(x^2 - 4) = x^3(x + 2)(x - 2).$$

Since x^3, $x + 2$, and $x - 2$ are factors, we conclude the x-intercepts are $(0,0)$, $(-2,0)$, and $(2,0)$. All of these intercepts are derived from factors with odd exponents, so the graph crosses the x-axis at each of these points. By using this information and observing that large values of x produce large values of y, we draw the graph in Figure 4.19(a). Keep in mind that our current methods produce only a rough sketch of the graph. A more detailed graph of $y = x^5 - 4x^3$ that is obtained by methods from calculus is given in Figure 4.19(b). Note that this graph indicates turning points when $x = \pm 2\sqrt{15}/5$ and an unanticipated wavy pattern between the turning points. Determining such subtleties is best left to a calculus course. You may, however, find it useful to remember that the number of turning points is, at most, one less than the degree of the polynomial function.

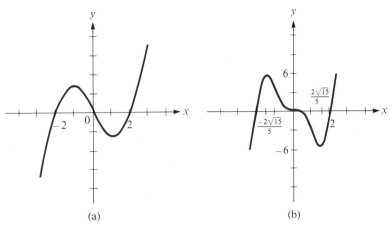

(a)   (b)

Figure 4.19   ☐

We now discuss some theorems that are fundamental to the theory of finding zeros of polynomial functions. You need a good grasp of number systems to really understand these statements, and it might be helpful if you review the relationships among the various sets of numbers, which are given in Sections 1.1 and 2.2.

Fundamental Theorem of Algebra	Every polynomial function of degree $n \geq 1$ with complex number coefficients has at least one complex zero.

Although the proof of this theorem is beyond the scope of this book, we can prove an important corollary of this theorem that answers a basic question: How many zeros are there for a polynomial function of degree n?

Number of Zeros Theorem	Every polynomial function of degree $n \geq 1$ has exactly n complex zeros, where zeros of multiplicity k are counted k times.

For instance, in Example 3 we saw that $P(x) = x^4(x - 2)(x + 7)^3$ is a polynomial function of degree 8 with distinct zeros of 0, 2, and −7. However, the multiplicities of these zeros are 4, 1, and 3, respectively, so that if we take into account the idea of multiplicity, then this 8th degree polynomial function has exactly eight zeros. More generally, we can prove the above theorem by first noting that if $y = P(x)$ is a polynomial function of degree $n \geq 1$ with leading coefficient a_n, then the fundamental theorem of algebra guarantees that $y = P(x)$ has at least one complex zero, say c_1. By the factor theorem, since c_1 is a zero, $x - c_1$ is a factor of $P(x)$, so

$$P(x) = (x - c_1)Q_1(x),$$

where polynomial $Q_1(x)$ has degree $n - 1$ and leading coefficient a_n. If the degree of $Q_1(x)$ is at least 1, then once again the fundamental theorem of algebra guarantees that $y = Q_1(x)$ has at least one zero, say c_2, so as above

$$Q_1(x) = (x - c_2)Q_2(x),$$

where polynomial $Q_2(x)$ has degree $n - 2$ and leading coefficient a_n. Then combining results, we have

$$P(x) = (x - c_1)(x - c_2)Q_2(x).$$

This process is continued for a total of n times until $Q_n(x) = a_n$. Thus, $P(x)$ can be factored into n linear factors and written as

$$P(x) = a_n(x - c_1)(x - c_2) \cdots (x - c_n)$$

so that, by the factor theorem, $y = P(x)$ has n zeros: $c_1, c_2, \ldots, c_n$. Furthermore, no other number, say c, distinct from $c_1, c_2, \ldots, c_n$ can be a zero since

$$P(c) = a_n(c - c_1)(c - c_2) \cdots (c - c_n) \neq 0$$

because none of the factors is zero. Thus, every polynomial function of degree $n \geq 1$ has exactly n (not necessarily distinct) complex zeros.

EXAMPLE 6   If 2 is a zero of $P(x) = x^3 - 2x^2 + 3x - 6$, find the other zeros.

Solution   Since $y = P(x)$ is a polynomial function of degree 3, there are exactly three (not necessarily distinct) zeros. We are given that 2 is one of these zeros, and if 2 is a zero of $y = P(x)$, then by the factor theorem, $x - 2$ is a factor of the polynomial. We can then use synthetic division to find another factor of $P(x)$ as follows:

$$
\begin{array}{r|rrrr}
2\rfloor & 1 & -2 & 3 & -6 \\
         &   & 2 & 0 & 6 \\
\hline
         & 1 & 0 & 3 & 0.
\end{array}
$$

The coefficients of another factor of $P(x)$ are given in the bottom row of this synthetic division. Thus,

$$P(x) = (x - 2)(x^2 + 3).$$

Setting the factor $x^2 + 3$ equal to 0 gives us the other two zeros of $i\sqrt{3}$ and $-i\sqrt{3}$.   □

Note in Example 6 that the zeros $i\sqrt{3}$ and $-i\sqrt{3}$ are conjugates of each other. We can anticipate that the two remaining zeros are conjugate pairs from our work with quadratic equations in Section 2.3, and in general, the following theorems show that when certain restrictions are placed on the coefficients of a polynomial, then zeros of polynomial functions always occur in conjugate pairs.

Conjugate-Pair Theorems

1. If a complex number $a + bi$ is a zero of a polynomial function of degree $n \geq 1$ with *real* number coefficients, then its conjugate $a - bi$ is also a zero.
2. Let a, b, and c be rational numbers. If an irrational number of the form $a + b\sqrt{c}$ is a zero of a polynomial function of degree $n \geq 1$ with *rational* number coefficients, then $a - b\sqrt{c}$ is also a zero.

The restrictions on the coefficients of the polynomial are crucial conditions in these theorems. For example, $P(x) = x - \sqrt{2}$ has only one zero, namely $\sqrt{2}$. Note that the conjugate $-\sqrt{2}$ is not also a zero, and that the second conjugate-pair theorem does not apply to $P(x) = x - \sqrt{2}$ because the polynomial does not have rational number coefficients.

We can prove the first conjugate-pair theorem by using the properties of complex conjugates discussed in Section 2.2 (which you may need to review). We start with a polynomial function of degree $n \geq 1$ with real number coefficients that we may write as

$$P(x) = a_n x^n + a_{n-1} x^{n-1} + \cdots + a_1 x + a_0.$$

Now if $z = a + bi$ is a zero of $y = P(x)$, then $P(z) = 0$, so

$$0 = a_n z^n + a_{n-1} z^{n-1} + \cdots + a_1 z + a_0.$$

We then take the conjugate of both sides of the equation (noting that the conjugate of 0 is 0) and proceed as follows:

$$
\begin{aligned}
0 &= \overline{a_n z^n + a_{n-1} z^{n-1} + \cdots + a_1 z + a_0} \\
&= \overline{a_n z^n} + \overline{a_{n-1} z^{n-1}} + \cdots + \overline{a_1 z} + \overline{a_0} && \text{(Conjugate of a sum} \\
& && \text{= Sum of conjugates)} \\
&= \overline{a_n}\,\overline{z^n} + \overline{a_{n-1}}\,\overline{z^{n-1}} + \cdots + \overline{a_1}\,\overline{z} + \overline{a_0} && \text{(Conjugate of a product} \\
& && \text{= Product of conjugates)} \\
&= a_n \overline{z^n} + a_{n-1} \overline{z^{n-1}} + \cdots + a_1 \overline{z} + a_0 && \text{(Conjugate of a real number} \\
& && \text{= The real number)} \\
&= a_n (\overline{z})^n + a_{n-1}(\overline{z})^{n-1} + \cdots + a_1 \overline{z} + a_0. && \text{(Conjugate of a power} \\
& && \text{= Power of the conjugate)}
\end{aligned}
$$

This last equation tells us that $P(\overline{z}) = 0$, so $\overline{z}$ is also a zero of $y = P(x)$, which proves the theorem.

EXAMPLE 7   If $-2i$ and $2 + \sqrt{7}$ are zeros of the polynomial function $P(x) = x^4 - 4x^3 + x^2 - 16x - 12$, find the other zeros.

Solution $P(x)$ has rational number coefficients. Thus, both conjugate-pair theorems apply, and if $-2i$ and $2 + \sqrt{7}$ are zeros, then $2i$ and $2 - \sqrt{7}$ are also zeros. There are no other zeros since the polynomial has degree 4.   □

EXAMPLE 8   Write a polynomial function (in factored form) of the lowest possible degree that has rational coefficients and zeros of 0, $1 - i$, and $\sqrt{2}$.

Solution   Since the polynomial has rational coefficients, we utilize both conjugate-pair theorems. Thus, in addition to the three zeros given, $1 + i$ and $-\sqrt{2}$ are also zeros. The lowest possible degree for the polynomial is then degree 5, and in factored form one possibility is

$$ P(x) = x(x - \sqrt{2})(x + \sqrt{2})(x - (1 + i))(x - (1 - i)). \qquad \square $$

In conclusion, let us discuss briefly why the fundamental theorem of algebra is so central to the theory of polynomial equations. During the historical growth of equation solving, the solution to certain types of equations continued to force mathematicians to develop new types of numbers. For example, if an equation uses just positive integers (like $x + 5 = 1$), it might be necessary to leave the positive integers for a solution. In the case of $x + 5 = 1$ an extension to -4 and the negative integers is needed. Integers were not enough for equations like $2x + 5 = 0$, so rational numbers were required. Similarly, $x^2 - 5 = 0$ brings an extension to irrational (or real) numbers, while $x^2 + 5 = 0$ can be solved only by introducing imaginary (or complex) numbers. However, the process stops here. The fundamental theorem of algebra guarantees that every polynomial equation involving complex numbers can be solved using only complex numbers. No extension of the number system is necessary. The mathematician Carl Friedrich Gauss was the first to prove this theorem. He did so in his doctoral thesis in 1799 when he was 20.

EXERCISES 4.4

In Exercises 1–4 find the zeros of each polynomial function. In each case state the degree of the polynomial function and the multiplicity of each zero.

1. $P(x) = (x + 3)^2$ **2.** $P(x) = x(x - 2)^3$

3. $P(x) = (x + 4)(x + \sqrt{2})(x - \sqrt{2})$

4. $P(x) = 4x^3(x - 1)^5(x + 6)$

In Exercises 5–10 write a polynomial function (in factored form) with the given degree and zeros.

5. degree 3; zeros are 1, 2, and 3
6. degree 3; zeros are $\sqrt{2}$, $-\sqrt{2}$, and 0
7. degree 4; zeros are -4, 0, i, and $-i$
8. degree 5; zeros are 5, $2 \pm i$, and $1 \pm \sqrt{3}$
9. degree 4; 1 and 5 are both zeros of multiplicity 2
10. degree 6; -2 is a zero of multiplicity 1, 0 is a zero of multiplicity 2, and 5 is a zero of multiplicity 3

In Exercises 11–20 graph the functions based on information obtained from the intercepts.

11. $y = (x - 2)(x + 1)$ **12.** $y = (x - 2)(x + 1)^2$

13. $y = (x - 2)^2(x + 1)$ **14.** $y = (x - 2)^2(x + 1)^2$

15. $y = (x^2 - 1)(x^2 - 4)$ **16.** $y = x^4 - 3x^2 + 2$

17. $y = x^3 - 2x^2 - 3x$ **18.** $y = x^3 - 4x$

19. $y = x^2 - x^3$ **20.** $y = x^3 + x^2$

In Exercises 21–26 write a polynomial function (in factored form) with rational coefficients of the lowest possible degree with the given zeros.

21. 2 and $\sqrt{3}$ **22.** i and 0

23. 0, $2 + 3i$, and $4 - \sqrt{3}$ **24.** -5, $-\sqrt{5}$, and $5i$

25. 3, -3, and $2i$ **26.** $3i$ and $1 + \sqrt{2}$

In Exercises 27–40 one or more zeros (is/are) given for each of the following polynomial functions. Find the other zeros.

27. $P(x) = x^2 - 4x - 3$; $2 + \sqrt{7}$

28. $P(x) = 3x^2 + 4x - 2$; $\dfrac{-2 - \sqrt{10}}{3}$

29. $P(x) = x^2 + 2$; $i\sqrt{2}$

30. $P(x) = 3x^2 - 2x + 1$; $\dfrac{1}{3} - \dfrac{i\sqrt{2}}{3}$

31. $P(x) = 2x^3 - 11x^2 + 28x - 24$; $\frac{3}{2}$ and $2 + 2i$

32. $P(x) = x^4 + 13x^2 - 48$; $\sqrt{3}$ and $-4i$

33. $P(x) = 2x^3 + 5x^2 + 4x + 1$; -1

34. $P(x) = 3x^3 - 2x^2 - 10x + 4$; 2

35. $P(x) = x^4 - 6x^3 + 7x^2 + 12x - 18$; 3 is a zero of multiplicity 2

36. $P(x) = 2x^4 - 5x^3 + 11x^2 - 3x - 5$; 1 and $-\frac{1}{2}$

37. $P(x) = x^4 - 4$; $\sqrt{2}$

38. $P(x) = x^4 - 14x^2 + 45$; $-\sqrt{5}$

39. $P(x) = x^4 - 1$; i

40. $P(x) = x^4 - 16$; $2i$

41. If -1 is a zero of $y = x^3 - x^2 - 10x - 8$, find all intercepts and then graph the function.

42. Graph the function in Exercise 35 based on information obtained from the intercepts.

In Exercises 43–46 $y = P(x)$ is a polynomial function with real coefficients. Answer true or false.

43. Every polynomial function of degree 3 has at least one real zero.

44. Every polynomial function of degree 4 has at least one real zero.

45. Every polynomial function of degree $n \geq 1$ has exactly n (not necessarily different) real zeros.

46. Every polynomial function of degree $n \geq 1$ has exactly n (not necessarily different) complex zeros.

47. Show that if $x - b$ is a factor of the polynomial $P(x)$, then b is a zero of $y = P(x)$.

4.5 RATIONAL ZEROS AND THE LOCATION THEOREM

The following theorem is used to obtain a list of *possible* rational zeros of a polynomial function with integer coefficients. We emphasize the word *possible*. The zeros may all be either irrational or complex. This theorem states only that if there are rational zeros, they satisfy the requirement indicated.

Rational Zero Theorem   If p/q, a rational number in lowest terms, is a zero of the polynomial function with integer coefficients,

$$P(x) = a_n x^n + a_{n-1} x^{n-1} + \cdots + a_1 x + a_0 \qquad (a_n \neq 0)$$

p is an integral factor of the constant term a_0 and q is an integral factor of the leading coefficient a_n.

To prove this theorem, we begin by noting that if p/q is a zero of $y = P(x)$, then

$$a_n (p/q)^n + a_{n-1} (p/q)^{n-1} + \cdots + a_1 (p/q) + a_0 = 0.$$

If we multiply both sides of this equation by q^n, we have

$$a_n p^n + a_{n-1} p^{n-1} q + \cdots + a_1 p q^{n-1} + a_0 q^n = 0.$$

Next, we subtract $a_0 q^n$ from both sides of the equation yielding

$$a_n p^n + a_{n-1} p^{n-1} q + \cdots + a_1 p q^{n-1} = -a_0 q^n,$$

and finally, we factor the common factor p from each term on the left to obtain

$$p(a_n p^{n-1} + a_{n-1} p^{n-2} q + \cdots + a_1 q^{n-1}) = -a_0 q^n.$$

Since p is a factor of the left-hand side of the equation, p must also be a factor of $-a_0 q^n$. But by hypothesis, p/q is in lowest terms, so p and q have no common factor, and p is not a factor of q^n. Thus, p is a factor of a_0.

To prove that q is a factor of a_n, we rewrite our second equation as

$$q(a_{n-1} p^{n-1} + a_{n-2} p^{n-2} q + \cdots + a_0 q^{n-1}) = -a_n p^n$$

and reason as above. In practice, the rational zero theorem is not difficult to apply.

EXAMPLE 1   List the possible rational zeros of the function

$$P(x) = 2x^3 - x^2 - 6x - 3.$$

Solution   The constant term a_0 is -3. The possibilities for p are the integers that are factors of -3. Thus,

$$p = \pm 3, \pm 1.$$

The leading coefficient a_n is 2. The possibilities for q are the integers that are factors of 2. Thus,

$$q = \pm 2, \pm 1.$$

The possible rational zeros p/q are then

$$3, \tfrac{3}{2}, 1, \tfrac{1}{2}, -\tfrac{1}{2}, -1, -\tfrac{3}{2}, -3. \qquad \square$$

Since we are often faced with a large list of possible rational zeros, we need some help in narrowing down the possibilities. Consider the theorem that follows, which helps us chip away at the problem. Note that it is a statement about real zeros. Thus, it applies to zeros that are rational or irrational.

Descartes' Rule of Signs	The maximum number of positive real zeros of the polynomial function $y = P(x)$ is the number of changes in sign of the coefficients in $P(x)$. The number of changes in sign of the coefficients in $P(-x)$ is the maximum number of negative real zeros. In both cases, if the number of zeros is not the maximum number, then it is less than this number by a multiple of 2.

We illustrate this theorem in the following example.

EXAMPLE 2   Find the zeros of $P(x) = 2x^3 + 5x^2 + 4x + 1$.

Solution
a. Use Descartes' rule of signs to determine the maximum number of positive and negative real zeros.

$$P(x) = 2x^3 + 5x^2 + 4x + 1$$

All the coefficients in $P(x)$ are positive. There are no positive real zeros since there are no changes in sign.

$$P(-x) = 2(-x)^3 + 5(-x)^2 + 4(-x) + 1$$
$$= -2x^3 + 5x^2 - 4x + 1$$

There are three sign changes in $P(-x)$. At most, there may be three negative real zeros. By decreasing this number by two, we determine that one negative real zero is the only other possibility.

b. Determine the possible rational zeros. The constant term a_0 is 1. The possibilities for p are the integers that are factors of 1. Thus,

$$p = \pm 1.$$

The leading coefficient a_n is 2. The integers that are factors of 2 give the possibilities for q. Thus,

$$q = \pm 2, \pm 1.$$

The possible rational zeros p/q are then

$$1, \tfrac{1}{2}, -\tfrac{1}{2}, -1.$$

From step 1 we eliminate the positive possibilities, leaving

$$-\tfrac{1}{2}, -1.$$

c. Use synthetic division to test whether one of the possibilities, say -1, is a zero.

$$
\begin{array}{r|rrrr}
-1 & 2 & 5 & 4 & 1 \\
   &   & -2 & -3 & -1 \\
\hline
   & 2 & 3 & 1 & 0
\end{array}
$$

Since the remainder is zero, $P(-1) = 0$. Thus, -1 is a zero of the function.

d. The factor theorem states that if -1 is a zero, $x - (-1)$ or $x + 1$ is a factor. The coefficients of the other factor are given in the bottom row of the synthetic division. That is,

$$2x^3 + 5x^2 + 4x + 1 = (x + 1)(2x^2 + 3x + 1).$$

We now find other zeros by setting the factor $2x^2 + 3x + 1$ equal to zero. The big advantage is that after finding a zero we may lower by 1 the degree of the equation we are solving. When we reach a second-degree equation, as in this case, we may apply the quadratic formula or (sometimes) the factoring method.

$$
\begin{aligned}
2x^2 + 3x + 1 &= 0 \\
(2x + 1)(x + 1) &= 0 \\
2x + 1 = 0 \qquad & \quad x + 1 = 0 \\
x = -\tfrac{1}{2} \qquad & \qquad x = -1
\end{aligned}
$$

Thus, the zeros are -1, -1, and $-\frac{1}{2}$. Remember that it is possible for a number to be counted as a zero more than once, and in this case we say -1 is a zero of multiplicity two. Based on the x- and y-intercepts and the multiplicity of each zero, the function is graphed in Figure 4.20.

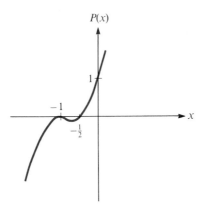

Figure 4.20

For the next example we add another useful rule, which permits us to establish upper and lower bounds on the zeros of a polynomial function.

Upper and Lower Bounds Theorem	**1.** If we divide a polynomial $P(x)$ synthetically by $x - b$, where $b > 0$, and all the numbers in the bottom row have the same sign, then there is no zero greater than b. We say b is an **upper bound** for the zeros of $y = P(x)$.
	2. If we divide a polynomial $P(x)$ synthetically by $x - c$, where $c < 0$, and the numbers in the bottom row alternate in sign, then there is no zero less than c. We say c is a **lower bound** for the zeros of $y = P(x)$.
	For the purposes of these tests, zero may be denoted as $+0$ or -0.

A rationale for these tests is contained in the next example.

EXAMPLE 3   Find the zeros of $P(x) = 2x^4 - 5x^3 + 11x^2 - 3x - 5$.

Solution
   a. Use Descartes' rule of signs to determine the maximum number of positive and negative real zeros.

$$P(x) = 2x^4 - 5x^3 + 11x^2 - 3x - 5$$

There are three sign changes in $P(x)$. Thus, the number of positive real zeros is three or one.

$$P(-x) = 2(-x)^4 - 5(-x)^3 + 11(-x)^2 - 3(-x) - 5$$
$$= 2x^4 + 5x^3 + 11x^2 + 3x - 5$$

There is one sign change in $P(-x)$, so Descartes' rule of signs guarantees exactly one negative *real* zero in this case.

b. Determine the possible rational zeros. The constant term a_0 is -5. The possibilities for p are the integers that are factors of -5. Thus,

$$p = \pm5, \pm1.$$

The leading coefficient a_n is 2. The integral factors of 2 give the possibilities for q. Thus,

$$q = \pm2, \pm1.$$

The possible rational zeros are then

$$5, \tfrac{5}{2}, 1, \tfrac{1}{2}, -\tfrac{1}{2}, -1, -\tfrac{5}{2}, -5.$$

c. We test negative zeros first, because step 1 indicated that there is, at most, one negative zero. If we find a negative zero, we then switch to the positive possibilities. Pick -1, a number in the middle of the negative choices. If it is not a zero, it may be a lower bound that eliminates other possibilities.

$$\begin{array}{r|rrrrr} -1 & 2 & -5 & 11 & -3 & -5 \\ & & -2 & 7 & -18 & 21 \\ \hline & 2 & -7 & 18 & -21 & 16 \end{array}$$

The remainder is 16, so -1 is not a zero. We tested a negative possibility and the numbers in the bottom row alternate in sign. Thus, there is no zero less than -1 as specified in statement 2 of the above theorem. To see why, note that if we test a negative choice, say c, greater in absolute value than -1, then the numbers in the bottom row of the synthetic division will continue to alternate in sign with respective numbers of greater absolute value (after the first entry) than our previous bottom row $(2, -7, 18, -21, 16)$. Thus $P(c) > 16$ when $c < -1$, so -1 is a lower bound for the zeros of $y = P(x)$. This eliminates $-\tfrac{5}{2}$ and -5, so try $-\tfrac{1}{2}$.

$$\begin{array}{r|rrrrr} -\tfrac{1}{2} & 2 & -5 & 11 & -3 & -5 \\ & & -1 & 3 & -7 & 5 \\ \hline & 2 & -6 & 14 & -10 & 0 \end{array}$$

Since the remainder is zero, $P(-\tfrac{1}{2}) = 0$. Thus, $-\tfrac{1}{2}$ is a zero of the function.

d. Since $-\frac{1}{2}$ is a zero, $x - (-\frac{1}{2})$ is a factor and we write

$$2x^4 - 5x^3 + 11x^2 - 3x - 5$$
$$= (x + \tfrac{1}{2})(2x^3 - 6x^2 + 14x - 10).$$

We now try to find the values of x for which $2x^3 - 6x^2 + 14x - 10$ equals zero. We found the one negative zero, so we switch to the positive possibilities. Pick 1. If it is not a zero, the numbers in the bottom row in the synthetic division may have the same sign. Such a result would establish 1 as an upper bound for the zeros (thereby eliminating $\frac{5}{2}$ and 5), since any synthetic division by $x - b$, where $b > 1$, would continue to produce bottom row entries that have the same sign and even larger absolute values (after the first entry) than those that result from the division involving $b = 1$.

$$
\begin{array}{r|rrrr}
1 & 2 & -6 & 14 & -10 \\
  &   & 2 & -4 & 10 \\
\hline
  & 2 & -4 & 10 & 0 \\
\end{array}
$$

Therefore, 1 is a zero of the function.

e. We now use the quadratic formula to complete the solution.

$$2x^2 - 4x + 10 = 0$$
$$x^2 - 2x + 5 = 0$$
$$x = \frac{-(-2) \pm \sqrt{(-2)^2 - 4(1)(5)}}{2(1)} = \frac{2 \pm \sqrt{-16}}{2}$$
$$= \frac{2 \pm 4i}{2} = 1 \pm 2i$$

The zeros are $-\frac{1}{2}$, 1, $1 + 2i$, and $1 - 2i$.   □

Up to this point we have limited our discussion to finding rational zeros of polynomial functions of degree greater than 2. Unfortunately, irrational zeros of higher degree polynomial functions are usually very difficult to find. Complicated formulas are available for polynomial functions of degrees 3 and 4, and it can be shown that no formula exists for degree 5 and greater. In such cases, numerical methods that utilize a computer are employed to approximate irrational zeros, and this topic is usually considered in detail in a course in numerical analysis. For our purposes, we need only illustrate the following key theorem for approximating zeros.

Location Theorem   Let $y = P(x)$ be a polynomial function. If $P(a)$ and $P(b)$ have opposite signs, then $y = P(x)$ has at least one real zero between a and b.

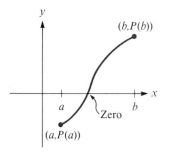

Figure 4.21

This result is easy to see geometrically. Consider Figure 4.21 and keep in mind that the graph of a polynomial function is an unbroken curve. Since the graph is on different sides of the x-axis at a and b, and no breaks are possible, the graph must cross the x-axis between a and b, guaranteeing at least one real zero in this interval.

EXAMPLE 4   Verify that $P(x) = 3x^3 + 16x^2 - 8$ has a zero between -5 and -6.

Solution   By direct substitution or synthetic division, we determine that $P(-5) = 17$ and $P(-6) = -80$. Since $P(-5)$ and $P(-6)$ have opposite signs, the location theorem assures us of at least one real zero between -5 and -6.   □

Once we have determined an interval that contains a zero, the job is to keep narrowing the interval until we obtain a sufficient degree of accuracy. The most efficient methods for narrowing the interval are developed in calculus. A simple and relatively effective alternative is to basically halve the interval each time until we attain a specified accuracy.

EXAMPLE 5   Find the zero of $P(x) = 3x^3 + 16x^2 - 8$ that is between -5 and -6 to the nearest tenth.

Solution   By using the location theorem and a calculator, and by successively halving the intervals, we determine the following.

Calculation	Comment: A Zero Is Between
$P(-5) = 17$	
$P(-6) = -80$	-5 and -6
$P(-5.5) \approx -23$	-5 and -5.5
$P(-5.25) \approx -1.1$	-5 and -5.25
$P(-5.13) \approx 8.1$	-5.13 and -5.25
$P(-5.19) \approx 3.6$	-5.19 and -5.25

Thus, to the nearest tenth, a zero is -5.2. (*Note:* Exercise 14 in this section asks you to find the exact value of this zero by using the previous methods of this section.)   □

EXERCISES 4.5

In Exercises 1–4 list the possible rational zeros of the function.

1. $P(x) = x^3 + 8x^2 - 10x - 20$
2. $P(x) = x^4 - 3x^3 + x^2 - x - 6$
3. $P(x) = 4x^4 - 3x^3 + x^2 - x - 6$
4. $P(x) = 3x^3 + 7x^2 + 8$

In Exercises 5–8 use Descartes' rule of signs to determine the maximum number of positive and negative real zeros.

5. $P(x) = x^3 - 2x^2 + x - 3$
6. $P(x) = 5x^3 + x^2 + 4x + 1$
7. $P(x) = 2x^4 + 5x^3 - x^2 - 3x - 7$
8. $P(x) = x^8 + 1$

In Exercises 9–18 find the zeros of the function.

9. $P(x) = x^3 - x^2 - 10x - 8$
10. $P(x) = x^3 + 6x^2 + 11x + 6$
11. $P(x) = 3x^3 + x^2 + 15x + 5$
12. $P(x) = 4x^3 - x^2 - 28x + 7$
13. $P(x) = 8x^3 - 12x^2 + 6x - 1$
14. $P(x) = 3x^3 + 16x^2 - 8$
15. $P(x) = 4x^4 - 5x^3 - 2x^2 - 3x - 10$
16. $P(x) = x^4 - 6x^3 + 7x^2 + 12x - 18$
17. $P(x) = 2x^4 + 7x^3 + 25x^2 + 47x + 18$
18. $P(x) = 6x^4 + 11x^3 - 63x^2 - 7x + 5$

In Exercises 19–22 use the location theorem and verify each statement.

19. $P(x) = 3x^3 + 16x^2 - 8$ has a zero between 0 and -1.
20. $P(x) = 4x^3 - x^2 - 28x + 7$ has a zero between 2 and 3.
21. $P(x) = x^4 - 6x^3 + 7x^2 + 12x - 18$ has a zero between 1 and 2 and another zero between -1 and -2.
22. $P(x) = 6x^4 + 11x^3 - 63x^2 - 7x + 5$ has a zero between 0 and 1 and another zero between -4 and -5.

In Exercises 23–28 approximate the zero of the function in the given interval to the nearest tenth.

23. $P(x) = x^3 + x - 1$; a zero is between 0.6 and 0.7.
24. $P(x) = x^3 + x^2 - 10x - 10$; a zero is between -3.1 and -3.2.
25. $P(x) = x^3 - 3x + 1$; a zero is between 0 and 1.
26. $P(x) = x^3 - 3x + 1$; a zero is between 1 and 2.
27. $P(x) = 2x^3 - 5x^2 - 3x + 9$; a zero is between -1 and -2.
28. $P(x) = 2x^3 - 5x^2 - 3x + 9$; a zero is between 2 and 3.
29. Use the rational zero theorem and the function $P(x) = x^2 - 2$ to show that $\sqrt{2}$ is not a rational number.
30. The volume of a rectangular box is 30 m³. If the length and width are respectively 3 m and 1 m greater than the height, show that only one set of dimensions is possible for the given box. What are these dimensions?

4.6 RATIONAL FUNCTIONS

When we add, subtract, or multiply two polynomials, the result is another polynomial. However, a function of the form

$$y = \frac{P(x)}{Q(x)} \qquad Q(x) \neq 0,$$

where $P(x)$ and $Q(x)$ are polynomials, is a different type of function called a **rational function.** For example,

$$y = \frac{1}{x}, \quad y = \frac{x-4}{x^2-1}, \quad \text{and} \quad y = \frac{3x^5+2}{x^3}$$

are rational functions. In our statements about rational functions we will assume that $P(x)$ and $Q(x)$ have no common factors, so the rational function is in lowest terms.

The behavior of a rational function often differs dramatically from that of a polynomial function. This difference may be easily seen by comparing the graph of a rational function to the smooth unbroken curves that characterize a polynomial.

Consider the rational function $y = 1/x$. Since division by zero is undefined, x cannot equal zero. Thus, the graph of this function does not intersect the line $x = 0$ (the y-axis). We can, however, let x approach zero and consider x values as close to zero as we wish. From the following tables note that as the x values squeeze in on zero, $|y|$ becomes larger and larger.

x	1	0.5	0.1	0.01	0.001
$y = \dfrac{1}{x}$	1	2	10	100	1,000

x	-1	-0.5	-0.1	-0.01	-0.001
$y = \dfrac{1}{x}$	-1	-2	-10	-100	$-1,000$

Figure 4.22 shows the behavior of $y = 1/x$ in the interval $-1 \le x \le 1$. The vertical line $x = 0$ that the curve approaches, but never touches, is called a **vertical asymptote.** We may use the following rule to determine if the graph of a rational function has any vertical asymptotes.

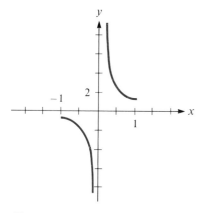

Figure 4.22

Vertical Asymptotes   If $P(x)$ and $Q(x)$ have no common factors, then the graph of the rational function $y = P(x)/Q(x)$ has as vertical asymptote the line $x = a$ for each value a at which $Q(a) = 0$.

EXAMPLE 1   Find any vertical asymptotes of the function

$$y = \frac{x+1}{(x-3)(x+2)}.$$

Figure 4.23

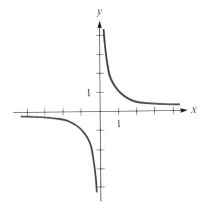

Figure 4.24

Solution   The polynomial in the denominator

$$Q(x) = (x - 3)(x + 2)$$

equals 0 when x is 3 or -2. Thus, there are two vertical asymptotes: $x = 3$ and $x = -2$. (*Note:* As a visual aid, the graph of this function is shown in Figure 4.23. By the end of this section, we will have developed sufficient techniques for you to be able to draw this graph.)   □

To complete the graph of $y = 1/x$, we must consider the behavior of the function as $|x|$ becomes larger. It is not difficult to see that $1/x$ squeezes in on zero from the positive side when x takes on larger and larger positive values. Similarly, $1/x$ squeezes in on zero from the negative side when x increases in magnitude in the negative direction. Thus, the curve gets closer and closer to the line $y = 0$ and the x-axis is a **horizontal asymptote**. Figure 4.24 shows the graph of $y = 1/x$.

Identifying any vertical and horizontal asymptotes is very important. These lines, together with plotting the intercepts and a few points, enable us to graph a rational function. A technique for determining horizontal asymptotes is included in the following examples.

EXAMPLE 2   If $y = \dfrac{x - 1}{x + 1}$, determine any vertical or horizontal asymptotes and graph the function.

Solution
  a. The polynomial in the denominator $x + 1$ equals zero when x is -1. Thus, $x = -1$ is a vertical asymptote.
  b. To determine any horizontal asymptotes, we change the form of the function by dividing each term in the numerator and the denominator by the highest power of x in the expression.

$$y = \frac{x - 1}{x + 1} = \frac{(x/x) - (1/x)}{(x/x) + (1/x)}$$
$$= \frac{1 - (1/x)}{1 + (1/x)} \qquad \text{(assuming that } x \neq 0\text{)}$$

Now as $|x|$ gets larger, $1/x$ approaches zero. Thus, y approaches $(1 - 0)/(1 + 0)$ and $y = 1$ is a horizontal asymptote.
  c. By setting $x = 0$, we may determine $(0, -1)$ is the y-intercept. Similarly, by setting $y = 0$, we may determine that $(1, 0)$ is the x-intercept.

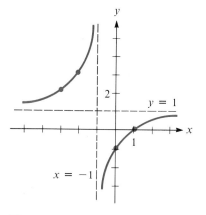

Figure 4.25

d. The vertical asymptote divides the *x*-axis into two regions. The intercepts are two points to the right of $x = -1$. If we plot a couple of points to the left of the vertical asymptote, say $(-2,3)$ and $(-3,2)$, we may complete the graph (see Figure 4.25).   □

In Example 2 an important question is unanswered. We know that as |*x*| becomes large, *y* approaches, but never quite reaches, 1. How do we know that *y* is not 1 when |*x*| is small? We answer this question by determining if there is any value of *x* for which

$$\frac{x-1}{x+1} = 1.$$

Solving this equation, we have

$$x - 1 = x + 1$$
$$-1 = 1 \qquad \text{false.}$$

The equation has no solution, and we may conclude that the curve never crosses the horizontal asymptote. Example 3 illustrates why this possibility must be considered.

EXAMPLE 3   If

$$y = \frac{2x^2 + x - 3}{x^2},$$

determine any vertical and horizontal asymptotes and graph the function.

Solution
a. The polynomial in the denominator equals zero when *x* is zero. Thus, $x = 0$ is a vertical asymptote.
b. To determine any horizontal asymptotes, we use the procedure from Example 2. Dividing each term in the numerator and the denominator by x^2 gives

$$y = \frac{(2x^2/x^2) + (x/x^2) - (3/x^2)}{x^2/x^2}$$

$$= \frac{2 + (1/x) - (3/x^2)}{1} \qquad (x \neq 0).$$

As |*x*| gets larger, $1/x$ and $-3/x^2$ approach 0. Thus, *y* approaches 2 and $y = 2$ is a horizontal asymptote.

c. To determine if the curve ever crosses the horizontal asymptote, find if there are any values of x for which

$$\frac{2x^2 + x - 3}{x^2} = 2.$$

Solving this equation for x, we have

$$2x^2 + x - 3 = 2x^2$$
$$x - 3 = 0$$
$$x = 3.$$

The curve crosses the asymptote at $(3,2)$.

d. Since x cannot be zero there is no y-intercept. After setting $y = 0$, we solve the equation $2x^2 + x - 3 = 0$ to find the x-intercepts $(1,0)$ and $(-\frac{3}{2},0)$.

e. To determine the behavior of the curve before it drops and starts approaching 2, we plot a couple of points to the right of $x = 3$, say $(4,\frac{33}{16})$ and $(5,\frac{52}{25})$. Additional points may always be plotted. Figure 4.26 shows the graph of the function. ▢

Figure 4.26

There are two theorems about asymptotes that can simplify our work. The first theorem indicates the behavior of the graph near any vertical asymptote and is based on the fact that even powers of positive and negative numbers have the same sign (positive) while odd powers of positive and negative numbers are opposite in sign.

Graph of $y = P(x)/Q(x)$ near Vertical Asymptotes

If $P(x)$ and $Q(x)$ have no common factors, and if $(x - a)^n$ is a factor of $Q(x)$, where n is the largest positive integer for which this statement is true, then:

1. The graph of $y = P(x)/Q(x)$ goes in opposite directions about the vertical asymptote $x = a$ when n is odd.

2. The graph of $y = P(x)/Q(x)$ goes in the same direction about the vertical asymptote $x = a$ when n is even.

For instance, in Example 2 the graph goes to positive infinity on the left side of $x = -1$ and to negative infinity on the right side (see Figure 4.25). We can predict this type of behavior since $Q(x) = (x + 1)^1$ and the odd exponent indicates the graph goes in opposite directions about the vertical asymptote.

In Example 3, however, $Q(x) = x^2$ and the even exponent indicates the graph goes in the same direction (to negative infinity) about the vertical asymptote ($x = 0$) as shown in Figure 4.26.

The second theorem about asymptotes enables us to pick out horizontal asymptotes almost by inspection. This theorem may be derived by using the procedure shown in Examples 2 and 3 for determining a horizontal asymptote and applying it to the general form for a rational function given below.

Horizontal Asymptote Theorem

The graph of the rational function

$$y = \frac{P(x)}{Q(x)} = \frac{a_n x^n + a_{n-1} x^{n-1} + \cdots + a_0}{b_m x^m + b_{m-1} x^{m-1} + \cdots + b_0},$$

where $a_n, b_m \neq 0$, has

1. a horizontal asymptote at $y = 0$ (the x-axis) if $n < m$
2. a horizontal asymptote at $y = a_n/b_m$ if $n = m$
3. no horizontal asymptote if $n > m$

For instance, the graph of $y = 1/x$ satisfies the first case since the numerator is a zero degree polynomial while the denominator is a first degree polynomial. Because the higher degree polynomial is in the denominator, $y = 0$ is a horizontal asymptote. Examples 2 and 3 illustrate the second case in which both polynomials are of the same degree so the horizontal asymptotes for

$$y = \frac{1x - 1}{1x + 1} \quad \text{and} \quad y = \frac{2x^2 + x - 3}{1x^2}$$

are $y = 1/1 = 1$ and $y = 2/1 = 2$, respectively. When there are no horizontal asymptotes as described in case 3, other techniques may be employed. In particular, if $n = m + 1$, that is, if the degree of the numerator is one more than the degree of the denominator, then the graph has a slant or oblique asymptote as discussed in the next example.

EXAMPLE 4   Graph the function $f(x) = \dfrac{x^2 - 1}{x + 2}$.

Solution

a. The polynomial in the denominator equals zero when x is -2. Thus, $x = -2$ is a vertical asymptote. Also, since $Q(x) = (x + 2)^1$, $Q(x)$ is an odd power of $x + 2$ so the graph of f goes in opposite directions about $x = -2$.

b. Because the higher degree polynomial is in the numerator, there is no horizontal asymptote. However, we can change the form of the equation by dividing $x^2 - 1$ by $x + 2$ as follows:

$$
\begin{array}{r}
x - 2 \\
x + 2 \overline{\smash{)}\,x^2 - 1} \\
\underline{x^2 + 2x } \\
-2x - 1 \\
\underline{-2x - 4} \\
3.
\end{array}
$$

Thus,

$$
f(x) = \frac{x^2 - 1}{x + 2} = x - 2 + \frac{3}{x + 2}.
$$

Now as $|x|$ gets larger, $3/(x + 2)$ approaches 0 and y approaches $x - 2$. Thus, the graph of f approaches the oblique (neither horizontal nor vertical) line $y = x - 2$, and we call this line an **oblique asymptote** of the graph of f.

c. By setting $x = 0$, we may determine that $(0, -\frac{1}{2})$ is the y-intercept. Similarly, by setting $f(x) = 0$, we may determine that $(1,0)$ and $(-1,0)$ are x-intercepts.

d. Using the above information and plotting one point to the left of the vertical asymptote, say $(-3, -8)$, we draw the graph in Figure 4.27.

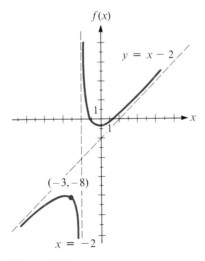

Figure 4.27

In our statements about rational functions we have assumed that $P(x)$ and $Q(x)$ had no common factors. If the same factor does appear in the numerator and the denominator, we can divide it out as long as we keep track of all values that make the factor zero. The next example illustrates such a case.

EXAMPLE 5   Graph the function $f(x) = \dfrac{x^2 - 4}{x - 2}$.

Solution   By factoring the numerator, we rewrite the expression as

$$f(x) = \frac{x^2 - 4}{x - 2} = \frac{(x + 2)(x - 2)}{x - 2}.$$

If $x = 2$, the expression becomes $0/0$, which is undefined. If $x \neq 2$, we can divide out common factors and obtain $x + 2$. Thus, we can rewrite the expression as follows:

$$f(x) = \frac{x^2 - 4}{x - 2} = \frac{(x + 2)(x - 2)}{x - 2} = \begin{cases} x + 2 & \text{if } x \neq 2 \\ \text{undefined} & \text{if } x = 2. \end{cases}$$

As shown in Figure 4.28, the graph is the set of all points on the line given by $f(x) = x + 2$ except $(2,4)$.   □

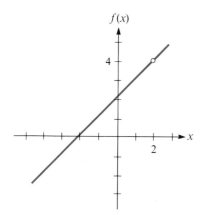

$f(x)$

Figure 4.28

Finally, here is an outline of the procedure for graphing rational functions with the form $y = P(x)/Q(x)$.

1. If $P(x)$ and $Q(x)$ have common factors, we can divide them out on the following conditions:
   a. If the degree of the common factor in the numerator is greater than or equal to the degree of the common factor in the denominator, then any value of x that makes the common factor zero produces a hole in the graph as shown in Example 5.
   b. If the degree of the common factor in the numerator is less than the degree of the common factor in the denominator, then any value of x that makes the common factor zero produces a vertical asymptote.
2. If $P(x)$ and $Q(x)$ have no common factors, find any vertical asymptotes by solving $Q(x) = 0$. Determine the behavior of the graph near any vertical asymptote by using the theorem given in this section.

3. Find any horizontal asymptotes by using the horizontal asymptote theorem or by dividing each term by the highest power of x and determining the value approached by y as $|x|$ gets larger. If the degree of the numerator is one more than the degree of the denominator, then determine an oblique asymptote by dividing $P(x)$ by $Q(x)$ and setting y equal to the quotient.

4. Check if the curve crosses the horizontal asymptote for small values of x by finding if there are any values of x for which $P(x)/Q(x)$ equals the value of the horizontal asymptote. There's no need to check for crossings of vertical asymptotes—they can't happen.

5. Find any x-intercepts by (setting $y = 0$ and) solving $P(x) = 0$. Find any y-intercepts by setting $x = 0$ and evaluating the expression.

6. Draw the graph. Plot additional points as needed (particularly if you have a calculator). You may wish to check for symmetries as discussed in Section 3.3.

EXERCISES 4.6

In Exercises 1–4 find any vertical asymptotes of the function.

1. $y = \dfrac{x + 7}{2x - 3}$

2. $y = \dfrac{x^2 + 1}{x(x + 4)}$

3. $y = \dfrac{2}{x^2 - 5x - 6}$

4. $y = \dfrac{x}{x^2 + x}$

In Exercises 5–24 determine any asymptotes and graph the function.

5. $y = \dfrac{-1}{x}$

6. $y = \dfrac{3}{x^2}$

7. $y = \dfrac{2}{x + 1}$

8. $y = \dfrac{-5}{3x + 2}$

9. $y = \dfrac{x - 2}{x + 2}$

10. $y = \dfrac{x + 3}{x - 5}$

11. $y = \dfrac{2x - 5}{3x + 5}$

12. $y = \dfrac{4x + 1}{2x + 7}$

13. $y = \dfrac{2}{(x - 1)^2}$

14. $y = \dfrac{2}{x^2 - 1}$

15. $y = \dfrac{1}{(x + 1)(x - 4)}$

16. $y = \dfrac{x - 1}{x^2 - 4}$

17. $y = \dfrac{x^2 - 1}{x^3}$

18. $y = \dfrac{x}{x^2 + 1}$

19. $y = \dfrac{x^2 - x - 6}{x^2}$

20. $y = \dfrac{3x^2 + x - 2}{2x^2}$

21. $y = \dfrac{x^2 + 1}{x}$

22. $y = \dfrac{(x - 1)^2}{x}$

23. $y = \dfrac{x^2 - 9}{x + 2}$

24. $y = \dfrac{x^2 - 2x - 3}{x + 3}$

In Exercises 25–33 graph the given function, with particular emphasis on any value of x for which $P(x) = Q(x) = 0$.

25. $h(x) = \dfrac{x(x + 1)}{x}$

26. $y = \dfrac{x(x + 1)}{x + 1}$

27. $y = \dfrac{x^2 - 4}{x + 2}$

28. $y = \dfrac{x^2 - x}{x}$

29. $g(x) = \dfrac{(2x + 1)(x - 3)(x + 2)}{(x - 3)(x + 2)}$

30. $f(x) = \dfrac{(x - 1)(x - 2)(x - 3)}{(x - 1)(x - 2)}$

31. $f(x) = \dfrac{x}{x(x + 1)}$

32. $y = \dfrac{x}{x^2 - 2x}$

33. $y = \dfrac{x - 1}{(x - 1)^2}$

CHAPTER OVERVIEW

Section	Key Concepts to Review

4.1	• Definitions of polynomial function and linear function
	• Point-slope equation: $y - y_1 = m(x - x_1)$
	• Slope-intercept equation: $y = mx + b$
	• For two lines with slopes m_1 and m_2:
	  parallel lines: $m_1 = m_2$
	  perpendicular lines: $m_1 \cdot m_2 = -1$

4.2	• Definitions of quadratic function, quadratic inequality, and axis of symmetry • Methods to graph a quadratic function • Axis of symmetry formula: $x = -b/2a$ • Vertex formula: $(-b/2a, f(-b/2a))$ • The graph of $f(x) = a(x - h)^2 + k$ (with $a \neq 0$) is the graph of $y = ax^2$ (a parabola) shifted so the vertex is (h,k) and the axis of symmetry is $x = h$. • Methods to solve a quadratic inequality
4.3	• Long division procedure for the division of polynomials • Synthetic division procedure for the division of a polynomial by $x - b$ • Remainder theorem
4.4	• Definitions of a zero of a function and the multiplicity of a zero • Factor theorem • Fundamental theorem of algebra • Number of zeros theorem • Theorem about when complex zeros come in conjugate pairs • Theorem about when irrational zeros of the form $a \pm b\sqrt{c}$ come in conjugate pairs • Theorem about graphing polynomial functions by knowing the multiplicity of real number zeros
4.5	• Rational zero theorem • Descartes' rule of signs • Upper and lower bounds theorem • Location theorem
4.6	• Definition of a rational function • Methods to determine vertical, horizontal, or oblique asymptotes • Horizontal asymptote theorem • Theorem about the behavior of a graph near any vertical asymptote(s) • Outline of the procedure for graphing rational functions

CHAPTER REVIEW EXERCISES

In Exercises 1–6 indicate if the given function is a polynomial function.

1. $f(x) = (2x - 1)/3$

2. $y = (2x - 1)/x$

3. $y = x^2 + x^{-1} + 2$

4. $y = x^2 + x + 2^{-1}$

5. $y = \sqrt{x}$

6. $y = \sqrt{2}$

In Exercises 7–12 find the zeros of the function.

7. $y = 2x^2 - 3x$

8. $f(x) = 3 - 5x$

9. $g(x) = x^3 - 2x^2 + x - 2$

10. $y = x^2(x + 7)(x - 4)$

11. $f(x) = (2x^2 - 5x + 1)/x^3$

12. $y = 1/(x^2 - 1)$

In Exercises 13–20 graph each function.

13. $y = x - x^2$

14. $y = \dfrac{3}{2x - 3}$

15. $y = \dfrac{x + 1}{x}$

16. $y = \dfrac{x + 1}{2}$

17. $y = \dfrac{x^2 + 2x}{x}$

18. $y = \dfrac{5x^2 + 4}{4x^2 - 1}$

19. $y = x^3 - 3x^2 - 4x$

20. $y = x^2(x + 7)(x - 4)$

In Exercises 21 and 22 solve each inequality.

21. $x^2 + x \geq 6$

22. $x^2 > x - 1$

23. What is the domain of $y = \dfrac{3x - 1}{4x + 2}$?

24. What is the range of $y = 2x - x^2$?

25. Determine the vertex and axis of symmetry of the graph of $f(x) = 3(x + 5)^2 - 1$.

26. Determine the vertical asymptotes for the function $y = \dfrac{3x + 1}{x^2 + 2x}$.

27. Determine the horizontal asymptote for the function $y = \dfrac{2x^2 + 3}{3x^2 + x - 2}$. Does the graph cross the horizontal asymptote? If yes, find the point at which it crosses.

28. What is the difference between the graphs of $y = \dfrac{x^2 - 4}{x - 2}$ and $y = x + 2$?

29. Write the equation of the line with an x-intercept of 3 and a y-intercept of -2.

30. Find the equation that defines the linear function f if $f(-1) = 6$ and $f(2) = -3$.

31. Line L_1 passes through $(-2, -2)$ and $(0, 3)$ and line L_2 passes through $(-4, k)$ and $(2, -1)$.
  a. Find k if L_1 is parallel to L_2.
  b. Find k if L_1 is perpendicular to L_2.

32. Write a polynomial function (in factored form) with rational coefficients of the lowest possible degree with zeros of $-i$, $\sqrt{2}$, and 0.

33. Write a polynomial function (in factored form) of degree 4 with zeros of 5, -2, 0, and 3.

34. If $f(x) = 2x^4 - x + 7$, find $f(-2)$ by (a) direct substitution and (b) the remainder theorem.

35. If the parabola $y = x^2 - 8x + k$ has its vertex on the x-axis, what is the value of k?

36. Divide $2x^4 - x + 7$ by $x^2 + 2$. Express the result in the form dividend = (divisor)(quotient) + remainder.

37. The dividend is x^2, the quotient is $x + 3$, and the remainder is 9. What is the divisor?

38. List the possible rational zeros of the function $P(x) = 3x^3 - 7x^2 + x - 2$.

39. Use Descartes' rule of signs to determine the maximum number of (a) positive and (b) negative real zeros of $P(x) = x^4 - 3x^3 + x^2 - x + 1$.

40. If 2 is a zero of $P(x) = 2x^3 - 4x^2 - 6x + 12$, find the other zeros.

41. Consider the following synthetic division:

$$\begin{array}{r|rrrr} 2 & 1 & -3 & 3 & -5 \\ & & 2 & -2 & 2 \\ \hline & 1 & -1 & 1 & -3. \end{array}$$

On the basis of this division can the function $P(x) = x^3 - 3x^2 + 3x - 5$ have any real zeros greater than 2? Justify your answer.

42. If $P(x) = x^3 + px + q$, determine the nature of the zeros when (a) p and q are both positive and (b) p is positive and q is negative.

43. If 1 is a root of $x^3 - 2x^2 - 5x + 6 = 0$, what is the quadratic equation that can be used to find the other two roots?

44. If b is a zero of $P(x) = x^3 + ax^2 + ax + 1$, show that $1/b$ is also a zero.

45. Find all the rational zeros of $y = 2x^{11} - x - 1$.

46. Verify that $P(x) = x^3 - 3x^2 - 1$ has a real zero between 3 and 4 by using the location theorem.

47. Approximate the zero of $P(x) = x^3 - 2x - 7$ that is between 2 and 3 to the nearest tenth.

48. Show that the largest rectangular field that can be enclosed by a given length of fence is a square.

49. A businessperson owns a parking lot that she wants to enclose with a fence. One side of the lot is bounded by a building, so no fencing is required. There is to be a 16-ft opening in the front of the fence, which is opposite the building. If side fencing costs $1 per foot and front fencing costs $1.50 per foot, what are the dimensions of the largest rectangular lot that can be fenced for $300?

50. A salesperson receives a monthly salary plus a commission on sales. She earned $1,050 during a month in which her sales totaled $5,000; the next month she sold $6,000 worth of merchandise and earned $1,150.
   a. Find the equation that defines this linear relationship.
   b. How much will she earn if she sells $9,000 worth of merchandise for the month?
   c. What is her monthly salary?
   d. What is her rate of commission?

51. A projectile shot vertically upward has a velocity of 192 ft/second at $t = 4$ seconds and 96 ft/second when $t = 7$ seconds.
   a. Find the equation that defines this linear relationship.
   b. Find the velocity of the projectile when $t = 9$ seconds.
   c. What is the initial velocity of the projectile (velocity when $t = 0$)?
   d. When is the velocity 0? What is the physical significance of this?

52. A computer manufacturer sells 6,000 units per week of its $3,000 model. Because of a backlog of computers a rebate program is introduced. It is estimated that each $100 decrease in price will result in the sale of 300 more units. The weekly revenue R is then given by the formula $R = (6{,}000 + 300x)(3{,}000 - 100x)$, where x is the number of $100 discounts offered. What rebate should be given to produce maximum revenue?

In Exercises 53–62 select the choice that answers the question or completes the statement.

53. The degree of $P(x) = x^2(x + 3)(x - 3)^4$ is
   a. 6   **b.** 7   **c.** 8   **d.** 16

54. If $P(x) = x^2(x + 3)(x - 3)^4$, what is the multiplicity of the zero of -3?
   a. 1   **b.** 2   **c.** 4   **d.** 5

55. A quadratic function with zeros of $2 \pm i$ is
   a. $y = x^2 - 4x + 5$ **b.** $y = x^2 - 5x + 4$
   c. $y = x^2 + 5x - 4$ **d.** $y = x^2 + 4x - 5$

56. The graph of $y = (x^2 + 1)/x^3$ lies in quadrants
   a. 1 and 2   **b.** 1 and 3
   c. 2 and 4   **d.** 3 and 4

57. If $x^3 - 2x^2 + ax + 9$ is exactly divisible by $x + 3$, then a equals
   a. -10 **b.** -12 **c.** 7   **d.** 16

58. The axis of symmetry for $y = 2x^2 + 4x - 1$ is
   a. $y = 1$ **b.** $x = 2$ **c.** $x = -1$ **d.** $y = 4$

59. The slope of the line $3x - 2y = 6$ is
   a. 3   **b.** -3 **c.** $\frac{3}{2}$ **d.** $-\frac{3}{2}$

60. The equation of the line that passes through the point $(-3,1)$ and is perpendicular to $y = \frac{1}{2}x + 4$ is
   a. $y = \frac{1}{2}x - \frac{1}{4}$ **b.** $y = \frac{1}{2}x + \frac{5}{2}$
   c. $y = 2x + 7$ **d.** $y = -2x - 5$

61. If $\frac{2}{5}$ is a zero of $y = ax^3 + bx^2 + cx + d$, in which $a, b, c,$ and d are integers with $a \neq 0$, then 5 must be a factor of
   a. a **b.** b **c.** c **d.** d

62. The function $y = x^3 - 3x - 3$ has exactly one positive real zero in which one of the following intervals?
   a. $[0,1]$ **b.** $[1,2]$ **c.** $[2,3]$ **d.** $[3,4]$

5 Exponential and Logarithmic Functions

If you win the state lottery and are offered $1 million or a penny that doubles in value on each day in the month of November, which option would you accept? What is the value of the penny by the end of the month? (See Example 2 of Section 5.1.)

E xponential functions and their inverses, the logarithmic functions, are important rules for analyzing a wide variety of relationships. Compound interest, population growth, radioactive decay, pH, decibels, and heat loss (from a home or a murder victim) are but some of the applications we consider in this chapter. Whereas a typical polynomial function looks like $y = x^2$, a typical exponential function looks like $y = 2^x$. The explosive growth of exponential functions like $y = 2^x$ can be quite amazing, and the introductory problem given above addresses this issue. Throughout this chapter, you will find that a scientific calculator is particularly useful, while a knowledge of exponents is indispensible.

5.1 EXPONENTIAL FUNCTIONS

In engineering, biology, economics, psychology, and other fields there are important phenomena whose behavior is described by functions in which the independent variable appears as the exponent of a fixed base. These functions are called exponential functions as shown in the following definition.

Exponential Function   The function f defined by

$$f(x) = b^x,$$

where $b > 0$ with $b \neq 1$, is called the **exponential function with base b.**

We eliminate a nonpositive base in the defining equation because such expressions as $(-1)^{1/2}$ and 0^{-2} are not real numbers. Since $1^x = 1$ for all values of x, we also eliminate 1.

EXAMPLE 1   In the biological reproduction of cells by division, one cell divides into two cells after a certain period of time. Each of these cells repeats the process and divides into two more, and so on. To be specific, assume that on a given day we start with one cell and the number doubles each day. Then there are 2 cells present after 1 day; $2 \cdot 2$, or 2^2, after 2 days; $2 \cdot 2^2$, or 2^3, after 3 days; and so on. Therefore, there are 2^x cells present after x days. This phenomenon of cell reproduction is described by the function

$$y = 2^x,$$

where x represents the number of complete days from the start of the experiment.   □

EXAMPLE 2   Solve the problem in the chapter introduction on page 235.

Solution   As in Example 1, the doubling process produces a base 2 exponential function. Since the initial value of a penny is 0.01, the value (V) of the penny after t days is given by

$$V = (0.01)2^t$$

and the penny's value increases as follows:

$$V = 0.01 \qquad\qquad\qquad \text{(when } t = 0\text{)}$$
$$V = (0.01)2 \quad\; = \$0.02 \qquad\qquad \text{(after 1 day)}$$
$$\vdots \qquad\qquad\qquad\qquad\qquad \vdots$$
$$V = (0.01)2^{10} = \$10.24 \qquad\qquad \text{(after 10 days)}$$
$$\vdots \qquad\qquad\qquad\qquad\qquad \vdots$$
$$V = (0.01)2^{20} = \$10,485.76 \qquad \text{(after 20 days)}$$
$$\vdots \qquad\qquad\qquad\qquad\qquad \vdots$$
$$V = (0.01)2^{25} = \$335,544.32 \qquad \text{(after 25 days)}$$
$$\vdots \qquad\qquad\qquad\qquad\qquad \vdots$$
$$V = (0.01)2^{30} = \$10,737,418 \qquad \text{(after 30 days)}.$$

Thus, by the end of November the penny is worth \$10,737,418 and is the better option. (*Note:* In this example we evaluated $(0.01)2^{30}$ by using the $\boxed{y^x}$ key on a calculator as follows:

$$0.01 \;\boxed{\times}\; 2 \;\boxed{y^x}\; 30 \;\boxed{=} \qquad 10{,}737{,}418.$$

Throughout this chapter, you will find that a scientific calculator is an essential aid. Exponential expressions as above [and later logarithmic expressions] are evaluated quickly and accurately on a calculator; otherwise, such evaluations require a cumbersome logarithmic approach that is discussed in Appendix A.3. So, at least for this chapter, try to obtain a calculator.) □

At present we are able to interpret exponents that are rational numbers. For example, if $f(x) = 2^x$, then

$$f(3) = 2^3 = 8$$
$$f(0) = 2^0 = 1$$
$$f(-2) = 2^{-2} = \frac{1}{2^2} = \frac{1}{4}$$
$$f(\tfrac{3}{2}) = 2^{3/2} = \sqrt{2^3} = \sqrt{8} \approx 2.83.$$

To interpret 2^x, where x is irrational, it is important to understand that real number exponents are defined so that the following "betweenness" property applies.

Property of Real Number Exponents

For $b > 0$, $b \neq 1$, and any real number x, if x lies between the rational numbers r and s, then b^x lies between b^r and b^s.

For example, to interpret $2^{\sqrt{2}}$ by using this property, we first note that $\sqrt{2} = 1.4142135\ldots$ may be written to any desired accuracy. Then, since

$$1.4 < \sqrt{2} < 1.5 \text{ and } 2^{1.4} < 2^{1.5}, \text{ we know } 2^{1.4} < 2^{\sqrt{2}} < 2^{1.5},$$
$$\text{so } 2.639\ldots < 2^{\sqrt{2}} < 2.828.\ldots$$

Similarly,

$$1.41 < \sqrt{2} < 1.42 \text{ and } 2^{1.41} < 2^{1.42},$$
$$\text{so } 2^{1.41} < 2^{\sqrt{2}} < 2^{1.42}, \text{ or } 2.657\ldots < 2^{\sqrt{2}} < 2.675\ldots,$$
$$1.414 < \sqrt{2} < 1.415 \text{ and } 2^{1.414} < 2^{1.415},$$
$$\text{so } 2^{1.414} < 2^{\sqrt{2}} < 2^{1.415}, \text{ or } 2.664\ldots < 2^{\sqrt{2}} < 2.666\ldots,$$

and so on. By continuing in this manner, we can make the difference between the endpoints of the interval containing $2^{\sqrt{2}}$ as small as we wish and thereby approximate the unique real number $2^{\sqrt{2}}$ to any desired accuracy. In actual practice, we usually obtain a direct estimate for $2^{\sqrt{2}}$ by using the $\boxed{y^x}$ key on a calculator as follows:

$$2 \ \boxed{y^x} \ 2 \ \boxed{\sqrt{\ }} \ \boxed{=} \qquad 2.6651441.$$

Thus, although a precise definition for irrational exponents can only be given in more advanced mathematics, we are now able to interpret expressions like 2^x for all real numbers x, and it can be shown that all previous laws of exponents are valid for all real number exponents. In the next example we use the results of our discussion to graph two typical exponential functions:

$$y = 2^x \quad \text{in which } b > 1$$
$$y = (\tfrac{1}{2})^x \quad \text{in which } 0 < b < 1.$$

EXAMPLE 3   Sketch the graphs of $g(x) = 2^x$ and $h(x) = (\tfrac{1}{2})^x$. In each case indicate the domain and range of the function on the graph.

Solution   By picking convenient integer values for x and substituting them in the given equations, we generate the following table.

x		-3	-2	-1	0	1	2	3
2^x		$\tfrac{1}{8}$	$\tfrac{1}{4}$	$\tfrac{1}{2}$	1	2	4	8
$(\tfrac{1}{2})^x$		8	4	2	1	$\tfrac{1}{2}$	$\tfrac{1}{4}$	$\tfrac{1}{8}$

Based on these points and the "betweenness" property of real number exponents, we then sketch the graphs of g and h as the smooth curves in Figure 5.1(a) and (b), respectively. In both cases the domain is the set of all real numbers and the range is the set of positive real numbers as shown on the graphs.   □

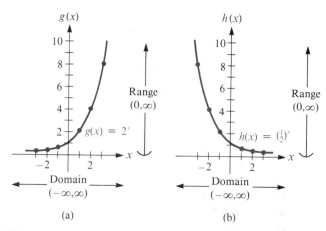

(a)

(b)

Figure 5.1

The graphs in Figure 5.1(a) and (b) illustrate the following important features of exponential functions of the form $f(x) = b^x$ (with $b > 0$, $b \neq 1$) and their graphs.

Properties of $f(x) = b^x$ **(with $b > 0, b \neq 1$)** **and Its Graph**	**1.** The domain of f is $(-\infty,\infty)$ and the range of f is $(0,\infty)$. **2.** The x-axis is a horizontal asymptote for the graph of f. **3.** The y-intercept is always $(0,1)$ and there are no x-intercepts. **4.** If $b > 1$, then as x increases, y increases. **5.** If $0 < b < 1$, then as x increases, y decreases.

We can graph certain variations of the function $y = b^x$ by using the graphing techniques given in Section 3.3. Example 4 illustrates such a case.

EXAMPLE 4   Graph $y = 1 - 2^{-x}$. What is the range of this function?

Solution   We begin by noting that 2^{-x} is equivalent to $(\frac{1}{2})^x$, so we start with the graph of $y = (\frac{1}{2})^x$ in Figure 5.1(b). We also note that $1 - 2^{-x} = -2^{-x} + 1$. To graph $y = -2^{-x} + 1$, we first reflect the graph of $y = 2^{-x}$ about the x-axis to obtain the graph of $y = -2^{-x}$ [see Figure 5.2(a)]. Then we raise the graph in Figure 5.2(a) up 1 unit, since the constant 1 is added after the exponential term. The completed graph is shown in Figure 5.2(b). From this graph we read that the range is the interval $(-\infty,1)$.

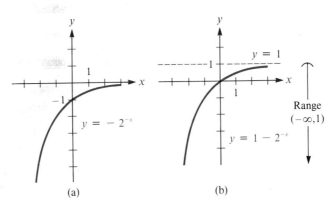

(a)   (b)

Figure 5.2

In this chapter we will frequently solve certain equations containing exponential expressions by using the principle that since exponential functions are one-to-one, if b is a positive number other than 1, then

$$b^x = b^y \text{ implies } x = y.$$

The next example shows where we use this principle.

EXAMPLE 5   If $(x, \frac{1}{27})$ is a solution of the equation $y = 9^x$, then find x.

Solution   We need to find the value of x that satisfies the equation $9^x = \frac{1}{27}$. In many cases it is hoped that through an understanding of exponents, you can determine the solution by inspection. If not, then try to rewrite the expressions in terms of a common base. In this case since $9 = 3^2$ and $27 = 3^3$ we solve the equation as follows:

$$9^x = \frac{1}{27}$$

$$(3^2)^x = \frac{1}{3^3}$$

$$3^{2x} = 3^{-3}$$

$$2x = -3 \qquad \text{Since } b^x = b^y \text{ implies } x = y$$

$$x = -\frac{3}{2}.$$

Thus, $-\frac{3}{2}$ is the x value in the ordered pair. (*Note:* In Section 5.4 we will consider equations like $3^x = 8$, in which it is difficult to rewrite the expressions in terms of a common base.)   ☐

Finally, let us consider two more applications involving exponential functions.

EXAMPLE 6   A company purchases a new machine for $3,000. The value of the machine depreciates at a rate of 10 percent each year.
a. Find a formula (rule) showing the value (y) of the machine at the end of x years.
b. What is the machine worth after 4 years?

Solution
a. If the machine depreciates 10 percent each year, at the end of a year the machine is worth 90 percent of the value at which it began the year. Thus, the value is 3,000(0.9) after 1 year; 3,000(0.9)(0.9) or 3,000(0.9)^2 after 2 years; and

$$y = 3,000(0.9)^x \qquad \text{after } x \text{ years.}$$

b. Evaluating the function when $x = 4$, we have

$$y = 3,000(0.9)^4$$
$$= \$1,968.30.$$ ☐

EXAMPLE 7 $1,000 is invested at 6 percent interest compounded annually.

a. Describe in functional notation the value of the investment at the end of t years.

b. How much is the investment worth after 3 years?

Solution

a. The $1,000 amounts to $1,000(1 + 0.06) or $1,060 at the end of 1 year. The principal during the second year is 1,000(1 + 0.06). Thus, the amount at the end of 2 years is 1,000(1 + 0.06)(1 + 0.06) or 1,000(1 + 0.06)². In general, the value of the investment after t years is

$$f(t) = 1,000(1 + 0.06)^t.$$

b. Evaluating the function when $t = 3$, we have

$$f(3) = 1,000(1 + 0.06)^3$$
$$= \$1,191.02.$$

(*Note:* Compound interest, which produces exponential growth, is a veritable gold mine when compared to the linear growth of simple interest. For example, if $1 had been invested and compounded annually at 3 percent interest since the birth of Christ the compounded amount in 1989 would be about $34,000,000,000,000,000,000,000,000,000. At simple interest the same investment would amount to $60.67!)   □

In this section the applications describe situations in which no change is considered to have occurred until an entire time period has elapsed. Thus, the $1,000 investment in Example 7 is compounded annually and does not change until the end of the year. For this reason, the domain of these functions is the set of nonnegative integers. In Section 5.5 we discuss phenomena that change continuously. The domain of the function in these applications is usually the set of nonnegative real numbers.

EXERCISES 5.1

In Exercises 1–10 fill in the missing component of each of the ordered pairs that make the pair a solution of the given equation. Also, choose a convenient scale for the vertical axis and graph each function.

1. $y = 2^x$ $(-3, \ ), (\ , 1), (\ , \frac{1}{2}), (2, \ )$

2. $y = (\frac{1}{2})^x$ $(0, \ ), (2, \ ), (\ , 2), (\ , \frac{1}{2})$

3. $y = (0.1)^x$ $(\ , 0.01), (\ , 1), (-1, \ ), (-2, \ )$

4. $g(x) = 4^{-x}$ $(2, \ ), (\ , 0.5), (-0.5, \ ), (\ , 4)$

5. $y = (0.25)^{-x}$ $(\frac{1}{2}, \ ), (\ , 1), (1, \ ), (\ , \frac{1}{2})$

6. $y = -5^x$ $(0, \ ), (1, \ ), (-1, \ ), (\ , -0.04)$

7. $h(x) = (\sqrt{2})^x$ $(0, \ ), (-2, \ ), (1, \ ), (\ , 2)$

8. $f(x) = 3(2)^x$ $(-1, \ ), (0, \ ), (1, \ ), (2, \ )$

9. $f(x) = 100(1.06)^x$ $(-1, \ ), (0, \ ), (1, \ ), (2, \ )$

10. $y = 3^x + 3^{-x}$ $(-1, \ ), (0, \ ), (1, \ ), (2, \ )$

In Exercises 11–20 graph each function. Also, determine the range of the function in each case.

11. $f(x) = 3^{-x}$

12. $y = -3^x$

13. $y = 1 - 3^{-x}$

14. $y = 3^x - 1$

15. $y = 2^{x-2}$

16. $y = 2^{2-x}$

17. $h(x) = -(\frac{1}{2})^{-x}$

18. $y = (\frac{1}{2})^{x+3}$

19. $y = 2^x + 2^{-x}$

20. $f(x) = 2^{|x|}$

21. Find the base of the exponential function $y = b^x$ that contains the given point.

  a. $(1,4)$ **b.** $(2,9)$ **c.** $(-1,3)$

  d. $(-2,\frac{1}{25})$ **e.** $(\frac{1}{2},4)$ **f.** $(0.5,10)$

  g. $(\frac{2}{3},\frac{1}{4})$ **h.** $(\frac{3}{2},27)$ **i.** $(-\frac{1}{3},2)$

  j. $(-\frac{2}{3},\frac{1}{9})$

22. Determine an integer n such that $n < x < n + 1$.

  a. $10^x = 451$ **b.** $10^x = 45,100$

  c. $10^x = 0.451$ **d.** $10^x = 0.00451$

23. Show that if $f(x) = b^x$ is the exponential function with base b, then $f(x_1 + x_2) = f(x_1) \cdot f(x_2)$.

24. A biologist has 500 cells in a culture at the start of an experiment. Hourly readings indicate that the number of cells is doubling every hour.

  a. Find a formula (rule) showing the number of cells present at the end of t hours.

  b. How many cells are present at the end of 4 hours?

  c. At the end of how many hours are 32,000 cells present?

25. $100 is invested at 5 percent compounded annually.

  a. Find a formula showing the value of the investment at the end of t years.

  b. How much is the investment worth at the end of 3 years?

26. The market value of a company's investment is $100,000. This value increases at a rate of 10 percent each year.

a. Describe in functional notation the market value of the investment t years from now.

b. What should be the market value of the investment 2 years from now?

27. A company purchases a machine for $10,000. The value of the machine depreciates at a rate of 20 percent each year.

  a. Find a formula to show the value of the machine after t years.

  b. What will be the value of the machine 3 years from now?

28. A biologist grows a colony of a certain kind of bacteria. It is found experimentally that $N = N_0 3^t$, where N represents the number of bacteria present at the end of t days. N_0 is the number of bacteria present at the start of the experiment. Suppose that there are 153,000 bacteria present at the end of 2 days.

  a. How many bacteria were present at the start of the experiment?

  b. How many bacteria are present at the end of 4 days?

  c. At the end of how many days are 459,000 bacteria present?

29. A new element has a half-life of 30 minutes; that is, if x oz of the element exist at a given time, $x/2$ oz exist 30 minutes later. The other half disintegrates into another element. Suppose that 1 g of the element is present 2 hours after the start of an experiment.

  a. Find a formula showing the amount of the element present t hours after the start of the experiment.

  b. How much of the element was present 1 hour after the start of the experiment?

30. You purchase a home for $150,000 and take out a 30-year mortgage. If the annual inflation rate remains fixed at 5 percent and the value of the home keeps pace with inflation, then what is the value of your home by the time the mortgage is paid off?

5.2 LOGARITHMIC FUNCTIONS

The **logarithm** (or log) of a number is the exponent to which a fixed base is raised to obtain the number. In the statement $10^2 = 100$, we call 10 the base, 100 the number, and 2 the exponent or logarithm. We say 2 is the log to the base 10 of 100. In logarithmic form $10^2 = 100$ is written as $\log_{10} 100 = 2$.

In general, the key relation between exponential form and logarithmic form is

$$\log_b N = L \text{ is equivalent to } b^L = N,$$

where b and N are positive numbers with $b \neq 1$. Examples 1 and 2 illustrate this definition.

EXAMPLE 1   Write $4^2 = 16$ and $b^r = s$ in logarithmic form.

Solution   Since $b^L = N$ implies $\log_b N = L$,

$4^2 = 16$ may be written as $\log_4 16 = 2$

$b^r = s$ may be written as $\log_b s = r$.   □

EXAMPLE 2   Write $\log_2 \frac{1}{8} = -3$ and $\log_a u = v$ in exponential form.

Solution   Since $\log_b N = L$ implies $b^L = N$,

$\log_2 \frac{1}{8} = -3$ may be written as $2^{-3} = \frac{1}{8}$

$\log_a u = v$ may be written as $a^v = u$.   □

Now that we have defined a logarithm, we introduce the concept of a logarithmic function. To begin, we first need to recall three facts about inverse functions from Section 3.5.

1. Two functions with exactly reverse assignments are inverse functions.
2. The domain of a function f is the range of its inverse function, and the range of f is the domain of its inverse.
3. A function is one-to-one when each x value in the domain is assigned a different y value so that no two ordered pairs have the same second component. We define the inverse function of f, denoted f^{-1}, only when f is a one-to-one function.

The logarithmic function is defined by a rule that reverses the assignments of the exponential function which is a one-to-one function. We start with the exponential function $y = b^x$ and obtain its inverse by interchanging the variables x and y. The resulting equation $x = b^y$ is a rule that assigns to each positive number exactly one exponent. Since $x = b^y$ is equivalent to $y = \log_b x$, this rule is written in logarithmic form as $y = \log_b x$ and we have the following definition.

Logarithmic Function   For $b > 0$, $b \neq 1$, and $x > 0$,

$$y = \log_b x \quad \text{if and only if} \quad x = b^y.$$

The function f defined by

$$f(x) = \log_b x \quad \text{for } x > 0$$

is called the **logarithmic function with base b.**

Note that the fixed base (b), as in the exponential function, may be any positive real number except 1. Also, since a positive base raised to any power is positive, only positive numbers may be substituted for x, so the domain of $f(x) = \log_b x$ is $(0,\infty)$. We expect this result because the domain and range of inverse functions are interchanged. Thus, the domain of $y = \log_b x$ is the range of $y = b^x$ (set of positive real numbers), and the range of $y = \log_b x$ is the domain of $y = b^x$ (set of all real numbers).

To illustrate that a logarithmic function reverses the assignments of an exponential function, consider Figure 5.3 in which $y = 2^x$ and $y = \log_2 x$ are sketched on the same axes.

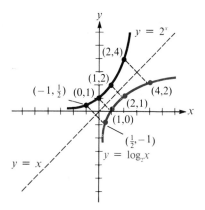

Figure 5.3

$y = 2^x$	x	-3	-2	-1	0	1	2	3		
	y	$\frac{1}{8}$	$\frac{1}{4}$	$\frac{1}{2}$	1	2	4	8		

$y = \log_2 x$ or $x = 2^y$	x	$\frac{1}{8}$	$\frac{1}{4}$	$\frac{1}{2}$	1	2	4	8
	y	-3	-2	-1	0	1	2	3

An examination of the tables and the two graphs shows that the functions have reverse assignments. Note in Figure 5.3 that the graphs are related in that each one is the reflection of the other across the line $y = x$ as expected.

Once again, we can graph variations of a familiar function $y = \log_b x$ by using the graphing techniques given in Section 3.3. Example 3 illustrates the case of a horizontal shift.

EXAMPLE 3   Graph $y = \log_2(x + 3)$. What is the domain of the function?

Solution   We start with the graph of $y = \log_2 x$ in Figure 5.3. To graph $y = \log_2(x + 3)$, we note that x has been replaced by $x + 3$, and the logarithmic rule is then applied to $x + 3$. Such a replacement means the graph of $y = \log_2(x + 3)$ is the graph of $y = \log_2 x$ shifted 3 units to the left. The completed graph is shown in Figure 5.4, and we read from this graph that the domain of $y = \log_2(x + 3)$ is $(-3, \infty)$. Without the graph, we may determine the domain by noting that logarithms are only defined for positive numbers. Thus, $x + 3 > 0$, so $x > -3$.   □

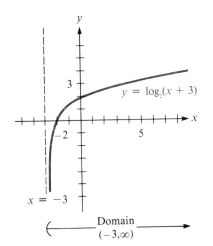

Figure 5.4

There are two specific bases associated with applications of logarithmic functions: base 10 or **common logarithms** and base e or **natural logarithms.** Logarithms to the base e are used extensively in calculus and will be considered in Section 5.5. Here we discuss common logarithms that are used to express many important formulas in science. As examples we have the formulas for pH (measure of the acidity of a solution), decibel (basic unit of loudness), and Richter scale ratings (measure of the magnitude of an earthquake). Before considering the pH formula, we first discuss how to evaluate common logarithm statements on a calculator.

It is standard notation to write a common logarithm (or $\log_{10} N$) as log N with base 10 being understood. Scientific calculators have a ⌷log⌷ key and to find log N, simply enter N and press ⌷log⌷. Examples 4 and 5 illustrate this.

EXAMPLE 4   log 12.3   12.3 ⌷log⌷   1.0899051 □

EXAMPLE 5 $\log(3.2 \times 10^{-5})$ 3.2 ⌷EE⌷ 5 ⌷+/−⌷ ⌷log⌷   ⌷−4.4949 00⌷ □

To reverse this process and find N if we know log N, we must remember that $\log_{10} x$ and 10^x are inverse functions. Therefore,

⌷INV⌷ ⌷log⌷ is equivalent to ⌷10^x⌷.

Some calculators write 10^x above the ⌷log⌷ key to specify that it is the inverse or second function associated with this key. We then find N as shown in Examples 6 and 7.

N equals

EXAMPLE 6   log $N = 1.13$ 1.13 ⌷INV⌷ ⌷log⌷   13.489629 □

EXAMPLE 7   log $N = -7.4$ 7.4 ⌷+/−⌷ ⌷INV⌷ ⌷log⌷   ⌷3.9811 −08⌷ □

In summary, here are a few notes.

1. You must understand scientific notation and how it is displayed on a calculator, so you may need to study Appendix A.1. In the last example the display $\boxed{3.9811 \ -08}$ means $N = 3.9811 \times 10^{-8}$.

2. In science it is common to refer to an inverse logarithm as an antilogarithm. Thus, antilog $x = \text{INV} \log x = 10^x$.

3. If you cannot obtain a calculator, logarithms and inverse logarithms can be determined as shown in Appendix A.3.

4. Remember that calculators replace log *evaluations,* not log *functions.* Analyzing relationships with logarithmic functions remains the major idea here.

We now consider the logarithmic function that defines pH.

You are probably aware of the concept of an acid, since it is associated with such familiar and diverse topics as indigestion, shampoo, swimming pools, and car batteries. In chemistry a logarithm is used to define pH (hydrogen potential), which is a convenient measure of the acidity of a solution. Briefly, let's see why pH is defined as a logarithm.

When an atom of hydrogen loses an electron, it is called a hydrogen ion. Since this atom is positively charged, the hydrogen ion is symbolized H^+. The electrical imbalance in an ion greatly affects chemical reactions, and the concentration of hydrogen ions (symbolized $[H^+]$) in a solution determines its acidity. However, hydrogen ion concentrations are very small numbers. For example, a weak acid solution might have a concentration (measured in moles/liter) of only 1 part H^+ in 10,000, or 1/10,000, or 10^{-4}. When writing such small numbers, the exponential form is obviously helpful. In 1909 a further simplification, pH notation, was introduced. With this notation only the exponent is considered, as follows:

$$\text{If } [H^+] = 10^{-4} \quad \text{then} \quad pH = 4$$
$$[H^+] = 10^{-8.3} \quad \text{then} \quad pH = 8.3$$
$$[H^+] = 10^{-x} \quad \text{then} \quad pH = x.$$

Since pH is defined as an exponent, we determine pH by finding a logarithm. That is, from

$$[H^+] = 10^{-pH}$$

we may write

$$-pH = \log_{10}[H^+]$$
$$pH = -\log[H^+].$$

Thus, pH is defined as the negative of the logarithm of the hydrogen ion concentration. A solution with $pH = 7$ is called neutral. In acids $pH < 7$; bases or alkalies have $pH > 7$.

EXAMPLE 8 The hydrogen ion concentration in a sample of human blood is 4.4×10^{-8}. Determine the pH of the sample.

Solution

$$pH = -\log[\text{H}^+]$$
$$= -\log(4.4 \times 10^{-8})$$

The calculation is as follows:

$4.4 \boxed{\text{EE}} \, 8 \, \boxed{+/-} \, \boxed{\log} \qquad \boxed{-7.3565\ 00}$.

Thus, pH ≈ 7.4. ◻

EXAMPLE 9 The pH of tomato juice is about 4.5. Determine $[\text{H}^+]$.

Solution

$$pH = -\log[\text{H}^+]$$

Replacing pH by 4.5 and solving for $\log[\text{H}^+]$, we have

$$\log[\text{H}^+] = -4.5.$$

We now compute an inverse logarithm as follows:

$4.5 \, \boxed{+/-} \, \boxed{\text{INV}} \, \boxed{\log} \qquad \boxed{3.1623\ -05}$.

Thus, $[\text{H}^+] \approx 3.2 \times 10^{-5}$. ◻

EXERCISES 5.2

In Exercises 1–10 express in logarithmic form.

1. $3^2 = 9$ **2.** $2^3 = 8$ **3.** $(\frac{1}{2})^2 = \frac{1}{4}$

4. $(\frac{1}{3})^3 = \frac{1}{27}$ **5.** $4^{-2} = \frac{1}{16}$ **6.** $(\frac{1}{2})^{-3} = 8$

7. $25^{1/2} = 5$ **8.** $8^{-1/3} = \frac{1}{2}$ **9.** $7^0 = 1$

10. $10^{-4} = 0.0001$

In Exercises 11–20 express in exponential form.

11. $\log_5 5 = 1$ **12.** $\log_2 32 = 5$

13. $\log_{1/3} \frac{1}{9} = 2$ **14.** $\log_{1/2} \frac{1}{16} = 4$

15. $\log_2 \frac{1}{4} = -2$ **16.** $\log_{1/4} 4 = -1$

17. $\log_{49} 7 = \frac{1}{2}$ **18.** $\log_{27} \frac{1}{3} = -\frac{1}{3}$

19. $\log_{100} 1 = 0$ **20.** $\log_{10} 0.001 = -3$

In Exercises 21–30 find the value of each expression.

21. $\log_3 9$ **22.** $\log_2 16$ **23.** $\log_4 4$

24. $\log_5 1$ **25.** $\log_3 \frac{1}{3}$ **26.** $\log_{10} 0.01$

27. $\log_3 \sqrt{3}$ **28.** $\log_9 3$ **29.** $\log_{10}(\log_5 5)$

30. $\log_2(\log_4 2)$

In Exercises 31–40 determine the value of the unknown by inspection or by writing in exponential form.

31. $\log_2 8 = y$ **32.** $\log_4 2 = y$

33. $\log_5 x = -1$ **34.** $\log_{10} x = -5$

35. $\log_b 125 = 3$ **36.** $\log_b 10 = \frac{1}{2}$

37. $\log_{10} 10^3 = y$ **38.** $\log_{10} 10^{2.4} = y$

39. $\log_b b = 1$ **40.** $\log_b 1 = 0$

In Exercises 41–43 evaluate each function for the given element of the domain.

41. $f(x) = \log_2 x$; $f(8)$; $f(0.5)$; $f(1)$

42. $g(x) = \log_9 x$; $g(3)$; $g(27)$; $g(\frac{1}{81})$

43. $h(x) = \log_{10} x$; $h(10,000)$; $h(1)$; $h(0.0000001)$

In Exercises 44–55 graph the function.

44. $y = \log_3 x$

45. $y = \log_{10} x$

46. $y = \log_{1/2} x$

47. $y = \log_{1/4} x$

48. $y = -\log_{1/4} x$

49. $y = -\log_4 x$

50. $y = \log_2(-x)$

51. $y = 1 - \log_2 x$

52. $y = 1 + \log_2 x$

53. $y = \log_2(1 + x)$

54. $y = \log_{10}|x|$

55. $y = |\log_{10} x|$

In Exercises 56–60 what is the domain of each function?

56. $f(x) = \log_{10}(2x - 1)$

57. $g(x) = \log_2(x^2 - 9)$

58. $h(x) = \log_5(-x)$

59. $f(x) = \log_{10}|x - 1|$

60. $f(x) = \log_{10} \sqrt{x + 1}$

In Exercises 61–64 determine the pH of the solution with the given hydrogen ion concentration.

61. $[H^+] = 10^{-7}$ (water)

62. $[H^+] = 4.2 \times 10^{-3}$ (5 percent vinegar)

63. $[H^+] = 2.5 \times 10^{-4}$ (orange juice)

64. $[H^+] = 3.3 \times 10^{-8}$ (swimming pool water)

In Exercises 65–68 determine the hydrogen ion concentration of the solution with the given pH.

65. pH = 0   (pure sulfuric acid)

66. pH = 8.9   (seawater)

67. pH = 2.1   (stomach juices)

68. pH = 11.3   (household ammonia)

The most widely used unit in communications engineering is the **decibel.** This name is derived from the basic unit of loudness, the bel (in honor of Alexander Graham Bell), with decibel meaning one-tenth bel. The decibel is defined as a logarithm because the ear perceives changes in sound levels logarithmically. That is, the ear is very perceptive to changes in loudness at low sound levels; at high sound levels the intensity of the sound must be greatly increased for the ear to perceive change. The formula for decibels (dB) is

$$dB = 10 \log_{10} \frac{P_2}{P_1},$$

where P_1 and P_2 represent the powers required to produce sound levels 1 and 2. Use this formula in Exercises 69–72. (*Note:* Many other important measures are defined as the logarithm of a ratio. The Richter scale, which is a measure of the magnitude of an earthquake, and the formula for the magnitude or brightness of a star are common examples.)

69. Determine the difference in loudness (in decibels) between sounds 1 and 2 if the power ratio (P_2/P_1) between the sounds is
   a. 1   **b.** 10   **c.** 100   **d.** 1,000
   As the power ratio is progressively multiplied by 10, how does the loudness in decibels increase?

70. What is the decibel gain of an amplifier with a 3-watt input and a 30-watt output?

71. In a normal conversation the power ratio (P_2/P_1) between the highest and lowest sound volumes is about 300 to 1. What is the range of speech in decibels? (*Note:* The decibel scale for loudness ranges from 0 dB for a sound at the threshold of hearing to about 140 dB for an airplane engine. Normal conversation is rated at about 60 dB.)

72. When rating an amplifier from a radio or public address system, P_1 assumes the arbitrary reference level value of 6 milliwatts (or 0.006 watt). What is the gain in decibels of a 60-watt amplifier? That is, find dB when $P_2 = 60$ watts and $P_1 = 0.006$ watt.

5.3 PROPERTIES OF LOGARITHMS

Since a logarithm is an exponent, the properties of logarithms may be derived from the following properties of exponents.

1. $b^y \cdot b^z = b^{y+z}$
2. $b^y/b^z = b^{y-z}$ $(b \neq 0)$
3. $(b^y)^z = b^{yz}$
4. $b^1 = b$
5. $b^0 = 1$ $(b \neq 0)$

The logarithm properties that correspond to each of these laws are now given below.

Properties of Logarithms   If b, x_1, x_2 are positive, with $b \neq 1$ and k any real number, then

1. $\log_b x_1 x_2 = \log_b x_1 + \log_b x_2$
2. $\log_b \dfrac{x_1}{x_2} = \log_b x_1 - \log_b x_2$
3. $\log_b (x_1)^k = k \log_b x_1$
4. $\log_b b = 1$
5. $\log_b 1 = 0$

To establish our first property, consider the product of two positive numbers x_1 and x_2. Let

$$x_1 = b^y \quad \text{or} \quad y = \log_b x_1$$

and

$$x_2 = b^z \quad \text{or} \quad z = \log_b x_2,$$

then

$$x_1 \cdot x_2 = b^y \cdot b^z = b^{y+z}.$$

Writing the last equation in logarithmic form, we have

$$\log_b x_1 x_2 = y + z,$$

so

$$\log_b x_1 x_2 = \log_b x_1 + \log_b x_2.$$

Through similar procedures we may establish properties 2 and 3 (and these proofs are requested in Exercise 45), while properties 4 and 5 are the direct result of the exponent laws $b^1 = b$ and $b^0 = 1$ and the definition of logarithm. An understanding of these properties is essential to the study of the theory and applications of logarithms, and you should consider carefully how these properties are used in the following examples.

EXAMPLE 1   Express each statement as the sum or difference of simpler logarithmic statements.

Statement   **Solution**

a. $\log_{10}(4 \cdot 5)$ $= \log_{10} 4 + \log_{10} 5$

b. $\log_2(\frac{3}{7})$ $= \log_2 3 - \log_2 7$

c. $\log_4 \sqrt{5}$ $= \log_4(5)^{1/2} = \frac{1}{2} \log_4 5$

d. $\log_{10}(xy/z)$ $= (\log_{10} x + \log_{10} y) - \log_{10} z$

e. $\log_{10} 2\pi \sqrt{L/g} = \log_{10} 2 + \log_{10} \pi + \log_{10} \sqrt{L/g}$

$\qquad\qquad = \log_{10} 2 + \log_{10} \pi + \frac{1}{2} \log_{10}(L/g)$

$\qquad\qquad = \log_{10} 2 + \log_{10} \pi + \frac{1}{2}(\log_{10} L - \log_{10} g)$ □

EXAMPLE 2   Express as a single logarithm with coefficient 1.

Statement   **Solution**

a. $\log_{10} 9 + \log_{10} 2$ $= \log_{10}(9 \cdot 2) = \log_{10} 18$

b. $\log_5 35 - \log_5 7$ $= \log_5(\frac{35}{7}) = \log_5 5$ (or 1)

c. $2 \log_4 5$ $= \log_4(5)^2 = \log_4 25$

d. $3 \log_b x + 2 \log_b y$ $= \log_b x^3 + \log_b y^2 = \log_b(x^3 \cdot y^2)$

e. $\frac{1}{2}(\log_{10} x - 3 \log_{10} y)$ $= \frac{1}{2}(\log_{10} x - \log_{10} y^3) = \frac{1}{2}(\log_{10}(x/y^3))$

$\qquad\qquad = \log_{10}(x/y^3)^{1/2} = \log_{10} \sqrt{x/y^3}$ □

EXAMPLE 3   If $\log_{10} 2 = 0.3010$ and $\log_{10} 3 = 0.4771$, simplify the expressions given.

Expression   **Solution**

a. $\log_{10} 9$ $= \log_{10} 3^2 = 2 \log_{10} 3 = 2(0.4771) = 0.9542$

b. $\log_{10} 12$ $= \log_{10}(2^2 \cdot 3) = \log_{10} 2^2 + \log_{10} 3$

$\qquad = 2 \log_{10} 2 + \log_{10} 3$

$\qquad = 2(0.3010) + 0.4771 = 1.0791$

c. $\log_{10} \sqrt{2} = \log_{10}(2)^{1/2} = \frac{1}{2} \log_{10} 2 = \frac{1}{2}(0.3010) = 0.1505$ □

There are two more important properties of logarithms that are often useful. We have been saying all along that a logarithm is an exponent. One way to express this idea is to say that

$$b^{\log_b x} = x$$

for all positive values of b and x. This property is merely a convenient way of restating the definition that $\log_b x$ is the exponent to which b is raised to obtain x. Similarly, since x is the exponent to which b is raised to obtain b^x, it follows that

$$\log_b b^x = x.$$

From a different viewpoint, these properties are a direct consequence of the inverse relation between exponential and logarithmic functions. That is, if we let $f(x) = b^x$, then $f^{-1}(x) = \log_b x$. From the definition of inverse functions in Section 3.5 we have

$$f[f^{-1}(x)] = f^{-1}[f(x)] = x.$$

Thus,

$$f[f^{-1}(x)] = b^{\log_b x} = x$$

and

$$f^{-1}[f(x)] = \log_b b^x = x.$$

The next example shows how these properties make it easy to simplify certain expressions that initially appear to be complicated.

EXAMPLE 4   Simplify.

a. $2^{\log_2 8}$ 　　　　　　　　　　　　　　　　**b.** $\log_{10} 10^p$

Solution
a. As a direct result of the above property, $2^{\log_2 8} = 8$. We can verify this answer since $\log_2 8 = 3$ so that $2^{\log_2 8} = 2^3 = 8$.
b. Since $\log_b b^x = x$, we have $\log_{10} 10^p = p$. We also could reason $\log_{10} 10^p = p \log_{10} 10 = p(1) = p$.   □

EXERCISES 5.3

In Exercises 1–18 express each logarithm as the sum or difference of simpler logarithms.

1. $\log_{10}(7 \cdot 5)$

2. $\log_b xyz$

3. $\log_6(\frac{3}{5})$

4. $\log_b(x/y)$

5. $\log_5(11)^{16}$

6. $\log_b x^{16}$

7. $\log_2 \sqrt{3}$

8. $\log_b \sqrt{x}$

9. $\log_{10} \sqrt[5]{16}$

10. $\log_b \sqrt[5]{x}$

11. $\log_4(4^2 \cdot 3^3)$

12. $\log_b x^2 y^3$

13. $\log_b \sqrt{xy}$

14. $\log_b \sqrt{x/y}$

15. $\log_b \sqrt[3]{x^5}$ **16.** $\log_b(x)^\pi$

17. $\log_b \sqrt[4]{xy^2/z^3}$

18. $\log_{10} \sqrt{s(s-a)(s-b)(s-c)}$

In Exercises 19–32 express each statement as a single logarithm with coefficient 1.

19. $\log_2 3 + \log_2 4$ **20.** $\log_b x + \log_b y$

21. $\log_4 20 - \log_4 5$ **22.** $\log_b y - \log_b x$

23. $2 \log_7 3$ **24.** $3 \log_b z$

25. $\frac{1}{2} \log_{10} 9 - 3 \log_{10} 2$ **26.** $3 \log_b x - \frac{1}{2} \log_b z$

27. $\frac{1}{3} \log_b x + \frac{2}{3} \log_b y$

28. $2 \log_b x + \log_b(x + y)$

29. $\log_b(x^2 - 1) - \log_b(x + 1)$

30. $\log_b(y - 2) + \log_b y - 2 \log_b x$

31. $\frac{1}{2}[\log_b x - (5 \log_b y - 3 \log_b z)]$

32. $\frac{1}{2}[(\log_b x - 5 \log_b y) - 3 \log_b z]$

In Exercises 33–40 simplify each expression.

33. $\log_{10} 10^5$ **34.** $\log_2(4^{1/3})$

35. $\log_{10}(m \times 10^k)$ **36.** $3^{\log_3 2}$

37. $2^{\log_2 15 - \log_2 5}$ **38.** $b^{r \log_b N}$

39. $\frac{n}{x} b^{n \log_b x}$ **40.** $b^{\log_b x + \log_b y}$

41. If $\log_{10} 2 = 0.3010$ and $\log_{10} 3 = 0.4771$, find

  a. $\log_{10} 6$ **b.** $\log_{10} 4$ **c.** $\log_{10} \sqrt{3}$

  d. $\log_{10} \frac{2}{3}$ **e.** $\log_{10}(0.5)$ **f.** $\log_{10} 5$

42. If $\log_a 2 = x$ and $\log_a 3 = y$, express each of the following in terms of x and y.

  a. $\log_a 27$ **b.** $\log_a 72$

  c. $\log_a \sqrt{\frac{1}{3}}$ **d.** $\log_a \sqrt[5]{\frac{2}{3}}$

43. Show that $\log_b(1/a) = -\log_b a$.

44. Give specific counterexamples to *disprove* each of the following statements.

  a. $\log_b xy = (\log_b x)(\log_b y)$

  b. $(\log_b x)(\log_b y) = \log_b x + \log_b y$

  c. $\log_b \frac{x}{y} = \frac{\log_b x}{\log_b y}$

  d. $\frac{\log_b x}{\log_b y} = \log_b x - \log_b y$

  e. $\log_b x^k = (\log_b x)^k$

  f. $(\log_b x)^k = k \log_b x$

45. Prove the following properties of logarithms.

  a. $\log_b \frac{x_1}{x_2} = \log_b x_1 - \log_b x_2$

  b. $\log_b(x_1)^k = k \log_b x_1$

46. What restrictions are placed on x_1, x_2, b, and k in Exercise 45?

47. If $f(x) = 10^x$ and $f^{-1}(x) = \log_{10} x$, verify that $f^{-1}[f(x)] = x$ for all x in the domain of f.

48. Examine the following line of reasoning: $3 > 2$. If we multiply both sides of the inequality by $\log_{10}(\frac{1}{2})$, we have

$$3 \log_{10}(\tfrac{1}{2}) > 2 \log_{10}(\tfrac{1}{2})$$

$$\log_{10}(\tfrac{1}{2})^3 > \log_{10}(\tfrac{1}{2})^2$$

$$\log_{10}(\tfrac{1}{8}) > \log_{10}(\tfrac{1}{4}).$$

Thus,

$$\tfrac{1}{8} > \tfrac{1}{4}.$$

Our conclusion is incorrect. What went wrong?

5.4 EXPONENTIAL AND LOGARITHMIC EQUATIONS

In Section 5.1 we solved exponential equations in which it was not hard to rewrite the expressions in terms of a common base. For instance, $9^x = 1/27$ was written as $3^{2x} = 3^{-3}$, so $x = -3/2$. Since this procedure is limited, we now consider a general approach that uses the following principle which is based on the fact that a logarithmic correspondence is a one-to-one function.

Equation-Solving Principle	If x, y, and b are positive real numbers with $b \neq 1$, then

1. $x = y$ implies $\log_b x = \log_b y$; and conversely,

2. $\log_b x = \log_b y$ implies $x = y$.

We can therefore solve **exponential equations** (the unknown is in the exponent) by taking the logarithm of both sides of the equation, as shown in the following examples.

EXAMPLE 1   Solve for x: $3^x = 8$.

Solution

$$3^x = 8$$
$$\log 3^x = \log 8 \qquad \text{Apply common logarithms to each side.}$$
$$x \log 3 = \log 8 \qquad \text{Property of logarithms}$$
$$x = \frac{\log 8}{\log 3}$$

If an approximation is desired, compute

$$8\ \boxed{\log}\ \boxed{\div}\ 3\ \boxed{\log}\ \boxed{=} \qquad 1.8927893.$$

Thus, $x \approx 1.893$, and the solution set is $\{1.893\}$ when rounded to four significant digits. □

EXAMPLE 2   Solve for x: $3^{x+2} = 5^{2x-1}$.

Solution

$$3^{x+2} = 5^{2x-1}$$
$$\log 3^{x+2} = \log 5^{2x-1} \qquad \text{Apply common logarithms to each side.}$$

$$(x + 2)\log 3 = (2x - 1)\log 5 \qquad \text{Property of logarithms}$$
$$x \log 3 + 2 \log 3 = 2x \log 5 - \log 5 \qquad \text{Distributive property}$$
$$x \log 3 - 2x \log 5 = -2 \log 3 - \log 5 \qquad \text{Equivalent equation grouping } x \text{ terms}$$

$$x(\log 3 - 2 \log 5) = -2 \log 3 - \log 5 \qquad \text{Factoring}$$

$$x = \frac{-2 \log 3 - \log 5}{\log 3 - 2 \log 5} \approx 1.795$$

To four significant digits, the solution set is $\{1.795\}$. □

The next example is not an exponential equation, but it is difficult to solve because the exponent is not a rational number. When an exponent is troublesome, a logarithmic solution is not far behind.

EXAMPLE 3   Solve for x: $x^{\sqrt{2}} = 23$.

Solution

$$x^{\sqrt{2}} = 23$$
$$\log x^{\sqrt{2}} = \log 23$$
$$\sqrt{2} \log x = \log 23$$
$$\log x = \frac{\log 23}{\sqrt{2}}$$

To solve for x, evaluate $(\log 23)/\sqrt{2}$ and then compute an inverse logarithm.

$$23 \boxed{\log} \boxed{\div} 2 \boxed{\sqrt{}} \boxed{=} \boxed{\text{INV}} \boxed{\log} \qquad 9.1809366$$

Thus, $x \approx 9.181$, so to four significant digits, the solution set is $\{9.181\}$.   □

We may change a logarithm in one base to a logarithm in another base by the above procedure. Suppose that $y = \log_b x$ and we wish to change to base a. This expression in exponential form is

$$b^y = x.$$

Taking the logarithm to the base a of both sides of the equation, we have

$$\log_a b^y = \log_a x$$
$$y \log_a b = \log_a x$$
$$y = \frac{\log_a x}{\log_a b}.$$

Since $y = \log_b x$, we have

$$\log_b x = \frac{\log_a x}{\log_a b}.$$

EXAMPLE 4   Use logarithms to the base 10 to determine

a. $\log_2 7$ **b.** $\log_5 0.043$

Solution

a. $\log_2 7 = \dfrac{\log 7}{\log 2} \approx 2.807$

b. $\log_5 0.043 = \dfrac{\log 0.043}{\log 5} \approx -1.955$ □

A **logarithmic equation** (the unknown is in the log statement) is sometimes solved by using the principle that $\log_b x = \log_b y$ implies $x = y$. For instance, if $\log(3x - 8) = \log x$, then $3x - 8 = x$, and $x = 4$. If the equation contains only one log statement, we often solve by changing to exponential form. Thus, if $\log (x + 7) = 2$, then $x + 7 = 10^2$, and $x = 93$. In some cases we obtain the single log statement by applying a property of logarithms as shown in the next example. One note of caution here—always check your answers in the original equation and accept only solutions that result in the logarithms of positive numbers.

EXAMPLE 5   Solve for x: $\log_3(x^2 - 4) - \log_3(x + 2) = 2$.

Solution

$$\log_3(x^2 - 4) - \log_3(x + 2) = 2$$
$$\log_3\left(\frac{x^2 - 4}{x + 2}\right) = 2$$
$$\log_3(x - 2) = 2$$
$$3^2 = x - 2$$
$$11 = x$$

If we substitute 11 in our original equation, both log expressions are defined. Thus, the solution set is $\{11\}$.   □

EXERCISES 5.4

In Exercises 1–20 use logarithms to solve the equation.

1. $2^x = 10$

2. $5^x = 100$

3. $4^{x+1} = 17$

4. $8^{x-1} = 25$

5. $2^{-x} = 9$

6. $3^{2x} = 5$

7. $(0.7)^t = 0.3$

8. $(1.05)^t = 3$

9. $2^x = 3^{2x+1}$

10. $3^{2x-1} = 4^{x+2}$

11. $x^{1.7} = 20$

12. $x^{2.5} = 1,000$

13. $x^{-0.6} = 12$

14. $x^{-4.1} = 47.6$

15. $x^{\sqrt{2}} = 8$

16. $(\sqrt{2})^x = 8$

17. $(1 - r)^8 = \dfrac{1}{2}$

18. $\left(1 + \dfrac{r}{2}\right)^4 = 2$

19. $6,000 = 2,000(1.03)^{2t}$

20. $1,000 = 100\left(1 + \dfrac{0.05}{4}\right)^{4t}$

In Exercises 21–40 solve the logarithmic equation.

21. $\log x = 2$

22. $\log x = 0$

23. $5 \log x = 5$

24. $2 \log x = \log 2$

25. $\log(1 - x) = -1$

26. $\log(-x) = \dfrac{1}{2}$

27. $\log(2x - 5) = \log x$

28. $\log x = \log(1 - x)$

29. $2 \log x = \log 8$

30. $3 \log x = \log(3x)$

31. $1 + \log x = \log 5$

32. $\log 2 + \log 3 = \log x$

33. $\log_4 x + \log_4 2 = 1$

34. $\log_3 x - \log_3 4 = 2$

35. $\log_2(x^2 - 1) = 3$

36. $\log_5(x + 1) - \log_5 x = 1$

37. $\log_6(x + 1) + \log_6 x = 1$

38. $\log_2(x + 1) + \log_2(x + 4) = 2$

39. $\log_b 2x = \log_b 4x - \log_b 2$

40. $\log_7(x^2 - x) = \log_7 x + \log_7(x - 1)$

In Exercises 41–48 use logarithms to the base 10 to determine each logarithm.

41. $\log_2 9$ **42.** $\log_4 31$ **43.** $\log_3 5$

44. $\log_5 3$ **45.** $\log_{1/2} 19$ **46.** $\log_{1/3} \frac{2}{3}$

47. $\log_4 0.012$ **48.** $\log_6 0.735$

49. If $\log_b x = 3$, find $\log_{1/b} x$.

50. Simplify $\log_x 5 \cdot \log_5 x$.

51. Simplify $(\log_2 10)(\log_{10} 12 - \log_{10} 3)$.

52. How long will it take for money invested at 5 percent compounded annually to double?

53. At what interest rate (compounded annually) must a sum of money be invested if it is to double in 4 years?

5.5 MORE APPLICATIONS AND THE NUMBER *e*

In Section 5.1 the applications describe situations in which no change is considered to have occurred until an entire time period has elapsed. We now wish to discuss phenomena that change continuously. To illustrate this type of change, we study more extensively the concept of compound interest.

EXAMPLE 1 $2,000 is invested at 6 percent compounded semiannually.

a. Find a formula showing the value of the investment at the end of t years.

b. How much is the investment worth after 2 years?

Solution

a. Since the amount is compounded semiannually, there are two conversions each year at a 3 percent interest rate. The $2,000 amounts to $2,000(1 + 0.06/2)$ after $\frac{1}{2}$ year; $2,000(1 + 0.06/2)^2$ after 1 year; and $2,000(1 + 0.06/2)^{2t}$ after t years. Thus,

$$A = 2,000(1.03)^{2t}.$$

b. Evaluating the function when $t = 2$, we have

$$A = 2,000(1.03)^{2(2)}$$
$$= 2,000(1.03)^4$$
$$= \$2,251.02. \qquad \square$$

We can generalize the procedure from Example 1 to obtain the following formula for the amount (including principal and interest) of an investment in terms of four items: (1) the original principal, P; (2) the rate of interest, r; (3) the number of conversions per year, n; and (4) the number of years, t.

$$A = P\left(1 + \frac{r}{n}\right)^{nt}$$

EXAMPLE 2 $100 is invested at 5 percent interest compounded quarterly. How much will it amount to in 3 years?

Solution   Substituting $P = 100$, $r = 0.05$, $n = 4$, and $t = 3$ in the formula above, we have

$$A = 100\left(1 + \frac{0.05}{4}\right)^{4(3)}$$
$$= 100(1.0125)^{12}$$
$$= \$116.08. \qquad \square$$

You should note that if the interest is computed frequently and added to the principal, the amount grows at a faster rate. Thus, it is more profitable to an investor for the conversion periods to be shorter. For example, $10,000 amounts to $10,600 when compounded annually at 6 percent; $10,609 when compounded semiannually; and $10,613.64 when compounded quarterly. An important consideration is to determine the effect of compounding more frequently, such as every day, or even every minute. Does the compounded amount become astronomical or is the growth limited?

To determine the result of continuous growth, let us examine the return on a $1 investment for 1 year at an interest rate of 100 percent. Progressively, we will compound our investment more frequently and observe the result. (*Note:* A $1 investment and a 100 percent interest rate are very unlikely, but we chose them to illustrate the situation with the easiest numbers.)

$$P = \$1 \qquad r = 100\% \quad \text{or} \quad 1 \qquad t = 1$$

Substituting in the formula

$$A = P\left(1 + \frac{r}{n}\right)^{nt},$$

we have

$$A = 1\left(1 + \frac{1}{n}\right)^{n(1)},$$

which simplifies to

$$A = \left(1 + \frac{1}{n}\right)^{n}.$$

If compounded

annually ($n = 1$):   then $A = (1 + \frac{1}{1})^1 = 2.00$

semiannually ($n = 2$):   then $A = (1 + \frac{1}{2})^2 = (1.5)^2 = 2.25$

quarterly ($n = 4$):   then $A = (1 + \frac{1}{4})^4 = (1.25)^4$
$$\approx 2.441 \ldots$$

monthly ($n = 12$):   then $A = (1 + \frac{1}{12})^{12} \approx (1.08333)^{12}$
$$\approx 2.613 \ldots$$

daily ($n = 365$):   then $A = (1 + \frac{1}{365})^{365} \approx (1.00274)^{365}$
$$\approx 2.714 \ldots$$

hourly ($n = 8{,}760$):   then $A = (1 + \frac{1}{8{,}760})^{8{,}760}$
$$\approx (1.000114)^{8{,}760}$$
$$\approx 2.718 \ldots$$

Note the small difference between compounding daily and compounding hourly. More frequent conversions result in even smaller changes in the amount, and it can be shown that as n gets larger and larger, $(1 + 1/n)^n$ gets closer and closer to an irrational number that is denoted by the letter e (to honor the Swiss mathematician Leonhard Euler). This number is extremely important in mathematics and to six significant digits

$$e \approx 2.71828.$$

Thus, a $1 investment compounded every instant (or continuously) for 1 year at an interest rate of 100 percent amounts to $2.71828 \ldots$, or e. In the next year each dollar again grows to e, so at the end of 2 years the amount is e^2. We can therefore find the amount to which the $1 has grown in t years by the formula

$$A = e^t,$$

where the base in the exponential function is the irrational number e.

This formula has to be generalized to handle more practical principals and interest rates as well as other physical situations that change in a similar manner. The generalized formula is

$$A = A_0 e^{kt},$$

where

A = amount at time t
A_0 = initial amount
k = growth or decay constant.

The important idea is that an exponential function with base e is used in a situation where a quantity is *continuously* either growing or decaying. The rate of increase or decrease is proportional to the amount present at any instant. k is positive if the quantity is increasing and negative if the quantity is decreasing.

EXAMPLE 3 $2,000 is invested at 6 percent compounded continuously. How much is the investment worth in 10 years?

Solution $A_0 = \$2,000$, $k = 0.06$, and $t = 10$. Substituting in the formula

$$A = A_0 e^{kt},$$

we have

$$A = 2,000 e^{(0.06)(10)}$$
$$= 2,000 e^{0.6}.$$

We compute $2,000 e^{0.6}$ by pressing

$$2,000 \boxed{\times} 0.6 \boxed{e^x} \boxed{=} \qquad 3,644.2376$$

although on some calculators you need the following sequence, which is explained shortly:

$$2,000 \boxed{\times} 0.6 \boxed{\text{INV}} \boxed{\ln} \boxed{=} \qquad 3,644.2376.$$

Thus, $A \approx \$3,644.24$. (*Note:* If we refer to Table 4 at the end of the book, we find that $e^{0.6} \approx 1.8221$ so that $A \approx 2,000(1.8221) \approx \$3,644.20$).   □

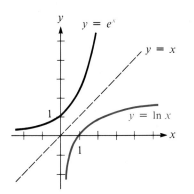

Figure 5.5

The calculator evaluation of $e^{0.6}$ in Example 3 once again forces us to consider the inverse relation between exponential and logarithmic functions. Just as 10^x and $\log x$ are inverse functions, so are e^x and $\log_e x$. Logarithms to the base e are called **natural logarithms** and $\log_e x$ is usually written as $\ln x$. As with all inverse functions, the graphs of $y = e^x$ and $y = \ln x$ are symmetric about the line $y = x$ as shown in Figure 5.5. Thus,

$$\boxed{\text{INV}} \boxed{\ln} \text{ is equivalent to } \boxed{e^x},$$

which explains the second keystroke sequence in Example 3. All the properties of logarithms and the equation-solving techniques from Section 5.4 apply to natural logarithms (or ln statements). In particular, note that the properties $\log_b b^x = x$ and $b^{\log_b x} = x$ (if $x > 0$) lead directly to

$$\ln e^x = x \text{ and } e^{\ln x} = x \text{ (if } x > 0)$$

when we use base e. Before continuing our discussion of exponential functions, it is useful to consider a few problems using these ideas.

EXAMPLE 4   Find x in decimal form.

a. $x = e^3$ **b.** $\ln x = 3$ **c.** $x = e^{\ln 3}$

d. $x = \ln e^3$ **e.** $e^x = 3$

Solution

a. Here we simply compute e^3 with one possible calculator sequence being

3 $\boxed{\text{INV}}$ $\boxed{\ln}$ 20.085537.

b. $\ln x = 3$ is equivalent to $x = e^3$, so parts **a** and **b** have the same solution.

c. Although we could use a calculator, it is not necessary. Since $e^{\ln x} = x$ (if $x > 0$), we know $e^{\ln 3} = 3$.

d. It follows from $\ln e^x = x$ that $\ln e^3 = 3$.

e. As in Section 5.4, proceed as follows:

$$e^x = 3$$
$$\ln e^x = \ln 3 \qquad \text{Apply natural logs to each side.}$$
$$x = \ln 3 \qquad \text{Property of logarithms}$$
$$x \approx 1.0986123. \qquad \text{Calculator or Table 3}$$

(*Note:* If you cannot obtain a calculator, you can evaluate natural logarithms by using Table 3 at the end of the book. Appendix A.3 discusses the method.)   □

Now that we can solve equations involving e, we continue our discussion of exponential functions.

EXAMPLE 5   An amount is invested at 7 percent compounded continuously. In how many years does the amount double?

Solution $A = 2A_0$ and $k = 0.07$. Substituting in the formula

$$A = A_0 e^{kt},$$

we have

$$2A_0 = A_0 e^{0.07t}$$
$$2 = e^{0.07t}$$
$$\ln 2 = \ln e^{0.07t} \qquad \text{Apply natural logs to each side.}$$
$$\ln 2 = 0.07t \qquad \text{Property of logarithms}$$
$$t = \frac{\ln 2}{0.07}.$$

To evaluate t, one possible calculator sequence is

$$2 \boxed{\text{ln}} \boxed{\div} 0.07 \boxed{=} \qquad 9.9021026.$$

Thus, $t \approx 9.9$ years. ☐

EXAMPLE 6 Physicists tell us that the half-life of radium is 1,620 years; that is, a given amount (A) of radium decomposes to the amount $A/2$ in 1,620 years. Radium (like all radioactive substances) decays at a rate proportional to the amount present. Thus, the formula $A = A_0 e^{kt}$ is appropriate. If a research center owns 1 g of radium, how much radium will it have in 100 years?

Solution We know that $A_0 = 1$ g; thus, our formula

$$A = A_0 e^{kt}$$

becomes

$$A = 1e^{kt}.$$

We can find k since we know that $A = 0.5$ when $t = 1,620$.

$$0.5 = e^{k(1,620)}$$
$$\ln 0.5 = \ln e^{1,620k} \qquad \text{Apply natural logs to each side.}$$
$$\ln 0.5 = 1,620k \qquad \text{Property of logarithms}$$
$$k = \frac{\ln 0.5}{1,620} \approx -0.00043$$

The formula is then

$$A = e^{-0.00043t}.$$

Substituting $t = 100$, we have

$$A = e^{-0.00043(100)}$$
$$= e^{-0.043}$$
$$A \approx 0.96 \text{ g.} \qquad ☐$$

EXAMPLE 7 If there is no restriction on food and living space, the rate of growth of a population of living organisms is proportional to the size of the population. Thus, the formula $A = A_0 e^{kt}$ is appropriate. Assume that in the absence of hunters, the deer population in New York State triples every 11 years.

a. Find the size of a herd of 500 deer after 5 years.
b. In how many years will the herd number 5,000?

Solution

a. We know that $A_0 = 500$. Thus, our formula

$$A = A_0 e^{kt}$$

becomes

$$A = 500 e^{kt}.$$

We can find k since we know that $A = 1,500$ when $t = 11$.

$$
\begin{aligned}
1,500 &= 500 e^{k(11)} \\
3 &= e^{11k} \\
\ln 3 &= \ln e^{11k} \\
\ln 3 &= 11k \\
k &= \frac{\ln 3}{11} \approx 0.10
\end{aligned}
$$

The formula is then

$$A = 500 e^{0.1t}.$$

Substituting $t = 5$, we have

$$
\begin{aligned}
A &= 500 e^{0.1(5)} \\
&= 500 e^{0.5} \\
&= 500(1.6487) \\
A &\approx 824 \text{ deer.}
\end{aligned}
$$

b. Substituting $A = 5,000$ in our formula, we have

$$
\begin{aligned}
5,000 &= 500 e^{0.1t} \\
10 &= e^{0.1t} \\
\ln 10 &= \ln e^{0.1t} \\
\ln 10 &= 0.1t \\
t &= \frac{\ln 10}{0.1} \approx 23 \text{ years.}
\end{aligned}
$$

□

EXAMPLE 8   Newton's law of cooling states that when a warm body is placed in colder surroundings at temperature t_α, the temperature (T) of the body at time t is given by

$$T - t_\alpha = D_0 e^{kt},$$

where D_0 is the initial difference in temperature and k is a constant. Because of an ice storm, there is a loss of power in a home heated to 68° F. If the outside temperature is 28° F and the temperature in the house drops from 68 to 64° F in 1 hour, when will the temperature in the house be down to 50° F?

Solution   We know that $t_\alpha = 28$ and $D_0 = 68 - 28 = 40$. Thus, our formula becomes

$$T - 28 = 40e^{kt}.$$

We can find k since we know that $T = 64$ when $t = 1$.

$$64 - 28 = 40e^{k(1)}$$
$$0.9 = e^k$$
$$\ln 0.9 = \ln e^k$$
$$\ln 0.9 = k$$
$$k \approx -0.105$$

(*Note:* k depends on the insulation in the house.) The formula is then

$$T - 28 = 40e^{-0.105t}.$$

Substituting $T = 50$, we have

$$50 - 28 = 40e^{-0.105t}$$
$$0.55 = e^{-0.105t}$$
$$\ln 0.55 = \ln e^{-0.105t}$$
$$\ln 0.55 = -0.105t$$
$$t = \frac{\ln 0.55}{-0.105}$$
$$t \approx 5.7 \text{ hours.} \qquad \square$$

In this section we discuss in detail only a few of the basic applications of the number e. However, keep in mind that the scope and power of the number e is immense. For example, consider the following relationships.

1. When the power is turned off, the electric current in a circuit does not vanish instantly but rather decreases exponentially.
2. The intensity of sunlight decreases exponentially as a diver descends further into ocean depths.
3. When a drug is injected into the bloodstream, the drug drains out of the body exponentially. Analyzing this relationship is important in medicine for it helps the doctor plan a series of injections that will maintain the prescribed level of some drug (such as an antibiotic) in a patient's bloodstream.
4. Atmospheric pressure decreases exponentially as altitude above sea level increases.

The list could easily be continued. In calculus you will probably consider some of these applications.

EXERCISES 5.5

In Exercises 1–20 write the indicated letter in decimal form or solve for the indicated letter.

1. $e^{0.16} = a;$ for a

2. $e^{-0.35} = a;$ for a

3. $e^a = 51;$ for a

4. $e^a = 0.8906;$ for a

5. $\ln x = -5.2;$ for x

6. $\ln x = 2.18;$ for x

7. $\ln e^x = 5;$ for x

8. $\ln e^{4x} = 12;$ for x

9. $e^{\ln 2} = k;$ for k

10. $e^{2 \ln 3} = b;$ for b

11. $10^x = e;$ for x

12. $10^x = e^2;$ for x

13. $\ln x = 2 \ln 3;$ for x

14. $\ln x = 2 \log 3;$ for x

15. $e^{2k} = 2;$ for k

16. $e^{-k} = 0.25;$ for k

17. $e^{-0.11t} = 0.45;$ for t

18. $e^{5,600k} = 0.5;$ for k

19. $e^{kt} = 1/2;$ for t

20. $y = ce^{kt};$ for t

21. Find the compounded amount for the following investments.
   a. $1,000 compounded semiannually for 3 years at 6 percent
   b. $5,000 compounded quarterly for 2 years at 12 percent
   c. $3,000 compounded monthly for 1 year at 5.4 percent

22. An amount is invested at 6 percent compounded monthly. In how many years does the amount triple?

23. How long will it take for $20,000 to grow to $30,000 at 10 percent compounded quarterly?

24. You are going to invest $10,000 for 3 years. Which offer is best?
   a. 10 percent compounded monthly
   b. 10.7 percent compounded annually
   c. 9.8 percent compounded continuously

25. Find the compounded amount for the following investments.
   a. $1,000 compounded continuously for 3 years at 6 percent
   b. $30,000 compounded continuously for 5 years at 13 percent
   c. $10,000 compounded continuously for 40 years at 8 percent
   d. $200,000 compounded continuously for 9 months at 10.4 percent

26. An amount is invested at 6 percent compounded continuously. In how many years does the amount triple?

27. At what rate of interest compounded continuously must money be deposited if the amount is to double in 7 years?

28. How much would a person have to invest today at 6 percent compounded continuously if he wants to have $1,000 1 year from today? (*Note:* An important consideration of an economist is how much a dollar x years from now is worth today.)

29. Assume that in a specific culture the number of bacteria (A) present after t hours is given by $A = 1,000e^{0.15t}$.
   a. How many bacteria are present after $\frac{1}{2}$ hour?
   b. How long will it take to have 40,000 bacteria?

30. The number of bacteria in a culture at the start of an experiment is about 10^5. Four hours later the population has increased to about 10^7.
   a. Approximate the size of the population after 6 hours.
   b. When was the number of bacteria 250,000?

31. In 1960 the population of a certain town was 1,000, and in 1970 it was 4,000. What population can the city planning commission expect in 1994 if this growth rate continues?

32. The amount (A) of a certain radioactive element remaining after t minutes is given by $A = A_0e^{-0.03t}$. How much of the element would remain after 1 hour if 50 g were present initially?

33. Physicists tell us that the half-life of an isotope of strontium is 25 years. If a research center owns 10 g of this isotope, how much will it have in 10 years?

34. A radioactive substance decays from 10 g to 6 g in 5 days. Find the half-life of the substance.

35. Specimens in geology and archaeology are dated by considering the exponential decay of a particular radioactive element found in the specimen. This technique is very reliable because radioactive disintegration is not affected by conditions such as pressure and temperature, which change the rate of ordinary chemical reactions. For dating to an age of about 60,000 years, the researcher considers the concentration of carbon 14, an isotope of carbon. In the cells of all living plants and animals, there is a fixed ratio of carbon 14 to ordinary stable carbon. However, when the plant or animal dies, the carbon 14 decreases according to the law of exponential decay.

a. The half-life of carbon 14 is about 5,600 years. Determine k, the decay constant, for this element.

b. What percentage of the carbon 14 is left in a specimen of bone that is 2,000 years old?

(*Note:* The radiocarbon dating method was developed in 1947 by Dr. Willard Libby, who received the Nobel prize in chemistry for his discovery. For specimens older than 60,000 years, too much carbon 14 has disintegrated for an accurate measurement. In such cases the researcher considers the concentration of other radioactive substances. For example, long periods of geologic time are frequently determined by measuring the presence of uranium 238, which has a half-life of about 4.5 billion years. Findings based on these reliable radioactive "clocks" are upsetting many of the traditional theories about the development of man, which were based on previous dating techniques.)

36. The temperature of a six-pack of beer (bought from a distributor on a hot summer day) is 90° F. The beer is placed in a refrigerator with a constant temperature of 40° F. If the beer cools to 60° F in 1 hour, when will the beer reach the more thirst-quenching temperature of 45° F?

37. A coroner examines the body of a murder victim at 9 A.M. and determines its temperature to be 88° F. An hour later the body temperature is down to 84° F. If the temperature of the room in which the body was found is 68° F and if the victim's body temperature was 98° F when she died, approximate the time of the murder.

38. The following formula (derived in calculus) is used to compute the values of e^x in Table 4 at the end of the book.

$$e^x = 1 + x + \frac{x^2}{2!} + \frac{x^3}{3!} + \cdots + \frac{x^n}{n!} + \cdots$$

where $n! = 1 \cdot 2 \cdot 3 \cdots n$ (for example, $4! = 1 \cdot 2 \cdot 3 \cdot 4$).

a. Use the first four terms of this formula to approximate $e^{0.2}$. Compare your result with the value in Table 4.

b. Use the first five terms of this formula to approximate e to the nearest hundredth.

In Exercises 39–41 express each logarithm as the sum or difference of simpler logarithms.

39. $\ln(1/x)$ **40.** $\ln\sqrt{x^2 + 1}$ **41.** $\ln\dfrac{x(x - 1)}{x + 2}$

42. Show that if $y = \dfrac{r}{1 + ce^{-at}}$, then $t = \dfrac{1}{a} \ln\dfrac{cy}{r - y}$.

CHAPTER OVERVIEW

Section	Key Concepts to Review
5.1	• Definition of the exponential function with base b
	• For $b > 0$, $b \neq 1$, and any real number x, if x lies between the rational numbers r and s, then b^x lies between b^r and b^s.
	• All previous laws of exponents hold for real number exponents.
	• For $f(x) = b^x$ with $b > 0$, $b \neq 1$: Domain: $(-\infty,\infty)$ Horizontal asymptote: x-axis Range: $(0,\infty)$ y-intercept: $(0,1)$
	• If $b > 0$, $b \neq 1$, then $b^x = b^y$ implies $x = y$.
	• Methods from Section 3.3 to graph variations of $f(x) = b^x$

Section	Key Concepts to Review

5.2	• Definitions of a logarithm, a common logarithm, and the logarithmic function with base b
	• $\log_b N = L$ is equivalent to $b^L = N$.
	• For $b > 0$, $b \neq 1$, and $x > 0$,
	$\quad y = \log_b x$ if and only if $x = b^y$
	and for $f(x) = \log_b x$:
	Domain: $(0,\infty)$ Vertical asymptote: y-axis
	Range: $(-\infty,\infty)$ x-intercept: $(1,0)$
	• The logarithmic function $y = \log_b x$ and the exponential function $y = b^x$ are inverse functions. Consequently,
	  a. The functions interchange their domain and range.
	  b. The graphs of the functions are symmetric about the line $y = x$.
	  c. Calculator evaluations of exponential and log expressions often involve the $\boxed{\text{INV}}$ key.
	• Methods from Section 3.3 to graph variations of $f(x) = \log_b x$

5.3	• Properties of logarithms (for b, x, $y > 0$, $b \neq 1$, and k any real number): **1.** $\log_b xy = \log_b x + \log_b y$ **2.** $\log_b(x/y) = \log_b x - \log_b y$ **3.** $\log_b x^k = k \log_b x$ **4.** $\log_b b = 1$ **5.** $\log_b 1 = 0$ **6.** $\log_b b^x = x$ **7.** $b^{\log_b x} = x$
5.4	• If x, y, $b > 0$, with $b \neq 1$, then $\quad x = y$ implies $\log_b x = \log_b y$; and $\log_b x = \log_b y$ implies $x = y$. • Change of base formula: $\log_b x = \dfrac{\log_a x}{\log_a b}$ • Methods to solve exponential and logarithmic equations
5.5	• Definition of a natural logarithm • As n gets larger and larger, $(1 + 1/n)^n$ gets closer and closer to an irrational number that is denoted by the letter e. To six significant digits, $e \approx 2.71828$. • Compound interest formula: $A = P\left(1 + \dfrac{r}{n}\right)^{nt}$ • Continuous growth or decay formula: $A = A_0 e^{kt}$ • $y = \ln x$ and $y = e^x$ are inverse functions. • Graphs of $y = \ln x$ and $y = e^x$ • $\ln e^x = x$ and $e^{\ln x} = x$ (if $x > 0$)

CHAPTER REVIEW EXERCISES

In Exercises 1–10 solve each equation for x. Use a calculator or tables where necessary.

1. $\log_5 5^7 = x$

2. $2^{-x} = 3$

3. $\ln 8 = \ln x - \ln 4$

4. $\log_{10} 0.123 = x$

5. $x^{0.1} = 5.67$

6. $x = e^{\ln a}$

7. $\log_{10} x = -2.4157$

8. $5^{x-3} = 1$

9. $3^{\log_x 4} = 4$

10. $\log_{10}(3x - 2) - \log_{10}(x + 4) = 1$

In Exercises 11 and 12 express the sum or difference as a single logarithm with coefficient 1.

11. $2 \ln x + \frac{1}{3} \ln y$

12. $\log_a(x + h) - \log_a x$

In Exercises 13 and 14 express each as the sum or difference of simpler logarithms.

13. $\log_b \sqrt{x^3/y}$

14. $\ln \sqrt[3]{xy^2}$

15. What is the domain of $f(x) = \ln(3x - 7)$?

16. Express in logarithmic form $4^{-1} = \frac{1}{4}$.

17. Express in exponential form $\log_2 8 = 3$.

18. Simplify $\dfrac{n}{x} e^{n \ln x}$.

19. Sketch on the same axes the graphs of $f(x) = (\frac{1}{2})^x$ and $g(x) = \log_{1/2} x$. What is the domain and range of each function?

20. True or false: if false, give a specific counterexample.

$$\frac{\log_b x}{\log_b y} = \log_b x - \log_b y$$

21. If $y = (\frac{1}{4})^x$, fill in the missing component of each of the following ordered pairs:

$(3,\ ),\ (\ ,4),\ (\ ,1),\ (\frac{3}{2},\ ),\ (-2,\ )$.

22. If \$500 is invested at 6 percent compounded annually, find a formula for the value (y) of the investment at the end of t years.

23. \$3,000 is invested at 5 percent compounded continuously. How much is the investment worth in 4 years?

24. A radioactive substance decays from 10 g to 8 g in 11 days. Find the half-life of the substance.

25. If $f = \{(3,-1), (4,-2)\}$, find f^{-1}. Determine the domain and range of each function.

26. If $\ln p^3 = b$, find $\ln p\sqrt{p}$ in terms of b.

27. If $\log_{10} M = K$, find 100^K in terms of M.

28. If $\ln 4 = 1.386$, solve $e^{3x} = 4$ for x.

29. Use $\log_{10} 2 = x$ and $\log_{10} 3 = y$ to find $\log_{10} \sqrt[3]{6}$ in terms of x and y.

30. Use the table of natural logarithms to determine $\ln 13.2$.

31. If $f(x) = \log_4 x$, find $f(1)$.

32. If $g(x) = -\ln x$, find (a) $g^{-1}(x)$; (b) $g^{-1}(-1)$.

33. Compute by means of logarithms $\sqrt{78.4}$. (See Appendix A.3.)

34. Use logarithms to base 10 to evaluate $\log_4 9$.

35. Show that if $y = r + ce^{-kt}$, then $t = \dfrac{1}{k} \ln \dfrac{c}{y - r}$.

36. Graph $y = 10 - 10^{-x}$.

37. Graph $y = -\log_3(x - 2)$.

38. A common formula for pH is $pH = \log_{10} \dfrac{1}{[H^+]}$. Show that this formula is equivalent to the one given in this chapter. That is, show $\log_{10} \dfrac{1}{[H^+]} = -\log_{10}[H^+]$.

39. Liquid X has 10 times the concentration of hydrogen ions as liquid Y. What is the relation of the pH of X to the pH of Y?

40. If $\log_{10} 9! = x$, find $\log_{10} 10!$.

In Exercises 41 and 42 simplify each expression algebraically.

41. $\left(\dfrac{e^x + e^{-x}}{2}\right)^2 - \left(\dfrac{e^x - e^{-x}}{2}\right)^2$

42. $\left(\dfrac{e^2 - 1}{4e} + \dfrac{e^2 + 1}{2e} - e\right)$ $- \left(\dfrac{e^2 - 1}{4e} - \dfrac{e^2 + 1}{2e} - \dfrac{1}{e}\right)$

43. Evaluate $\log_9 3 - \frac{1}{2} \log_5 1 + 2 \log_3 27$.

44. Evaluate $\log_2 36 + \log_2 \frac{4}{9}$.

45. Solve $50 = 35 + 40e^{10k}$ for k by using natural logarithms.

In Exercises 46–53 select the choice that completes the statement.

46. The graph of $f(x) = \log_3 x$ lies in quadrants

    a. 1 and 2   **b.** 1 and 3

    c. 1 and 4   **d.** 2 and 4

47. If $\log_{10} x^2 = a$, then $\log_{10} 10x$ equals

    a. $(a + 2)/2$ **b.** $10a$ **c.** $a + 1$ **d.** a^2

48. An expression equivalent to $(\ln 27)/3$ is

    a. $\ln 9$ **b.** $-\ln 27$ **c.** $\ln 3$ **d.** $\ln 24$

49. An expression equivalent to $\dfrac{\log_{10} x}{\log_5 x}$ is

    a. $\log_2 x$ **b.** 2 **c.** $\log_5 10$ **d.** $\log_{10} 5$

50. If $4^x = \sqrt{8}$, then x equals

    a. $\frac{3}{4}$ **b.** $-\frac{2}{3}$ **c.** $\frac{3}{2}$ **d.** $\frac{1}{3}$

51. If $x^a = y^{a+1}$, and x and y are positive numbers other than 1, then a equals

    a. $\dfrac{\log x}{\log x - \log y}$ **b.** $\dfrac{\log y}{\log x - \log y}$

    c. $\dfrac{\log y - \log x}{\log x}$ **d.** $\dfrac{\log x + \log y}{\log y}$

52. If $f(x) = \ln x$, then $f(ab)$ equals

    a. $f(a) \cdot f(b)$ **b.** $f(a + b)$

    c. $f(a) + f(b)$ **d.** $f(a) - f(b)$

53. If $A/2 = Ae^{-kt}$, then k equals

    a. $\dfrac{\ln \frac{1}{2}}{t}$ **b.** $t \cdot \ln \frac{1}{2}$ **c.** $\ln \dfrac{2}{t}$ **d.** $\dfrac{\ln 2}{t}$

6 *The Trigonometric Functions*

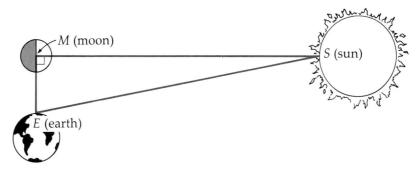

The first major attempt to measure the relative distances of the sun and the moon from the earth was made in about 280 B.C. by the Alexandrian astronomer Aristarchus. Basically he reasoned that the moon has no light of its own since we do not always see a "full moon." Therefore, the moon, like the earth, must receive its light from the sun and then reflect this light toward earth. As illustrated in the figure, he also correctly reasoned that at a quarter phase of the moon, when it is half light and half dark, angle *EMS* is a right angle. Using primitive instruments, Aristarchus measured angle *SEM* to be 29/30 of a right angle (or 87°). About how many times more distant is the sun than the moon in Aristarchus's estimate? (See Example 12 of Section 6.1.)

I n this chapter and in Chapter 7 we turn to trigonometry. This topic can be approached from different viewpoints, and in this chapter we provide three definitions of the trigonometric functions. These definitions are presented in order of abstraction and sophistication. In this section we begin with the rather concrete study of right triangle trigonometry, and we follow this in Section 6.2 with a general angle approach. Finally, in Section 6.4 we discuss the modern concept of the trigonometric functions of real numbers that is the approach most useful in calculus. In the end you must merge these definitions for a thorough understanding of trigonometry.

Historically, trigonometry involved the study of triangles for the purpose of measuring angles and distances in astronomy (as illustrated above). In early times this science was the primary concern of scholars who sought to understand God's design of the universe aesthetically and, practically, to obtain a more accurate system of navigation. As is often the case in mathematics, the ideas developed in this study later proved useful in analyzing a wide variety of situations.

6.1 TRIGONOMETRIC FUNCTIONS OF ACUTE ANGLES

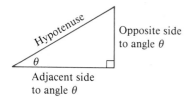

Figure 6.1

To define the trigonometric functions with respect to a triangle, we start with a right triangle. A **right triangle** is a triangle that contains a 90 degree (or right) angle. The two other angles in the triangle are **acute angles,** which by definition measure between 0° and 90°. The side opposite the right angle is called the **hypotenuse.** In a right triangle six different ratios can be formed using the measures of the three sides of the triangle, and these ratios are related to the measure of an acute angle in the following definitions.

Definition of the Trigonometric Functions

If θ (theta) is an acute angle in a right triangle as shown in Figure 6.1, then

Name of Function	Abbreviation	Ratio	
sine of angle θ	$\sin \theta$	$\dfrac{\text{opposite}}{\text{hypotenuse}}$	reciprocal functions
cosecant of angle θ	$\csc \theta$	$\dfrac{\text{hypotenuse}}{\text{opposite}}$	
cosine of angle θ	$\cos \theta$	$\dfrac{\text{adjacent}}{\text{hypotenuse}}$	reciprocal functions
secant of angle θ	$\sec \theta$	$\dfrac{\text{hypotenuse}}{\text{adjacent}}$	
tangent of angle θ	$\tan \theta$	$\dfrac{\text{opposite}}{\text{adjacent}}$	reciprocal functions
cotangent of angle θ	$\cot \theta$	$\dfrac{\text{adjacent}}{\text{opposite}}$	

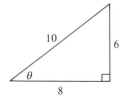

Figure 6.2

EXAMPLE 1   Find the values of the six trigonometric functions of angle θ in Figure 6.2.

Solution   The length of the side opposite angle θ is 6, the adjacent side length is 8, and the length of the hypotenuse is 10. Substituting these numbers in the above definitions gives

$$\sin\theta = \frac{\text{opp}}{\text{hyp}} = \frac{6}{10} = \frac{3}{5} \;\leftarrow \text{reciprocals} \rightarrow\; \csc\theta = \frac{\text{hyp}}{\text{opp}} = \frac{10}{6} = \frac{5}{3}$$

$$\cos\theta = \frac{\text{adj}}{\text{hyp}} = \frac{8}{10} = \frac{4}{5} \;\leftarrow \text{reciprocals} \rightarrow\; \sec\theta = \frac{\text{hyp}}{\text{adj}} = \frac{10}{8} = \frac{5}{4}$$

$$\tan\theta = \frac{\text{opp}}{\text{adj}} = \frac{6}{8} = \frac{3}{4} \;\leftarrow \text{reciprocals} \rightarrow\; \cot\theta = \frac{\text{adj}}{\text{opp}} = \frac{8}{6} = \frac{4}{3}.$$

Note that if we say $\tan\theta = \frac{3}{4}$, we do not necessarily mean that the opposite side length is 3 and the adjacent side length is 4.   ☐

EXAMPLE 2   Find the values of the remaining trigonometric functions of acute angle θ if $\tan\theta = \frac{2}{5}$.

Solution   First, draw a right triangle as in Figure 6.3 and label one acute angle θ. Since $\tan\theta = \frac{2}{5}$, the ratio of the opposite side length to the adjacent side length is 2:5. Although many choices are possible, it is easiest to label the opposite side 2 and the adjacent side 5. The hypotenuse in the triangle is found by the Pythagorean relationship

$$(\text{hypotenuse})^2 = 2^2 + 5^2 = 29$$
$$\text{hypotenuse} = \sqrt{29}.$$

Figure 6.3

Then

$$\sin\theta = \frac{2}{\sqrt{29}} \;\leftarrow \text{reciprocals} \rightarrow\; \csc\theta = \frac{\sqrt{29}}{2}$$

$$\cos\theta = \frac{5}{\sqrt{29}} \;\leftarrow \text{reciprocals} \rightarrow\; \sec\theta = \frac{\sqrt{29}}{5}$$

$$\cot\theta = \frac{5}{2}.$$

Note that $\cot\theta = \frac{5}{2}$ can be determined from its reciprocal relation to $\tan\theta = \frac{2}{5}$ without constructing the triangle.   ☐

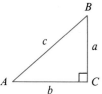

Figure 6.4

It is common notation to label a right triangle as in Figure 6.4. Capital letters such as *A*, *B*, and *C* denote angles, while the lengths of the sides opposite these angles are labeled with the corresponding lowercase letters *a*, *b*, and *c*. The next example illustrates this notation.

EXAMPLE 3   In right triangle *ABC* with $C = 90°$, if $\cos B = \frac{5}{13}$ and $c = 39$, find *a*.

Solution   From Figure 6.4 we have

$$\cos B = \frac{\text{adj}}{\text{hyp}} = \frac{a}{c} = \frac{5}{13}.$$

Substituting $c = 39$ gives

$$\frac{a}{39} = \frac{5}{13}.$$

Then

$$a = 39\left(\tfrac{5}{13}\right)$$
$$= 15. \qquad \square$$

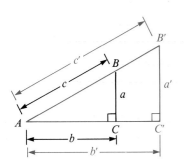

Figure 6.5

To demonstrate why the trigonometric relationships define functions, consider Figure 6.5. Right triangles *ABC* and *AB′C′* contain the same angle measures. Two triangles with the same angle measures are called **similar triangles,** and one of the properties of similar triangles is that the lengths of corresponding sides are proportional. Therefore,

$$\frac{a}{c} = \frac{a'}{c'}, \qquad \frac{b}{c} = \frac{b'}{c'}, \qquad \frac{a}{b} = \frac{a'}{b'}, \qquad \text{and so on.}$$

This means the six trigonometric ratios depend only on angle *A* and not on the size of the right triangle that contains *A*. Since there is exactly one number associated with a particular trigonometric ratio of angle *A*, the term *function* applies.

So far we have determined the values of the trigonometric functions by applying their definitions to triangles whose side measures were known. A common problem is determining these values if we know the measure of angle *θ*. First, let us consider the special angles whose trigonometric values are known exactly. Figure 6.6 shows the acute angles 30°, 45°, and 60° contained in two right triangles. The diagonal in a square with a side length of 1 unit forms the

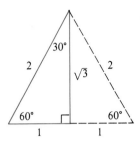

Figure 6.6

right triangle with a 45° angle; the altitude in an equilateral triangle with a side length of 2 units forms the right triangle with angles of 30° and 60°. Using these two triangles, we may determine the value of any of their trigonometric functions. The results are tabulated in the following table.

θ	$\sin \theta$	$\csc \theta$	$\cos \theta$	$\sec \theta$	$\tan \theta$	$\cot \theta$
30°	$\dfrac{1}{2}$	2	$\dfrac{\sqrt{3}}{2}$	$\dfrac{2}{\sqrt{3}}$	$\dfrac{1}{\sqrt{3}}$	$\sqrt{3}$
60°	$\dfrac{\sqrt{3}}{2}$	$\dfrac{2}{\sqrt{3}}$	$\dfrac{1}{2}$	2	$\sqrt{3}$	$\dfrac{1}{\sqrt{3}}$
45°	$\dfrac{1}{\sqrt{2}}$	$\sqrt{2}$	$\dfrac{1}{\sqrt{2}}$	$\sqrt{2}$	1	1

EXAMPLE 4   If the length of the short leg of a 30–60–90 triangle is 5, determine the lengths of the other two sides in the triangle.

Solution   The side lengths in a 30–60–90 triangle are in the ratio of $1:\sqrt{3}:2$. If the length of the short leg is 5, then the longer leg length is $5\sqrt{3}$ and the length of the hypotenuse is 10.   □

Note in the table of exact values for special angles the similarity between the values of the trigonometric functions for 30° and 60°. For example:

$$\sin 30° = \frac{1}{2} \quad \text{and} \quad \cos 60° = \frac{1}{2}$$

$$\tan 30° = \frac{1}{\sqrt{3}} \quad \text{and} \quad \cot 60° = \frac{1}{\sqrt{3}}$$

$$\sec 30° = \frac{2}{\sqrt{3}} \quad \text{and} \quad \csc 60° = \frac{2}{\sqrt{3}}.$$

This similarity results from the fact that in the right triangle, the side opposite the 30° angle is adjacent to the 60° angle. Thus,

$$\sin 30° = \frac{\text{side opposite 30° angle}}{\text{hypotenuse}} = \frac{1}{2}$$

$$= \frac{\text{side adjacent to 60° angle}}{\text{hypotenuse}} = \cos 60°$$

$$\tan 30° = \frac{\text{side opposite 30° angle}}{\text{side adjacent to 30° angle}} = \frac{1}{\sqrt{3}}$$

$$= \frac{\text{side adjacent to 60° angle}}{\text{side opposite 60° angle}} = \cot 60°$$

$$\sec 30° = \frac{\text{hypotenuse}}{\text{side adjacent to 30° angle}} = \frac{2}{\sqrt{3}}$$

$$= \frac{\text{hypotenuse}}{\text{side opposite 60° angle}} = \csc 60°.$$

Observe that in each case a trigonometric function of 30° is equal to the corresponding cofunction of 60°. The corresponding cofunction is easy to remember since the *co*function of the sine is the *co*sine, the cofunction of the tangent is the *co*tangent, and the cofunction of the secant is the *co*secant. We can generalize from these examples concerning 30° and 60° to any two angles *A* and *B* that are **complementary** (that is, *A* + *B* = 90°), since in any right triangle the side opposite angle *A* is adjacent to angle *B*. Thus, a trigonometric function of any acute angle is equal to the corresponding cofunction of the complementary angle.

EXAMPLE 5   cot 15° = tan 75° since 15° and 75° are complementary angles (that is, 15° + 75° = 90°) and the cofunction of the cotangent is the tangent.   □

Except for a few special angles such as 30°, 45°, and 60°, we evaluate trigonometric functions for acute angles by a calculator or a table. First consider Table 5 at the end of the book. Because of the relationship just mentioned between cofunctions of complementary angles, Table 5 lists angles from 0° to 45° on the left side of the page and their complementary angles from 45° to 90° on the right side. Functions listed at the top and bottom of each column are cofunctions. To use the table to find the approximate value of a particular trigonometric function for an acute angle, do the following:

To Evaluate Trigonometric Functions Using Table 5

1. Locate the row containing the angle measure.
2. Read at the intersection of this row and the column corresponding to the particular trigonometric function. The function names are taken from the top row if the angle is between 0° and 45°. Function names in the bottom row apply to angles between 45° and 90°.

Table 5 lists angle measures in 10 minute increments where one **minute** (written 1′) equals $\frac{1}{60}$ of one degree. One degree may also be subdivided by using decimal degrees, which is a unit more convenient for evaluation with a calculator. Table 5 values are usually accurate to four significant digits, and greater accuracy may be obtained with a calculator.

EXAMPLE 6   Use Table 5 to approximate sin 22° and cos 68°.

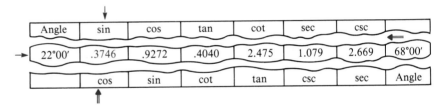

Figure 6.7

Solution   Consider the excerpt from Table 5 in Figure 6.7. Thus,

sin 22° ≈ 0.3746
cos 68° ≈ 0.3746.   □

EXAMPLE 7   Use Table 5 to approximate cot 4°, cos 17° 10′, tan 72° 50′, sec 89° 50′, and csc 0°.

Solution   From Table 5 we read the values

cot 4° ≈ 14.30
cos 17° 10′ ≈ 0.9555
tan 72° 50′ ≈ 3.237
sec 89° 50′ ≈ 343.8
csc 0° is undefined.   □

A scientific calculator provides the most direct method for evaluating trigonometric functions, but you must be careful and understand a few relationships in trigonometry. You need to be careful because calculators usually work with three units of angle measure—degrees, radians, and grads. Radian measure will be discussed in Section 6.3, while grad measure is used in road construction and will not be considered. So first consult the owner's manual to your calculator and learn the procedure for operating in degree mode. Choosing the correct angular mode is easy to do but is a common source of error because it's easy to forget. Most calculators are in the degree mode when first turned on, so you may have no problem here. Now you need only enter the angle and press $\boxed{\text{sin}}$, $\boxed{\text{cos}}$, or $\boxed{\text{tan}}$. For example,

Table 5	Calculator
$\sin 22° \approx 0.3746$	22 $\boxed{\text{sin}}$ 0.3746066

Many calculators do not have a system for entering angle measure in minutes but instead express angles in decimal degrees. If the problem is stated in degrees-minutes, you can change to decimal degrees by dividing the minutes by 60. Thus,

Table 5	Calculator
$\cos 17°10' \approx 0.9555$	17 $\boxed{+}$ 10 $\boxed{÷}$ 60 $\boxed{=}$ $\boxed{\text{cos}}$ 0.9554502

Cotangent, secant, and cosecant are usually computed as reciprocal functions of tangent, cosine, and sine by using the $\boxed{1/x}$ key as follows:

Table 5	Calculator
$\cot 4° \approx 14.30$	4 $\boxed{\text{tan}}$ $\boxed{1/x}$ 14.300666
$\sec 89°50' \approx 343.8$	89 $\boxed{+}$ 50 $\boxed{÷}$ 60 $\boxed{=}$ $\boxed{\text{cos}}$ $\boxed{1/x}$ 343.77516
$\csc 0°$ is undefined	0 $\boxed{\text{sin}}$ $\boxed{1/x}$ Error

To find an angle when one of the trigonometric functions is known, we reverse the assignments of the trig functions and press $\boxed{\text{INV}}$ $\boxed{\text{sin}}$ (or $\boxed{\text{sin}^{-1}}$ or $\boxed{\text{Arcsin}}$) and so on. This sequence calculates the acute angle whose trig function is the displayed positive value.

Table 5	Calculator
$\sin 22° \approx 0.3746$	0.3746 $\boxed{\text{INV}}$ $\boxed{\text{sin}}$ 21.999593

Once again, cotangent, secant, and cosecant problems are solved by using the $\boxed{1/x}$ key in conjunction with the corresponding reciprocal function. For example, the following sequence is used because $\cot \theta = 14.30$ is equivalent to $\tan \theta = 1/14.30$.

Table 5	Calculator
$\cot 4° \approx 14.30$	$14.30 \boxed{1/x} \boxed{\text{INV}} \boxed{\tan} 4.0001858$

The sine or cosine of an angle cannot be greater than 1; correspondingly, their reciprocals, cosecant and secant, cannot be positive numbers less than 1. An error message appears in the display when you attempt to solve such problems. The tangent and cotangent functions range over all real numbers.

A calculator is usually accurate to at least eight decimal places in contrast to four for Table 5. However, in most cases the extra digits are really "overkill." In the exercises and in the remainder of the chapter we will usually round off answers for trigonometric values to four significant digits and angle measures to the nearest tenth of a degree.

Right Triangle Applications

To solve a right triangle means to find the measures of the two acute angles and the lengths of the three sides of the triangle. To accomplish this, at least two of these five values must be known and one or more must be a side length. In this section we follow the standard practice of simplifying the notation by always labeling the angles of the triangle as A, B, and C, with C designating the right angle. Before attempting these problems, it is suggested that you study the topic of significant digits in the Appendix. (*Note:* As discussed in the Appendix, a bar above a zero, $\bar{0}$, is used to avoid ambiguity and indicates a zero that is a significant digit.)

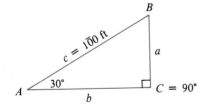

Figure 6.8

EXAMPLE 8   Solve the right triangle ABC in which $A = 30°$ and $c = 1\bar{0}0$ ft.

Solution   First, sketch Figure 6.8. We can find angle B since angles A and B are complementary.

$$A + B = 90°$$
$$30° + B = 90°$$
$$B = 60°$$

Second, we can find side length a by using the sine function.

$$\sin A = \frac{a}{c}$$

$$\sin 30° = \frac{a}{100}$$

$$100(\sin 30°) = a$$

$$100(0.5000) = a$$

$$5\overline{0} \text{ ft} = a \qquad \text{(two significant digits)}$$

Third, we can find b by using the cosine function.

$$\cos A = \frac{b}{c}$$

$$\cos 30° = \frac{b}{100}$$

$$100(\cos 30°) = b$$

$$100(0.8660) = b$$

$$87 \text{ ft} = b \qquad \text{(two significant digits)}$$

To summarize, we found that $B = 60°$, $a = 5\overline{0}$ ft, and $b = 87$ ft.   □

In applications of trigonometry our values are only as accurate as the devices we use to measure the data. However, although all our answers are approximations, the symbol for equality ($=$) is generally used instead of the more precise symbol for approximation ($\approx$). Computed results should not be used to determine other parts of a triangle since the given data produce more accurate answers. The results are usually rounded off as follows:

Accuracy of Sides	Accuracy of Angle
Two significant digits	Nearest degree
Three significant digits	Nearest 10 minutes or tenth of a degree
Four significant digits	Nearest minute or hundredth of a degree

EXAMPLE 9   Solve the right triangle ABC in which $a = 11.0$ ft and $b = 5.00$ ft.

Solution   First, sketch Figure 6.9. Now find the length of the hypotenuse by the Pythagorean theorem.

$$c^2 = a^2 + b^2$$

$$= 11^2 + 5^2$$

$$= 146$$

$$c = \sqrt{146}$$

$$= 12.1 \text{ ft} \qquad \text{(three significant digits)}$$

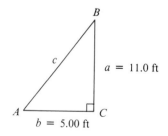

Figure 6.9

Second, we can find angle A by using the tangent function.

$$\tan A = \frac{a}{b}$$

$$= \frac{11}{5} = 2.2$$

$$A = 65°30'   \text{(from Table 5 to the nearest 10 minutes)}$$

Third, we can find angle B since angle A and angle B are complementary.

$$A + B = 90°$$
$$65°30' + B = 90°$$
$$B = 24°30'$$

To summarize, we found that $c = 12.1$ ft, $A = 65°30'$, and $B = 24°30'$. □

In many practical applications of right triangles, an angle is measured with respect to a horizontal line. This measurement is accomplished by use of a transit. (By centering a bubble of air in a water chamber, the table of this instrument may be horizontally leveled.) The sighting tube of the transit is then tilted upward or downward until the desired object is sighted. This measuring technique will result in an angle that is described as either an angle of elevation or an angle of depression (see Figure 6.10). The measurement results in an **angle of elevation** if the object being sighted is above the observer, and the measurement results in an **angle of depression** if the object being sighted is below the observer.

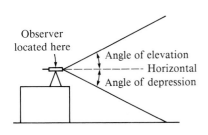

Figure 6.10

EXAMPLE 10   It is necessary to determine the height of a smokestack to estimate the cost of painting it. At a point 225 ft from the base of the stack, the angle of elevation is 33.0°. How high is the smokestack?

Solution   First, sketch Figure 6.11. In right triangle ABC we can find x, the height of the smokestack, by using the tangent function.

$$\tan A = \frac{\text{opposite}}{\text{adjacent}}$$

$$\tan 33.0° = \frac{x}{225}$$

$$225(\tan 33.0°) = x$$
$$225(0.6494) = x$$
$$146 \text{ ft} = x   \text{(three significant digits)} \quad □$$

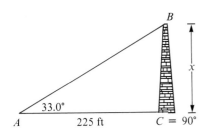

Figure 6.11

EXAMPLE 11   The measure of the angle of depression of a buoy from the platform of a radar tower that is 85 ft above the ocean is 15°. Find the distance of the buoy from the base of the tower. (See Figure 6.12.)

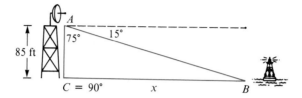

Figure 6.12

Solution   In right triangle ABC we can find the measure of angle A since angle A and the angle of depression are complementary.

$$A + 15° = 90°$$
$$A = 75°$$

We know the side length adjacent to angle A and we need to find x, which is the opposite side length. Thus,

$$\tan 75° = \frac{x}{85}$$
$$85(\tan 75°) = x$$
$$85(3.732) = x$$
$$320 \text{ ft} = x \qquad \text{(two significant digits)}.$$ □

EXAMPLE 12   Solve the problem in the chapter introduction on page 269.

Solution   Consider Figure 6.13. The length of the hypotenuse $(\overline{ES})$ represents the distance from the earth to the sun, while the length of the adjacent side $(\overline{EM})$ represents the distance from the earth to the moon. The ratio $\overline{ES}/\overline{EM}$ represents how many times more distant is the sun than the moon from the earth. Then

$$\frac{\overline{ES}}{\overline{EM}} = \frac{\text{hypotenuse}}{\text{adjacent}} = \sec 87° \approx 19.11.$$

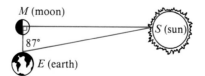

Figure 6.13

Thus, Aristarchus estimated that the sun is about 19 times farther away than the moon. (*Note:* Actually, Aristarchus computed his estimate by using Euclidean geometry. Trigonometry was developed later to solve precisely this type of problem. In reality, Aristarchus's estimate is not close to the true ratio. The error is caused by his measurement for angle *SEM*, which should be about 89°50'. On the basis of the correct measurement for angle *SEM*, about how many times farther away is the sun than the moon?)   ☐

EXERCISES 6.1

In Exercises 1–4 find the six trigonometric functions of the acute angles in the right triangle.

1.

2.

3.

4.
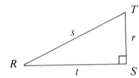

In Exercises 5–12 use these two triangles.

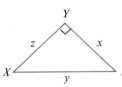

5. Find cos Z.
6. Find sec θ.
7. Find cot α.
8. Find sin α.
9. Find csc X.
10. Find tan θ.
11. Represent the ratio y/z as a trigonometric function of an acute angle.
12. Represent the ratio $9/40$ as a trigonometric function of an acute angle.

In Exercises 13–18 refer to this triangle.

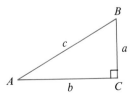

13. If $\cos B = \frac{4}{5}$ and $a = 16$, find c.
14. If $\sin A = \frac{12}{13}$ and $c = 39$, find a.
15. If $\tan A = \frac{7}{12}$ and $a = 21$, find b.
16. If $\csc B = 2$ and $c = 8$, find a.
17. If $a = 5$, $b = 12$, and $c = 13$, find the cosine of the smaller acute angle.
18. If $c = 17$ and $a = 8$, find the tangent of the larger acute angle.

In Exercises 19–24 use the given trigonometric ratio to find the other trigonometric functions of θ. Use the definitions of the functions, not a calculator.

19. $\sin \theta = \frac{1}{2}$ **20.** $\csc \theta = 2$ **21.** $\cot \theta = \frac{3}{5}$
22. $\cos \theta = \frac{8}{17}$ **23.** $\tan \theta = 2$ **24.** $\sec \theta = 5$

In Exercises 25–30 express each term as a function of the angle complementary to the given angle.

25. $\sin 17°$ **26.** $\cos 64°$ **27.** $\tan 81°30'$
28. $\sec 33°58'$ **29.** $\csc 68.1°$ **30.** $\cot 0.5°$

In Exercises 31–38 use the exact trigonometric values of 30°, 45°, or 60°.

31. Does $2 \sin 30° = \sin 60°$?
32. Does $\sin 60° \csc 30° = \tan 60°$?
33. Does $\sin 45° \cos 45° = 1$?
34. Does $\sin 30° = \sqrt{\dfrac{1 - \cos 60°}{2}}$?
35. Does $(\tan 45°)^2 + 1 = (\sec 45°)^2$?
36. Does $\cos 60° = 1 - 2(\sin 30°)^2$?
37. If the length of the short leg of a 30–60–90 triangle is 2, determine the lengths of the other two sides of the triangle.
38. If the length of a leg of a 45–45–90 triangle is 5, determine the length of the hypotenuse of the triangle.

In Exercises 39–48 approximate each using a calculator or Table 5. If the angle is given in decimal degrees, use a calculator.

39. cos 7°   **40.** sin 42°   **41.** cot 83°

42. csc 54°   **43.** tan 79°30′   **44.** sec 65°10′

45. csc 51°00′   **46.** cot 16°40′   **47.** sin 12.3°

48. tan 47.6°

In Exercises 49–58 approximate the measure of angle θ. Write solutions to the nearest 10 minutes from Table 5 or to the nearest tenth of a degree by calculator.

49. sin $\theta \approx 0.7071$ **50.** cos $\theta \approx 0.8660$

51. tan $\theta \approx 0.7907$ **52.** cot $\theta \approx 2.699$

53. sec $\theta \approx 1.781$ **54.** csc $\theta \approx 49.11$

55. cot $\theta \approx 0.7651$ **56.** tan $\theta \approx 0.0402$

57. csc $\theta \approx 16.00$ **58.** sec $\theta \approx 1.549$

59. Explain why the values of sin θ range between 0 and 1 when θ is an acute angle.

60. Explain why the values of csc θ are always greater than 1 when θ is an acute angle.

In Exercises 61–79 solve each right triangle ABC ($C = 90°$) for the given data. Round off answers using the guidelines given in this section. (*Note:* As discussed in the Appendix, a bar above a zero, $\bar{0}$, is used to avoid ambiguity and indicates a zero that is a significant digit.)

61. $A = 30°$, $a = 5\bar{0}$ ft   **62.** $B = 45°$, $a = 85$ ft

63. $A = 60°$, $c = 15$ ft   **64.** $A = 22°$, $b = 62$ ft

65. $B = 71°$, $c = 25$ ft   **66.** $A = 19°$, $a = 17$ ft

67. $A = 55°$, $c = 25$ ft   **68.** $B = 10.3°$, $a = 24.5$ ft

69. $A = 45.5°$, $a = 86.6$ ft   **70.** $A = 84°50′$, $c = 12.4$ ft

71. $B = 52°40′$, $c = 625$ ft   **72.** $A = 31.5°$, $b = 29.7$ ft

73. $B = 88°10′$, $a = 31.2$ ft **74.** $a = 5\bar{0}$ ft, $b = 120$ ft

75. $a = 6.0$ ft, $c = 15$ ft   **76.** $b = 1.0$ ft, $c = 2.0$ ft

77. $a = 1.0$ ft, $b = 1.0$ ft   **78.** $a = 7.00$ ft, $c = 11.0$ ft

79. $b = 12.0$ ft, $c = 26.0$ ft

In Exercises 80–99 solve each problem by making a careful diagram and using right triangles.

80. A ladder leans against the side of a building and makes an angle of 60° with the ground. If the foot of the ladder is $1\bar{0}$ ft from the building, find the height the ladder reaches on the building.

81. An escalator from the first floor to the second floor of a building is $5\bar{0}$ ft long and makes an angle of 30° with the floor. Find the vertical distance between the floors.

82. If the angle of elevation of the sun at a certain time is 40°, find the height of a tree that casts a shadow of 45 ft.

83. A road has a uniform elevation of 6°. Find the increase in elevation in driving $5\bar{0}0$ yd along the road.

84. The distance from ground level to the underside of a cloud is called the "ceiling." At an airport, a ceiling light projector throws a spotlight vertically on the underside of a cloud. At a distance of $6\bar{0}0$ ft from the projector, the angle of elevation of the spot of light on the cloud is 58°. What is the ceiling?

85. To find the width of a river, a surveyor sets up her transit at C and sights across the river to point B (both B and C are at the water's edge, as shown below). She then measures off $2\bar{0}0$ ft from C to A such that C is a right angle. If she determines that angle A is 24°, how wide is the river?

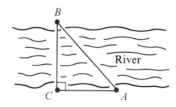

86. A pilot in an airplane at an altitude of $4,\bar{0}00$ ft observes the angle of depression of an airport to be 12°. How far is the airport from the point on the ground directly below the plane?

87. A lighthouse built at sea level is 180 ft high. From its top, the angle of depression of a buoy is 24°. Find the distance from the buoy to the foot of the lighthouse.

88. A surveyor stands on a cliff $5\bar{0}$ ft above the water of the river below. If the angle of depression to the water's edge on the opposite bank is 10°, how wide is the river at this point?

89. At an airport, cars drive down a ramp 85 ft long to reach the lower baggage claim area 15 ft below the main level. What angle does the ramp make with the ground at the lower level?

90. For maximum safety the distance between the base of a ladder and a building should be one-fourth of the length of the ladder. Find the angle that the ladder makes with the ground when it is set up in the safest position.

91. In building a warehouse, a carpenter checks the drawings and finds the roof span to be $4\overline{0}$ ft, as shown in the sketch. If the slope of the roof is 17°, what length of 2- by 6-in. stock will he need to make rafters if a 12-in. overhang is desired?

92. A machinist is given a 5.00 in. diameter steel rod with instructions to make a tapered pin 12.0 in. long. The pin must have diameters of 4.00 in. and 2.00 in. What angle of taper should he use to obtain the right dimensions? (*Hint:* Make use of the dashed line parallel to the center axis in the diagram.)

93. A circular disc 24.0 in. in diameter is to have five equally spaced holes as shown. Determine the correct setting (x) for the dividers to space these holes.

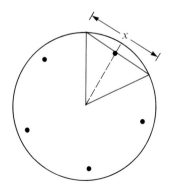

94. In an isosceles triangle each of the equal sides is $7\overline{0}$ in. and the base is $8\overline{0}$ in. Find the angles of the triangle.

95. The sides of a rectangle are 18 and 31 ft. Find the angle that the diagonal makes with the shorter side.

96. An observer on the third floor of a building determines that the angle of depression of the foot of a building across the street is 28° and the angle of elevation of the top of the same building is 51°. If the distance between the two buildings is $5\overline{0}$ ft, find the height of the observed building.

97. An observer on the top of a hill 350 ft above the level of a road spots two cars due east of her. Find the distance between the cars if the angles of depression noted by the observer were 16° and 27°.

98. A right triangle, called an **impedance triangle**, is used to analyze alternating current (a.c.) circuits. Consider diagram (a), which shows a resistor and an inductor in series. As current flows through these components it encounters some resistance, and the total effective resistance is called the **impedance.** We determine the impedance by making the resistances of the two circuit components the measures of the legs of a right triangle. As shown in diagram (b), the hypotenuse of the triangle then represents the impedance. The degree to which the voltage and current are in phase is given by angle θ, which is called the **phase angle.** From the data in this figure determine the impedance and the phase angle.

(a)

(b)

99. An important principle in the mathematical analysis of light is the **law of reflection.** This law states that a ray of light that strikes a reflecting surface is reflected so the angle of incidence (i) equals the angle of reflection (r). In the diagram below a photographer positions his flash at A. For a better lighting effect he aims this flash at position P on a reflecting surface to take a picture of a subject at B. If $\overline{AC} = 4.0$ ft, $\overline{CD} = 12$ ft, and $i = 41°$, find the length of BD.

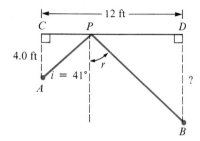

100. A major principle in the theory of light is the **law of refraction.** Refraction is the bending of light as it passes from one medium to another. For example, consider the following diagram, which shows a ray of light bending toward the perpendicular as it passes from air to water. The bending is caused by the change in speed of the light ray as it slows down in the water, which is the denser medium. The mathematical relation between the angle of incidence (i) and the angle of refraction (r) is a trigonometric ratio called **Snell's law.** The law is

$$\frac{\sin i}{\sin r} = \frac{v_i}{v_r},$$

where v_i/v_r is the ratio between the velocities of light in the two mediums.

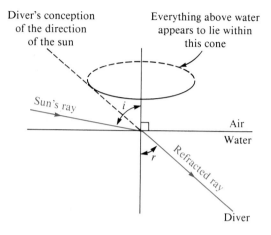

a. As light passes from air to water, v_i/v_r is about $\frac{4}{3}$. Find the angle of refraction if $i = 28.5°$.

b. When the sun is near the horizon, the angle of incidence approaches 90° and the angle at which the sun's rays penetrate the water approaches a limiting value. Determine this value of r. It is interesting to note that this restriction on the angle of refraction causes an optical illusion for a diver under water, as shown in the diagram. As he looks up, the world above the surface appears to be in the shape of a cone. This distorted perspective is called the "fish-eye view of the world."

6.2 TRIGONOMETRIC FUNCTIONS OF GENERAL ANGLES

Right triangle trigonometry is useful, but limited. To solve triangles that might not contain a right angle, we need trigonometric functions that are defined for $0° < \theta < 180°$. Also, although triangle trigonometry is important, an approach emphasizing the repetitive characteristics of the trigonometric functions is very useful today. The rhythmic motion of the heart, the sound waves produced by musical instruments, and alternating electric current are only a few of the physical phenomena that occur in cycles and can be studied through the aid of the trigonometric functions. To begin this general study, we need to be more precise about angles and their measure.

A **ray** is a half-line that begins at a point and extends indefinitely in some direction. Two rays that share a common end point (or vertex) form an angle. If we designate one ray as the **initial ray** and the other ray as the **terminal ray** (see Figure 6.14), the **measure of the angle** is the amount of rotation needed to make the initial ray coincide with the terminal ray. Notice that there are many rotations that will make the rays coincide, since there is no limitation on the number of revolutions made by the initial ray. In fact, it is useful to allow the initial ray to rotate through many revolutions, since the rotating initial ray will demonstrate a cyclic behavior that can serve as a model to simulate physical phenomena that occur in cycles. Also, the initial ray can rotate in two possible directions, as shown in Figure 6.14. To show the direction of the rotation, we define the measure of an angle to be positive if the rotation is counterclockwise and negative if the rotation is clockwise.

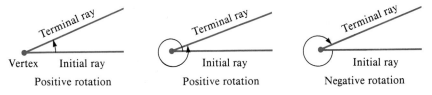

Figure 6.14

The frame of reference that we use to define the trigonometric functions is the Cartesian coordinate system. First, we put an angle θ (theta) in **standard position** (see Figure 6.15).

1. We place the vertex of the angle at the origin (0,0).
2. We place the initial ray of the angle along the positive x-axis.

Figure 6.15

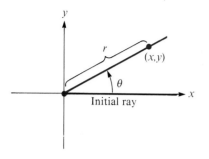

Figure 6.16

When we have the angle in standard position, we can define the trigonometric functions of an angle by considering any point on the terminal ray of θ [except (0,0)]. Three numbers can be associated with the location of this point (see Figure 6.16):

1. The x-coordinate of the point
2. The y-coordinate of the point
3. The distance r between the point and the origin

Since r represents the distance from the origin (0,0) to the point (x,y), we can find the relationship between x, y, and r by using the distance formula.

$$r = \sqrt{(x - 0)^2 + (y - 0)^2}$$

$$r = \sqrt{x^2 + y^2}$$

or

$$r^2 = x^2 + y^2$$

If we consider the number of ratios that can be obtained from the three variables x, y, and r, we find that there are six. It is these six ratios that define the six *trigonometric functions*.

Definition of the Trigonometric Functions

If θ is an angle in standard position and if (x,y) is any point on the terminal ray of θ [except (0,0)], then

Name of Function	Abbreviation	Ratio	
sine of angle θ	sin θ	$= \dfrac{y}{r}$	reciprocal functions
cosecant of angle θ	csc θ	$= \dfrac{r}{y}$ $(y \neq 0)$	
cosine of angle θ	cos θ	$= \dfrac{x}{r}$	reciprocal functions
secant of angle θ	sec θ	$= \dfrac{r}{x}$ $(x \neq 0)$	
tangent of angle θ	tan θ	$= \dfrac{y}{x}$ $(x \neq 0)$	reciprocal functions
cotangent of angle θ	cot θ	$= \dfrac{x}{y}$ $(y \neq 0)$	

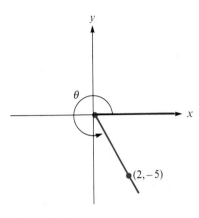

Figure 6.17

EXAMPLE 1 Find the value of the six trigonometric functions of an angle θ if $(2, -5)$ is a point on the terminal ray of θ (see Figure 6.17).

Solution

$$x = 2 \qquad y = -5$$
$$r = \sqrt{x^2 + y^2}$$
$$= \sqrt{(2)^2 + (-5)^2}$$
$$= \sqrt{4 + 25} = \sqrt{29} \qquad (\textit{Note: } \sqrt{4 + 25} \neq \sqrt{4} + \sqrt{25}.)$$

$$\sin \theta = \frac{y}{r} = \frac{-5}{\sqrt{29}} \quad \leftarrow \text{reciprocals} \rightarrow \quad \csc \theta = \frac{r}{y} = \frac{\sqrt{29}}{-5}$$

$$\cos \theta = \frac{x}{r} = \frac{2}{\sqrt{29}} \quad \leftarrow \text{reciprocals} \rightarrow \quad \sec \theta = \frac{r}{x} = \frac{\sqrt{29}}{2}$$

$$\tan \theta = \frac{y}{x} = \frac{-5}{2} \quad \leftarrow \text{reciprocals} \rightarrow \quad \cot \theta = \frac{x}{y} = \frac{2}{-5} \qquad \square$$

Note two important features of the definition above. First, (x,y) may be any point (except the origin) on the terminal ray of θ. Figure 6.18 shows that if we pick different points on the terminal ray of θ, say (x_1,y_1) and (x_2,y_2), they determine similar triangles. It follows as in Section 6.1 that corresponding side lengths are proportional, so the trigonometric ratios are uniquely determined by the terminal ray of θ. This means that in addition to picking any point on the terminal ray of θ, we may also choose any of the angle measures associated with the angle.

Second, note in Figure 6.19 that if θ is a positive acute angle in standard position with the point (x,y) on its terminal ray, then a right triangle may be formed with opposite side of length y, adjacent side of length x, and hypotenuse of length r. Thus, right triangle trigonometry is a special case of the more general definition of the trigonometric functions for it arises when we are dealing with an acute angle whose terminal ray lies in the first quadrant.

We now consider the problem of evaluating trigonometric functions for any angle. At present we have discussed how a calculator or Table 5 can be used to approximate trigonometric values for acute angles. The extension to evaluating any angle on a scientific calculator is simple. Many calculators compute trigonometric values for all angles (positive or negative), so you need only enter the angle and press the appropriate function keys as in Section 6.1. Although some calculators limit the size of the angle you can enter (under 1,440°, for instance, on Casio fx-910), you will rarely exceed this input range; if you do, we discuss how to reduce large angles later in this section. Thus, a calculator should easily give you a number associated with a trigonometric evaluation.

Figure 6.18

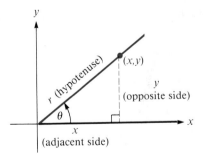

Figure 6.19

What is missing in this simple calculator method, however, is an understanding of some important concepts in trigonometry. Reference and coterminal angles, the signs of the functions in the various quadrants, and special angles with exact trigonometric values that are easily found and often used are just some of the ideas associated with a noncalculator approach. So keep in mind that a final numerical answer is only part of the objective of this section. We begin by using the definition of the trigonometric functions in the following example to evaluate functions for a special angle.

EXAMPLE 2   Find the six trigonometric functions of 90°.

Solution   We have shown that we may pick any point on the terminal ray of the angle when applying the definition. The terminal ray of a 90° angle is the positive y-axis and a choice of $r = 1$ determines the point $(0,1)$ as shown in Figure 6.20. Then since $x = 0$, $y = 1$, and $r = 1$, we have

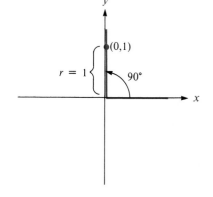

Figure 6.20

$$\sin 90° = \frac{y}{r} = \frac{1}{1} = 1 \;\leftarrow \text{reciprocals} \rightarrow\; \csc 90° = \frac{r}{y} = \frac{1}{1} = 1$$

$$\cos 90° = \frac{x}{r} = \frac{0}{1} = 0 \;\leftarrow \text{reciprocals} \rightarrow\; \sec 90° = \frac{r}{x} = \frac{1}{0} \text{ undefined}$$

$$\tan 90° = \frac{y}{x} = \frac{1}{0} \text{ undefined} \;\leftarrow \text{reciprocals} \rightarrow\; \cot 90° = \frac{x}{y} = \frac{0}{1} = 0.$$

☐

An angle of 90° is "special" because its terminal ray coincides with one of the axes. Such angles are called **quadrantal angles,** and any quadrantal angle can be expressed as the product of 90° and some integer. For example, 270° = 90° · 3 and −180° = 90° · (−2). We can repeat the procedure from Example 2 when evaluating functions for any quadrantal angle, and Figure 6.21 shows the four possible positions for the terminal ray of such angles. There are other quadrantal angles besides 0°, 90°, 180°, and 270° (which are shown in the figure), but their trigonometric values must be the same as one of the

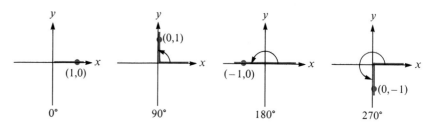

Figure 6.21

four listed. For example, the trigonometric values for 360° will be the same as the trigonometric values for 0°, since the terminal ray for both angles is the same (positive x-axis). In general, if two angles have the same terminal ray, they are called **coterminal** and the trigonometric functions of coterminal angles are equal. The following table summarizes the values of the trigonometric functions for quadrantal angles.

θ	$\sin \theta$	$\csc \theta$	$\cos \theta$	$\sec \theta$	$\tan \theta$	$\cot \theta$
0°	0	undefined	1	1	0	undefined
90°	1	1	0	undefined	undefined	0
180°	0	undefined	−1	−1	0	undefined
270°	−1	−1	0	undefined	undefined	0

EXAMPLE 3   Find the exact value of sin 540°.

Solution   540° is a quadrantal angle that is coterminal with 180° (see Figure 6.22). Therefore, sin 540° = sin 180° = 0.   □

EXAMPLE 4   Find the exact value of cos(−270°).

Solution   −270° is a quadrantal angle that is coterminal with 90° (see Figure 6.23). Therefore, cos(−270°) = cos 90° = 0. (*Note:* Remember that a negative angle means a rotation in a clockwise direction.)

Figure 6.22

Figure 6.23   □

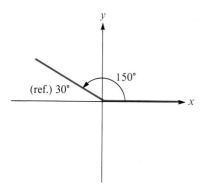

Figure 6.24

To evaluate trigonometric functions for angles that are not quadrantal angles, we introduce the concept of a **reference angle.** The reference angle for an angle θ is defined to be the positive acute angle formed by the terminal ray of θ and the horizontal axis. For example, the reference angle for 150° is 30°, since the closest segment of the horizontal axis (negative x-axis) may correspond to a rotation of 180° and $|180° - 150°| = 30°$, as shown in Figure 6.24.

Now consider the angles 30°, 150°, 210°, and 330°, where the reference angle for each of these angles is 30°. Note that the points on the terminal ray for each of these angles differ only in the sign of the x-coordinate or the y-coordinate. Therefore, the values of the trigonometric functions (which are ratios among x, y, and r) of these angles can differ only in their sign. For instance, by considering the appropriate points in Figure 6.25, we can determine $\sin 30° = \frac{1}{2}$, $\sin 150° = \frac{1}{2}$, $\sin 210° = -\frac{1}{2}$, and $\sin 330° = -\frac{1}{2}$. Thus, the reference angle provides us with a method for relating an angle in any quadrant to some acute angle whose trigonometric values can be found. Although the trigonometric values of an angle and its reference angle may differ in sign, we can determine the correct sign according to the quadrant

Figure 6.25

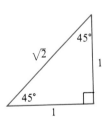

Figure 6.26

containing the terminal ray of the angle. By considering the definition of the trigonometric functions and the signs of x and y in the various quadrants, we construct the chart in Figure 6.26. To illustrate, since $\cos \theta = x/r$ with $r > 0$, the sign in the cosine ratio depends on the sign of x. Thus, $\cos \theta$ is positive in Q_1 and Q_4 and negative in Q_2 and Q_3. As a memory aid, the underlined first letters in the chart in Figure 6.26 correspond to the first letters in the mnemonic "*A*ll *s*tudents *t*ake *c*alculus."

EXAMPLE 5   Find the value of $\tan 135°$.

Solution   First, determine the reference angle.

$$|180° - 135°| = 45°   \text{(See Figure 6.27.)}$$

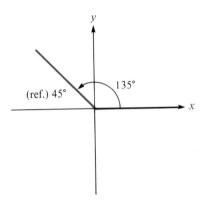

Figure 6.27

Second, determine $\tan 45°$.

$$\tan 45° = 1   \text{(See Figure 6.28.)}$$

Third, determine the correct sign.

$135°$ is in the second quadrant where the value of the tangent function is negative.

Therefore, $\tan 135° = -1$.   □

Figure 6.28

In general, we can find the trigonometric value of nonquadrantal angles by doing the following.

To Evaluate Trigonometric Functions

For nonquadrantal angles
1. Find the reference angle for the given angle.
2. Find the trigonometric value of the reference angle using the appropriate function. Since reference angles are always positive acute angles, this value can always be determined from Table 5. (*Note:* If the reference angle is 30°, 45°, or 60°, the exact answer is preferable.)
3. Determine the correct sign according to the terminal ray of the angle and the chart in Figure 6.26.

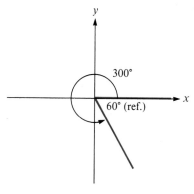

Figure 6.29

EXAMPLE 6   Find the exact value of cos 300°.

Solution   First, determine the reference angle.

$$|360° - 300°| = 60°\qquad \text{(See Figure 6.29.)}$$

Second, determine cos 60°.

$$\cos 60° = \tfrac{1}{2}\qquad \text{(See Figure 6.30.)}$$

Figure 6.30

Third, determine the correct sign.

300° is in Q_4, where the cosine function is positive.

Therefore, cos 300° = cos 60° = $\tfrac{1}{2}$.   □

EXAMPLE 7   Evaluate sec 510°.

Solution   First, determine the reference angle.

$$|540° - 510°| = 30°\qquad \text{(See Figure 6.31.)}$$

Figure 6.31

Second, determine sec 30°.

$$\sec 30° = \frac{2}{\sqrt{3}} \qquad \text{(See Figure 6.30.)}$$

Third, determine the correct sign.

510° is in Q_2, where the secant function is negative.

Therefore, sec 510° = −sec 30° = −$2/\sqrt{3}$.   □

EXAMPLE 8   Approximate cos 227° 20′.

Solution   First, determine the reference angle.

|227° 20′ − 180°| = 47° 20′   (See Figure 6.32.)

Second, determine cos 47° 20′.

cos 47° 20′ ≈ 0.6777   (Table 5)

Third, determine the correct sign.

227° 20′ is in Q_3, where the cosine function is negative.

Therefore, cos 227° 20′ = −cos 47° 20′ ≈ −0.6777.   □

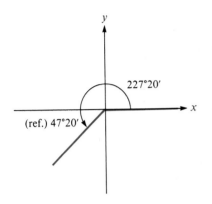

Figure 6.32

EXAMPLE 9   Approximate csc(−412° 50′).

Solution   First, determine the reference angle.

|−360 − (−412° 50′)| = 52° 50′   (See Figure 6.33.)

Second, determine csc 52° 50′.

csc 52° 50′ ≈ 1.255   (Table 5)

Third, determine the correct sign.

−412° 50′ is in Q_4, where the cosecant function is negative.

Therefore, csc(−412° 50′) = −csc 52° 50′ ≈ −1.255.   □

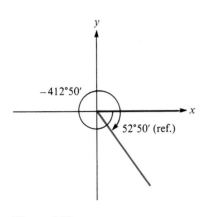

Figure 6.33

Finally, consider the next example which illustrates another use for the chart in Figure 6.26.

EXAMPLE 10   Find the value of the remaining trigonometric functions if $\tan \theta = \frac{2}{3}$ and $\sec \theta < 0$.

Solution   First, determine the quadrant that contains the terminal ray of θ. $\tan \theta = \frac{2}{3}$, which is positive. Therefore, θ is in Q_1 or Q_3; $\sec \theta < 0$ means that $\sec \theta$ is negative and θ is in Q_2 or Q_3. Only Q_3 satisfies both conditions. Therefore, the terminal ray of θ is in Q_3, where x is negative and y is negative.

Second, determine appropriate values for x, y, and r.

$$\tan \theta = \frac{y}{x} = \frac{2}{3}$$

Since both x and y are negative in Q_3, let $x = -3$ and $y = -2$; then find r.

$$\begin{aligned} r &= \sqrt{x^2 + y^2} \\ &= \sqrt{(-3)^2 + (-2)^2} \\ &= \sqrt{9 + 4} \\ &= \sqrt{13} \end{aligned}$$

Third, calculate the values of the different trigonometric functions. If $x = -3$, $y = -2$, and $r = \sqrt{13}$, then

$$\sin \theta = \frac{y}{r} = \frac{-2}{\sqrt{13}} \quad \leftarrow \text{reciprocals} \rightarrow \quad \csc \theta = \frac{r}{y} = \frac{\sqrt{13}}{-2}$$

$$\cos \theta = \frac{x}{r} = \frac{-3}{\sqrt{13}} \quad \leftarrow \text{reciprocals} \rightarrow \quad \sec \theta = \frac{r}{x} = \frac{\sqrt{13}}{-3}$$

$$\tan \theta = \frac{y}{x} = \frac{-2}{-3} = \frac{2}{3} \quad \leftarrow \text{reciprocals} \rightarrow \quad \cot \theta = \frac{x}{y} = \frac{-3}{-2} = \frac{3}{2}.$$

Note once again that the value of a trigonometric function is a ratio. $\tan \theta = \frac{2}{3}$ did not mean that $y = 2$ and $x = 3$. Also, in the second step we could have proceeded by drawing a right triangle with opposite side 2 and adjacent side 3. With this method we then determine the hypotenuse, set up the trigonometric ratios, and affix the correct sign depending on the function value in Q_3.   □

EXERCISES 6.2

In Exercises 1–10 find the values of the six trigonometric functions of the angle θ which is in standard position and satisfies the given condition.

1. $(3, -4)$ is on the terminal ray of θ.
2. $(-1, -1)$ is on the terminal ray of θ.
3. $(-12, 5)$ is on the terminal ray of θ.
4. $(2, 4)$ is on the terminal ray of θ.
5. $(9, -12)$ is on the terminal ray of θ.
6. $(-7, -13)$ is on the terminal ray of θ.
7. The terminal ray of θ lies in Q_1 on the line $y = x$.
8. The terminal ray of θ lies in Q_3 on the line $y = x$.
9. The terminal ray of θ lies on the line $y = -2x$ and θ is in Q_2.
10. The terminal ray of θ lies on the line $y = -2x$ and θ is in Q_4.

In Exercises 11–16 determine in which quadrant the terminal ray of θ lies.

11. $\sin \theta$ is negative, $\cos \theta$ is positive.
12. $\cot \theta$ is positive, $\csc \theta$ is positive.
13. $\tan \theta$ is positive, $\sec \theta$ is negative.
14. $\sec \theta > 0$, $\csc \theta < 0$.
15. $\tan \theta < 0$, $\sin \theta > 0$.
16. $\cos \theta < 0$, $\cot \theta < 0$.

In Exercises 17–24 find the values of the remaining trigonometric functions of θ.

17. $\sin \theta = -\frac{3}{5}$, terminal ray of θ is in Q_3.
18. $\sec \theta = \frac{13}{12}$, terminal ray of θ is in Q_4.
19. $\cos \theta = \frac{1}{2}$, $\tan \theta$ is negative.
20. $\tan \theta = \frac{2}{3}$, $\sin \theta$ is negative.
21. $\csc \theta = -\frac{4}{3}$, $\cot \theta$ is positive.
22. $\cot \theta = 1$, $\csc \theta < 0$.
23. $\sin \theta = \frac{1}{3}$, $\sec \theta > 0$.
24. $\csc \theta = 2$, $\cos \theta < 0$.

In Exercises 25–32 find the reference angle for the given angle.

25. $205°$ **26.** $181°$ **27.** $60°$ **28.** $-30°$
29. $-97°$ **30.** $111.1°$ **31.** $385°30'$ **32.** $1,000°$

In Exercises 33–64 find the *exact* value of each expression.

33. $\sin 210°$ **34.** $\tan 225°$
35. $\sec 330°$ **36.** $\cos 135°$
37. $\cot 315°$ **38.** $\csc 120°$
39. $\sin 150°$ **40.** $\tan 300°$
41. $\cos 225°$ **42.** $\cot 240°$

43. $\cos 450°$ **44.** $\sin 630°$
45. $\sec 390°$ **46.** $\tan 420°$
47. $\cot 690°$ **48.** $\cos 840°$
49. $\sin 1,035°$ **50.** $\csc 675°$
51. $\cos 495°$ **52.** $\sin 570°$
53. $\sin(-60°)$ **54.** $\cos(-45°)$
55. $\tan(-120°)$ **56.** $\sec(-225°)$
57. $\cot(-315°)$ **58.** $\csc(-210°)$
59. $\sec(-330°)$ **60.** $\cos(-480°)$
61. $\cot(-495°)$ **62.** $\sin(-1,050°)$
63. $\sin(-90°)$ **64.** $\cot(-540°)$

In Exercises 65–94 find the approximate value of each expression. Use Table 5 or a calculator.

65. $\sin 212°$ **66.** $\cos 307°$
67. $\tan 254°$ **68.** $\cot 115°$
69. $\sec 301°20'$ **70.** $\csc 163.4°$
71. $\cos 148°50'$ **72.** $\sin 354.5°$
73. $\cot 298°10'$ **74.** $\tan 190°40'$
75. $\sin 177°10'$ **76.** $\cos 252°20'$
77. $\csc 672°40'$ **78.** $\cot 392.1°$
79. $\sin 626°40'$ **80.** $\sin 531.5°$
81. $\cos 952°20'$ **82.** $\sec 452°20'$
83. $\tan 738°30'$ **84.** $\sin 521°50'$
85. $\cos(-81°)$ **86.** $\sin(-25°)$
87. $\tan(-131°)$ **88.** $\csc(-322°40')$
89. $\cos(-61°)$ **90.** $\sin(-251.5°)$
91. $\tan(-214°10')$ **92.** $\sec(-400°)$
93. $\cot(-512°)$ **94.** $\csc(-938°20')$

6.3 RADIANS

Up to now we have measured angles in degrees. However, in many applications of these functions and in calculus a different angle measure, called a **radian,** is more useful. We define the radian measure of an angle by first placing the vertex of the angle at the center of a circle. Let s be the length of the intercepted arc and let r be the radius. Then

$$\theta = \frac{s}{r}$$

is the radian measure of the angle (see Figure 6.34). Equivalently, an angle measuring 1 radian intercepts an arc equal in length to the radius of the circle (see Figure 6.35). In plane geometry it is shown that *s varies directly as r,* so that we can find the radian measure of θ by using a circle of *any* radius.

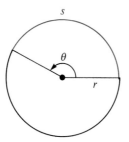

Figure 6.34   **Figure 6.35**

EXAMPLE 1   A central angle in a circle of radius 3 in. intercepts an arc of 12 in. Find the radian measure of the angle.

Solution   Substituting $r = 3$ in. and $s = 12$ in. in the formula

$$\theta = \frac{s}{r},$$

we have

$$\theta = \frac{12 \text{ in.}}{3 \text{ in.}} = 4.$$

Thus, the radian measure of the angle is 4. Note that r and s are measured in the same unit, which divides out in the ratio. Thus, the radian measure of an angle is a number without dimension. Although the word "radian" is often added, an angle measure with no units means radian measure.   □

We may find the relation between degrees and radians by considering the measure of an angle that makes one complete rotation. In degrees the measure of the angle is 360. Since the circumference of a circle is $2\pi r$, the radian measure is

$$\theta = \frac{s}{r} = \frac{2\pi r}{r} = 2\pi.$$

Thus,

$$360° = 2\pi \text{ radians.}$$

From this relation we derive the following conversion rules between degrees and radians.

Degree-Radian Conversion Rules	Degrees to radians formula: $$1° = \frac{\pi}{180} \text{ radian} \approx 0.0175 \text{ radians}$$ Radians to degrees formula: $$1 \text{ radian} = \frac{180°}{\pi} \approx 57.3°$$

EXAMPLE 2   Express 30°, 45°, and 270° in terms of radians.

Solution   Using the degrees to radians formula, we have

$$30° = 30 \cdot 1° = 30 \cdot \frac{\pi}{180} = \frac{\pi}{6}$$

$$45° = 45 \cdot 1° = 45 \cdot \frac{\pi}{180} = \frac{\pi}{4}$$

$$270° = 270 \cdot 1° = 270 \cdot \frac{\pi}{180} = \frac{3\pi}{2}. \qquad \square$$

EXAMPLE 3   Express $\pi/3$, π, and $7\pi/5$ radians in terms of degrees.

Solution   Using the radians to degrees formula, we have

$$\frac{\pi}{3} = \frac{\pi}{3} \cdot 1 = \frac{\pi}{3} \cdot \frac{180°}{\pi} = 60°$$

$$\pi = \pi \cdot 1 = \pi \cdot \frac{180°}{\pi} = 180°$$

$$\frac{7\pi}{5} = \frac{7\pi}{5} \cdot 1 = \frac{7\pi}{5} \cdot \frac{180°}{\pi} = 252°. \qquad \square$$

To illustrate an application of radians, consider the problem of determining the area of the shaded section in Figure 6.36. If you do not remember from geometry, notice intuitively that the area (A) varies directly as the central angle (θ). That is,

$$A = k\theta, \text{ where } k \text{ is a constant.}$$

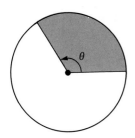

Figure 6.36

We may find k since we know that $A = \pi r^2$ when the angle (θ) is a complete rotation of 2π radians. Thus,

$$\pi r^2 = k \cdot 2\pi$$
$$\tfrac{1}{2}r^2 = k.$$

The formula (when θ is expressed in radians) is then

$$A = \tfrac{1}{2}r^2\theta.$$

EXAMPLE 4   In a circle of radius 12 in., find the area of a sector whose central angle is 120°.

Solution   To use the formula we must first convert the angle to radians.

$$120° = 120 \cdot \frac{\pi}{180} = \frac{2\pi}{3}$$

Now, substituting $r = 12$ and $\theta = 2\pi/3$ in the formula

$$A = \tfrac{1}{2}r^2\theta,$$

we have

$$A = \frac{1}{2}(12)^2 \cdot \frac{2\pi}{3}$$
$$= 48\pi \text{ in.}^2.$$

One of the most important applications of radians concerns linear and angular velocity. If a point P moves along the circumference of a circle at a constant speed, then the linear velocity, v, is given by the formula

$$v = \frac{s}{t},$$

where s is the distance traveled by the point (or the length of the arc traversed) and t is the time required to travel this distance. During the same time interval, the radius of the circle connected to point P swings through θ angular units. Thus, the angular velocity, ω (Greek lowercase omega), is given by

$$\omega = \frac{\theta}{t}.$$

We can relate linear and angular velocity through the familiar formula

$$s = \theta r.$$

Dividing both sides of this equation by t, we obtain

$$\frac{s}{t} = \frac{\theta}{t} r.$$

Substituting v for s/t and ω for θ/t yields

$$v = \omega r.$$

Thus, the linear velocity is equal to the product of the angular velocity and the radius.

Since $s = \theta r$ is valid only when θ is measured in radians, the angular velocity ω must be expressed in radians per unit of time when using the formula $v = \omega r$. However, in many common applications the angular velocity is expressed in revolutions per minute. For these problems remember to convert ω to radians (rad) per unit of time through the relationship 1 rpm $= 2\pi$ rad/minute before using the formula.

EXAMPLE 5   A circular saw blade 12 in. in diameter rotates at 1,600 rpm. Find (a) the angular velocity in radians per second and (b) the linear velocity in inches per second at which the teeth would strike a piece of wood.

Solution

a. Since 1 rpm $= 2\pi$ rad/minute, we have

$$\omega = 1{,}600 \text{ rpm} = 1{,}600\left(2\pi\frac{\text{rad}}{\text{minute}}\right) = 3{,}200\pi\frac{\text{rad}}{\text{minute}}.$$

To convert to rad/second we use 1 minute $= 60$ seconds, as follows:

$$\omega = 3{,}200\pi\frac{\text{rad}}{\text{minute}}\frac{1 \text{ minute}}{60 \text{ seconds}} = \frac{160\pi}{3}\frac{\text{rad}}{\text{second}}.$$

b. Since ω is expressed in radians per unit time, we find v by the formula $v = \omega r$, with $r = 6$ in.

$$v = \omega r$$
$$= \frac{160\pi}{3}\frac{\text{rad}}{\text{second}} \cdot 6 \text{ in.}$$
$$= 320\pi\frac{\text{in.}}{\text{second}}$$

Remember that the radian measure of an angle is a number without dimension, so the word "radian" does not appear in the units for linear velocity.   □

EXERCISES 6.3

In Exercises 1–8 complete the table by replacing each question mark with the appropriate number.

The Radius Is	The Intercepted Arc Is	The Central Angle Is
1. 20 ft	100 ft	?
2. 8 in.	?	3
3. ?	56 yd	7
4. 5.2 meters	5.2 meters	?
5. ?	4.8 meters	$\frac{1}{2}$
6. 11 ft	?	3.2
7. 1 unit	5 units	?
8. 1 unit	π units	?

9. The radius of a wheel is 20 in. When the wheel moves 110 in., through how many radians does a point on the wheel turn? How many revolutions are made by the wheel?

10. The radius of a wheel is 16 in. Find the number of radians through which a point on the circumference turns when the wheel moves a distance of 2 ft. How many revolutions are made by the wheel?

11. Find the area of a sector of a circle whose central angle is $\pi/4$ and radius is 10 in.

12. Find the area of a sector of a circle whose central angle is 120° and radius is 5 in.

13. In a circle of radius 2 ft the arc length of a sector is 8 ft. Find the area of the sector.

14. In a circle of radius 5 ft the arc length of a sector is 3 ft. Find the area of the sector.

In Exercises 15–35 express each angle in radian measure.

15. 30°	**16.** 45°	**17.** 60°	**18.** 90°
19. 120°	**20.** 135°	**21.** 150°	**22.** 180°
23. 210°	**24.** 225°	**25.** 240°	**26.** 270°
27. 300°	**28.** 315°	**29.** 330°	**30.** 360°
31. 200°	**32.** 75°	**33.** 20°	**34.** 162°
35. 305°			

In Exercises 36–47 express each angle in degree measure.

36. $\frac{\pi}{4}$	**37.** $\frac{\pi}{3}$	**38.** $\frac{\pi}{6}$	**39.** $\frac{2\pi}{9}$
40. $\frac{11\pi}{36}$	**41.** $\frac{2\pi}{3}$	**42.** $\frac{13\pi}{10}$	**43.** $\frac{7\pi}{9}$
44. $\frac{2\pi}{15}$	**45.** $\frac{12\pi}{5}$	**46.** $\frac{\pi}{12}$	**47.** $\frac{7\pi}{18}$

In Exercises 48–53 find, to the nearest degree, the number of degrees in each angle. (*Note:* Use $\pi \approx 3.14$.)

48. 1   **49.** 2   **50.** 6   **51.** 4   **52.** 3.5   **53.** 5.8

54. What is the linear velocity in inches/second of an object that moves along the circumference of a circle of radius 12 in. with an angular velocity of 7 rad/second?

55. What is the angular velocity in radians/second of an object that moves along the circumference of a circle of radius 9 in. with a linear velocity of 54 in./second?

56. What is the angular velocity in radians/second of the minute hand of a clock?

57. A phonograph record 12 in. in diameter is being played at $33\frac{1}{3}$ rpm.
  a. Find the angular velocity in radians/minute.
  b. Find the linear velocity in inches/minute of a point on the circumference of the record.

58. A pulley 20 in. in diameter makes 400 rpm.
  a. Find the angular velocity in radians/second.
  b. Find the speed in inches/second of the belt that drives the pulley.
  (*Note:* The speed of a point on the circumference of the pulley is the same as the speed of the belt.)

59. In a dynamo an armature 12 in. in diameter makes 1,500 rpm. Find the linear velocity in inches/second of the tip of the armature.

60. A car is traveling at 60 mi/hour (88 ft/second) on tires 30 in. in diameter. Find the angular velocity of the tires in radians/second.

61. Arcs of circles are used in many styles of molding that can be found in homes. Consider the illustration shown here. If $\overline{AB} = \overline{DE} = \frac{1}{2}\overline{BC} = \frac{1}{2}\overline{CD}$ and $\overline{AB} = \frac{1}{2}$ in., what is the length of the curved portion of the molding?

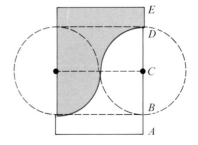

62. In architecture, arcs of circles are basic to the construction of many beautiful designs. For example, consider the quatrefoil in this illustration. If the side of the square measures 4 in., determine (a) the perimeter of the quatrefoil and (b) the area of the quatrefoil (the color portion of the figure).

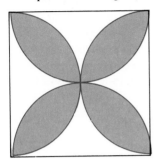

63. The formula $s = \theta r$ may be used to estimate certain distances when θ is a small angle (to about 10°). For example, consider the illustration below. An observer measures the angle from earth to opposite ends of the sun's diameter to be 0.53°. Since θ is a small angle, the arc length s is a good approximation for the diameter d. If the distance from the earth to the sun is 93,000,000 mi, determine the diameter of the sun to two significant digits.

6.4 TRIGONOMETRIC FUNCTIONS OF REAL NUMBERS

Many natural phenomena repeat over definite periods of time. The waves broadcast by a radio station; the rhythmic motion of the heart; alternating electric current; weather-related issues such as air pollution; and the economic pattern of expansion, retrenchment, recession, and recovery—all these occur in cycles. The trigonometric functions, also repetitive, are very useful in analyzing such periodic phenomena. However, the independent variable in these applications is the time and not the angle. Thus, we need to define the trigonometric functions in terms of real numbers and not degrees. Since radians measure angles in terms of real numbers, they are the starting point for our discussion of the trigonometry of real numbers.

Consider Figure 6.37, in which the central angle θ is measured in radians. It is convenient to label the radius of the circle as 1 unit so that $r = 1$. Then

$$\theta = \frac{s}{r}$$
$$= \frac{s}{1}$$
$$\theta = s.$$

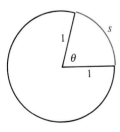

Figure 6.37

In a unit circle the same real number measures both the central angle θ and the intercepted arc s. Thus, we may base the definitions of the trigonometric functions on either an angle or an arc length. The results are valid for both interpretations. Since angles are meaningless in periodic phenomena, it is useful to emphasize the interpretation of a real number as the measure of an arc length *s*. We liberally interpret arc length as the distance traveled by a point as it moves around the unit circle repeating its behavior every 2π units.

To illustrate more forcefully the correspondence between real numbers and arc lengths, consider a unit circle with its center at the origin of the Cartesian coordinate system. The equation of this circle is $x^2 + y^2 = 1$. Through the point (1,0) and parallel to the *y*-axis we draw a real number line, labeled *s*. The zero point of *s* coincides with the point (1,0) on the circle. Units are marked off in the same scale as the *y*-axis (see Figure 6.38).

If the positive half of *s* is wrapped around the circle counterclockwise and the negative half of *s* is wrapped around the circle clockwise, we establish a one-to-one correspondence between real numbers and arc lengths of the circle (see Figure 6.39). Thus, each numbered point on *s* coincides with exactly one point on the circle, and the length of an arc may be read from this curved *s*-axis. We now relate this discussion to trigonometry.

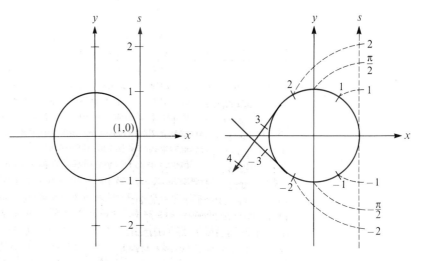

Figure 6.38   Figure 6.39

Definition of the Sine and Cosine Functions	Consider a point (x,y) on the unit circle $x^2 + y^2 = 1$ at arc length *s* from (1,0). We define the cosine of *s* to be the *x*-coordinate of the point and the sine of *s* to be the *y*-coordinate.

$$\cos s = x$$
$$\sin s = y$$

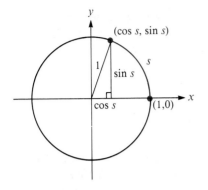

Figure 6.40

Note in Figure 6.40 that our definitions are consistent with the definitions of sine and cosine in terms of the sides and hypotenuse of a right triangle. The domain of both functions is the set of all real numbers, since the arc length s is determined by wrapping a real number line around the unit circle. The x- and y-coordinates in a unit circle vary between -1 and 1 (inclusive). Thus, for the range we have

$$-1 \leq \cos s \leq 1$$
$$-1 \leq \sin s \leq 1.$$

The remaining trigonometric functions are defined as follows:

Name of Function	Abbreviation	Ratio	
tangent of s	tan s	$y/x = \sin s/\cos s$	$(x \neq 0)$
cotangent of s	cot s	$x/y = \cos s/\sin s$	$(y \neq 0)$
secant of s	sec s	$1/x = 1/\cos s$	$(x \neq 0)$
cosecant of s	csc s	$1/y = 1/\sin s$	$(y \neq 0)$

We evaluate trigonometric functions of real numbers by using these definitions or by using a scientific calculator. Once again, the calculator method is most efficient for evaluating a trigonometric expression. Just set the calculator for radian mode, enter the number, and press the appropriate function keys as discussed in Section 6.1. However, this shortcut bypasses many important ideas in trigonometry. So let us consider how to evaluate the trigonometric functions from their definitions as we did in Section 6.2.

Using the above definitions, we find the values of the trigonometric functions by determining the rectangular coordinates (x,y) of points on the unit circle. For certain real numbers these coordinates are easy to find. For example, let us determine the values of the trigonometric functions of zero. The coordinates for an arc length of zero are $(1,0)$ (see Figure 6.41). Thus,

$$\sin 0 = y = 0 \leftarrow \text{reciprocals} \rightarrow \csc 0 = \frac{1}{y} = \frac{1}{0} \text{ undefined}$$

$$\cos 0 = x = 1 \leftarrow \text{reciprocals} \rightarrow \sec 0 = \frac{1}{x} = \frac{1}{1} = 1$$

$$\tan 0 = \frac{y}{x} = \frac{0}{1} = 0 \leftarrow \text{reciprocals} \rightarrow \cot 0 = \frac{x}{y} = \frac{1}{0} \text{ undefined.}$$

Figure 6.41

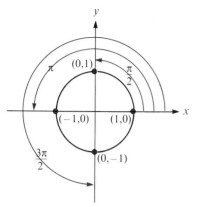

Figure 6.42

We can repeat this procedure for other arc lengths that terminate at one of the axes. Since the circumference of a circle of radius r is $2\pi r$, the circumference of a unit circle is 2π. The x- or y-axes may then intersect the unit circle at arc lengths of 0, $\pi/2$, π, and $3\pi/2$. Consider Figure 6.42, which illustrates the points of intersection for these arc lengths.

Using these coordinates, we may determine the trigonometric values of these numbers; the results are summarized in the following table and repeated on the endpaper at the back of the book. Note that tabular values match our results for $0°$, $90°$, $180°$, and $270°$ from Section 6.2.

s	$\sin s$	$\csc s$	$\cos s$	$\sec s$	$\tan s$	$\cot s$
0	0	undefined	1	1	0	undefined
$\pi/2$	1	1	0	undefined	undefined	0
π	0	undefined	-1	-1	0	undefined
$3\pi/2$	-1	-1	0	undefined	undefined	0

Other numbers terminate at one of the axes, but their trigonometric values are the same as one of the four listed. For example, the trigonometric values of 2π are the same as the trigonometric values of 0 since both numbers are assigned the point $(1,0)$ on the unit circle. The numbers 4π, 6π, -2π, and -4π are also assigned this point. A basic fact in our development is that in laying off a length 2π, we pass around the circle and return to our original point. Thus, the x- and y-coordinates repeat themselves at intervals of length 2π, and for any trigonometric function f we have

$$f(s + 2\pi k) = f(s), \text{ where } k \text{ is an integer.}$$

For example,

$$f(4\pi) = f(0 + 2\pi(2)) = f(0)$$
$$f(-2\pi) = f(0 + 2\pi(-1)) = f(0).$$

This observation is very important because it means that if we determine the values of the trigonometric functions in the interval $[0,2\pi)$, we know their values for all real s.

EXAMPLE 1   Find $\sin 7\pi$, $\cos 12\pi$, $\tan(11\pi/2)$, $\cot(17\pi/2)$, $\sec(-5\pi)$, and $\csc(-5\pi/2)$.

Solution   Since the values of the trigonometric functions repeat themselves at multiples of 2π, we have

$$\sin 7\pi = \sin(\pi + 6\pi) = \sin[\pi + 2\pi(3)] = \sin \pi = 0$$

$$\cos 12\pi = \cos(0 + 12\pi) = \cos[0 + 2\pi(6)] = \cos 0 = 1$$

$$\tan \frac{11\pi}{2} = \tan 5\tfrac{1}{2}\pi = \tan\left(\frac{3\pi}{2} + 4\pi\right) = \tan\left[\frac{3\pi}{2} + 2\pi(2)\right]$$

$$= \tan \frac{3\pi}{2} \qquad \text{undefined}$$

$$\cot \frac{17\pi}{2} = \cot 8\tfrac{1}{2}\pi = \cot\left(\frac{\pi}{2} + 8\pi\right) = \cot\left[\frac{\pi}{2} + 2\pi(4)\right]$$

$$= \cot \frac{\pi}{2} = 0$$

$$\sec(-5\pi) = \sec[\pi + (-6\pi)] = \sec[\pi + 2\pi(-3)] = \sec \pi = -1$$

$$\csc\left(\frac{-5\pi}{2}\right) = \csc(-2\tfrac{1}{2}\pi) = \csc\left[\frac{3\pi}{2} + (-4\pi)\right]$$

$$= \csc\left[\frac{3\pi}{2} + 2\pi(-2)\right] = \csc \frac{3\pi}{2} = -1. \qquad \square$$

A **trigonometric identity** is a statement that is true for all real numbers for which the expressions are defined. We simplify our work by developing identities that relate a trigonometric function of a negative number to the same function of a positive number. The symmetry of the unit circle makes these identities easy to derive. Consider Figures 6.43 and 6.44, which illustrate the symmetry for two possible values of s. Note that the numbers s and $-s$ are assigned the same x-coordinate, so

$$\cos(-s) = \cos s.$$

The y-coordinates differ only in their sign. Thus,

$$\sin(-s) = -\sin s.$$

The remaining functions are ratios of the sine and cosine, so it follows that

$$\tan(-s) = -\tan s$$
$$\cot(-s) = -\cot s$$
$$\sec(-s) = \sec s$$
$$\csc(-s) = -\csc s.$$

Figure 6.43

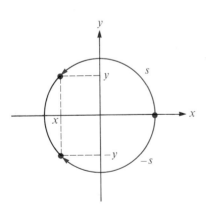

Figure 6.44

EXAMPLE 2   Find $\cos(-3\pi/2)$ and $\csc(-5\pi/2)$.

Solution

$$\cos\left(\frac{-3\pi}{2}\right) = \cos\frac{3\pi}{2} = 0$$

$$\csc\left(\frac{-5\pi}{2}\right) = -\csc\frac{5\pi}{2} = -\csc 2\tfrac{1}{2}\pi = -\csc\left(\frac{\pi}{2} + 2\pi\right)$$

$$= -\csc\frac{\pi}{2} = -1 \qquad\qquad \square$$

Besides arc lengths that terminate at one of the axes, the numbers $\pi/4$, $\pi/3$, and $\pi/6$ are also considered special numbers. These numbers correspond to angles of 45°, 60°, and 30°, respectively, and in Section 6.1 we derived their exact trigonometric values. These results are given in the following table and repeated on the endpaper at the back of the book.

s	$\sin s$	$\csc s$	$\cos s$	$\sec s$	$\tan s$	$\cot s$
$\dfrac{\pi}{3}$	$\dfrac{\sqrt{3}}{2}$	$\dfrac{2}{\sqrt{3}}$	$\dfrac{1}{2}$	2	$\sqrt{3}$	$\dfrac{1}{\sqrt{3}}$
$\dfrac{\pi}{4}$	$\dfrac{1}{\sqrt{2}}$	$\sqrt{2}$	$\dfrac{1}{\sqrt{2}}$	$\sqrt{2}$	1	1
$\dfrac{\pi}{6}$	$\dfrac{1}{2}$	2	$\dfrac{\sqrt{3}}{2}$	$\dfrac{2}{\sqrt{3}}$	$\dfrac{1}{\sqrt{3}}$	$\sqrt{3}$

The above results can also be obtained by unit circle derivations. Let us illustrate in the case of $s = \pi/4$. Since $\pi/4$ is halfway between 0 and $\pi/2$, $\pi/4$ is assigned the midpoint of the arc joining the points (1,0) and (0,1) on the unit circle (see Figure 6.45). Thus, the x- and y-coordinates are equal (that is, $x = y$). Since the coordinates of any point on the circle satisfy the equation $x^2 + y^2 = 1$, we have

$$x^2 + x^2 = 1$$

$$x^2 = \frac{1}{2}$$

$$x = \frac{1}{\sqrt{2}} \quad \text{or} \quad -\frac{1}{\sqrt{2}}$$

and

$$y = \frac{1}{\sqrt{2}} \text{ or } -\frac{1}{\sqrt{2}}.$$

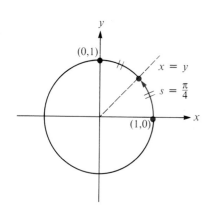

Figure 6.45

Since x and y are positive in the first quadrant, we have

$$x = y = \frac{1}{\sqrt{2}}.$$

The trigonometric values of $\pi/4$ are then

$$\sin\frac{\pi}{4} = y = \frac{1}{\sqrt{2}} \leftarrow \text{reciprocals} \rightarrow \csc\frac{\pi}{4} = \frac{1}{\sin(\pi/4)} = \sqrt{2}$$

$$\cos\frac{\pi}{4} = x = \frac{1}{\sqrt{2}} \leftarrow \text{reciprocals} \rightarrow \sec\frac{\pi}{4} = \frac{1}{\cos(\pi/4)} = \sqrt{2}$$

$$\tan\frac{\pi}{4} = \frac{\sin(\pi/4)}{\cos(\pi/4)} = 1 \leftarrow \text{reciprocals} \Rightarrow \cot\frac{\pi}{4} = \frac{\cos(\pi/4)}{\sin(\pi/4)} = 1.$$

EXAMPLE 3   Find $\sin(-\pi/3)$, $\cos(-\pi/4)$, $\sin(9\pi/4)$, and $\tan(13\pi/3)$.

Solution   Using the table on page 306 and our previous procedure, we have

$$\sin\left(-\frac{\pi}{3}\right) = -\sin\frac{\pi}{3} = -\frac{\sqrt{3}}{2}$$

$$\cos\left(-\frac{\pi}{4}\right) = \cos\frac{\pi}{4} = \frac{1}{\sqrt{2}}$$

$$\sin\frac{9\pi}{4} = \sin 2\tfrac{1}{4}\pi = \sin\left(\frac{\pi}{4} + 2\pi\right) = \sin\frac{\pi}{4} = \frac{1}{\sqrt{2}}$$

$$\tan\frac{13\pi}{3} = \tan 4\tfrac{1}{3}\pi = \tan\left(\frac{\pi}{3} + 4\pi\right) = \tan\frac{\pi}{3} = \sqrt{3}. \qquad \square$$

A basic fact in our development has been that in laying off a length 2π, we pass around the unit circle and return to our original point. Thus, if we determine the values of the trigonometric functions in the interval $[0,2\pi)$, we know their values for all real s. Consider Figure 6.46, which illustrates how the symmetry of the circle may be used to further simplify the evaluation.

Note that the coordinates of the point assigned to $\pi/6$ differ only in sign from the coordinates assigned to $\pi - \pi/6 = 5\pi/6$, $\pi + \pi/6 = 7\pi/6$, and $2\pi - \pi/6 = 11\pi/6$. Thus, for a specific trigonometric function, say sine, we have

$$\sin\frac{\pi}{6} = \left|\sin\frac{5\pi}{6}\right| = \left|\sin\frac{7\pi}{6}\right| = \left|\sin\frac{11\pi}{6}\right| = \frac{1}{2}.$$

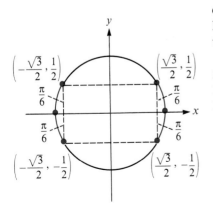

Figure 6.46

The number $\pi/6$ is called the **reference number** for $5\pi/6$, $7\pi/6$, and $11\pi/6$. In general, we determine the reference number for s (denoted by s_R) by finding the shortest positive arc length between the point on the circle assigned to s and the x-axis. This discussion indicates that the trigonometric values of s and s_R are related as follows:

Trigonometric Values of s and s_R	Any trigonometric function of s is equal in absolute value to the same named function of its reference number s_R.

We determine the correct sign by considering the function definitions together with the sign of x and y in the four quadrants. The chart in Figure 6.47 indicates the signs of the functions in the various quadrants (and this diagram corresponds to our previous chart in Figure 6.26).

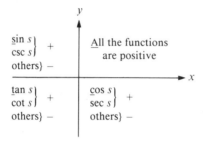

Figure 6.47

EXAMPLE 4   Find $\sin(7\pi/4)$.

Solution   First, determine the reference number.

$$s_R = 2\pi - \frac{7\pi}{4} = \frac{\pi}{4} \qquad \text{(See Figure 6.48.)}$$

Second, determine $\sin(\pi/4)$.

$$\sin \frac{\pi}{4} = \frac{1}{\sqrt{2}}$$

Third, determine the correct sign. The point assigned to $7\pi/4$ is in Q_4, where the value of the sine function is negative. Therefore,

$$\sin \frac{7\pi}{4} = -\frac{1}{\sqrt{2}}. \qquad \square$$

Figure 6.48

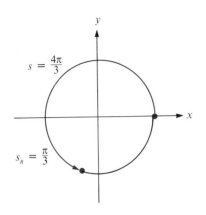

$s = \dfrac{4\pi}{3}$

$s_R = \dfrac{\pi}{3}$

Figure 6.49

EXAMPLE 5   Find $\cos(-10\pi/3)$.

Solution   First, simplify the expression to the form $\cos s$, where $0 \le s < 2\pi$.

$$\cos\left(\frac{-10\pi}{3}\right) = \cos\frac{10\pi}{3} = \cos 3\tfrac{1}{3}\pi = \cos\left(\frac{4\pi}{3} + 2\pi\right) = \cos\frac{4\pi}{3}$$

Second, determine s_R.

$$s_R = \frac{4\pi}{3} - \pi = \frac{\pi}{3} \qquad \text{(See Figure 6.49.)}$$

Third, determine $\cos(\pi/3)$.

$$\cos\frac{\pi}{3} = \frac{1}{2}$$

Fourth, determine the correct sign. The point assigned to $4\pi/3$ is in Q_3, where the value of the cosine function is negative. Therefore,

$$\cos\left(\frac{-10\pi}{3}\right) = -\frac{1}{2}. \qquad\qquad \Box$$

Until now we have been considering special numbers in that their trigonometric values may be found *exactly* by geometric means. To approximate the trigonometric values of other numbers we use the same procedure, together with Table 6 at the end of the book. This table lists the trigonometric values for numbers between 0 and 1.57 (or about $\pi/2$). These values are computed through the aid of calculus, and it should be understood that they are *approximations* that are rounded off to four significant digits. In these problems we use 3.14 as an approximation for π.

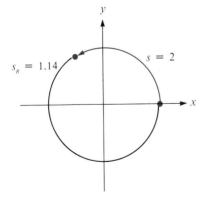

$s_R = 1.14$

$s = 2$

Figure 6.50

EXAMPLE 6   Approximate $\sin 2$.

Solution   First, determine the reference number.

$$s_R = \pi - 2 \approx 3.14 - 2 = 1.14 \qquad \text{(See Figure 6.50.)}$$

Second, determine $\sin 1.14$.

$$\sin 1.14 \approx 0.9086 \qquad \text{(Table 6)}$$

Third, determine the correct sign. The point assigned to 2 is in Q_2, where the value of the sine function is positive. Therefore,

$$\sin 2 \approx 0.9086. \qquad\qquad \Box$$

EXAMPLE 7   Approximate $\tan(-7.12)$.

Solution   First, simplify the expression to the form $\tan s$, where $0 \le s < 6.28$ (or $0 \le s < 2\pi$).

$$\tan(-7.12) = -\tan 7.12 = -\tan(0.84 + 6.28) = -\tan 0.84$$

Second, determine $\tan 0.84$.

$$\tan 0.84 \approx 1.116   \text{(Table 6)}$$

Therefore,

$$\tan(-7.12) = -\tan 0.84 \approx -1.116. \qquad \square$$

EXAMPLE 8   Approximate $\sec(8\pi/5)$.

Solution   First, determine the reference number.

$$s_R = 2\pi - \frac{8\pi}{5} = \frac{2\pi}{5} \approx \frac{2(3.14)}{5} = 1.26 \qquad \text{(See Figure 6.51.)}$$

Second, determine $\sec 1.26$.

$$\sec 1.26 \approx 3.270 \qquad \text{(Table 6)}$$

Third, determine the correct sign. The point assigned to $8\pi/5$ is in Q_4, where the value of the secant function is positive. Therefore,

$$\sec \frac{8\pi}{5} \approx 3.270. \qquad \square$$

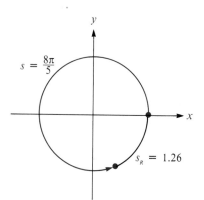

Figure 6.51

Here are a few notes on calculator usage.

1. Be sure to set the calculator for radian mode. Most calculators are in degree mode when first turned on, and forgetting this step is a very common error.
2. Some calculators limit the size of the number you can enter (under 8π, for instance, on Casio fx-910). If you exceed this input range, the calculator displays an error message. To evaluate such numbers you must reduce the number to an acceptable range by subtracting a multiple of 2π as discussed in this section.
3. Remember that cotangent, secant, and cosecant are usually computed by first evaluating either tangent, cosine, or sine, respectively, and then pressing the reciprocal key $\boxed{1/x}$.
4. Although a calculator is usually correct to at least eight decimal places, we will be displaying four decimal places in our solutions.

5. In Example 8 we determined $\sec(8\pi/5) \approx 3.270$. A calculator solution is

(radian mode) $8\boxed{\times}\pi\boxed{\div}5\boxed{=}\boxed{\cos}\boxed{1/x}$ 3.236068.

The calculator solution is much more accurate. The example solution is only accurate to two significant digits because 3.14 is used as an estimate for π.

EXERCISES 6.4

In Exercises 1–30 find the *exact* function value.

1. $\cos 4\pi$

2. $\sin 100\pi$

3. $\sin 5\pi$

4. $\cos 11\pi$

5. $\tan(-2\pi)$

6. $\cot(-36\pi)$

7. $\cos\left(-\dfrac{\pi}{2}\right)$

8. $\sin\left(-\dfrac{3\pi}{2}\right)$

9. $\sec\dfrac{5\pi}{2}$

10. $\tan\dfrac{19\pi}{2}$

11. $\csc\left(\dfrac{-7\pi}{2}\right)$

12. $\cos\left(\dfrac{-9\pi}{2}\right)$

13. $\sin\left(\dfrac{11\pi}{2}\right)$

14. $\sin\left(\dfrac{-97\pi}{2}\right)$

15. $\cos\left(-\dfrac{\pi}{3}\right)$

16. $\sin\left(-\dfrac{\pi}{4}\right)$

17. $\tan\dfrac{9\pi}{4}$

18. $\cos\dfrac{9\pi}{4}$

19. $\sin\dfrac{13\pi}{6}$

20. $\sec\dfrac{17\pi}{4}$

21. $\cot\dfrac{7\pi}{3}$

22. $\sin\dfrac{19\pi}{3}$

23. $\cos\left(\dfrac{-13\pi}{6}\right)$

24. $\sin\left(\dfrac{-13\pi}{3}\right)$

25. $\csc\left(\dfrac{-31\pi}{3}\right)$

26. $\cos\left(\dfrac{-25\pi}{4}\right)$

27. $\cot\dfrac{55\pi}{3}$

28. $\tan\left(\dfrac{-73\pi}{6}\right)$

29. $\sin\left(\pi-\dfrac{4\pi}{3}\right)$

30. $\cos\left(2\pi-\dfrac{13\pi}{6}\right)$

In Exercises 31–50 find the *exact* function value. Use reference numbers.

31. $\cos\dfrac{7\pi}{6}$

32. $\sin\dfrac{5\pi}{3}$

33. $\tan\dfrac{4\pi}{3}$

34. $\sec\dfrac{11\pi}{6}$

35. $\sin\dfrac{7\pi}{4}$

36. $\cos\dfrac{2\pi}{3}$

37. $\csc\dfrac{5\pi}{6}$

38. $\sin\dfrac{3\pi}{4}$

39. $\cos\dfrac{5\pi}{3}$

40. $\cot\dfrac{5\pi}{4}$

41. $\sin\left(\dfrac{-7\pi}{6}\right)$

42. $\cos\left(\dfrac{-4\pi}{3}\right)$

43. $\sec\left(\dfrac{-2\pi}{3}\right)$

44. $\tan\left(\dfrac{-7\pi}{4}\right)$

45. $\cos\dfrac{11\pi}{4}$

46. $\sin\dfrac{8\pi}{3}$

47. $\cot\dfrac{23\pi}{6}$

48. $\csc\dfrac{15\pi}{4}$

49. $\sin\left(\dfrac{-20\pi}{3}\right)$

50. $\cos\left(\dfrac{-23\pi}{6}\right)$

In Exercises 51–70 find the approximate function value. Use a calculator or reference numbers and Table 6.

51. $\cos 2$

52. $\sin 3$

53. $\tan 4$

54. $\sec 5$

55. $\sin 5.41$

56. $\csc 2.23$

57. $\cot 3.71$

58. $\cos 1.84$

59. $\sin(-6.07)$

60. $\tan(-1.69)$

61. $\cos 11.73$

62. $\cot 9.61$

63. $\csc(-7.57)$

64. $\sin(-10.25)$

65. $\sin\dfrac{2\pi}{5}$

66. $\tan\dfrac{9\pi}{5}$

67. $\cos\dfrac{3\pi}{7}$

68. $\sin\dfrac{4\pi}{9}$

69. $\sec\left(\dfrac{-7\pi}{8}\right)$

70. $\cos\left(\dfrac{-7\pi}{10}\right)$

71. The following formulas (derived from calculus) are used to compute the values of $\cos x$ and $\sin x$ in Table 6.

$$\cos x = 1 - \frac{x^2}{2!} + \frac{x^4}{4!} - \frac{x^6}{6!} + \cdots$$

$$\sin x = x - \frac{x^3}{3!} + \frac{x^5}{5!} - \frac{x^7}{7!} + \cdots,$$

where $n! = 1 \cdot 2 \cdots n$ (for example, $3! = 1 \cdot 2 \cdot 3$). Use the first three terms of these formulas to approximate the following expressions. Compare your results with the values in Table 6.

a. $\sin 1$ **b.** $\cos 1$ **c.** $\cos 0$ **d.** $\sin 0.5$

72. Consider the unit circle in the following figure, in which each of the six trigonometric functions can be represented as a line segment. For example, since $\overline{OC} = 1$ in right triangle OAC, we have

$$\sin \theta = \frac{\overline{AC}}{\overline{OC}} = \frac{\overline{AC}}{1} = \overline{AC}.$$

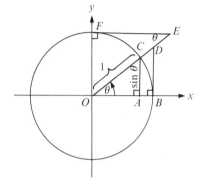

Notice we obtained the desired line segment by selecting a right triangle where the denominator in the defining ratio is 1. Determine the line segments representing the five remaining trigonometric functions in the figure.

73. Determine the coordinates of the point on the unit circle assigned to $\pi/6$. (*Hint:* The method is similar to the one given for $\pi/4$. The arc in the circle from $-\pi/6$ to $\pi/6$ is equal to the arc from $\pi/6$ to $\pi/2$.)

74. Classify each of the six trigonometric functions as either an even function or an odd function.

6.5 GRAPHS OF SINE AND COSINE FUNCTIONS

A picture or graph of the sine and cosine functions helps us understand their cyclic behavior. We use the Cartesian coordinate system and associate the arc length values with points on the horizontal or x-axis. This means we are now using x, instead of s, to represent the independent variable, which is an arc length. The vertical axis, labeled the y-axis, is used to represent the function values.

We begin by considering the values of the sine function as we lay off on the x-axis a length 2π and pass once around the unit circle. The following table indicates some values of the sine function on this typical interval.

$y = \sin x$													
x	0	$\frac{\pi}{6}$	$\frac{\pi}{3}$	$\frac{\pi}{2}$	$\frac{2\pi}{3}$	$\frac{5\pi}{6}$	π	$\frac{7\pi}{6}$	$\frac{4\pi}{3}$	$\frac{3\pi}{2}$	$\frac{5\pi}{3}$	$\frac{11\pi}{6}$	2π
y	0	0.5	0.87	1	0.87	0.5	0	-0.5	-0.87	-1	-0.87	-0.5	0

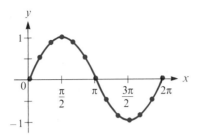

Figure 6.52

If we plot these points and join them with a smooth curve, we obtain the graph in Figure 6.52, which describes the essential characteristics of the sine function during one cycle. The plot of $y = \sin x$ starts at the origin, attains a maximum at one-fourth of the cycle length, returns to zero halfway through the cycle, attains a minimum at the three-quarter point, and returns to zero at the end of the cycle. Each time we lay off a length 2π we pass around the circle and repeat this behavior. Thus, the graph of $y = \sin x$ weaves continuously through cycles in both directions, as illustrated in Figure 6.53.

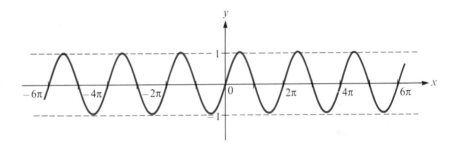

Figure 6.53

When we look at the graph of the sine function, we see that the function repeats its values on intervals of length 2π. For this reason, the sine function is said to be *periodic*. We define a periodic function as follows:

Periodic Function   A function f is periodic if

$$f(x) = f(x + p)$$

for all x in the domain of f. The smallest positive number p for which this is true is called the **period** of the function.

This definition applies to the sine function since

$$\sin x = \sin(x + 2\pi).$$

Thus, the sine function is periodic with period 2π.

We can use this information to sketch functions of the form $y = \sin bx$, where b represents a positive real number. These functions are similar to $y = \sin x$ in that they have the same basic shape, but they may differ by having different periods. For example, to obtain the graph of $y = \sin 2x$ we multiply x by 2 and take the sine of the resulting number. This means that when $x = \pi$, we evaluate the sine of $2 \cdot \pi$ or $\sin 2\pi$. Thus, when we substitute numbers from 0 to π for x, we evaluate the function from $\sin 0$ to $\sin 2\pi$ and complete one cycle. The period for $y = \sin 2x$ is therefore π. In general, the period of $y = \sin bx$ $(b > 0)$ is $2\pi/b$. Figure 6.54 compares the graphs of $y = \sin x$ and $y = \sin 2x$ for $0 \le x \le 2\pi$.

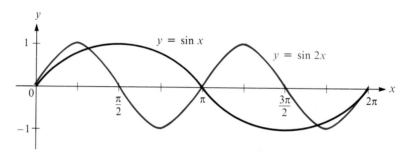

Figure 6.54

EXAMPLE 1   Sketch the graph of $y = \sin \frac{1}{2}x$ for $0 \le x \le 2\pi$.

Solution   Determine the period by substituting $\frac{1}{2}$ for b.

$$\text{Period} = \frac{2\pi}{b} = \frac{2\pi}{1/2} = 4\pi$$

If the curve completes one cycle for $0 \le x \le 4\pi$, the curve completes one-half of a sine wave for $0 \le x \le 2\pi$ as shown in Figure 6.55.   □

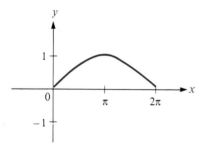

Figure 6.55

A second important characteristic of the graph of $y = \sin x$ is that the greatest y value is 1. When the graph of a periodic function is centered about the x-axis, the maximum y value is called the **amplitude** of the function. Thus, the amplitude of $y = \sin x$ is 1.

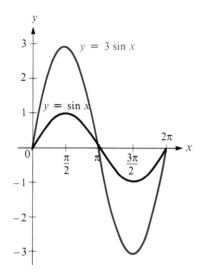

Figure 6.56

We can use this information to sketch functions of the form $y = a \sin bx$, where a represents a real number. These functions are similar to $y = \sin bx$ in that they have the same basic shape and period, but they may differ by having different amplitudes. For example, to sketch the graph of $y = 3 \sin x$ we obtain values for $\sin x$ and multiply these values by 3. Since the greatest y value that $\sin x$ attains is 1, the greatest y value that $3 \sin x$ attains is 3. Thus, the amplitude of $y = 3 \sin x$ is 3.

In general, since the greatest value that $\sin x$ attains is 1, the greatest value that $a \sin x$ attains is $|a|$. Thus, the amplitude of $y = a \sin bx$ is $|a|$. Figure 6.56 compares the graphs of $y = \sin x$ and $y = 3 \sin x$ on $[0,2\pi]$.

EXAMPLE 2   Sketch the graph of $y = 2 \sin 3x$ for $0 \le x \le 2\pi$.

Solution   First, determine the amplitude and period.

$$\text{Amplitude} = |a| = |2| = 2$$

$$\text{Period} = \frac{2\pi}{b} = \frac{2\pi}{3}$$

If the curve completes one cycle on the interval $[0,2\pi/3]$, the curve completes three cycles on the given interval. Note that b gives the number of cycles on the interval $[0,2\pi]$. Since the amplitude is 2, the curve oscillates between a maximum value of 2 and a minimum value of -2. This graph is sketched in Figure 6.57.

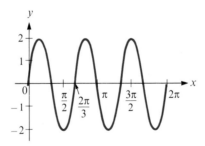

Figure 6.57   □

EXAMPLE 3   Sketch the graph of $y = -\sin \pi x$ for $0 \le x \le 2\pi$.

Solution   First, determine the amplitude and period.

$$\text{Amplitude} = |a| = |-1| = 1$$

$$\text{Period} = \frac{2\pi}{b} = \frac{2\pi}{\pi} = 2$$

If the curve completes one cycle every 2 units, the curve completes slightly more than three cycles on the interval $[0,2\pi]$. Since the amplitude is 1, the curve oscillates between a maximum value of 1 and a minimum value of -1. Because a is negative, we obtain the graph by reflecting the graph of $y = \sin \pi x$ about the x-axis as shown in Figure 6.58.

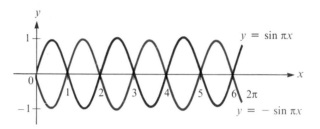

Figure 6.58 □

The graph of the cosine function has the same essential characteristics as the graph of the sine function. That is, the amplitude of the cosine function is 1 and the period is 2π. This is evidenced by the following table, which indicates some values of the cosine function on the interval $[0,2\pi]$.

$y = \cos x$													(for $0 \le x \le 2\pi$)
x	0	$\dfrac{\pi}{6}$	$\dfrac{\pi}{3}$	$\dfrac{\pi}{2}$	$\dfrac{2\pi}{3}$	$\dfrac{5\pi}{6}$	π	$\dfrac{7\pi}{6}$	$\dfrac{4\pi}{3}$	$\dfrac{3\pi}{2}$	$\dfrac{5\pi}{3}$	$\dfrac{11\pi}{6}$	2π
y	1	0.87	0.5	0	-0.5	-0.87	-1	-0.87	-0.5	0	0.5	0.87	1

If we plot these points and join them with a smooth curve, we obtain the graph shown in Figure 6.59. This graph demonstrates that the cosine function completes one cycle on the interval $[0,2\pi]$ and attains a maximum value of 1. Like that of the sine function, this graph can be reproduced indefinitely in both directions to obtain as much of the graph of the cosine function as desired (see Figure 6.60).

Figure 6.59

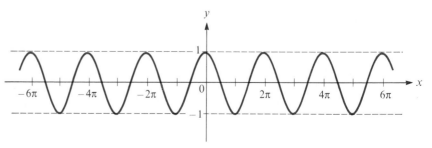

Figure 6.60

The great similarity between the graph of the sine and cosine functions should be apparent. In fact, if we shift the graph of the cosine function $\pi/2$ units to the right, the resulting graph is the sine function. Thus, the only difference between the two graphs is that one curve leads the other by $\pi/2$. That is, $\cos(x - \pi/2) = \sin x$.

We graph functions of the form $y = a \cos bx$ in a manner similar to graphing $y = a \sin bx$; that is, we find the amplitude by computing $|a|$ and the period by evaluating $2\pi/b$. The difference is that the graph of $y = a \cos bx$ attains a maximum or minimum height at $x = 0$. A common error in these problems is to draw the graph of a sine wave, which has a height of 0 when $x = 0$.

EXAMPLE 4   Sketch the graph of $y = \frac{1}{2} \cos 2x$ for $0 \le x \le 2\pi$.

Solution   First, determine the amplitude and period.

$$\text{Amplitude} = |a| = \left| \frac{1}{2} \right| = \frac{1}{2}$$

$$\text{Period} = \frac{2\pi}{b} = \frac{2\pi}{2} = \pi$$

If the curve completes one cycle on $[0,\pi]$, the curve will complete two cycles on $[0,2\pi]$. Since the amplitude is $\frac{1}{2}$, the curve oscillates between a maximum value of $\frac{1}{2}$ and a minimum value of $-\frac{1}{2}$ as shown in Figure 6.61.

Figure 6.61

EXAMPLE 5   Sketch one cycle of the graph of $y = -4 \cos 10x$.

Solution   First, determine the amplitude and period.

$$\text{Amplitude} = |a| = |-4| = 4$$

$$\text{Period} = \frac{2\pi}{b} = \frac{2\pi}{10} = \frac{\pi}{5}$$

The curve completes one cycle on the interval $[0,\pi/5]$ and attains a maximum value of 4 and a minimum value of -4. Since a is a negative number, we obtain the graph shown in Figure 6.62 by starting and ending the graph at a minimum point.

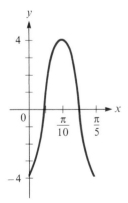

Figure 6.62

Finally, let us determine the effect of the constant c in a function of the form $y = a \sin(bx + c)$. Consider $y = \sin(x - \pi/2)$. As $x - \pi/2$ ranges from 0 to 2π, the curve completes one sine wave.

$$x - \frac{\pi}{2} = 0 \quad \text{when} \quad x = \frac{\pi}{2}$$

$$x - \frac{\pi}{2} = 2\pi \quad \text{when} \quad x = \frac{5\pi}{2}$$

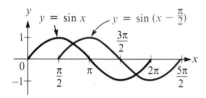

Figure 6.63

Thus, the function completes one cycle in the interval from $\pi/2$ to $5\pi/2$. The period of the function is 2π and the amplitude is 1. For comparison, one cycle of the graphs of $y = \sin x$ and $y = \sin(x - \pi/2)$ is given in Figure 6.63.

The complete graph of $y = \sin(x - \pi/2)$ is sketched by repeating the cycle shown in Figure 6.63 to the left and to the right. Notice that the graph of $y = \sin(x - \pi/2)$ may be obtained by shifting the graph of $y = \sin x$ to the right $\pi/2$ units. The sine wave then starts a cycle at $\pi/2$ instead of 0 and we call $\pi/2$ the **phase shift.**

In general, the constant c in the function $y = a \sin(bx + c)$ causes a shift of the graph of $y = a \sin bx$. The shift is of distance $|c/b|$ and is to the left if $c > 0$ and to the right if $c < 0$. The phase shift is given by $-c/b$. Similar remarks hold for functions of the form $y = a \cos(bx + c)$.

EXAMPLE 6   Graph one cycle of the function $y = 3 \cos(2x + \pi/2)$. Indicate the amplitude, period, and phase shift.

Solution   Determine the amplitude and period.

$$\text{Amplitude} = |a| = |3| = 3$$

$$\text{Period} = \frac{2\pi}{b} = \frac{2\pi}{2} = \pi$$

The function completes one cosine cycle as $2x + \pi/2$ varies from 0 to 2π.

$$2x + \frac{\pi}{2} = 0 \quad \text{when} \quad x = -\frac{\pi}{4}$$

$$2x + \frac{\pi}{2} = 2\pi \quad \text{when} \quad x = \frac{3\pi}{4}$$

Thus, the function completes one cycle in the interval from $-\pi/4$ to $3\pi/4$ (see Figure 6.64). This interval checks with the computed period since $3\pi/4 - (-\pi/4) = \pi$. A cycle starts at $-\pi/4$, so $-\pi/4$ is the phase shift. We may verify the phase shift since

$$\frac{-c}{b} = \frac{-\pi/2}{2} = -\frac{\pi}{4}.$$

Figure 6.64

Note produced by a tuning fork
(a)

Note produced by a tuning fork
with a higher pitch
(b)

Sound produced by a piano
(c)

Figure 6.65

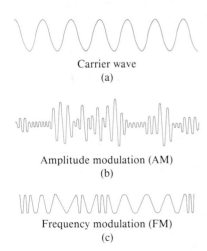

Carrier wave
(a)

Amplitude modulation (AM)
(b)

Frequency modulation (FM)
(c)

Figure 6.66

Amplitude, period, and phase shift are important considerations when analyzing any periodic phenomena. For example, let us briefly discuss two familiar concepts: musical sounds and radio waves. Musical sounds are caused by regular vibrations that have a definite period. On an electronic instrument called an **oscilloscope,** which changes sounds to electrical impulses and then to light waves, the sound from a tuning fork has the shape illustrated in Figure 6.65(a). The period of the wave depends on the pitch of the sound. With higher notes the pitch or frequency of the sound increases, producing a wave that has a smaller period [Figure 6.65(b)]. Humans can detect frequencies between about 50 and 15,000 vibrations (cycles) per second. However, some animals, such as the bat, can hear frequencies as high as 120,000 hertz (cycles per second). The amplitude of the wave depends on the intensity of the sound. Since humans hear best at a frequency of about 3,500 hertz, the loudness of a sound depends on both intensity and frequency. Although most musical sounds are very complex [such as the sound produced by the piano in Figure 6.65(c)], the French mathematician Joseph Fourier, in about 1800, showed that any periodic function is the sum of simple sine functions. Thus, all these sounds can be graphed and analyzed by some combination of sine waves. This analysis is indispensable in the design of sound recording and reproducing equipment.

Radio waves, used to transmit information at the speed of light (186,000 mi/second), are produced by oscillations in an electric current. Each current cycle produces a single radio wave. Special electronic equipment is needed to produce the alternating current since ordinary stations broadcast between 500,000 and 1,500,000 radio waves per second. Each station is licensed to broadcast a fixed number of radio waves per second, called its **frequency.** You receive the station's program by adjusting or tuning your set to accept this frequency.

Information is imposed on a radio wave as follows: A carrier wave [Figure 6.66(a)] is produced at a transmitting station, which makes the licensed number of cycles per second. This carrier wave is then modulated by the program current from the broadcasting site. In amplitude-modulated (AM) broadcasting, the amplitude of the carrier is made to vary according to the message; the wavelength remains constant [see Figure 6.66(b)]. With frequency modulation (FM), the amplitude of the wave remains constant and the wavelength varies [see Figure 6.66(c)].

FM broadcasting is superior to AM in that it produces better fidelity of sound and is relatively free from static and interference. However, AM stations are more numerous since they are less expensive and have a greater broadcasting range. In television an AM signal is used to transmit the picture but an FM signal carries the sound.

EXERCISES 6.5

In Exercises 1–10 state the amplitude and the period and sketch the curve for $-p \le x \le p$, where p is the period of the function.

1. $y = 2 \sin x$

2. $y = 3 \cos 2x$

3. $y = -3 \cos x$

4. $y = -4 \sin \frac{1}{3}x$

5. $y = 2 \sin 3x$

6. $y = \frac{1}{2} \cos 4x$

7. $y = -\cos 18x$

8. $y = -6 \sin \dfrac{x}{4}$

9. $y = 10 \cos \pi x$

10. $y = 110 \sin 120\pi x$

In Exercises 11–20 state the amplitude and the period and sketch the curve for $0 \le x \le 2\pi$.

11. $y = 3 \cos 4x$

12. $y = 2 \cos \dfrac{x}{4}$

13. $y = -\sin \dfrac{x}{2}$

14. $y = -3 \sin 2x$

15. $y = \frac{1}{2} \sin 3x$

16. $y = 1.5 \cos \frac{1}{3}x$

17. $y = \sin \dfrac{\pi}{2}x$

18. $y = 2 \cos \pi x$

19. $y = -2 \cos \dfrac{\pi}{3}x$

20. $y = -0.9 \sin \dfrac{\pi}{4}x$

In Exercises 21–30 state the amplitude, the period, and the phase shift and sketch one cycle of the function.

21. $y = \sin\left(x + \dfrac{\pi}{2}\right)$

22. $y = 2 \sin(x - \pi)$

23. $y = \cos\left(x - \dfrac{\pi}{4}\right)$

24. $y = 3 \cos\left(x + \dfrac{\pi}{3}\right)$

25. $y = \frac{1}{2} \cos\left(2x + \dfrac{\pi}{4}\right)$

26. $y = -\sin\left(\dfrac{x}{2} - \pi\right)$

27. $y = -\cos\left(\dfrac{x}{4} + \dfrac{\pi}{2}\right)$

28. $y = \sin(x - 1)$

29. $y = \sin(\pi x - \pi)$

30. $y = 1.2 \cos\left(2\pi x - \dfrac{\pi}{2}\right)$

In Exercises 31–40 find an equation for the curves with the given single cycle. The equations should be written in the form $y = a \sin bx$ or $y = a \cos bx$.

31.

32.

33.

34.

35.

36.

37.

38.

39.

40.

In Exercises 41–59 select the choice that completes the statement or answers the question.

41. The amplitude of the function $y = 4 \sin 3x$ is

    a. 1   **b.** 12   **c.** 3   **d.** 4

42. The maximum value of $y = 3 \cos \frac{1}{2}x$ is

    a. 1   **b.** $\frac{1}{2}$ **c.** 3   **d.** $\frac{3}{2}$

43. The minimum value of $y = 2 \cos 3x$ is

    a. 0   **b.** -3 **c.** 3   **d.** -2

44. The maximum value of $y = 2 + \sin x$ is

    a. 1   **b.** 2   **c.** 3   **d.** 4

45. The expression $y = 3 \sin \frac{1}{2}x$ reaches its maximum value when x equals

    a. 0   **b.** $\pi/2$ **c.** π **d.** $3\pi/2$

46. The expression $y = 2 \sin 3x$ reaches its maximum value when x equals

    a. 0   **b.** $\pi/6$ **c.** $\pi/4$ **d.** $2\pi/3$

47. The expression $y = 3 \cos \frac{1}{2}x$ reaches its minimum value when x equals

    a. 0   **b.** $\pi/2$ **c.** π **d.** 2π

48. The expression $y = 2 \cos 3x$ reaches its minimum value when x equals

    a. 0   **b.** $\pi/3$ **c.** $\pi/2$ **d.** $2\pi/3$

49. The period of the curve $y = 2 \sin x$ is

    a. 0   **b.** $\pi/2$ **c.** π **d.** 2π

50. The period of the curve $y = 3 \cos 2x$ is

    a. $\pi/2$ **b.** π **c.** 2π **d.** 4π

51. A function having the period π is

    a. $y = 2 \cos x$ **b.** $y = \cos \frac{1}{2}x$

    c. $y = \frac{1}{2} \cos x$ **d.** $y = \cos 2x$

52. A function having a period of $\pi/2$ is

    a. $y = \frac{1}{4} \sin x$ **b.** $y = 4 \sin 2x$

    c. $y = 2 \sin 4x$ **d.** $y = 4 \sin \frac{1}{4}x$

53. As x increases from $\pi/2$ to $3\pi/2$, then $y = \sin x$

    a. increases   **b.** decreases

    c. increases, then decreases

    d. decreases, then increases

54. As x increases from π to 2π, the cosine of x

    a. increases   **b.** decreases

    c. increases, then decreases

    d. decreases, then increases

55. As x increases in the interval from $\pi/2$ to $3\pi/2$, the value of $y = \cos x$ will

    a. increase   **b.** decrease

    c. increase, then decrease

    d. decrease, then increase

56. As x increases from $-\pi/2$ to 0, sin x

   a. increases from 0 to 1
   b. decreases from 0 to -1
   c. increases from -1 to 0
   d. decreases from 1 to 0

57. $y = \sin x$ and $y = \cos x$ both increase in

   a. quadrant 1   **b.** quadrant 2
   c. quadrant 3   **d.** quadrant 4

58. Between $x = 0$ and $x = 2\pi$, the graphs of $y = \sin x$ and $y = \cos x$ have in common

   a. no points   **b.** one point
   c. two points   **d.** four points

59. When graphs of $y = \sin x$ and $y = \cos x$ are drawn on the same axes, how many times do they intersect on the interval $[\pi/2,\pi]$?

   a. 0   **b.** 1   **c.** 2   **d.** 3

60. For any periodic function the amplitude is defined as $(M - m)/2$, where M is the maximum function value and m is the minimum function value. Use this definition to do the following problems.

   a. Show the amplitude of $y = \sin x$ is 1.
   b. Show the amplitude of $y = a \sin x$ is $|a|$.
   c. Show the amplitude of $y = 3 + \sin x$ is 1.
   d. What are the amplitude and the period of the function shown in this illustration?

6.6 GRAPHS OF THE OTHER TRIGONOMETRIC FUNCTIONS

Although the tangent function is periodic, its behavior differs dramatically from that of the sine and cosine. To see this difference, compare the graph of $y = \tan x$ to the smooth weaving curves of these functions. We begin by constructing the following table, which lists some values of the tangent function on the interval $[0,2\pi]$.

		$y = \tan x = \sin x/\cos x$			(for $0 \le x \le 2\pi$)							
x	0	$\dfrac{\pi}{6}$	$\dfrac{\pi}{3}$	$\dfrac{\pi}{2}$	$\dfrac{2\pi}{3}$	$\dfrac{5\pi}{6}$	π	$\dfrac{7\pi}{6}$	$\dfrac{4\pi}{3}$	$\dfrac{3\pi}{2}$	$\dfrac{5\pi}{3}$	$\dfrac{11\pi}{6}$	2π
y	0	0.6	1.7	und.	-1.7	-0.6	0	0.6	1.7	und.	-1.7	-0.6	0

Unlike the sine and cosine, the tangent function is not defined for all real numbers. That is, $y = \tan x = \sin x/\cos x$ is undefined when $\cos x = 0$. Thus, we must exclude from the domain of this function $\pi/2$, $3\pi/2$, and any x for which $x = (\pi/2) + k\pi$ (k any integer). As x approaches $\pi/2$, sin x approaches 1 and cos x approaches 0. This means that $|\tan x|$ becomes very large as x gets close to $\pi/2$. Consider Table 6 at the end of the book, which verifies this observation. On the basis of the preceding discussion and the fact that $\tan (-x) = -\tan x$, a portion of the graph of $y = \tan x$ is presented in Figure 6.67.

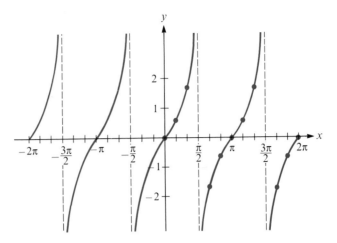

Figure 6.67

Using this graph we can see that the tangent function differs from the sine and cosine in the following important respects.

Comparison of the Sine, Cosine, and Tangent Functions

1. The domain of the sine and cosine functions is the set of all real numbers. The tangent function excludes $x = (\pi/2) + k\pi$ (k any integer).
2. The range of the sine and cosine functions is $[-1,1]$. The range of the tangent function is the set of all real numbers.
3. The sine and cosine are periodic with period 2π. Careful consideration of Figure 6.67 shows that the tangent function repeats its values every π units. Thus, the period of the tangent function is π.

From this discussion, the value of graphs in analyzing the behavior of various functions should be clear.

The other three trigonometric functions may be graphed as the reciprocals of the sine, cosine, or tangent. That is,

$$\csc x = \frac{1}{\sin x}, \qquad \sec x = \frac{1}{\cos x}, \qquad \tan x = \frac{1}{\cot x}.$$

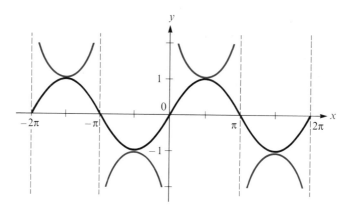

Figure 6.68 $y = \csc x$

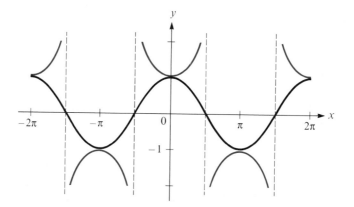

Figure 6.69 $y = \sec x$

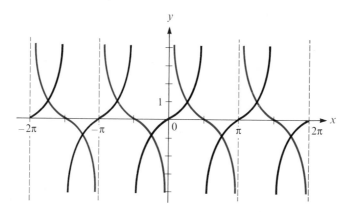

Figure 6.70 $y = \cot x$

Figure 6.71

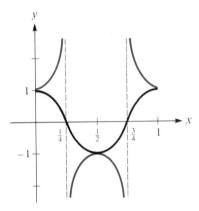

Figure 6.72

In Figures 6.68–6.70 we first graph the sine, cosine, and tangent as black curves. By obtaining the reciprocals of various y values, we then graph in color the cosecant, secant, and cotangent. Note the following about a function and its reciprocal function.

1. As one function increases, the other decreases, and vice versa.
2. The two functions always have the same sign.
3. When one function is zero, the other is undefined.
4. When the value of the function is 1 or -1, the reciprocal function has the same value.

The dashed lines in these figures are vertical asymptotes.

EXAMPLE 1   Graph one cycle of the function $y = \tan(x + \pi/4)$.

Solution   We start with the graph of $y = \tan x$ in Figure 6.67. To graph $y = \tan(x + \pi/4)$ we note that x has been replaced by $x + \pi/4$, so by our graphing techniques from Section 3.3, the graph of $y = \tan(x + \pi/4)$ is the graph of $y = \tan x$ shifted $\pi/4$ units to the left. Figure 6.71 shows the graph of one cycle on the interval that contains $x = 0$.   □

EXAMPLE 2   Graph one cycle of the function $y = \sec 2\pi x$.

Solution   Since $\sec 2\pi x = 1/\cos 2\pi x$, we first sketch one cycle of the graph of $y = \cos 2\pi x$ to use as a reference. The amplitude and period for $y = \cos 2\pi x$ are both 1, and the black curve in Figure 6.72 shows the graph of this function. Then by obtaining the reciprocals of the various y values, we graph one cycle of $y = \sec 2\pi x$ as shown in color in the figure.   □

EXERCISES 6.6

In Exercises 1–3 complete the table for the function and then sketch the curve from these points on the interval $[0,2\pi]$.

x	0	$\dfrac{\pi}{6}$	$\dfrac{\pi}{3}$	$\dfrac{\pi}{2}$	$\dfrac{2\pi}{3}$	$\dfrac{5\pi}{6}$	π	$\dfrac{7\pi}{6}$	$\dfrac{4\pi}{3}$	$\dfrac{3\pi}{2}$	$\dfrac{5\pi}{3}$	$\dfrac{11\pi}{6}$	2π
y													

1. $y = \cot x$ **2.** $y = \sec x$ **3.** $y = \csc x$

In Exercises 4–6 use Figures 6.68–6.70 to determine the domain, the range, and the period of the function.

4. $y = \cot x$ **5.** $y = \sec x$ **6.** $y = \csc x$

In Exercises 7–12 complete the table by determining if the function is increasing or decreasing in the interval. Use graphs to determine your answers.

	$\left(0,\dfrac{\pi}{2}\right)$	$\left(\dfrac{\pi}{2},\pi\right)$	$\left(\pi,\dfrac{3\pi}{2}\right)$	$\left(\dfrac{3\pi}{2},2\pi\right)$
7. $\sin x$				
8. $\cos x$				
9. $\tan x$				
10. $\cot x$				
11. $\sec x$				
12. $\csc x$				

13. Of the six trigonometric functions, only $y = \sin x$ and $y = \cos x$ have amplitudes. Why?

In Exercises 14–21 graph one cycle of the given functions.

14. $y = \tan\left(x - \dfrac{\pi}{4}\right)$

15. $y = -\cot\left(x + \dfrac{\pi}{2}\right)$

16. $y = \cot \frac{1}{2} x$

17. $y = \tan 2x$

18. $y = \sec 2x$

19. $y = 3 \csc \dfrac{x}{2}$

20. $y = 2 \csc \pi x$

21. $y = -\sec\left(x + \dfrac{\pi}{4}\right)$

6.7 TRIGONOMETRIC IDENTITIES

Trigonometry is characterized by many formulas that may be used to simplify trigonometric expressions. For example, the formula

$$\cos(-x) = \cos x$$

made it easier to evaluate the cosine of a negative number. In Section 6.6 we used the formula

$$\csc x = \frac{1}{\sin x}$$

to study the behavior of the cosecant function. These formulas are examples of **trigonometric identities.** That is, they are true for all values of x for which the expressions are defined.

Some basic trigonometric identities are easy to develop. Since all the trigonometric functions may be defined in terms of the sine and/or cosine, these functions are interrelated. This enables us to change the form of many trigonometric expressions to an expression that is either simpler or more useful for a particular problem. For convenience, we list below six fundamental identities that are used often. Except for identity 6, which is verified in Example 1, these statements follow directly from the definition of the trigonometric functions.

Six Fundamental Identities	**Identity 1**	$\csc x = \dfrac{1}{\sin x}$ or $\sin x = \dfrac{1}{\csc x}$ or $\sin x \csc x = 1$
	Identity 2	$\sec x = \dfrac{1}{\cos x}$ or $\cos x = \dfrac{1}{\sec x}$ or $\cos x \sec x = 1$
	Identity 3	$\cot x = \dfrac{1}{\tan x}$ or $\tan x = \dfrac{1}{\cot x}$ or $\tan x \cot x = 1$
	Identity 4	$\tan x = \dfrac{\sin x}{\cos x}$
	Identity 5	$\cot x = \dfrac{\cos x}{\sin x}$
	Identity 6	$\sin^2 x + \cos^2 x = 1$

EXAMPLE 1   Verify the identity $\sin^2 x + \cos^2 x = 1$.

Solution   To avoid confusion we return to the original notation of using s to represent the real numbers associated with arc lengths on the unit circle. Then by definition

$$x = \cos s \quad \text{and} \quad y = \sin s.$$

Any point (x,y) on the unit circle must satisfy the equation $x^2 + y^2 = 1$. Thus, by substitution we have

$$(\cos s)^2 + (\sin s)^2 = 1.$$

For convenience, this statement is usually written with x as the independent variable. Thus,

$$\sin^2 x + \cos^2 x = 1. \qquad \square$$

EXAMPLE 2   Show that the expression $\sin x \cot x \sec x$ is identical to 1.

Solution

$$\sin x \cot x \sec x = \sin x \left(\frac{\cos x}{\sin x} \right) \sec x \qquad \text{Identity 5}$$

$$= \sin x \, \frac{\cos x}{\sin x} \, \frac{1}{\cos x} \qquad \text{Identity 2}$$

$$= 1 \qquad \text{Simplify.}$$

Thus, $\sin x \cot x \sec x = 1$ for all values of x at which the expression is defined. $\qquad \square$

EXAMPLE 3   Prove the identity $\tan^2 x + 1 = \sec^2 x$.

Solution

$$\tan^2 x + 1 = \frac{\sin^2 x}{\cos^2 x} + 1 \qquad \text{Identity 4}$$

$$= \frac{\sin^2 x + \cos^2 x}{\cos^2 x}$$

$$= \frac{1}{\cos^2 x} \qquad \text{Identity 6}$$

$$= \sec^2 x \qquad \text{Identity 2}$$

Thus, $\tan^2 x + 1 = \sec^2 x$ is an identity.   □

In Example 3 we established the identity $\tan^2 x + 1 = \sec^2 x$, and in Exercise 16 you are asked to verify that $\cot^2 x + 1$ is identical to $\csc^2 x$. These two identities are used often and we need to add them to our list of basic identities.

> **Identity 7** $\tan^2 x + 1 = \sec^2 x$
>
> **Identity 8** $\cot^2 x + 1 = \csc^2 x$

The eight identities listed are commonly referred to as the **fundamental identities,** and among these, Identities 6, 7, and 8 are called the **Pythagorean identities.** In the next example we include an alternative solution that effectively employs Identity 7.

EXAMPLE 4   Show that the equation $\cos^2 x(1 + \tan^2 x) = 1$ is an identity.

Solution

$$\cos^2 x(1 + \tan^2 x) = \cos^2 x\left(1 + \frac{\sin^2 x}{\cos^2 x}\right) \qquad \text{Identity 4}$$

$$= \cos^2 x \cdot 1 + \cos^2 x \cdot \frac{\sin^2 x}{\cos^2 x} \qquad \text{Simplify.}$$

$$= \cos^2 x + \sin^2 x$$

$$= 1 \qquad \text{Identity 6}$$

Alternative Solution

$$\cos^2 x (1 + \tan^2 x) = \cos^2 x (\sec^2 x) \qquad \text{Identity 7}$$

$$= \cos^2 x \left(\frac{1}{\cos^2 x} \right) \qquad \text{Identity 2}$$

$$= 1$$

Thus, $\cos^2 x (1 + \tan^2 x) = 1$ is an identity.   ☐

In general, there is no standard procedure for working with identities. In fact, a given identity can usually be proved in several ways. However, these suggestions should be helpful.

Guidelines for Proving Identities

1. Change the more complicated expression in the identity to the same form as the less complicated expression. If both expressions are complicated, you might try to change them both to the same expression.
2. Until you become more familiar with identities, change all functions to sines and cosines. This procedure might necessitate more algebra in some instances, but it will provide a direct approach to the problem. Gradually try to make use of the other trigonometric functions.
3. Do *not* attempt to prove an identity by treating it as an equation and using the associated techniques, for this involves assuming what you want to prove.

EXAMPLE 5   Show that the expression $(\tan^2 x - 1)/(\tan^2 x + 1)$ is identical to the expression $\sin^2 x - \cos^2 x$.

Solution

$$\frac{\tan^2 x - 1}{\tan^2 x + 1} = \frac{(\sin^2 x / \cos^2 x) - 1}{(\sin^2 x / \cos^2 x) + 1} \qquad \text{Identity 4}$$

$$= \frac{\cos^2 x [(\sin^2 x / \cos^2 x) - 1]}{\cos^2 x [(\sin^2 x / \cos^2 x) + 1]} \left.\begin{array}{l} \\ \\ \end{array}\right\} \begin{array}{l} \text{Simplify the} \\ \text{complex fraction.} \end{array}$$

$$= \frac{\sin^2 x - \cos^2 x}{\sin^2 x + \cos^2 x}$$

$$= \frac{\sin^2 x - \cos^2 x}{1} \qquad \text{Identity 6}$$

$$= \sin^2 x - \cos^2 x$$

Thus,

$$\frac{\tan^2 x - 1}{\tan^2 x + 1} = \sin^2 x - \cos^2 x$$

is an identity.   ☐

EXAMPLE 6   Prove the identity $(\cos x \csc x)/\cot^2 x = \tan x$.

Solution

$$\frac{\cos x \csc x}{\cot^2 x} = \frac{\cos x (1/\sin x)}{\cos^2 x/\sin^2 x}$$ Identities 1 and 5

$$= \frac{\sin^2 x(\cos x/\sin x)}{\sin^2 x(\cos^2 x/\sin^2 x)}$$

$$= \frac{\sin x \cos x}{\cos^2 x}$$ Simplify the complex fraction.

$$= \frac{\sin x}{\cos x}$$

$$= \tan x$$ Identity 4

Thus, the given equation is an identity.   ☐

EXERCISES 6.7

In Exercises 1–20 transform each first expression and show that it is identical to the second expression.

1. $\sin x \sec x$; $\tan x$

2. $\tan x \csc x$; $\sec x$

3. $\sin^2 x \csc x$; $\sin x$

4. $\tan x \cot^2 x$; $\cot x$

5. $\cos x \tan x$; $1/\csc x$

6. $\sec x \cot x$; $1/\sin x$

7. $\cos x \tan x \csc x$; 1

8. $\cot x \sec x \sin x$; 1

9. $\sin^2 x \cot^2 x$; $\cos^2 x$

10. $\tan^2 x \csc^2 x$; $\sec^2 x$

11. $\sin^2 x \sec^2 x$; $\tan^2 x$

12. $\dfrac{\cot x}{\csc x}$; $\cos x$

13. $\dfrac{\cos^2 x}{\cot^2 x}$; $\sin^2 x$

14. $\dfrac{\sin x \sec x}{\tan x}$; 1

15. $\dfrac{\cot x \tan x}{\sec x}$; $\cos x$

16. $1 + \cot^2 x$; $\csc^2 x$

17. $(1 - \cos x)(1 + \cos x)$; $1/\csc^2 x$

18. $\sin x(\csc x - \sin x)$; $\cos^2 x$

19. $\dfrac{\cos^2 x}{1 + \sin x}$; $1 - \sin x$

20. $\sin^4 x - \cos^4 x$; $\sin^2 x - \cos^2 x$

In Exercises 21–40 prove that the equation is an identity.

21. $\dfrac{1}{\sin^2 x} - 1 = \cot^2 x$

22. $\csc x \sin x - \dfrac{1}{\sec^2 x} = \sin^2 x$

23. $\dfrac{1 + \tan^2 x}{\csc^2 x} = \tan^2 x$

24. $\dfrac{1 + \tan^2 x}{\tan^2 x} = \csc^2 x$

25. $\sin x \tan x + \cos x = \sec x$

26. $\tan x \csc^2 x - \tan x = \cot x$

27. $\dfrac{\tan^2 x - \sin^2 x}{\tan^2 x} = \sin^2 x$

28. $\dfrac{\sec^2 x + \csc^2 x}{\sec^2 x} = \csc^2 x$

29. $\dfrac{\csc x}{\tan x + \cot x} = \cos x$

30. $\dfrac{\sec x + \csc x}{1 + \tan x} = \csc x$

31. $(1 - \sin x)(\sec x + \tan x) = \cos x$

32. $\sec x \csc x - 2 \cos x \csc x + \cot x = \tan x$

33. $\sin x \cos^3 x + \cos x \sin^3 x = \sin x \cos x$

34. $\cos^4 x + 2 \cos^2 x \sin^2 x + \sin^4 x = 1$

35. $(\sin x + \cos x)^2 + (\sin x - \cos x)^2 = 2$

36. $1/(\sec^3 x \cos^4 x) = \sec x$

37. $\sec x - \cos x = \sin x \tan x$

38. $\sec^4 x - \tan^4 x = 2 \sec^2 x - 1$

39. $\dfrac{\sin x}{\csc x} + \dfrac{\cos x}{\sec x} = 1$

40. $\dfrac{\csc x}{\sin x} - \dfrac{\cot x}{\tan x} = 1$

6.8 MORE ON TRIGONOMETRIC IDENTITIES

We now develop other useful trigonometric identities. It is important to be able to manipulate these formulas to establish a desired result. In the following examples we first state, and then verify, the particular identity.

EXAMPLE 1   Verify the identity

$$\cos(x_1 + x_2) = \cos x_1 \cos x_2 - \sin x_1 \sin x_2.$$

Solution   As in Example 1 from Section 6.7, we first prove the identity by using s as the independent variable. The derivation is quite long, but once established, this formula yields many important results. Consider the coordinates of the points on the unit circle in Figure 6.73. The arc lengths from $(1,0)$ to $[\cos(s_1 + s_2), \sin(s_1 + s_2)]$ and from $[\cos(-s_1), \sin(-s_1)]$ to $(\cos s_2, \sin s_2)$ are both of length $s_1 + s_2$. Equal arcs in a circle subtend equal chords. Thus, by the distance formula we have

$$\sqrt{[\cos(s_1 + s_2) - 1]^2 + [\sin(s_1 + s_2) - 0]^2}$$
$$= \sqrt{(\cos s_2 - \cos s_1)^2 + [\sin s_2 - (-\sin s_1)]^2}.$$

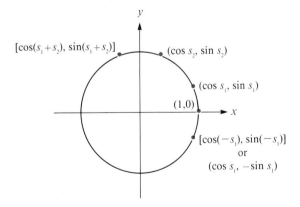

Figure 6.73

After squaring both sides to remove radicals and performing the indicated operations, we obtain

$$\cos^2(s_1 + s_2) - 2\cos(s_1 + s_2) + 1 + \sin^2(s_1 + s_2)$$
$$= \cos^2 s_2 - 2\cos s_1 \cos s_2 + \cos^2 s_1 + \sin^2 s_2$$
$$+ 2\sin s_1 \sin s_2 + \sin^2 s_1.$$

Because $\sin^2 s + \cos^2 s = 1$, we regroup the terms as follows:

$$[\sin^2(s_1 + s_2) + \cos^2(s_1 + s_2)] - 2\cos(s_1 + s_2) + 1$$
$$= (\sin^2 s_2 + \cos^2 s_2) + (\sin^2 s_1 + \cos^2 s_1) - 2\cos s_1 \cos s_2$$
$$+ 2\sin s_1 \sin s_2.$$

Substituting 1 for the appropriate expressions, we have

$$1 - 2\cos(s_1 + s_2) + 1 = 1 + 1 - 2\cos s_1 \cos s_2 + 2\sin s_1 \sin s_2$$
$$-2\cos(s_1 + s_2) = -2\cos s_1 \cos s_2 + 2\sin s_1 \sin s_2$$
$$\cos(s_1 + s_2) = \cos s_1 \cos s_2 - \sin s_1 \sin s_2.$$

For convenience, we let x represent the independent variable. Thus,

$$\cos(x_1 + x_2) = \cos x_1 \cos x_2 - \sin x_1 \sin x_2.$$

Note that

$$\cos(x_1 + x_2) \neq \cos x_1 + \cos x_2. \qquad \square$$

EXAMPLE 2   Verify the identity

$$\mathbf{\cos(x_1 - x_2) = \cos x_1 \cos x_2 + \sin x_1 \sin x_2.}$$

Solution   Rewrite $\cos(x_1 - x_2)$ as $\cos[x_1 + (-x_2)]$ and use the formula from Example 1.

$$\cos[x_1 + (-x_2)] = \cos x_1 \cos(-x_2) - \sin x_1 \sin(-x_2)$$

Since $\cos(-x_2) = \cos x_2$ and $\sin(-x_2) = -\sin x_2$, we have

$$\cos[x_1 + (-x_2)] = \cos x_1 \cos x_2 - \sin x_1(-\sin x_2)$$

or

$$\cos(x_1 - x_2) = \cos x_1 \cos x_2 + \sin x_1 \sin x_2. \qquad \square$$

EXAMPLE 3   Verify the identity $\mathbf{\cos[(\pi/2) - x] = \sin x.}$

Solution   Use the formula for the cosine of the difference of two numbers (Example 2).

$$\cos\left(\frac{\pi}{2} - x\right) = \cos\frac{\pi}{2}\cos x + \sin\frac{\pi}{2}\sin x$$

$$= 0 \cdot \cos x + 1 \cdot \sin x$$

$$= \sin x \qquad \qquad \square$$

EXAMPLE 4   Verify the identity $\cos 2x = \cos^2 x - \sin^2 x$.

Solution   Write $2x$ as $x + x$ and use the formula for the cosine of the sum of two numbers.

$$\cos 2x = \cos(x + x) = \cos x \cos x - \sin x \sin x$$

$$= \cos^2 x - \sin^2 x$$

There are two alternative forms for $\cos 2x$. From $\sin^2 x + \cos^2 x = 1$ it follows that $\sin^2 x = 1 - \cos^2 x$. Thus,

$$\cos 2x = \cos^2 x - \sin^2 x$$

$$= \cos^2 x - (1 - \cos^2 x)$$

$$= 2\cos^2 x - 1.$$

The remaining alternative form is

$$\cos 2x = 1 - 2\sin^2 x. \qquad \qquad \square$$

EXAMPLE 5   Verify the identity $\cos\dfrac{x}{2} = \pm\sqrt{\dfrac{1 + \cos x}{2}}$.

Solution   In Example 4 we obtained the identity

$$\cos 2x = 2\cos^2 x - 1.$$

If we replace x by $x/2$ in this identity, we have

$$\cos 2\left(\frac{x}{2}\right) = 2\cos^2\frac{x}{2} - 1$$

$$\cos x = 2\cos^2\frac{x}{2} - 1$$

$$\frac{1 + \cos x}{2} = \cos^2\frac{x}{2}$$

$$\pm\sqrt{\frac{1 + \cos x}{2}} = \cos\frac{x}{2}.$$

The correct sign is determined by the quadrant in which $x/2$ lies. $\square$

From the examples you should notice how new formulas are derived from previous results. To help you grasp this concept only identities concerned with the cosine function were developed. We now show how the formula for the sine of the sum of two numbers is usually derived. This formula can then be used to derive a list of identities concerning the sine function, and the exercises ask you to verify many of these formulas.

In Example 3 we proved

$$\cos\left(\frac{\pi}{2} - x\right) = \sin x.$$

Replacing x by $x_1 + x_2$ gives

$$\sin(x_1 + x_2) = \cos\left[\frac{\pi}{2} - (x_1 + x_2)\right] = \cos\left[\left(\frac{\pi}{2} - x_1\right) - x_2\right].$$

If we now use the formula for the cosine of the difference of two numbers, we obtain

$$\sin(x_1 + x_2) = \cos\left[\left(\frac{\pi}{2} - x_1\right) - x_2\right]$$
$$= \cos\left(\frac{\pi}{2} - x_1\right)\cos x_2 + \sin\left(\frac{\pi}{2} - x_1\right)\sin x_2.$$

By the identity in Example 3 we know $\cos(\pi/2 - x_1) = \sin x_1$. In Exercise 6 of this section you are asked to verify $\sin(\pi/2 - x_1) = \cos x_1$. The formula for the sine of the sum of two numbers is then

$$\sin(x_1 + x_2) = \sin x_1 \cos x_2 + \cos x_1 \sin x_2.$$

This formula leads directly to the following identities whose verification is left to the exercises.

$$\sin(x_1 - x_2) = \sin x_1 \cos x_2 - \cos x_1 \sin x_2$$
$$\sin 2x = 2 \sin x \cos x$$

We now consider some numerical examples using the various identities.

EXAMPLE 6   If $\sin x = \frac{3}{5}$ and $\pi/2 < x < \pi$, find

a. $\cos x$ **b.** $\tan x$ **c.** $\sec x$

d. $\cos 2x$ **e.** $\sin 2x$ **f.** $\cos x/2$

Solution

a. Since $\pi/2 < x < \pi$, where $\cos x < 0$, the identity $\sin^2 x + \cos^2 x = 1$ when solved for $\cos x$ becomes

$$\cos x = -\sqrt{1 - \sin^2 x} = -\sqrt{1 - \left(\frac{3}{5}\right)^2} = -\sqrt{\frac{16}{25}} = -\frac{4}{5}.$$

b. $\tan x = \dfrac{\sin x}{\cos x} = \dfrac{3/5}{-4/5} = \dfrac{3}{-4}$

c. $\sec x = \dfrac{1}{\cos x} = \dfrac{1}{-4/5} = \dfrac{5}{-4}$

d. $\cos 2x = 1 - 2\sin^2 x = 1 - 2\left(\frac{3}{5}\right)^2 = \dfrac{7}{25}$

e. $\sin 2x = 2\sin x \cos x = 2\left(\frac{3}{5}\right)\left(-\frac{4}{5}\right) = -\dfrac{24}{25}$

f. If $\pi/2 < x < \pi$, then $x/2$ is between $\pi/4$ and $\pi/2$ and $\cos(x/2)$ is positive. Thus,

$$\cos\frac{x}{2} = \sqrt{\frac{1 + \cos x}{2}} = \sqrt{\frac{1 + (-4/5)}{2}} = \sqrt{\frac{1}{10}} \text{ or } \frac{\sqrt{10}}{10}. \quad \square$$

EXAMPLE 7   If $\sin x_1 = -\frac{1}{2}$, where $\pi < x_1 < 3\pi/2$, and $\cos x_2 = \sqrt{3}/2$, where $0 < x_2 < \pi/2$, find

a. $\cos x_1$ **b.** $\sin x_2$ **c.** $\cos(x_1 + x_2)$ **d.** $\sin(x_1 - x_2)$

Solution

a. Since $\pi < x_1 < 3\pi/2$, where $\cos x_1$ is negative, we have

$$\cos x_1 = -\sqrt{1 - \sin^2 x_1} = -\sqrt{1 - \left(-\frac{1}{2}\right)^2}$$

$$= -\sqrt{\frac{3}{4}} = -\frac{\sqrt{3}}{2}.$$

b. Since $0 < x_2 < \pi/2$, where $\sin x_2$ is positive, we have

$$\sin x_2 = \sqrt{1 - \cos^2 x_2} = \sqrt{1 - \left(\frac{\sqrt{3}}{2}\right)^2} = \sqrt{\frac{1}{4}} = \frac{1}{2}.$$

c. $\cos(x_1 + x_2) = \cos x_1 \cos x_2 - \sin x_1 \sin x_2$

$$= \left(-\frac{\sqrt{3}}{2}\right)\left(\frac{\sqrt{3}}{2}\right) - \left(-\frac{1}{2}\right)\left(\frac{1}{2}\right)$$

$$= -\frac{3}{4} - \left(-\frac{1}{4}\right)$$

$$= -\frac{1}{2}$$

d. $\sin(x_1 - x_2) = \sin x_1 \cos x_2 - \cos x_1 \sin x_2$

$$= \left(-\frac{1}{2}\right)\left(\frac{\sqrt{3}}{2}\right) - \left(-\frac{\sqrt{3}}{2}\right)\left(\frac{1}{2}\right)$$

$$= -\frac{\sqrt{3}}{4} - \left(-\frac{\sqrt{3}}{4}\right)$$

$$= 0 \qquad \square$$

Finally, the major identities we have considered are summarized below. We have derived many of these formulas, while the verifications of others have been left to the exercises. For reference purposes, we include on this list the *product-to-sum formulas* and the *sum-to-product formulas* that have occasional use in higher mathematics. These formulas are derived from the sum or difference formulas (see Exercises 35 and 36) and you should be able to specify these formulas by consulting a reference source if the need arises.

Summary of Trigonometric Identities

Fundamental Identities

$$\csc x = \frac{1}{\sin x} \qquad \sec x = \frac{1}{\cos x} \qquad \cot x = \frac{1}{\tan x}$$

$$\tan x = \frac{\sin x}{\cos x} \qquad \cot x = \frac{\cos x}{\sin x}$$

$$\left.\begin{array}{l} \sin^2 x + \cos^2 x = 1 \\ \tan^2 x + 1 = \sec^2 x \\ \cot^2 x + 1 = \csc^2 x \end{array}\right\} \begin{array}{l} \text{Pythagorean} \\ \text{Identities} \end{array}$$

Negative Angle Formulas

$$\sin(-x) = -\sin x \qquad \cos(-x) = \cos x$$

$$\tan(-x) = -\tan x$$

Double-Angle Formulas

$$\sin 2x = 2 \sin x \cos x \qquad \cos 2x = \cos^2 x - \sin^2 x$$

$$\tan 2x = \frac{2 \tan x}{1 - \tan^2 x} \qquad\qquad = 2 \cos^2 x - 1$$

$$= 1 - 2 \sin^2 x$$

Half-Angle Formulas

$$\sin \frac{x}{2} = \pm \sqrt{\frac{1 - \cos x}{2}} \qquad \cos \frac{x}{2} = \pm \sqrt{\frac{1 + \cos x}{2}}$$

$$\tan \frac{x}{2} = \pm \sqrt{\frac{1 - \cos x}{1 + \cos x}}$$

Alternate Forms of cos 2x Formulas

$$\sin^2 x = \frac{1 - \cos 2x}{2}, \ \cos^2 x = \frac{1 + \cos 2x}{2}$$

Sum or Difference Formulas

$\sin(x_1 \pm x_2) = \sin x_1 \cos x_2 \pm \cos x_1 \sin x_2$

$\cos(x_1 \pm x_2) = \cos x_1 \cos x_2 \mp \sin x_1 \sin x_2$

$\tan(x_1 \pm x_2) = \dfrac{\tan x_1 \pm \tan x_2}{1 \mp \tan x_1 \tan x_2}$

(*Note:* Read the upper signs for the sum formula and the lower signs for the difference formula.)

Product-to-Sum Formulas

$\sin x \cos y = \frac{1}{2}[\sin(x - y) + \sin(x + y)]$

$\sin x \sin y = \frac{1}{2}[\cos(x - y) - \cos(x + y)]$

$\cos x \cos y = \frac{1}{2}[\cos(x - y) + \cos(x + y)]$

Sum-to-Product Formulas

$\sin A + \sin B = 2 \sin \dfrac{A + B}{2} \cos \dfrac{A - B}{2}$

$\sin A - \sin B = 2 \cos \dfrac{A + B}{2} \sin \dfrac{A - B}{2}$

$\cos A + \cos B = 2 \cos \dfrac{A + B}{2} \cos \dfrac{A - B}{2}$

$\cos A - \cos B = -2 \sin \dfrac{A + B}{2} \sin \dfrac{A - B}{2}$

EXERCISES 6.8

In Exercises 1–6 verify the identity. Use the formulas for the cosine of either the sum or difference of two numbers.

1. $\cos\left(\dfrac{\pi}{2} + x\right) = -\sin x$

2. $\cos(\pi + x) = -\cos x$

3. $\cos(\pi - x) = -\cos x$

4. $\cos\left(\dfrac{3\pi}{2} - x\right) = -\sin x$

5. $\cos\left(\dfrac{3\pi}{2} + x\right) = \sin x$

6. Show $\sin(\pi/2 - x) = \cos x$. (*Hint:* Write $\cos x$ as $\cos[\pi/2 - (\pi/2 - x)]$ and use the formula for the cosine of the difference of two numbers.)

In Exercises 7 and 8 verify the identity. Use the formula $\sin(x_1 + x_2) = \sin x_1 \cos x_2 + \cos x_1 \sin x_2$.

7. $\sin(x_1 - x_2) = \sin x_1 \cos x_2 - \cos x_1 \sin x_2$

8. $\sin 2x = 2 \sin x \cos x$

In Exercises 9–14 verify the identity. Use the formulas for the sine of either the sum or difference of two numbers.

9. $\sin\left(\dfrac{\pi}{2} - x\right) = \cos x$ 10. $\sin\left(\dfrac{\pi}{2} + x\right) = \cos x$

11. $\sin(\pi + x) = -\sin x$ 12. $\sin(\pi - x) = \sin x$

13. $\sin\left(\dfrac{3\pi}{2} - x\right) = -\cos x$

14. $\sin\left(\dfrac{3\pi}{2} + x\right) = -\cos x$

15. Verify the identity $\cos 2x = 1 - 2 \sin^2 x$.

16. Use Exercise 15 and Example 5 to verify the identity

$\sin \dfrac{x}{2} = \pm \sqrt{\dfrac{1 - \cos x}{2}}$.

In Exercises 17 and 18 use $\tan x = (\sin x / \cos x)$ to verify the identity.

17. $\tan(x_1 + x_2) = \dfrac{\tan x_1 + \tan x_2}{1 - \tan x_1 \tan x_2}$

18. $\tan \dfrac{x}{2} = \pm \sqrt{\dfrac{1 - \cos x}{1 + \cos x}}$

In Exercises 19 and 20 verify the identity. Use the formula for the tangent of the sum of two numbers.

19. $\tan(x_1 - x_2) = \dfrac{\tan x_1 - \tan x_2}{1 + \tan x_1 \tan x_2}$

20. $\tan 2x = \dfrac{2 \tan x}{1 - \tan^2 x}$

21. If $\sin x = \frac{12}{13}$ and $0 < x < \pi/2$, find

    a. $\cos x$ **b.** $\cot x$ **c.** $\csc x$ **d.** $\cos 2x$

    e. $\sin 2x$ **f.** $\cos(x/2)$ **g.** $\sin(x/2)$

22. If $\cos x = a$ and $\pi < x < 3\pi/2$, express each of the following in terms of a.

    a. $\sin x$ **b.** $\sec x$ **c.** $\cos 2x$ **d.** $\sin 2x$

    e. $\sin(x/2)$ **f.** $\cos(x/2)$

23. If $\sin x_1 = \frac{1}{2}$, where $0 < x_1 < \pi/2$, and $\cos x_2 = -\sqrt{3}/2$, where $\pi/2 < x_2 < \pi$, find

    a. $\cos x_1$ **b.** $\sin x_2$

    c. $\cos(x_1 + x_2)$ **d.** $\sin(x_1 - x_2)$

24. If $\cos x_1 = \frac{3}{5}$, where $3\pi/2 < x_1 < 2\pi$, and $\sin x_2 = -\frac{5}{13}$, where $3\pi/2 < x_2 < 2\pi$, find

    a. $\sin x_1$ **b.** $\cos x_2$

    c. $\cos(x_1 - x_2)$ **d.** $\sin(x_1 + x_2)$

25. Use the formula for the sine of the sum of two numbers and derive a formula for $\sin 3x$ in terms of $\sin x$.

26. Use the formula for the cosine of the sum of two numbers to verify the identity.

    a. $\cos 3x = 4 \cos^3 x - 3 \cos x$

    b. $\cos 4x = 8 \cos^4 x - 8 \cos^2 x + 1$

In Exercises 27–35 prove the equation is an identity.

27. $\csc x \sin 2x = 2 \cos x$ **28.** $\dfrac{\sin 2x}{2 \sin x} = \cos x$

29. $\cos^2 x = \dfrac{\cos 2x + 1}{2}$ **30.** $\sin^2 x = \dfrac{1 - \cos 2x}{2}$

31. $\cos 2x + 2 \sin^2 x = 1$ **32.** $\sin 2x = \dfrac{2 \cot x}{\csc^2 x}$

33. $\dfrac{1 + \cos 2x}{\sin 2x} = \cot x$ **34.** $\dfrac{\cos 2x}{\sin x} + \dfrac{\sin 2x}{\cos x} = \csc x$

35. $\sin(x + y) + \sin(x - y) = 2 \sin x \cos y$

36. Show $\sin A + \sin B = 2 \sin \frac{1}{2}(A + B) \cos \frac{1}{2}(A - B)$ is an identity. (*Hint:* Start with the identity in Exercise 35 and let $A = x + y$ and $B = x - y$. Then solve for x and y in terms of A and B.)

37. If $\tan(x + y) = 2$ and $\tan x = \mathrm{i}$, find $\tan y$.

38. Show that angle C equals the sum of angles A and B in this diagram by showing $\sin C = \sin(A + B)$.

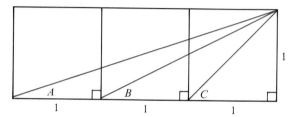

(*Note:* This problem can be approached in many interesting ways, and Charles Trigg published 54 different proofs of the result in the *Journal of Recreational Mathematics,* Vol. 4, April 1971.)

6.9 TRIGONOMETRIC EQUATIONS

In Sections 6.7 and 6.8 we considered **trigonometric identities,** which are equations like $\sin x \csc x = 1$ that are true for all values of x for which the expressions are defined. We now consider **conditional trigonometric equations** like $\sin x = -\frac{1}{2}$ that are true only for certain values of x. In solving such equations we are looking for the set of all values of the variable that satisfy the given equation. Before we attempt to find a procedure for writing all the solutions to a particular equation, let us first establish a method for finding solutions between 0 and 2π.

EXAMPLE 1   Find the exact values of x ($0 \le x < 2\pi$) for which the equation $\sin x = -\frac{1}{2}$ is a true statement.

Solution   First, determine the quadrant that contains the point assigned to x.

$$\sin x = -\tfrac{1}{2}, \qquad \text{which is a negative number.}$$

The point assigned to x could be in either Q_3 or Q_4, since the sine function is negative in both quadrants.

Second, determine the reference number (x_R).

$$\sin \frac{\pi}{6} = \frac{1}{2} \qquad \text{(from Section 6.4)}$$

Therefore, the reference number is $\pi/6$. (*Note:* When the reference number is $\pi/3$, $\pi/4$, or $\pi/6$, we may read exact values from the table in Section 6.4. Otherwise, we need Table 6 at the back of the book.)

Third, determine the appropriate values of x (Figure 6.74).

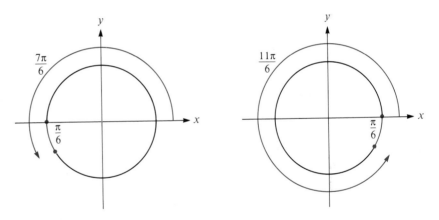

Figure 6.74

$\pi + \pi/6 = 7\pi/6$ is the number in Q_3 with a reference number of $\pi/6$.

$2\pi - \pi/6 = 11\pi/6$ is the number in Q_4 with a reference number of $\pi/6$.

Therefore, $7\pi/6$ and $11\pi/6$ make the equation a true statement, and the solution set in the interval $[0,2\pi)$ is $\{7\pi/6, 11\pi/6\}$. $\square$

EXAMPLE 2   Solve the equation $2 \sin x \cos x - \sin x = 0$ in the interval $[0,2\pi)$.

Solution

$$2 \sin x \cos x - \sin x = 0$$
$$\sin x (2 \cos x - 1) = 0$$

Note that we have found two factors whose product is zero. Hence, the original equation will be satisfied whenever either factor is zero, and we treat each factor separately from this point on.

First Factor $\sin x = 0$ The sine function is 0 when $x = 0$ and $x = \pi$.

Second Factor $2 \cos x - 1 = 0$

$$\cos x = \tfrac{1}{2}$$

1. Since $\cos x$ is positive, the point assigned to x could be in Q_1 or Q_4.
2. Since $\cos \pi/3 = \tfrac{1}{2}$, the reference number is $\pi/3$.
3. $\pi/3$ is the number in Q_1 with a $2\pi - \pi/3 = 5\pi/3$ is the number reference number of $\pi/3$ in Q_4 with a reference number of (Figure 6.75). $\pi/3$ (Figure 6.76).

Figure 6.75

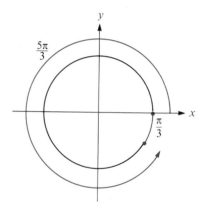

Figure 6.76

Thus, in the interval $[0,2\pi)$ the solution set is $\{0,\pi/3,\pi,5\pi/3\}$.   □

Once we have the solutions of a trigonometric equation that are between 0 and 2π, we can determine all the solutions, since the trigonometric functions are periodic. In laying off a length 2π we pass around the unit circle and return to our original point. Thus, we generate all the solutions to an equation by adding multiples of 2π to the solutions that are in the interval $[0,2\pi)$.

EXAMPLE 3   Approximate all the solutions to $4 \cos x + 1 = 0$.

Solution   First, solve the equation for $\cos x$.

$$4 \cos x + 1 = 0$$
$$\cos x = -\tfrac{1}{4} = -0.2500$$

Second, determine the quadrant that contains the point assigned to x.

$\cos x = -0.2500$, which is a negative number.

The point assigned to x could be in either Q_2 or Q_3, since the cosine function is negative in both quadrants.

Third, determine the reference number.

$\cos 1.32 \approx 0.2500$ (from Table 6)

Therefore, the reference number is 1.32.

Fourth, determine the appropriate values of x (Figure 6.77).

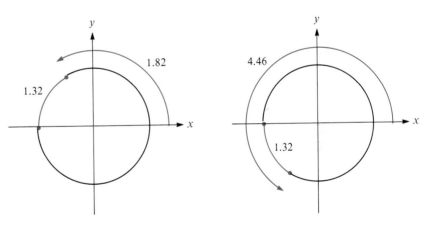

Figure 6.77

$3.14 - 1.32 = 1.82$ is the number in Q_2 with a reference number of 1.32.

$3.14 + 1.32 = 4.46$ is the number in Q_3 with a reference number of 1.32.

Thus, the formulas

$$1.82 + k2\pi \quad \text{and} \quad 4.46 + k2\pi,$$

where k is an integer, generate all the solutions to the equation, and the solution set is $\{x : x = 1.82 + k2\pi \text{ or } x = 4.46 + k2\pi, k \text{ any integer}\}$.   □

EXAMPLE 4   Find the values of x in the interval $[0,2\pi)$ for which the equation $\sin 3x = 1/\sqrt{2}$ is a true statement.

Solution   Solve the equation for $3x$.
a. Since $\sin 3x$ is positive, the point assigned to $3x$ could be in Q_1 or Q_2.
b. Since $\sin \pi/4 = 1/\sqrt{2}$, the reference number is $\pi/4$.
c. Determine the appropriate values of $3x$ and then of x (Figure 6.78).

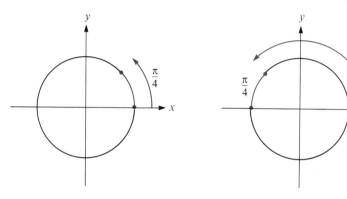

Figure 6.78

$\pi/4$ is the number in Q_1 with a reference number of $\pi/4$.

$\pi - \pi/4 = 3\pi/4$ is the number in Q_2 with a reference number of $\pi/4$.

Thus,

$$3x = \frac{\pi}{4} + k2\pi \quad \text{or} \quad 3x = \frac{3\pi}{4} + k2\pi,$$

from which we have

$$x = \frac{\pi}{12} + k\frac{2\pi}{3} \quad \text{or} \quad x = \frac{\pi}{4} + k\frac{2\pi}{3}.$$

d. To obtain solutions in the interval $[0,2\pi)$, use 0, 1, and 2 as replacements for k.

$$\frac{\pi}{12} + (0)\frac{2\pi}{3} = \frac{\pi}{12} \qquad \frac{\pi}{4} + (0)\frac{2\pi}{3} = \frac{\pi}{4}$$

$$\frac{\pi}{12} + (1)\frac{2\pi}{3} = \frac{3\pi}{4} \qquad \frac{\pi}{4} + (1)\frac{2\pi}{3} = \frac{11\pi}{12}$$

$$\frac{\pi}{12} + (2)\frac{2\pi}{3} = \frac{17\pi}{12} \qquad \frac{\pi}{4} + (2)\frac{2\pi}{3} = \frac{19\pi}{12}$$

Thus, $\{\pi/12, \pi/4, 3\pi/4, 11\pi/12, 17\pi/12, 19\pi/12\}$ is the solution set for $0 \le x < 2\pi$. ☐

The following example illustrates how the solution to a problem can depend on a knowledge of identities.

EXAMPLE 5   Solve the equation $\tan x + \cot x = 2 \csc x$ in the interval $[0, 2\pi)$.

Solution $\tan x + \cot x = 2 \csc x$. Using *fundamental identities* 4, 5, and 1, we rewrite an equivalent equation that contains only $\sin x$ and $\cos x$.

$$\frac{\sin x}{\cos x} + \frac{\cos x}{\sin x} = \frac{2}{\sin x}$$

Simplifying the fractional equation, we have

$$\sin x \cos x \left(\frac{\sin x}{\cos x} + \frac{\cos x}{\sin x} \right) = \sin x \cos x \frac{2}{\sin x}$$

$$\sin^2 x + \cos^2 x = 2 \cos x$$

$$1 = 2 \cos x \qquad \text{Identity 6}$$

$$\tfrac{1}{2} = \cos x.$$

As illustrated in Example 2 of this section, $\cos x = \frac{1}{2}$ when $x = \pi/3$ and $x = 5\pi/3$, so the solution set in $[0, 2\pi)$ is $\{\pi/3, 5\pi/3\}$. ☐

Calculators are not programmed to find all solutions to a trigonometric equation, or even the solutions between 0 and 2π. Instead, the $\boxed{\text{INV}}$ key in conjunction with the trig function keys produce the number closest to zero whose trig value is in the display. For this reason, calculators are probably best used here to obtain reference numbers by entering the magnitude of the trig value and using the inverse function keys. Solutions in the various quadrants can then be determined from the reference number as shown in this section. In the exercises remember the following points.

1. Set the calculator for radian mode.
2. Cotangent, secant, and cosecant problems are solved by using the $\boxed{1/x}$ key and then using the inverse of the corresponding reciprocal function.
3. The sine or cosine of an angle cannot be greater than 1; and, correspondingly, their reciprocals, cosecant and secant, cannot be positive numbers less than 1. An error message appears in the display when you attempt to solve such problems. The tangent and cotangent functions range over all real numbers.

EXERCISES 6.9

In Exercises 1–16 solve for x in the interval $[0,2\pi)$.

1. $\sin x = \dfrac{\sqrt{3}}{2}$ **2.** $\sin x = -\dfrac{\sqrt{3}}{2}$

3. $\cos x = -\frac{1}{2}$ **4.** $\tan x = 1$

5. $\tan x = -1$ **6.** $\sec x = 2$

7. $\sin x = 0.1219$ **8.** $\sin x = -0.1219$

9. $\tan x = -3.145$ **10.** $\tan x = 3.145$

11. $\sqrt{2} \cos x = 1$ **12.** $\csc x - \sqrt{2} = 0$

13. $\sin x + 2 = 0$ **14.** $\cos x - 3 = 0$

15. $3 \tan x - 1 = 0$ **16.** $2 \tan x + 7 = 0$

In Exercises 17–20 find all solutions to the equation.

17. $4 \sin x - 1 = 1$ **18.** $10 \cos x - 2 = 0$

19. $4 \csc x + 9 = 0$ **20.** $5 \cot x + 3 = 0$

In Exercises 21–36 solve for x in the interval $[0,2\pi)$.

21. $2 \sin^2 x - 1 = 0$ **22.** $3 \tan^2 x - 1 = 0$

23. $\cos^2 x = \cos x$ **24.** $\sin^3 x = \sin x$

25. $2 \cos x \sin x - \cos x = 0$

26. $\tan x \cos x = \tan x$

27. $\tan^2 x + 4 \tan x - 21 = 0$

28. $2 \sin^2 x - 5 \sin x = 3$

29. $\cos 3x = 0$ **30.** $\sin 4x = 1$

31. $\sin 2x = \frac{1}{2}$ **32.** $\sin \frac{1}{2}x = \frac{1}{2}$

33. $\tan \frac{1}{3}x = 1$ **34.** $\csc \frac{1}{4}x = 2$

35. $\sin^2 2x = \sin 2x$

36. $\cot^2 2x + 2 \cot 2x + 1 = 0$

In Exercises 37–40 solve for x in the interval $[0,2\pi)$. Use the quadratic formula.

37. $\cos^2 x + \cos x - 1 = 0$

38. $3 \sin^2 x - \sin x - 1 = 0$

39. $\sin^2 x + \sin x + 1 = 0$

40. $2 \tan^2 x + \tan x - 5 = 0$

In Exercises 41–54 solve each equation for $0 \le x < 2\pi$ through the aid of the *fundamental identities*.

41. $\sin x = \cos x$ **42.** $\sqrt{3} \sin x - \cos x = 0$

43. $\sin x = \tan x$ **44.** $2 \sin x - \tan x = 0$

45. $\sin^2 x + \cos^2 x = \tan x$

46. $\tan^2 x - \sec^2 x = \cos x$

47. $\sin x + \cos x \tan x = 1$

48. $\tan x + 4 \cot x = 5$

49. $2 \sin x - \csc x = 1$

50. $1 - \cos^2 x = \sin x$

51. $2 \cos^2 x = 3 \sin x + 3$

52. $2 \sin^2 x = \cos x + 1$

53. $\sec^2 x + 5 \tan x + 4 = 0$

54. $\tan^2 x - 3 \sec x = 9$

6.10 INVERSE TRIGONOMETRIC FUNCTIONS

The concept of an inverse function was introduced in Section 3.5. Recall these three facts.

1. Two functions with exactly reverse assignments are inverse functions.
2. The domain of a function f is the range of its inverse function and the range of f is the domain of its inverse.
3. A function is one-to-one when each x value in the domain is assigned a different y value so that no two ordered pairs have the same second component. We define the inverse of f, denoted f^{-1}, only when f is a one-to-one function.

Now consider the graph of $y = \sin x$ in Figure 6.79. Because the sine function is periodic, many x values are assigned the same y value. For example,

$$\sin 0 = \sin \pi = \sin 2\pi = \sin(-\pi) = 0.$$

Thus, the inverse of the sine function is not a function. The so-called **inverse sine function** is defined by restricting the domain of $y = \sin x$ to the interval $[-\pi/2,\pi/2]$. Note in Figure 6.79 that each x value is assigned a different y value in this limited version of the sine function.

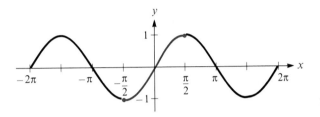

Figure 6.79

We find the rule for the inverse sine function by interchanging the x and y variables. Thus, the inverse of the function defined by

$$y = \sin x \qquad -\frac{\pi}{2} \leq x \leq \frac{\pi}{2}$$

is defined by

$$x = \sin y \qquad -\frac{\pi}{2} \leq y \leq \frac{\pi}{2},$$

which is written in inverse notation as

$$y = \arcsin x \quad \text{or} \quad y = \sin^{-1} x.$$

For example, since $\sin(\pi/2) = 1$, we write

$$\frac{\pi}{2} = \arcsin 1 \quad \text{or} \quad \frac{\pi}{2} = \sin^{-1} 1.$$

Both expressions are read "$\pi/2$ is the arc (or number) whose sin is 1," and note that the -1 in the function name $\sin^{-1}$ is not an exponent. Because a function and its inverse interchange their domain and range, for the inverse sine function the domain is the interval $[-1,1]$ and the range is $[-\pi/2,\pi/2]$. The following definition sums up the key aspects of our discussion.

Inverse Sine Function

The inverse sine function, denoted by **arcsin** or **sin⁻¹**, is defined by

$$y = \arcsin x \text{ if and only if } x = \sin y,$$

where $-1 \leq x \leq 1$ and $-\pi/2 \leq y \leq \pi/2$.

Consider Figure 6.80 which gives the graph of $y = \arcsin x$. Note that the rule in the inverse sine function basically selects the number *closest to zero* whose sine is x.

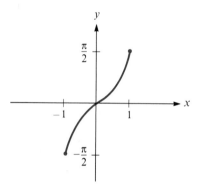

Figure 6.80

Through similar considerations we may define inverses for the other five trigonometric functions. For example, Figure 6.81 shows that $y = \cos x$ is one-to-one and assumes all its range values for $0 \leq x \leq \pi$. Thus, we may define the inverse function, $y = \arccos x$ as graphed in color, with domain $[-1,1]$ and range $[0,\pi]$ from this limited version of the cosine function. Similarly, Figure 6.82 shows that we use $y = \tan x$ for $-\pi/2 < x < \pi/2$ to define $y = \arctan x$ with the set of all real numbers for its domain and the interval $(-\pi/2, \pi/2)$ for its range. The remaining inverse functions are used less frequently, so we just list their respective domains and ranges in the following summary. A consideration of the graphs of $y = \cot x$, $y = \sec x$, and $y = \csc x$ would show the respective intervals to be suitable (though arbitrary) choices.

Figure 6.81

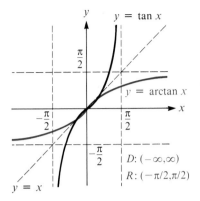

Figure 6.82

Inverse Trigonometric Functions	Function	Domain	Range
	$y = \arcsin x$	$\{x: -1 \le x \le 1\}$	$\left\{y: -\dfrac{\pi}{2} \le y \le \dfrac{\pi}{2}\right\}$
	$y = \arccos x$	$\{x: -1 \le x \le 1\}$	$\{y: 0 \le y \le \pi\}$
	$y = \arctan x$	Set of all real numbers	$\left\{y: -\dfrac{\pi}{2} < y < \dfrac{\pi}{2}\right\}$
	$y = \text{arccot } x$	Set of all real numbers	$\{y: 0 < y < \pi\}$
	$y = \text{arcsec } x$	$\{x: x \le -1 \quad \text{or} \quad x \ge 1\}$	$\left\{y: 0 \le y \le \pi, y \ne \dfrac{\pi}{2}\right\}$
	$y = \text{arccsc } x$	$\{x: x \le -1 \quad \text{or} \quad x \ge 1\}$	$\left\{y: -\dfrac{\pi}{2} \le y \le \dfrac{\pi}{2}, y \ne 0\right\}$

There are two areas concerning inverse trigonometric functions in which there is no general agreement, so take note.

1. In terms of notation an inverse trig function is sometimes capitalized and written as Arcsin x or $\text{Sin}^{-1} x$. With this convention arcsin x and $\sin^{-1} x$ do not represent functions but represent the set of all numbers whose sine is x with no restriction placed on the range values.

2. The inverse secant and cosecant functions are sometimes assigned different range intervals. Since these functions are seldom used, we will not consider the alternatives with their respective advantages.

EXAMPLE 1   Find arcsin $\frac{1}{2}$.

Solution   Let $y = $ arcsin $\frac{1}{2}$; then $\frac{1}{2} = \sin y$. We solve this equation and find the number y in the interval $[-\pi/2, \pi/2]$ whose sin is $\frac{1}{2}$. Since $\sin \pi/6 = \frac{1}{2}$, arcsin $\frac{1}{2} = \pi/6$.   □

EXAMPLE 2   Find $\cos^{-1}(-\frac{1}{2})$.

Solution   Let $y = \cos^{-1}(-\frac{1}{2})$; then $-\frac{1}{2} = \cos y$. We solve this equation and find the number y in the interval $[0, \pi]$ whose cosine is $-\frac{1}{2}$. Since $\cos(2\pi/3) = -\frac{1}{2}$, $\cos^{-1}(-\frac{1}{2}) = 2\pi/3$.   □

EXAMPLE 3   Find arctan(-1).

Solution   Let $y = $ arctan(-1); then $-1 = \tan y$. We solve this equation and find the number y in the interval $(-\pi/2, \pi/2)$ whose tangent is -1. Since $\tan(-\pi/4) = -1$, arctan$(-1) = -\pi/4$.   □

EXAMPLE 4   Find arccot 0.5312.

Solution   Let $y = $ arccot 0.5312; then $0.5312 = \cot y$. We solve this equation and find the number y in the interval $(0, \pi)$ whose cotangent is 0.5312. Table 6 at the end of the book indicates that $\cot 1.08 \approx 0.5312$. Thus, arccot $0.5312 \approx 1.08$.   □

EXAMPLE 5   Find sin(arccos 0).

Solution   First, determine arccos 0. Since $\cos(\pi/2) = 0$, we have

$$\frac{\pi}{2} = \text{arccos } 0.$$

Now replace arccos 0 by $\pi/2$ in the original expression.

$$\sin(\text{arccos } 0) = \sin\frac{\pi}{2} = 1 \qquad □$$

EXAMPLE 6   Find $\tan(\sin^{-1} \frac{4}{5})$.

Solution   It is useful to interpret $\sin^{-1} \frac{4}{5}$ as the measure of an angle in a right triangle. Let $\theta = \sin^{-1} \frac{4}{5}$ and sketch the triangle in Figure 6.83. The length of the side opposite θ is 4 and the hypotenuse has length 5. The remaining side length is found by the Pythagorean relationship.

$$x = \sqrt{5^2 - 4^2}$$
$$= 3$$

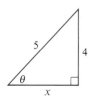

Figure 6.83 $\theta = \sin^{-1} \frac{4}{5}$

Thus,

$$\tan \theta = \tan(\sin^{-1} \tfrac{4}{5}) = \tfrac{4}{3}.$$ □

The method in Example 6 generalizes to a type of problem that occurs in calculus. The next two examples illustrate this important technique.

EXAMPLE 7   If $x = \sin \theta$, write $-\cot \theta - \theta$ as a function of x.

Solution   If $x = \sin \theta$, then $\theta = \arcsin x$. We can represent $\cot \theta$ (as in Example 6) by interpreting θ as the measure of an angle in a right triangle as shown in Figure 6.84. The length of the side opposite θ is x, the hypotenuse has length 1, and the remaining side length is found by the Pythagorean relationship to be $\sqrt{1 - x^2}$. Thus, $\cot \theta = \sqrt{1 - x^2}/x$ so

$$-\cot \theta - \theta = \frac{-\sqrt{1 - x^2}}{x} - \arcsin x.$$ □

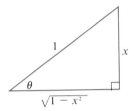

Figure 6.84

EXAMPLE 8   If $x = 4 \tan \theta$, write $\ln|\sec \theta + \tan \theta|$ as a function of x.

Solution   If $x = 4 \tan \theta$, then $\theta = \arctan (x/4)$. As shown in Figure 6.85, the length of the side opposite θ is x and the adjacent side length is 4. By the Pythagorean theorem the length of the hypotenuse is $\sqrt{x^2 + 16}$. Thus,

$$\sec \theta = \frac{\sqrt{x^2 + 16}}{4} \quad \text{and} \quad \tan \theta = \frac{x}{4}$$

so

$$\ln|\sec \theta + \tan \theta| = \ln \left| \frac{\sqrt{x^2 + 16}}{4} + \frac{x}{4} \right|.$$ □

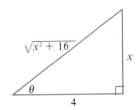

Figure 6.85

The first six example problems from this section may all be solved easily on a calculator. Just set the calculator for radian mode and proceed as follows:

EXAMPLE 1 $\arcsin \tfrac{1}{2}$ 1 $\boxed{\div}$ 2 $\boxed{=}$ $\boxed{\text{INV}}$ $\boxed{\sin}$ 0.5235988   □

EXAMPLE 2 $\cos^{-1}(-\tfrac{1}{2})$ 1 $\boxed{+/-}$ $\boxed{\div}$ 2 $\boxed{=}$ $\boxed{\text{INV}}$ $\boxed{\cos}$ 2.0943951   □

EXAMPLE 3 $\arctan(-1)$ 1 $\boxed{+/-}$ $\boxed{\text{INV}}$ $\boxed{\tan}$ -0.7853982 □

EXAMPLE 4   arccot 0.5312   0.5312 $\boxed{1/x}$ $\boxed{\text{INV}}$ $\boxed{\tan}$ 1.0825014   □

EXAMPLE 5   sin(arccos 0)   0 $\boxed{\text{INV}}$ $\boxed{\cos}$ $\boxed{\sin}$ 1   □

EXAMPLE 6   tan(sin^{-1} $\frac{4}{5}$)   4 $\boxed{\div}$ 5 $\boxed{=}$ $\boxed{\text{INV}}$ $\boxed{\sin}$ $\boxed{\tan}$ 1.3333333   □

Remember that, when possible, exact answers like $\pi/6$ and $2\pi/3$ are preferred.

One troublesome spot on a calculator is evaluating the inverse cotangent of a negative number. For this function we define $\pi/2 < y < \pi$ for $x < 0$. But the range of the inverse tangent function is defined and is programmed for $-\pi/2 < y < 0$ when $x < 0$. Since the two functions do not have the same range, you should use the calculator to find the reference number and then determine the solution between $\pi/2 < y < \pi$ as discussed in Section 6.9. For instance, to determine arccot(-0.5312) compute as in Example 4 above to find the reference number. Then the solution is

$$\pi - 1.0825014 = 2.0590913.$$

Fortunately, the inverse cotangent function is seldom used.

EXERCISES 6.10

In Exercises 1–30 evaluate the expression.

1. arccos $\frac{1}{2}$

2. arccos$\left(-\frac{1}{2}\right)$

3. sin$^{-1}\left(-\dfrac{\sqrt{3}}{2}\right)$

4. arcsin $\dfrac{\sqrt{3}}{2}$

5. arctan 1

6. arccot(-1)

7. arcsin(-1)

8. arcsec(-1)

9. arcsin $\dfrac{1}{\sqrt{2}}$

10. tan$^{-1}\sqrt{3}$

11. csc$^{-1}(-2)$

12. arccos$\left(-\dfrac{1}{\sqrt{2}}\right)$

13. arcsin 0.3124

14. sin$^{-1}(-0.3124)$

15. cos$^{-1}(-0.5509)$

16. arccos 0.5509

17. arctan 1.758

18. tan$^{-1}(-1.758)$

19. sec$^{-1}(-\frac{5}{4})$

20. arccot $\frac{2}{3}$

21. cos(arcsin 0)

22. sin(arccos 1)

23. sin$\left(\cos^{-1} \dfrac{\sqrt{3}}{2}\right)$

24. cos[sin$^{-1}(-\frac{1}{2})$]

25. cos(arccos 0)

26. cos^{-1}(cos 0)

27. arcsin(sin 1)

28. sin(arcsin 1)

29. tan[arcsin(-0.5518)]

30. csc(arccos 0.0129)

In Exercises 31–39 simplify the expression. Use right triangles.

31. sin(tan^{-1} $\frac{3}{4}$)

32. cos[arcsin($-\frac{1}{3}$)]

33. tan(arcsin $\frac{12}{13}$)

34. cot(cos^{-1} $\frac{8}{17}$)

35. cos(sin^{-1} $\frac{3}{5}$ $-$ cos^{-1} $\frac{24}{25}$)

36. sin[sin^{-1} $\frac{8}{17}$ $+$ cos$^{-1}(-\frac{12}{13})$]

37. cot(arcsin x)

38. sec$\left(\tan^{-1} \dfrac{x}{2}\right)$

39. tan$\left(\arcsin \dfrac{x-2}{3}\right)$

In Exercises 40–47 write the expression as a function of x. Use inverse functions and/or right triangles.

40. θ if $x = \tan \theta$

41. θ if $x - 2 = 3 \sec \theta$

42. ln|sec θ + tan θ| if $x = \tan \theta$

43. ln|csc θ $-$ cot θ| if $x = \sin \theta$

44. 2 sec θ tan θ if $x = 2 \sec \theta$

45. $\dfrac{-1}{4 \sin \theta}$ if $x = 2 \tan \theta$

46. $\theta + \sin \theta \cos \theta$ if $x - 1 = \sin \theta$

47. $\frac{13}{4}\theta - 2 \ln|\cos \theta|$ if $x - 5 = 4 \tan \theta$

In Exercises 48–53 solve the formula for θ or t.

48. $\tan \theta = \dfrac{b}{a}$

49. $m \cdot \sin \theta = 1$

50. $V_x = V \cos \theta$

51. $T = \dfrac{2V_0 \sin \theta}{g}$

52. $a = \sin \pi t$

53. $b = \cos(t - \pi)$

54. Consider the illustration below.

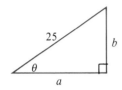

a. Write θ as a function of a.
b. Write θ as a function of b.

55. Consider the illustration below. Write θ as the difference between two inverse tangent expressions.

56. Sketch the graph of $y = \operatorname{arccot} x$.
57. Sketch the graph of $y = \arcsin(x - 1)$.

In Exercises 58–63 for what values of x is each statement true?

58. $\sin(\sin^{-1} x) = x$

59. $\sin^{-1}(\sin x) = x$

60. $\cos(\cos^{-1} x) = x$

61. $\cos^{-1}(\cos x) = x$

62. $\arctan(\tan x) = x$

63. $\tan(\arctan x) = x$.

CHAPTER OVERVIEW

Section	Key Concepts to Review
6.1	• Definitions of right triangle, acute angle, hypotenuse, similar triangles, cofunctions, complementary angles, one minute (written 1′), reciprocal functions, angle of elevation, and angle of depression
	• Definition of the trigonometric functions of an acute angle θ of a right triangle
	• The side lengths in a 30–60–90 triangle are in the ratio of $1 : \sqrt{3} : 2$.
	• The side lengths in a 45–45–90 triangle are in the ratio of $1 : 1 : \sqrt{2}$.
	• A trigonometric function of any acute angle is equal to the corresponding cofunction of the complementary angle.
	• Guidelines for using a scientific calculator or Table 5 to evaluate a trigonometric function of an acute angle
	• Methods to solve a right triangle
	• Guidelines for accuracy in computed results

Section	Key Concepts to Review
6.2	• Definitions of ray, measure of an angle, standard position of angle θ, quadrantal angle, coterminal angle, and reference angle • Definition of the trigonometric functions of an angle in standard position • Chart summarizing the signs of the trigonometric ratios • Methods to determine (if defined) approximate trigonometric values for any angle and exact values in special cases
6.3	• Definition of 1 radian • Radian measure formula: $\theta = s/r$ • Degrees to radians formula: $1° = \pi/180$ radians • Radians to degrees formula: 1 radian $= 180°/\pi$ • Area formula: $A = \frac{1}{2}r^2\theta$ (θ in radians) • Formula relating linear velocity (v) and angular velocity (ω): $v = \omega r$
6.4	• Definition of the trigonometric functions for a unit circle • Definitions of trigonometric identity and reference number • In a unit circle the same real number measures both the central angle θ and the intercepted arc s. (That is, $\theta = s$.) • For the sine and cosine functions, the domain is the set of all real numbers, and the range is the set of real numbers between -1 and 1, inclusive. • Methods to determine (if defined) approximate trigonometric values for any number and exact values in special cases • For any trigonometric function f, we have $f(s + 2\pi k) = f(s)$, where k is an integer. • Negative angle identities • When using a scientific calculator, be sure to set the calcuator for radian mode.
6.5	• Definitions of periodic function, period, amplitude, and phase shift • For $y = a \sin(bx + c)$ and $y = a \cos(bx + c)$, with $b > 0$, Amplitude $= |a|$, Period $= 2\pi/b$, Phase shift $= -c/b$
6.6	• Graphs of $y = \tan x$, $y = \cot x$, $y = \sec x$, and $y = \csc x$ • Domain, range, and period for $y = \tan x$, $y = \cot x$, $y = \sec x$, and $y = \csc x$ • Relations between a function and its reciprocal function
6.7	• Statements of the eight *fundamental identities* (See back endpaper 1–8.) • Guidelines for proving identities

Section	Key Concepts to Review
6.8	• Statements of the sum, difference, double-angle, and half-angle formulas for sine and cosine • Verification of the tangent formulas and the reduction formulas by using the above identities • Summary of the major trigonometric identities
6.9	• Methods to solve certain trigonometric equations (Reference numbers, identities, and a scientific calculator may be involved.) • Determine solutions between 0 and 2π as follows:

Quadrant	Solution
1	Reference number
2	π − reference number
3	π + reference number
4	2π − reference number

• We generate all the solutions to a trigonometric equation by adding multiples of 2π to the solutions that are in the interval $[0,2\pi)$.

6.10	• The inverse sine function is denoted by arcsin or $\sin^{-1}$. By definition, $\quad y = \arcsin x$ if and only if $x = \sin y$, where $-1 \le x \le 1$ and $-\pi/2 \le y \le \pi/2$. Similar remarks hold for the other inverse trigonometric functions. • Domain and range of the six inverse trigonometric functions • Graphs of $y = \arcsin x$, $y = \arccos x$, and $y = \arctan x$

CHAPTER REVIEW EXERCISES

In Exercises 1–10 evaluate the given expression.

1. $\sin 1$

2. $\sin^{-1} 1$

3. $\arccos(-1.11)$

4. $\cos 135°$

5. $\sin 99\pi$

6. $\tan(56\pi/3)$

7. $\sec 317°40'$

8. $\arcsin(-0.4439)$

9. $\sin[\cos^{-1}(-1)]$

10. $\tan(\sin^{-1} \frac{15}{17})$

In Exercises 11–14 state the amplitude, the period, and the phase shift and sketch one cycle of the function.

11. $y = \frac{1}{2} \cos 3x$

12. $y = 2 \sin \pi x$

13. $y = -\cos(2x + \pi/2)$

14. $y = \csc 2x$

In Exercises 15–18 solve for x in the interval $[0,2\pi)$.

15. $\csc x = -2$

16. $4 \sin^2 x - \sin x - 2 = 0$

17. $\sin 2x = \sin x$

18. $\tan 3x = 1$

In Exercises 19–22 verify the given identity.

19. $\dfrac{\sin x + \tan x}{1 + \sec x} = \sin x$

20. $\sec x \cdot \csc x = \tan x + \cot x$

21. $(\sin x - \cos x)^2 = 1 - \sin 2x$

22. $\cos(x + y) + \cos(x - y) = 2 \cos x \cos y$

23. Find the area of a sector of a circle of radius 10 in. that is subtended by a central angle of 60°.

24. In which quadrant do $y = \sin x$ and $y = \cos x$ both decrease?

25. Use the formula for the sine of the difference of two numbers and simplify the expression $\sin(3\pi - x)$.

26. What is the domain of $y = \csc x$?

27. Find an equation for the curve in the following illustration. The equation should be written in the form $y = a \sin bx$ or $y = a \cos bx$.

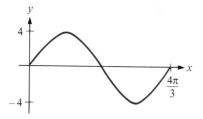

28. If $x = 3 \sin \theta$, write $\frac{9}{2}\theta - \frac{9}{4} \sin \theta \cos \theta$ as a function of x.

29. **a.** Change 210 degrees to radians.
  b. Change $2\pi/5$ radians to degrees.

30. If $\sin x = \frac{5}{13}$ and $\pi/2 < x < \pi$, find $\sin 2x$.

31. In a right triangle the length of the hypotenuse is 17 and the longer side length is 15. Determine the sine of the smaller acute angle.

32. Consider Snell's law (see Exercise 100 in Section 6.1). As light passes from air to water, v_i/v_r is about $\frac{4}{3}$. Find the angle of incidence of a ray of light if the angle of refraction is 25°40′.

33. In right triangle ABC with $C = 90°$, if $b = 5.00$ and $c = 8.00$, determine angle B.

34. If $y = \arccos(x/2)$, find x in terms of y.

35. If $\cos x = a$, express $\cos 2x$ in terms of a.

36. Match each of the graphs below with one of the following functions: $y = 2^x$, $y = x^2$, $y = \log_2 x$, and $y = \arcsin x$.

(a)   (b)

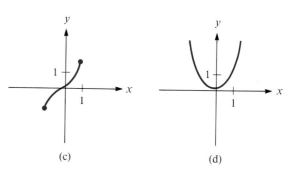

(c)   (d)

37. Express $\cos 22°$ as a function of the angle complementary to 22°.

38. Use the formula for the sine of the sum of two numbers and derive a formula for $\sin 4x$ in terms of $\sin x$ and $\cos x$.

39. In a circle a central angle intercepts an arc equal in length to the diameter of the circle. What is the measure of the central angle in radians?

40. Is $\sin(\pi/2 + x) = -\cos x$ a conditional equation or an identity?

41. Find all solutions to $\sin x = 1$ where (a) x is a real number, (b) x is an angle given in radian measure, (c) x is an angle given in degree measure.

42. If the point $(1, -1)$ is on the terminal ray of θ, find $\cos \theta$.

In Exercises 43–57 select the choice that completes the statement or answers the question.

43. The expression $\sin x + (\cos^2 x/\sin x)$ is identical to

　　a. sec x　　**b.** csc x　　**c.** cos x　　**d.** 1

44. If $x = \arccos(-\frac{1}{2})$, then x equals

　　a. $\pi/3$　　**b.** $\pi/6$　　**c.** $5\pi/6$　　**d.** $2\pi/3$

45. Which number is not in the range of $y = \sin x$?

　　a. 1　　**b.** $-\frac{1}{2}$　　**c.** 2　　**d.** 0

46. If $\sin x < 0$ and $\tan x > 0$, then the point assigned to x lies in quadrant

　　a. 1　　**b.** 2　　**c.** 3　　**d.** 4

47. The equation $\cos(-x) = -\cos x$ is true for

　　a. all values of x　　**b.** only certain values of x

　　c. no values of x

48. A function having the period π is

　　a. $y = 2 \sin x$　　**b.** $y = \frac{1}{2} \sin x$

　　c. $y = \sin \frac{1}{2}x$　　**d.** $y = \sin 2x$

49. To convert from radians to degrees, we multiply the number of radians by

　　a. $180°/\pi$　　**b.** $\pi/90°$　　**c.** $90°/\pi$　　**d.** $\pi/180°$

50. The expression $\cos(x - \pi)$ simplifies to

　　a. cos x　　**b.** $-\cos x$　　**c.** sin x　　**d.** $-\sin x$

51. If a central angle of a radians intercepts an arc of length b in a circle of radius c, then

　　a. $a = b/c$　　**b.** $a = \pi/b$

　　c. $a = c/b$　　**d.** $a = bc$

52. If $f(x) = \cos 3x + \tan 2x$, then $f(\pi/6)$ equals

　　a. $\sqrt{3}$　　**b.** $1 + \sqrt{3}$

　　c. $\sqrt{3}/3$　　**d.** $(3 + \sqrt{3})/3$

53. If $\log \tan x = a$, then $\log \cot x$ equals

　　a. $1/a$　　**b.** $-a$　　**c.** $1 - a$　　**d.** a

54. Which one of the following is an identity?

　　a. $\sin x + \cos x = 1$　　**b.** $\sec x \cdot \csc x = 1$

　　c. $\sin \frac{1}{2}x = \frac{1}{2} \sin x$　　**d.** $\cos^2 x = 1 - \sin^2 x$

55. Find sec R in the following right triangle.

　　a. $3/5$　　**b.** $4/3$　　**c.** $\sqrt{41}/5$　　**d.** $\sqrt{41}/4$

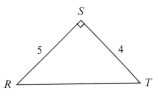

56. The value of $\sin 100°$ is equal to

　　a. $\sin 10°$　　**b.** $\sin 80°$

　　c. $-\sin 80°$　　**d.** $-\sin 10°$

57. If the terminal ray of θ lies on the line $y = 2x$ and θ is in Q_3, then one angle measure for θ is

　　a. $225°$　　**b.** $210°$　　**c.** $207°$　　**d.** $243°$

58. A road rises 25 ft in a horizontal distance of $4\overline{0}0$ ft. Find the angle that the road makes with the horizontal.

59. From the top of a building $3\overline{0}$ ft tall, the angle of depression to the foot of a building across the street is $60°$ and the angle of elevation to the top of the same building is $70°$. How tall is the building?

60. A carpenter has to build a stairway. The total rise is 8 ft 6 in. and the angle of rise is $30°$, as shown in the illustration. What is the shortest piece of 2- × -12-in. stock that can be used to make the stringer?

61. Consider the diagram below, which illustrates that twilight lasts until the sun is 18° below the horizon. From this, estimate the height (*h*) of the atmosphere.

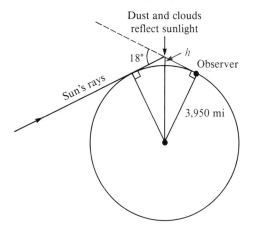

62. To determine the radius of the sun, an observer on earth at point *O* measures angle θ to be 16′, as illustrated by the diagram. If the distance from the earth to the sun is about 93,000,000 mi, what is the radius of the sun?

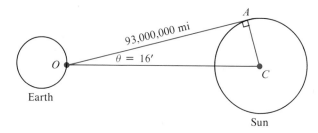

7 Trigonometry: *General Angle Applications*

On a ground ball to the first base side, the pitcher is expected to cover first base and beat the batter to this bag. If a baseball diamond is a square that is 90.0 ft long on each side, and if the pitcher's mound is 60.5 ft from home plate on the diagonal from home to second base, then how far is the pitcher's mound from first base? (See Example 4 of Section 7.2.)

60.5 ft

90.0 ft

I n this chapter we continue our discussion of trigonometry by considering some applications of the trigonometric functions with angle measures as domain elements. These applications involve the law of sines, the law of cosines, vectors, the trigonometric form of complex numbers, and polar coordinates.

7.1 LAW OF SINES

In Section 6.1 we learned to solve right triangles. We now wish to extend our ability to solve triangles by considering the solution of general triangles, which may or may not be right triangles. Remember that we "solve" a triangle by finding the measures of its three angles and three sides. To accomplish this, at least three of these six values must be known and one or more must be a side length.

The first technique that we use to solve general triangles is called the **law of sines.** We can derive this law by placing triangle *ABC* on a rectangular coordinate system so that angle *A* is in standard position. Figure 7.1(a) and (b) show the result when *A* is an acute angle and an obtuse angle, respectively. In both cases we draw the altitude of the triangle, *CD*, and note that its length is *y*.

(a)

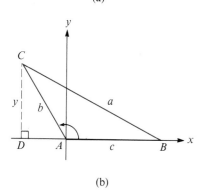

(b)

Figure 7.1 (a) A is an acute angle. (b) A is an obtuse angle.

Then by the general definition of the sine function,

$$\sin A = \frac{y}{b}, \text{ so } y = b \sin A.$$

Also, in right triangle BDC,

$$\sin B = \frac{y}{a}, \text{ so } y = a \sin B.$$

Setting these two expressions for y equal to each other, we have

$$b \sin A = a \sin B, \text{ so } \frac{\sin A}{a} = \frac{\sin B}{b}.$$

Similarly, by placing angle C in standard position we can show that

$$\frac{\sin A}{a} = \frac{\sin C}{c}.$$

Combining these results, we have

$$\frac{\sin A}{a} = \frac{\sin B}{b} = \frac{\sin C}{c}.$$

This relationship, the **law of sines,** states the following.

Law of Sines	The sines of the angles in a triangle are proportional to the lengths of the opposite sides.

Note that if C is a right angle, $\sin C = \sin 90° = 1$, and the law of sines yields the right triangle relationships

$$\sin A = \frac{a}{c} \text{ and } \sin B = \frac{b}{c}.$$

The law of sines can be used to solve any triangle in the following two cases:

1. If we know the measures for two angles and one side of the triangle
2. If we know the measures for two sides of the triangle and the angle opposite one of them

The following example illustrates how the law of sines can be used to solve a triangle in the first case in which the measures for two angles and one side of the triangle are known. Note that in computing results the symbol for equality (=) is generally used even though the symbol for approximation (≈) may be more appropriate.

EXAMPLE 1   Approximate the missing parts of triangle ABC in which $A = 35°$, $B = 50°$, and $a = 12$ ft.

Solution   First, sketch Figure 7.2. We can find angle C since the sum of the angles in a triangle is $180°$.

$$A + B + C = 180°$$
$$35° + 50° + C = 180°$$
$$C = 95°$$

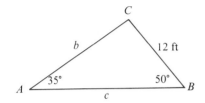

Figure 7.2

Second, we find side length b by applying the law of sines.

$$\frac{\sin A}{a} = \frac{\sin B}{b}$$

$$\frac{\sin 35°}{12} = \frac{\sin 50°}{b}$$

$$b = \frac{12 \sin 50°}{\sin 35°}$$

$$= \frac{12(0.7660)}{0.5736}$$

$$= 16 \text{ ft}$$

Third, we find c by applying the law of sines.

$$\frac{\sin A}{a} = \frac{\sin C}{c}$$

$$\frac{\sin 35°}{12} = \frac{\sin 95°}{c}$$

$$c = \frac{12 \sin 95°}{\sin 35°}$$

$$= \frac{12(0.9962)}{0.5736} \qquad (\textit{Note: } \sin 95° = \sin 85° = 0.9962.)$$

$$= 21 \text{ ft}$$

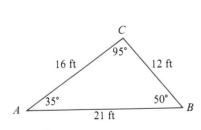

Figure 7.3

Thus, the solution to the triangle is shown in Figure 7.3.   □

(*Note:* Remember that our computed results cannot be more accurate than the data that are given. Guidelines for the desired accuracy in a solution can be found in Section 6.1. The arithmetic involved in these calculations is quite tedious, so you are encouraged to use a calculator.)

EXAMPLE 2   Two observation towers, *A* and *B*, are located $1\overline{0}$ mi apart. A fire is sighted at point *C* and the observer in tower *A* measures angle *CAB* to be 80°. At the same time the observer in tower *B* measures angle *CBA* to be 40°. How far is the fire from tower *A*?

Solution   First, draw a diagram picturing the data (Figure 7.4). Second, find angle *C*.

$$A + B + C = 180°$$
$$80° + 40° + C = 180°$$
$$C = 60°$$

Third, we find *b* by applying the law of sines.

$$\frac{\sin B}{b} = \frac{\sin C}{c}$$

$$\frac{\sin 40°}{b} = \frac{\sin 60°}{10}$$

$$b = \frac{10 \sin 40°}{\sin 60°}$$

$$= \frac{10(0.6428)}{0.8660}$$

$$= 7.4$$

Thus, the fire is about 7.4 mi from station *A*.   □

The following examples illustrate how the law of sines can be used to solve a triangle in the second case in which the measures for two sides of the triangle and the angle opposite one of them are known.

EXAMPLE 3   Solve the triangle *ABC* in which $B = 60°$, $b = 5\overline{0}$ ft, and $c = 3\overline{0}$ ft.

Solution   First, sketch Figure 7.5. We find angle *C* by applying the law of sines.

$$\frac{\sin B}{b} = \frac{\sin C}{c}$$

$$\frac{\sin 60°}{50} = \frac{\sin C}{30}$$

$$\frac{30(\sin 60°)}{50} = \sin C$$

$$\frac{30(0.8660)}{50} = \sin C$$

$$0.5196 = \sin C$$

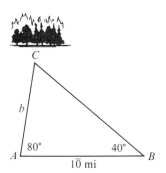

C

b

80°   40°

A ———— $1\overline{0}$ mi ———— *B*

Figure 7.4

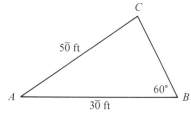

C

$5\overline{0}$ ft

60°

A ———— $3\overline{0}$ ft ———— *B*

Figure 7.5

We now have two possibilities for angle C since the sine of both first and second quadrant angles is positive.

Case 1   (acute angle in Q_1):   **Case 2**   (obtuse angle in Q_2):

   $\sin C = 0.5196$ $\sin C = 0.5196$

   reference angle $= 31°20'$ (from Table 5)

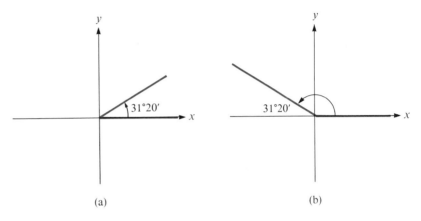

(a)   (b)

Figure 7.6

$31°20'$ is the angle in Q_1 with a reference angle of $31°20'$ [Figure 7.6(a)].

$148°40'$ is the angle in Q_2 with a reference angle of $31°20'$ [Figure 7.6(b)].

Therefore, $C = 31°20'$ or $C = 148°40'$.

Second, we find angle A in both of the above cases.

$$A + B + C = 180°$$
$$A + 60° + 31°20' = 180°$$
$$A = 88°40'$$

$$A + B + C = 180°$$
$$B = 60° \qquad C = 148°40'$$

Here we find $B + C = 208°40'$, so regardless of the value of A, the sum of the angles of the triangle exceeds $180°$. Therefore, we reject $C = 148°40'$ as a solution.

Third, we find side length a by applying the law of sines.

$$\frac{\sin A}{a} = \frac{\sin B}{b}$$

$$\frac{\sin 88°40'}{a} = \frac{\sin 60°}{50}$$

$$\frac{50(\sin 88°40')}{\sin 60°} = a$$

$$\frac{50(0.9997)}{0.8660} = a$$

$$58 \text{ ft} = a$$

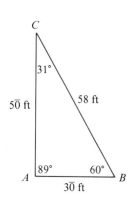

Figure 7.7

When we round off the angle measures to the nearest degree, the solution to the triangle is as shown in Figure 7.7.   □

EXAMPLE 4   Approximate the missing parts of triangle ABC in which $A = 37°20'$, $a = 12.5$ ft, and $c = 20.1$ ft.

Solution   First, sketch Figure 7.8. We find angle C by applying the law of sines.

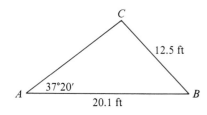

Figure 7.8

$$\frac{\sin A}{a} = \frac{\sin C}{c}$$

$$\frac{\sin 37°20'}{12.5} = \frac{\sin C}{20.1}$$

$$\frac{20.1(\sin 37°20')}{12.5} = \sin C$$

$$\frac{20.1(0.6065)}{12.5} = \sin C$$

$$0.9753 = \sin C$$

We now have two possibilities.

Case 1   (acute angle in Q_1):   **Case 2**   (obtuse angle in Q_2):

$\sin C = 0.9753$ $\sin C = 0.9753$

reference angle $= 77°10'$ (from Table 5)

77°10' is the angle in Q_1 with a reference angle of 77°10' [Figure 7.9(a)].   102°50' is the angle in Q_2 with a reference angle of 77°10' [Figure 7.9(b)].

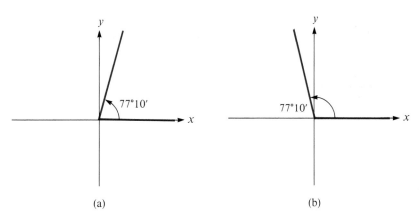

(a)   (b)

Figure 7.9

Therefore, $C = 77°10'$ or $C = 102°50'$.

Second, we find angle B in both of the above cases.

$$A + B + C = 180° \qquad\qquad A + B + C = 180°$$
$$37°20' + B + 77°10' = 180° \qquad 37°20' + B + 102°50' = 180°$$
$$B + 114°30' = 180° \qquad\qquad B + 140°10' = 180°$$
$$B = 65°30' \qquad\qquad\qquad B = 39°50'$$

Third, we find side length b by applying the law of sines.

$$\frac{\sin A}{a} = \frac{\sin B}{b} \qquad\qquad\qquad \frac{\sin A}{a} = \frac{\sin B}{b}$$

$$\frac{\sin 37°20'}{12.5} = \frac{\sin 65°30'}{b} \qquad\qquad \frac{\sin 37°20'}{12.5} = \frac{\sin 39°50'}{b}$$

$$b = \frac{12.5(\sin 65°30')}{\sin 37°20'} \qquad\qquad b = \frac{12.5(\sin 39°50')}{\sin 37°20'}$$

$$= \frac{12.5(0.9100)}{0.6065} \qquad\qquad = \frac{12.5(0.6406)}{0.6065}$$

$$= 18.8 \text{ ft} \qquad\qquad\qquad = 13.2 \text{ ft}$$

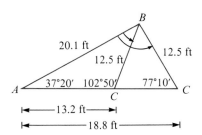

Figure 7.10

The two possible solutions from the given data are shown in Figure 7.10. □

Note that when we attempt to solve a triangle in which the measures for two sides of the triangle and the angle opposite one of them are given, there may be one triangle that fits the data (as in Example 3) or there may be two triangles that fit the data (as in Example 4). Sometimes no triangle can be constructed from the data; then we say the data is inconsistent.

EXERCISES 7.1

(*Note:* There are more problems on the law of sines in Exercises 7.2.) In Exercises 1–12 approximate the remaining parts of the triangle for the data given.

1. $A = 30°$, $a = 25$ ft, and $B = 45°$
2. $C = 60°$, $c = 4\overline{0}$ ft, and $A = 80°$
3. $B = 120°$, $C = 40°$, and $a = 55$ ft
4. $C = 135°$, $c = 98$ ft, and $B = 15°$
5. $A = 62°10'$, $a = 31.5$ ft, and $B = 76°30'$
6. $A = 98°30'$, $B = 6°10'$, and $a = 415$ ft
7. $A = 45°$, $a = 8\overline{0}$ ft, and $b = 5\overline{0}$ ft
8. $C = 60°$, $c = 75$ ft, and $a = 45$ ft
9. $B = 30°$, $b = 3\overline{0}$ ft, and $a = 4\overline{0}$ ft
10. $B = 22°$, $b = 78$ ft, and $a = 86$ ft
11. $C = 150°$, $c = 92$ ft, and $b = 69$ ft
12. $C = 105°30'$, $c = 46.1$ ft, and $b = 75.2$ ft

13. Two surveyors establish a baseline AB on a level field. The surveyor at point A is $20\overline{0}$ ft from the surveyor at point B. Each one sights a stake at point C. The surveyor at A measures angle CAB to be $72°30'$, while the surveyor at B measures angle CBA to be $81°20'$. Find the distance from B to C.

14. Engineers wish to build a bridge across a river to join point A on one side to either point B or point C on the other side. This distance from B to C is $40\overline{0}$ ft, angle ABC is $67°20'$, and angle ACB is $84°30'$. By how many feet does the distance from A to B exceed the distance from A to C?

15. Airport A is $30\overline{0}$ mi due north of airport B. Their radio stations receive a distress signal from a ship located at point C. It is determined that point C is located 54° south of east with respect to airport A and 76° north of east from airport B. How far is the ship from airport A?

16. Two engineers are located at points A and B on the opposite sides of a hill. They are both able to see a stake at point C, which is at a distance of $80\overline{0}$ ft from A and $70\overline{0}$ ft from B. If angle ABC is 25°, find the distance $\overline{AB}$ through the hill.

17. In the illustration below find acute angles θ, α, and β to the nearest degree and x and y to two significant digits.

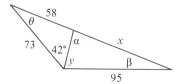

7.2 LAW OF COSINES

In Section 7.1 we found that the law of sines can be used to solve any triangle in the following two cases:

1. If we know the measures for **two angles** and **one side** of the triangle
2. If we know the measures for **two sides** of the triangle and the **angle opposite** one of them

However, there exist two other cases for which the law of sines cannot be applied. They are:

3. If we know the measures for **two sides** of the triangle and the **angle between** these two sides
4. If we know the measures for the **three sides** of the triangle

In these cases we use the **law of cosines,** which states the following.

Law of Cosines   In any triangle, the square of any side length equals the sum of the squares of the other two side lengths, minus twice the product of these other two side lengths and the cosine of their included angle.

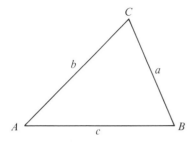

Figure 7.11

Thus, for triangle *ABC* in Figure 7.11 we have

$$a^2 = b^2 + c^2 - 2bc \cos A$$
$$b^2 = a^2 + c^2 - 2ac \cos B$$
$$c^2 = a^2 + b^2 - 2ab \cos C.$$

If we know the measures for two sides of the triangle and the included angle, we find the third side length by substituting in one of these formulas. After finding this part, we complete the solution by use of the law of sines. To obtain accuracy in the angle measures, the computed side length should be carried in the calculations to at least one more significant digit than stated in the solution.

If the three forms of the law of cosines are solved for the cosine of the angle, we have

$$\cos A = \frac{b^2 + c^2 - a^2}{2bc}$$

$$\cos B = \frac{a^2 + c^2 - b^2}{2ac}$$

$$\cos C = \frac{a^2 + b^2 - c^2}{2ab}.$$

These formulas are used to find the angle measures in a triangle when we know the three side lengths. In this case we do not use the law of sines to complete the solution because results are more accurate when they are computed from the data given.

Before starting the sample problems, let us first derive the law of cosines. Once again, we place triangle *ABC* on a rectangular coordinate system with angle *A* in standard position and consider both an acute and obtuse possibility for angle *A* as shown in Figure 7.12(a) and (b). In both cases vertex *B* obviously has coordinates $(c,0)$. Also, in general, $\cos A = x/b$ and $\sin A = y/b$, so the *x*-coordinate of vertex *C* is $b \cos A$ and the *y*-coordinate is $b \sin A$. If we now apply the distance formula to find the square of side length *a*, we have

(a)

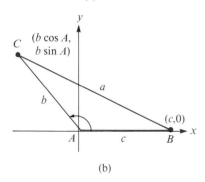

(b)

Figure 7.12

$$a^2 = (c - b \cos A)^2 + (0 - b \sin A)^2$$
$$= c^2 - 2bc \cos A + b^2 \cos^2 A + b^2 \sin^2 A$$
$$= b^2(\sin^2 A + \cos^2 A) + c^2 - 2bc \cos A.$$

Since $\sin^2 A + \cos^2 A = 1$ is an identity, the equation becomes

$$a^2 = b^2 + c^2 - 2bc \cos A,$$

which is one form of the law of cosines. A similar procedure with angles B and C, respectively, placed in standard position yields the other two forms.

EXAMPLE 1   Approximate the missing parts of triangle ABC in which $A = 60°$, $b = 25$ ft, and $c = 42$ ft.

Solution   First, sketch Figure 7.13. We find side length a by applying the law of cosines.

$$
\begin{aligned}
a^2 &= b^2 + c^2 - 2bc \cos A \\
&= (25)^2 + (42)^2 - 2(25)(42) \cos 60° \\
&= 625 + 1{,}764 - 2{,}100(0.5000) \\
&= 2{,}389 - 1{,}050 \\
&= 1{,}339 \\
a &= \sqrt{1{,}339} \approx 36.6 \\
a &= 37 \text{ ft}
\end{aligned}
$$

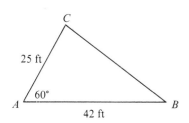

Figure 7.13

Second, we find the smaller of the remaining angles, angle B, by applying the law of sines. This angle must be acute. (Why?) We use 36.6 for a for better accuracy.

$$
\begin{aligned}
\frac{\sin A}{a} &= \frac{\sin B}{b} \\
\frac{\sin 60°}{36.6} &= \frac{\sin B}{25} \\
\frac{25 \sin 60°}{36.6} &= \sin B \\
\frac{25(0.8660)}{36.6} &= \sin B \\
0.5915 &= \sin B \\
36° &= B
\end{aligned}
$$

(*Note:* $\sin B = 0.5915$ is true if $B = 36°$ or if $B = 144°$. We eliminate 144° as a possible solution since we know that angle B must be acute.)

Third, we find angle C.

$$
\begin{aligned}
A + B + C &= 180° \\
60° + 36° + C &= 180° \\
C &= 84°
\end{aligned}
$$

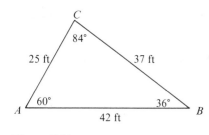

Figure 7.14

Thus, the solution to the triangle is as shown in Figure 7.14.   □

EXAMPLE 2   Approximate the missing parts of triangle ABC in which $a = 23.5$ ft, $b = 44.2$ ft, and $c = 30.1$ ft.

Solution   First, sketch Figure 7.15. We find angle A by applying the law of cosines.

$$\cos A = \frac{b^2 + c^2 - a^2}{2bc}$$

$$= \frac{(44.2)^2 + (30.1)^2 - (23.5)^2}{2(44.2)(30.1)}$$

$$= 0.8672$$

$$A = 29°50'$$

(*Note:* Remember that if the cosine of the angle is positive, the angle is acute; if the cosine of the angle is negative, the angle is in Q_2 and is obtuse.)

Second, we find the smaller of the remaining angles, acute angle C, by applying the law of cosines.

$$\cos C = \frac{a^2 + b^2 - c^2}{2ab}$$

$$= \frac{(44.2)^2 + (23.5)^2 - (30.1)^2}{2(44.2)(23.5)}$$

$$= 0.7701$$

$$C = 39°40'$$

Third, we find angle B.

$$A + B + C = 180°$$
$$29°50' + B + 39°40' = 180°$$
$$B = 110°30'$$

Thus, the solution to the triangle is as shown in Figure 7.16.   ☐

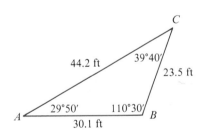

Figure 7.15

Figure 7.16

EXAMPLE 3   In parallelogram $ABCD$ the lengths of sides AB and AD are 12 m and 19 m, respectively. If $A = 38°$, find the length of the longer diagonal of the parallelogram.

Solution   First, sketch Figure 7.17 and note that the longer diagonal is AC. In the parallelogram $\overline{AB} = \overline{DC} = 12$ and the sum of angles A and D is 180°. Thus,

$$A + D = 180°$$
$$38° + D = 180°$$
$$D = 142°.$$

Figure 7.17

We find the length of the longer diagonal by applying the law of cosines.

$$(\overline{AC})^2 = (\overline{AD})^2 + (\overline{DC})^2 - 2(\overline{AD})(\overline{DC}) \cos D$$
$$(\overline{AC})^2 = (19)^2 + (12)^2 - 2(19)(12) \cos 142°$$

Since $\cos 142° = -\cos 38° = -0.7880$, we have

$$(\overline{AC})^2 = (19)^2 + (12)^2 - 2(19)(12)(-0.7880)$$
$$(\overline{AC})^2 = 864.3$$
$$\overline{AC} = 29.$$

Thus, the longer diagonal is about 29 m.   □

EXAMPLE 4   Solve the problem in the chapter introduction on page 357.

Solution   Consider the simplified sketch of the problem in Figure 7.18. We find x by applying the law of cosines to obtain

$$x^2 = (60.5)^2 + (90)^2 - 2(60.5)(90) \cos 45°$$
$$= 3,660.25 + 8,100 - 10,890(0.7071)$$
$$= 4,059.93$$
$$x = \sqrt{4,059.93} \approx 63.7.$$

Thus, the pitcher's mound is about 63.7 ft from first base, which should give the pitcher an adequate head start in a race to the bag.   □

Figure 7.18

EXERCISES 7.2

In Exercises 1–10 approximate the remaining parts of triangle ABC. The law of cosines will be needed in at least one of the steps.

1. $a = 12$ ft, $b = 15$ ft, and $C = 60°$

2. $a = 2\overline{0}$ ft, $c = 3\overline{0}$ ft, and $B = 30°$

3. $c = 19.2$ ft, $a = 46.1$ ft, and $B = 10°20'$

4. $b = 36$ ft, $c = 75$ ft, and $A = 98°$

5. $b = 11.1$ ft, $a = 19.2$ ft, and $C = 95°40'$

6. $a = 11$ ft, $b = 15$ ft, and $c = 19$ ft

7. $a = 12$ ft, $b = 5.2$ ft, and $c = 8.1$ ft

8. $a = 4.9$ ft, $b = 5.3$ ft, and $c = 2.6$ ft

9. $a = 34.4$ ft, $b = 56.1$ ft, and $c = 42.3$ ft

10. $a = 45.0$ ft, $b = 108$ ft, and $c = 117$ ft

In Exercises 11–14 use the law of cosines.

11. In parallelogram $ABCD$ the lengths of sides AB and AD are 21 m and 13 m, respectively. If $A = 52°$, find the length of the longer diagonal of the parallelogram.

12. In parallelogram $ABCD$ the length of sides AB and AD are 6.0 and 8.0 m, respectively. If the length of the shorter diagonal is 5.0 m, find angle A.

13. A surveyor at point C sights two points A and B on opposite sides of a lake. If C is 760 ft from A and 920 ft from B and angle ACB is 96°, how wide is the lake?

14. What is the distance between the two islands in the illustration below?

In Exercises 15–26 the problems are mixed, that is, some use the law of sines, some the law of cosines, and some use both. Solve each triangle.

15. $A = 15°$, $C = 87°$, and $b = 42$ ft

16. $B = 68°$, $C = 72°$, and $a = 18$ ft

17. $a = 4\overline{0}$ ft, $b = 5\overline{0}$ ft, and $C = 120°$

18. $b = 126$ ft, $c = 92.1$ ft, and $A = 72°50'$

19. $B = 111°20'$, $C = 35°40'$, and $a = 142$ ft

20. $c = 127$ ft, $b = 315$ ft, and $A = 162°30'$

21. $a = 4.0$ ft, $b = 2.0$ ft, and $c = 3.0$ ft

22. $A = 95°$, $a = 54$ ft, and $c = 38$ ft

23. $B = 7°10'$, $b = 74.8$ ft, and $c = 92.4$ ft

24. $a = 84.8$ ft, $b = 36.8$ ft, and $c = 76.5$ ft

25. $a = 15\overline{0}$ ft, $b = 175$ ft, and $c = 20\overline{0}$ ft

26. $B = 152°50'$, $b = 13\overline{0}$ ft, and $c = 45.0$ ft

In Exercises 27–31 the solution will require a decision as to whether the law of sines or the law of cosines is appropriate.

27. A ship sails due east for $4\overline{0}$ mi and then changes direction and sails 20° north of east for $6\overline{0}$ mi. How far is the ship from its starting point?

28. One gun is located at point A, while a second gun at point B is located 5.0 mi directly east of A. From point A the direction to the target is 27° north of east. From point B the direction to the target is 72° north of east. For what firing range should the guns be set?

29. In a parallelogram the shorter diagonal makes angles of 25° and 72° with the sides. If the length of the shorter side is 15 m, what is the length of the longer side?

30. Find the perimeter of the regular pentagon in the illustration below.

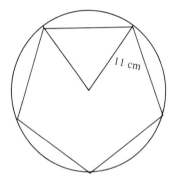

11 cm

31. A and B are two points located on opposite edges of a lake. A third point C is located so that $\overline{AC}$ is 421 ft and $\overline{BC}$ is 376 ft. Angle ABC is measured to be 65.5°. Compute the distance $\overline{AB}$ across the lake.

7.3 VECTORS

An application of trigonometry occurs with the study of physical quantities that act in a definite direction. For example, when meteorologists describe wind they mention both the speed of the wind and the direction from which the wind is blowing. Similarly, quantities such as forces, weights, and velocities must be described in such a way that both the strength (magnitude) and the direction of the quantity can be determined. Mathematically, we represent such a quantity by a line segment with an arrowhead at one end; this directed line segment is called a **vector.** The direction in which the arrowhead is pointing represents the direction in which the quantity is acting, while the length of the line segment is proportional to the magnitude of the quantity.

EXAMPLE 1   Use a vector to represent graphically a wind that is blowing due north at a speed of 40 mi/hour.

Solution   If 1 unit represents 10 mi/hour, the directed line segment **OA,** which is 4 units long and pointing due north, is the appropriate vector (Figure 7.19). (*Note:* A vector that starts at *O* and ends at *A* is labeled **OA;** a vector that starts at *A* and ends at *O* is a different vector and is labeled **AO.**)

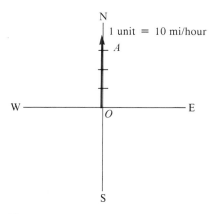

Figure 7.19   □

Frequently, there are two (or more) forces acting on a body from different directions, and their net effect is a third force with a new direction. This new force is called the **resultant,** or vector sum, of the given forces. For example, in Figure 7.20 two forces of 20 lb and 30 lb are both acting on a body with an angle of 30° between the two forces. It can be shown experimentally that the resultant of vectors **OA** and **OB** is vector **OC,** which is the diagonal of a parallelogram formed from the given vectors. The magnitude of the resultant **OC** can be determined by the length of the line segment from *O* to *C*. The direction of the resultant is the same as the direction of the arrowhead on vector **OC** and is described in terms of the original vectors by finding either angle *AOC* or angle *BOC*. The next three examples illustrate this procedure for finding the resultant of two forces. Examples 2 and 3 require only right triangle trigonometry, while Example 4 uses both the law of cosines and the law of sines.

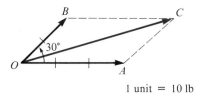

1 unit = 10 lb

Figure 7.20

EXAMPLE 2   A force of 3.0 lb and a force of 4.0 lb are acting on a body with an angle of 90° between the two forces. Find (a) the magnitude of the resultant and (b) the angle between the resultant and the larger force.

Figure 7.21

Solution   Let **OA** represent the 4.0-lb force and **OB** the 3.0-lb force. Then vector **OC** represents the resultant (see Figure 7.21).

a. We can find the length of **OC** by using the Pythagorean theorem in right triangle OAC. Notice that line segments OB and AC have the same length.

$$(\overline{OC})^2 = (4)^2 + (3)^2$$
$$(\overline{OC})^2 = 25$$
$$\overline{OC} = 5.0$$

Thus, the magnitude of the resultant is 5.0 lb.

b. We can find the angle between the resultant and the 4.0-lb force by using the tangent function in right triangle OAC.

$$\tan \theta = \frac{3}{4}$$
$$\theta = 37° \qquad \text{(to the nearest degree)}$$

Thus, the angle between the resultant and the larger force is 37°.   □

EXAMPLE 3   A ship is headed due east at $2\overline{0}$ knots (nautical miles per hour) while the current carries the ship due south at 5.0 knots. Find (a) the speed of the ship and (b) the direction (course) of the ship.

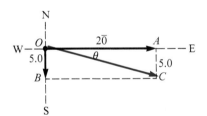

Figure 7.22

Solution   Let vector **OA** represent the velocity of the ship in the easterly direction. Let vector **OB** represent the velocity of the ship in the southerly direction due to the current. Let vector **OC** represent the actual velocity of the ship. (See Figure 7.22.)

a. We can find the length of **OC** by using the Pythagorean theorem in right triangle OAC.

$$(\overline{OC})^2 = (20)^2 + (5)^2$$
$$(\overline{OC})^2 = 425$$
$$\overline{OC} = \sqrt{425} \approx 21$$

Thus, the speed of the ship is 21 knots.

b. We can find the angle between the resultant and vector **OA,** which is due east, by using the tangent function in right triangle OAC.

$$\tan \theta = \frac{5}{20}$$
$$\theta = 14° \qquad \text{(to the nearest degree)}$$

Thus, the ship is heading in a direction that is 14° south of east.   □

Figure 7.23

EXAMPLE 4   A body is acted on by two forces with magnitudes of 78 lb and 42 lb that act at an angle of 25° with each other, as shown in Figure 7.23. Find (a) the magnitude of the resultant and (b) the angle between the resultant and the larger force.

Solution

a. The resultant of the two given forces is the diagonal of the parallelogram $ABCD$ formed from the given vectors. In the parallelogram $\overline{AB} = \overline{DC} = 42$ and the sum of angle A and angle D equals 180°. Thus,

$$A + D = 180°$$
$$25° + D = 180°$$
$$D = 155°.$$

We find the magnitude of resultant **AC** by applying the law of cosines.

$$(\overline{AC})^2 = (78)^2 + (42)^2 - 2(78)(42) \cos 155°$$

Since $\cos 155° = -\cos 25° = -0.9063$, we have

$$(\overline{AC})^2 = (78)^2 + (42)^2 - 2(78)(42)(-0.9063)$$
$$\overline{AC} = \sqrt{13{,}786} \approx 117$$
$$\overline{AC} = 120.$$

Thus, the magnitude of the resultant is 120 lb.

b. We find the angle between the resultant and the larger force, angle CAD, by applying the law of sines. Note that since angle D is greater than 90°, angle CAD must be acute. We use 117 for $\overline{AC}$ for better accuracy.

$$\frac{\sin CAD}{42} = \frac{\sin 155°}{117}$$
$$\sin CAD = \frac{42 \sin 155°}{117}$$

Since $\sin 155° = \sin 25° = 0.4226$, we have

$$\sin CAD = \frac{42(0.4226)}{117}$$
$$= 0.1517$$
$$CAD = 9°.$$

Thus, the angle between the resultant and the larger force is 9°.   □

In the previous examples we considered how the net effect of having two forces acting on a body produced a third force, the resultant. However, the reverse situation often arises where we are given a single force that we think of as the resultant and we need to calculate two forces, which are called **components,** that produce the resultant. This process of expressing a single force in terms of two components, which are usually at right angles to each other, is called **resolving a vector.** The following examples illustrate the usefulness of resolving a vector into components.

EXAMPLE 5   An airplane, pointed due west, is traveling 10° north of west at a rate of 4$\overline{0}$0 mi/hr. This resultant course is due to a wind blowing north. Find (a) the velocity of the plane if there were no wind (that is, the vector pointing due west) and (b) the velocity of the wind (that is, the vector pointing due north).

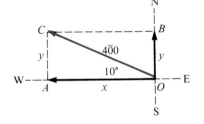

Figure 7.24

Solution   Let vector **OC** represent the resultant velocity of the plane. The resolution of vector **OC** results in a westerly component vector **OA** (the velocity of the plane if there were no wind) and a northerly component vector **OB** (the velocity of the wind) (see Figure 7.24).
  a. We can find the velocity of the plane if there were no wind by using the cosine function in right triangle *OAC.*

$$\cos 10° = \frac{x}{400}$$

$$400(\cos 10°) = x$$
$$400(0.9848) = x$$
$$390 = x \quad \text{(two significant digits)}$$

Thus, the velocity of the plane if there were no wind would be 390 mi/hour due west.
  b. We can find the velocity of the wind by using the sine function in right triangle *OAC.*

$$\sin 10° = \frac{y}{400}$$

$$400(\sin 10°) = y$$
$$400(0.1736) = y$$
$$69 = y \quad \text{(two significant digits)}$$

Thus, the velocity of the wind would be 69 mi/hour due north.   □

EXAMPLE 6   A man pulls with a force of $4\overline{0}$ lb on a window pole in an effort to lower a window. What part of the man's force lowers the window if the pole makes an angle of 20° with the window?

Solution   Let vector **OA** represent the pull on the window pole of $4\overline{0}$ lb. The resolution of vector **OA** results in the vertical component vector **OC** (the force lowering the window) and the horizontal component vector **OB** (wasted force) (see Figure 7.25).
    We can find the magnitude of vector **OC** by using the cosine function in right triangle *ACO*.

$$\cos 20° = \frac{\overline{OC}}{40}$$
$$40(\cos 20°) = \overline{OC}$$
$$40(0.9397) = \overline{OC}$$
$$38 = \overline{OC} \quad \text{(two significant digits)}$$

Figure 7.25

Thus, a force of 38 lb is lowering the window.   □

EXAMPLE 7   Find the horizontal and vertical components of a vector that has a magnitude of 25 lb and makes an angle of 40° with the positive *x*-axis.

Solution   Let vector **OA** be the given vector. The resolution of vector **OA** results in the horizontal component, vector **OB**, and the vertical component, vector **OC** (see Figure 7.26).
    We can find the magnitude of the horizontal component, vector **OB**, by using the cosine function in right triangle *OBA*.

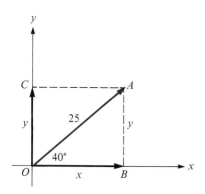

$$\cos 40° = \frac{x}{25}$$
$$25(\cos 40°) = x$$
$$25(0.7660) = x$$
$$19 = x \quad \text{(two significant digits)}$$

Figure 7.26

We can find the magnitude of the vertical component, vector **OC**, by using the sine function in right triangle *OBA*.

$$\sin 40° = \frac{y}{25}$$
$$25(\sin 40°) = y$$
$$25(0.6428) = y$$
$$16 = y \quad \text{(two significant digits)}$$

Thus, the magnitude of the horizontal component is 19 lb and the magnitude of the vertical component is 16 lb.   □

Example 7 shows how to resolve a vector into horizontal (or *x*) and vertical (or *y*) components. This breakdown can be very useful if we want to add two (or more) vectors. That is because the resultant of two (or more) vectors may be found by breaking each vector into its *x* and *y* components. The sum of the *x* components of each vector is the *x* component of the resultant (labeled R_x). Similarly, the sum of the *y* components of each vector is the *y* component of the resultant (labeled R_y). By the Pythagorean theorem the magnitude of the resultant is then

$$R = \sqrt{(R_x)^2 + (R_y)^2}$$

and the angle the resultant makes with the positive *x*-axis (labeled θ_R) is determined from the equation

$$\tan \theta_R = \frac{R_y}{R_x}.$$

Example 8 illustrates this approach to vector addition.

EXAMPLE 8   Vector **A** has a magnitude of 5.0 lb and makes an angle of 35° with the positive *x*-axis. Vector **B** has a magnitude of 9.0 lb and makes an angle of 160° with the positive *x*-axis. Find the resultant (or vector sum) of **A** and **B**.

Solution   We organize the procedure described above with the following table.

Vector	Magnitude	Direction	*x* Component	*y* Component
A	5.0	35°	5.0 cos 35° = 4.096	5.0 sin 35° = 2.868
B	9.0	160°	9.0 cos 160° = −8.457	9.0 sin 160° = 3.078
R			R_x = −4.361	R_y = 5.946

The magnitude of the resultant is

$$R = \sqrt{(R_x)^2 + (R_y)^2}$$
$$= \sqrt{(-4.361)^2 + (5.946)^2}$$
$$= \sqrt{54.37}$$
$$= 7.4.$$

The angle the resultant makes with the positive x-axis is found by solving

$$\tan \theta_R = \frac{R_y}{R_x} = \frac{5.946}{-4.361} = -1.363.$$

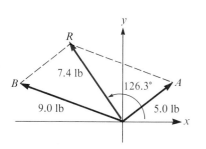

Figure 7.27

Since the y component is positive and the x component is negative, θ_R is in Q_2. By calculator (or Table 5) we determine the reference angle is 53.7°. Then

$$\theta_R = 180° - 53.7°$$
$$= 126.3°.$$

Thus, the resultant has magnitude 7.4 lb and makes an angle of 126.3° with the positive x-axis. (See Figure 7.27.)   □

EXERCISES 7.3

In Exercises 1–10 find the magnitude of all forces to two significant digits and all angles to the nearest degree.

1. A force of 5.0 lb and a force of 12 lb are acting on a body with an angle of 90° between the two forces. Find
   a. the magnitude of the resultant
   b. the angle between the resultant and the larger force

2. A force of 12 lb and a force of 16 lb are acting on a body with an angle of 90° between the two forces. Find
   a. the magnitude of the resultant
   b. the angle between the resultant and the smaller force

3. Two forces, one $1\overline{0}$ lb and the other 15 lb, act on the same object at right angles to each other. Find
   a. the magnitude of the resultant
   b. the angle between the resultant and the larger force

4. Two velocities, one $2\overline{0}$ mi/hour north and the other $3\overline{0}$ mi/hour east, are acting on the same body. Find
   a. the speed of the resultant velocity
   b. the angle of the resultant velocity with respect to a direction of due east

5. Two velocities, one 5.0 mi/hour south and the other 15 mi/hour west, are acting on the same body. Find
   a. the speed of the resultant velocity
   b. the angle of the resultant velocity with respect to a direction of due west

6. A ship is headed due south at 18 knots (nautical miles per hour) while the current carries the ship due west at 5.0 knots. Find
   a. the speed of the ship
   b. the direction (course) of the ship

7. An airplane can fly $5\overline{0}0$ mi/hour in still air. If it is heading due north in a wind that is blowing due east at a rate of $8\overline{0}$ mi/hour, find
   a. the distance the plane can fly in 1 hour
   b. the angle that the resultant velocity will make with respect to due east

8. An object is dropped from a plane that is moving horizontally at a speed of $3\overline{0}0$ ft/second. If the vertical velocity of the object in terms of time is given by the formula $v = 32t$, 5 seconds later:
   a. What is the speed of the object?
   b. What angle does the direction of the object make with the horizontal?

9. A pilot wishes to fly due east at $3\overline{0}0$ mi/hour when a $7\overline{0}$ mi/hour wind is blowing due north.
   a. How many degrees south of east should the pilot point the plane to attain the desired course?
   b. What airspeed should the pilot maintain?

10. A ship wishes to travel due south at 25 knots in an easterly current of 7.0 knots.
    a. How many degrees west of south should the navigator direct the ship?
    b. What speed must the ship maintain?

In Exercises 11–14 find the horizontal and vertical component of each vector.

11. A magnitude of $1\bar{0}0$ lb; makes an angle of 30° with the positive x-axis.
12. A magnitude of 75 lb; makes an angle of 45° with the positive x-axis.
13. A magnitude of 18 lb; makes an angle of 27° with the positive x-axis.
14. A magnitude of 125 lb; makes an angle of 72° with the positive x-axis.

In Exercises 15–20 find the resultant (or vector sum) of the given vectors by resolving each vector into its x and y components.

	Vector	Magnitude	Direction (with respect to positive x-axis)
15.	A	5.0 lb	26°
	B	3.0 lb	84°
16.	C	6.0 lb	95°
	D	9.0 lb	15°
17.	C	12 lb	110°
	D	15 lb	180°
18.	A	4.0 lb	90°
	B	3.0 lb	190°
19.	A	$1\bar{0}$ lb	42°
	B	$2\bar{0}$ lb	140°
	C	$3\bar{0}$ lb	240°
20.	A	5.0 lb	60°
	B	2.0 lb	210°
	C	1.0 lb	270°

21. An airplane pointed due east is traveling 15° north of east at a rate of 350 mi/hour. The resultant course is due to a wind blowing north. Find
    a. the velocity of the plane if there were no wind (that is, the vector pointing due east)
    b. the velocity of the wind (that is, the vector pointing due north)
22. A man pulls with a force of 25 lb on a window pole in an effort to lower a window. How much force is wasted if the pole makes an angle of 15° with the window?
23. A woman pushes with a force of $4\bar{0}$ lb on the handle of a lawn mower that makes an angle of 33° with the ground. How much force pushes the lawn mower forward?

24. A car weighing 3,500 lb is parked in a driveway that makes an angle of 5° with the horizontal. Find the minimum brake force that is needed to keep the car from rolling down the driveway. (Use the illustration below and assume no friction.)

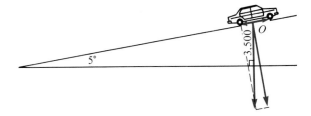

25. Find the force needed to keep a barrel weighing $1\bar{0}0$ lb from rolling down a ramp that makes an angle of 15° with the horizontal (assume no friction).
26. A box is resting on a ramp that makes an angle of 18° with the horizontal. What is the force of friction between the box and the ramp if the box weighs $8\bar{0}$ lb?
27. A body is acted on by two forces with magnitudes of 16 lb and 25 lb which act at an angle of 30° with each other. Find
    a. the magnitude of the resultant
    b. the angle between the resultant and the larger force
28. Two forces with magnitudes of 45 lb and $9\bar{0}$ lb are applied to the same point. If the angle between them measures 72°, find
    a. the magnitude of the resultant
    b. the angle between the resultant and the smaller force
29. Forces with magnitudes of 126 lb and 198 lb act simultaneously on a body in such a way that the angle between the forces is 14°50′. Find
    a. the magnitude of the resultant
    b. the angle between the resultant and the larger force
30. Two forces with magnitudes of 15 lb and $2\bar{0}$ lb act on a body in such a way that the magnitude of the resultant is 28 lb. Find the angles that the three forces make with each other.
31. Two forces act on a body to produce a resultant of 75 lb. If the angle between the two forces is 56° and one of the forces is $6\bar{0}$ lb, find the magnitude of the other force.

32. City *B* is located 50° north of east of city *A*. There is a 3̄0-mi/hour wind from the west and the pilot wishes to maintain an airspeed of 450 mi/hour. How many degrees north of east should the pilot head the plane to arrive directly at city *B* from city *A?*

33. Two forces with magnitudes of 42 lb and 71 lb are applied to the same object. If the magnitude of the resultant is 85 lb, find the angles that the three forces make with each other.

34. A force **A** of 4̄0̄0 lb and a force **B** of 6̄0̄0 lb act at a point. Their resultant, **R,** makes an angle of 42° with **A**. Find the magnitude of **R**.

7.4 TRIGONOMETRIC FORM OF COMPLEX NUMBERS

Figure 7.28

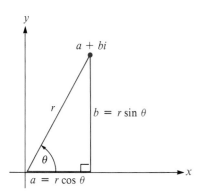

Figure 7.29

Each complex number $a + bi$ involves a pair of real numbers, a and b. Graphically, this means that we may represent a complex number as a point in a rectangular coordinate system. The values for a are plotted on the horizontal axis (x-axis) and the values for b are plotted on the vertical axis (y-axis). Thus, the complex number $a + bi$ is represented by the point (a,b) with x value a and y value b. The plane on which complex numbers are graphed is called the **complex plane,** and in this context the horizontal axis is called the **real axis** and the vertical axis is called the **imaginary axis.** In Figure 7.28 the geometric representations of several complex numbers are shown.

A useful new way of writing a complex number is to interpret the position of its geometric representation in terms of the trigonometric functions. Consider the point corresponding to the nonzero complex number $a + bi$ in Figure 7.29. When we use trigonometry, we obtain the following relationships.

$$\sin \theta = \frac{b}{r} \quad \text{therefore} \quad b = r \sin \theta$$

$$\cos \theta = \frac{a}{r} \quad \text{therefore} \quad a = r \cos \theta$$

$$r = \sqrt{a^2 + b^2}$$

$$\tan \theta = \frac{b}{a}$$

Thus, we may change the form of the complex number as follows:

$$a + bi = r \cos \theta + (r \sin \theta)i = r(\cos \theta + i \sin \theta).$$

In this trigonometric form, θ is called the **argument** of the complex number and r is called the **absolute value.** Note that the absolute value of a complex number $a + bi$ is interpreted geometrically as the distance from the origin to (a,b).

EXAMPLE 1   Write the number $3 - 3i$ in trigonometric form.

Solution   First, determine the absolute value, r. Since $a = 3$ and $b = -3$, we have

$$r = \sqrt{a^2 + b^2} = \sqrt{(3)^2 + (-3)^2} = \sqrt{18} \quad \text{or} \quad 3\sqrt{2}.$$

Now determine the argument.

$$\tan \theta = \frac{b}{a} = \frac{-3}{3} = -1$$

Since $(3, -3)$ is in Q_4 and the reference angle is $45°$, we conclude that

$$\theta = 315°.$$

The number is written in trigonometric form as

$$3\sqrt{2}(\cos 315° + i \sin 315°).$$

See Figure 7.30. (*Note:* Any angle coterminal with $315°$ may also be used. For example, we may write $3\sqrt{2}[\cos(-45°) + i \sin(-45°)]$.)   □

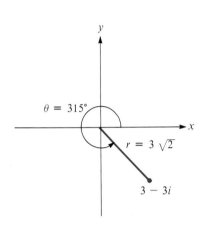

Figure 7.30

EXAMPLE 2   Write the number -2 in trigonometric form.

Solution   First, determine the absolute value, r. Since $a = -2$ and $b = 0$, we have

$$r = \sqrt{a^2 + b^2} = \sqrt{(-2)^2 + 0^2} = \sqrt{4} = 2.$$

The argument is $180°$ since -2 is a point on the negative portion of the x-axis. The number in trigonometric form is

$$2(\cos 180° + i \sin 180°). \qquad \text{(See Figure 7.31.)}$$ □

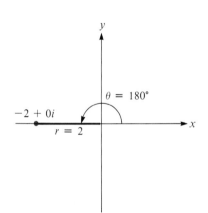

Figure 7.31

EXAMPLE 3   Write the number $2(\cos 150° + i \sin 150°)$ in the form $a + bi$.

Solution   Since $\cos 150° = -\sqrt{3}/2$ and $\sin 150° = 1/2$, we have

$$2(\cos 150° + i \sin 150°) = 2\left[\frac{-\sqrt{3}}{2} + i\left(\frac{1}{2}\right)\right]$$
$$= -\sqrt{3} + i.$$ □

Multiplication and division of complex numbers is easy when the numbers are in trigonometric form. We derive the product rule as follows:

$$r_1(\cos \theta_1 + i \sin \theta_1) \cdot r_2(\cos \theta_2 + i \sin \theta_2)$$
$$= r_1 r_2[\cos \theta_1 \cos \theta_2 + i \cos \theta_1 \sin \theta_2 + i \sin \theta_1 \cos \theta_2 + i^2 \sin \theta_1 \sin \theta_2].$$

Replacing i^2 by -1 and grouping the terms yields

$$r_1 r_2[(\cos \theta_1 \cos \theta_2 - \sin \theta_1 \sin \theta_2) + i(\sin \theta_1 \cos \theta_2 + \cos \theta_1 \sin \theta_2)].$$

We now simplify the expressions in the parentheses by applying the formulas for the sine and cosine of the sum of two angles (see Section 6.8) and obtain our final result.

$$r_1(\cos \theta_1 + i \sin \theta_1) \cdot r_2(\cos \theta_2 + i \sin \theta_2)$$
$$= r_1 r_2[\cos(\theta_1 + \theta_2) + i \sin(\theta_1 + \theta_2)]$$

In words, the product of two complex numbers is a third complex number. The absolute value is the product of the absolute values of the given numbers, and the argument is the sum of the arguments of the given numbers.

By a similar procedure we can obtain the following quotient rule.

$$\frac{r_1(\cos \theta_1 + i \sin \theta_1)}{r_2(\cos \theta_2 + i \sin \theta_2)} = \frac{r_1}{r_2}[\cos(\theta_1 - \theta_2) + i \sin(\theta_1 - \theta_2)], \text{ if } r_2 \neq 0$$

In words, the quotient of two complex numbers is a third complex number. The absolute value is the quotient of the absolute values of the given numbers, and the argument is the difference of the arguments of the given numbers.

EXAMPLE 4   Determine
a. the product $z_1 \cdot z_2$
b. the quotient z_1/z_2
c. the quotient z_2/z_1
of the following complex numbers.

$$z_1 = 3(\cos 72° + i \sin 72°), z_2 = 5(\cos 43° + i \sin 43°)$$

Solution

a. $3(\cos 72° + i \sin 72°) \cdot 5(\cos 43° + i \sin 43°)$
$= 3 \cdot 5[\cos(72° + 43°) + i \sin(72° + 43°)]$
$= 15(\cos 115° + i \sin 115°)$

b. $\dfrac{3(\cos 72° + i \sin 72°)}{5(\cos 43° + i \sin 43°)} = \dfrac{3}{5}[\cos(72° - 43°) + i \sin(72° - 43°)]$

$$= \frac{3}{5}(\cos 29° + i \sin 29°)$$

c. $\dfrac{5(\cos 43° + i \sin 43°)}{3(\cos 72° + i \sin 72°)} = \dfrac{5}{3}[\cos(43° - 72°) + i \sin(43° - 72°)]$

$$= \frac{5}{3}[\cos(-29°) + i \sin(-29°)]$$

or

$$= \frac{5}{3}(\cos 331° + i \sin 331°) \qquad \square$$

The trigonometric form is very useful when working with powers and roots of complex numbers. Consider the following statement, which is known as *De Moivre's theorem.*

De Moivre's Theorem   If $r(\cos \theta + i \sin \theta)$ is any complex number and n is any real number, then

$$[r(\cos \theta + i \sin \theta)]^n = r^n(\cos n\theta + i \sin n\theta).$$

We illustrate this theorem in the following example.

EXAMPLE 5   Find $(1 + i)^{12}$.

Solution   First, write the number in trigonometric form.

$$r = \sqrt{a^2 + b^2} = \sqrt{(1)^2 + (1)^2} = \sqrt{2}$$

$$\tan \theta = \frac{b}{a} = \frac{1}{1} = 1 \qquad \theta = 45°$$

Thus, $1 + i = \sqrt{2}(\cos 45° + i \sin 45°)$. Now De Moivre's theorem tells us that

$$\begin{aligned}
[\sqrt{2}(\cos 45° + i \sin 45°)]^{12} &= (\sqrt{2})^{12}[\cos(12 \cdot 45°) + i \sin(12 \cdot 45°)]\\
&= (2^{1/2})^{12}[\cos 540° + i \sin 540°]\\
&= 2^6[(-1) + i(0)]\\
&= -64. \qquad \square
\end{aligned}$$

Any complex number has two square roots, three cube roots, four fourth roots, and so on. To find these roots we use De Moivre's theorem. However, to find all of the roots you must remember that there are many trigonometric representations for the same complex number. For example, since the angle $\theta + k \cdot 360°$ (k any integer) is coterminal to θ, we have

$$1 + i = \sqrt{2}(\cos 45° + i \sin 45°) = \sqrt{2}(\cos 405° + i \sin 405°)$$
$$= \sqrt{2}(\cos 765° + i \sin 765°), \text{ and so on.}$$

Thus, the method is to find one root, add $360°$ to θ, use the new trigonometric representation to find another root, and repeat this procedure until all n roots are found.

EXAMPLE 6   Find the five fifth roots of 32.

Solution   First, we write 32 in trigonometric form as

$32(\cos 0° + i \sin 0°)$.

Now applying De Moivre's theorem, we have

$$[32(\cos 0° + i \sin 0°)]^{1/5} = 32^{1/5}\left(\cos \frac{0°}{5} + i \sin \frac{0°}{5}\right)$$
$$= 2(\cos 0° + i \sin 0°) = 2.$$

The first root is 2. To find another root, first add $360°$ to θ to obtain a different trigonometric representation for 32.

$32 = 32[\cos(0° + 360°) + i \sin(0° + 360°)]$

Then by De Moivre's theorem

$$\text{2nd root is } 32^{1/5}\left(\cos\frac{0° + 360°}{5} + i \sin\frac{0° + 360°}{5}\right)$$
$$= 2(\cos 72° + i \sin 72°).$$

Repeating this procedure, we have

$$\text{3rd root is } 32^{1/5}\left(\cos\frac{0° + 2 \cdot 360°}{5} + i \sin\frac{0° + 2 \cdot 360°}{5}\right)$$
$$= 2(\cos 144° + i \sin 144°)$$
$$\text{4th root is } 32^{1/5}\left(\cos\frac{0° + 3 \cdot 360°}{5} + i \sin\frac{0° + 3 \cdot 360°}{5}\right)$$
$$= 2(\cos 216° + i \sin 216°)$$
$$\text{5th root is } 32^{1/5}\left(\cos\frac{0° + 4 \cdot 360°}{5} + i \sin\frac{0° + 4 \cdot 360°}{5}\right)$$
$$= 2(\cos 288° + i \sin 288°).$$

Note that except for the first root, which is 2, these numbers are expressed more simply in trigonometric form.   □

In much of the literature dealing with complex numbers and their applications the expression

$$r(\cos \theta + i \sin \theta)$$

is written as

$$r \text{ cis } \theta \quad \text{or} \quad r \angle \theta.$$

These forms are merely convenient abbreviations and the expressions are interchangeable.

EXERCISES 7.4

In Exercises 1–14 write the number in trigonometric form.

1. 3   2. i 3. $-2i$

4. -4 5. $1 - i$ 6. $-1 + i$

7. $4 + 3i$ 8. $-5 - 12i$ 9. $-3 - 2i$

10. $7 - 4i$ 11. $-1 + \sqrt{3}i$ 12. $\sqrt{3} - i$

13. $\sqrt{2} - \sqrt{2}i$ 14. $-2 - 2\sqrt{3}i$

In Exercises 15–24 write the number in the form $a + bi$.

15. $3(\cos 90° + i \sin 90°)$

16. $5(\cos 0° + i \sin 0°)$

17. $4(\cos 180° + i \sin 180°)$

18. $\sqrt{2}(\cos 270° + i \sin 270°)$

19. $\sqrt{3}(\cos 120° + i \sin 120°)$

20. $2(\cos 210° + i \sin 210°)$

21. $2(\cos 225° + i \sin 225°)$

22. $\sqrt{6}(\cos 315° + i \sin 315°)$

23. $\cos 52° + i \sin 52°$

24. $10(\cos 115° + i \sin 115°)$

In Exercises 25–34 find (a) the product $z_1 \cdot z_2$, (b) the quotient z_1/z_2, and (c) the quotient z_2/z_1 of the given complex numbers.

25. $z_1 = 2(\cos 52° + i \sin 52°)$,
    $z_2 = 4(\cos 11° + i \sin 11°)$

26. $z_1 = 6(\cos 7° + i \sin 7°)$,
    $z_2 = 9(\cos 90° + i \sin 90°)$

27. $z_1 = \cos 90° + i \sin 90°$,
    $z_2 = \cos 180° + i \sin 180°$

28. $z_1 = \cos 33° + i \sin 33°$,
    $z_2 = 4(\cos 63° + i \sin 63°)$

29. $z_1 = 3(\cos 131° + i \sin 131°)$,
    $z_2 = 12(\cos 205° + i \sin 205°)$

30. $z_1 = \cos 270° + i \sin 270°$,
    $z_2 = 5(\cos 3° + i \sin 3°)$

31. $z_1 = 2(\cos 300° + i \sin 300°)$,
    $z_2 = \sqrt{2}(\cos 45° + i \sin 45°)$

32. $z_1 = 3\sqrt{2}(\cos 315° + i \sin 315°)$,
    $z_2 = \cos 0° + i \sin 0°$

33. $z_1 = \cos(-20°) + i \sin(-20°)$,
    $z_2 = \cos(-45°) + i \sin(-45°)$

34. $z_1 = 3(\cos 0° + i \sin 0°)$,
    $z_2 = 4[\cos(-90°) + i \sin(-90°)]$

In Exercises 35–40 use De Moivre's theorem and express the result in the form $a + bi$.

35. $(1 + i)^8$ 36. $(-1 + i)^6$

37. $(1 - \sqrt{3}i)^5$ 38. $(-\sqrt{3} - i)^7$

39. $\left(\dfrac{\sqrt{2}}{2} + \dfrac{\sqrt{2}}{2}i\right)^{16}$ 40. $\left(\dfrac{\sqrt{2}}{2} - \dfrac{\sqrt{2}}{2}i\right)^{14}$

In Exercises 41–46 use De Moivre's theorem to find all roots. Leave the answers in trigonometric form when Table 5 at the end of the book is required.

41. $\sqrt[3]{1}$ 42. $\sqrt[4]{-1}$ 43. $\sqrt[4]{-16i}$

44. $\sqrt[3]{8i}$ 45. $\sqrt[5]{1 + i\sqrt{3}}$ 46. $\sqrt[6]{1 + i}$

In Exercises 47–50 find all the solutions of the equation. Leave the answers in trigonometric form when Table 5 is required.

47. $x^3 + 8 = 0$ 48. $x^3 - 27i = 0$

49. $x^4 - 1 = 0$ 50. $x^5 + 32 = 0$

7.5 POLAR COORDINATES

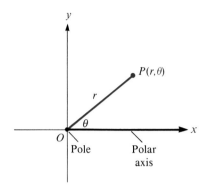

Figure 7.32

Up to now we have specified points in the plane by using the Cartesian (or rectangular) coordinate system. Another important system that can be used to obtain a picture of a relationship is the **polar coordinate system.** In this system we use a fixed point O, called the **pole** or **origin,** and a ray that extends from O and forms an axis, called the **polar axis.** The pole and polar axis correspond to the origin and positive x-axis of the rectangular coordinate system as shown in Figure 7.32. Now any point P in the plane may be represented by an ordered pair of numbers (r,θ), called **polar coordinates,** where r gives the distance from O to P and θ gives an angle measure from the polar axis to ray OP. Example 1 graphs a few points that are given in polar coordinate form.

EXAMPLE 1   Plot the points $(3,45°)$, $(1,7\pi/6)$, $(2,-60°)$, and $(0,\pi)$.

Solution   The points are shown in Figure 7.33. Note that polar coordinates may use degree or radian angle measures that may be positive or negative. Also, the point $(0,\pi)$ is the pole. In general, $(0,\theta)$ is the pole for all values of θ. □

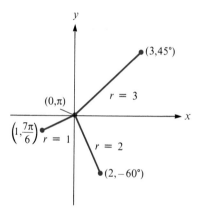

Figure 7.33

Negative values for r often arise when graphing equations like $r = \sin \theta$ in polar coordinates, since negative values for the trigonometric functions are common. If $r < 0$, then we plot (r,θ) by measuring $|r|$ units in a direction opposite to the terminal ray of θ. Example 2 illustrates this interpretation.

EXAMPLE 2   Plot the points $(-2,60°)$ and $(-1,-\pi/4)$.

Solution   The point $(-2,60°)$ is located 2 units in the opposite direction from the terminal ray of a $60°$ angle, while $(-1,-\pi/4)$ is located 1 unit along the extension of the terminal ray of $-\pi/4$. Both points are shown in Figure 7.34. □

There are many ways to represent a given point in polar coordinates, and we need to adjust to this. For instance, the point at $(1,30°)$ may also be described as

$$(1,390°), (1,750°), (1,-330°), (-1,210°), \text{ and so on.}$$

In general, different representations for a given point are obtained as follows:

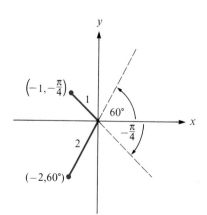

Figure 7.34

$(r,\theta) = (r,\theta + n \cdot 360°)$, where n is an integer
$(-r,\theta) = (r,\theta + n \cdot 180°)$, where n is an odd integer.

Basically, these rules tell us that when the angle changes by $360°$, r remains the same; but when the angle changes by $180°$, r changes sign.

EXAMPLE 3   Express the points $(2,-45°)$, $(-1,2\pi/3)$, and $(-3,7\pi/6)$ in polar coordinates with $r \geq 0$ and $0° \leq \theta < 360°$ (or $0 \leq \theta < 2\pi$).

Solution   We satisfy the given conditions as follows:

$$(2,-45°) = (2,-45° + 360°) = (2,315°)$$
$$\left(-1,\frac{2\pi}{3}\right) = \left(-(-1),\frac{2\pi}{3} + \pi\right) = \left(1,\frac{5\pi}{3}\right)$$
$$\left(-3,\frac{7\pi}{6}\right) = \left(-(-3),\frac{7\pi}{6} - \pi\right) = \left(3,\frac{\pi}{6}\right).$$

Polar equations relating r and θ, and graphed in polar coordinates, provide an efficient vehicle for analyzing some basic curves. For example, consider the problem of graphing $r = \sin\theta$ in polar coordinates. First, let's construct the following table and try point-by-point plotting.

θ	0°	30°	60°	90°	120°	150°	180°	210°	240°	270°	300°	330°	360°
$r = \sin\theta$	0	0.50	0.87	1	0.87	0.50	0	−0.50	−0.87	−1	−0.87	−0.50	0

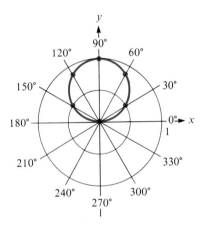

Figure 7.35

Note in Figure 7.35 that $r = \sin\theta$ graphs as a circle and that the complete circle is traced out for $0° \leq \theta < 180°$. When $r < 0$, for $180° < \theta < 360°$ we merely obtain the same points a second time, and the sine function will continue this pattern if other values of θ are considered.

Point-by-point plotting is not very efficient and we usually try to draw graphs by recognizing that certain types of curves come from certain types of equations. In the case of the circle we have the following important special cases.

Equation	Graph	Comment
$r = a$		A circle of radius a and centered at the origin
$r = 2a \sin \theta$		A circle of radius a, centered on the positive y-axis, and passing through the origin
$r = -2a \sin \theta$		A circle of radius a, centered on the negative y-axis, and passing through the origin
$r = 2a \cos \theta$		A circle of radius a, centered on the positive x-axis, and passing through the origin
$r = -2a \cos \theta$		A circle of radius a, centered on the negative x-axis, and passing through the origin

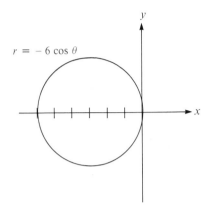

$r = -6 \cos \theta$

Figure 7.36

EXAMPLE 4   Graph $r = -6 \cos \theta$ in polar coordinates.

Solution   By matching $r = -6 \cos \theta$ to $r = -2a \cos \theta$ and using the above chart, we determine that the graph is a circle of radius 3, centered on the negative x-axis, and passing through the origin. From this information it is easy to sketch the graph in Figure 7.36.   □

Besides recognizing a curve from a given equation, we also want to make use of symmetry. We can test for x- and y-axis symmetry as follows:

Substitution	Symmetry
Replacing θ by $-\theta$ produces an equivalent equation.	x-axis
Replacing θ by $180° - \theta$ produces an equivalent equation.	y-axis

Since $\cos(-\theta) = \cos \theta$ and $\sin(180° - \theta) = \sin \theta$, we look for x-axis symmetry for polar equations involving cosine and y-axis symmetry for polar curves involving sine.

EXAMPLE 5   Graph $r = 2 - 2 \sin \theta$ in polar coordinates.

Solution   Since $r = 2 - 2 \sin(180° - \theta)$ and $r = 2 - 2 \sin \theta$ are equivalent equations, we have symmetry about the y-axis. Therefore, we first graph the curve to the right of the y-axis and then reflect this portion about the y-axis to obtain the entire graph. We can obtain the first portion by plotting the points in the following table as shown in Figure 7.37(a).

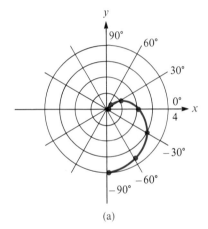

(a)

θ	90°	60°	30°	0°	−30°	−60°	−90°
$r = 2 - 2 \sin \theta$	0	0.27	1	2	3	3.7	4

Now we just reflect the curve in Figure 7.37(a) about the y-axis to obtain the entire graph [Figure 7.37(b)].   □

Because the curve in Figure 7.37(b) is heart shaped, it is called a **cardioid.** We can anticipate cardioids as graphs of equations of the form

$$r = a + a \sin \theta \qquad r = a - a \sin \theta$$
$$r = a + a \cos \theta \qquad r = a - a \cos \theta.$$

Each case produces the same heart-shaped figure but the orientation of the cardioid relative to the axes varies, with the sine versions being symmetric with respect to the y-axis and the cosine versions being symmetric with respect to the x-axis.

(b)

Figure 7.37

Three loops ($n = 3$)

Four loops ($n = 2$)

Figure 7.38

In our final illustration of graphing we wish to consider the graph of $r = \sin 2\theta$. We have seen that $r = 2 \sin \theta$ graphs as a circle, but doubling θ, instead of $\sin \theta$, is totally different. In general, equations of the form

$$r = a \sin n\theta$$
$$r = a \cos n\theta$$

graph as curves called **roses** (see Figure 7.38), because they contain loops (or petals) that produce a flower-shaped appearance. The rose has n loops when n is odd and $2n$ loops when n is even. Example 6 shows how to draw the rose corresponding to $r = \sin 2\theta$.

EXAMPLE 6   Graph $r = \sin 2\theta$ in polar coordinates.

Solution   The equation fits the form $r = a \sin n\theta$, with $n = 2$, so we know it graphs as a rose with 4 (or $2n$) loops. To draw the rose we need to determine the values of θ for which $|r|$ is greatest, since these points may be endpoints for the loops. By drawing the graph in rectangular coordinates of $r = \sin 2\theta$ in Figure 7.39, we see that 45°, 135°, 225°, and 315° produce the greatest values for $|r|$. We can now draw the rose for $r = \sin 2\theta$ as in Figure 7.40. The Q_1 and Q_3 loops are straightforward, but note that the Q_4 loop is formed for 90° < θ < 180° and the Q_2 loop is formed for 270° < θ < 360°, because r is negative in these intervals. It is useful to know that when n is even, we have $2n$ loops with endpoints at *both* maximum and minimum r values. When n is odd, however, we need only consider maximum r values, because the minimum r values give no new points.

Figure 7.39

Figure 7.40   ☐

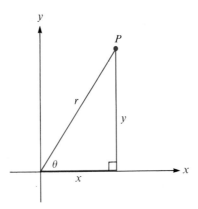

Figure 7.41

We often analyze a problem by using both polar and rectangular coordinates, so it is important we understand the basic relationships between the two systems. These may be found by considering Figure 7.41 and using our knowledge of trigonometry.

$$\sin \theta = \frac{y}{r}, \text{ so } y = r \sin \theta$$

$$\cos \theta = \frac{x}{r}, \text{ so } x = r \cos \theta$$

$$\tan \theta = \frac{y}{x}$$

$$x^2 + y^2 = r^2$$

The top two formulas show us how to switch from polar coordinates to rectangular coordinates, while the bottom two formulas let us go the other way.

EXAMPLE 7   Express $(2, 210°)$ in rectanglar coordinates.

Solution   We substitute $r = 2$ and $\theta = 210°$ in the given formulas.

$$x = r \cos \theta = 2 \cos 210° = 2\left(\frac{-\sqrt{3}}{2}\right) = -\sqrt{3}$$

$$y = r \sin \theta = 2 \sin 210° = 2\left(-\frac{1}{2}\right) = -1$$

Thus, $(2, 210°)$ is written in rectangular coordinates as $(-\sqrt{3}, -1)$.   □

EXAMPLE 8   Express $(-1,1)$ in polar coordinates with $r \geq 0$ and $0° \leq \theta < 360°$.

Solution   We are given $x = -1$ and $y = 1$. Then

$$r^2 = x^2 + y^2 = (-1)^2 + (1)^2 = 2.$$

Since $r \geq 0$, we know $r = \sqrt{2}$. Also,

$$\tan \theta = \frac{y}{x} = \frac{1}{-1} = -1.$$

Since $(-1,1)$ lies in Q_2 and $r \geq 0$, we conclude $\theta = 135°$. Thus, $(-1,1)$ is written as $(\sqrt{2},135°)$ in polar coordinates. (*Note:* Without the stated restrictions on r and θ, many representations are possible.)   □

The four basic relationships may also be used to change a polar equation to rectangular form and vice versa.

EXAMPLE 9   Transform $x^2 + y^2 + 6x = 0$ to an equation in polar form.

Solution   Since $x^2 + y^2 = r^2$ and $x = r \cos \theta$, we proceed as follows:

$$x^2 + y^2 + 6x = 0$$
$$r^2 + 6r \cos \theta = 0$$
$$r(r + 6 \cos \theta) = 0$$
$$r = 0 \qquad r + 6 \cos \theta = 0$$
$$r = -6 \cos \theta.$$

Since $r = -6 \cos \theta$ includes the pole, or $r = 0$, the complete relation is expressed by $r = -6 \cos \theta$. Note that both equations fit the form for circles in their respective systems.   □

EXAMPLE 10   Transform $\theta = 3\pi/4$ to an equation in rectangular form.

Solution   When θ is a constant, we only need $\tan \theta = y/x$. Since $\theta = 3\pi/4$,

$$\tan \theta = \frac{y}{x}$$

$$\tan \frac{3\pi}{4} = \frac{y}{x}$$

$$-1 = \frac{y}{x}$$

$$y = -x.$$

Thus, $\theta = 3\pi/4$ may be written in rectangular form as $y = -x$.   □

EXERCISES 7.5

1. Plot the following points in polar coordinates.
   a. $(2,60°)$ b. $(3,\pi)$ c. $(1,-30°)$
   d. $(0,150°)$ e. $(0.5,-2\pi/3)$ f. $(2,500°)$

2. Plot the following points in polar coordinates.
   a. $(-1,90°)$ b. $(-2,-\pi/6)$ c. $(-\frac{3}{2},135°)$
   d. $(-2,210°)$ e. $(-3,-\pi)$ f. $(-1,2\pi/3)$

3. Express the following points in polar coordinates with $r \geq 0$ and $0° \leq \theta < 360°$ (or $0 \leq \theta < 2\pi$).
   a. $(1,-30°)$ b. $(-2,5\pi/6)$ c. $(-1,4\pi/3)$
   d. $(3,585°)$ e. $(-1,-7\pi/4)$ f. $(2,-180°)$

4. List three other polar representations for $(3,210°)$ with one having $r < 0$.

In Exercises 5–20 graph each equation in polar coordinates and give its name.

5. $r = 3$ 6. $\theta = \pi/4$
7. $r = \cos \theta$ 8. $r = 3 \sin \theta$
9. $r = -4 \sin \theta$ 10. $r = -2 \cos \theta$
11. $r = 1 + \sin \theta$ 12. $r = 1 - \sin \theta$
13. $r = \sin \theta - 1$ 14. $r = 1 + \cos \theta$
15. $r = 2(1 - \cos \theta)$ 16. $r = \cos \theta - 1$
17. $r = \cos 2\theta$ 18. $r = 3 \sin 2\theta$
19. $r = 2 \sin 3\theta$ 20. $r = \cos 4\theta$

In Exercises 21–28 graph each equation in polar coordinates.

21. $r = 2 + \cos \theta$ (limacon)
22. $r = 3 - \sin \theta$ (limacon)
23. $r = 1 - 2 \sin \theta$ (limacon, inner loop)
24. $r = 1 + 2 \cos \theta$ (limacon, inner loop)
25. $r^2 = 4 \cos 2\theta$ (lemniscate)
26. $r^2 = \sin 2\theta$ (lemniscate)
27. $r = \theta, \theta \geq 0$ (spiral)
28. $r = \theta, \theta \leq 0$ (spiral)

In Exercises 29–34 express each point in rectangular coordinates.

29. $(1,\pi)$ 30. $(2,60°)$ 31. $(-3,3\pi)$
32. $(-6,0°)$ 33. $(\sqrt{2},5\pi/4)$ 34. $(-\sqrt{3},-120°)$

In Exercises 35–40 express each point in polar coordinates with $r \geq 0$ and $0° \leq \theta < 360°$.

35. $(-1,0)$ 36. $(0,-2)$ 37. $(1,-1)$
38. $(-3,-3)$ 39. $(-\sqrt{3},1)$ 40. $(-2,2\sqrt{3})$

In Exercises 41–50 transform each equation to polar form.

41. $x = 2$ 42. $y = -2$
43. $x^2 + y^2 = 9$ 44. $y = x$
45. $x^2 + y^2 = x$ 46. $x^2 + y^2 - 4y = 0$
47. $y = x^2$ $y = x^3$
49. $x^2 - y^2 = 4$ 50. $x^2 + y^2 + x = \sqrt{x^2 + y^2}$

In Exercises 51–60 transform each equation to rectangular form.

51. $r = 3$ 52. $\theta = -\pi/6$
53. $r \cos \theta = 2$ 54. $r \sin \theta = -3$
55. $r = -3 \sin \theta$ 56. $r = 3 \cos \theta$
57. $r = 2 \csc \theta$ 58. $r = \sec \theta$
59. $r(3 \sin \theta + 2 \cos \theta) = 1$ 60. $r = \sin 2\theta$

CHAPTER OVERVIEW

Section	Key Concepts to Review
7.1	• Law of sines: $\dfrac{\sin A}{a} = \dfrac{\sin B}{b} = \dfrac{\sin C}{c}$
	• Guidelines on when to use the law of sines
	• When given the measures for two sides of a triangle and the angle opposite one of them, there may be one triangle that fits the data (see Example 3) or there may be two triangles that fit the data (see Example 4). Sometimes no triangle can be constructed from the data; then we say the data is inconsistent.

Section	Key Concepts to Review
7.2	• Law of cosines: $a^2 = b^2 + c^2 - 2bc \cos A$ $\qquad\qquad b^2 = a^2 + c^2 - 2ac \cos B$ $\qquad\qquad c^2 = a^2 + b^2 - 2ab \cos C$ • Guidelines on when to use the law of cosines
7.3	• Definitions of vector, resultant of two (or more) vectors, and components of a vector • Methods to determine the resultant of two (or more) vectors • Methods to determine the horizontal and vertical components of a vector • Resultant formulas: $R = \sqrt{(R_x)^2 + (R_y)^2}$ and $\tan \theta = R_y/R_x$
7.4	• Definitions of complex plane, real axis, imaginary axis, and the argument and absolute value of a complex number • Graph of a complex number • In trigonometric form $a + bi$ is written as $r(\cos \theta + i \sin \theta)$. • The absolute value, r, is given by $r = \sqrt{a^2 + b^2}$. • The argument, θ, is given by $\tan \theta = b/a$. • Methods to multiply and divide two complex numbers in trigonometric form • De Moivre's theorem
7.5	• Definitions of pole, polar axis, and polar coordinates • If $r < 0$, we plot (r,θ) by measuring $\lvert r \rvert$ units in a direction opposite to the terminal ray of θ. • Formulas for different representations of a given point: $\qquad (r,\theta) = (r,\theta + n \cdot 360°)$, where n is an integer $\qquad (-r,\theta) = (r,\theta + n \cdot 180°)$, where n is an odd integer • Two tests for x- and y-axis symmetry • Basic graphs: $\qquad r = a, r = \pm 2a \sin \theta, r = \pm 2a \cos \theta$: circle $\qquad\quad r = a \pm a \sin \theta, r = a \pm a \cos \theta$: cardioid $\qquad\quad r = a \sin n\theta, r = a \cos n\theta$: rose $\begin{cases} n \text{ loops, if } n \text{ is odd} \\ 2n \text{ loops, if } n \text{ is even} \end{cases}$ • Formulas relating rectangular coordinates and polar coordinates: $\qquad x = r \cos \theta, y = r \sin \theta, x^2 + y^2 = r^2, \tan \theta = y/x$

CHAPTER REVIEW EXERCISES

In Exercises 1–6 find the indicated part in triangle *ABC*.

1. Determine *B* if $a = 5.0$, $b = 9.0$, and $c = 6.0$.
2. Determine *C* if $A = 81°$, $a = 11$, and $c = 35$.
3. Determine *A* if $C = 44°50'$, $B = 86°20'$, and $a = 62.7$.
4. Determine *B* if $C = 90°$, $b = 5.00$, and $c = 8.00$.
5. Determine *c* if $C = 120°$, $a = 3.0$, and $b = 4.0$.
6. Determine *b* if $A = 40°$, $B = 60°$, and $c = 6.0$.

In Exercises 7–10 perform the indicated operations. Write the answers in the form $a + bi$ and in trigonometric form.

7. $8(\cos 90° + i \sin 90°) \div (\cos 0° + i \sin 0°)$
8. $2(\cos 73° + i \sin 73°) \cdot 3(\cos 107° + i \sin 107°)$
9. $\left(\dfrac{\sqrt{2}}{2} + \dfrac{\sqrt{2}}{2}i\right)^{10}$ 10. $\sqrt[3]{-1}$ (all roots)

In Exercises 11–14 graph each equation in polar coordinates.

11. $r = -2 \sin \theta$ 12. $r = 2$
13. $r = 2 + 2 \cos \theta$ 14. $r = 2 \cos 3\theta$

15. Express the point $(-1, -1)$ in polar coordinates with $r \geq 0$ and $0° \leq \theta < 360°$.
16. Express the point $(-2, 7\pi/4)$ in rectangular coordinates.
17. What is the vertical component of a vector with a magnitude of $5\overline{0}$ lb which makes an angle of $25°$ with the positive *x*-axis?
18. Find all solutions for the equation $x^4 + 4 = 0$.
19. If *ABC* is a right triangle ($C = 90°$), show that the law of cosines simplifies to the relationship $c^2 = a^2 + b^2$.
20. In triangle *ABC*, $b = 5$, $c = 6$, and $\cos A = -\frac{1}{3}$. Find side length *a*.
21. Transform the equation $r(2 \cos \theta - \sin \theta) = 3$ to rectangular form.
22. Transform the equation $x^2 + y^2 = y$ to polar form.
23. Express the roots of $x^3 + 27 = 0$ in the form $r(\cos \theta + i \sin \theta)$.
24. Express the number $-1 - i$ in trigonometric form.

25. If the outfielders are positioned as shown below, then how far is each of them from third base in case they need to throw out a runner at that base?

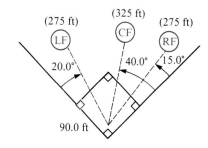

26. Two points, *A* and *B*, are $1\overline{0}0$ yd apart. Point *C* across a canyon is located so that angle *CAB* is $70°$ and angle *CBA* is $80°$. Compute the distance $\overline{BC}$ across the canyon.
27. A draftsman drew to scale (1 in. = $5\overline{0}$ yd) a map of a development that includes a triangular recreation area with sides of lengths 75 yd, 125 yd, and 150 yd. What are the angles of the triangle representing the recreation area on the map?
28. A force of 12 lb and a force of 15 lb are acting on a body with an angle of $90°$ between the two forces. Find the magnitude of the resultant and the angle between the resultant and the larger force.
29. A force of $4\overline{0}$ lb and a force of $3\overline{0}$ lb act on a body so that their resultant is a force of 38 lb. Find the angle between the two original forces.
30. Vector **A** has a magnitude of 6.0 lb and makes an angle of $55°$ with the positive *x*-axis. Vector **B** has a magnitude of 2.0 lb and makes an angle of $110°$ with the positive *x*-axis. Find the resultant (or vector sum) of **A** and **B**.

In Exercises 31–40 select the choice that completes the statement or answers the question.

31. The number of triangles satisfying the conditions that $B = 30°$, $a = 57$ ft, and $b = 39$ ft is

   a. two   **b.** one   **c.** none

32. We write $2(\cos 120° + i \sin 120°)$ in the form $a + bi$ as

    a. $1 - \sqrt{3}i$ **b.** $-\sqrt{3} + i$

    c. $\sqrt{3} - i$ **d.** $-1 + \sqrt{3}i$

33. The graph of $r = 2 + 2 \cos \theta$ is a

    a. circle   **b.** cardioid

    c. rose   **d.** line

34. The x component of a vector with a magnitude of 25 lb which makes an angle of 67° with the x-axis is

    a. 9.8 lb   **b.** 12 lb   **c.** 17 lb   **d.** 23 lb

35. An alternate polar representation for $(-1, -45°)$ is

    a. $(1, 45°)$ **b.** $(1, 135°)$

    c. $(-1, 135°)$ **d.** $(1, 315°)$

36. To solve a triangle when we know the measures for two sides of the triangle and the angle between these two sides, we first apply the

    a. law of sines   **b.** law of cosines

37. $(1 + i)^6$ simplifies to

    a. 8   **b.** -8 **c.** $-8i$ **d.** $8i$

38. In triangle ABC, if $\sin A = \frac{3}{4}$ and $\sin B = \frac{1}{2}$, then the ratio of side length a to side length b is

    a. 3:2   **b.** 8:3   **c.** 3:1   **d.** 4:3

39. Which illustration shows the graph of $r = \sin 5\theta$?

(a)   (b)

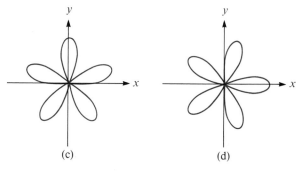

(c)   (d)

40. The absolute value of the complex number $-2 + 2i$ is

    a. $2\sqrt{2}$ **b.** 2   **c.** 4   **d.** $\sqrt{2}$

8 Analytic Geometry: *Conic Sections*

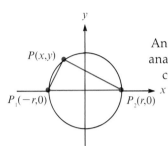

Analytic geometry allows us to use the powerful methods of algebra to analyze geometry problems. To illustrate this, use the given figure and the concept of slope to show that if a triangle is inscribed in a circle with the diameter as one of its sides, then the triangle is a right angle. (See Example 3 of Section 8.1.)

A nalytic geometry bridges the gap between algebra and geometry. By representing the ordered pairs that satisfy some algebraic equation as points in the Cartesian coordinate system, we generate a geometric picture or graph. The fundamental relationship between an equation and its graph is as follows: Every ordered pair that satisfies the equation corresponds to a point in its graph, and every point in the graph corresponds to an ordered pair that satisfies the equation.

We have considered some techniques for sketching the graph of a given equation. However, the reverse question is of major importance. If we are given in geometric terms some object or motion from the physical world, can we describe it with some equation? If so, the powerful methods of algebra can be used to analyze them. Thus, we now consider the problem of determining an equation that corresponds to a given geometric condition.

8.1 INTRODUCTION TO ANALYTIC GEOMETRY

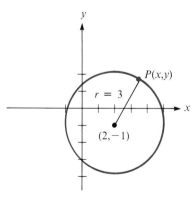

Suppose the graph of an equation is the set of all points in a plane that are 3 units from the point $(2, -1)$. It is not hard to tell that the graph is a circle of radius 3. Let us consider the more difficult problem of determining an equation for this circle.

We start by making the sketch in Figure 8.1, where we let $P(x,y)$ be any point on the given circle. By the distance formula, the distance from $(2, -1)$ to P is

$$d = \sqrt{(x - 2)^2 + [y - (-1)]^2}.$$

Figure 8.1

This distance is the radius of the circle, which is given as 3. Thus,

$$3 = \sqrt{(x-2)^2 + [y - (-1)]^2}.$$

Squaring both sides of this equation yields

$$9 = (x-2)^2 + (y+1)^2.$$

We now simplify as follows:

$$9 = x^2 - 4x + 4 + y^2 + 2y + 1$$
$$0 = x^2 + y^2 - 4x + 2y - 4.$$

This equation defines the circle formed by the given geometric condition.

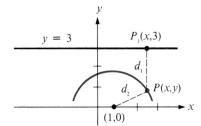

Figure 8.2

EXAMPLE 1   Find an equation for the graph that is the set of all points in a plane equidistant from the line $y = 3$ and the point $(1,0)$.

Solution   Let $P(x,y)$ represent any point satisfying the given geometric condition. Then draw the sketch in Figure 8.2 that includes a portion of the given graph. P_1 lies on the line $y = 3$, so we represent the point as $(x,3)$. Since P_1 and P have the same x-coordinate, the distance d_1 is given by

$$d_1 = |y - 3|.$$

The distance from P to $(1,0)$ is given by

$$d_2 = \sqrt{(x-1)^2 + (y-0)^2}.$$

The given geometric condition states that

$$d_1 = d_2.$$

Thus,

$$|y - 3| = \sqrt{(x-1)^2 + (y-0)^2}.$$

Squaring both sides of the equation yields

$$(y-3)^2 = (x-1)^2 + (y-0)^2.$$

We simplify this equation as follows:

$$y^2 - 6y + 9 = x^2 - 2x + 1 + y^2$$
$$0 = x^2 - 2x + 6y - 8.$$

The graph of this equation satisfies the given geometric condition.   □

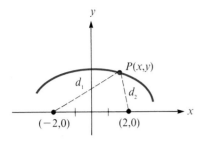

Figure 8.3

EXAMPLE 2   Find an equation for the graph that is the set of all points in a plane the sum of whose distances from (2,0) and (−2,0) is 10.

Solution   Let $P(x,y)$ represent any point satisfying the given geometric condition. Then draw the sketch in Figure 8.3 that includes a portion of the given graph. By the distance formula, we have

$$d_1 = \sqrt{[x - (-2)]^2 + (y - 0)^2}$$
$$d_2 = \sqrt{(x - 2)^2 + (y - 0)^2}.$$

The given geometric condition states that

$$d_1 + d_2 = 10.$$

Thus,

$$\sqrt{(x + 2)^2 + y^2} + \sqrt{(x - 2)^2 + y^2} = 10$$

or

$$\sqrt{(x + 2)^2 + y^2} = 10 - \sqrt{(x - 2)^2 + y^2}.$$

Square both sides of the equation and simplify.

$$(x + 2)^2 + y^2 = 100 - 20\sqrt{(x - 2)^2 + y^2} + (x - 2)^2 + y^2$$
$$8x = 100 - 20\sqrt{(x - 2)^2 + y^2}$$
$$2x - 25 = -5\sqrt{(x - 2)^2 + y^2}$$

Square both sides of the equation and simplify.

$$4x^2 - 100x + 625 = 25[(x - 2)^2 + y^2]$$
$$4x^2 - 100x + 625 = 25x^2 - 100x + 100 + 25y^2$$
$$0 = 21x^2 + 25y^2 - 525$$

The graph of this equation satisfies the given geometric condition.   □

Another aspect of analytic geometry is the proof of geometric theorems by using a coordinate system and algebraic methods as illustrated by the next example.

EXAMPLE 3   Solve the problem in the chapter introduction on page 395.

Solution   If the product of the slopes of line segments PP_1 and PP_2 is −1, then these sides are perpendicular and the triangle is a right triangle. Then

$$m_1m_2 = \frac{y - 0}{x - (-r)} \cdot \frac{y - 0}{x - r} = \frac{y^2}{x^2 - r^2}.$$

Since $x^2 + y^2 = r^2$ is an equation of a circle of radius r with center at the origin (as shown in Section 3.2), we know $y^2 = r^2 - x^2$, so

$$m_1 m_2 = \frac{r^2 - x^2}{x^2 - r^2} = -1.$$

Thus, the product of the slopes is -1, so the triangle is a right triangle. □

Up to now we have relied on distance and slope formulas in our analysis. Another useful formula that is simple to apply is the formula for finding the coordinates of the midpoint of a line segment. To derive this formula, let $P_1(x_1, y_1)$ and $P_2(x_2, y_2)$ be two points in a plane with $x_1 < x_2$ and $y_1 < y_2$, and let $P(x, y)$ be any other point on line segment $P_1 P_2$. We then construct similar triangles $P_1 R P$ and $P S P_2$ as shown in Figure 8.4, and since corresponding side lengths are proportional in such triangles, we know

$$\frac{x - x_1}{x_2 - x} = \frac{y - y_1}{y_2 - y} = \frac{\overline{P_1 P}}{\overline{P P_2}}.$$

If $P(x, y)$ is the midpoint of the segment, then $\overline{P_1 P}/\overline{P P_2} = 1$, and we may solve for x and y as follows:

$$\frac{x - x_1}{x_2 - x} = 1 \qquad\qquad \frac{y - y_1}{y_2 - y} = 1$$
$$x - x_1 = x_2 - x \qquad\qquad y - y_1 = y_2 - y$$
$$2x = x_1 + x_2 \qquad\qquad 2y = y_1 + y_2$$
$$x = \frac{x_1 + x_2}{2} \qquad\qquad y = \frac{y_1 + y_2}{2}.$$

These results are valid regardless of the locations of P_1 and P_2 and we have the following result.

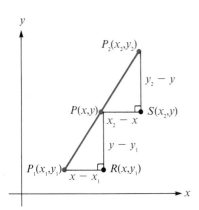

Figure 8.4

Midpoint Formula   The midpoint of the line segment joining (x_1, y_1) and (x_2, y_2) is

$$\left(\frac{x_1 + x_2}{2}, \frac{y_1 + y_2}{2} \right).$$

Note that the x-coordinate of the midpoint is simply the average of the x-coordinates of the two endpoints, and a similar relation applies for the y-coordinate of the midpoint. Also, in the above derivation of the midpoint formula, it is easy to extend our result by letting $\overline{P_1 P}/\overline{P P_2}$ equal any positive number r and thereby develop a formula for the coordinates of the point that divides a line segment in any given ratio. We pursue this idea in Exercise 32.

EXAMPLE 4   Find the midpoint of the line segment joining $(-3,4)$ and $(7,-1)$.

Solution   Using the midpoint formula, the midpoint is

$$\left(\frac{-3+7}{2}, \frac{4+(-1)}{2}\right) = \left(2, \frac{3}{2}\right).$$

EXAMPLE 5   Find an equation for the line that is the perpendicular bisector of the line segment whose endpoints are $(3,5)$ and $(11,-7)$.

Solution   (See Figure 8.5.)

a. The perpendicular bisector of a line segment passes through the midpoint of the segment. Substituting the coordinates of the endpoints $(3,5)$ and $(11,-7)$ in the midpoint formula gives

$$\left(\frac{3+11}{2}, \frac{5+(-7)}{2}\right) = (7,-1)$$

as the midpoint of the segment.

b. The slope of the given line segment is

$$m_1 = \frac{y_2 - y_1}{x_2 - x_1} = \frac{-7-5}{11-3} = \frac{-12}{8} = \frac{-3}{2}.$$

The slope m_2 of the perpendicular bisector must satisfy $-\frac{3}{2}m_2 = -1$, so $m_2 = \frac{2}{3}$.

c. We find an equation for the line through $(7,-1)$ with slope $\frac{2}{3}$ as follows:

$$y - y_1 = m(x - x_1)$$
$$y - (-1) = \tfrac{2}{3}(x - 7)$$
$$y = \tfrac{2}{3}x - \tfrac{17}{3}.$$

Thus, an equation for the specified line is $y = \frac{2}{3}x - \frac{17}{3}$.

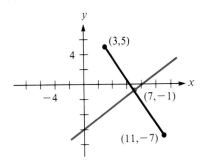

Figure 8.5

EXAMPLE 6   Use analytic geometry to prove that the midpoint of the hypotenuse of any right triangle is equidistant from each of the three vertices.

Solution   We are free to place the specified figure in any convenient position on a coordinate system, so we start by placing the legs of a right triangle on the x- and y-axes as shown in Figure 8.6. The vertex of the right angle is at the origin, while the other vertices are at the points $A(a,0)$ and $B(0,b)$. By the midpoint formula, the midpoint M of the hypotenuse is

$$\left(\frac{a+0}{2}, \frac{0+b}{2}\right) = \left(\frac{a}{2}, \frac{b}{2}\right).$$

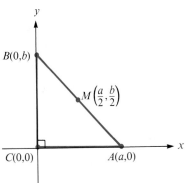

Figure 8.6

Then by the distance formula, the distance between the midpoint and each vertex is as follows:

$$\overline{AM} = \sqrt{\left(a - \frac{a}{2}\right)^2 + \left(0 - \frac{b}{2}\right)^2} = \sqrt{\frac{a^2}{4} + \frac{b^2}{4}}$$

$$\overline{BM} = \sqrt{\left(0 - \frac{a}{2}\right)^2 + \left(b - \frac{b}{2}\right)^2} = \sqrt{\frac{a^2}{4} + \frac{b^2}{4}}$$

$$\overline{CM} = \sqrt{\left(0 - \frac{a}{2}\right)^2 + \left(0 - \frac{b}{2}\right)^2} = \sqrt{\frac{a^2}{4} + \frac{b^2}{4}}.$$

Therefore, $\overline{AM} = \overline{BM} = \overline{CM}$, which establishes the given theorem. □

EXERCISES 8.1

In Exercises 1–10 find an equation for the graph in a plane that is defined by the given geometric condition.

1. The set of all points equidistant from $(0,2)$ and $(-1,5)$
2. The set of all points equidistant from $(1,-3)$ and $(-4,2)$
3. The set of all points that are 4 units from the point $(-2,-3)$
4. The set of all points that are 2 units from the point $(4,0)$
5. The set of all points equidistant from the line $x = 4$ and the point $(0,-2)$
6. The set of all points equidistant from the x-axis and the point $(3,-1)$
7. The set of all points the sum of whose distances from $(1,0)$ and $(-1,0)$ is 4
8. The set of all points the sum of whose distances from $(0,3)$ and $(0,-3)$ is 8
9. The set of all points the difference of whose distances from $(0,3)$ and $(0,-3)$ is 3
10. The set of all points the difference of whose distances from $(4,0)$ and $(-4,0)$ is 6

In Exercises 11–16 find the midpoint of the line segment joining the given pair of points.

11. $(-2,3)$, $(6,-1)$
12. $(5,-4)$, $(-9,0)$
13. $(7,2)$, $(-7,-9)$
14. $(-1,1)$, $(1,-1)$
15. $(0,0)$, $(0,3)$
16. $(8,-2)$, $(5,-2)$

In Exercises 17–24 find an equation for the line that is the perpendicular bisector of the line segment whose endpoints are given.

17. $(1,2)$, $(7,6)$
18. $(1,3)$, $(-3,-1)$
19. $(-4,-9)$, $(0,-5)$
20. $(-3,5)$, $(1,-1)$
21. $(2,-3)$, $(-3,4)$
22. $(1,0)$, $(-2,7)$
23. $(6,-1)$, $(-1,2)$
24. $(-4,4)$, $(3,-5)$

In Exercises 25–28 use the illustration below to prove the given theorem.

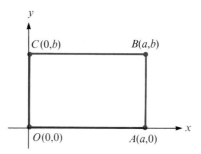

25. The diagonals of a rectangle are equal in length.
26. The diagonals of a rectangle bisect each other.
27. If the diagonals of a rectangle are perpendicular, then the rectangle is a square.
28. The midpoints of the sides of a rectangle are the vertices of a rhombus (that is, a parallelogram with all sides equal in length).

In Exercises 29–31 prove each theorem by the methods of analytic geometry.

29. The diagonals of a parallelogram bisect each other.
30. The line segments joining the midpoints of the opposite sides of any quadrilateral bisect each other.
31. The line segment joining the midpoints of two sides of a triangle is parallel to the third side and one-half its length.
32. a. Consider the derivation of the midpoint formula. If $P(x,y)$ is a point that divides the line segment from P_1 to P_2 so that $\overline{P_1P}/\overline{PP_2}$ equals a given positive ratio r, then show the coordinates of $P(x,y)$ are

$$\left(\frac{x_1 + rx_2}{1 + r}, \frac{y_1 + ry_2}{1 + r}\right).$$

This formula is called the *point-of-division formula.*

b. Derive the midpoint formula from the point-of-division formula.
c. Find the coordinates of the point that divides the line segment from $P_1(1,3)$ to $P_2(8,-4)$ in the ratio $\frac{3}{4}$.
d. Determine the two points of trisection of the line segment joining $(0,-4)$ and $(6,4)$.

8.2 THE CIRCLE

Circles, ellipses, hyperbolas, and parabolas are referred to as **conic sections.** This name was first used by Greek mathematicians who discovered that these curves result from the intersections of a cone with an appropriate plane as shown in Figure 8.7. If the cutting plane does not contain the vertex, then the important conic sections are obtained as follows:

Circle   The intersecting plane is parallel to the base of the cone.
Ellipse   The intersecting plane is parallel to neither the base nor the side and intersects only one nappe of the cone.
Hyperbola   The intersecting plane cuts both nappes of the cone.
Parabola   The intersecting plane is parallel to the side of the cone.

Three degenerate cases occur when the cutting plane passes through the vertex. These degenerate conic sections are a point, a line, and a pair of intersecting lines.

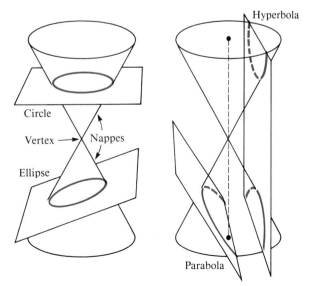

Figure 8.7

The conic sections may be analyzed from different viewpoints, and in this chapter we take the analytic geometry approach in which we derive equations for the conic sections by starting with their geometric definitions. To illustrate, we begin with the definition of a circle that is given in plane geometry.

Definition of a Circle	A **circle** is the set of all points in a plane at a given distance from a fixed point.

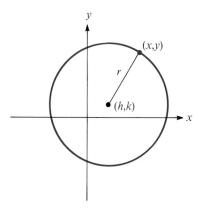

Figure 8.8

From this definition we derive an equation of a circle as follows (see also Figure 8.8):

Let (h,k) be the fixed point, the center of the circle.
Let r be the given distance, the radius of the circle.
Let (x,y) be any point on the circle.

Then by the distance formula, we have

$$\sqrt{(x - h)^2 + (y - k)^2} = r$$

or, after squaring both sides of the equation,

$$(x - h)^2 + (y - k)^2 = r^2.$$

This equation is the **standard form of the equation of a circle of radius r with center (h,k).**

EXAMPLE 1   Find the equation in standard form of the circle with center at $(2,-1)$ and radius 4.

Solution   Substituting $h = 2$, $k = -1$, and $r = 4$ in the equation

$$(x - h)^2 + (y - k)^2 = r^2,$$

we have

$$(x - 2)^2 + [y - (-1)]^2 = (4)^2$$
$$(x - 2)^2 + (y + 1)^2 = 16. \qquad \square$$

EXAMPLE 2   Find the equation in standard form of the circle with center at $(4,1)$ that passes through the origin.

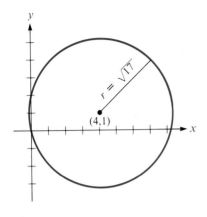

Figure 8.9

Solution   Substituting $h = 4$ and $k = 1$ into the standard equation for a circle, we have

$$(x - 4)^2 + (y - 1)^2 = r^2.$$

Since the circle passes through the origin, we find r^2 by substituting the coordinates of the point $(0,0)$ in the equation.

$$(0 - 4)^2 + (0 - 1)^2 = r^2$$
$$17 = r^2$$

The equation of the circle in standard form is then

$$(x - 4)^2 + (y - 1)^2 = 17 \quad \text{(see Figure 8.9)}. \qquad \square$$

EXAMPLE 3   Find the center and radius of the circle given by

$$x^2 + y^2 + 4x - 6y = 12.$$

Solution   We must transform the given equation to the standard form.

$$(x - h)^2 + (y - k)^2 = r^2$$

We start by completing the square in the x terms and the y terms. To do this, first group the equation as

$$(x^2 + 4x \quad) + (y^2 - 6y \quad) = 12.$$

Now find one-half of the coefficient of x (2). Square it (4) and add the result to both sides of the equation. Similarly, find one-half of the coefficient of y (-3). Square it (9) and add the result to both sides of the equation. Thus, we have

$$(x^2 + 4x + 4) + (y^2 - 6y + 9) = 12 + 4 + 9$$
$$(x + 2)^2 + (y - 3)^2 = 25.$$

By comparing this equation to the standard form, we determine that the center of the circle is at $(-2,3)$ and the radius is $\sqrt{25}$, or 5 (see Figure 8.10). $\square$

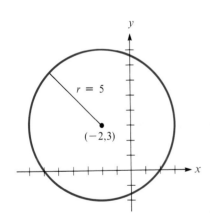

Figure 8.10

EXERCISES 8.2

In Exercises 1–10 find the equation in standard form for each of the circles from the given information.

1. Center at $(-3,4)$, radius 2
2. Center at $(-2,1)$, radius 3
3. Center at $(0,0)$, radius 1
4. Center at $(0,2)$, radius 5

5. Center at $(-4,-1)$, tangent to the line $x = -1$
   (*Note:* A line is tangent to a circle if the two intersect at exactly one point. The tangent is perpendicular to the radius at the point of intersection.)
6. Center at $(1,4)$, tangent to the line $y = -3$
7. Center at $(3,-3)$, passes through the origin

8. Center at $(2,-5)$, passes through $(1,0)$
9. Center at $(-1,-1)$, passes through $(6,2)$
10. Center at $(-4,-3)$, passes through $(-1,-5)$

In Exercises 11–14 find the center and radius of each circle.

11. $(x - 2)^2 + (y - 5)^2 = 16$
12. $(x + 1)^2 + (y + 4)^2 = 49$
13. $(x - 3)^2 + y^2 = 20$
14. $x^2 + (y + 2)^2 = 5$

In Exercises 15–24 find the center and radius of each circle.

15. $x^2 + y^2 = 9$
16. $x^2 + y^2 - 4 = 0$
17. $x^2 + y^2 - 10x - 6y = 15$
18. $x^2 + y^2 - 4x + 6y = 23$
19. $x^2 + y^2 + 8x - 2y - 1 = 0$
20. $x^2 + y^2 + 4x - 9 = 0$
21. $x^2 + y^2 + 7x + 3y + 4 = 0$
22. $x^2 + y^2 - 5x - y - 3 = 0$
23. $4x^2 + 4y^2 + 8x - 16y - 29 = 0$
24. $2x^2 + 2y^2 - 10x - 2y - 31 = 0$

8.3 THE ELLIPSE

The next conic section we consider is the ellipse, which is basically an oval-shaped curve. Once again, we begin our study with a geometric definition.

Definition of an Ellipse

An **ellipse** is the set of all points in a plane the sum of whose distances from two fixed points (**foci**) is a constant.

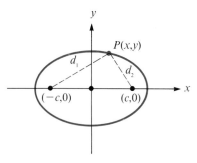

Figure 8.11

Equations for ellipses may be derived from this definition in the following way. For simplicity, let the foci be positioned at $(-c,0)$ and $(c,0)$ and let $2a$ represent the constant sum. If we let $P(x,y)$ be any point on the ellipse as shown in Figure 8.11, then by definition $d_1 + d_2 = 2a$, so it follows from the distance formula that

$$\sqrt{(x + c)^2 + (y - 0)^2} + \sqrt{(x - c)^2 + (y - 0)^2} = 2a$$

or

$$\sqrt{(x + c)^2 + y^2} = 2a - \sqrt{(x - c)^2 + y^2}.$$

Now square both sides of the equation and simplify.

$$(x + c)^2 + y^2 = 4a^2 - 4a\sqrt{(x - c)^2 + y^2} + (x - c)^2 + y^2$$
$$xc = a^2 - a\sqrt{(x - c)^2 + y^2}$$
$$a\sqrt{(x - c)^2 + y^2} = a^2 - cx$$

Again square both sides of the equation and simplify.

$$a^2[(x - c)^2 + y^2] = a^4 - 2a^2cx + c^2x^2$$
$$a^2x^2 - 2a^2cx + a^2c^2 + a^2y^2 = a^4 - 2a^2cx + c^2x^2$$
$$a^2x^2 - c^2x^2 + a^2y^2 = a^4 - a^2c^2$$
$$(a^2 - c^2)x^2 + a^2y^2 = a^2(a^2 - c^2)$$
$$\frac{x^2}{a^2} + \frac{y^2}{a^2 - c^2} = 1$$

Since $d_1 + d_2$ is greater than the distance between the foci, it follows that $a > c$ and $a^2 - c^2 > 0$. If we now define

$$b^2 = a^2 - c^2 \quad \text{or} \quad a^2 = b^2 + c^2,$$

we obtain the **standard form** of the equation of an ellipse with center at the origin and foci on the x-axis.

$$\frac{x^2}{a^2} + \frac{y^2}{b^2} = 1$$

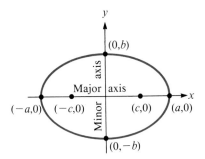

Foci: $(-c,0)$, $(c,0)$
Vertices: $(-a,0)$, $(a,0)$

Figure 8.12

Consider Figure 8.12. By setting $y = 0$, we find that the x-intercepts are $(-a,0)$ and $(a,0)$. By setting $x = 0$, we find that the y-intercepts are $(0,-b)$ and $(0,b)$. The larger segment from $(-a,0)$ to $(a,0)$ is called the **major axis**; the **minor axis** is the segment from $(0,-b)$ to $(0,b)$. The endpoints of the major axis are called the **vertices of the ellipse.**

Foci: $(-c,0)$, $(c,0)$
Vertices: $(-a,0)$, $(a,0)$

If the foci are placed on the y-axis at $(0,-c)$ and $(0,c)$, then the **standard form** of the equation of an ellipse is

$$\frac{x^2}{b^2} + \frac{y^2}{a^2} = 1,$$

where the larger denominator is denoted by a^2. The major axis is then along the y-axis as shown in Figure 8.13. Note that the foci always lie on the major axis.

Foci: $(0,c)$, $(0,-c)$
Vertices: $(0,a)$, $(0,-a)$

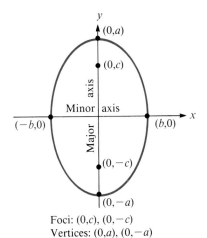

Foci: $(0,c)$, $(0,-c)$
Vertices: $(0,a)$, $(0,-a)$

Figure 8.13

EXAMPLE 1   Find the coordinates of the foci and the endpoints of the major and minor axes of the ellipse given by $4x^2 + y^2 = 36$.

Solution   First, divide both sides of the equation by 36 to obtain the standard form.

$$\frac{x^2}{9} + \frac{y^2}{36} = 1$$

Then

$$a^2 = 36 \qquad a = 6$$
$$b^2 = 9 \qquad b = 3.$$

To find c, we replace a^2 by 36 and b^2 by 9 in the formula

$$a^2 = b^2 + c^2$$
$$36 = 9 + c^2$$
$$27 = c^2 \quad \text{or} \quad c = \sqrt{27}.$$

The major axis is along the y-axis since the larger denominator appears in the y term. Thus,

endpoints of major axis: $(0,-6), (0,6)$
endpoints of minor axis: $(-3,0), (3,0)$
coordinates of foci: $(0,-\sqrt{27}), (0,\sqrt{27})$.

These points are shown in the graph of the ellipse in Figure 8.14.

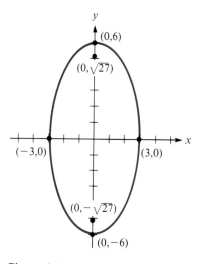

Figure 8.14

EXAMPLE 2   Find the equation in standard form of the ellipse with center at the origin, one focus at (3,0), and one vertex at (4,0).

Solution   Since c is the distance from the center to a focus, c is 3. Since a is the distance from the center to a vertex, a is 4. To find b^2, replace c by 3 and a by 4 in the formula

$$a^2 = b^2 + c^2$$
$$(4)^2 = b^2 + (3)^2$$
$$7 = b^2.$$

Since the major axis is along the x-axis, the standard form is

$$\frac{x^2}{a^2} + \frac{y^2}{b^2} = 1.$$

Thus,

$$\frac{x^2}{16} + \frac{y^2}{7} = 1$$

is the standard equation of the ellipse.   □

If the ellipse is centered at the point (h,k), the **standard form** is

$$\frac{(x - h)^2}{a^2} + \frac{(y - k)^2}{b^2} = 1$$

when the major axis is parallel to the x-axis and

$$\frac{(x - h)^2}{b^2} + \frac{(y - k)^2}{a^2} = 1$$

when the major axis is parallel to the y-axis. As in the case of the circle, the standard form is obtained by completing the square.

EXAMPLE 3   Find the coordinates of the foci and the endpoints of the major and minor axes of the ellipse given by

$$9x^2 + 4y^2 + 18x - 8y - 23 = 0.$$

Also, sketch the curve.

Solution   First, put the equation in standard form by completing the square.

$$(9x^2 + 18x \quad) + (4y^2 - 8y \quad) = 23$$
$$9(x^2 + 2x \quad) + 4(y^2 - 2y \quad) = 23$$
$$9(x^2 + 2x + 1) + 4(y^2 - 2y + 1) = 23 + 9(1) + 4(1) = 36$$

Be careful to add 9(1) and 4(1) to the right side of the equation. Now divide both sides of the equation by 36.

$$\frac{(x + 1)^2}{4} + \frac{(y - 1)^2}{9} = 1$$

From this equation in standard form we conclude that the center is at $(-1,1)$.

$$a^2 = 9 \qquad a = 3$$
$$b^2 = 4 \qquad b = 2$$

To find c, we replace a^2 by 9 and b^2 by 4 in the formula

$$a^2 = b^2 + c^2$$
$$9 = 4 + c^2$$
$$5 = c^2 \quad \text{or} \quad c = \sqrt{5}.$$

The major axis is parallel to the y-axis since the larger denominator appears in the y term; a is the distance from the center to the vertices. Thus,

endpoints of major axis: $(-1,-2), (-1,4)$.

The distance from the center to the endpoints of the minor axis is b. Thus,

endpoints of minor axis: $(-3,1), (1,1)$.

The distance from the center to the foci is c. Thus,

coordinates of the foci: $(-1,1 - \sqrt{5}), (-1,1 + \sqrt{5})$.

The ellipse is graphed in Figure 8.15.   □

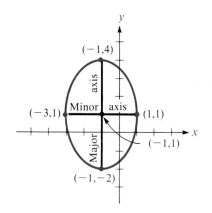

Figure 8.15

Finally, let us consider a few of the ways the ellipse appears in the physical world.

1. In our solar system many bodies revolve in elliptical orbits around a larger body that is located at one focus. For instance, all planets and some comets travel in elliptical paths with the sun at one focus (see Figure 8.16). The moon and artificial satellites follow a similar path

Figure 8.16

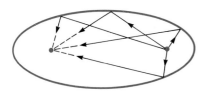

Figure 8.17

around the earth. The ellipse was first analyzed in detail by the Greek mathematician Apollonius in about 230 B.C. for its aesthetic, not practical, value. However, if this study had not been available, it is likely that Johannes Kepler (A.D. 1600) would have been unable to explain planetary motion scientifically, which would have caused a long delay in the development of astronomy and subsequent sciences. This episode is just one of many cases in which a purely mathematical investigation later proved to be of great practical value.

2. The ellipse has a reflection property that causes any ray or wave that originates at one focus to strike the ellipse and pass through the other focus. Acoustically, this property means that in a room with an elliptical ceiling, a slight noise made at one focus can be heard at the other focus (see Figure 8.17). However, if you are standing between the foci, you hear nothing. Such rooms are called whispering galleries, and a famous one is located in the Capitol in Washington, D.C.

3. Arches whose main purpose is beauty and not strength are often elliptical in shape. These ornamental arches are usually large structures made of masonry.

EXERCISES 8.3

In Exercises 1–14 find the coordinates of the foci and the endpoints of the major and minor axes. Sketch each ellipse.

1. $\dfrac{x^2}{25} + \dfrac{y^2}{16} = 1$

2. $\dfrac{x^2}{36} + \dfrac{y^2}{100} = 1$

3. $\dfrac{x^2}{1} + \dfrac{y^2}{4} = 1$

4. $\dfrac{x^2}{49} + \dfrac{y^2}{9} = 1$

5. $x^2 + 9y^2 = 36$

6. $9x^2 + 4y^2 = 144$

7. $4x^2 + y^2 = 1$

8. $x^2 + 9y^2 = 25$

9. $\dfrac{(x-1)^2}{9} + \dfrac{(y+3)^2}{25} = 1$

10. $\dfrac{(x+2)^2}{100} + \dfrac{(y-2)^2}{64} = 1$

11. $4(x-3)^2 + 25(y+1)^2 = 100$

12. $25(x+4)^2 + 16y^2 = 400$

13. $4x^2 + 9y^2 + 8x - 54y + 49 = 0$

14. $9x^2 + y^2 - 36x + 2y + 1 = 0$

In Exercises 15–24 find the equation in standard form of the ellipse satisfying the conditions.

15. Center at origin, vertex (0,5), focus (0,3)
16. Center at origin, vertex (−10,0), focus (−8,0)
17. Center at origin, focus (4,0), length of major axis 12
18. Center at origin, focus (0,−2), length of minor axis 2
19. Center at origin, vertex (3,0), length of minor axis 4
20. Center at origin, vertex (0,3), passes through (2,1)
21. Center at (2,2), focus (2,−1), vertex (2,−3)
22. Center at (−1,3), vertex (3,3), length of minor axis 6
23. Foci at (5,3) and (−1,3), length of major axis 10
24. Vertices at (4,1) and (−8,1), focus (−4,1)
25. The earth travels in an elliptical orbit around the sun, which is located at one of the foci. The longest distance between the earth and the sun is 94,500,000 mi and the shortest distance is 91,500,000 mi.
    a. Find the length of the major axis of the ellipse.
    b. Find the distance between the foci of the orbit.
26. An elliptical arch has a height of 20 ft and a span of 50 ft. How high is the arch 15 ft each side of center?

8.4 THE HYPERBOLA

The geometric definition of a hyperbola resembles the definition of an ellipse and is stated below.

Definition of a Hyperbola	A **hyperbola** is the set of all points in a plane the difference of whose distances from two fixed points (**foci**) is a positive constant.

Thus, the distances between the foci and a point on the figure maintain a *constant difference* for a hyperbola and a *constant sum* for an ellipse. To derive the standard form of the equation of a hyperbola with foci on the *x*-axis and center at the origin, we position the foci at $(-c,0)$ and $(c,0)$. Let $2a$ represent the positive constant of the definition and let $P(x,y)$ be any point on the hyperbola as shown in Figure 8.18.

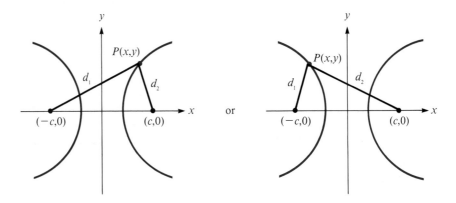

Figure 8.18

Since $d_1 < d_2$ or $d_1 > d_2$, by definition,

$$|d_1 - d_2| = 2a \quad \text{or} \quad d_1 - d_2 = \pm 2a.$$

Then by the distance formula, we have

$$\sqrt{(x + c)^2 + (y - 0)^2} - \sqrt{(x - c)^2 + (y - 0)^2} = \pm 2a$$

or

$$\sqrt{(x + c)^2 + y^2} = \pm 2a + \sqrt{(x - c)^2 + y^2}.$$

Now square both sides of the equation and simplify.

$$(x + c)^2 + y^2 = 4a^2 \pm 4a\sqrt{(x - c)^2 + y^2} + (x - c)^2 + y^2$$
$$4cx = 4a^2 \pm 4a\sqrt{(x - c)^2 + y^2}$$
$$cx - a^2 = \pm a\sqrt{(x - c)^2 + y^2}$$

Again square both sides and simplify.

$$c^2x^2 - 2cxa^2 + a^4 = a^2[(x - c)^2 + y^2]$$
$$c^2x^2 - 2cxa^2 + a^4 = a^2x^2 - 2cxa^2 + a^2c^2 + a^2y^2$$
$$c^2x^2 - a^2x^2 - a^2y^2 = a^2c^2 - a^4$$
$$(c^2 - a^2)x^2 - a^2y^2 = a^2(c^2 - a^2)$$
$$\frac{x^2}{a^2} - \frac{y^2}{c^2 - a^2} = 1$$

If we now define

$$b^2 = c^2 - a^2 \quad \text{or} \quad c^2 = a^2 + b^2,$$

we obtain the **standard form** of a hyperbola with center at the origin and foci on the x-axis.

$$\frac{x^2}{a^2} - \frac{y^2}{b^2} = 1$$

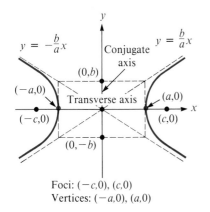

Foci: $(-c,0), (c,0)$
Vertices: $(-a,0), (a,0)$

Figure 8.19

Consider Figure 8.19. By setting $y = 0$, we find that the x-intercepts are $(-a,0)$ and $(a,0)$. The line segment joining these two points is called the **transverse axis.** The endpoints of the transverse axis are called the **vertices of the hyperbola.** By setting $x = 0$, we find that there are no y-intercepts. The line segment from $(0,b)$ to $(0,-b)$ is called the **conjugate axis.** To determine the significance of b, we rewrite $(x^2/a^2) - (y^2/b^2) = 1$ as

$$y = \frac{\pm bx}{a}\sqrt{1 - \frac{a^2}{x^2}}.$$

As $|x|$ gets very large, $1 - a^2/x^2$ approaches 1. Thus, the graph of the hyperbola approaches the lines

$$y = \pm\frac{b}{a}x.$$

These lines are called the **asymptotes of the hyperbola.** They are a great aid in sketching the curve. As shown in Figure 8.19, the asymptotes are the diagonals of a rectangle of dimensions $2a$ by $2b$.

EXAMPLE 1   For the hyperbola given by $4x^2 - y^2 = 36$, find the coordinates of the vertices and the foci. Also, determine the asymptotes and sketch the curve.

Solution   First, divide both sides of the equation by 36 to obtain the standard form.

$$\frac{x^2}{9} - \frac{y^2}{36} = 1$$

Then

$$a^2 = 9 \qquad a = 3$$
$$b^2 = 36 \qquad b = 6.$$

Note that a^2 is not necessarily the larger denominator. Since a is the distance from the center to the vertices, we have

coordinates of the vertices: $(-3,0), (3,0)$.

To find c, we replace a^2 by 9 and b^2 by 36 in the formula

$$c^2 = a^2 + b^2$$
$$= 9 + 36$$
$$c^2 = 45 \quad \text{or} \quad c = \sqrt{45}.$$

The distance from the center to the foci is c. Thus,

coordinates of the foci: $(-\sqrt{45},0), (\sqrt{45},0)$.

For the asymptotes we have

$$y = \pm\frac{b}{a}x = \pm\frac{6}{3}x = \pm 2x.$$

The hyperbola is sketched in Figure 8.20.

Figure 8.20

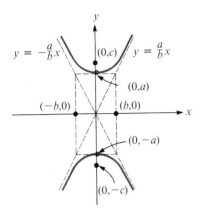

$y = -\dfrac{a}{b}x$ $(0,c)$ $y = \dfrac{a}{b}x$

$(0,a)$

$(-b,0)$ $(b,0)$

$(0,-a)$

$(0,-c)$

Figure 8.21

If the foci are positioned on the y-axis at $(0,-c)$ and $(0,c)$, the **standard form** of the equation of a hyperbola is

$$\frac{y^2}{a^2} - \frac{x^2}{b^2} = 1.$$

In this case the asymptotes are given by $y = \pm ax/b$, as shown in Figure 8.21.

EXAMPLE 2   For the hyperbola given by $25y^2 - 4x^2 = 100$, find the coordinates of the vertices and the foci. Also, determine the asymptotes and sketch the curve.

Solution   First, divide both sides of the equation by 100 to obtain the standard form.

$$\frac{y^2}{4} - \frac{x^2}{25} = 1$$

Then

$$a^2 = 4 \qquad a = 2$$
$$b^2 = 25 \qquad b = 5.$$

Since a is the distance from the center to the vertices, we have

coordinates of the vertices: $(0,2),\ (0,-2)$.

To find c, we replace a^2 by 4 and b^2 by 25 in the formula

$$c^2 = a^2 + b^2$$
$$= 4 + 25$$
$$c^2 = 29 \quad \text{or} \quad c = \sqrt{29}.$$

The distance from the center to the foci is c. Thus,

coordinates of foci: $(0,\sqrt{29}),\ (0,-\sqrt{29})$.

For the asymptotes we have

$$y = \pm\frac{a}{b}x = \pm\frac{2}{5}x.$$

The hyperbola is sketched in Figure 8.22.   □

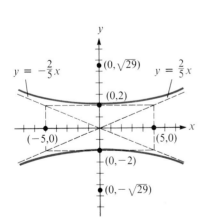

$y = -\dfrac{2}{5}x$ $(0,\sqrt{29})$ $y = \dfrac{2}{5}x$

$(0,2)$

$(-5,0)$ $(5,0)$

$(0,-2)$

$(0,-\sqrt{29})$

Figure 8.22

EXAMPLE 3   Find the equation in standard form of the hyperbola with center at the origin, one focus at $(0,5)$, and one vertex at $(0,3)$.

Solution   Since c is the distance from the center to a focus, c is 5. Since a is the distance from the center to a vertex, a is 3. To find b^2, we replace c by 5 and a by 3 in the formula

$$c^2 = a^2 + b^2$$
$$(5)^2 = (3)^2 + b^2$$
$$16 = b^2.$$

Since the foci are along the y-axis, the standard form is

$$\frac{y^2}{a^2} - \frac{x^2}{b^2} = 1.$$

Thus, the standard equation of the hyperbola is

$$\frac{y^2}{9} - \frac{x^2}{16} = 1. \qquad \square$$

If the hyperbola is centered at the point (h,k), the **standard form** is

$$\frac{(x - h)^2}{a^2} - \frac{(y - k)^2}{b^2} = 1$$

when the transverse axis is parallel to the x-axis. In this case the asymptotes are given by $y - k = (\pm b/a)(x - h)$. If the transverse axis is parallel to the y-axis, the **standard form** is

$$\frac{(y - k)^2}{a^2} - \frac{(x - h)^2}{b^2} = 1$$

and the asymptotes are given by $y - k = (\pm a/b)(x - h)$.

EXAMPLE 4   Find the coordinates of the vertices and the foci of the hyperbola given by $16y^2 - x^2 - 32y + 4x - 4 = 0$. Also, determine the asymptotes and sketch the curve.

Solution   First, put the equation in standard form by completing the square.

$$(16y^2 - 32y \quad) - (x^2 - 4x \quad) = 4$$
$$16(y^2 - 2y \quad) - (x^2 - 4x \quad) = 4$$
$$16(y^2 - 2y + 1) - (x^2 - 4x + 4) = 4 + 16(1) + (-1)4$$
$$16(y - 1)^2 - (x - 2)^2 = 16$$

Dividing both sides of the equation by 16, we have

$$\frac{(y-1)^2}{1} - \frac{(x-2)^2}{16} = 1.$$

From this equation in standard form we conclude that the center is at $(2,1)$.

$$a^2 = 1 \qquad a = 1$$
$$b^2 = 16 \qquad b = 4$$

The transverse axis is parallel to the y-axis since the y term is positive; a is the distance from the center to the vertices. Thus,

coordinates of vertices: $(2,0)$, $(2,2)$.

To find c, we replace a^2 by 1 and b^2 by 16 in the formula

$$c^2 = a^2 + b^2$$
$$= 1 + 16$$
$$c^2 = 17 \quad \text{or} \quad c = \sqrt{17}.$$

The distance from the center to the foci is c. Thus,

coordinates of foci: $(2,1 - \sqrt{17})$, $(2,1 + \sqrt{17})$.

The asymptotes are the diagonals of the rectangle of dimensions $2a$ by $2b$ that is centered at $(2,1)$. The equations of these asymptotes are

$$y - k = \pm\frac{a}{b}(x - h)$$

$$y - 1 = \pm\frac{1}{4}(x - 2).$$

The hyperbola is graphed in Figure 8.23.

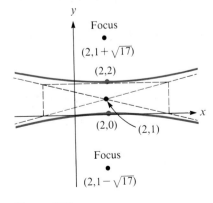

Figure 8.23

EXERCISES 8.4

In Exercises 1–14 find the coordinates of the vertices and the foci, determine the asymptotes, and sketch each curve.

1. $\dfrac{x^2}{16} - \dfrac{y^2}{9} = 1$ **2.** $\dfrac{x^2}{36} - \dfrac{y^2}{100} = 1$

3. $\dfrac{y^2}{25} - \dfrac{x^2}{16} = 1$ **4.** $\dfrac{y^2}{1} - \dfrac{x^2}{4} = 1$

5. $x^2 - 9y^2 = 36$ **6.** $9x^2 - 4y^2 = 144$

7. $4y^2 - x^2 = 1$ **8.** $y^2 - 9x^2 = 25$

9. $\dfrac{(x + 2)^2}{9} - \dfrac{(y - 3)^2}{25} = 1$

10. $\dfrac{(y - 1)^2}{64} - \dfrac{(x - 2)^2}{36} = 1$

11. $4(y + 1)^2 - 25(x + 3)^2 = 100$

12. $25(x + 4)^2 - 16y^2 = 400$

13. $4x^2 - 9y^2 + 8x - 54y - 113 = 0$

14. $9y^2 - x^2 - 36y + 2x - 1 = 0$

In Exercises 15–24 find the equation in standard form of the hyperbola satisfying the conditions.

15. Center at origin, focus (0,5), vertex (0,4)

16. Center at origin, focus (−10,0), vertex (−8,0)

17. Center at origin, focus (4,0), length of transverse axis 6

18. Center at origin, focus (0,−2), length of conjugate axis 2

19. Center at origin, vertex (3,0), length of conjugate axis 10

20. Center at origin, vertex (0,2), passes through (3,4)

21. Center at (−3,−4), focus (2,−4), vertex (0,−4)

22. Center at (5,0), vertex (5,6), length of conjugate axis 8

23. Foci at (3,4) and (3,−2), length of transverse axis 4

24. Vertices at (4,1) and (−10,1), focus (7,1)

8.5 THE PARABOLA

We have already done some work with parabolas in connection with graphing quadratic functions in Section 4.2. We now expand our coverage by considering some geometric properties of this curve and developing the standard forms of equations of parabolas with a horizontal or vertical axis of symmetry. As in the preceding sections, a geometric definition is our starting point.

Definition of a Parabola   A **parabola** is the set of all points in a plane equidistant from a fixed line (**directrix**) and a fixed point (**focus**) not on the line.

From this definition we derive a general equation for a parabola as follows. For simplicity, let the directrix be the line $x = -p$ and position the focus at $(p,0)$. $P(x,y)$ represents any point on the parabola (see Figure 8.24). Since P_1 and P have the same y-coordinate, the distance d_1 is given by

$$d_1 = |x - (-p)| = |x + p|.$$

The distance from P to $(p,0)$ is given by

$$d_2 = \sqrt{(x - p)^2 + (y - 0)^2}.$$

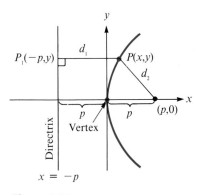

$P_1(-p,y)$

d_1

$P(x,y)$

d_2

$(p,0)$

x

y

p p

Vertex

Directrix

$x = -p$

Figure 8.24

The geometric condition for a parabola states that

$$d_1 = d_2.$$

Thus,

$$|x + p| = \sqrt{(x - p)^2 + y^2}.$$

Squaring both sides of the equation yields

$$(x + p)^2 = (x - p)^2 + y^2.$$

Simplifying this equation, we have

$$x^2 + 2px + p^2 = x^2 - 2px + p^2 + y^2$$

$$4px = y^2.$$

This equation is the **standard form** of a parabola with directrix $x = -p$ and focus at $(p,0)$. The line through the focus that is perpendicular to the directrix is called the **axis of symmetry.** In this case the axis of symmetry is the x-axis. The point on the axis of symmetry that is midway between the focus and the directrix is called the **vertex.** The vertex is the turning point of the parabola. In this case the vertex is the origin. Note that p is the distance from the vertex to the focus and from the vertex to the directrix.

EXAMPLE 1   Find the focus and the directrix of the parabola given by $y^2 = 12x$.

Solution   Matching the equation $y^2 = 12x$ to the standard form

$$y^2 = 4px,$$

we write

$$y^2 = 12x = 4(3)x.$$

Thus,

$$p = 3.$$

The focus is on the axis of symmetry (the x-axis) p units to the right of the vertex (the origin). Thus,

focus: $(3,0)$.

The directrix is the line p units to the left of the vertex. Thus,

directrix: $x = -3$.   □

There are three other possibilities for a parabola with a directrix that is parallel to the *x*- or *y*-axis and whose vertex is the origin. These cases are illustrated in Figures 8.25–8.27. In all cases $p > 0$.

To summarize, the graphs of the equations $y^2 = \pm 4px$ and $x^2 = \pm 4py$ are parabolas. If the *y* term is squared, the axis of symmetry is the *x*-axis. The parabola opens to the right when the coefficient of *x* is positive and to the left when this coefficient is negative. When the *x* term is squared, the axis of symmetry is the *y*-axis. The parabola opens upward when the coefficient of *y* is positive and downward when this coefficient is negative. In all cases *p* gives the distance from the vertex to the focus and from the vertex to the directrix.

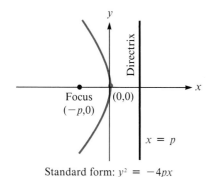

Standard form: $y^2 = -4px$
Axis of symmetry: *x*-axis
opens to the left

Figure 8.25   Case 2

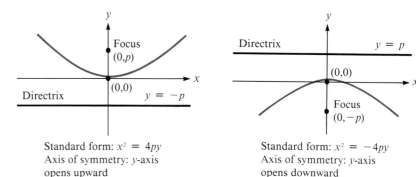

Standard form: $x^2 = 4py$
Axis of symmetry: *y*-axis
opens upward

Figure 8.26   Case 3

Standard form: $x^2 = -4py$
Axis of symmetry: *y*-axis
opens downward

Figure 8.27   Case 4

EXAMPLE 2   Find the focus and directrix of the parabola given by $x^2 = -6y$.

Solution   Matching the equation $x^2 = -6y$ to the standard form in case 4,

$$x^2 = -4py,$$

we write

$$x^2 = -6y = -4(\tfrac{3}{2})y.$$

Thus,

$$p = \tfrac{3}{2}.$$

The focus is on the axis of symmetry (the y-axis) p units down from the vertex (the origin). Thus,

focus: $(0, -\frac{3}{2})$.

The directrix is the line p units up from the vertex. Thus,

directrix: $y = \frac{3}{2}$.   □

EXAMPLE 3   A parabola with its vertex at the origin has its focus at $(-5,0)$. Find the directrix and the equation in standard form of the parabola and sketch the curve.

Solution   This is an example of case 2. The term p, which represents the distance from the vertex $(0,0)$ to the focus $(-5,0)$, is 5. Thus,

directrix: $x = 5$.

To find the standard equation of the parabola, replace p by 5 in the form

$$y^2 = -4px$$

to obtain

$$y^2 = -20x.$$

The parabola opens to the left and is graphed in Figure 8.28.   □

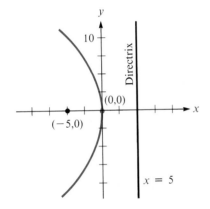

Figure 8.28

If the vertex of the parabola is at the point (h,k) and the directrix is parallel to the x- or y-axis, then there are four possible standard forms for the parabola. These cases are illustrated in Figure 8.29. In all cases $p > 0$.

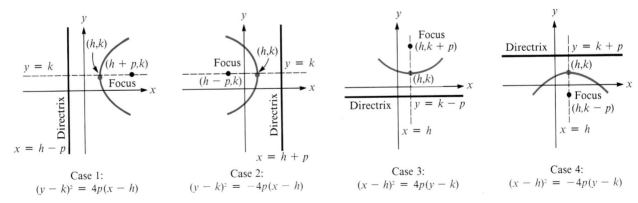

Figure 8.29

EXAMPLE 4   For the parabola given by $y = x^2 - 2x$, find the vertex, the focus, and the directrix. Also, sketch the curve.

Solution   First, put the equation in standard form by completing the square.

$$y = (x^2 - 2x \qquad)$$
$$y + 1 = (x^2 - 2x + 1)$$
$$y + 1 = (x - 1)^2$$

Matching this equation to the form in case 3,

$$(x - h)^2 = 4p(y - k),$$

we conclude that

$$h = 1, \qquad k = -1, \qquad 4p = 1, \quad \text{or} \quad p = \tfrac{1}{4}.$$

Thus,

vertex: $(1, -1)$.

The focus is on the axis of symmetry $(x = 1)$ p units up from the vertex $(1, -1)$. Thus,

focus: $(1, -\tfrac{3}{4})$.

The directrix is the line p units down from the vertex. Thus,

directrix: $y = -\tfrac{5}{4}$.

The parabola opens upward and is graphed in Figure 8.30.

Figure 8.30

Figure 8.31

Figure 8.32

In conclusion, let us consider a few of the applications of the parabola.

1. The parabola has an important reflection property. Any ray or wave that originates at the focus and strikes the parabola is reflected parallel to the axis of symmetry (see Figure 8.31). For this reason, an instrument such as a flashlight or a searchlight uses a parabolic reflector with the bulb located at the focus. The reflector redirects light that would otherwise be wasted parallel to the axis so that a straight beam of light is formed. In an automobile headlight a bulb located at the focus produces the high beam. For the low beam the bulb is placed slightly ahead of and above the focus. In reverse, the reflection property of a parabola causes any ray or wave that comes into a parabolic reflector parallel to the axis of symmetry to be directed to the focus point. Radar, radio antennas, and reflecting telescopes work on this principle (see Figure 8.32).
2. Arches whose main purpose is strength are usually parabolic in shape and constructed of steel.
3. In a suspension bridge the cables are designed for engineering purposes to hang in a shape that is almost parabolic.
4. Every comet in our solar system travels in an orbit that is either an ellipse, a parabola, or a branch of a hyperbola. In each case the sun is located at one focus. Only comets with an elliptical orbit reappear.
5. Projectiles, such as a baseball or the water that streams from a fountain, follow a parabolic path.

EXERCISES 8.5

In Exercises 1–14 find the vertex, the focus, and the equation of the directrix and sketch each curve.

1. $y^2 = 4x$
2. $y^2 = -8x$
3. $x^2 = -2y$
4. $x^2 = 5y$
5. $y^2 + 12x = 0$
6. $y^2 - 10x = 0$
7. $4x^2 - 3y = 0$
8. $2x^2 + 3y = 0$
9. $(y + 2)^2 = 4(x + 1)$
10. $(x - 1)^2 = -6y$
11. $y = x^2 - 4x$
12. $x = y^2 + 2y$
13. $x^2 - 2x - 4y - 7 = 0$
14. $y^2 + 4y + 3x - 8 = 0$

In Exercises 15–24 find the equation in standard form of the parabola satisfying the conditions.

15. Vertex at origin, focus (0,3)
16. Vertex at origin, focus (−2,0)
17. Vertex at origin, directrix $x = -\frac{1}{2}$
18. Vertex at origin, directrix $y = \frac{3}{2}$
19. Focus (0,4), directrix $y = -4$
20. Focus $\left(-\frac{5}{2},0\right)$, directrix $x = \frac{5}{2}$
21. Focus (2,1), directrix $x = -4$
22. Focus (−1,−3), directrix $y = 5$
23. Vertex (3,0), focus (3,3)
24. Vertex (1,4), directrix $x = \frac{7}{2}$

25. Consider the parabolic reflector shown. Choose axes in a convenient position and determine (a) an equation of the parabola and (b) the location of the focus point.

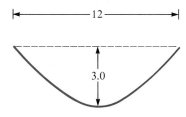

26. Answer the questions in Exercise 25 for the parabolic reflector shown below.

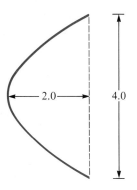

27. A parabolic arch has a height of 25 ft and a span of 40 ft. How high is the arch 8 ft each side of the center?

8.6 CLASSIFYING CONIC SECTIONS

All of the standard forms for conic sections that we have developed may be converted through basic algebra to an equation of the form

$$Ax^2 + Cy^2 + Dx + Ey + F = 0,$$

where A and C are not both zero. This equation is called the **general form** of an equation of a conic section with axis or axes parallel to the coordinate axes. If a graph is a circle, an ellipse, a hyperbola, or a parabola, then in general form A and C must satisfy certain conditions that are apparent from the standard forms of these conics. We summarize these conditions below, along with the degenerate possibilities for each case.

Conditions on A and C	Conic Section	Degenerate Possibilities
$A = C \neq 0$	Circle	A point or no gráph at all
$A \neq C$, A and C have the same sign.	Ellipse	A point or no graph at all
A and C have opposite signs.	Hyperbola	Two intersecting straight lines
$A = 0$ or $C = 0$ (but not both)	Parabola	A line, a pair of parallel lines, or no graph at all

Example 1 illustrates the main application of this chart.

EXAMPLE 1   Identify the graph of the following equations if the graph is either a circle, an ellipse, a hyperbola, or a parabola.

a. $4x^2 - 9y^2 + 8x - 54y - 113 = 0$ **b.** $2x^2 - x = -3y^2$

Solution   Since the degenerate possibilities have been eliminated, we proceed as follows:

a. The equation $4x^2 - 9y^2 + 8x - 54y - 113 = 0$ is in general form with $A = 4$ and $C = -9$. Since A and C have different signs the graph of the equation is a hyperbola.

b. To determine the conic given by $2x^2 - x = -3y^2$, we first write the equation in general form as

$$2x^2 + 3y^2 - x = 0.$$

In this case A and C have the same sign with $A \neq C$. Thus, the graph of the equation is an ellipse.   □

The next example shows the analysis of two equations that lead to degenerate cases.

EXAMPLE 2   Identify the graph of the following equations.

a. $2x^2 + y^2 + 1 = 0$ **b.** $x^2 + y^2 + 4x - 6y + 13 = 0$

Solution

a. The equation $2x^2 + y^2 + 1 = 0$ does not define an ellipse, even though A and C have the same sign with $A \neq C$. There is no ordered pair of real numbers that satisfy $2x^2 + y^2 = -1$ since $2x^2$ and y^2 are never negative, so their sum can never be -1. Thus, the equation has no graph.

b. On first inspection, this equation seems to define a circle. To be sure, we need to complete the square and convert the equation to standard form. By doing so, in this case we obtain

$$(x + 2)^2 + (y - 3)^2 = 0.$$

Only $x = -2$, $y = 3$ satisfies the equation, so the graph is the point $(-2,3)$.   □

EXERCISES 8.6

In Exercises 1–10 identify the graph of each equation if the graph is either a circle, an ellipse, a hyperbola, or a parabola.

1. $y^2 - 3x + 2y + 1 = 0$
2. $9y^2 - x^2 - 36x + 2y - 1 = 0$
3. $4x^2 + 9y^2 + 8x - 54y - 49 = 0$
4. $x^2 + y^2 + 7x + 3y - 4 = 0$
5. $x^2 = y^2 - 4$ 6. $3y^2 = 4x - x^2$
7. $3y = 4x - x^2$ 8. $5x^2 = 1 - 5y^2$
9. $2y^2 = 9 - 2x^2$ 10. $5x - 2y = y^2$

In Exercises 11–20 identify the graph of each equation. Degenerate cases are possible here.

11. $x^2 - y^2 - 4y - 4 = 0$
12. $x^2 + y^2 - 4y + 4 = 0$
13. $x^2 + 2x = 1 - y^2$ 14. $x^2 + 2x = y^2 - 1$
15. $x^2 + 2x = 1 - y$ 16. $x^2 + 2x = y^2 - 4$
17. $9x^2 + 4y^2 + 18x - 8y + 13 = 0$
18. $4x^2 + 9y^2 + 2x - 4 = 0$

19. $4x^2 + 9y^2 - 2x + 4 = 0$
20. $4y^2 + 3 = 8y$
21. Match each graph with the equation that illustrates that case.

Graph	Equation
1. Two distinct lines through the origin	a. $x^2 + y^2 = 0$
2. Two distinct parallel lines	b. $x^2 + y^2 = -1$
3. One line through the origin	c. $x^2 - y^2 = 0$
4. A point (the origin)	d. $x^2 = 1$
5. No graph	e. $x^2 = 0$

22. Generalize from the result in Exercise 21 and specify conditions for A, C, and k in the equation

$$Ax^2 + Cy^2 = k, \text{ with } A^2 + C^2 \neq 0$$

that result in a graph that is a circle, an ellipse, a hyperbola, or each of the special cases listed above.

CHAPTER OVERVIEW

Section	Key Concepts to Review
8.1	• Fundamental relationship between an equation and its graph: Every ordered pair that satisfies the equation corresponds to a point in its graph, and every point in the graph corresponds to an ordered pair that satisfies the equation.
	• Procedures for finding an equation that corresponds to a given geometric condition
	• Procedures for proving certain theorems in plane geometry by using a coordinate system and algebraic methods
	• Midpoint formula: $\left(\dfrac{x_1 + x_2}{2}, \dfrac{y_1 + y_2}{2}\right)$
8.2	• Definition of circle
	• Standard form of a circle of radius r with center (h,k): $(x - h)^2 + (y - k)^2 = r^2$

Section	Key Concepts to Review

8.3
- Definitions of ellipse, foci, major axis, minor axis, and vertices
- Summary for an ellipse:

Foci	Center	Standard Form $(a > b)$
On x-axis at $(\pm c, 0)$	Origin	$\dfrac{x^2}{a^2} + \dfrac{y^2}{b^2} = 1$
On y-axis at $(0, \pm c)$	Origin	$\dfrac{x^2}{b^2} + \dfrac{y^2}{a^2} = 1$
On major axis parallel to x-axis	(h,k)	$\dfrac{(x-h)^2}{a^2} + \dfrac{(y-k)^2}{b^2} = 1$
On major axis parallel to y-axis	(h,k)	$\dfrac{(x-h)^2}{b^2} + \dfrac{(y-k)^2}{a^2} = 1$

- In general, $a^2 = b^2 + c^2$, where a, b, and c represent the following distances:

  a is the distance from the center to the endpoints on the major axis
  b is the distance from the center to the endpoints on the minor axis
  c is the distance from the center to the foci

8.4
- Definitions of hyperbola, foci, transverse axis, vertices, conjugate axis, and asymptotes of the hyperbola
- Summary for a hyperbola:

Foci	Center	Standard Form	Asymptotes
On x-axis at $(\pm c, 0)$	Origin	$\dfrac{x^2}{a^2} - \dfrac{y^2}{b^2} = 1$	$y = \pm\dfrac{b}{a}x$
On y-axis at $(0, \pm c)$	Origin	$\dfrac{y^2}{a^2} - \dfrac{x^2}{b^2} = 1$	$y = \pm\dfrac{a}{b}x$
On transverse axis parallel to x-axis	(h,k)	$\dfrac{(x-h)^2}{a^2} - \dfrac{(y-k)^2}{b^2} = 1$	$y - k = \pm\dfrac{b}{a}(x-h)$
On transverse axis parallel to y-axis	(h,k)	$\dfrac{(y-k)^2}{a^2} - \dfrac{(x-h)^2}{b^2} = 1$	$y - k = \pm\dfrac{a}{b}(x-h)$

- In general, $c^2 = a^2 + b^2$, where a, b, and c represent the following distances:

  a is the distance from the center to the endpoints on the transverse axis
  b is the distance from the center to the endpoints on the conjugate axis
  c is the distance from the center to the foci

Section	Key Concepts to Review
8.5	• Definitions of parabola, directrix, focus, axis of symmetry, and vertex
	• Summary for a parabola with vertex at the origin: (In all cases p gives the distance from the vertex to the focus and from the vertex to the directrix. See Figures 8.24–8.27.)

Standard Form	Opens	Axis of Symmetry	Focus	Directrix
$y^2 = 4px$	Right	x-axis	$(p,0)$	$x = -p$
$y^2 = -4px$	Left	x-axis	$(-p,0)$	$x = p$
$x^2 = 4py$	Upward	y-axis	$(0,p)$	$y = -p$
$x^2 = -4py$	Downward	y-axis	$(0,-p)$	$y = p$

	• Figure 8.29 summarizes the cases when the vertex of the parabola is at the point (h,k) and the directrix is parallel to the x- or y-axis.
8.6	• The general form of an equation of a conic section with axis or axes parallel to the coordinate axes is $$Ax^2 + Cy^2 + Dx + Ey + F = 0,$$ where A and C are not both zero.
	• Chart summarizing the graphing possibilities for the above equation

CHAPTER REVIEW EXERCISES

1. Find an equation for the graph that is the set of all points in a plane equidistant from the line $y = 5$ and the point $(-1,-3)$.

2. Classify the graph defined by each of the following equations.

   a. $9x^2 = 4y^2 - 54x - 16y + 61$

   b. $x^2 + 4x + 8y = 4$

   c. $7x - 2y = 3$

   d. $16x^2 + 25y^2 - 32x - 284 = 0$

   e. $x^2 + 6x = 3 - y^2$

3. Determine the center and radius of the circle given by $x^2 + y^2 + 4x - 6y = 12$.

4. Determine the standard equation of the circle with center at $(4,-3)$ that passes through $(0,-1)$.

5. Find the coordinates of the foci of the ellipse given by $25x^2 + 16y^2 = 400$.

6. What is the equation in standard form of the ellipse with vertices at $(3,2)$ and $(-7,2)$ and a minor axis of length 6?

7. Find the equations of the asymptotes of the hyperbola given by $y^2/64 - x^2/100 = 1$.

8. Determine the standard equation of the hyperbola whose center is at the origin with one focus at $(-8,0)$ and a transverse axis of length 14.

9. What are the coordinates of the focus point of the parabola given by $x^2 = -7y$?

10. Find the standard equation of the parabola with focus at $(5,1)$ and directrix $x = -1$.

11. Find the standard equation of the circle whose diameter extends from $(-2,2)$ to $(4,2)$.

12. Show by the methods of analytic geometry that the midpoints of the sides of a rectangle are the vertices of a quadrilateral whose perimeter is equal to the sum of the lengths of the diagonals of the rectangle.

13. Graph $16(y - 2)^2 - 25(x + 3)^2 = 400$.

14. If the line $y = mx - 1$ passes through the center of the circle $(x - 2)^2 + (y + 3)^2 = 25$, then find the value of m.

15. What are the coordinates of the vertices of the hyperbola given by $9(x - 3)^2 - 4(y + 1)^2 = 144$?

In Exercises 16–20 select the choice that completes the statement or answers the question.

16. The graph of $x^2 = 1 - 2y^2$ is
    a. a circle   b. an ellipse
    c. a parabola   d. a hyperbola

17. Which conic section is defined as the set of all points in a plane equidistant from a fixed line and a fixed point not on the line?
    a. circle   b. ellipse
    c. parabola   d. hyperbola

18. If s varies directly as the square of t, then the graph of this relation is a
    a. line   b. circle
    c. parabola   d. hyperbola

19. Which of the following is the equation of a hyperbola?
    a. $2x^2 + y^2 = 5$ b. $x^2 = 1 - y^2$
    c. $x = y^2 - 4$ d. $x^2 = y^2 - 4$

20. The standard equation of the circle with center at $(1, -3)$ and radius 5 is
    a. $(x + 1)^2 + (y - 3)^2 = 5$
    b. $(x - 1)^2 + (y + 3)^2 = 25$
    c. $(x - 1)^2 + (y + 3)^2 = 5$
    d. $(x + 1)^2 + (y - 3)^2 = 25$

9 Systems of Equations and Inequalities

1 cube, 2 bolts   8 nails   1 cube   2 nails, 1 bolt   1 cube   ?

Consider the figure. How many nails are needed to balance the cube? (See Example 8 of Section 9.1.)

In the analysis of a problem we often must take into account many variables and many relationships among the variables. This situation leads to the topic of setting up and solving systems of equations and inequalities. In most cases we deal with **linear equations in *n* variables,** which are equations of the form

$$a_1x_1 + a_2x_2 + \cdots + a_nx_n = c,$$

where $a_1, a_2, \ldots, a_n$, and c are real numbers and $x_1, x_2, \ldots, x_n$ are variables. A set of linear equations is called a **linear system.** In this chapter we first review the substitution and addition-elimination methods for solving simple linear systems. Then we show how to solve more complicated linear systems by using determinants, Gaussian elimination, and matrix algebra. Finally, we discuss nonlinear systems of equations and systems of linear inequalities. Partial fractions and linear programming are considered as applications of systems of equations and inequalities, respectively.

9.1 SYSTEMS OF LINEAR EQUATIONS IN TWO VARIABLES

More than one linear equation is often needed to describe a situation adequately. For example, suppose that you have to decide between two positions as a salesperson. The first company offers a straight 20 percent commission, whereas the second company offers a salary of $70 per week plus a 10 percent commission. Before you choose your job, it is important to determine how much you would have to sell per week before the two job offers would produce the same income. For the purpose of comparison, we have graphed the two job offers in Figure 9.1.

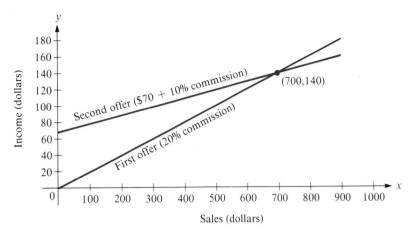

Figure 9.1

First offer (20% commission): $y = 0.2x$

Sales (in dollars) x	100	200	300	400	500	600	700	800	900	1,000
Income (in dollars) y	20	40	60	80	100	120	140	160	180	200

Second offer ($70 + 10% commission): $y = 70 + 0.1x$

Sales (in dollars) x	100	200	300	400	500	600	700	800	900	1,000
Income (in dollars) y	80	90	100	110	120	130	140	150	160	170

By examining the graph, we can see that the solution is the point (700,140), where the two lines intersect. That is, you earn an income of $140 from either job when you sell $700 worth of merchandise. Above $700 the larger commission is the better offer. Notice in our example that there are many ordered pairs that satisfy the first job offer and many ordered pairs that satisfy the second offer, but our solution is the only ordered pair common to the two equations. In general, the solution set of a system of linear equations is the set of all the ordered pairs that satisfy both equations. Graphically this corresponds to the collection of points where the lines intersect. Usually, as in our example, the two lines will intersect at only one point and, consequently, our solution is only one ordered pair. However, the next two examples illustrate two other possibilities.

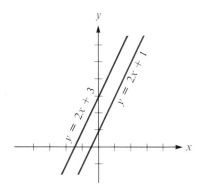

EXAMPLE 1   Find all the ordered pairs that satisfy the pair of equations

$$y = 2x + 1$$
$$y = 2x + 3.$$

Solution   These two lines are parallel and do not intersect, as shown in Figure 9.2. Thus, there is no ordered pair that satisfies both equations and the solution set is $\emptyset$. This situation will arise whenever two lines have the same slope and different y-intercepts. Such a system of equations is called **inconsistent.**   □

EXAMPLE 2   Find all the ordered pairs that satisfy the pair of equations

$$y = 2x + 3$$
$$2y = 4x + 6.$$

Solution   When we attempt to graph both equations (see Figure 9.3), we find that any ordered pair that satisfies the first equation also satisfies the second equation. Thus, the two equations are equivalent. This means that the same line is the graph of both equations and that graphically the solution set is the set of all points on that line. This situation occurred because the second equation is obtained if we multiply both sides of the first equation by 2. In more complicated situations it is often difficult to determine if two equations are equivalent. A system containing equivalent equations is called **dependent.**   □

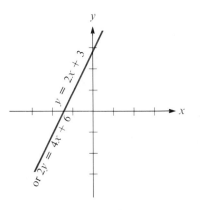

Figure 9.3

By far the most useful situation is when the graphs of the two linear equations intersect at one point so there is only one ordered pair that satisfies both equations. We have been finding this solution graphically. This technique is good for illustrating the principle involved and for obtaining an approximate solution. However, if an exact solution is required, we must use algebraic methods. The first method we discuss makes use of the **substitution method,** as in the following example.

EXAMPLE 3   Find all the ordered pairs that satisfy the pair of equations

$$y = x + 2$$
$$y = 2x - 3.$$

Solution   We wish to find the value for x that makes both y values the same or, equivalently, we want to know the value for x that makes $x + 2$ equal to $2x - 3$.

$$x + 2 = 2x - 3$$
$$2 = x - 3$$
$$5 = x$$

Thus, the x-coordinate of the solution is 5. To find the y-coordinate, substitute 5 for x in either of the given equations.

$$y = x + 2 \quad \text{or} \quad y = 2x - 3$$
$$= (5) + 2 \qquad\qquad = 2(5) - 3$$
$$= 7 \qquad\qquad\qquad = 7$$

Thus, the solution is (5,7). (*Note:* A good check of your result is to find the y-coordinate by substituting the x-coordinate in both equations.)   □

EXAMPLE 4   The total cost of producing gadgets consists of $300 per month for rent plus $4 per unit for material. If the selling price for a gadget is $9, how many units must be made and sold per month for the company to break even?

Solution   If x represents the number of gadgets made and sold, then

$$\text{Total cost} = 300 + 4x$$
$$\text{Total revenue} = 9x.$$

We wish to find the value for x that makes the total cost equal to the revenue or, equivalently, we want to know the value for x that makes $300 + 4x$ equal to $9x$.

$$300 + 4x = 9x$$
$$300 = 5x$$
$$60 = x$$

Thus, the company will break even when it makes and sells 60 gadgets. To check the result, substitute 60 in the original equations (see Figure 9.4).

$$\text{Total cost} = 300 + 4x \quad \text{or} \quad \text{Total revenue} = 9x$$
$$= 300 + 4(60) \qquad\qquad\qquad = 9(60)$$
$$= 540 \qquad\qquad\qquad\qquad = 540$$

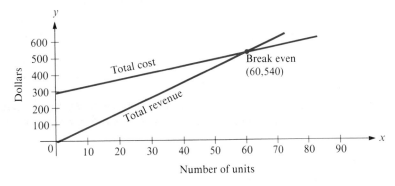

Figure 9.4   □

In these examples the substitution method has been appropriate since all the linear equations have been given in the form $y = mx + b$. However, if one of the linear equations is given as $3x - 5y - 4 = 0$, we would have to change the equation to $y = \frac{3}{5}x - \frac{4}{5}$ and continue with this equation. This procedure is awkward algebraically. Thus, a different method is used. The following example illustrates the **addition-elimination method.** It is very important because it can be used to solve complicated systems of equations with more variables.

EXAMPLE 5   Find all the ordered pairs that satisfy the pair of equations

$$x + y = 9$$
$$x - y = 5.$$

Solution   This method attempts to eliminate one of the variables by adding the two equations together. In this example the coefficients for y are $+1$ and -1. If we add the equations, the result will be an equation that contains only x.

$$x + y = 9$$
$$\underline{x - y = 5}$$
$$2x \quad = 14 \qquad \text{Add the equations.}$$
$$x \quad = 7$$

Thus, the x-coordinate of the solution is 7.

To find the y-coordinate, substitute 7 for x in either of the given equations.

$$x + y = 9 \quad \text{or} \quad x - y = 5$$
$$(7) + y = 9 \qquad\quad 7 - y = 5$$
$$y = 2 \qquad\qquad\quad y = 2$$

Thus, the solution is (7,2).   □

EXAMPLE 6   Find all the ordered pairs that satisfy the pair of equations

$$3x - 2y = 27$$
$$2x + 5y = -1.$$

Solution   If we form equivalent equations by multiplying the top equation by -2 and the bottom equation by 3, we can eliminate the x variable.

$$-6x + 4y = -54$$
$$\underline{6x + 15y = -3}$$
$$19y = -57 \qquad \text{Add the equations.}$$
$$y = -3$$

Thus, the y-coordinate of the solution is -3.

To find the x-coordinate, substitute -3 for y in either of the given equations.

$$3x - 2y = 27 \quad \text{or} \quad 2x + 5y = -1$$
$$3x - 2(-3) = 27 \qquad\qquad 2x + 5(-3) = -1$$
$$x = 7 \qquad\qquad\qquad x = 7$$

Thus, the solution is $(7, -3)$.   □

EXAMPLE 7   Find the forces F_1 and F_2 that achieve equilibrium for the beam in Figure 9.5.

Figure 9.5

Solution   If we consider two of the laws of physics, we can determine two equations that contain F_1 and F_2. First, since the system is in equilibrium, the sum of the forces pointing down must be equal to the sum of the forces pointing up.

$$F_1 + F_2 = 14$$

Second, the tendency of the lumber to rotate about point A in a clockwise direction is the product of F_1 and the distance (2 ft) between F_1 and the turning point A; the tendency of the lumber to rotate in a counterclockwise direction is the product of F_2 and the distance (5 ft) between F_2 and the turning point A. Since the system is in equilibrium, we have

$$\text{clockwise turning effect} = \text{counterclockwise turning effect}$$
$$2 \cdot F_1 \qquad = \qquad 5 \cdot F_2.$$

Thus, we can determine F_1 and F_2 by finding the ordered pair that satisfies these two equations.

$$F_1 + F_2 = 14$$
$$2F_1 - 5F_2 = 0 \quad \text{or} \quad (2F_1 = 5F_2)$$

We can eliminate F_2 by multiplying the top equation by 5 and then adding the two equations.

$$5F_1 + 5F_2 = 70$$
$$\underline{2F_1 - 5F_2 = 0}$$
$$7F_1 = 70$$
$$F_1 = 10$$

Thus, F_1 is 10 lb. To find F_2, substitute 10 for F_1 in either of the given equations.

$$F_1 + F_2 = 14 \quad \text{or} \quad 2F_1 = 5F_2$$
$$(10) + F_2 = 14 \qquad\qquad 2(10) = 5F_2$$
$$F_2 = 4 \qquad\qquad\quad 4 = F_2$$

Thus,

$$F_1 = 10 \text{ lb and } F_2 = 4 \text{ lb.}$$

EXAMPLE 8   Solve the problem in the chapter introduction on page 428.

Solution   If we let b, c, and n represent the weights of a bolt, a cube, and a nail, respectively, then by inspection of the scales, we determine the following system of equations.

$$c + 2b = 8n$$
$$c = 2n + b$$

One way to find c in terms of n is to solve the second equation for b to determine that $b = c - 2n$ and then substitute the result in the first equation.

$$c + 2(c - 2n) = 8n$$
$$3c - 4n = 8n$$
$$3c = 12n$$
$$c = 4n$$

Thus, 4 nails are needed to balance the cube.

In economics an analysis of the law of supply and demand involves the intersection of two curves. Basically, as the price for an item increases, the quantity of the product that is supplied increases while the quantity that is demanded decreases. The point at which the supply and demand curves intersect is called the **point of market equilibrium.** This principle is shown in Figure 9.6, where, for illustrative purposes, we assume the supply and demand equations to be linear.

Figure 9.6

The equilibrium price is the value for p at which supply equals demand. The law of supply and demand states that in a system of free enterprise (a big assumption!) a product will sell near its equilibrium price. The theory asserts that if the price is above the equilibrium price, more of the product will be put on the market to capitalize on the higher price. A surplus results, causing the price of the product to lower. When the price is below the equilibrium price, the demand for the product will exceed the supply, so the sellers can raise their prices.

EXERCISES 9.1

In Exercises 1–20 find all the ordered pairs that satisfy the pair of equations. In 1–10 first estimate the solution graphically.

1. $y = x - 2$
   $y = 5x + 6$

2. $y = -x - 3$
   $y = 2x + 3$

3. $y = 3x + 4$
   $y = -x - 2$

4. $y = 4x + 7$
   $y = 3x + 5$

5. $y = x + 4$
   $y = 2x + 4$

6. $y = -5x + 2$
   $y = 4x - 7$

7. $y = x$
   $x - 2y = 6$

8. $y = -3x$
   $2x + 3y = -21$

9. $y = \frac{1}{3}x - 5$
   $x - 3y - 15 = 0$

10. $y = -\frac{5}{2}x + 3$
   $5x + 2y + 6 = 0$

11. $x + y = 25$
   $6x - y = 3$

12. $2x + 3y = 8$
   $2x - 7y = -32$

13. $5x + 3y = -2$
   $x - 2y = -3$

14. $-2x - y = -5$
   $5x + 2y = -17$

15. $3x - 2y = 1$
   $6x - 4y = 5$

16. $2x - 3y = 1$
   $6x - 9y = 3$

17. $2x - 5y = 5$
   $4x + 3y = 23$

18. $4x + 2y = 2$
   $6x - 5y = 27$

19. $7x - 2y - 19 = 0$
   $3x + 5y + 14 = 0$

20. $6x + 10y - 7 = 0$
   $15x - 4y - 3 = 0$

In Exercises 21–40 set up a system of linear equations and solve by an appropriate method illustrated in this section.

21. The sum of two numbers is 70, their difference is 22. What are the numbers?

22. Find two complementary angles whose difference is 20°.

23. Find two supplementary angles whose difference is 100°.

24. A piece of lumber is 120 in. long. Where must it be cut for one piece to be four times longer than the other piece?

25. A container holding a liquid weighs 500 g. If one-half the liquid is poured out, the weight is 350 g. What is the weight of the empty container?

26. If four black metal balls and one red metal ball are placed on a scale, they balance a weight of 100 g. A weight of 90 g will balance two black and three red balls. Find the weight of each kind of metal ball.

27. The velocity of a particle that accelerates at a uniform rate is linearly related to the elapsed time by the equation $v = v_0 + at$, where v_0 is the initial velocity and a the acceleration. If $v = 36$ ft/second when $t = 2$ seconds and $v = 4$ ft/second when $t = 3$ seconds, find values for v_0 and a.

28. In Exercise 27 find the values of v_0 and a if $v = 45$ ft/second when $t = 2$ seconds and if $v = 72$ ft/second when $t = 5$ seconds.

29. Find the forces F_1 and F_2 that achieve equilibrium for the beam in the following diagram.

30. Find the forces F_1 and F_2 that achieve equilibrium for the beam in the following diagram.

31. It takes a boat 1 hour to go 6 mi upstream (against the current); the return trip downstream (with the current) takes $\frac{1}{2}$ hour. What is the rate of the current? What would be the speed of the boat if there were no current?

32. A chemist has a 10 percent solution and a 25 percent solution of alcohol. How much of each should be mixed together to obtain 100 gal of a 20 percent solution?

33. A college wants to invest $10,000 to have an annual income of $700 for a scholarship. The college plans to invest part of the money in a bank that yields 6 percent interest and the remainder in a speculative fund that promises to yield 11 percent interest. How much should be invested in each to obtain the desired income?

34. You are trying to decide between two positions as a salesperson. The first offer is a straight 15 percent commission while the second offer pays a salary of $60 per week plus a 10 percent commission. How much must you sell each week for the two jobs to pay the same?

35. A manufacturer wants to know whether it will pay him to buy and install a special machine to turn out widgets that, until now, he has been purchasing from an outside supplier for $1 each. The machine will cost him $500 per year and will be able to produce widgets for 50 cents each. How many widgets would he have to make and use each year to justify purchasing this new machine?

36. A company is trying to decide between two machines for packaging its new product. Machine A will cost $5,000 per year plus $2 to package each unit; machine B will cost $8,000 per year plus $1 to package each unit.

   a. How many units must be produced for the cost of the two machines to be the same? If the company plans to produce more units, which machine should they purchase?

   b. If packaging can be subcontracted to another firm at a cost of $5 per unit, how many units would have to be made before the purchase of machine A would be worthwhile?

37. Consider Figure 9.6 and the accompanying discussion of the law of supply and demand. Determine the equilibrium price (p) in terms of the constants $a, b, c,$ and d. Which of these constants are positive? Which are negative?

Exercises 38–39 are problems that are usually considered in the third semester of accounting. In these problems we find several important business considerations defined in terms of each other. Thus, they have interlocking solutions that lead naturally to simultaneous equations.

38. A company operates in a state that levies a 10 percent tax on the income that remains after paying the federal tax. Meanwhile, the federal tax is 50 percent of the income that remains after paying the state tax. If during the current year a corporation has $400,000 in taxable income, determine the state and federal income taxes.

39. A company is to give a performance bonus to one of its top managers. The income in the manager's division is $100,000 and the bonus rate is 10 percent. The bonus is based on profit, so it is based on income after both taxes and the bonus have been subtracted from the $100,000. The taxes amount to 30 percent of the taxable income, with the bonus counting as a tax deduction. What is the manager's bonus?

40. In electronics the analysis of a circuit often utilizes several laws that apply at the same time. This situation leads naturally to simultaneous equations. By applying Kirchhoff's laws to the circuit shown in the diagram, we obtain the following equations:

$$R_1I_1 + R_3(I_1 + I_2) = E_1$$
$$R_2I_2 + R_3(I_1 + I_2) = E_2.$$

Determine the value of the currents I_1 and I_2 if $E_1 = 8$ volts, $E_2 = 5$ volts, $R_1 = 6$ ohms, $R_2 = 10$ ohms, and $R_3 = 3$ ohms.

9.2 DETERMINANTS

A system of linear equations may be solved by formula if we find the solution of the general system

$$a_1x + b_1y = c_1$$
$$a_2x + b_2y = c_2.$$

Once we do the hard work of solving this system, we need only substitute values for the constants to solve a specific problem. Also, a problem that may be solved by formula is done easily on a computer.

To solve for x, we multiply each member of the first equation by b_2 and each member of the second equation by b_1 to obtain

$$a_1b_2x + b_1b_2y = c_1b_2$$
$$b_1a_2x + b_1b_2y = b_1c_2.$$

Now we subtract the second equation from the first to obtain

$$a_1b_2x - b_1a_2x = c_1b_2 - b_1c_2.$$

Factoring the left side of the equation, we have

$$(a_1b_2 - b_1a_2)x = c_1b_2 - b_1c_2.$$

Thus,

$$x = \frac{c_1b_2 - b_1c_2}{a_1b_2 - b_1a_2}.$$

In a similar manner, if we multiply each member of the first equation by a_2 and each member of the second equation by a_1 and then subtract, we have

$$y = \frac{a_1c_2 - c_1a_2}{a_1b_2 - b_1a_2}.$$

These formulas may now be used to find x and y whenever $a_1b_2 - b_1a_2 \neq 0$. If $a_1b_2 - b_1a_2 = 0$, the system is either dependent (in which case the numerators of x and y are zero) or inconsistent (in which case the numerators of x and y are not zero).

We do not memorize the formulas in this form since they may be obtained by defining what is called a **determinant.** Consider the expression $a_1b_2 - b_1a_2$, which is the denominator in the formulas for both x and y. We define a determinant to be a square array of numbers enclosed by vertical bars, such as

$$\begin{vmatrix} a_1 & b_1 \\ a_2 & b_2 \end{vmatrix},$$

and we define the value of this determinant to be $a_1b_2 - b_1a_2$.

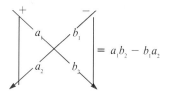

Figure 9.7

The numbers a_1 and b_2 are elements of the principal diagonal and the numbers b_1 and a_2 are elements of the secondary diagonal. Note in Figure 9.7 that the value of the determinant is the product of the elements of the principal diagonal minus the product of the elements of the secondary diagonal.

EXAMPLE 1

$$\begin{vmatrix} 1 & 2 \\ 3 & 4 \end{vmatrix} = 1(4) - 2(3) = 4 - 6 = -2 \qquad \square$$

EXAMPLE 2

$$\begin{vmatrix} 1 & -5 \\ 4 & -7 \end{vmatrix} = 1(-7) - (-5)(4) = -7 + 20 = 13 \qquad \square$$

We now state Cramer's rule, which shows how determinants can be used to solve a system of equations.

Cramer's Rule   (Two equations with two variables) The solution to the system

$$a_1 x + b_1 y = c_1$$
$$a_2 x + b_2 y = c_2$$

with $a_1 b_2 - b_1 a_2 \neq 0$ is

$$x = \frac{D_x}{D} = \frac{\begin{vmatrix} c_1 & b_1 \\ c_2 & b_2 \end{vmatrix}}{\begin{vmatrix} a_1 & b_1 \\ a_2 & b_2 \end{vmatrix}} = \frac{c_1 b_2 - b_1 c_2}{a_1 b_2 - b_1 a_2}$$

$$y = \frac{D_y}{D} = \frac{\begin{vmatrix} a_1 & c_1 \\ a_2 & c_2 \end{vmatrix}}{\begin{vmatrix} a_1 & b_1 \\ a_2 & b_2 \end{vmatrix}} = \frac{a_1 c_2 - c_1 a_2}{a_1 b_2 - b_1 a_2}.$$

In Cramer's rule the formulas for x and y are not difficult to remember. In both cases the determinant in the denominator is formed from the coefficients of x and y. Different determinants are used in the numerators. When solving for x, the coefficients of x, which are a_1 and a_2, are replaced by the constants c_1 and c_2. Similarly, when solving for y, the coefficients of y, which are b_1 and b_2, are replaced by the constants c_1 and c_2.

EXAMPLE 3   Use Cramer's rule to solve the system of equations

$$3x - 2y = 27$$
$$2x + 5y = -1.$$

Solution   First, evaluate the determinant in the denominator, which consists of the coefficients of x and y.

$$D = \begin{vmatrix} 3 & -2 \\ 2 & 5 \end{vmatrix} = 3(5) - (-2)(2) = 15 + 4 = 19$$

Second, evaluate the determinant in the numerator of the formula for x. Replace the column containing the coefficients of x by the column with the constants on the right side of the equations.

$$D_x = \begin{vmatrix} 27 & -2 \\ -1 & 5 \end{vmatrix} = 27(5) - (-2)(-1) = 135 - 2 = 133$$

Third, evaluate the determinant in the numerator of the formula for y. We replace the column containing the coefficients of y by the column with the constants on the right side of the equations.

$$D_y = \begin{vmatrix} 3 & 27 \\ 2 & -1 \end{vmatrix} = 3(-1) - 27(2) = -3 - 54 = -57$$

Fourth, substitute the value of the determinants in the formulas.

$$x = \frac{D_x}{D} = \frac{\begin{vmatrix} 27 & -2 \\ -1 & 5 \end{vmatrix}}{\begin{vmatrix} 3 & -2 \\ 2 & 5 \end{vmatrix}} = \frac{133}{19} = 7$$

$$y = \frac{D_y}{D} = \frac{\begin{vmatrix} 3 & 27 \\ 2 & -1 \end{vmatrix}}{\begin{vmatrix} 3 & -2 \\ 2 & 5 \end{vmatrix}} = \frac{-57}{19} = -3$$

Thus, the solution is $(7, -3)$.   □

EXAMPLE 4   Use determinants to solve the system of equations

$$5x + 2y - 1 = 0$$
$$x - 3y - 7 = 0.$$

Solution   First, change the equations to the form from which the formulas were derived.

$$5x + 2y = 1$$
$$x - 3y = 7$$

$$x = \frac{D_x}{D} = \frac{\begin{vmatrix} 1 & 2 \\ 7 & -3 \end{vmatrix}}{\begin{vmatrix} 5 & 2 \\ 1 & -3 \end{vmatrix}} = \frac{1(-3) - 2(7)}{5(-3) - 2(1)} = \frac{-17}{-17} = 1$$

$$y = \frac{D_y}{D} = \frac{\begin{vmatrix} 5 & 1 \\ 1 & 7 \end{vmatrix}}{\begin{vmatrix} 5 & 2 \\ 1 & -3 \end{vmatrix}} = \frac{5(7) - 1(1)}{-17} = \frac{34}{-17} = -2$$

Thus, the solution is $(1, -2)$.   □

Determinants may be used to solve more complicated systems, such as

$$a_1x + b_1y + c_1z = d_1$$
$$a_2x + b_2y + c_2z = d_2$$
$$a_3x + b_3y + c_3z = d_3.$$

This system involves three equations and three variables. Its solution is the collection of all values for x, y, and z that satisfy all three equations simultaneously. To solve this system, we must learn to evaluate determinants with three rows and three columns.

We first define the **minor** of an element to be the determinant formed by deleting the row and column containing the given element. For example,

$$\text{minor for } a_1 = \begin{vmatrix} a_1 & b_1 & c_1 \\ a_2 & b_2 & c_2 \\ a_3 & b_3 & c_3 \end{vmatrix} = \begin{vmatrix} b_2 & c_2 \\ b_3 & c_3 \end{vmatrix}$$

$$\text{minor for } b_2 = \begin{vmatrix} a_1 & b_1 & c_1 \\ a_2 & b_2 & c_2 \\ a_3 & b_3 & c_3 \end{vmatrix} = \begin{vmatrix} a_1 & c_1 \\ a_3 & c_3 \end{vmatrix}.$$

Now we define the **cofactor** of an element by considering the row and column position of that element. If an element is in the ith row and the jth column, then the cofactor of the element is given by the product

$$(-1)^{i+j} \cdot (\text{minor of the element}).$$

Figure 9.8

In other words, if $i + j$ is even, the cofactor of the element is the same as the minor of the element; but if $i + j$ is odd, then the cofactor equals the negative of the minor of the element. In the case of a 3 by 3 determinant, this definition means that the cofactor of an element is found by attaching the sign from the pattern in Figure 9.8 to the minor of that element. Now we can evaluate 3 by 3 (and more complicated) determinants by following these steps.

To Evaluate a Determinant	**1.** Pick any row or any column of the determinant. **2.** Multiply each entry in that row or column by its cofactor. **3.** Add the results. This sum is defined to be the value of the determinant.

Note that to each determinant there corresponds exactly one number as the value of that determinant. This determinant value is independent of the row or column that is chosen in step 1.

EXAMPLE 5   Evaluate the following determinant by (a) expansion along the first row of the determinant and (b) expansion along the second column of the determinant.

$$\begin{vmatrix} 1 & 1 & 3 \\ 5 & 0 & 2 \\ -2 & 3 & -1 \end{vmatrix}$$

Solution

a. The sign pattern in Figure 9.8 for row 1 is $+, -, +$. Using row 1 and following the steps given above, we have

$$\begin{vmatrix} 1 & 1 & 3 \\ 5 & 0 & 2 \\ -2 & 3 & -1 \end{vmatrix}$$

$$= 1\begin{vmatrix} 1 & 1 & 3 \\ 5 & 0 & 2 \\ -2 & 3 & -1 \end{vmatrix} - 1\begin{vmatrix} 1 & 1 & 3 \\ 5 & 0 & 2 \\ -2 & 3 & -1 \end{vmatrix} + 3\begin{vmatrix} 1 & 1 & 3 \\ 5 & 0 & 2 \\ -2 & 3 & -1 \end{vmatrix}$$

$$= 1\begin{vmatrix} 0 & 2 \\ 3 & -1 \end{vmatrix} - 1\begin{vmatrix} 5 & 2 \\ -2 & -1 \end{vmatrix} + 3\begin{vmatrix} 5 & 0 \\ -2 & 3 \end{vmatrix}$$

$$= 1[0(-1) - 2(3)] - 1[5(-1) - 2(-2)] + 3[5(3) - 0(-2)]$$

$$= 1(-6) - 1(-1) + 3(15)$$

$$= -6 + 1 + 45$$

$$= 40.$$

b. The sign pattern in Figure 9.8 for column 2 is $-, +, -$. Expanding the determinant by this column gives

$$\begin{vmatrix} 1 & 1 & 3 \\ 5 & 0 & 2 \\ -2 & 3 & -1 \end{vmatrix}$$

$$= -1 \begin{vmatrix} 1 & 1 & 3 \\ 5 & 0 & 2 \\ -2 & 3 & -1 \end{vmatrix} + 0 \begin{vmatrix} 1 & 1 & 3 \\ 5 & 0 & 2 \\ -2 & 3 & -1 \end{vmatrix} - 3 \begin{vmatrix} 1 & 1 & 3 \\ 5 & 0 & 2 \\ -2 & 3 & -1 \end{vmatrix}$$

$$= -1 \begin{vmatrix} 5 & 2 \\ -2 & -1 \end{vmatrix} + 0 \begin{vmatrix} 1 & 3 \\ -2 & -1 \end{vmatrix} - 3 \begin{vmatrix} 1 & 3 \\ 5 & 2 \end{vmatrix}$$

$$= -1[5(-1) - 2(-2)] + 0[\text{not needed}] - 3[1(2) - 3(5)]$$
$$= -1(-1) + 0 - 3(-13)$$
$$= 1 + 39$$
$$= 40.$$

Thus, the value of the determinant is 40. As expected, the answer in parts **a** and **b** is the same. Since you have a choice, you may want to evaluate a determinant by picking the row or column with the most 0's. As shown in part **b**, it is unnecessary to evaluate the minor of 0 elements. □

Cramer's rule extends to systems with three equations and three variables, and the formulas for x, y, and z follow an arrangement similar to that of the system with two equations and two variables. In each case the determinant in the denominator is formed from the coefficients of x, y, and z. Different determinants are used in the numerators. When solving for x, the coefficients of x are replaced by the constants d_1, d_2, and d_3. Similarly, when solving for y or z, the constants d_1, d_2, and d_3 replace the coefficients of the desired variable.

EXAMPLE 6   Use determinants to solve the system of equations

$$3x - y + 6z = 1$$
$$x + 2y - 3z = 0$$
$$2x - 3y - z = -9.$$

Solution   The determinant in the denominator is formed from the coefficients of x, y, and z. To find the determinant in the numerator for x, we replace the column containing the coefficients of x by the column containing the constants on the right side of the equations.

$$x = \frac{D_x}{D} = \frac{\begin{vmatrix} 1 & -1 & 6 \\ 0 & 2 & -3 \\ -9 & -3 & -1 \end{vmatrix}}{\begin{vmatrix} 3 & -1 & 6 \\ 1 & 2 & -3 \\ 2 & -3 & -1 \end{vmatrix}}$$

$$= \frac{1 \begin{vmatrix} 2 & -3 \\ -3 & -1 \end{vmatrix} - 0 \begin{vmatrix} -1 & 6 \\ -3 & -1 \end{vmatrix} + (-9) \begin{vmatrix} -1 & 6 \\ 2 & -3 \end{vmatrix}}{3 \begin{vmatrix} 2 & -3 \\ -3 & -1 \end{vmatrix} - \begin{vmatrix} -1 & 6 \\ -3 & -1 \end{vmatrix} + 2 \begin{vmatrix} -1 & 6 \\ 2 & -3 \end{vmatrix}}$$

expansion by column 1 because of 0 element

$$= \frac{70}{-70} = -1$$

The value of the determinant in the denominator for both y and z is -70. The column containing the constants on the right side of the equations first replaces the coefficients of y (to find y) and then the coefficients of z (to find z).

$$y = \frac{D_y}{D} = \frac{\begin{vmatrix} 3 & 1 & 6 \\ 1 & 0 & -3 \\ 2 & -9 & -1 \end{vmatrix}}{-70}$$

expansion by column 2 because of 0 element

$$= \frac{-1 \begin{vmatrix} 1 & -3 \\ 2 & -1 \end{vmatrix} + 0 \begin{vmatrix} 3 & 6 \\ 2 & -1 \end{vmatrix} - (-9) \begin{vmatrix} 3 & 6 \\ 1 & -3 \end{vmatrix}}{-70} = \frac{-140}{-70} = 2$$

$$z = \frac{D_z}{D} = \frac{\begin{vmatrix} 3 & -1 & 1 \\ 1 & 2 & 0 \\ 2 & -3 & -9 \end{vmatrix}}{-70}$$

expansion by column 3 because of 0 element

$$= \frac{1 \begin{vmatrix} 1 & 2 \\ 2 & -3 \end{vmatrix} - 0 \begin{vmatrix} 3 & -1 \\ 2 & -3 \end{vmatrix} + (-9) \begin{vmatrix} 3 & -1 \\ 1 & 2 \end{vmatrix}}{-70} = \frac{-70}{-70} = 1$$

The solution is $x = -1$, $y = 2$, $z = 1$. □

There are two additional points to mention about Cramer's rule. First, if the determinant in the denominator is zero, merely write that there is no unique solution to the problem. Such systems are either inconsistent or dependent. Second, although Cramer's rule extends to systems beyond three equations with three variables, the determinants associated with such systems are usually hard to evaluate. An easier approach, which introduces matrix theory, is discussed in the next section for use in such cases.

EXERCISES 9.2

In Exercises 1–20 evaluate the determinant.

1. $\begin{vmatrix} 1 & 3 \\ 4 & 2 \end{vmatrix}$

2. $\begin{vmatrix} -2 & 5 \\ 1 & 3 \end{vmatrix}$

3. $\begin{vmatrix} 1 & -5 \\ 2 & -8 \end{vmatrix}$

4. $\begin{vmatrix} 2 & -3 \\ 2 & -5 \end{vmatrix}$

5. $\begin{vmatrix} 5 & -3 \\ 2 & 4 \end{vmatrix}$

6. $\begin{vmatrix} -2 & 14 \\ 3 & -3 \end{vmatrix}$

7. $\begin{vmatrix} -3 & 0 \\ 4 & 1 \end{vmatrix}$

8. $\begin{vmatrix} 5 & -28 \\ 4 & -10 \end{vmatrix}$

9. $\begin{vmatrix} 3 & 8 \\ -8 & 11 \end{vmatrix}$

10. $\begin{vmatrix} -7 & -9 \\ -10 & -6 \end{vmatrix}$

11. $\begin{vmatrix} 2 & 5 \\ -4 & -10 \end{vmatrix}$

12. $\begin{vmatrix} 4 & -8 \\ 6 & -12 \end{vmatrix}$

13. $\begin{vmatrix} -2 & 3 & 2 \\ 0 & 7 & 4 \\ 1 & -1 & 3 \end{vmatrix}$

14. $\begin{vmatrix} 0 & 2 & -1 \\ 5 & 1 & 0 \\ -3 & 0 & -4 \end{vmatrix}$

15. $\begin{vmatrix} 2 & 1 & 1 \\ 1 & -2 & -1 \\ 3 & 3 & -1 \end{vmatrix}$

16. $\begin{vmatrix} 2 & 3 & 2 \\ 1 & -3 & 3 \\ 5 & 1 & -4 \end{vmatrix}$

17. $\begin{vmatrix} 2 & -7 & 1 \\ 4 & 1 & 2 \\ 6 & -3 & 3 \end{vmatrix}$

18. $\begin{vmatrix} 9 & 10 & -2 \\ 4 & 1 & 8 \\ -1 & -7 & 3 \end{vmatrix}$

19. $\begin{vmatrix} 6 & 3 & 11 \\ -3 & 4 & 5 \\ -1 & -2 & 3 \end{vmatrix}$

20. $\begin{vmatrix} -1 & 2 & -1 \\ 5 & -2 & -3 \\ 2 & -4 & 2 \end{vmatrix}$

In Exercises 21–30 use determinants to solve the systems of equations in Exercises 11–20 of Section 9.1.

In Exercises 31–40 use determinants to solve the given systems of equations. If a variable is missing from an equation, enter zero in the determinant as its coefficient.

31. $x - y + z = -1$
      $x + y - 3z = 3$
    $2x + y - 2z = 4$

32. $x + y + z = 1$
     $x + y - z = 3$
     $x - y - z = 5$

33. $-2x + y - z = 3$
     $2x + 3y + 2z = 5$
     $3x + 2y + z = -3$

34. $x + y - 3z = -4$
    $-x - y + 5z = 8$
     $2x + 3y - z = 3$

35. $x - y + z = 3$
    $3x + y + z = 1$
    $2x + 2y - z = -3$

36. $x + y + 3z = 1$
    $2x + 5y + 2z = 0$
    $3x - 2y - z = 3$

37. $2x + 3y = 10$
    $3x + 2z = 2$
    $4y + z = 6$

38. $x + y + z = 0$
    $2x - 5y = 3$
    $-5y + 3z = 7$

39. $2x - y + 3z = 1$
    $-x + y - z = -1$
    $-4x + 2y - 6z = -2$

40. $x + y + z = 1$
    $2x - 3y - 2z = -4$
    $3x - 2y - z = -2$

In Exercises 41–44 solve the equations simultaneously to determine the value of the currents I_1, I_2, and I_3 shown in the diagram.

$$I_1 + I_2 + I_3 = 0$$
$$R_1 I_1 - R_3 I_3 = E_1$$
$$R_2 I_2 - R_3 I_3 = E_2$$

E_1 (volts)	E_2 (volts)	R_1 (ohms)	R_2 (ohms)	R_3 (ohms)
41. 3	10	2	9	5
42. 8	5	6	10	3
43. 7	2	5	4	9
44. 10	20	50	12	100

45. An equation for the line that passes through points (x_1, y_1) and (x_2, y_2) may be written in determinant form as

$$\begin{vmatrix} x & y & 1 \\ x_1 & y_1 & 1 \\ x_2 & y_2 & 1 \end{vmatrix} = 0.$$

Use this rule and find the equation of the line that passes through $(-1, 6)$ and $(2, -3)$.

9.3 TRIANGULAR FORM AND MATRICES

Systems of three linear equations with three variables, and more complicated linear systems, can always be solved by using a technique known as **Gaussian elimination.** This method is basically the addition-elimination method from Section 9.1, and its systematic nature can be programmed to allow effective computer solutions. To understand this method, first consider the following system of equations that is said to be in **triangular form.**

$$\begin{aligned} 4x + 2y - z &= 2 \\ 3y - 3z &= 6 \\ 5z &= 10 \end{aligned}$$

This system is easy to solve. The third equation, $5z = 10$, tells us $z = 2$, then back substitution yields

$$\begin{aligned} 3y - 3(2) &= 6 && 4x + 2(4) - (2) = 2 \\ 3y &= 12 && \quad\text{and}\quad 4x = -4 \\ y &= 4 && x = -1. \end{aligned}$$

Thus, the solution is $x = -1$, $y = 4$, and $z = 2$. However, systems of equations rarely start out in triangular form, so we need some procedure for obtaining this form for any given system. The three operations that follow are called the **elementary operations,** and they are used in Gaussian elimination to produce **equivalent systems** (ones with the same solution) until we reach triangular form.

Operations that Produce Equivalent Systems	**1.** Multiply both sides of an equation by a nonzero number. **2.** Add a multiple of one equation to another equation. **3.** Interchange the order in which two equations of a system are listed.

Note that we have already used the first two operations in Section 9.1, while the third operation clearly affects only the form of the system and not the solution. Now consider carefully Example 1, which shows how we can use these operations to change a system into triangular form.

EXAMPLE 1   Solve the system of equations

$$\begin{aligned} 3x - y + 6z &= 1 \\ x + 2y - 3z &= 0 \\ 2x - 3y - z &= -9. \end{aligned}$$

Solution   We want x to appear only in the first equation. It is easier to eliminate x from the other equations when the coefficient of x in the first equation is 1. So, first, we change the order of the equations to

$$\begin{aligned} x + 2y - 3z &= 0 \\ 3x - y + 6z &= 1 \\ 2x - 3y - z &= -9. \end{aligned}$$

We now add -3 times the first equation to the second equation, and we also add -2 times the first equation to the third equation. The result is

$$\begin{aligned} x + 2y - 3z &= 0 \\ -7y + 15z &= 1 \\ -7y + 5z &= -9. \end{aligned}$$

Finally, to eliminate y in the third equation, we add -1 times the second equation to the third equation to obtain

$$\begin{aligned} x + 2y - 3z &= 0 \\ -7y + 15z &= 1 \\ -10z &= -10. \end{aligned}$$

The system is now in triangular form. From the third equation we know $z = 1$, while back substitution gives

$$\begin{array}{ccc} -7y + 15(1) = 1 & & x + 2(2) - 3(1) = 0 \\ -7y = -14 & \text{and} & x = -1. \\ y = 2 & & \end{array}$$

Thus, the solution is $x = -1$, $y = 2$, and $z = 1$. □

We can improve the procedure in our current method by keeping track of only the constants in the equations and not writing down the variables. The standard notation for such an abbreviation utilizes matrices. A **matrix** is a rectangular array of numbers that is enclosed in brackets (or parentheses). Each number in the matrix is called an **entry** or **element** of the matrix. There are two matrices associated with the system

$$\begin{aligned} a_1x + b_1y + c_1z &= d_1 \\ a_2x + b_2y + c_2z &= d_2 \\ a_3x + b_3y + c_3z &= d_3. \end{aligned}$$

The **coefficient matrix** consists of the coefficients of x, y, and z and is written as

$$\begin{bmatrix} a_1 & b_1 & c_1 \\ a_2 & b_2 & c_2 \\ a_3 & b_3 & c_3 \end{bmatrix},$$

while the **augmented matrix** that follows includes these coefficients and an additional column (usually separated by a dashed line) that contains the constants on the right side of the equals sign.

$$\begin{bmatrix} a_1 & b_1 & c_1 & \vdots & d_1 \\ a_2 & b_2 & c_2 & \vdots & d_2 \\ a_3 & b_3 & c_3 & \vdots & d_3 \end{bmatrix}$$

We now restate in the language of matrices the operations that produce equivalent equations. These operations are called **elementary row operations** and they can be used to obtain a matrix solution to a system of equations.

Corresponding Operations for Solving a Linear System	Elementary Operations on Equations	Elementary Row Operations on Matrices
	1. Multiply both sides of an equation by a nonzero number.	1. Multiply each entry in a row by a nonzero number.
	2. Add a multiple of one equation to another.	2. Add a multiple of the entries in one row to another row.
	3. Interchange two equations.	3. Interchange two rows.

We display both of these methods in the next example to reinforce the similarities in the methods.

EXAMPLE 2   Solve the system

$$\begin{aligned} 4x - 2y + z &= 11 \\ x - y + 3z &= 6 \\ x + y + z &= 2 \end{aligned}$$

and show both the matrix form of the system and the corresponding equations.

Solution   We use the elementary operations above and proceed as follows:

Equation Form

$$\begin{aligned} 4x - 2y + z &= 11 \\ x - y + 3z &= 6 \\ x + y + z &= 2 \end{aligned}$$

↓ Interchange equations 1 and 3.

$$\begin{aligned} x + y + z &= 2 \\ x - y + 3z &= 6 \\ 4x - 2y + z &= 11 \end{aligned}$$

Matrix Form

$$\begin{bmatrix} 4 & -2 & 1 & \vdots & 11 \\ 1 & -1 & 3 & \vdots & 6 \\ 1 & 1 & 1 & \vdots & 2 \end{bmatrix}$$

↓ Interchange rows 1 and 3.

$$\begin{bmatrix} 1 & 1 & 1 & \vdots & 2 \\ 1 & -1 & 3 & \vdots & 6 \\ 4 & -2 & 1 & \vdots & 11 \end{bmatrix}$$

↓ Add -1 times the first equation to the second equation.

$$\begin{aligned} x + y + z &= 2 \\ -2y + 2z &= 4 \\ 4x - 2y + z &= 11 \end{aligned}$$

↓ Add -1 times each entry in row 1 to the corresponding entry in row 2.

$$\begin{bmatrix} 1 & 1 & 1 & \vdots & 2 \\ 0 & -2 & 2 & \vdots & 4 \\ 4 & -2 & 1 & \vdots & 11 \end{bmatrix}$$

↓ Add -4 times the first equation to the third equation.

$$\begin{aligned} x + y + z &= 2 \\ -2y + 2z &= 4 \\ -6y - 3z &= 3 \end{aligned}$$

↓ Add -4 times each entry in row 1 to the corresponding entry in row 3.

$$\begin{bmatrix} 1 & 1 & 1 & \vdots & 2 \\ 0 & -2 & 2 & \vdots & 4 \\ 0 & -6 & -3 & \vdots & 3 \end{bmatrix}$$

↓ Add -3 times the second equation to the third equation.

$$\begin{aligned} x + y + z &= 2 \\ -2y + 2z &= 4 \\ -9z &= -9 \end{aligned}$$

↓ Add -3 times each entry in row 2 to the corresponding entry in row 3.

$$\begin{bmatrix} 1 & 1 & 1 & \vdots & 2 \\ 0 & -2 & 2 & \vdots & 4 \\ 0 & 0 & -9 & \vdots & -9 \end{bmatrix}$$

The last row or last equation tells us $-9z = -9$, so $z = 1$. Then

$$\begin{aligned} -2y + 2(1) &= 4 \\ y &= -1 \end{aligned} \quad \text{and} \quad \begin{aligned} x + (-1) + 1 &= 2 \\ x &= 2. \end{aligned}$$

Thus, the solution is $x = 2$, $y = -1$, and $z = 1$.   □

EXAMPLE 3   Use matrix form to solve the system

$$\begin{aligned} 7x - 2y &= 19 \\ 3x + 5y &= -14. \end{aligned}$$

Solution   The augmented matrix for the system is

$$\begin{bmatrix} 7 & -2 & \vdots & 19 \\ 3 & 5 & \vdots & -14 \end{bmatrix}.$$

To obtain 0 for the second entry in column 1, we first multiply each entry in row 1 by 3 and each entry in row 2 by -7.

$$\begin{bmatrix} 21 & -6 & \vdots & 57 \\ -21 & -35 & \vdots & 98 \end{bmatrix}$$

Now add the first row to the second row.

$$\begin{bmatrix} 21 & -6 & \vdots & 57 \\ 0 & -41 & \vdots & 155 \end{bmatrix}$$

The last row tells us that $-41y = 155$, so $y = -\frac{155}{41}$. Then

$$7x - 2(-\tfrac{155}{41}) = 19$$
$$7x = \tfrac{469}{41}$$
$$x = \tfrac{67}{41}.$$

Thus, the solution is $(\frac{67}{41}, -\frac{155}{41})$.   □

EXAMPLE 4   Use matrix form to solve the system

$$a + b + c + d = 1$$
$$a - b + c + d = 1$$
$$b + c - d = 1$$
$$b \quad\ + d = 1.$$

Solution   The augmented matrix for the system is

$$\begin{bmatrix} 1 & 1 & 1 & 1 & \vdots & 1 \\ 1 & -1 & 1 & 1 & \vdots & 1 \\ 0 & 1 & 1 & -1 & \vdots & 1 \\ 0 & 1 & 0 & 1 & \vdots & 1 \end{bmatrix}.$$

To obtain 0's in the first column after row 1, we need only add -1 times the first row to the second row.

$$\begin{bmatrix} 1 & 1 & 1 & 1 & \vdots & 1 \\ 0 & -2 & 0 & 0 & \vdots & 0 \\ 0 & 1 & 1 & -1 & \vdots & 1 \\ 0 & 1 & 0 & 1 & \vdots & 1 \end{bmatrix}$$

The second row represents the equation $-2b = 0$, so we already know $b = 0$. To obtain 0's in the second column after row 2, we first multiply each entry in row 2 by $-\frac{1}{2}$ to make the coefficient of b in the second equation a 1.

$$\begin{bmatrix} 1 & 1 & 1 & 1 & \vdots & 1 \\ 0 & 1 & 0 & 0 & \vdots & 0 \\ 0 & 1 & 1 & -1 & \vdots & 1 \\ 0 & 1 & 0 & 1 & \vdots & 1 \end{bmatrix}$$

Now add -1 times the second row to the third row as well as to the fourth row.

$$\begin{bmatrix} 1 & 1 & 1 & 1 & \vdots & 1 \\ 0 & 1 & 0 & 0 & \vdots & 0 \\ 0 & 0 & 1 & -1 & \vdots & 1 \\ 0 & 0 & 0 & 1 & \vdots & 1 \end{bmatrix}$$

Row 2 and row 4 tell us that $b = 0$ and $d = 1$, while substitution of these values into the equations corresponding to row 3 and row 1 gives

$$c - d = 1$$
$$c - (1) = 1 \quad \text{and}$$
$$c = 2$$

$$a + b + c + d = 1$$
$$a + 0 + 2 + 1 = 1$$
$$a = -2.$$

Thus, the solution is $a = -2$, $b = 0$, $c = 2$, and $d = 1$. ☐

There are two additional points to mention about Gaussian elimination. First, if you come up with a row of zeros in the coefficient portion in any matrix, merely write that there is no unique solution to the problem. Such systems are either dependent (if the corresponding last column entry is also zero) or inconsistent (if the last column entry is not zero). Second, in each of our example problems we stopped when we reached triangular form and completed the solution by back substitution. An alternative method is to continue to produce equivalent matrices (or systems) until we reach a form like

$$\begin{bmatrix} 1 & 0 & 0 & \vdots & a \\ 0 & 1 & 0 & \vdots & b \\ 0 & 0 & 1 & \vdots & c \end{bmatrix}.$$

From this final form of the matrix, we directly read off that $x = a$, $y = b$, and $z = c$.

EXERCISES 9.3

In Exercises 1–4 solve the given systems of equations and show both the matrix form of the systems and the corresponding equations.

1. $x + y = 7$
$x - y = -2$

2. $2a + 5b = 1$
$3a - 4b = 13$

3. $a + b + c = 2$
$3a - b - c = -1$
$2a + 2b - c = 1$

4. $4x - 2y + 2z = 0$
$3x \qquad + 2z = 0$
$x - 2y \qquad = 0$

In Exercises 5–20 use matrix form to solve the given systems of equations.

5. $5x + 2y + 17 = 0$
$-2x - y + 5 = 0$

6. $x - y = 3$
$-x - y = -3$

7. $2x - 3y = 1$
$-6x + 9y = -3$

8. $x - y = 3$
$-x + y = -5$

9. $x + y \qquad = 4$
$\quad y + z = -8$
$x \qquad + z = 2$

10. $-a + b \qquad = -1$
$\quad b - c = 3$
$a \qquad + c = -12$

11. $x + y + z = 1$
$x + y - z = 3$
$x - y - z = 5$

12. $x + y + 3z = 1$
$2x + 5y + 2z = 0$
$3x - 2y - z = 3$

13. $x_1 - x_2 + x_3 = 2$
$2x_1 - 3x_2 + 2x_3 = 6$
$3x_1 + x_2 + x_3 = 2$

14. $x_1 - x_2 + x_3 = 3$
$3x_1 + x_2 + x_3 = 1$
$2x_1 + 2x_2 - x_3 = -3$

15. $6A + 3B + 2C = 1$
$5A + 4B + 3C = 0$
$A + B + C = 0$

16. $2a + 3b - c = 3$
$a + b - 3c = -4$
$-a - b + 5c = 8$

17. $3x - 2y - z = -2$
$2x + 2y + 2z = 2$
$2x - 3y - 2z = -4$

18. $x + y = 2$
$y + z = 1$
$z + w = -1$
$x + w = 0$

19. $a + b + c + d = 0$
$a - b + 2c + d = 1$
$4a + b + 2c \qquad = 5$
$5a \qquad + 4c + 2d = 6$

20. $x_1 + x_2 + x_3 + x_4 = 1$
$x_1 - x_2 + x_3 - x_4 = -1$
$-2x_1 + x_2 - x_3 - 2x_4 = 1$
$2x_1 - 2x_2 + 2x_3 + x_4 = 1$

21. The points $(-1,-8)$, $(1,2)$, $(2,4)$ lie on the parabola $y = ax^2 + bx + c$. Find a, b, and c.

22. Let $f(x) = ax^2 + bx + c$. Find values of a, b, and c such that $f(1) = 0$, $f(2) = 8$, and $f(3) = 22$.

23. If the graph of $y = ax^3 + bx^2 + cx + d$ passes through the points $(1,1)$, $(2,1)$, $(3,-11)$, and $(-1,-11)$, find a, b, c, and d.

24. The points $(3,3)$, $(-2,-2)$, and $(1,-1)$ lie on the circle $x^2 + y^2 + Dx + Ey + F = 0$. Find D, E, and F. What are the center and radius of this circle?

25. Find the radius of each circle in the diagram below.

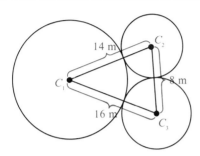

26. Solve for x_1, x_2, and x_3 in terms of the constants a, b, and c.

$$\begin{aligned} x_1 + x_2 \qquad &= a \\ x_2 + x_3 &= b \\ x_1 \qquad + x_3 &= c \end{aligned}$$

27. A 2,000-seat theater was sold out for a weekend concert. Total receipts on Friday night were \$31,500 when seats in the orchestra, mezzanine, and balcony sold for \$20, \$15, and \$10, respectively. On Saturday, prices were raised to \$25, \$20, and \$10 and total receipts increased to \$39,000. How many seats are in the mezzanine?

9.4 SOLVING SYSTEMS BY MATRIX ALGEBRA

There is one other method for solving a system of linear equations that merits our attention. In the last section we used elementary row operations on matrices to solve the system. A different matrix method is to develop a matrix algebra that defines how to perform various operations with matrices and includes some properties that hold in this system. This approach is beneficial in two ways.

1. Matrices have many applications independent of systems of equations, and the ideas discussed here can serve as an introduction to this important topic. Because of this we develop a little more matrix algebra than is needed to solve a system of equations. More detailed coverage of this topic is usually given in a course in linear algebra.

2. Although the matrix algebra approach usually requires more work than our other methods to solve a given system, it is an efficient technique for solving systems such as

$$\begin{aligned} 2x + 3y &= b_1 \\ 3x + 5y &= b_2, \end{aligned}$$

where the coefficients of x and y remain fixed but different pairs of constants replace the b's. We consider this problem at the end of this section.

To begin, we first recall that a **matrix** is a rectangular array of numbers enclosed in brackets (or parentheses) and that each number in the matrix is called an **entry** or **element** of the matrix. We now add some other basic definitions.

Matrix Definitions

1. A matrix with *m* rows and *n* columns is of **dimension *m* × *n*.**
2. A **square matrix** is a matrix having the same number of rows as columns.
3. A **zero matrix** is a matrix containing only zero elements. We often denote a zero matrix by a boldface zero: **0.**
4. Two matrices are **equal** if and only if the elements in corresponding positions are equal.
5. To **add** (or **subtract**) two matrices of the same dimension, add (or subtract) elements in corresponding positions. Only matrices of the same dimension can be added (or subtracted).
6. To multiply a matrix by a real number, multiply each element in the matrix by that number. In this context we call the real number a **scalar,** and we call this operation **scalar multiplication.**
7. $A - B = A + (-B)$, where $-B = (-1)B$.

Now consider Example 1, which illustrates these definitions.

EXAMPLE 1   Answer the following questions given

$$A = \begin{bmatrix} 2 & -1 & 3 \\ 0 & 1 & -2 \end{bmatrix}, B = \begin{bmatrix} 0 & 0 \\ 0 & 0 \end{bmatrix},$$

$$C = \begin{bmatrix} 1 & 0 \\ 0 & 1 \end{bmatrix}, \text{ and } D = \begin{bmatrix} 1 & 3 & 6 \\ 2 & 5 & 7 \end{bmatrix}.$$

a. What is the dimension of each matrix?
b. Find (if possible) $A + D$.
c. Find (if possible) $A + C$.
d. Find $-A$.
e. If $A + 2X = D$, find X.

Solution
a. *A* and *D* are matrices of dimension 2 × 3 since both have 2 rows and 3 columns. *B* and *C* are square matrices with 2 rows and 2 columns and are of dimension 2 × 2. Matrix *B* is an example of a zero matrix.
b. *A* and *D* are both of dimension 2 × 3, so $A + D$ is defined and is found as follows:

$$\begin{bmatrix} 2 & -1 & 3 \\ 0 & 1 & -2 \end{bmatrix} + \begin{bmatrix} 1 & 3 & 6 \\ 2 & 5 & 7 \end{bmatrix} = \begin{bmatrix} 2+1 & -1+3 & 3+6 \\ 0+2 & 1+5 & -2+7 \end{bmatrix}$$

$$= \begin{bmatrix} 3 & 2 & 9 \\ 2 & 6 & 5 \end{bmatrix}.$$

c. A and C are not of the same dimension, so $A + C$ is undefined.

d. $-A = (-1)A$, so to find $-A$ we multiply each element in A by -1. Then

$$-A = -1 \begin{bmatrix} 2 & -1 & 3 \\ 0 & 1 & -2 \end{bmatrix} = \begin{bmatrix} -2 & 1 & -3 \\ 0 & -1 & 2 \end{bmatrix}.$$

e. Since A and D are of dimension 2×3, matrix X must also have 2 rows and 3 columns. If we let

$$X = \begin{bmatrix} a & b & c \\ d & e & f \end{bmatrix}, \text{ then } 2X = \begin{bmatrix} 2a & 2b & 2c \\ 2d & 2e & 2f \end{bmatrix}, \text{ so}$$

$$\begin{array}{ccccc} A & + & 2X & = & D \end{array}$$

$$\begin{bmatrix} 2 & -1 & 3 \\ 0 & 1 & -2 \end{bmatrix} + \begin{bmatrix} 2a & 2b & 2c \\ 2d & 2e & 2f \end{bmatrix} = \begin{bmatrix} 1 & 3 & 6 \\ 2 & 5 & 7 \end{bmatrix}$$

$$\begin{bmatrix} 2 + 2a & -1 + 2b & 3 + 2c \\ 0 + 2d & 1 + 2e & -2 + 2f \end{bmatrix} = \begin{bmatrix} 1 & 3 & 6 \\ 2 & 5 & 7 \end{bmatrix}.$$

For two matrices to be equal, elements in corresponding positions must be the same. Thus,

$$\begin{array}{ll} 2 + 2a = 1, \text{ so } a = -\tfrac{1}{2}; & -1 + 2b = 3, \text{ so } b = 2; \\ 3 + 2c = 6, \text{ so } c = \tfrac{3}{2}; & 0 + 2d = 2, \text{ so } d = 1; \\ 1 + 2e = 5, \text{ so } e = 2; & -2 + 2f = 7, \text{ so } f = \tfrac{9}{2}. \end{array}$$

So

$$X = \begin{bmatrix} -\tfrac{1}{2} & 2 & \tfrac{3}{2} \\ 1 & 2 & \tfrac{9}{2} \end{bmatrix}. \qquad \square$$

Many of the properties from the algebra of real numbers carry over to the algebra of matrices. That is because we have defined our operations to this point in terms of operating on elements that are real numbers and that are in corresponding positions. Some of the more basic properties are as follows:

Properties of Matrices If A, B, and C are matrices and c and k are scalars (real numbers), then

1. $(A + B) + C = A + (B + C)$
2. $A + B = B + A$
3. $A + 0 = 0 + A = A$
4. $A + (-A) = (-A) + A = 0$
5. $c(kA) = (ck)A$
6. $c(A + B) = cA + cB$

These properties provide us with an alternative method for solving the problem in Example 1e that is similar to our previous work with equations. That is, we start with $A + 2X = D$ and solve for X as follows:

$$A + 2X = D$$
$$A + (-A) + 2X = D + (-A) \quad \text{Add } -A \text{ to both sides.}$$
$$\mathbf{0} + 2X = D - A \quad \text{Property 4 and subtraction definition}$$
$$2X = D - A \quad \text{Property 3}$$
$$X = \tfrac{1}{2}(D - A). \quad \text{Multiply both sides by the scalar } \tfrac{1}{2}.$$

Then

$$X = \tfrac{1}{2}\left(\begin{bmatrix} 1 & 3 & 6 \\ 2 & 5 & 7 \end{bmatrix} - \begin{bmatrix} 2 & -1 & 3 \\ 0 & 1 & -2 \end{bmatrix}\right)$$

$$= \tfrac{1}{2}\begin{bmatrix} -1 & 4 & 3 \\ 2 & 4 & 9 \end{bmatrix}$$

$$= \begin{bmatrix} -\tfrac{1}{2} & 2 & \tfrac{3}{2} \\ 1 & 2 & \tfrac{9}{2} \end{bmatrix}.$$

Note that this answer agrees with our previous result.

To this point, our definitions and properties have been straightforward and predictable. Matrix multiplication, however, is unusual and we do *not* multiply matrices by simply multiplying elements in corresponding positions. To understand the usefulness of the different type of product that is defined in matrix multiplication, consider the following matrices.

$$\begin{array}{c} \quad\quad \text{Printers} \quad \text{Monitors} \\ \begin{array}{c} \text{Plant A} \\ \text{Plant B} \\ \text{Plant C} \end{array} \begin{bmatrix} 70 & 30 \\ 50 & 90 \\ 80 & 10 \end{bmatrix} = A \end{array} \qquad \begin{array}{c} \text{Value} \\ \begin{array}{c} \text{Printers} \\ \text{Monitors} \end{array} \begin{bmatrix} 300 \\ 200 \end{bmatrix} = B \end{array}$$

Matrix A summarizes information about the location of computer equipment, while matrix B indicates the value of that equipment. We can find the total value of the equipment at each plant as follows:

$$\left(\begin{array}{c}\text{number}\\\text{of}\\\text{printers}\end{array}\right)\left(\begin{array}{c}\text{value}\\\text{per}\\\text{printer}\end{array}\right) + \left(\begin{array}{c}\text{number}\\\text{of}\\\text{monitors}\end{array}\right)\left(\begin{array}{c}\text{value}\\\text{per}\\\text{monitor}\end{array}\right) = \begin{array}{c}\text{total}\\\text{value}\end{array}$$

Plant A	70	$\cdot$	300	$+$	30 $\cdot$ 200	$= 27{,}000$
Plant B	50	$\cdot$	300	$+$	90 $\cdot$ 200	$= 33{,}000$
Plant C	80	$\cdot$	300	$+$	10 $\cdot$ 200	$= 26{,}000$

and we can represent this result in the following matrix.

$$
\begin{array}{c}
\text{Value} \\
\begin{array}{l}
\text{Plant A} \\
\text{Plant B} \\
\text{Plant C}
\end{array}
\begin{bmatrix}
27{,}000 \\
33{,}000 \\
26{,}000
\end{bmatrix} = C
\end{array}
$$

We wish to define matrix multiplication so that $AB = C$. First, note that for this product to make sense the columns in matrix A (type of computer equipment) must match the rows of matrix B. Thus, the product AB of two matrices is defined only when the number of columns in A is the same as the number of rows in B. Also note that the product AB (or matrix C) gets its row description from the rows of A (plant location) and its column description from the column(s) of matrix B (value). Thus, the product AB of two matrices has as many rows as A and as many columns as B. We summarize these results in this diagram.

$$
\text{product } AB \text{ is of dimension } m \times p
$$

Finally, the element in row 1, column 1 in AB is found by multiplying the entries in the first row of A by the corresponding entries in the first column of B and adding the results. Similarly, the element in row 2, column 1 is found by multiplying the entries in the second row of A by the corresponding entries in the first column of B. This pattern continues and our observations suggest the following definition for matrix multiplication.

Matrix Multiplication	If A is an $m \times n$ matrix and B is an $n \times p$ matrix, then the product AB is an $m \times p$ matrix in which the ith row, jth column element of AB is found by multiplying each element in the ith row of A by the corresponding element in the jth column of B and adding the results.

Examples 2–4 illustrate the definition.

EXAMPLE 2   Let

$$
A = \begin{bmatrix} 3 & 0 & 1 \\ 4 & 4 & 2 \end{bmatrix} \quad \text{and} \quad B = \begin{bmatrix} 1 & 2 \\ 4 & 0 \\ 7 & 3 \end{bmatrix}.
$$

Find (if possible) the product AB.

Solution A is of dimension 2×3 and B is of dimension 3×2. Thus, the product is defined and AB is a 2×2 matrix. Such a matrix has 4 elements that are determined as follows:

Step 1 (row 1, column 1)

$$\begin{bmatrix} 3 & 0 & 1 \\ 4 & 4 & 2 \end{bmatrix} \begin{bmatrix} 1 & 2 \\ 4 & 0 \\ 7 & 3 \end{bmatrix}$$

Computation

$3 \cdot 1 + 0 \cdot 4 + 1 \cdot 7 = 10$

Step 2 (row 1, column 2)

$$\begin{bmatrix} 3 & 0 & 1 \\ 4 & 4 & 2 \end{bmatrix} \begin{bmatrix} 1 & 2 \\ 4 & 0 \\ 7 & 3 \end{bmatrix}$$

$3 \cdot 2 + 0 \cdot 0 + 1 \cdot 3 = 9$

Step 3 (row 2, column 1)

$$\begin{bmatrix} 3 & 0 & 1 \\ 4 & 4 & 2 \end{bmatrix} \begin{bmatrix} 1 & 2 \\ 4 & 0 \\ 7 & 3 \end{bmatrix}$$

$4 \cdot 1 + 4 \cdot 4 + 2 \cdot 7 = 34$

Step 4 (row 2, column 2)

$$\begin{bmatrix} 3 & 0 & 1 \\ 4 & 4 & 2 \end{bmatrix} \begin{bmatrix} 1 & 2 \\ 4 & 0 \\ 7 & 3 \end{bmatrix}$$

$4 \cdot 2 + 4 \cdot 0 + 2 \cdot 3 = 14$

Thus,

$$AB = \begin{bmatrix} 10 & 9 \\ 34 & 14 \end{bmatrix}.$$

□

EXAMPLE 3   Let

$$A = [1 \quad 2 \quad 3 \quad 4] \quad \text{and} \quad B = \begin{bmatrix} -1 & 4 \\ 0 & 1 \\ 5 & -3 \\ -2 & 0 \end{bmatrix}.$$

Find (if possible) (a) the product AB and (b) the product BA.

Solution

a. A is a 1×4 matrix and B is a 4×2 matrix. Thus, the product is defined and is a 1×2 matrix. We determine these two elements as follows:

Entry in row 1, column 1: $1(-1) + 2(0) + 3(5) + 4(-2) = 6$
Entry in row 1, column 2: $1(4) + 2(1) + 3(-3) + 4(0) = -3$.

Thus,

$$AB = [6 \quad -3].$$

b. Since B is a 4×2 matrix and A is a 1×4 matrix, the number of columns in B does not equal the number of rows in A, so the product BA is undefined.

[*Note:* This example shows that matrix multiplication is not commutative (that is, in general $AB \neq BA$), so we must be careful about the order in which we write matrices when expressing a product. For square matrices of the same order we do, however, have an associative property $(AB)C = A(BC)$ and a distributive property $A(B + C) = AB + AC$.] □

EXAMPLE 4 Let

$$A = \begin{bmatrix} 2 & 3 \\ 3 & 5 \end{bmatrix}, \quad X = \begin{bmatrix} x \\ y \end{bmatrix}, \quad \text{and} \quad B = \begin{bmatrix} 5 \\ 9 \end{bmatrix}.$$

Write the system of linear equations represented by $AX = B$.

Solution A is a 2×2 matrix and X is a 2×1 matrix, so AX is defined and is a 2×1 matrix. The two elements in the product matrix are

Entry in row 1, column 1: $2x + 3y$ so $AX = \begin{bmatrix} 2x + 3y \\ 3x + 5y \end{bmatrix}.$
Entry in row 2, column 1: $3x + 5y$

Since $AX = B$, we know

$$\begin{bmatrix} 2x + 3y \\ 3x + 5y \end{bmatrix} = \begin{bmatrix} 5 \\ 9 \end{bmatrix},$$

so from the definition of equal matrices we obtain the system

$$2x + 3y = 5$$
$$3x + 5y = 9. \qquad\qquad □$$

Example 4 illustrates that a system of linear equations can be represented as a matrix equation $AX = B$, so if we can solve this matrix equation for X, the result will give us the solution to the system of equations. To solve $AX = B$ for X, you might suggest we divide both sides by A, since that is one approach to solving equations. However, division of matrices is not defined, so this approach doesn't work here. A second approach is to multiply both sides of the equation by $1/A$ (or the multiplicative inverse of A), since in the algebra of real numbers

$$(1/A)Ax = (1/A)B$$
$$1 \cdot x = (1/A)B$$
$$x = (1/A)B.$$

This approach can be extended to matrix algebra if we define matrices that can play the parts of $1/A$ and 1 from the equations above. The necessary definitions for the part of 1 are as follows:

1. The **principal diagonal** of a square matrix consists of the elements in the diagonal extending from the upper left corner to the lower right corner.
2. The $n \times n$ **identity matrix** is the square matrix with n rows and n columns with 1's on the principal diagonal and 0's elsewhere. An identity matrix is symbolized by I.

An identity matrix is important because

$$AI = IA = A$$

for any square matrix A. Example 5 illustrates this property.

EXAMPLE 5   If $A = \begin{bmatrix} 3 & 1 \\ 2 & 4 \end{bmatrix}$ verify that $AI = IA = A$.

Solution   Since A is a 2×2 matrix, we use $I = \begin{bmatrix} 1 & 0 \\ 0 & 1 \end{bmatrix}$, which is the 2×2 identity matrix. Then

$$AI = \begin{bmatrix} 3 & 1 \\ 2 & 4 \end{bmatrix}\begin{bmatrix} 1 & 0 \\ 0 & 1 \end{bmatrix}$$

$$= \begin{bmatrix} 3 \cdot 1 + 1 \cdot 0 & 3 \cdot 0 + 1 \cdot 1 \\ 2 \cdot 1 + 4 \cdot 0 & 2 \cdot 0 + 4 \cdot 1 \end{bmatrix}$$

$$= \begin{bmatrix} 3 & 1 \\ 2 & 4 \end{bmatrix} = A.$$

Also,

$$IA = \begin{bmatrix} 1 & 0 \\ 0 & 1 \end{bmatrix}\begin{bmatrix} 3 & 1 \\ 2 & 4 \end{bmatrix}$$

$$= \begin{bmatrix} 1 \cdot 3 + 0 \cdot 2 & 1 \cdot 1 + 0 \cdot 4 \\ 0 \cdot 3 + 1 \cdot 2 & 0 \cdot 1 + 1 \cdot 4 \end{bmatrix}$$

$$= \begin{bmatrix} 3 & 1 \\ 2 & 4 \end{bmatrix} = A. \qquad \square$$

To play the part of $1/A$, we now wish to find a square matrix A^{-1}, called the **multiplicative inverse of A,** such that

$$A^{-1}A = AA^{-1} = I$$

It can be shown that A has an inverse if and only if the determinant of A is not zero, and that when A^{-1} does exist, it is unique and can be obtained in the following way.

To Find A^{-1}

1. Write the augmented matrix $[A \mid I]$, where I is the identity matrix with the same dimension as A.
2. Use the elementary row operations shown in Section 9.3 to replace matrix $[A \mid I]$ with a matrix of the form $[I \mid B]$.
3. Then A^{-1} is matrix B.

When it is not possible to obtain $[I \mid B]$ (for instance, you may obtain all 0's in a row in the left portion of the matrix), then A has no inverse. Example 6 illustrates the procedure. Note that A^{-1} denotes the multiplicative inverse of matrix A and does not equal $1/A$ in matrix algebra.

EXAMPLE 6   If $A = \begin{bmatrix} 2 & 3 \\ 3 & 5 \end{bmatrix}$, find A^{-1} (if it exists).

Solution   First, we form the augmented matrix $[A \mid I]$.

$$\begin{bmatrix} 2 & 3 & \vdots & 1 & 0 \\ 3 & 5 & \vdots & 0 & 1 \end{bmatrix}$$

We then convert $[A \mid I]$ to the form $[I \mid B]$ using elementary row operations. To obtain 1 for the row 1, column 1 entry, we multiply each element in row 1 by $\frac{1}{2}$.

$$\begin{bmatrix} 1 & \frac{3}{2} & \vdots & \frac{1}{2} & 0 \\ 3 & 5 & \vdots & 0 & 1 \end{bmatrix}$$

To obtain 0 for the row 2, column 1 entry, we add -3 times each entry in row 1 to the corresponding entry in row 2.

$$\begin{bmatrix} 1 & \frac{3}{2} & \vdots & \frac{1}{2} & 0 \\ 0 & \frac{1}{2} & \vdots & -\frac{3}{2} & 1 \end{bmatrix}$$

Next, multiply each entry in row 2 by 2 so the row 2, column 2 entry is 1.

$$\begin{bmatrix} 1 & \frac{3}{2} & \vdots & \frac{1}{2} & 0 \\ 0 & 1 & \vdots & -3 & 2 \end{bmatrix}$$

Finally, add $-\frac{3}{2}$ times each entry in row 2 to the corresponding entry in row 1 to obtain 0 for the row 1, column 2 entry.

$$\begin{bmatrix} 1 & 0 & \vdots & 5 & -3 \\ 0 & 1 & \vdots & -3 & 2 \end{bmatrix}$$

The matrix is now in the form $[I \mid B]$, so by the given rule, B is the inverse of A. Thus,

$$A^{-1} = \begin{bmatrix} 5 & -3 \\ -3 & 2 \end{bmatrix}.$$ $\square$

If A^{-1} exists, the matrix equation $AX = B$ can now be solved by multiplying both sides of the equation by A^{-1} and simplifying as follows:

$$\begin{aligned} A^{-1}AX &= A^{-1}B \\ IX &= A^{-1}B \quad \text{(since } A^{-1}A = I) \\ X &= A^{-1}B \quad \text{(since } IX = X). \end{aligned}$$

Thus, the product $A^{-1}B$ gives the solution to a system of linear equations. In practice, it is usually easier to use row operations to solve a particular linear system than to determine A^{-1}. However, Example 7 considers the problem mentioned in the introduction to this section in which the matrix algebra method is an efficient approach.

EXAMPLE 7   Solve the system

$$\begin{aligned} 2x + 3y &= b_1 \\ 3x + 5y &= b_2 \end{aligned}$$

when (a) $b_1 = 5$, $b_2 = 9$ and (b) $b_1 = -3$, $b_2 = 0$.

Solution

a. The given system can be represented by $AX = B$, where

$$A = \begin{bmatrix} 2 & 3 \\ 3 & 5 \end{bmatrix}, \qquad X = \begin{bmatrix} x \\ y \end{bmatrix}, \qquad \text{and} \qquad B = \begin{bmatrix} 5 \\ 9 \end{bmatrix}.$$

If A^{-1} exists, the solution to $AX = B$ is $X = A^{-1}B$. We know from Example 6 that

$$A^{-1} = \begin{bmatrix} 5 & -3 \\ -3 & 2 \end{bmatrix},$$

so

$$\begin{bmatrix} x \\ y \end{bmatrix} = \begin{bmatrix} 5 & -3 \\ -3 & 2 \end{bmatrix} \begin{bmatrix} 5 \\ 9 \end{bmatrix} = \begin{bmatrix} 5(5) + (-3)(9) \\ -3(5) + 2(9) \end{bmatrix} = \begin{bmatrix} -2 \\ 3 \end{bmatrix}.$$

Thus, $x = -2$ and $y = 3$.

b. We simply proceed as before but with different elements in matrix B.

$$\begin{bmatrix} x \\ y \end{bmatrix} = \begin{bmatrix} 5 & -3 \\ -3 & 2 \end{bmatrix} \begin{bmatrix} -3 \\ 0 \end{bmatrix} = \begin{bmatrix} 5(-3) + (-3)(0) \\ -3(-3) + 2(0) \end{bmatrix} = \begin{bmatrix} -15 \\ 9 \end{bmatrix}$$

Thus, $x = -15$ and $y = 9$. Note that once we know A^{-1}, it is easy to solve the system for any pair of values for b_1 and b_2.   □

EXERCISES 9.4

In Exercises 1–4 specify the dimension of each matrix.

1. $\begin{bmatrix} 3 & 0 & 1 & -5 \\ -2 & 8 & 4 & 7 \end{bmatrix}$

2. $\begin{bmatrix} 0 & 1 \\ -1 & 0 \end{bmatrix}$

3. $\begin{bmatrix} 6 \\ 4 \end{bmatrix}$

4. $\begin{bmatrix} 1 & -3 & 5 \end{bmatrix}$

In Exercises 5–8 find the values of the variables.

5. $\begin{bmatrix} a & b \\ 5 & 4 \end{bmatrix} = \begin{bmatrix} 10 & -3 \\ c & d \end{bmatrix}$

6. $\begin{bmatrix} x \\ y \end{bmatrix} = \begin{bmatrix} -1 \\ 4 \end{bmatrix}$

7. $\begin{bmatrix} 1 + a & -1 - b & 3 + c \end{bmatrix} = \begin{bmatrix} 3 & -1 & 0 \end{bmatrix}$

8. $\begin{bmatrix} 2 + 2a & -1 + 2b & 3 + 2c \\ 2d & 1 + 2e & -2 + 2f \end{bmatrix}$

$= \begin{bmatrix} 4 & 0 & 1 \\ -2 & 7 & 2 \end{bmatrix}$

In Exercises 9–14 perform the indicated operations.

9. $\begin{bmatrix} 1 & 2 \\ 3 & 1 \end{bmatrix} + \begin{bmatrix} -2 & -5 \\ 1 & 9 \end{bmatrix}$

10. $\begin{bmatrix} \frac{1}{8} \\ \frac{5}{8} \\ \frac{7}{8} \end{bmatrix} - \begin{bmatrix} \frac{3}{8} \\ -\frac{1}{8} \\ \frac{5}{8} \end{bmatrix}$

11. $\frac{1}{2} \begin{bmatrix} 0 & 4 & -2 \\ 8 & -6 & 2 \end{bmatrix}$

12. $-3 \begin{bmatrix} 1 & 0 \\ 0 & 1 \end{bmatrix}$

13. $2 \begin{bmatrix} 3 & 2 \\ 1 & 0 \end{bmatrix} - \begin{bmatrix} -4 & 2 \\ 3 & 5 \end{bmatrix}$

14. $-\frac{1}{2} \begin{bmatrix} 2 & -4 & 6 \end{bmatrix} + \frac{1}{3} \begin{bmatrix} -9 & 3 & 6 \end{bmatrix}$

In Exercises 15–20 find the dimensions of the product AB and the product BA if they are defined.

15. A is 2×2, B is 2×2.

16. A is 2×3, B is 2×3.

17. A is 2×3, B is 3×2.

18. A is 4×1, B is 1×4.

19. A is 4×1, B is 3×2.

20. A is 3×3, B is 1×3.

In Exercises 21–26 determine the given product.

21. $\begin{bmatrix} 1 & 3 \\ -1 & 0 \end{bmatrix} \begin{bmatrix} 2 & -3 \\ 3 & 1 \end{bmatrix}$

22. $\begin{bmatrix} 2 & -3 \\ 3 & 1 \end{bmatrix} \begin{bmatrix} 1 & 3 \\ -1 & 0 \end{bmatrix}$

23. $\begin{bmatrix} 1 & 3 \\ -1 & 0 \end{bmatrix} \begin{bmatrix} 1 & 0 \\ 0 & 1 \end{bmatrix}$

24. $\begin{bmatrix} 1 & 3 \\ -1 & 0 \end{bmatrix} \begin{bmatrix} 2 \\ 3 \end{bmatrix}$

25. $\begin{bmatrix} 3 & -4 & 1 \\ 0 & 5 & -3 \\ 2 & -1 & 4 \end{bmatrix} \begin{bmatrix} -1 \\ 2 \\ 1 \end{bmatrix}$

26. $\begin{bmatrix} 1 & 3 & -2 \\ -1 & 0 & 2 \end{bmatrix} \begin{bmatrix} -1 & 3 \\ 2 & 0 \\ 1 & 4 \end{bmatrix}$

In Exercises 27–38 perform the indicated operations (if they are defined) for matrices A, B, C, D, and I as defined below.

$$A = \begin{bmatrix} 2 & -1 & 3 \\ 0 & 1 & -2 \end{bmatrix}, B = \begin{bmatrix} 1 & 2 \\ -3 & -4 \end{bmatrix},$$

$$C = \begin{bmatrix} -1 & 0 \\ 0 & -1 \end{bmatrix}, D = \begin{bmatrix} 1 & 3 & 6 \\ 2 & 5 & 7 \end{bmatrix},$$

$$I = \begin{bmatrix} 1 & 0 \\ 0 & 1 \end{bmatrix}.$$

27. $A - D$ **28.** $A - B$ **29.** AB

30. BA **31.** BI **32.** IB

33. $2(B + C)$ **34.** $3D - 2A$ **35.** II

36. $(BC)A$ **37.** $C(A + D)$ **38.** $(A + D)C$

In Exercises 39 and 40 write the system of linear equations represented by $AX = B$.

39. $A = \begin{bmatrix} 1 & 1 \\ 5 & -2 \end{bmatrix}$, $X = \begin{bmatrix} x \\ y \end{bmatrix}$, and $B = \begin{bmatrix} 0 \\ 3 \end{bmatrix}$.

40. $A = \begin{bmatrix} 3 & -1 & 5 \\ 1 & 0 & -2 \\ -1 & 1 & -1 \end{bmatrix}$,

$X = \begin{bmatrix} x \\ y \\ z \end{bmatrix}$, and $B = \begin{bmatrix} 1 \\ -1 \\ 1 \end{bmatrix}$.

In Exercises 41–49 find the inverse, if it exists, for each matrix.

41. $\begin{bmatrix} -2 & -1 \\ 7 & 3 \end{bmatrix}$ **42.** $\begin{bmatrix} 1 & 1 \\ 6 & -1 \end{bmatrix}$

43. $\begin{bmatrix} 5 & 3 \\ 1 & -2 \end{bmatrix}$ **44.** $\begin{bmatrix} -2 & -1 \\ 5 & 2 \end{bmatrix}$

45. $\begin{bmatrix} 3 & -2 \\ 6 & -4 \end{bmatrix}$ **46.** $\begin{bmatrix} 7 & -2 \\ 3 & 5 \end{bmatrix}$

47. $\begin{bmatrix} 1 & -1 & 1 \\ -1 & 1 & 0 \\ 0 & -1 & 1 \end{bmatrix}$ **48.** $\begin{bmatrix} 2 & 3 & 0 \\ 3 & 0 & 2 \\ 0 & 4 & 1 \end{bmatrix}$

49. $\begin{bmatrix} 2 & -1 & 3 \\ -1 & 1 & -1 \\ -4 & 2 & -6 \end{bmatrix}$

In Exercises 50–58 write the given system in the form $AX = B$ and use A^{-1} (if it exists) to solve the system. (*Note:* The inverses for these problems are found in Exercises 41–49.)

50. $\begin{aligned} -2x - y &= 2 \\ 7x + 3y &= -1 \end{aligned}$ **51.** $\begin{aligned} x + y &= 25 \\ 6x - y &= 3 \end{aligned}$

52. $\begin{aligned} 5x + 3y &= -2 \\ x - 2y &= -3 \end{aligned}$ **53.** $\begin{aligned} -2x - y &= -5 \\ 5x + 2y &= -17 \end{aligned}$

54. $\begin{aligned} 3x - 2y &= 1 \\ 6x - 4y &= 5 \end{aligned}$ **55.** $\begin{aligned} 7x - 2y &= 19 \\ 3x + 5y &= -14 \end{aligned}$

56. $\begin{aligned} x - y + z &= 1 \\ -x + y &= -1 \\ -y + z &= 2 \end{aligned}$ **57.** $\begin{aligned} 2x + 3y &= 10 \\ 3x + 2z &= 2 \\ 4y + z &= 6 \end{aligned}$

58. $\begin{aligned} 2x - y + 3z &= 1 \\ -x + y - z &= -1 \\ -4x + 2y - 6z &= -2 \end{aligned}$

59. If the inverse of $\begin{bmatrix} 3 & -5 \\ -4 & 7 \end{bmatrix}$ is $\begin{bmatrix} 7 & 5 \\ 4 & 3 \end{bmatrix}$, then solve

$$\begin{aligned} 3x - 5y &= b_1 \\ -4x + 7y &= b_2 \end{aligned}$$

when (a) $b_1 = 2$, $b_2 = 3$ and (b) $b_1 = 0$, $b_2 = -5$.

60. If the inverse of $\begin{bmatrix} 6 & -1 & -5 \\ -7 & 1 & 5 \\ -10 & 2 & 11 \end{bmatrix}$

is $\begin{bmatrix} -1 & -1 & 0 \\ -27 & -16 & -5 \\ 4 & 2 & 1 \end{bmatrix}$, then solve

$$\begin{aligned} 6x - y - 5z &= b_1 \\ -7x + y + 5z &= b_2 \\ -10x + 2y + 11z &= b_3 \end{aligned}$$

when (a) $b_1 = 1$, $b_2 = 0$, $b_3 = 0$ and (b) $b_1 = 2$, $b_2 = -1$, $b_3 = -2$.

9.5 PARTIAL FRACTIONS

Solving a system of linear equations is a useful component for determining algebraic identities that enable us to analyze rational functions. Recall from Section 4.6 that rational functions fit the form

$$y = \frac{P(x)}{Q(x)} \qquad Q(x) \neq 0,$$

where $P(x)$ and $Q(x)$ are polynomials. In certain situations (especially in calculus), if $P(x)/Q(x)$ is complicated, we try to write this expression as the sum of simpler fractions that are easier to analyze. For example, from the usual addition of fractions we know that

$$\frac{2}{x + 2} + \frac{3}{x - 1}$$

is identical to

$$\frac{5x + 4}{(x + 2)(x - 1)}.$$

The question is how to work this problem backward and split the more complicated fraction into the sum of the simpler fractions. This sum is called the **partial fraction decomposition** of the expression, and each term in the sum is called a **partial fraction.**

$$\frac{5x + 4}{(x + 2)(x - 1)} \underset{\text{adding}}{\overset{\text{decomposing}}{=}} \underbrace{\frac{2}{x + 2} + \frac{3}{x - 1}}_{\text{partial fractions}}$$

In the above problem, determining the partial fraction decomposition (assume it is unknown) amounts to finding constants A and B such that

$$\frac{5x + 4}{(x + 2)(x - 1)} = \frac{A}{x + 2} + \frac{B}{x - 1}$$

is an identity. To do this, we first clear the expression of fractions by multiplying both sides of the equation by $(x + 2)(x - 1)$ to obtain

$$5x + 4 = A(x - 1) + B(x + 2).$$

Now since this equation is an identity, the coefficients of like powers of x must be equal on both sides of the equation. By rewriting the above equation as

$$5x + 4 = (A + B)x - A + 2B$$

and equating coefficients, we obtain the linear system

$$A + B = 5$$
$$-A + 2B = 4.$$

Adding these two equations yields $3B = 9$, so $B = 3$. Then it is easy to determine that $A = 2$ and that our results check with the original problem.

In this example we have illustrated the method of equating coefficients to determine the constants A and B because it is a powerful method that extends to more complicated cases, and it also gives us additional exposure to systems of linear equations. A shortcut to finding A and B in this particular example is to realize that

$$5x + 4 = A(x - 1) + B(x + 2)$$

is an identity and therefore true when $x = 1$ and $x = -2$. Substitution of $x = 1$ readily yields $9 = 3B$, so $B = 3$; while replacing x by -2 gives $-6 = -3A$, so $A = 2$. However, in the long run this method is limited, and it is most useful only when A and B are coupled with factors that are linear expressions.

EXAMPLE 1   Determine the constants A, B, and C so that

$$\frac{16}{(x + 2)(x^2 + 4)} = \frac{A}{x + 2} + \frac{Bx + C}{x^2 + 4}$$

is an identity.

Solution   First, multiply both sides of the given equation by $(x + 2)(x^2 + 4)$ to obtain

$$16 = A(x^2 + 4) + (Bx + C)(x + 2).$$

Then expand the right side of this equation.

$$16 = Ax^2 + 4A + Bx^2 + 2Bx + Cx + 2C$$

Next, group together like powers of x and then factor.

$$16 = (A + B)x^2 + (2B + C)x + 4A + 2C$$

Now equate coefficients of like powers of x and set up a system of linear equations. (Note that $16 = 0x^2 + 0x + 16$.)

Equating coefficients of x^2:	$A + B = 0$
Equating coefficients of x:	$2B + C = 0$
Equating constant terms:	$4A + 2C = 16$

Finally, we solve this system by the methods in this chapter to determine that $A = 2$, $B = -2$, and $C = 4$.   □

To this point, the form of the partial fraction decomposition has been given, and our job has been to determine constants so the given equation is an identity. In practice, however, you will start only with a rational function

$$y = \frac{P(x)}{Q(x)} \qquad Q(x) \neq 0$$

and you need to supply the form of the decomposition. To accomplish this, first divide out any common factors of $P(x)$ and $Q(x)$ and then (if necessary) do the following.

1. The degree of $P(x)$ must be less than the degree of $Q(x)$. If it is not, then through long division rewrite the fraction as

$$\frac{P(x)}{Q(x)} = \text{quotient} + \frac{\text{remainder}}{Q(x)},$$

where the remainder term is in proper form.

2. The denominator, $Q(x)$, must be factored into linear factors and/or quadratic factors. The quadratic factors should be **irreducible,** which means they cannot be factored into linear factors (with real coefficients). Factoring $Q(x)$ in this manner is always possible (theoretically), but it may be very difficult.

Now consider Example 1, which illustrates the form used for distinct linear and quadratic factors. For a linear factor like $x + 2$, we merely put a constant, say A, in the numerator of the partial fraction. For an irreducible quadratic factor like $x^2 + 4$, we just put a linear expression, say $Bx + C$, in the numerator of the partial fraction. In general, we state these rules as follows:

Nonrepeated Linear or Quadratic Factors

Distinct Linear Factors Each linear factor $ax + b$ of $Q(x)$ that is not repeated produces in the decomposition a term of the form

$$\frac{A}{ax + b}.$$

Distinct Quadratic Factors Each irreducible quadratic factor $ax^2 + bx + c$ of $Q(x)$ that is not repeated produces in the decomposition a term of the form

$$\frac{Ax + B}{ax^2 + bx + c}.$$

We illustrate our ideas to this point in the following example.

EXAMPLE 2   Express $\dfrac{x^3 + 8}{x^3 + 4x}$ as a sum of partial fractions.

Solution   Since the degree of the numerator is not less than the degree of the denominator, we need long division to determine

$$\frac{x^3 + 8}{x^3 + 4x} = 1 + \frac{-4x + 8}{x^3 + 4x}.$$

Now we work with the remainder term. First, factor the denominator and then write the partial fraction decomposition in the form determined by the above rules.

$$\frac{-4x + 8}{x^3 + 4x} = \frac{-4x + 8}{x(x^2 + 4)} = \frac{A}{x} + \frac{Bx + C}{x^2 + 4}$$

Then multiplying both sides of this equation by $x(x^2 + 4)$ gives

$$\begin{aligned}
-4x + 8 &= A(x^2 + 4) + (Bx + C)x \\
&= Ax^2 + 4A + Bx^2 + Cx \\
&= (A + B)x^2 + Cx + 4A.
\end{aligned}$$

Equating the coefficients, we have

$$\begin{aligned}
A + B &= 0 \\
C &= -4 \\
4A &= 8
\end{aligned}$$

so

$$A = 2, B = -2, \text{ and } C = -4.$$

Thus,

$$\frac{x^3 + 8}{x^3 + 4x} = 1 + \frac{2}{x} + \frac{-2x - 4}{x^2 + 4}. \qquad \Box$$

To handle the cases in which linear or quadratic factors may be repeated in the denominator, we need a more general version of our methods for partial fraction decomposition. These more powerful procedures are as follows:

Partial Fraction Decomposition of $P(x)/Q(x)$

Linear Factors Each linear factor of $Q(x)$ of the form $(ax + b)^n$ produces in the decomposition a sum of n terms of the form

$$\frac{A_1}{ax + b} + \frac{A_2}{(ax + b)^2} + \cdots + \frac{A_n}{(ax + b)^n}.$$

Quadratic Factors Each irreducible quadratic factor of $Q(x)$ of the form $(ax^2 + bx + c)^n$ produces in the decomposition a sum of n terms of the form

$$\frac{A_1x + B_1}{ax^2 + bx + c} + \frac{A_2x + B_2}{(ax^2 + bx + c)^2} + \cdots + \frac{A_nx + B_n}{(ax^2 + bx + c)^n}.$$

Note that when a linear or quadratic factor appears just once, then $n = 1$ and the above procedures simplify to our previous methods. Also, in practice, we usually denote the constants in the numerators as A, B, C, and so on, instead of using subscript notation.

EXAMPLE 3   Express $\dfrac{1}{x^4 - 2x^3 + x^2}$ as a sum of partial fractions.

Solution   No division is necessary here, so first factor the denominator.

$$x^4 - 2x^3 + x^2 = x^2(x^2 - 2x + 1) = x^2(x - 1)^2$$

Now $(x - 1)^2$ is a repeated linear factor, and x^2 is also analyzed with the repeated linear factor procedure. Each of these factors produces two terms in the decomposition as follows:

$$\frac{1}{x^2(x - 1)^2} = \frac{A}{x} + \frac{B}{x^2} + \frac{C}{x - 1} + \frac{D}{(x - 1)^2}.$$

Multiplying both sides of this equation by $x^2(x - 1)^2$ gives

$$\begin{aligned}
1 &= Ax(x - 1)^2 + B(x - 1)^2 + Cx^2(x - 1) + Dx^2 \\
&= Ax^3 - 2Ax^2 + Ax + Bx^2 - 2Bx + B + Cx^3 - Cx^2 + Dx^2 \\
&= (A + C)x^3 + (-2A + B - C + D)x^2 + (A - 2B)x + B.
\end{aligned}$$

Equating the coefficients, we have

$$\begin{aligned}
A + C &= 0 \\
-2A + B - C + D &= 0 \\
A - 2B &= 0 \\
B &= 1,
\end{aligned}$$

so

$$A = 2, B = 1, C = -2, \text{ and } D = 1.$$

Thus,

$$\frac{1}{x^4 - 2x^3 + x^2} = \frac{2}{x} + \frac{1}{x^2} + \frac{-2}{x - 1} + \frac{1}{(x - 1)^2}. \qquad \square$$

EXERCISES 9.5

In Exercises 1–10 determine the constants A, B, C, and/or D so that the equation is an identity.

1. $\dfrac{1}{(x + 2)(x - 2)} = \dfrac{A}{x + 2} + \dfrac{B}{x - 2}$

2. $\dfrac{1}{(x - 1)(x + 2)} = \dfrac{A}{x - 1} + \dfrac{B}{x + 2}$

3. $\dfrac{6}{(x + 1)(x^2 + 1)} = \dfrac{A}{x + 1} + \dfrac{Bx + C}{x^2 + 1}$

4. $\dfrac{2x^3 + 5x + 2}{(x^2 + 1)(x^2 + 2)} = \dfrac{Ax + B}{x^2 + 1} + \dfrac{Cx + D}{x^2 + 2}$

5. $\dfrac{2x}{(x + 2)^2} = \dfrac{A}{x + 2} + \dfrac{B}{(x + 2)^2}$

6. $\dfrac{3x - 10}{(x - 1)^2} = \dfrac{A}{x - 1} + \dfrac{B}{(x - 1)^2}$

7. $\dfrac{x^3 - 2x^2 - 4x + 3}{(x^2 + 1)^2} = \dfrac{Ax + B}{x^2 + 1} + \dfrac{Cx + D}{(x^2 + 1)^2}$

8. $\dfrac{4x^2 + 3x - 1}{(x^2 + x + 1)^2} = \dfrac{Ax + B}{x^2 + x + 1} + \dfrac{Cx + D}{(x^2 + x + 1)^2}$

9. $\dfrac{-2x + 4}{(x - 1)^2(x^2 + 1)} = \dfrac{A}{x - 1} + \dfrac{B}{(x - 1)^2} + \dfrac{Cx + D}{x^2 + 1}$

10. $\dfrac{2x^3 - 5x^2 + 4x - 3}{x^2(x^2 + 1)} = \dfrac{A}{x} + \dfrac{B}{x^2} + \dfrac{Cx + D}{x^2 + 1}$

In Exercises 11–24 find the partial fraction decomposition of the given expression.

11. $\dfrac{1}{1 - x^2}$

12. $\dfrac{x + 2}{x^2 - 5x + 4}$

13. $\dfrac{x}{x^2 + 2x + 1}$

14. $\dfrac{5x + 6}{x^3 - 2x^2}$

15. $\dfrac{x^3}{x^2 - 4}$

16. $\dfrac{x^2}{x^2 - 2x + 1}$

17. $\dfrac{-12x - 6}{x^3 + x^2 - 6x}$

18. $\dfrac{1}{(x + 1)(x + 2)(x + 3)}$

19. $\dfrac{1}{x^3 + x}$

20. $\dfrac{x^5 - x^3 + x - 1}{x^4 + x^2}$

21. $\dfrac{4x}{x^4 - 1}$

22. $\dfrac{1}{x^3 - 1}$

23. $\dfrac{4x^2 + x + 8}{x^4 + 3x^2 + 2}$

24. $\dfrac{4x^3 - 7x^2 + 5x - 2}{x^4 + 2x^2 + 1}$

9.6 NONLINEAR SYSTEMS OF EQUATIONS

We started this chapter by showing how to determine any intersection points for two straight lines. A more general problem is to find intersection points for any two curves. For instance, in calculus it is often necessary to find any intersection points of one of the common conic sections (circle, ellipse, hyperbola, and parabola) with a line or another conic section. The substitution method and the addition-elimination method from Section 9.1 are sufficient in many cases to solve systems even when one or more of the equations is not linear. The following examples illustrate a few such cases.

EXAMPLE 1   Find all the ordered pairs that satisfy the pair of equations

$$y = x^2 - 6x + 8$$
$$y = x + 2.$$

Solution   We wish to find the value(s) for x that make both y values the same or, equivalently, we want to know the value(s) for x that make $x^2 - 6x + 8$ equal to $x + 2$.

$$x^2 - 6x + 8 = x + 2$$
$$x^2 - 7x + 6 = 0$$
$$(x - 6)(x - 1) = 0$$
$$x - 6 = 0 \qquad x - 1 = 0$$
$$x = 6 \qquad\quad x = 1$$

Thus, the parabola and the line intersect at $x = 1$ and $x = 6$. To find the y-coordinates, substitute these numbers in the simpler equation $y = x + 2$.

$$y = x + 2 \qquad \text{and} \qquad y = x + 2$$
$$= (1) + 2 \qquad\qquad\qquad = (6) + 2$$
$$= 3 \qquad\qquad\qquad\qquad = 8$$

Thus, the solutions are $(1,3)$ and $(6,8)$. Consider Figure 9.9, which shows the parabola and the line intersecting at these points. □

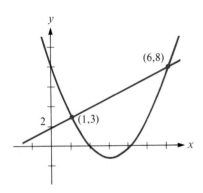

Figure 9.9

EXAMPLE 2   Find all intersection points of the ellipse $2x^2 + 3y^2 = 5$ and the hyperbola $4y^2 - 3x^2 = 1$.

Solution   The system we need to solve is

$$2x^2 + 3y^2 = 5$$
$$-3x^2 + 4y^2 = 1.$$

If we form equivalent equations by multiplying the top equation by 3 and the bottom equation by 2, we can eliminate the x variable.

$$6x^2 + 9y^2 = 15$$
$$\underline{-6x^2 + 8y^2 = 2}$$
$$17y^2 = 17 \qquad \text{Add the equations.}$$
$$y^2 = 1$$
$$y = \pm 1$$

Now substituting $y = \pm 1$ into $2x^2 + 3y^2 = 5$ gives

$$2x^2 + 3(\pm 1)^2 = 5$$
$$x^2 = 1$$
$$x = \pm 1.$$

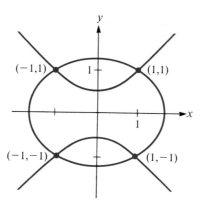

Figure 9.10

Thus, the solutions are $(1,1)$, $(1,-1)$, $(-1,1)$, and $(-1,-1)$. Figure 9.10 shows the ellipse and the hyperbola intersecting at these points. $\quad\square$

EXAMPLE 3   Solve the system

$$\frac{5}{x} + \frac{2}{y} - 1 = 0$$

$$\frac{1}{x} - \frac{3}{y} - 7 = 0.$$

Solution   If we rewrite the above system as

$$5\left(\frac{1}{x}\right) + 2\left(\frac{1}{y}\right) = 1$$

$$\left(\frac{1}{x}\right) - 3\left(\frac{1}{y}\right) = 7$$

and let $a = 1/x$ and $b = 1/y$, we obtain

$$5a + 2b = 1$$
$$a - 3b = 7,$$

which is a relatively simple system of linear equations. If we now add -5 times the second equation to the first equation, we find

$$17b = -34 \quad \text{so} \quad b = -2.$$

Then substituting $b = -2$ into equation 2 gives

$$a - 3(-2) = 7$$
$$a = 1.$$

Finally, from the definition of a and b we have

$$\frac{1}{x} = 1 \qquad \text{and} \qquad \frac{1}{y} = -2$$

$$x = 1 \qquad \text{and} \qquad y = -\tfrac{1}{2}.$$

Thus, the solution is $(1, -\tfrac{1}{2})$.   □

EXAMPLE 4   Solve the system

$$x^2 + y^2 = 1$$
$$y = x^2.$$

Solution   Substituting $y = x^2$ into the top equation gives

$$x^2 + (x^2)^2 = 1 \quad \text{or} \quad (x^2)^2 + x^2 - 1 = 0,$$

which is an equation with quadratic form. As discussed in Section 2.4, if we let $t = x^2$, the equation becomes

$$t^2 + t - 1 = 0.$$

By the quadratic formula,

$$t = \frac{-1 \pm \sqrt{(1)^2 - 4(1)(-1)}}{2(1)} = \frac{-1 \pm \sqrt{5}}{2},$$

so

$$x^2 = \frac{-1 \pm \sqrt{5}}{2} \quad \text{and} \quad x = \pm\sqrt{\frac{-1 \pm \sqrt{5}}{2}}.$$

Since x must be a real number we eliminate $\pm\sqrt{\dfrac{-1 - \sqrt{5}}{2}}$ and conclude that

$$x = \pm\sqrt{\frac{-1 + \sqrt{5}}{2}}.$$

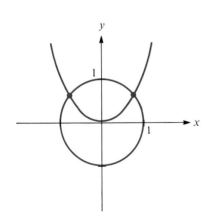

Figure 9.11

Finally, substituting these two values of x into $y = x^2$ gives us solutions of

$$\left(\sqrt{\frac{-1 + \sqrt{5}}{2}}, \frac{-1 + \sqrt{5}}{2}\right) \quad \text{and} \quad \left(-\sqrt{\frac{-1 + \sqrt{5}}{2}}, \frac{-1 + \sqrt{5}}{2}\right).$$

Figure 9.11 shows the circle $x^2 + y^2 = 1$ and the parabola $y = x^2$ intersecting at these points, which are approximately $(0.8, 0.6)$ and $(-0.8, 0.6)$.   □

EXERCISES 9.6

In Exercises 1–30 find all the ordered pairs that satisfy the pairs of equations.

1. $y = x^2$
$y = x$

2. $f(x) = 2 - x^2$
$g(x) = -x$

3. $g(x) = 10 - x^2$
$f(x) = 1$

4. $y = x^2 - 2x - 3$
$y = 2x + 2$

5. $y = x^2 - 4$
$y = 4 - x^2$

6. $f(x) = x^2$
$g(x) = 2 - x^2$

7. $x^2 + y^2 = 100$
$3x + y = 10$

8. $x^2 + y^2 = 1$
$y = x + 1$

9. $x^2 + y^2 = 8$
$2x^2 - y^2 = 4$

10. $x^2 - y^2 = 8$
$2x^2 + y^2 = 19$

11. $x^2 + 2y^2 = 40$
$2x^2 + y^2 = 32$

12. $2x^2 + 3y^2 = 5$
$3x^2 = 4y^2 - 1$

13. $xy = 15$
$x + y = 8$

14. $xy = 1$
$x - 2y = 1$

15. $xy = 2$
$x^2 + 2y^2 = 9$

16. $xy = 4$
$x^2 + y^2 = 8$

17. $x^2 + y^2 = 2$
$x = y^2$

18. $x^2 + y^2 = 3$
$x = y^2$

19. $x - y = 2$
$x = y^2 - 4$

20. $y^2 = x + 9$
$x = 3 - 2y - y^2$

21. $3x + y = 7$
$x^2 y = 4$

22. $x = y^3 - 3y^2$
$x - y = -3$

23. $x^2 + y^2 = 5$
$y = x^2$

24. $(x - 1)^2 + (y + 2)^2 = 5$
$y + 2x = 0$

25. $\dfrac{x^2}{4} + \dfrac{y^2}{9} = 1$
$\dfrac{x^2}{4} - \dfrac{y^2}{9} = 1$

26. $\dfrac{x^2}{4} + y^2 = 1$
$\dfrac{x^2}{4} - \dfrac{y^2}{2} = 1$

27. $\dfrac{3}{x} - \dfrac{2}{y} = 27$
$\dfrac{2}{x} + \dfrac{5}{y} = -1$

28. $\dfrac{5}{x} + \dfrac{3}{y} - 5 = 0$
$\dfrac{4}{x} + \dfrac{1}{y} - 11 = 0$

29. $\dfrac{2}{x} + \dfrac{2}{y} - \dfrac{1}{z} = 0$
$\dfrac{2}{x} - \dfrac{1}{y} + \dfrac{2}{z} = 3$
$\dfrac{1}{x} - \dfrac{2}{y} - \dfrac{2}{z} = 12$

30. $\dfrac{1}{x} + \dfrac{2}{y} - \dfrac{3}{z} = 0$
$\dfrac{3}{x} - \dfrac{1}{y} + \dfrac{6}{z} = 1$
$\dfrac{2}{x} - \dfrac{3}{y} - \dfrac{1}{z} = -9$

31. Find all intersection points of the ellipse $x^2 + 2y^2 = 3$ and the hyperbola $3x^2 - y^2 = 2$.

32. Find all intersection points of the semicircle $y = -\sqrt{36 - x^2}$ and the line $y = -x$.

33. Find to the nearest tenth all points where the circle $x^2 + y^2 = 9$ and the ellipse $(x^2/25) + (y^2/4) = 1$ intercept.

34. Find to the nearest hundredth all intersection points of the circle $x^2 + y^2 = 1$ and the parabola $x = y^2$.

35. Show that a hyperbola defined by the equation $(x^2/a^2) - (y^2/b^2) = 1$ does not intersect its asymptotes $y = \pm(b/a)x$.

36. A rectangle is inscribed in a circle. If the areas of the rectangle and circle are 4 m² and $\frac{5}{2}\pi$ m², respectively, then find the dimensions of the rectangle.

37. Find the length of the legs of a right triangle if the area is 36 ft² and the hypotenuse is $5\sqrt{6}$ ft.

38. Find the radius of each circle in the illustration below if the combined areas of the two circles is 48.5π square inches.

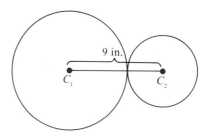

39. Both the parallelogram and the triangle in the illustration below have areas of 36m². Find b and h.

40. If the area and perimeter of a rectangle are denoted by a and p, respectively, then show the length is given by $(p + \sqrt{p^2 - 16a})/4$.

41. Three pipes *A*, *B*, and *C* can fill a tank in 2 hours. If only *A* and *B* are used it takes 3 hours, while *B* and *C* can do the job in 4 hours. How long does it take each pipe alone to fill the tank?

42. Pumps *A* and *B* working together can fill a tank in 7 hours and 30 minutes. This same tank may also be filled by running pumps *A* and *C* for 6 hours or pumps *B* and *C* for 10 hours. How long does it take each pump alone to fill the tank? How long does it take to fill the tank using all three pumps?

9.7 SYSTEMS OF LINEAR INEQUALITIES AND LINEAR PROGRAMMING

The analysis of certain problems often requires a system of inequalities instead of a system of equations. Example 1 illustrates this point.

EXAMPLE 1   A company manufactures two products, *A* and *B*. Product *A* costs $20 per unit to produce and product *B* costs $60 per unit. The company has $90,000 to spend on the production. If *x* represents the number of units produced of product *A* and *y* represents the number of units produced of product *B*, write an inequality that expresses the restriction placed on the company because of available funds.

Solution   *x* represents the number of units produced of product *A* and each unit costs $20 to produce. Thus, the company spends $20x$ dollars to produce *x* units of product *A*. Similarly, the company spends $60y$ dollars to produce *y* units of product *B*. Since the total cost cannot exceed $90,000, we have

$$20x + 60y \leq 90,000.$$

Since the number of units produced cannot be negative, we also have

$$x \geq 0 \text{ and } y \geq 0.$$

Thus, the restriction placed on the company because of available funds is given by

$$20x + 60y \leq 90,000 \text{ with } x \geq 0 \text{ and } y \geq 0. \qquad \square$$

The question in Example 1 is designed to show the usefulness of inequalities in describing practical situations. Capital is only one of many restrictions that a company must consider because of limited resources. For example, machine time, labor hours, and storage space are other considerations. Subject to these restrictions, the company must determine the number of units of each product that should be manufactured for maximum profit. The solution to such problems is found by techniques from a branch of applied mathematics called **linear programming.** To explore this topic further, we first need to develop methods for graphing the solution set to a system of inequalities.

An inequality in two variables such as $y < x$ has infinitely many solutions. To illustrate these solutions graphically, we note that the line $y = x$ separates the plane into two regions. Each region consists of the set of points on one side of the line and is called a **half-plane.** The solution set to $y < x$ is given by the half-plane *below* the line $y = x$, while the solution set to $y > x$ is given by the half-plane *above* $y = x$. We usually indicate the half-plane in the solution set through shading as shown in Figure 9.12, and the sketch of the solution set gives the graph of the inequality. Note that we use either a solid line or a dashed line in the graph, depending on whether the line is included in the solution set.

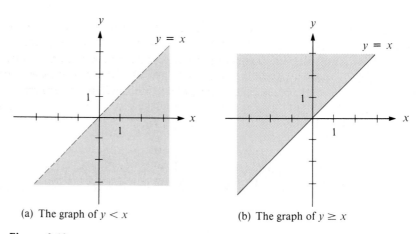

(a) The graph of $y < x$ (b) The graph of $y \geq x$

Figure 9.12

EXAMPLE 2   Graph each inequality.

a. $y \leq 2x - 1$ **b.** $y > -3x$

Solution
a. See Figure 9.13.
b. See Figure 9.14.

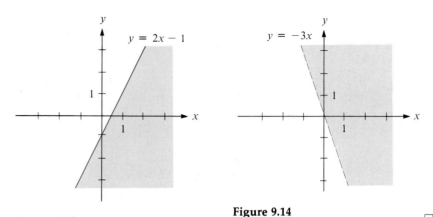

Figure 9.14

Figure 9.13

When the inequality is not solved for y, you may have difficulty choosing which half-plane to shade in your answer. The easiest way to decide in such cases is usually to pick some **test point** that is not on the line and substitute the coordinates of this point into the inequality. If the resulting statement is true, then shade the half-plane containing the test point. Otherwise, shade the half-plane on the other side of the line from the test point. The origin $(0,0)$ is often a convenient point to use in this test.

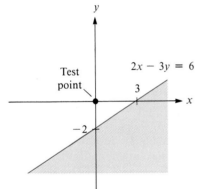

Figure 9.15

EXAMPLE 3 Graph $2x - 3y \geq 6$.

Solution The origin is not on the line $2x - 3y = 6$, so substitute 0 for x and y in the above inequality.

$$2(0) - 3(0) \geq 6$$
$$0 \geq 6$$

Since this statement is false we shade the half-plane not containing the origin, as shown in Figure 9.15. If you feel uncomfortable with the test point method, an alternative is to solve the given inequality for y. The result, $y \leq \frac{2}{3}x - 2$, tells us to shade the half-plane below the line. ☐

The solution set of a system of inequalities is the intersection of the solution sets of all the individual inequalities in the system. The best way to give the solution set is in a graph. To do this, we graph the solutions to each inequality in the system and then shade in the overlap (intersection) of these half-planes.

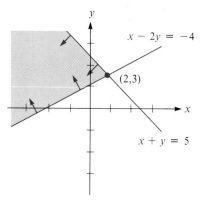

Figure 9.16

EXAMPLE 4   Graph the solution set of the system

$$x + y \le 5$$
$$x - 2y \le -4.$$

Solution   By using the origin in both cases as a test point, we determine that $x + y \le 5$ is satisfied by the points on or below the line $x + y = 5$, while $x - 2y \le -4$ is satisfied by the points on or above the line given by $x - 2y = -4$. The intersection of these two half-planes is shown in Figure 9.16.   □

The **vertices** or **corners** of a region are the intersection points of the bounding sides in the region. For example, the corner in the region defined by the system of inequalities in Example 4 is (2,3) as shown in Figure 9.16. Such corners are important in the solution to linear programming problems, and they can be found by any of the methods we have considered for finding the intersection point for two lines. The next example includes a discussion of this concept.

EXAMPLE 5   Graph the solution set of the system

$$x + y \le 6, \qquad x - 2y \le 3,$$
$$y - x \le 2, \qquad x \ge 0, \qquad y \ge 0.$$

Specify the coordinates of any corner in the graph.

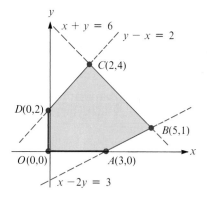

Figure 9.17

Solution   The inequalities $x \ge 0$ and $y \ge 0$ restrict the graph of the solution set to the first quadrant. By using the origin as a test point, we determine that the inequalities specify the region on or below the lines $x + y = 6$ and $y - x = 2$ as well as the region on or above the line $x - 2y = 3$. Figure 9.17 shows the graph of the solution set. To determine the coordinates of the corners O, A, B, C, and D, we note that O is the origin (0,0); A is the x-intercept of $x - 2y = 3$, which is (3,0); and D is the y-intercept of $y - x = 2$, which is (0,2). To find point B, we solve the system

$$x + y = 6$$
$$x - 2y = 3$$

by adding -1 times the bottom equation to the top equation to get $3y = 3$. Then $y = 1$, so $x = 5$, and point B is (5,1). Finally, the system

$$x + y = 6$$
$$y - x = 2$$

solves by adding the two equations together to get $2y = 8$. Then $y = 4$, so $x = 2$, and point C is (2,4).   □

We now return to the topic of linear programming. Basically, linear programming involves finding the maximum or minimum value of a linear function $F = ax + by$ which is called the **objective function.** The variables x and y must satisfy certain inequalities, called **constraints,** that impose restrictions on the solution. Only an ordered pair (x,y) that satisfies all the constraints is a possible or **feasible solution,** and the set of all feasible solutions is given by the graph of the given system of inequalities (or constraints). Although the set of feasible solutions may seem overwhelming, the corners are the crucial points as shown in the following theorem.

Corner Point Theorem	If a linear function $F = ax + by$ assumes a maximum or minimum value subject to a system of linear inequalities, then it does so at the corners or vertices of that system of constraints.

Furthermore, we are guaranteed that a maximum or minimum exists as long as the set of feasible solutions is a closed convex polygon. The condition that a polygon be closed merely means that the region include its boundary lines, while a region is said to be **convex** provided the region contains the line segment joining *any* two points in the region. Figure 9.18(a) shows a set of points that is convex, while Figure 9.18(b) shows a nonconvex region.

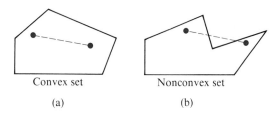

Convex set   Nonconvex set

(a)   (b)

Figure 9.18

Since the solution set of a system of linear inequalities is always a convex region, this condition poses no problem. Thus, the only time we may not be able to find a maximum or minimum value occurs when the set of feasible solutions is unbounded and does not form a polygon.

We can now consider a typical problem in linear programming.

EXAMPLE 6   Find the maximum value of $F = 3x - 2y$ subject to the constraints

$$x + y \le 6$$
$$y - x \le 2$$
$$x - 2y \le 3$$
$$x \ge 0$$
$$y \ge 0.$$

Solution   The constraints are the system of inequalities in Example 5, so the set of all feasible solutions including all vertices is shown in Figure 9.17. Since the set of feasible solutions is a closed convex polygon, we know that a maximum value exists at one of the corners. We need only substitute the coordinates of these vertices into the objective function to determine the correct corner.

Vertices	$F = 3x - 2y$
O (0,0)	$3(0) - 2(0) = 0$
A (3,0)	$3(3) - 2(0) = 9$
B (5,1)	$3(5) - 2(1) = 13$
C (2,4)	$3(2) - 2(4) = -2$
D (0,2)	$3(0) - 2(2) = -4$

From the table we see that when $x = 5$ and $y = 1$, the function attains a maximum value of 13. Note that we can also determine from the table that the minimum function value of -4 occurs at (0,2).   □

If we look a little closer at Example 6, we can see why any maximum or minimum occurs at a corner. Consider Figure 9.19, which plots the line $F = 3x - 2y$ for various values of F. Note that as F increases, the y-intercepts of the line change but the slopes remain fixed at $\frac{3}{2}$. The result is a family of parallel lines. The line with the maximum or minimum value of F will then intersect the set of feasible solutions at a corner as shown in the figure.

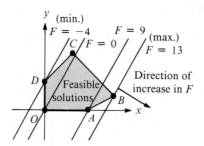

Figure 9.19

The next example considers a linear programming problem in a more practical setting.

EXAMPLE 7   A manufacturer of personal computers makes two types of printers, A and B. To comply with contracts, the company must produce at least 100 type A printers and 200 type B printers each week. A total of 4 labor hours is required to assemble printer A and a total of 2 labor hours is required to assemble printer B with a maximum of 1,000 labor hours available each week. If the company can sell all their printers for a profit of $70 on printer A and $40 on printer B, then how many of each type should be produced weekly for the maximum profit?

Solution   If we let x and y represent the number of assembled type A and type B printers, respectively, then the profit function is

$$P = 70x + 40y$$

and the constraints are

$$x \geq 100$$
$$y \geq 200$$
$$4x + 2y \leq 1,000.$$

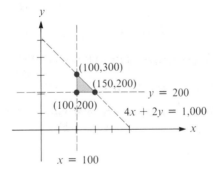

Figure 9.20

The constraints determine the set of feasible solutions shown in Figure 9.20 with vertices (100,200), (150,200), and (100,300). Since the set of feasible solutions is a closed convex polygon, we know that a maximum value exists at one of these corners. The value of the profit function at each of these vertices is as follows:

Vertices	$P = 70x + 40y$
(100,200)	$70(100) + 40(200) = 15,000$
(150,200)	$70(150) + 40(200) = 18,500$
(100,300)	$70(100) + 40(300) = 19,000$

Thus, the profit function reaches a maximum value of $19,000 when 100 type A printers and 300 type B printers are produced each week.   □

The geometrical method we have discussed is intended as an introduction to linear programming. In most cases the situation being analyzed is quite complicated and an algebraic method, called the **Simplex method,** is used to solve the problem using computers. Although we have only considered this topic in the context of maximizing profit, other applications are common. Some of these include:

1. Determining the most economical mixture of ingredients that will result in a product with certain minimum requirements.
2. Determining the quickest and most economical route in distributing a product.
3. Determining the most efficient use of industrial machinery.
4. Determining the most effective production schedule.

EXERCISES 9.7

In Exercises 1–8 graph the given inequalities. Use shading to indicate the graph.

1. $y \leq 2x$

2. $y > x - 2$

3. $y > 1 - 2x$

4. $y \geq 2$

5. $x \leq 4$

6. $x - y > 0$

7. $2x - 4y \geq 8$

8. $x + 2y \leq 5$

In Exercises 9–16 graph the solution set of the given system of inequalities. Use shading to indicate the graph and specify the coordinates of any corner in the graph.

9. $x + y \leq 7$
$x - 2y \leq -8$

10. $x + 2y \geq -12$
$2x - y \geq 1$

11. $y \leq 3 - 4x$
$x - y \geq 0$

12. $y \geq 3x + 4$
$x + y \leq -2$

13. $x + y \leq 4$
$x \geq 0$
$y \geq 0$

14. $x + y \geq 1$
$x - y \leq 1$
$x + 2y \leq 2$

15. $y \leq x + 1$
$y \geq -x + 1$
$y \geq x - 1$
$y \leq -x + 3$

16. $x + y \leq 8$
$2x + y \leq 11$
$-2x + y \leq 5$
$x \geq 0$
$y \geq 0$

17. Find the maximum and minimum values of $F = 2x - 3y$ subject to the constraints given in Exercise 13.

18. Find the maximum and minimum values of $F = 6x - 12y$ subject to the constraints given in Exercise 14.

19. Find the values of x and y that yield the maximum value of $F = \frac{1}{2}x + \frac{1}{4}y$ subject to the constraints given in Exercise 15.

20. Find the values of x and y that yield the minimum value of $F = 5x - y$ subject to the constraints given in Exercise 16.

In Exercises 21–22 find the maximum value of P subject to the given constraints.

21. $P = 60x + 40y$
$4x + 2y \leq 2{,}000$
$x \geq 100$
$y \geq 200$

22. $P = 120x + 20y$
$15x + 3y \leq 12{,}000$
$x \geq 300$
$y \geq 600$

23. Minimize $C = -2x - 3y + 200$ subject to
$3x + 4y \leq 180$
$15 \leq x \leq 40$
$10 \leq y \leq 30$.

24. Maximize $P = 15x + 8y + 9$ subject to
$x - y \leq 7$
$x - y \geq 1$
$8 \leq x \leq 13$.

25. A woman has $20,000 to invest in the stock market. Stock A sells for $90 per share, stock B costs $50 per share, and stock C costs $30 per share. If x, y, and z represent the number of shares that she bought of each stock, respectively, find an inequality that expresses this relationship.

26. A TV manufacturer has available 10,000 labor hours per month. A console requires 7 labor hours and a portable requires 5 labor hours. If x represents the number of consoles produced and y represents the number of portables produced, find an inequality that expresses this relationship.

27. Government regulations limit the amount of a pollutant that a company can discharge into the water to A gal/day. If product 1 produces B gal of pollutant per unit and product 2 produces C gal of pollutant per unit, find an inequality that expresses the restriction placed on this company because of the government regulation.

28. Sad Sam has been informed by his doctor that he would be less sad if he obtained at least the minimum adult requirements of thiamine (2 mg) and of niacin (25 mg) per day. A trip to the supermarket reveals the following facts about his favorite cereals.

Cereal	Thiamine per Ounce (mg)	Niacin per Ounce (mg)
A	1.0	10.0
B	0.7	12.0

If x represents the number of ounces he eats of cereal A and y represents the number of ounces he eats of cereal B, find two inequalities that express what must be done if he is to satisfy his minimum daily requirement.

29. A manufacturer of personal computers makes two types of disc drives, A and B. To comply with contracts, the company must produce at least 80 type A and 150 type B disc drives each week. A total of 5 labor hours is required to assemble drive A and a total of 4 labor hours is required to assemble drive B with a maximum of 2,000 labor hours available each week. If the company can sell all their disc drives for a profit of $90 on type A and $75 on type B, then how many of each should be produced weekly for the maximum profit?

30. Chris needs at least 9 units of vitamin A and 12 units of vitamin C per day. The table below summarizes information about two of her favorite foods.

Food	Vitamin A per Ounce	Vitamin C per Ounce	Cost per Ounce
A	3 units	3 units	25 cents
B	2 units	4 units	20 cents

How many ounces of each food should Chris eat to satisfy at least cost her minimum requirements for vitamins A and C?

31. Repeat Exercise 30 with food A costing 28 cents per ounce and food B costing 18 cents per ounce.

32. Use the information given in Exercise 28 and determine the combination of cereals Sam should eat to satisfy his doctor's recommendation at least cost if cereal A costs 15 cents per ounce and cereal B costs 13 cents per ounce.

CHAPTER OVERVIEW

Section	Key Concepts to Review
9.1	• Definitions of linear equations with n variables, linear system, inconsistent system, and dependent system
	• The solution set of a system of linear equations in two variables is the set of all the ordered pairs that satisfy both equations. Graphically this corresponds to the collection of points where the lines intersect.
	• Methods to solve a system of equations by the substitution method and by the addition-elimination method

Section	Key Concepts to Review
9.2	• Definitions of determinant, minor of an element, and cofactor of an element
	• To evaluate a 2 by 2 determinant, use $\begin{vmatrix} a_1 & b_1 \\ a_2 & b_2 \end{vmatrix} = a_1b_2 - b_1a_2$.
	• Cramer's rule
	• Method to evaluate 3 by 3 (and more complicated) determinants
9.3	• Definitions of triangular form, equivalent systems, matrix, entry or element of a matrix, coefficient matrix, and augmented matrix
	• The following operations are used in Gaussian elimination to solve a linear system.

Elementary Operations on Equations	Elementary Row Operations on Matrices
1. Multiply both sides of an equation by a nonzero number.	1. Multiply each entry in a row by a nonzero number.
2. Add a multiple of one equation to another.	2. Add a multiple of the entries in one row to another row.
3. Interchange two equations.	3. Interchange two rows.

Section	Key Concepts to Review
9.4	• Definitions of dimension, square matrix, zero matrix, equal matrices, scalar, $n \times n$ identity matrix, and the multiplicative inverse of A (symbolized A^{-1})
	• Definitions for scalar multiplication and addition, subtraction, and multiplication of matrices
	• Six basic properties of matrices
	• Only matrices of the same dimension can be added or subtracted.
	• The product AB of two matrices is defined only when the number of columns in A matches the number of rows in B.
	• Methods to determine if A^{-1} exists, and if it does, how to find it
	• If A^{-1} exists, the solution to the linear system $AX = B$ is given by $X = A^{-1}B$.
9.5	• Definitions of partial fraction decomposition, partial fraction, and irreducible quadratic factor
	• Methods to determine the partial fraction decomposition of $P(x)/Q(x)$, where $P(x)$ and $Q(x)$ are (nonzero) polynomials (Note that the denominator in each partial fraction in the decomposition is either a linear factor, a repeated linear factor, an irreducible quadratic factor, or a repeated irreducible quadratic factor.)
9.6	• Methods to solve nonlinear systems of equations by the substitution method and by the addition-elimination method

Section	Key Concepts to Review
9.7	• Definitions of half-plane, vertices or corners of a region, objective function, constraints, feasible solution, and convex region
	• Methods to graph inequalities in two variables
	• Methods to solve a system of inequalities
	• The solution set of a system of inequalities is the intersection of the solution sets of all the individual inequalities in the system.
	• Methods to find the maximum and/or minimum value of an objective function subject to given constraints
	• Corner point theorem

CHAPTER REVIEW EXERCISES

In Exercises 1–10 solve the given system of linear equations using the method indicated.

1. $2x + 3y = -17$ (addition-elimination)
$3x - y = 2$

2. $y = -3x + 2$ (substitution)
$y = 2x - 3$

3. $y = \frac{1}{2}x - 3$ (graphing)
$3x - 7y - 18 = 0$

4. $3x + y = 2$ (Cramer's rule)
$-3x + y = -2$

5. $x + y = 10$ (row operations on matrices)
$-2x - 5y = -11$

6. $5x + 3y = 1$ (matrix algebra)
$3x + 2y = 0$

7. $x - y + z = 5$ (substitution)
$y - z = 0$
$z = -1$

8. $x + y + z = 2$ (row operations on matrices)
$x + y - z = 4$
$2x - y - 4z = -5$

9. $x + z = 10$ (Cramer's rule)
$y + 2z = 3$
$-x + y = -2$

10. $x + 3y + 3z = 5$ (matrix algebra)
$x + 4y + 3z = 2$
$x + 3y + 4z = 8$

In Exercises 11–22 solve each system of equations.

11. $x = 3y - 5$
$y = 4x - 2$

12. $5x - 4y - 11 = 0$
$2x + 7y + 13 = 0$

13. $\dfrac{2}{x} - \dfrac{1}{y} = 1$

$\dfrac{3}{x} + \dfrac{2}{y} = 12$

14. $x^2 + y^2 = 10$
$2x^2 - y^2 = 17$

15. $x + y = 6$
$xy = 4$

16. $3^{x + y} = 27$
$2^{x - y} = 4$

17. $x + y = 6$
$y = \sqrt{x}$

18. $y = x^3 - 6x^2 + 8x$
$y = x^2 - 4x$

19. $2x - y + z = 3$
$x - 3y + 5z = 4$
$x + y - 2z = 0$

20. $x + y - z = 2$
$2x - y + 5z = 3$
$-x - y + z = -1$

21. $-x + y + z = 0$
$x - y + z = 0$
$x + y - z = 0$

22. $\dfrac{1}{x} - \dfrac{1}{y} + \dfrac{1}{z} = 2$

$\dfrac{1}{x} + \dfrac{2}{y} - \dfrac{1}{z} = 7$

$\dfrac{2}{x} - \dfrac{6}{y} + \dfrac{1}{z} = 1$

In Exercises 23–26 evaluate each determinant.

23. $\begin{vmatrix} -2 & 3 \\ 1 & 4 \end{vmatrix}$ **24.** $\begin{vmatrix} 1 & -6 & 2 \\ 2 & 2 & -3 \\ 3 & -4 & 1 \end{vmatrix}$

25. $\begin{vmatrix} 3 & -3 & 1 \\ -1 & 9 & -3 \\ 5 & -6 & 2 \end{vmatrix}$ **26.** $\begin{vmatrix} 1 & 0 & 1 & 2 \\ 0 & 0 & -1 & 0 \\ -2 & 3 & 0 & -1 \\ 1 & 2 & 3 & 0 \end{vmatrix}$

In Exercises 27 and 28 verify the given identity.

27. $\begin{vmatrix} a & b \\ c & d \end{vmatrix} = -\begin{vmatrix} c & d \\ a & b \end{vmatrix}$

28. $\begin{vmatrix} ka & b \\ kc & d \end{vmatrix} = k\begin{vmatrix} a & b \\ c & d \end{vmatrix}$

In Exercises 29–36 find (if defined) the specified matrix given

$$A = \begin{bmatrix} 1 & -2 & 3 \\ -1 & 0 & 4 \end{bmatrix}, B = \begin{bmatrix} -3 & 0 \\ 0 & -3 \end{bmatrix},$$

$$C = \begin{bmatrix} 4 & 5 \\ 3 & 4 \end{bmatrix}, \text{ and } D = \begin{bmatrix} 3 & -1 \\ 0 & 1 \\ 2 & 5 \end{bmatrix}.$$

29. $B - C$ **30.** BC **31.** CD

32. DC **33.** $3C$ **34.** C^{-1}

35. $AD + C$ **36.** $2A - 3D$

37. Graph $2x - y \le 7$. Use shading to indicate the graph.

38. Graph the solution set of the system

$$x + 4y \le 17$$
$$4x + 3y \ge 16.$$

39. Maximize $P = 50x + 30y$ subject to

$$5x + 2y \le 1,000$$
$$x \ge 50$$
$$y \ge 100.$$

40. Minimize $C = 3x - 2y + 20$ subject to

$$x + y \le 8$$
$$y - x \le 6$$
$$x - 2y \le 2$$
$$x \ge 0$$
$$y \ge 0.$$

41. Find a, b, and c so the parabola $y = ax^2 + bx + c$ passes through the points $(1,3)$, $(-1,9)$, and $(2,6)$.

42. Find the intersection point of the line $y = \frac{3}{4}x - \frac{1}{4}$ and the line whose x- and y-intercepts are $(6,0)$ and $(0,4)$.

43. Find in terms of r all intersection points of the circle $x^2 + y^2 = r^2$ and the hyperbola $x^2 - y^2 = 5$. For what values of r do the curves intersect?

44. The area and perimeter of a rectangle are 341 m² and 84 m, respectively. Find the length and width of the rectangle.

45. If the vertices of a triangle are at points (x_1,y_1), (x_2,y_2), and (x_3,y_3), then the area of the triangle is given by the absolute value of the following determinant.

$$\frac{1}{2}\begin{vmatrix} x_1 & y_1 & 1 \\ x_2 & y_2 & 1 \\ x_3 & y_3 & 1 \end{vmatrix}$$

Use this rule to find the area of the triangle whose vertices are at $(0,-3)$, $(-2,-1)$, and $(1,4)$.

46. One of the classic stories from the history of mathematics concerns Archimedes (250 B.C.), a great mathematician and scientist of ancient times. Archimedes lived in the Greek city-state of Syracuse, where the king, Hieron, suspected a goldsmith of giving him a gold crown that contained hidden silver. The king referred the problem to Archimedes and asked him to determine, without destroying the crown, the percentage of pure gold in the crown. While taking a bath, Archimedes found the principle needed to solve this problem. What Archimedes noticed in his bath was the obvious fact that when a body is immersed in water, it displaces a volume of water that is equal to the volume of the body. He also knew that bodies of the same weight do not necessarily have the same volume. Using these principles, Archimedes then filled a bucket of water to the brim. Suppose the crown weighed 10 lb and displaced 18 cubic in. of water and that 10 lb of pure gold and 10 lb of pure silver displaced 15 and 30 in.³ of water, respectively. What percentage of the crown would you tell King Hieron is made of gold?

47. Multiple choice: Which of the figures would be used to solve graphically the system

$$5x + 3y + 2 = 0$$
$$x - 2y + 3 = 0?$$

(a)   (b)

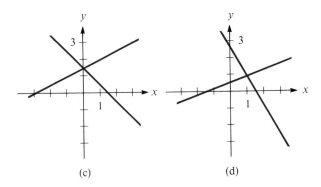

(c)   (d)

48. A company manufactures two types of fertilizer, A and B. A bag of fertilizer A is made from 4 units of nitrogen and 1 unit of potassium, while a bag of fertilizer B is made from 2 units of nitrogen and 3 units of potassium. Raw materials in stock are limited to 600 units of nitrogen and 260 units of potassium. If the company can sell all their fertilizer for profits of $5 per bag of fertilizer A and $4 per bag of fertilizer B, then how many of each type should be produced for the maximum profit?

49. True or False: $\begin{bmatrix} 1 & 2 \\ 3 & 4 \end{bmatrix} = \begin{bmatrix} 3 & 4 \\ 1 & 2 \end{bmatrix}$

50. True or False: $\begin{vmatrix} 1 & 2 \\ 3 & 4 \end{vmatrix} = -\begin{vmatrix} 3 & 4 \\ 1 & 2 \end{vmatrix}$

In Exercises 51 and 52 determine the constants A, B, C, and/or D so that the equation is an identity.

51. $\dfrac{-4x - 5}{(x + 3)^2} = \dfrac{A}{x + 3} + \dfrac{B}{(x + 3)^2}$

52. $\dfrac{-x^3 - x^2 - 8x + 2}{(x^2 + 3)(x^2 - 2)} = \dfrac{Ax + B}{x^2 + 3} + \dfrac{Cx + D}{x^2 - 2}$

In Exercises 53 and 54 find the partial fraction decomposition of the given expression.

53. $\dfrac{x^2}{x^2 - 9}$

54. $\dfrac{2x^3 + 8x}{x^4 + 4x^2 + 4}$

55. In Exercise 63 of Section 2.1 we considered a problem from the *Mathematical Puzzles of Sam Loyd*, a two-volume series edited by Martin Gardner and published by Dover Publications, © 1960. Here's another one for you to try.

How much does each jar hold?

MRS. HUBBARD has invented a clever system for keeping tabs on her jars of blackberry jam. She has arranged the jars in her cupboard so that she has twenty quarts of jam on each shelf. The jars are in three sizes. Can you tell how much each size contains?

10 Discrete Algebra and Probability

If 512 players enter a single elimination tennis tournament (one loss and you're out), determine the number of matches needed to declare a winner by considering the following two approaches:

a. The number of matches from each round form a geometric progression. Write down this sequence and determine its sum.

b. Instead of reasoning from the first round to the end of the tournament, try a favorite technique in problem solving—reason from the end result. Write an equivalent question to "How many matches were played?" that transforms this problem into one that is really simple. Now, how many matches are needed in this tournament? (See Example 3 of Section 10.2.)

U p to now we have dealt primarily with situations that are continuous in nature and that are analyzed by using the set of real numbers. However, in this chapter we discuss topics in which the positive integers play an important defining role because of the step-by-step or discrete nature of the processes in these problems. Throughout this chapter, as we discuss sequences, series, mathematical induction, the binomial theorem, and counting techniques, it is important to note that in our formulas, n is restricted to the positive integers. Finally, we conclude this chapter with an introduction to probability, since it is a natural outgrowth of our discussion of counting techniques.

10.1 SEQUENCES

A **sequence** is a set of numbers arranged in a definite order. For example, the collection of even positive integers

$$2, 4, 6, 8, \ldots$$

is a sequence. Each number in the sequence is called a **term** of the sequence. We usually write a sequence as

$$a_1, a_2, a_3, \ldots, a_n, \ldots,$$

where the subscript gives the term number and a_n represents the general or nth term. We also denote a sequence by $\{a_n\}$. In the above sequence, 2 is the first term, a_1, 4 is the second term, a_2, and so on. A sequence with a first and last term is called a **finite sequence;** a sequence with an infinite number of terms is called an **infinite sequence.**

Any term in a sequence is an assignment of a_n to n. Thus, we can think of a sequence as a function. For example, the sequence of even positive integers is given by

$$a(n) = 2n, \qquad n = 1, 2, 3, \ldots.$$

Note that a is a function name (just like f) and the domain of a is the set of positive integers. We can convert from this form and list the terms of the sequence by substituting the positive integers for n in the given rule.

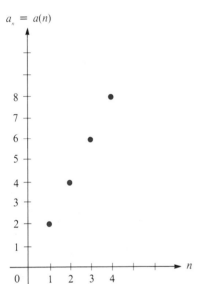

$a_n = a(n)$

Figure 10.1

	Subscript Notation		Functional Notation		Terms of Sequence or Range Elements
First Term	=	a_1	=	$a(1)$	$= 2(1) =$	2
Second Term	=	a_2	=	$a(2)$	$= 2(2) =$	4
Third Term	=	a_3	=	$a(3)$	$= 2(3) =$	6
.		.		.	.	.
.		.		.	.	.
.		.		.	.	.
General or nth Term	=	a_n	=	$a(n)$	$= 2(n) =$	$2n$
.		.		.	.	.
.		.		.	.	.
.		.		.	.	.

The graph of this function is given in Figure 10.1 and consists of the points

$$(1,2), (2,4), (3,6), \ldots.$$

Note that we do not connect the points and that we are limited to plotting a finite number of points that suggest the behavior of the function.

The functional interpretation of a sequence leads to the following definitions.

Definition of a Sequence   A **sequence** is a function whose domain is the set of positive integers 1, 2, 3, ... The functional values or range elements are called the **terms** of the sequence.

The word *sequence* normally means infinite sequence and this definition is consistent with that understanding. In the examples and exercises that follow we stick with the custom of using subscript notation, instead of functional notation, when giving a sequence.

EXAMPLE 1 Write the first four terms of the sequence given by $a_n = 3n - 1$; also find a_{25}.

Solution Substituting $n = 1, 2, 3, 4$ in the rule for a_n gives

$$a_1 = 3(1) - 1 = 2, \quad a_2 = 3(2) - 1 = 5,$$
$$a_3 = 3(3) - 1 = 8, \quad a_4 = 3(4) - 1 = 11.$$

For the 25th term, a_{25}, we have

$$a_{25} = 3(25) - 1 = 74. \qquad \square$$

In this book we discuss in detail two specific kinds of sequences: arithmetic progressions and geometric progressions. A famous prediction involving these sequences appeared in an essay on population written by the English economist Thomas Malthus in 1798. In this essay he was the first to express concern about the rapid growth of the earth's population and claimed that population increases in a geometric sequence while subsistence (food) increases in an arithmetic sequence. As we investigate these sequences, consider the consequences of this theory for the future. We begin with the definition of an arithmetic sequence.

Definition of an Arithmetic Progression

An **arithmetic progression** is a sequence of numbers in which each term after the first is found by adding a constant to the preceding term. This constant is called the **common difference** and is symbolized by d.

For example, the following sequences are arithmetic progressions.

$7, 10, 13, 16, \ldots$ common difference: 3

$5, 1, -3, -7, \ldots$ common difference: -4

$1, 3/2, 2, 5/2, \ldots$ common difference: 1/2

We find the rule for the general term in an arithmetic progression with first term a_1 and common difference d by observing the following pattern in such a sequence.

1st term, 2nd term, 3rd term, . . . , *n*th term

$$a_1, \quad a_1 + 1 \cdot d, \quad a_1 + 2 \cdot d, \ldots, a_1 + (n - 1)d$$

Thus, the formula for the nth term is

$$a_n = a_1 + (n - 1)d.$$

EXAMPLE 2   Find the formula for the general term a_n in the arithmetic progression

$$3, 7, 11, \ldots .$$

What is the 17th term in the sequence?

Solution   The first term is $a_1 = 3$ and the common difference is $d = 7 - 3 = 4$. Substituting these numbers in the above formula gives

$$a_n = 3 + (n - 1)4$$
$$= 4n - 1.$$

To find the 17th term, a_{17}, replace n by 17. Thus,

$$a_{17} = 4(17) - 1 = 67. \qquad \square$$

Next we consider geometric progressions.

Definition of a Geometric Progression	A **geometric progression** is a sequence of numbers in which each number after the first is found by multiplying the preceding term by a constant. This constant is called the **common ratio** and is symbolized by r.

For example, the following sequences are geometric progressions.

$$2, 4, 8, 16, \ldots \qquad \text{constant ratio: } \frac{4}{2} = \frac{8}{4} = \frac{16}{8} = 2$$

$$\frac{1}{2}, -\frac{1}{4}, \frac{1}{8}, -\frac{1}{16}, \ldots \qquad \text{constant ratio: } -\frac{1}{2}$$

$$1, 1.06, (1.06)^2, (1.06)^3, \ldots \qquad \text{constant ratio: } 1.06$$

We find the rule for the general term in a geometric progression with first term a_1 and common ratio r by observing the following pattern in such a sequence.

1st term, 2nd term, 3rd term, . . . , nth term

$$a_1, \qquad a_1 r^1, \qquad a_1 r^2, \ldots, a_1 r^{n-1}$$

Thus, the formula for the nth term is

$$a_n = a_1 r^{n-1}.$$

EXAMPLE 3   Find the formula for the general term a_n in the geometric progression

$$24, -12, 6, \ldots .$$

What is the 7th term in the sequence?

Solution   The first term is $a_1 = 24$ and the common ratio is $r = -\frac{12}{24} = -\frac{1}{2}$. Substituting these values in the above formula gives

$$a_n = 24(-\tfrac{1}{2})^{n-1}.$$

To find the 7th term a_7, replace n by 7. Thus,

$$
\begin{aligned}
a_7 &= 24(-\tfrac{1}{2})^{7-1} \\
&= 24(-\tfrac{1}{2})^6 \\
&= 24(\tfrac{1}{64}) = \tfrac{3}{8}.
\end{aligned}
$$

$\square$

Arithmetic and geometric progressions by no means exhaust the list of important sequences. As an example consider the set of numbers

$$1, 1, 2, 3, 5, 8, 13, 21, 34, 55, 89, \ldots ,$$

which is called the **Fibonacci sequence.** The rule that generates these numbers is as follows: Start with 1 and 1 and thereafter each term is the sum of the two previous terms. In symbols we write

$$a_1 = 1, \quad a_2 = 1, \quad a_n = a_{n-1} + a_{n-2} \qquad \text{for } n \geq 3.$$

Leonardo Fibonacci, who was a mathematician in the Middle Ages, introduced this sequence innocently enough in his book on arithmetic and algebra as the answer to the following problem (paraphrased): A pair of rabbits produces a pair of baby rabbits by the end of their second month and every month thereafter. Each new pair of rabbits does the same. If none of the rabbits die, how many pairs of rabbits are there at the beginning of each month? The answer for the first seven months (and take the time to analyze why) is

Beginning of month number	1	2	3	4	5	6	7
Numbers of pairs	1	1	2	3	5	8	13

What is special about this sequence? Why is a mathematics journal, *The Fibonacci Quarterly,* devoted exclusively to issues connected with these numbers? Well, the numbers in the Fibonacci sequence keep appearing in the analysis of topics in such diverse areas as mathematics (naturally), physics, chemistry, art, and music, as well as appearing with surprising regularity in nature. Most of these applications are a bit involved, but it is not difficult to consider some of the fascinating ways in which Fibonacci numbers occur in nature.

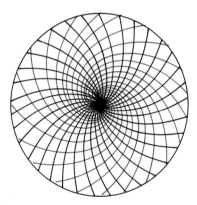

Figure 10.2

On pine cones, pineapples, sunflowers, and other growths that appear on the tip of a stem there are two distinct sets of spirals. As shown in the diagram of a small sunflower in Figure 10.2, one set radiates clockwise; the other, counterclockwise. The number of spirals in each set is a Fibonacci number. In a small sunflower (as in this figure) there are 21 clockwise and 34 counterclockwise spirals. Some common counts for such growths are

pine cone	5, 8	small sunflower	21, 34
pineapple	8, 13	large sunflower	34, 55
large pineapple	13, 21	giant sunflower	55, 89.

Note that each pair of numbers gives consecutive terms in the Fibonacci sequence.

These numbers also occur when considering the spiral leaf growth pattern of many trees. For example, consider the cyclic arrangements in the positioning of the leaf buds on the sample from an oak tree in Figure 10.3. Note that there are 5 buds to a complete cycle. In addition, each cycle contains 2 revolutions of the spiral. Both numbers are terms of the Fibonacci sequence. Similar remarks are true of other trees (such as elm and birch trees) and certain types of bushes. These examples just begin to scratch the surface of the immense volume of material dealing with the Fibonacci sequence.

     1   2   3   4   5   6

Figure 10.3

EXERCISES 10.1

In Exercises 1–10 write the first four terms in the sequence with general terms as given; also find the indicated term.

1. $a_n = 2n - 1$; a_{50} **2.** $a_n = 7n + 2$; a_{25}

3. $a_n = n^2$; a_7 **4.** $a_n = 1/n$; a_{33}

5. $a_n = \dfrac{(-1)^{n+1}}{n}$; a_{10} **6.** $a_n = \dfrac{(-1)^n}{n}$; a_{10}

7. $a_n = \dfrac{1}{2^n}$; a_5 **8.** $a_n = \dfrac{3^{n-1}}{4^{n-1}}$; a_5

9. $a_n = \dfrac{n(n+1)}{2}$; a_8

10. $a_n = \frac{1}{6}n(n+1)(2n+1)$; a_6

In Exercises 11–20 state whether the sequence is an arithmetic progression, geometric progression, or neither. For any arithmetic progression state the common difference and write the next two terms. For any geometric progression state the common ratio and write the next two terms.

11. $2, 7, 12, 17, 22, \ldots$

12. $-2, 4, -8, 16, -32, \ldots$

13. $1, \frac{1}{2}, \frac{1}{3}, \frac{1}{4}, \frac{1}{5}, \ldots$

14. $1, -4, -9, -14, -19, \ldots$

15. $1, 1.02, (1.02)^2, (1.02)^3, (1.02)^4, \ldots$

16. $\frac{1}{2}, \frac{2}{3}, \frac{3}{4}, \frac{4}{5}, \frac{5}{6}, \ldots$

17. $3, 0.3, 0.03, 0.003, 0.0003, \ldots$

18. $1.01, 1.02, 1.03, 1.04, 1.05, \ldots$

19. $1, \dfrac{1}{2^2}, \dfrac{1}{3^2}, \dfrac{1}{4^2}, \dfrac{1}{5^2}, \ldots$

20. $1, \dfrac{1}{2}, \dfrac{1}{2^2}, \dfrac{1}{2^3}, \dfrac{1}{2^4}, \ldots$

In Exercises 21–30 the sequences are arithmetic progressions. Find the formula for the general term a_n and the value of the indicated term in the sequence.

21. $3, 7, 11, \ldots$; 40th term

22. $2, 9, 16, \ldots$; 50th term

23. $9, 4, -1, \ldots$; 5th term

24. $2, -2, -6, \ldots$; 10th term

25. $1.1, 1.4, 1.7, \ldots$; a_{25}

26. $3.15, 3.10, 3.05, \ldots$; a_{18}

27. $\frac{1}{3}, \frac{5}{3}, 3, \ldots$; a_{30} **28.** $\frac{10}{7}, \frac{1}{7}, -\frac{8}{7}, \ldots$; a_9

29. $4, \frac{35}{8}, \frac{19}{4}, \ldots$; a_8 **30.** $\frac{13}{4}, \frac{9}{2}, \frac{23}{4}, \ldots$; a_{20}

In Exercises 31–40 the sequences are geometric progressions. Find the formula for the general term a_n and the value of the indicated term in the sequence.

31. $2, 6, 18, \ldots$; 7th term

32. $2, -6, 18, \ldots$; 7th term

33. $1, -\frac{1}{2}, \frac{1}{4}, \ldots$; 6th term

34. $1, \frac{1}{2}, \frac{1}{4}, \ldots$; 6th term

35. $\sqrt{2}, 2, 2\sqrt{2}, \ldots$; a_8

36. $1, 1.04, (1.04)^2, \ldots$; a_{10}

37. $6, 4, \frac{8}{3}, \ldots$; a_7

38. $1, 0.1, 0.01, \ldots$; a_{10}

39. $1, -\frac{1}{3}, \frac{1}{9}, \ldots$; a_6

40. $1, \frac{3}{4}, \frac{9}{16}, \ldots$; a_5

41. Write the first four terms in the arithmetic progression in which $a_1 = 4$ and $d = \frac{1}{2}$.

42. Write the first four terms in the geometric progression in which $a_1 = 4$ and $r = \frac{1}{2}$.

43. Which term of the geometric progression $18, 9, \ldots$ is $\frac{9}{16}$?

44. Which term of the arithmetic progression $7, 4, \ldots$ is -68?

45. Find the first term in an arithmetic progression in which $a_{17} = 39$ and $d = 2$.

46. The first term of a geometric progression is 8 and the fourth term is 27. Find the common ratio.

47. Find two different values of x so that $-\frac{2}{3}, x, -\frac{25}{54}$ are in geometric progression.

48. If, in an arithmetic progression, $a_1 = 80$, $d = -3$, and $a_n = 26$, find n.

49. If, in an arithmetic progression, $a_6 = 15$ and $a_{12} = 24$, find a_1 and d.

50. If, in an arithmetic progression, $a_{15} = \frac{22}{3}$ and $a_{30} = \frac{91}{6}$, find a_{40}.

51. Write the first fifteen terms of the Fibonacci sequence.

52. The Fibonacci sequence appears in the family tree of a male bee. A male bee develops from an unfertilized egg and has only one parent—his mother. A female bee has both a mother and a father. The number of ancestors of a male bee in each successive generation is therefore the numbers of the Fibonacci sequence as shown below.

Number of Ancestors

1

2

3

Continue the diagram for the next three generations and verify that the number of ancestors in these generations is given by the next three terms in the Fibonacci sequence.

10.2 SERIES

Associated with any sequence

$$a_1, a_2, \ldots, a_n, \ldots$$

is a **series**

$$a_1 + a_2 + \cdots + a_n + \cdots$$

which is the sum of all the terms in the sequence. The series associated with arithmetic and geometric progressions are called, respectively, an **arithmetic series** and a **geometric series.** In this section we consider only series with a *finite* number of terms. The meaning of a series with an infinite number of terms is much more complex. Infinite series are discussed briefly in the next section and in detail in calculus.

We begin with a classic story from the history of mathematics that suggests the procedure for finding the sum of an arithmetic series. Carl Friedrich Gauss dominated mathematics in the nineteenth century and is considered, along with Archimedes and Newton, in a special class among the great mathematicians. He contributed profoundly to astronomy, physics (especially electromagnetic theory), non-Euclidean geometry, number theory, probability and statistics, and function theory. Reportedly, he demonstrated his imaginative insight to problem solving at the early age of ten when his class at school was assigned some "busy work." Their task—add all the numbers from 1 to 100. To the teacher's amazement Gauss quickly scribbled the correct result 5,050. How did he do it? Well, as is often the case when an ingenious method is used, the answer is easily obtained. Gauss noticed that the 100 numbers could be arranged in 50 pairs that all add up to 101 as follows:

$$
\begin{array}{ccccccccc}
1 + & 2 + & 3 + & 4 + & \cdots + & 48 + & 49 + & 50 + \\
100 + & 99 + & 98 + & 97 + & \cdots + & 53 + & 52 + & 51 \\
\hline
101 + & 101 + & 101 + & 101 + & \cdots + & 101 + & 101 + & 101.
\end{array}
$$

Thus, the result is simply 50(101) or 5,050.

By a similar method of adding pairs of terms with the same sum, we can develop the formula for the sum, denoted S_n, of the first n terms of an arithmetic progression. In general form an arithmetic series can be written as

$$S_n = a_1 + (a_1 + d) + (a_1 + 2d) + \cdots + (a_n - 2d) + (a_n - d) + a_n.$$

By reversing the order of the terms, we can also write this series as

$$S_n = a_n + (a_n - d) + (a_n - 2d) + \cdots + (a_1 + 2d) + (a_1 + d) + a_1.$$

If we now add term by term the equivalent expressions for S_n, we have n pairs, which all add up to $a_1 + a_n$. Thus,

$$2S_n = n(a_1 + a_n)$$

so that the general formula is

$$S_n = \frac{n}{2}(a_1 + a_n).$$

This formula is used when we know a_1 and a_n. However, a_n is often not given. If we replace a_n by $a_1 + (n - 1)d$, the above formula becomes

$$S_n = \frac{n}{2}\{a_1 + [a_1 + (n - 1)d]\}$$

$$= \frac{n}{2}[2a_1 + (n - 1)d].$$

We summarize our results as follows.

Arithmetic Series Formulas   The sum of the first n terms of an arithmetic series is given by

$$S_n = \frac{n}{2}(a_1 + a_n)$$

or

$$S_n = \frac{n}{2}[2a_1 + (n-1)d].$$

EXAMPLE 1   Find the sum of the first 21 terms of the arithmetic progression 4, 7, 10, ...

Solution   Here $a_1 = 4$, $d = 3$, and $n = 21$. Thus,

$$S_n = \frac{n}{2}[2a_1 + (n-1)d]$$

$$= \frac{21}{2}[2(4) + (21-1)3]$$

$$= \frac{21}{2}(68) = 714. \qquad \square$$

We now develop a formula for the sum of a given number of terms in a geometric progression. In general form a geometric series with n terms can be written as

$$S_n = a_1 + a_1 r + a_1 r^2 + \cdots + a_1 r^{n-1}.$$

Multiplying both sides of this equation by r gives

$$r S_n = a_1 r + a_1 r^2 + \cdots + a_1 r^{n-1} + a_1 r^n.$$

Subtracting the second equation from the first, we obtain

$$S_n - r S_n = a_1 - a_1 r^n$$
$$S_n(1 - r) = a_1 - a_1 r^n$$
$$S_n = \frac{a_1 - a_1 r^n}{1 - r}.$$

This formula is used when we know a_1, r, and n. If we know the nth or last term in the series, a different formula is used. Since $a_n = a_1 r^{n-1}$, we have $a_1 = a_n/r^{n-1}$. This replacement in the above formula gives

$$S_n = \frac{a_1 - (a_n/r^{n-1})r^n}{1 - r} = \frac{a_1 - a_n r}{1 - r}.$$

To summarize, we have derived the following formulas.

Geometric Series Formulas

The sum of the first n terms of a geometric series is given by

$$S_n = \frac{a_1 - a_1 r^n}{1 - r}$$

or

$$S_n = \frac{a_1 - a_n r}{1 - r}.$$

EXAMPLE 2   Find the sum of the first seven terms of the geometric progression $1, \frac{1}{2}, \frac{1}{4}, \ldots$.

Solution   Here $a_1 = 1$, $r = \frac{1}{2}$, and $n = 7$. Thus,

$$S_n = \frac{a_1 - a_1 r^n}{1 - r}$$

$$= \frac{1 - 1(\frac{1}{2})^7}{1 - \frac{1}{2}}$$

$$= \frac{1 - (1/128)}{1/2} = \frac{127/128}{1/2} = \frac{127}{64}.$$ □

EXAMPLE 3   Solve the problem in the chapter introduction on page 486.

Solution

a. In the first round the 512 players are paired up two at a time, so 256 matches are needed. The 256 winners then advance to the second round which will require 128 matches. Each subsequent round continues to require half the number of matches as the preceding round, so the number of matches from each round forms the following geometric progression.

256, 128, 64, 32, 16, 8, 4, 2, 1

It is easy enough to just add up these numbers, but we can also find the sum by using the second of our formulas for the sum of a geometric series.

$$S_n = \frac{a_1 - a_n r}{1 - r} = \frac{256 - 1(\frac{1}{2})}{1 - \frac{1}{2}} = \frac{512 - 1}{2 - 1} = 511$$

Thus, 511 matches are needed in the tournament.

b. When the tournament is *over* there are 511 losers and 1 winner. Each loser lost exactly one match so the question "How many matches are played?" can be found by answering "How many losers are there?" With this approach it is easy to see that 511 matches are needed. □

A series is a sum, so it is convenient to use the Greek letter Σ, read "sigma," to mean ADD. By this convention, called **sigma notation,** we write

$$S_n = a_1 + a_2 + \cdots + a_n$$

in compact form as

$$S_n = \sum_{i=1}^{n} a_i$$

so that $\displaystyle\sum_{i=1}^{n} a_i$ means to add the terms that result from replacing i by 1, then 2, . . . , then n. The use of the letter i in this notation is arbitrary so that

$$\sum_{i=1}^{n} a_i, \qquad \sum_{j=1}^{n} a_j, \qquad \sum_{k=1}^{n} a_k$$

all represent the same series.

EXAMPLE 4   Write the series $\displaystyle\sum_{i=1}^{4} (2i + 3)$ in expanded form and determine the sum.

Solution

$$\sum_{i=1}^{4}(2i + 3) = (2 \cdot 1 + 3) + (2 \cdot 2 + 3) + (2 \cdot 3 + 3) + (2 \cdot 4 + 3)$$
$$= \quad 5 \quad + \quad 7 \quad + \quad 9 \quad + \quad 11$$
$$= 32 \qquad\qquad\qquad\qquad\qquad\qquad\qquad □$$

EXAMPLE 5   Write the series $\sum_{j=2}^{7} 3^j$ in expanded form and determine the sum.

Solution

$$\sum_{j=2}^{7} 3^j = 3^2 + 3^3 + 3^4 + 3^5 + 3^6 + 3^7$$

This series is a geometric series with $a_1 = 3^2$, $a_n = 3^7$, and $r = 3$. Thus,

$$S_n = \frac{a_1 - a_n r}{1 - r}$$

$$= \frac{3^2 - (3^7)3}{1 - 3}$$

$$= \frac{9 - 6{,}561}{-2} = 3{,}276.$$

Therefore $\sum_{j=2}^{7} 3^j = 3{,}276.$ □

EXAMPLE 6   Find an expression for the general term and write the series $2 + 5 + 8 + 11 + 14 + 17$ in sigma notation.

Solution   In most cases it is hoped that you can guess the general term by inspection. For this series it is

$$3i - 1$$

and since there are six terms, we have

$$\sum_{i=1}^{6} (3i - 1) = 2 + 5 + 8 + 11 + 14 + 17.$$

If you could not guess the general term in this case, you could note the series is an arithmetic series with $a_1 = 2$ and $d = 3$ so that

$$a_n = a_1 + (n - 1)d$$
$$= 2 + (n - 1)3$$
$$= 3n - 1.$$ □

We now consider one of the important applications of geometric series. Sooner than you might imagine one (or more) of your former classmates will probably call you up and ask you to invest in your future. The plan—purchase an annuity. An **annuity** is any series of equal payments made at equal time intervals. There are different types of annuities. Here we discuss an **ordinary annuity,** in which payments are made at the end of each time period that coincides with a point at which interest is converted. Here's how it goes.

Suppose that at the end of each year for 5 years you invest $100, as shown below, in an account that pays 6 percent interest compounded annually.

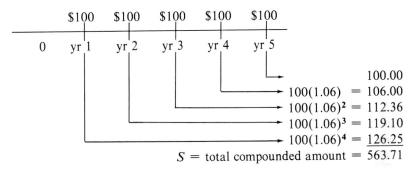

Each deposit earns interest for a different period of time. Thus, we compute separately the compounded amount of each $100 investment by the methods from Section 5.1 and add the results to obtain the total compounded amount. This sum is called the amount of the annuity and denoted S. The amount of this annuity (or the value of this series of equal periodic payments) is $563.71.

Treating each $100 investment separately is cumbersome. It is more efficient to note that

$$S = 100 + 100(1.06) + 100(1.06)^2 + 100(1.06)^3 + 100(1.06)^4$$

is a geometric series with $a_1 = 100$, $r = 1.06$, and $n = 5$. The formula for the sum of a geometric series

$$S = \frac{a_1 - a_1 r^n}{1 - r}$$

is usually written for these problems as

$$S = \frac{a_1(r^n - 1)}{r - 1}$$

so that for this series we have

$$S = \frac{100[(1.06)^5 - 1]}{1.06 - 1}$$

$$= \frac{100[1.338226 - 1]}{0.06}$$

$$= \frac{33.8226}{0.06} = \$563.71.$$

This answer coincides with our previous result and is easier to obtain.

Finally, let us generalize our discussion and derive the formula for the amount of an annuity—the most efficient way to determine S. Refer to the specific example just given to clarify any steps you do not understand. If we let

$p =$ periodic payment of the annuity
$i =$ interest rate per period
$n =$ number of payment periods
$S =$ amount of the annuity

then

$$S = p + p(1 + i) + p(1 + i)^2 + \cdots + p(1 + i)^{n-1}.$$

This series is a geometric series with n terms in which $r = 1 + i$ and $a_1 = p$. Substituting these values in the formula

$$S = \frac{a_1(r^n - 1)}{r - 1}$$

gives

$$S = \frac{p[(1 + i)^n - 1]}{(1 + i) - 1} = \frac{p[(1 + i)^n - 1]}{i}.$$

This formula is usually written as

$$S = p\left[\frac{(1 + i)^n - 1}{i}\right].$$

To simplify the evaluation of this expression, tables (which may be found in most business math books) give the values of

$$\frac{(1 + i)^n - 1}{i}$$

for common values of i and n. Then we need only multiply the appropriate table value by p (the periodic payment) to determine S (the amount of the annuity).

EXERCISES 10.2

In Exercises 1–10 find the sum of the indicated number of terms of the following arithmetic and geometric progressions.

1. 3, 7, 11, . . . ; 10 terms
2. 9, 4, −1, . . . ; 12 terms
3. 2, 6, 18, . . . ; 7 terms
4. 2, −6, 18, . . . ; 7 terms
5. 1, 1.04, $(1.04)^2$, . . . ; 5 terms
6. 1, 0.1, 0.01, . . . ; 9 terms
7. $1, \frac{3}{4}, \frac{9}{16}, \ldots$; 6 terms
8. 1.1, 1.4, 1.7, . . . ; 8 terms
9. 3.15, 3.10, 3.05, . . . ; 11 terms
10. $\frac{13}{4}, \frac{9}{2}, \frac{23}{4}, \ldots$; 10 terms

In Exercises 11–20 write the series in expanded form and determine the sum.

11. $\displaystyle\sum_{i=1}^{10} i$

12. $\displaystyle\sum_{i=1}^{4} i^2$

13. $\displaystyle\sum_{j=1}^{6} (-1)^j$

14. $\displaystyle\sum_{j=1}^{3} \frac{(-1)^{j+1}}{j}$

15. $\displaystyle\sum_{k=1}^{12} (2k-1)$

16. $\displaystyle\sum_{k=1}^{20} (6k+1)$

17. $\displaystyle\sum_{i=2}^{5} (\tfrac{1}{2}i + 3)$

18. $\displaystyle\sum_{j=2}^{8} 2^j$

19. $\displaystyle\sum_{k=1}^{5} (\tfrac{1}{3})^{k-1}$

20. $\displaystyle\sum_{i=1}^{4} (2 + 3^i)$

In Exercises 21–30 write the series in sigma notation. (*Note:* A given series can be expressed in sigma notation in more than one way.)

21. $3 + 4 + 5 + \cdots + 10$
22. $1 + 4 + 9 + 16 + 25$
23. $\frac{1}{2} + \frac{1}{4} + \frac{1}{8} + \frac{1}{16}$
24. $\frac{1}{2} - \frac{1}{4} + \frac{1}{8} - \frac{1}{16}$
25. $\frac{1}{2} + \frac{2}{3} + \frac{3}{4} + \frac{4}{5} + \frac{5}{6}$
26. $\frac{2}{3} + \frac{3}{5} + \frac{4}{7} + \frac{5}{9} + \frac{6}{11}$
27. $x + x^2 + x^3 + x^4 + x^5 + x^6$

28. $x + x^3 + x^5 + x^7$
29. $1 + 2^3 + 3^3 + \cdots + n^3$
30. $1 + 2^3 + 3^3 + \cdots + (n-1)^3$
31. What is the sum of the first n even positive integers?
32. A grandfather clock chimes as many times as the hour. How many times does it chime in striking the hours 1 through 12?
33. A free-falling body that starts from rest drops about 16 ft the first second, 48 ft the second second, 80 ft the third second, and so on. How many feet does the object fall in 10 seconds?
34. You accept a position at a salary of $10,000 for the first year with an increase of $1,000 per year each year thereafter. How many years will you have to work for your total earnings to equal $231,000?
35. What is the total number of your ancestors in the seven generations that immediately precede you? (*Note:* One generation back is your parents, two generations back is your grandparents, and so on.)
36. A certain ball always rebounds $\frac{1}{2}$ as far as it falls. If the ball is thrown 10 ft into the air, how far up and down has it traveled when it hits the ground for the sixth time?
37. A certain ball always rebounds $\frac{2}{3}$ as far as it falls. If the ball is dropped from a height of 9 ft, how far up and down has it traveled when it hits the ground for the seventh time?
38. At the end of each year a woman deposits $1,000 in an account that pays 6 percent interest compounded annually. How much is in the account after 8 deposits have been made?
39. The principle of an annuity might help one to stop smoking. Suppose one saves the price of a package of cigarettes (let's say $1.50) each day of the year (use 365 days). At the end of the year the money is deposited in an account paying 9 percent interest compounded annually. How much is in the account after 10 deposits have been made?
40. Let $a_1, a_2, \ldots$ be the Fibonacci sequence. For successive values of n, find $a_1 + a_2 + \cdots + a_n$. Compare the sums with the terms of the sequence and determine the formula for the sum of the first n terms in the Fibonacci sequence.

10.3 INFINITE GEOMETRIC SERIES

Consider the series

$$\tfrac{1}{2} + \tfrac{1}{4} + \tfrac{1}{8} + \tfrac{1}{16} + \cdots + (\tfrac{1}{2})^n + \cdots .$$

This series contains infinitely many terms and raises a fundamental question: Can we ever add infinitely many numbers? In the usual sense the operation of addition has meaning only with respect to a finite number of terms so, in the usual sense, the answer to the question quite simply is no. However, instead of abandoning this train of thought let us look at the behavior of S_n, the sum of the first n terms in the series, as n increases from 1 to 6.

$$S_1 = \tfrac{1}{2}$$
$$S_2 = \tfrac{1}{2} + \tfrac{1}{4} = \tfrac{3}{4}$$
$$S_3 = \tfrac{1}{2} + \tfrac{1}{4} + \tfrac{1}{8} = \tfrac{7}{8}$$
$$S_4 = \tfrac{1}{2} + \tfrac{1}{4} + \tfrac{1}{8} + \tfrac{1}{16} = \tfrac{15}{16}$$
$$S_5 = \tfrac{1}{2} + \tfrac{1}{4} + \tfrac{1}{8} + \tfrac{1}{16} + \tfrac{1}{32} = \tfrac{31}{32}$$
$$S_6 = \tfrac{1}{2} + \tfrac{1}{4} + \tfrac{1}{8} + \tfrac{1}{16} + \tfrac{1}{32} + \tfrac{1}{64} = \tfrac{63}{64}$$

It appears that as n gets larger, S_n approaches, but never equals, 1. In fact, we can get S_n as close to 1 as we wish merely by taking a sufficiently large value for n. It is in this sense that we are going to assign the number 1 as the "sum" of the series. Note we are bending the usual meaning of sum to suit our purposes. To say that 1 is the sum of the above infinite series does not mean we add up all the numbers and get 1; it means that as n gets larger and larger, S_n converges or closes in on 1.

Even with our new meaning for sum, many infinite series are still meaningless. Infinite arithmetic series and many infinite geometric series do not close in on some number. Intuitively, you can probably see that if the series is to converge to some number, then the terms in the series as n gets larger and larger must approach 0. This condition occurs in infinite geometric series with $|r| < 1$ (or, equivalently, with $-1 < r < 1$), so we investigate this type of series further.

The formula for the sum of the first n terms of a geometric series is

$$S_n = \frac{a_1 - a_1 r^n}{1 - r} .$$

Now if $|r| < 1$, as n gets larger and larger $a_1 r^n$ approaches 0, so that S_n closes in on $a_1/(1 - r)$. Thus, we have the following important result.

Sum of an Infinite Geometric Series

An infinite geometric series with $|r| < 1$ converges to the value or sum

$$S = \frac{a_1}{1 - r}.$$

EXAMPLE 1   Find the sum of the infinite geometric series

$$1 + \tfrac{1}{3} + \tfrac{1}{9} + \tfrac{1}{27} + \cdots.$$

Solution   Here $a_1 = 1$ and $r = \tfrac{1}{3}$. Since the common ratio is between -1 and 1 we can assign a sum to the series. Substituting in the formula

$$S = \frac{a_1}{1 - r},$$

we have

$$S = \frac{1}{1 - \tfrac{1}{3}} = \frac{1}{\tfrac{2}{3}} = \frac{3}{2}. \qquad \square$$

In Section 1.1 we saw that every repeating decimal is a rational number and can be expressed as the quotient of two integers. One method for finding the appropriate fraction is to use an infinite geometric series. For example, the repeating decimal $0.\overline{3}$ (or equivalently $0.333\ldots$) can be written as

$$0.3 + 0.03 + 0.003 + \cdots.$$

This series is an infinite geometric series with $r = (0.03/0.3) = 0.1$ and $a_1 = 0.3$, so

$$S = \frac{a_1}{1 - r} = \frac{0.3}{1 - (0.1)} = \frac{0.3}{0.9} = \frac{1}{3}.$$

Thus, $0.\overline{3}$ is equivalent to $\tfrac{1}{3}$.

EXAMPLE 2   Express the repeating decimal $7.\overline{54}$ as the ratio of two integers.

Solution   The repeating decimal $7.\overline{54}$ can be written as

$$7 + 0.54 + 0.0054 + 0.000054 + \cdots.$$

The series

$$0.54 + 0.0054 + 0.000054 + \cdots$$

is an infinite geometric series with $a_1 = 0.54$ and $r = 0.01$, so

$$S = \frac{a_1}{1 - r} = \frac{0.54}{1 - (0.01)} = \frac{0.54}{0.99} = \frac{54}{99} = \frac{6}{11}.$$

Adding the first term in the original series, 7, to $\frac{6}{11}$ gives

$$7 + \tfrac{6}{11} = \tfrac{77}{11} + \tfrac{6}{11} = \tfrac{83}{11}.$$

Thus, $7.\overline{54}$ is equivalent to $\frac{83}{11}$.   □

EXERCISES 10.3

In Exercises 1–10 find the sum of the infinite geometric series.

1. $1 + \frac{1}{2} + \frac{1}{4} + \frac{1}{8} + \cdots$

2. $1 + \frac{1}{5} + \frac{1}{25} + \frac{1}{125} + \cdots$

3. $\frac{1}{3} - \frac{1}{9} + \frac{1}{27} - \frac{1}{81} + \cdots$

4. $2 - 1 + \frac{1}{2} - \frac{1}{4} + \cdots$

5. $3 + \frac{3}{10} + \frac{3}{100} + \frac{3}{1,000} + \cdots$

6. $5 + \frac{5}{3} + \frac{5}{9} + \frac{5}{27} + \cdots$

7. $\frac{3}{2} - 1 + \frac{2}{3} - \frac{4}{9} + \cdots$

8. $\frac{1}{2} - \frac{1}{3} + \frac{2}{9} - \frac{4}{27} + \cdots$

9. $5 + 0.5 + 0.05 + 0.005 + \cdots$

10. $11 + 1.1 + 0.11 + \cdots$

In Exercises 11–20 express each repeating decimal as the ratio of two integers.

11. $0.\overline{2}$ **12.** $0.\overline{9}$ **13.** $0.\overline{07}$

14. $0.0\overline{7}$ **15.** $0.3\overline{21}$ **16.** $0.6\overline{332}$

17. $5.\overline{9}$ **18.** $4.8\overline{1}$ **19.** $2.14\overline{3}$

20. $2.1\overline{43}$

21. A certain ball always rebounds $\frac{2}{3}$ as far as it falls. If the ball is dropped from a height of 9 ft, how far up and down has it traveled before it comes to rest?

22. For what values of x does the series

$$1 + x + x^2 + x^3 + \cdots$$

converge to a sum? What is this sum?

23. For what values of x does the series

$$(x + 1) + 2(x + 1)^2 + 4(x + 1)^3 + \cdots$$

converge to a sum? What is this sum?

24. Show that

$$1 + \sin^2 x + \sin^4 x + \cdots = \sec^2 x.$$

For what values of x is the statement valid?

25. Consider the series

$$1 - 1 + 1 - 1 + 1 - 1 + \cdots.$$

What is the sum of this series? The problem can be attacked from three viewpoints.

Method 1   The sum is 0 since

$$(1 - 1) + (1 - 1) + (1 - 1) + \cdots$$
$$= 0 + 0 + 0 + \cdots = 0.$$

Method 2   The sum is 1 since

$$1 + (-1 + 1) + (-1 + 1) + \cdots$$
$$= 1 + 0 + 0 + \cdots = 1.$$

Method 3   The sum is $\frac{1}{2}$ since the series is a geometric series with $r = -1$, so

$$S = \frac{a_1}{1 - r} = \frac{1}{1 - (-1)} = \frac{1}{2}.$$

Therefore $0 = 1 = \frac{1}{2}$. What is wrong?

10.4 MATHEMATICAL INDUCTION

Certain statements are true for every positive integer, and these results are quite useful in mathematics. We usually prove these statements about positive integers by a method called *mathematical induction*. The word *induction* is used because in many cases we can guess the general formula by looking at some specific cases. For example, consider the following sums of odd integers.

$$1 \qquad \text{or} \quad 1^2$$
$$1 + 3 = 4 \quad \text{or} \quad 2^2$$
$$1 + 3 + 5 = 9 \quad \text{or} \quad 3^2$$
$$1 + 3 + 5 + 7 = 16 \quad \text{or} \quad 4^2$$
$$1 + 3 + 5 + 7 + 9 = 25 \quad \text{or} \quad 5^2$$

From these additions we strongly suspect (or inductively conclude) that the sum of the first n odd integers always equals n^2 or, equivalently, that

$$1 + 3 + 5 + \cdots + (2n - 1) = n^2$$

for every positive integer value of n. However, the validity of this formula in some specific cases does not guarantee that the rule will not break down for some higher number of odd integers. We can never prove the principle for *all* positive integers by checking a finite number of cases. Thus, mathematicians developed a technique called "proof by mathematical induction" that is based on the following principle.

Principle of Mathematical Induction	If a given statement S_n concerning positive integer n is true for $n = 1$, and if its truth for $n = k$ implies its truth for $n = k + 1$, then S_n is true for every positive integer n.

According to this principle, we can prove that a statement (or formula) is true for every positive integer value of n by showing two things.

1. Prove the statement is true for $n = 1$.
2. Prove that if the statement is true for any positive integer k, then it is also true for the next highest integer $k + 1$.

Proof by mathematical induction is similar to lining up a row of dominoes so that when the first is knocked over it causes the second domino to fall, the second knocks down the third, and so on. In Step 1 we knock down the first domino, and by proving Step 2 we show the desired chain reaction then takes place.

By using mathematical induction, we can now continue our discussion concerning the sum of the first *n* odd integers, and we can prove that this sum *always* equals n^2. We show how in Example 1.

EXAMPLE 1   Use mathematical induction to prove that

$$1 + 3 + 5 + \cdots + (2n - 1) = n^2$$

is true for every positive integer *n*.

Solution   We prove the statement by mathematical induction with the following two steps.

Step 1   Replacing *n* by 1 in the formula gives

$$2(1) - 1 = 1^2$$

which is true. Step 1 is the easy part of the proof.

Step 2   Replace *n* by *k* and write what it means for the formula to work for any positive integer *k*. Call this equation (1). It is the induction assumption.

$$1 + 3 + 5 + \cdots + (2k - 1) = k^2 \tag{1}$$

Now replace *n* by $k + 1$ and write what it means for the formula to work for the next highest integer $k + 1$. Call this equation (2).

$$1 + 3 + 5 + \cdots + (2k - 1) + [2(k + 1) - 1] = (k + 1)^2 \tag{2}$$

We now show that equation (1) implies equation (2). That is, we assume equation (1) is true and use this information to prove equation (2) is true. We accomplish this in either of the following ways.

Method 1   We start with equation (1) and add the $k + 1$ term to both sides of this equation. Then

$$
\begin{aligned}
1 + 3 + \cdots + (2k - 1) + [2(k + 1) - 1] &= k^2 + [2(k + 1) - 1] \\
&= k^2 + 2k + 1 \\
&= (k + 1)^2
\end{aligned}
$$

which is equation (2). Thus, equation (1) implies equation (2).

Method 2   We start with the left side of equation (2) and derive the other side with the aid of equation (1). Then

$$1 + 3 + \cdots + (2k - 1) + [2(k + 1) - 1]$$
$$= [1 + 3 + \cdots + (2k - 1)] + [2(k + 1) - 1]$$

$$\underbrace{\text{induction assumption}}$$

$$= \quad\quad k^2 \quad\quad\quad\quad + [2(k + 1) - 1]$$
$$= k^2 + 2k + 1$$
$$= (k + 1)^2$$

which shows equation (1) implies equation (2).

The proof is complete. Step 1 shows the statement is true for $n = 1$. Step 2 guarantees the statement is true for the next highest integer 2, then 3, and so on, with no positive integer escaping the chain reaction.   □

EXAMPLE 2   Prove that if a and b are real numbers, then $(ab)^n = a^n b^n$ for every positive integer n.

Solution   We prove the statement by mathematical induction with the following two steps.

Step 1   Replacing n by 1 in the statement gives

$$(ab)^1 = a^1 b^1$$

which is true.

Step 2   When $n = k$, the statement becomes

$$(ab)^k = a^k b^k. \tag{1}$$

When $n = k + 1$, the statement becomes

$$(ab)^{k + 1} = a^{k + 1} b^{k + 1}. \tag{2}$$

We now use equation (1) to establish equation (2) as follows:

$$(ab)^{k + 1} = (ab)^k \cdot (ab) \quad\quad \text{1st exponent law: } x^m \cdot x^n = x^{m + n}$$
$$= (a^k b^k) \cdot (ab) \quad\quad \text{By equation (1)}$$
$$= (a^k \cdot a) \cdot (b^k \cdot b)$$
$$= a^{k + 1} \cdot b^{k + 1}.$$

The proof is complete. Note that it is important to state clearly what you need to prove (the "$k + 1$" statement) and what you assume (the "k" statement) so you can prove it.   □

Some statements are not true for *all* positive integers but instead are valid only from some particular integer on up. We prove a statement for all integers greater than or equal to some particular integer (say q) as follows:

1. By direct substitution show the statement is true for $n = q$.
2. Show that if the statement is true for any positive integer $k \geq q$, then it is also true for the next highest integer $k + 1$.

Example 3 illustrates this case.

EXAMPLE 3   Prove that $2^n > n + 1$ for every positive integer $n \geq 2$.

Solution   To prove the statement by mathematical induction, we need two steps. In Step 1, however, we cannot start at $n = 1$ because 2^1 is not greater than $1 + 1$. Thus, as indicated in the problem, we start at $n = 2$.

Step 1   Replacing n by 2 in the inequality gives

$$2^2 > 2 + 1$$

which is true.

Step 2   We need to show that

$$2^k > k + 1 \text{ implies } 2^{k+1} > (k + 1) + 1$$

or, equivalently, that

$$2^k > k + 1 \text{ implies } 2^{k+1} > k + 2.$$

Since $2^k \cdot 2 = 2^{k+1}$, we start with $2^k > k + 1$ and multiply both sides by 2 as follows:

$$2^k > k + 1$$
$$2 \cdot 2^k > 2(k + 1)$$
$$2^{k+1} > 2k + 2.$$

Since $k \geq 2$, it follows that $2k + 2$ is greater than $k + 2$. Thus,

$$2^{k+1} > 2k + 2 > k + 2$$

and the proof is complete.   □

The principle of mathematical induction is a very powerful method of proof that can be applied to a wide variety of mathematical statements. The main difficulty is that some ingenuity is often needed in deciding how to establish the "$k + 1$" statement from the *assumption* of the "k" statement. Note that we accomplished this with different approaches in the example problems after illustrating these approaches in Example 1. To be specific, in Example 2 we started with one side of the "$k + 1$" statement and derived the other side with the aid of the "k" statement. However, in Example 3 we started with the "k" statement and did something to it (we multiplied both sides by 2) to establish the "$k + 1$" statement. It will be easier for you to decide how to proceed in Step 2 if you are aware of both methods. Eventually, Step 2 becomes manageable through the experience gained in considering many problems.

EXERCISES 10.4

In Exercises 1–10 prove by mathematical induction that the following formulas are true for every positive integer value of n.

1. $1 + 2 + 3 + \cdots + n = \dfrac{n(n + 1)}{2}$

2. $\frac{1}{2} + \frac{2}{2} + \frac{3}{2} + \cdots + \dfrac{n}{2} = \dfrac{n(n + 1)}{4}$

3. $2 + 4 + 6 + \cdots + 2n = n(n + 1)$

4. $3 + 6 + 9 + \cdots + 3n = \dfrac{3n(n + 1)}{2}$

5. $4 + 8 + 12 + \cdots + 4n = 2n(n + 1)$

6. $2 + 6 + 10 + \cdots + (4n - 2) = 2n^2$

7. $2 + 2^2 + 2^3 + \cdots + 2^n = 2(2^n - 1)$

8. $3 + 3^2 + 3^3 + \cdots + 3^n = \dfrac{3(3^n - 1)}{2}$

9. $1^3 + 2^3 + 3^3 + \cdots + n^3 = \dfrac{n^2(n + 1)^2}{4}$

10. $1^2 + 2^2 + 3^2 + \cdots + n^2 = \dfrac{n(n + 1)(2n + 1)}{6}$

In Exercises 11–20 prove the statement by mathematical induction.

11. $2^n > n$ for all positive integers n.

12. $3^n < 3^{n + 1}$ for all positive integers n.

13. $n^2 < 2^n$ for all positive integers $n \geq 5$.

14. If a is a real number, then $(a^2)^n = a^{2n}$ for all positive integers n.

15. If the x's are positive real numbers, then
$\log(x_1 x_2 \cdots x_n) = \log x_1 + \log x_2 + \cdots + \log x_n$
for all positive integers $n \geq 2$.

16. If P dollars are invested at r percent compounded annually, then at the end of n years the compounded amount is given by $A_n = P(1 + r)^n$.

17. If a is a real number such that $0 < a < 1$, then $0 < a^n < 1$ for all positive integers n.

18. $n^2 + n$ is divisible by 2 for all positive integers n.

19. If $x \neq a$, then $x^n - a^n$ is divisible by $x - a$ for all positive integers n. (*Hint:* Add and subtract xa^k to $x^{k + 1} - a^{k + 1}$, then group terms and factor.)

20. If $x \neq y$, then $x^{2n} - y^{2n}$ is divisible by $x - y$ for all positive integers n. (*Note:* See Exercise 19.)

10.5 BINOMIAL THEOREM

A leading candidate for the most common mistake in algebra appears when students are asked to find the product or expansion of $(x + y)^2$. Too frequently they write the expansion as $x^2 + y^2$, which leaves out the middle term $2xy$. The problem recurs when expanding other expressions of the form $(x + y)^n$, where n is a positive integer, since there are usually many more terms than just x^n and y^n. We now discuss a method for finding these middle terms that is much easier than repeated multiplication.

We start by trying to find some patterns in the following expansions of the powers of $a + b$. You can verify each expansion by direct multiplication.

$$(a + b)^1 = a + b \qquad\qquad \text{(2 terms)}$$
$$(a + b)^2 = a^2 + 2ab + b^2 \qquad\qquad \text{(3 terms)}$$
$$(a + b)^3 = a^3 + 3a^2b + 3ab^2 + b^3 \qquad\qquad \text{(4 terms)}$$
$$(a + b)^4 = a^4 + 4a^3b + 6a^2b^2 + 4ab^3 + b^4 \qquad\qquad \text{(5 terms)}$$
$$(a + b)^5 = a^5 + 5a^4b + 10a^3b^2 + 10a^2b^3 + 5ab^4 + b^5 \qquad \text{(6 terms)}$$

Observe that in all the above cases the expansion of $(a + b)^n$ behaved as follows:

1. The number of terms in the expansion is $n + 1$. For example, in the expression $(a + b)^5$ we have $n = 5$ and there are six terms in the expansion.
2. The first term is a^n and the last term is b^n.
3. The second term is $na^{n-1}b$ and the nth term is nab^{n-1}.
4. The exponent of a decreases by 1 in each successive term while the exponent of b increases by 1.
5. The sum of the exponents of a and b in any term is n.

Thus, if we assume that this pattern continues when n is a positive integer greater than 5 (and it can be proved that it does), we need only find a method for determining the constant coefficients if we wish to expand such expressions. Our observations to this point may be summarized in the following expansion formula for positive integral powers of $a + b$.

$$(a + b)^n = a^n + na^{n-1}b + (\text{constant})a^{n-2}b^2$$
$$+ (\text{constant})a^{n-3}b^3 + \cdots + nab^{n-1} + b^n$$

You may think of this expansion formula as an informal version of the binomial theorem. A complete statement of the binomial theorem includes a method for obtaining the constant coefficients, and we will do this later. However, for our present purpose, we have arranged the constant coefficients in the above expansions of $(a + b)^n$ for $n \leq 5$ in the triangular array shown in the following chart.

Powers of $a + b$	Constant Coefficients (Pascal's Triangle)

$(a + b)^0$	Row 0
$(a + b)^1$	Row 1
$(a + b)^2$	Row 2
$(a + b)^3$	Row 3
$(a + b)^4$	Row 4
$(a + b)^5$	Row 5

The triangular array of numbers that specifies the constant coefficients in the expansions of $(a + b)^n$ for $n = 0, 1, 2, \ldots$ is called **Pascal's triangle.** Note that except for the 1's, each entry in Pascal's triangle is the sum of the two numbers on either side of it in the preceding row as diagrammed in the chart. We are now ready to try some problems.

EXAMPLE 1   Expand $(x + y)^8$ using Pascal's triangle.

Solution   In our expansion formula we substitute x for a, y for b, and 8 for n. To determine the constant coefficients, we must continue the pattern in the above chart until we have specified the entries in row 8 of Pascal's triangle. To do this, we start at row 5 and proceed as follows:

Row 5	1	5	10	10	5	1	
Row 6	1	6	15	20	15	6	1
Row 7	1	7	21	35	35	21	7	1
Row 8	1	8	28	56	70	56	28	8	1

Thus, the expansion of $(x + y)^8$ is

$$(x + y)^8 = x^8 + 8x^7y + 28x^6y^2 + 56x^5y^3$$
$$+ 70x^4y^4 + 56x^3y^5 + 28x^2y^6 + 8xy^7 + y^8. \qquad \square$$

EXAMPLE 2   Expand $(x + 2y)^4$ using Pascal's triangle.

Solution   We substitute x for a, $2y$ for b, and 4 for n in our expansion formula. From Pascal's triangle we determine the coefficients of the five terms in the expansion as 1, 4, 6, 4, and 1, respectively. Thus, we have

$$(x + 2y)^4 = x^4 + 4x^3(2y) + 6x^2(2y)^2 + 4x(2y)^3 + (2y)^4$$
$$= x^4 + 8x^3y + 24x^2y^2 + 32xy^3 + 16y^4. \qquad \square$$

EXAMPLE 3 Expand $(2x - y)^3$ using Pascal's triangle.

Solution First, rewrite $(2x - y)^3$ as $[2x + (-y)]^3$. In our expansion formula we now substitute $2x$ for a, $-y$ for b, and 3 for n. From Pascal's triangle we determine the coefficients of the four terms in the expansion as 1, 3, 3, and 1, respectively. Thus, we have

$$(2x - y)^3 = (2x)^3 + 3(2x)^2(-y) + 3(2x)(-y)^2 + (-y)^3$$
$$= 8x^3 - 12x^2y + 6xy^2 - y^3.$$

Note that when the binomial is the difference of two terms, the terms in the expansion alternate in sign. □

By now you should have the feel of the basic patterns in the binomial expansion, so let's add the final step—a formula for the constant coefficients. We have been using Pascal's triangle to generate these constants. However, this approach is often impractical since constructing the triangle can be time consuming, particularly for large values of n. The formula for the coefficients can be quite efficient and follows a nice pattern. First, though, it is helpful to introduce **factorial** notation. For any positive integer n, the symbol $n!$ (read "n factorial") means the product $n \cdot (n - 1) \cdot (n - 2) \cdot \ldots \cdot 3 \cdot 2 \cdot 1$. For example,

$$6! = 6 \cdot 5 \cdot 4 \cdot 3 \cdot 2 \cdot 1 = 720.$$

We also define $0! = 1$. Using factorials we now state the binomial theorem.

Binomial Theorem For any positive integer n,

$$(a + b)^n = a^n + \frac{n}{1!}a^{n-1}b + \frac{n(n-1)}{2!}a^{n-2}b^2 + \cdots$$
$$+ \frac{n(n-1)(n-2)\cdots(n-r+1)a^{n-r}b^r}{r!} + \cdots + b^n.$$

The binomial theorem is proved by mathematical induction, and we consider such a proof in Exercise 52. You will find the application of this theorem to be very systematic if you consider the following example carefully.

EXAMPLE 4 Expand $(x + y)^5$ by the binomial theorem.

Solution   In our statement of the binomial theorem we substitute x for a and y for b. We also note that in this case $n = 5$. The expansion goes as follows:

$$(x + y)^5 = x^5 + \frac{5}{1!}x^4y^1 + \frac{5 \cdot 4}{2!}x^3y^2 + \frac{5 \cdot 4 \cdot 3}{3!}x^2y^3$$

$$+ \frac{5 \cdot 4 \cdot 3 \cdot 2}{4!}xy^4 + y^5$$

$$= x^5 + 5x^4y + 10x^3y^2 + 10x^2y^3 + 5xy^4 + y^5. \qquad \square$$

In Example 4 there are a few patterns in determining the coefficients that you should note.

1. The coefficients 1, 5, 10, 10, 5, 1 are symmetrical; that is, they increase and then decrease in the same manner. This symmetry cuts our job in half since we need determine the coefficients only as far as the middle term to obtain the entire list.

2. The coefficients go

$$1, \frac{5}{1!}, \frac{5 \cdot 4}{2!}, \frac{5 \cdot 4 \cdot 3}{3!}, \quad \text{etc.}$$

or, in general, as

$$1, \frac{n}{1!}, \frac{n(n-1)}{2!}, \frac{n(n-1)(n-2)}{3!}, \quad \text{etc.}$$

Each new coefficient (after n) can be generated from the previous coefficient. Just insert the next lowest integer as a factor in the numerator and use the next highest factorial in the denominator. It might also help if you note that in each coefficient the denominator is the factorial of the exponent for y in that term.

3. When computing each coefficient it is possible to divide out common factors so that the expression is easy to evaluate. For example,

$$\frac{5 \cdot 4 \cdot 3 \cdot 2}{4!} = \frac{5 \cdot 4 \cdot 3 \cdot 2}{4 \cdot 3 \cdot 2 \cdot 1} = \frac{5}{1} = 5.$$

With a little practice you will find this method to be quite straightforward—the idea is to balance the different benefits that Pascal's triangle and the formula method have to offer. Here is a problem in which the formula method is a lot easier to use than constructing Pascal's triangle.

EXAMPLE 5   Write the first four terms in the expansion of $(x - 2)^{14}$.

Solution   First, rewrite $(x - 2)^{14}$ as $[x + (-2)]^{14}$. Here $n = 14$, $a = x$, and $b = -2$. Then by the binomial theorem the first four terms are

$$x^{14} + \frac{14}{1!}x^{13}(-2)^1 + \frac{14 \cdot 13}{2!}x^{12}(-2)^2 + \frac{14 \cdot 13 \cdot 12}{3!}x^{11}(-2)^3$$

which simplifies to

$$x^{14} - 28x^{13} + 364x^{12} - 2{,}912x^{11}. \qquad \square$$

To state the binomial theorem in its most efficient form, we need one more refinement in our methods. Note in the statement of the binomial theorem that the constant coefficient of the term containing b^r for $0 < r \leq n$ is

$$\frac{n(n - 1) \cdots (n - r + 1)}{r!}.$$

For instance, in the expansion of $(x + y)^5$ in Example 4 the constant coefficient of the term containing y^4 is

$$\frac{5 \cdot 4 \cdot 3 \cdot 2}{4!} \text{ or } \frac{5(5 - 1)(5 - 2)(5 - 4 + 1)}{4!}.$$

In its present form this formula is awkward to use and we may restate the formula more efficiently as follows:

$$\frac{n(n - 1) \cdots (n - r + 1)}{r!} = \frac{n(n - 1) \cdots (n - r + 1)}{r!} \cdot \frac{(n - r)!}{(n - r)!}$$

$$= \frac{n!}{r!(n - r)!}.$$

If we now follow customary notation and denote the constant coefficient of the term containing b^r by $\binom{n}{r}$, then we have developed that a formula for $\binom{n}{r}$ is

$$\binom{n}{r} = \frac{n!}{r!(n - r)!}.$$

In addition, $\binom{n}{0} = \binom{n}{n} = 1$, and we may obtain these results with our formula since by definition $0! = 1$. Our discussion has led us to the definition of a binomial coefficient.

Binomial Coefficient

Let r and n be nonnegative integers with $r \leq n$. Then the symbol $\binom{n}{r}$ is defined by

$$\binom{n}{r} = \frac{n!}{r!(n-r)!}.$$

Each of the numbers $\binom{n}{r}$ is called a **binomial coefficient.**

As you might expect, binomial coefficients are entries in Pascal's triangle, and the next example illustrates this point.

EXAMPLE 6   Evaluate the following binomial coefficients and compare the results to the entries in Pascal's triangle:

$$\binom{4}{0}, \binom{4}{1}, \binom{4}{2}, \binom{4}{3}, \binom{4}{4}.$$

Solution   By applying the formula in the definition, we have

$$\binom{4}{0} = \frac{4!}{0!(4-0)!} = \frac{4!}{0!4!} = 1$$

$$\binom{4}{1} = \frac{4!}{1!(4-1)!} = \frac{4!}{1!3!} = \frac{4 \cdot 3 \cdot 2 \cdot 1}{1 \cdot (3 \cdot 2 \cdot 1)} = 4$$

$$\binom{4}{2} = \frac{4!}{2!(4-2)!} = \frac{4!}{2!2!} = \frac{4 \cdot 3 \cdot 2 \cdot 1}{(2 \cdot 1)(2 \cdot 1)} = 6$$

$$\binom{4}{3} = \frac{4!}{3!(4-3)!} = \frac{4!}{3!1!} = \frac{4 \cdot 3 \cdot 2 \cdot 1}{(3 \cdot 2 \cdot 1) \cdot 1} = 4$$

$$\binom{4}{4} = \frac{4!}{4!(4-4)!} = \frac{4!}{4!0!} = 1.$$

By direct comparison, we note that the binomial coefficients 1, 4, 6, 4, and 1 match the entries in row 4 of Pascal's triangle. This example illustrates the general result that the binomial coefficients

$$\binom{n}{0}, \binom{n}{1}, \binom{n}{2}, \ldots, \binom{n}{n}$$

are precisely the numbers in the nth row of Pascal's triangle. Note also that by using a scientific calculator with a factorial key $\boxed{x!}$, it is easy to evaluate binomial coefficients. For instance, since $\binom{18}{7} = \dfrac{18!}{7!11!}$, we may evaluate $\binom{18}{7}$ by calculator as follows:

$$18 \boxed{x!} \boxed{\div} 7 \boxed{x!} \boxed{\div} 11 \boxed{x!} \boxed{=} 31{,}824.$$ ☐

By using binomial coefficients and sigma notation, the binomial theorem may be stated in a very efficient way.

Binomial Theorem   For any positive integer n,

$$(a + b)^n = \sum_{r=0}^{n} \binom{n}{r} a^{n-r} b^r$$

$$= \binom{n}{0} a^n + \binom{n}{1} a^{n-1} b + \binom{n}{2} a^{n-2} b^2 + \cdots + \binom{n}{n} b^n.$$

We now redo the expansion from Example 1 using this version of the theorem.

EXAMPLE 7   Expand $(x + y)^8$ by the binomial theorem.

Solution   Applying the binomial theorem with $x = a$, $y = b$, and $n = 8$ yields

$$(x + y)^8 = \binom{8}{0} x^8 + \binom{8}{1} x^7 y + \binom{8}{2} x^6 y^2 + \binom{8}{3} x^5 y^3 + \binom{8}{4} x^4 y^4$$

$$+ \binom{8}{5} x^3 y^5 + \binom{8}{6} x^2 y^6 + \binom{8}{7} xy^7 + \binom{8}{8} y^8.$$

To evaluate the binomial coefficients, we first note that $\binom{8}{0} = \binom{8}{8} = 1$. Then we use the symmetry in the coefficients and the given formula to obtain

$$\binom{8}{1} = \binom{8}{7} = \frac{8!}{1!7!} = 8, \qquad \binom{8}{2} = \binom{8}{6} = \frac{8!}{2!6!} = 28,$$

$$\binom{8}{3} = \binom{8}{5} = \frac{8!}{3!5!} = 56, \qquad \binom{8}{4} = \frac{8!}{4!4!} = 70.$$

Thus, the desired expansion is

$$(x + y)^8 = x^8 + 8x^7 y + 28x^6 y^2 + 56x^5 y^3 + 70x^4 y^4$$
$$+ 56x^3 y^5 + 28x^2 y^6 + 8xy^7 + y^8.$$

(*Note:* The symmetry in the binomial coefficients follows from the identity

$$\binom{n}{r} = \binom{n}{n-r}.$$

You are asked to verify this property in Exercise 48.) □

With our current methods it is now easy to write any single term of a binomial expansion without producing the rest of the terms. Note in the binomial theorem that the exponent of b is 1 in the second term, 2 in the third term, and, in general, $r - 1$ in the rth term. Since the sum of the exponents of a and b is always n, the exponent above a in the rth term must be $n - (r - 1)$. Finally, the binomial coefficient of the term containing b^{r-1} is $\binom{n}{r-1}$. These observations lead to the following formula.

---

rth Term of the Binomial Expansion

The rth term of the binomial expansion of $(a + b)^n$ is

$$\binom{n}{r-1} a^{n-(r-1)} b^{r-1}.$$

EXAMPLE 8   Find the 14th term in the expansion of $(3x - y)^{15}$.

Solution   In this case $n = 15$, $r = 14$, $a = 3x$, and $b = -y$. By the above formula, the 14th term is

$$\binom{15}{13}(3x)^2(-y)^{13} = 105(9x^2)(-y)^{13}$$
$$= -945x^2y^{13}.$$ □

EXERCISES 10.5

In Exercises 1–12 expand each expression by the binomial theorem. Use both Pascal's triangle and the formula methods in determining the constant coefficients.

1. $(x + y)^6$
2. $(x + y)^7$
3. $(x - y)^5$
4. $(x - y)^4$
5. $(x + h)^4$
6. $(x + h)^3$
7. $(x - 1)^7$
8. $(y + 1)^5$
9. $(2x + y)^3$
10. $(x - 2y)^6$
11. $(3c - 4d)^4$
12. $(4c + 3d)^3$

In Exercises 13–16 write the first four terms in the expansion of the given expression.

13. $(x + y)^{15}$
14. $(x - y)^{10}$
15. $(x - 3y)^{12}$
16. $(x + 3)^{17}$

In Exercises 17 and 18 evaluate each of the binomial coefficients and compare the results to the entries in Pascal's triangle.

17. $\binom{3}{0}, \binom{3}{1}, \binom{3}{2}, \binom{3}{3}$

18. $\binom{5}{0}, \binom{5}{1}, \binom{5}{2}, \binom{5}{3}, \binom{5}{4}, \binom{5}{5}$

In Exercises 19 and 20 express the entries in the given row of Pascal's triangle using the notation $\binom{n}{r}$.

19. Row 6   **20.** Row 2

In Exercises 21–30 evaluate each of the binomial coefficients.

21. $\binom{5}{3}$ **22.** $\binom{6}{4}$ **23.** $\binom{9}{4}$

24. $\binom{10}{3}$ **25.** $\binom{10}{0}$ **26.** $\binom{9}{9}$

27. $\binom{20}{2}$ **28.** $\binom{25}{24}$ **29.** $\binom{18}{8}$

30. $\binom{15}{5}$

In Exercises 31–40 write the indicated term of the expansion.

31. 2nd term of $(x + y)^{18}$
32. 5th term of $(x - y)^7$
33. 12th term of $(3x - y)^{13}$
34. 6th term of $(x + 3y)^{10}$
35. 6th term of $(x^2 + 1)^9$
36. 11th term of $(2x^3 - 1)^{14}$
37. 3rd term of $(2 - \sqrt{x})^5$
38. 4th term of $(\sqrt{x} + 3)^6$
39. Middle term of $(3x^2 + 2y^3)^6$
40. Middle term of $(x^{-1} + y^{-1})^{10}$
41. Find the coefficient of x^7y^4 in the expansion of $(x + y)^{11}$.

42. Find the coefficient of $x^{15}y^5$ in the expansion of $(x - y)^{20}$.
43. Find the coefficient of the term containing x^7 in the expansion of $(x - 2y)^{10}$.
44. Find the term with the variable factor x^5 in the expansion of $(2x - y)^{15}$.
45. Find the value of $(1.2)^4$ by writing it in the form $(1 + 0.2)^4$ and using the binomial theorem.
46. Find the value of $(0.97)^3$ by writing it in the form $(1 - 0.03)^3$ and using the binomial theorem.

47. Show that $\binom{n}{n-1} = n$.

48. Show that $\binom{n}{r} = \binom{n}{n-r}$.

49. **a.** Use the binomial expansion of $(1 + 1)^n$ to show that

$$\binom{n}{0} + \binom{n}{1} + \binom{n}{2} + \cdots + \binom{n}{n} = 2^n.$$

b. What is the sum of the entries in row 12 of Pascal's triangle?

50. If $f(x) = x^n$, where n is a positive integer, show that

$$\frac{f(x + h) - f(x)}{h}$$

$$= nx^{n-1} + \frac{n(n-1)}{2!} x^{n-2}h + \cdots + h^{n-1}.$$

51. **a.** Show that $\binom{4}{3} + \binom{4}{2} = \binom{5}{3}$.

b. Consider the positions in Pascal's triangle of the binomial coefficients in part **a.** What feature of Pascal's triangle does the statement illustrate?

c. The property we have been considering states that for all positive integers k and r with $r \le k$,

$$\binom{k}{r} + \binom{k}{r-1} = \binom{k+1}{r}.$$

Prove this property by using the binomial coefficient formula.

52. Below we use mathematical induction and the property shown in Exercise 51 to prove the binomial theorem (that is, $(a + b)^n = \sum_{r=0}^{n} \binom{n}{r} a^{n-r} b^r$). Fill in the missing steps (denoted by the letters **a–i**) along with appropriate justifications.

Step 1   Replacing n by 1 gives **a,** which is true.

Step 2   When $n = k$ the statement becomes

$$(a + b)^k = \binom{k}{0} a^k + \binom{k}{1} a^{k-1} b + \cdots \qquad (1)$$
$$+ \binom{k}{k-1} ab^{k-1} + \binom{k}{k} b^k.$$

When $n = k + 1$ the statement becomes

b. $\qquad\qquad\qquad\qquad\qquad (2)$

We now use equation (1) to establish equation (2). To begin, multiply both sides of equation (1) by $(a + b)$ and then continue as follows:

$$(a + b)^{k+1} = a\left[\binom{k}{0} a^k + \binom{k}{1} a^{k-1} b + \cdots\right.$$
$$\left. + \binom{k}{k-1} ab^{k-1} + \binom{k}{k} b^k\right] + b[\mathbf{c}]$$
$$= \binom{k}{0} a^{k+1} + \binom{k}{1} a^k b + \cdots$$
$$+ \binom{k}{k-1} a^2 b^{k-1} + \binom{k}{k} ab^k + \mathbf{d}$$
$$= \binom{k}{0} a^{k+1} + \left[\binom{k}{1} + \binom{k}{0}\right] a^k b$$
$$+ \cdots + [\mathbf{e}] ab^k + \binom{k}{k} b^{k+1}$$
$$= (\mathbf{f}) a^{k+1} + (\mathbf{g}) a^k b + \cdots$$
$$+ (\mathbf{h}) ab^k + (\mathbf{i}) b^{k+1}$$

which is equation (2). Thus, equation (1) implies equation (2) and the proof by induction is complete.

10.6 COUNTING TECHNIQUES

When making certain decisions, it is often important to analyze the possibilities associated with the situation. For example, if your college offers 2 sections of a required math course and 3 sections of a required English course, then how many schedules are possible in enrolling for these courses? We can answer this question by constructing a **tree diagram** as shown in Figure 10.4.

Figure 10.4

We start by listing the 2 math sections and then we branch out from these points to the 3 English sections. Reading each branch from left to right, we determine there are 6 schedules as shown in Figure 10.4. Note that the number of possible schedules is the product of the number of math choices and the number of English choices. This example illustrates the following key principle in counting.

Fundamental Counting Principle	If event *A* can occur in *a* ways, and following this, event *B* can occur in *b* ways, then the two events taken together can occur in *a · b* ways.

Furthermore, this principle extends to three or more events as considered in Example 1.

EXAMPLE 1   In how many ways can a three-item true–false test be answered? Include a tree diagram listing the possibilities.

Solution   There are three questions, so we start by making three boxes.

☐   ☐   ☐

Since there are two choices (true or false) for each question, we now enter a 2 in each box and then multiply as indicated in the fundamental counting principle.

$$\boxed{2} \cdot \boxed{2} \cdot \boxed{2} = 8$$

Thus, there are 8 ways to answer the test. As shown in Figure 10.5, we can systematically list these choices by writing T and F under the first question and then branching to T and F for each successive question. Following each branch from left to right then gives the possibilities shown.

First question	Second question	Third question	Possible answer
		T	TTT
	T	F	TTF
T		T	TFT
	F	F	TFF
		T	FTT
	T	F	FTF
F		T	FFT
	F	F	FFF

Figure 10.5

EXAMPLE 2   Find the number of different batting orders that are possible for a baseball team that starts 9 players. (Assume the team has only 9 players.)

Solution   The leadoff batter may be any of the 9 players. However, there are only 8 choices for the second position (since the leadoff batter cannot be used again), 7 choices for the third position, and so on. By the fundamental counting principle, this gives

$$9 \cdot 8 \cdot 7 \cdot 6 \cdot 5 \cdot 4 \cdot 3 \cdot 2 \cdot 1 = 362{,}880$$

different batting orders.   □

The question in Example 2 illustrates an important type of counting problem. A **permutation** of a collection of objects or symbols is an arrangement without repetition of these objects in which order is important. Thus, in Example 2 we say there are 362,880 permutations or orderings of the 9 players on a baseball team. In general, when ordering n objects we have n choices for the first position, $n - 1$ choices for the second position, $n - 2$ choices for the third position, and so on. By using the fundamental counting principle and recalling factorial notation (see Section 10.5), we then conclude the following:

Permutations of a Set of n Objects   The number of permutations of n different objects using all of them is

$$n(n - 1)(n - 2) \cdots 3 \cdot 2 \cdot 1 = n!$$

EXAMPLE 3   In how many ways can four candidates be arranged on a ballot?

Solution   Since each candidate is listed only once and the order or arrangement is significant, we need to determine the number of permutations of the four candidates. From the rule above, this number is

$$4! = 4 \cdot 3 \cdot 2 \cdot 1 = 24.$$ □

In many cases it is useful to compute the number of permutations of n objects using only some of the objects in each permutation. For instance, by the fundamental counting principle, the number of four-letter permutations of the 26 letters in the alphabet is $26 \cdot 25 \cdot 24 \cdot 23 = 358{,}800$. We refer to this problem as the number of permutations of 26 objects taken 4 at a time and symbolize this as $_{26}P_4$. Note that

$$_{26}P_4 = \underbrace{26 \cdot 25 \cdot 24 \cdot 23}_{4 \text{ factors}}$$

so that to evaluate $_{26}P_4$ we multiply four consecutive integers that start at 26 and then decrease. This example suggests the following rule.

Permutation Formula   If $_nP_r$ represents the number of permutations of n objects taken r at a time, then

$$_nP_r = \underbrace{n(n - 1)(n - 2)\cdots(n - r + 1)}_{r \text{ factors}} \qquad r \leq n.$$

Note in this formula that n is the starting factor while r indicates the number of factors. When the number of factors is large, it is convenient to evaluate $_nP_r$ by applying the formula derived below and using a scientific calculator with a factorial key.

$$\begin{aligned}
_nP_r &= n(n - 1)(n - 2)\cdots(n - r + 1) \\
&= \frac{n(n - 1)(n - 2)\cdots(n - r + 1)[(n - r)\cdots 2 \cdot 1]}{[(n - r)\cdots 2 \cdot 1]}
\end{aligned}$$

so

$$_nP_r = \frac{n!}{(n - r)!}$$

EXAMPLE 4   In how many ways can nine members from a hockey team be assigned to six different starting positions?

Solution   Since position is important and repetition cannot occur, the number is

$$_9P_6 = \underbrace{9 \cdot 8 \cdot 7 \cdot 6 \cdot 5 \cdot 4}_{6 \text{ factors}} = 60{,}480.$$

Using our alternative formula, the evaluation is

$$_9P_6 = \frac{9!}{(9 - 6)!} = \frac{9!}{3!} = \frac{9 \cdot 8 \cdot 7 \cdot 6 \cdot 5 \cdot 4 \cdot 3 \cdot 2 \cdot 1}{3 \cdot 2 \cdot 1} = 60{,}480. \quad \square$$

When we cannot distinguish between certain members in our collection of objects, we must revise our permutation formula. For instance, how many distinct permutations can be made from the letters of the word ERROR? If we could distinguish among the R's by calling them say R_1, R_2, and R_3, then there would be 5! or 120 permutations of the five letters. Since there are 3! or 6 orderings of the three R's, we could group the 120 possibilities into 20 groups in which the 6 arrangements in each group are not distinguishable when we drop the subscripts. Thus, the number of distinct arrangements of the letters in ERROR is 20, which is $5! \div 3!$. This example suggests the following rule for analyzing this type of problem.

Distinguishable Permutations Formula	The number of distinguishable permutations of n objects taken all at a time when n_1 are of one kind, n_2 are of a second kind, . . . , and n_k are of a kth kind, where $n_1 + n_2 + \cdots + n_k = n,$ is $$\frac{n!}{n_1! n_2! \cdots n_k!}.$$

EXAMPLE 5   How many distinct permutations can be made from the letters of the word BEGINNING?

Solution   The word BEGINNING has 9 letters with duplications of 3 N's, 2 I's, and 2 G's. By the given theorem the number of distinct permutations is

$$\frac{9!}{3!2!2!} = 15,120. \qquad \Box$$

We now consider one last, but very important, type of counting problem. A distinguishing characteristic of a permutation is that the order of the objects is significant. However, in many situations order does not matter. For example, if we want to know the number of ways in which 2 people can be hired from 5 job applicants, then selecting individuals A and B is the same as selecting B and A. What is important here is only who gets the jobs. A **combination** of a collection of objects or symbols is a selection without repetition in which the order of selection does not matter. In our example problem we therefore need to determine the number of combinations of 5 applicants when taken 2 at a time. To answer this, we first note that if order does matter, there are

$$_5P_2 = 5 \cdot 4 = 20$$

possibilities. However, since the two positions can be ordered in 2! or 2 ways, only half of these possibilities are different when we disregard order. Thus, there are $20 \div 2$, or 10, combinations of applicants who can be hired. This example illustrates that we analyze combinations as follows:

Combination Formula   If $_nC_r$ represents the number of combinations of n objects taken r at a time, then

$$_nC_r = \frac{_nP_r}{r!} = \frac{n!}{(n-r)!\,r!}.$$

Note that the formula for $_nC_r$ matches the formula for the binomial coefficient $\binom{n}{r}$, which we considered in Section 10.5. For this reason, $\binom{n}{r}$ is often used in place of $_nC_r$ to denote the number of combinations of n objects taken r at a time.

EXAMPLE 6   From a class of 16 students an instructor asks 3 students to write their homework on the board. How many student selections are possible?

Solution   Order does not matter and repetition is not possible, so we need to determine the number of combinations of 16 students taken 3 at a time.

$$_{16}C_3 = \frac{_{16}P_3}{3!} = \frac{\overbrace{16 \cdot 15 \cdot 14}^{3\text{ factors}}}{3!} = 560$$

An alternative evaluation is

$$_{16}C_3 = \frac{16!}{(16-3)!\,3!} = \frac{16!}{13!\,3!} = \frac{16 \cdot 15 \cdot 14}{3!} = 560. \qquad \square$$

EXAMPLE 7   How many different 5-card poker hands can be dealt from a deck of 52 cards?

Solution   The number of combinations of 52 cards taken 5 at a time is

$$_{52}C_5 = \frac{_{52}P_5}{5!} = \frac{52 \cdot 51 \cdot 50 \cdot 49 \cdot 48}{5 \cdot 4 \cdot 3 \cdot 2 \cdot 1} = 2{,}598{,}960. \qquad \square$$

If you use the formula $_nC_r = {_nP_r}/r!$ and r is a large number, it is useful to know that

$$_nC_r = {_nC_{n-r}}.$$

For example, this formula enables us to evaluate $_{16}C_{13}$ by replacing it with $_{16}C_3$ and proceeding as in Example 6. If you consider the alternative evaluation given in this example, you should be able to see why these two combinations are equal.

EXERCISES 10.6

In Exercises 1–10 evaluate the expression.

1. 6!

2. $(6 - 3)!$

3. $\dfrac{9!}{5!4!}$

4. $\dfrac{100!}{98!}$

5. $_5P_4$

6. $_5C_4$

7. $_8P_5$

8. $_4P_2$

9. $_{10}C_1$

10. $_{25}C_{22}$

11. One reason 0! is defined as 1 is that this definition helps validate our permutation and combination formulas when r equals n or 0. Use $0! = 1$ when evaluating each of the following.

    a. $_5P_5$ **b.** $_5C_5$ **c.** $_5C_0$

In Exercises 12–15 answer the question using the fundamental counting principle.

12. In how many ways can a 2-item true–false test be answered? Include a tree diagram listing the possibilities.

13. How many head–tail orderings are possible when a coin is flipped 3 times? Include a tree diagram listing the possibilities.

14. In how many ways can a student choose 1 of 4 math courses and 1 of 3 business courses?

15. In how many ways can a student choose 2 of 3 computer courses and 3 of 4 elective courses?

In Exercises 16–19 answer the question by using the permutation theorems.

16. How many distinct permutations can be made from the letters of the word NUMBER?

17. How many distinct permutations can be made from the letters of the word MISSISSIPPI?

18. In how many ways can 8 members from a basketball team be assigned to 5 different starting positions?

19. How many orders of finish (excluding ties) are possible in a race among 6 people?

In Exercises 20–23 answer the question by using the combination theorems.

20. In how many ways can we obtain 4 volunteers from 10 people?

21. In how many ways can 3 elective courses be selected from 8 choices?

22. How many different 13-card bridge hands can be dealt from a deck of 52 cards?

23. How many subcommittees of 4 can be made from a 12-person committee?

In Exercises 24–40 consider carefully whether order is significant and repetition is possible. Then answer the question using an appropriate theorem.

24. In how many ways can the 12 members of a jury be seated in the jury box?

25. How many 12-member juries can be selected from 18 potential jurors?

26. In how many ways can 2 cards be selected from a deck of 52 cards if the first card is replaced before the second card is chosen? How many ways are possible if the first card is not replaced?

27. In how many ways can 4 red flags and 2 white flags be arranged one above the other?

28. Find the number of different batting orders that are possible for a baseball team that starts 9 players if the pitcher must bat last. (Assume the team has only 9 players.)

29. In a group of 10 people how many handshakes are possible if each person shakes hands once with the other members of the group?

30. In how many ways can the 4 quadrants in the Cartesian coordinate system be colored if 4 different colors are used?

31. In how many ways can 3 of the same math book and 2 of the same English book be arranged on a shelf?

32. In how many ways can a 5-item multiple-choice test with choices a, b, c, and d be answered?

33. How many subcommittees of 3 Republicans and 3 Democrats can be selected from a 10-member committee that includes 6 Republicans and 4 Democrats?

34. In how many ways can we divide 10 people into 2 groups of 5?

35. In how many ways can we divide 10 people into 5 groups of 2?

36. On a test you are asked to answer 4 problems out of 7. How many choices are possible?

37. How many 3-digit odd numbers can be written from the digits 1, 2, 3, 4, and 5 without repetition in the digits?

38. A soccer league consists of 6 teams. If each team plays 2 games with each of the other teams, then find the total number of games played.

39. In how many ways can we introduce, one by one, the 5 nominees for an award?

40. How many line segments can be drawn joining 7 points if no 3 of the points lie on the same line?

41. Solve for n: $_nP_2 = 20$.

42. Solve for n: $_nC_2 = 66$.

43. Use $_nC_r = \dfrac{n!}{(n-r)!r!}$ and prove that $_nC_r = {_nC_{n-r}}$.

44. Use $_nC_r = {_nC_{n-r}}$ and solve $_nC_{16} = {_nC_4}$ for n.

45. If $_nC_a = {_nC_b}$, where $a \neq b$ and a, b, and n are positive integers, then
    - **a.** $n = a + b$
    - **b.** $n = a - b$
    - **c.** $n = ab$
    - **d.** $n = a/b$

46. If n is a positive integer greater than 1, then
    - **a.** $_nC_n > {_nC_{n-1}}$
    - **b.** $_nC_n = {_nC_{n-1}}$
    - **c.** $_nC_n < {_nC_{n-1}}$

47. By using sigma notation and combinations the binomial theorem may be efficiently stated as follows:

    For any positive integer n,

    $$(a + b)^n = \sum_{r=0}^{n} {_nC_r}\, a^{n-r} b^r.$$

    Expand $(a + b)^4$ using this formula.

48. Expand $(x - 2y)^5$ using the formula in Exercise 47.

49. The rth term in the expansion of $(a + b)^n$ is $_nC_{r-1} a^{n-(r-1)} b^{r-1}$. Use this formula and find the 3rd term in the expansion of $(x + y)^6$.

50. Find in simplest form the 5th term in the expansion of $(x^2 + x^{-1})^8$ by using the formula in Exercise 49.

10.7 PROBABILITY

Once we can analyze the number of ways in which certain events can occur, a natural and important question is to ask what the chances are of these events taking place. Probability theory is used to assign a number between 0 and 1, inclusive, that indicates how certain we are that an event will occur. On this scale, impossible events are assigned a probability of 0 and events that must take place are assigned a probability of 1. Before we begin to develop some of the laws for determining probabilities, it is important to understand that there are two varieties of probability statements. To illustrate this, consider the following predictions.

There is an 80 percent chance interest rates will go down.
There is a 60 percent chance the Bears will win.
There is a 50 percent chance a flipped coin will land heads.
There is a 25 percent chance of correctly guessing both answers on a two-item true–false test.

The first two statements are examples of subjective statements in that the stated number is a reflection of (hopefully) an expert's opinion and cannot really be verified. In such cases a different expert will often predict a different number. However, the last two statements are examples of **theoretical probability** in that the numbers are determined by the laws of probability and everyone should obtain the same theoretical result. When we say there is a 50 percent chance a flipped coin will land heads, we note that in practice we rarely obtain 50 heads in 100 flips of a coin, but theoretically

$$\frac{\text{number of heads}}{n \text{ tosses}} \xrightarrow{\text{approaches}} 50 \text{ percent}$$

as n gets larger and larger. This idea that probability is a ratio suggests the following definition for the probability of an event.

Definition of the Probability of Event E

If an event E can occur in s ways out of a total of T equally likely outcomes, then the probability of E, denoted $P(E)$, is

$$P(E) = \frac{s}{T} = \frac{\text{number of ways } E \text{ can occur}}{\text{total number of equally likely outcomes}}.$$

Note that this definition assures us that if E cannot occur, $P(E) = 0$, while if E is certain, then $P(E) = 1$. Now consider Examples 1–4 which illustrate this definition.

EXAMPLE 1   Find the probability of selecting a spade from a regular deck of 52 cards.

Solution   There are 52 cards with the same chance of being picked. Among these are 13 spades. Thus,

$$P(\text{spade}) = \frac{13}{52}, \text{ or } \frac{1}{4}, \text{ or } 0.25, \text{ or } 25 \text{ percent.}$$

Note that fractional answers are simplified (if possible) and that probabilities often take the form of decimals and percents.   □

EXAMPLE 2 Two dice are rolled. What is the probability that the sum of the outcomes on the dice is 7?

```
1,1   1,2   1,3   1,4   1,5  (1,6)
2,1   2,2   2,3   2,4  (2,5)  2,6
3,1   3,2   3,3  (3,4)  3,5   3,6
4,1   4,2  (4,3)  4,4   4,5   4,6
5,1  (5,2)  5,3   5,4   5,5   5,6
(6,1)  6,2   6,3   6,4   6,5   6,6
```

Figure 10.6

Solution There are 6 equally likely outcomes on each die, so by the fundamental counting principle, there are 6 × 6 or 36 outcomes to consider. By using a tree diagram and associating each outcome on one die with each possible outcome on the other, we obtain the possibilities in Figure 10.6. As encircled in Figure 10.6, there are 6 outcomes that produce a sum of 7, so

$$P(\text{sum } 7) = \frac{6}{36}, \text{ or } \frac{1}{6}.$$

EXAMPLE 3 From a group of 8 students, 3 are selected randomly for prizes. If Ellen, Michael, and Ilene are among the group, then what is the probability they are the winners?

Solution The prizes may be awarded to $_8C_3$ or 56 equally likely groups. Only one of these combinations is Ellen, Michael, and Ilene. Thus, the chances are $\frac{1}{56}$.

EXAMPLE 4 From a committee of 6 men and 6 women, a 3-person subcommittee is randomly chosen. What is the probability that the subcommittee contains 2 men and 1 woman?

Solution There are $_{12}C_3$ or 220 equally likely subcommittees. Among these are $_6C_2 \cdot _6C_1$ or 90 subcommittees with 2 of 6 men and 1 of 6 women. Thus, the chances are $\frac{90}{220}$, or $\frac{9}{22}$.

To determine probabilities involving more than one event, it is useful to add to our current methods probability formulas that enable us to analyze "and," "or," and "not" statements. We first illustrate the "not" formula. On one roll of a die it is easy to determine that

$$P(4) = \frac{1}{6} \text{ while } P(\text{not } 4) = \frac{5}{6}.$$

Note that $P(4) + P(\text{not } 4) = 1$ or, equivalently, $P(\text{not } 4) = 1 - P(4)$. We generalize this observation as follows:

Formula for *P*(not *E*)   The probability that event *E* will not occur is given by

$$P(\text{not } E) = 1 - P(E).$$

EXAMPLE 5   If you guess randomly at the answers on a two-item true–false test, what is the probability you do not get both answers right?

Solution   From a tree diagram the possibilities are

$$R,R \qquad R,W \qquad W,R \qquad W,W,$$

where the question one outcome is listed as the first component in each pair. Since there are only two answer choices, $P(R) = P(W)$ so these outcomes are equally likely. Then

$$P(\text{both right}) = \frac{1}{4} \text{ while } P(\text{not both right}) = 1 - \frac{1}{4} = \frac{3}{4}. \qquad \square$$

In Example 5 we can determine the chances of being right on both questions without listing all the outcomes. Note that

$$P(\text{question one right}) = \frac{1}{2}, \text{ and } P(\text{question two right}) = \frac{1}{2},$$

$$\text{while } P(\text{both right}) = \frac{1}{4}.$$

Thus,

$$P(\text{both right}) = P(\text{question one right}) \cdot P(\text{question two right}).$$

In general, we say two events are **independent** if the outcome of one does not affect the outcome of the other, and we have the following theorem for independent events.

Probability of Independent Events   If *A* and *B* are independent events, then

$$P(A \text{ and } B) = P(A) \cdot P(B).$$

Furthermore, this rule extends to three or more independent events.

EXAMPLE 6   Three pills and five capsules are in a box. Two of the pills cause drowsiness, while four of the capsules remedy this side effect. What is the probability that a person who swallows a pill and a capsule will be drowsy? What are the chances of not being drowsy?

Solution   For a person to be drowsy, two independent events must occur as follows:

P(drowsy) = P(pill causes drowsiness *and* capsule is not a remedy)

$$= \quad \frac{2}{3} \qquad \times \qquad \frac{1}{5}$$

$$= \quad \frac{2}{15}.$$

Also,

$$P(\text{not drowsy}) = 1 - P(\text{drowsy}) = 1 - \frac{2}{15} = \frac{13}{15}. \qquad \square$$

Finally, we need a probability formula for "or" statements. Two events that cannot occur at the same time are called **mutually exclusive,** and for such events the "or" formula is simply

$$P(A \text{ or } B) = P(A) + P(B).$$

For example, on one selection from a regular deck of cards

$$P(\text{jack or queen}) = P(\text{jack}) \quad + P(\text{queen})$$

$$= \frac{4}{52} \qquad + \frac{4}{52}$$

$$= \frac{8}{52}, \text{ or } \frac{2}{13}.$$

When A and B have outcomes in common, we must subtract this overlapping from the sum so we don't count these outcomes twice. Thus, the general "or" formula is

$$P(A \text{ or } B) = P(A) + P(B) - P(A \text{ and } B).$$

For example,

$$\overbrace{\text{jack of spades}}$$

$$P(\text{jack or spade}) = P(\text{jack}) + P(\text{spade}) - P(\text{jack and spade})$$

$$= \frac{4}{52} + \frac{13}{52} - \frac{1}{52}$$

$$= \frac{16}{52}, \quad \text{or} \quad \frac{4}{13}.$$

To summarize, the "or" formulas are as follows:

Formulas for *P(A or B)*   For any events A and B,

$$P(A \text{ or } B) = P(A) + P(B) - P(A \text{ and } B).$$

When A and B are mutually exclusive, this formula simplifies to

$$P(A \text{ or } B) = P(A) + P(B).$$

EXAMPLE 7   We flip a coin and roll a die. Find the following probabilities by both listing the equally likely outcomes and also using the "and" and "or" formulas.
a. tails on the coin *and* 5 on the die
b. tails on the coin *or* 5 on the die

Solution
a. There are 2 equally likely outcomes for the coin and 6 for the die. So by the fundamental counting principle we must list 2 × 6 or 12 outcomes. By using a tree diagram these outcomes are

H1	H2	H3	H4	H5	H6
T1	T2	T3	T4	T5	T6.

The outcome *T*5 represents tails on the coin *and* 5 on the die, so the chances for this event are $\frac{1}{12}$. We obtain this answer by formula as follows:

$$P(T \text{ and } 5) = P(T) \cdot P(5) = \frac{1}{2} \cdot \frac{1}{6} = \frac{1}{12}.$$

b. The outcomes $T1$, $T2$, $T3$, $T4$, $T5$, $T6$, and $H5$ all contain either a T or a 5, so the probability of tails on the coin *or* 5 on the die is $\frac{7}{12}$. This answer is obtained by formula as follows:

$$P(T \text{ or } 5) = P(T) + P(5) - P(T \text{ and } 5)$$
$$= \frac{1}{2} + \frac{1}{6} - \frac{1}{12}$$
$$= \frac{7}{12}.$$

EXERCISES 10.7

1. A committee consists of 7 men and 5 women. If the chairperson is selected at random, find the probability the chairperson is a woman.

2. If a fair die is rolled once, find the probability of obtaining each of the following.
  a. a 3   **b.** an outcome less than 3
  c. a 7   **d.** an outcome less than 7

3. A card is drawn from a regular deck of 52 cards. Find the probability that the card drawn is
  a. the queen of spades **b.** a heart
  c. a picture card   **d.** not a red queen
  e. an ace or a king   **f.** neither a jack nor a queen
  g. a spade or a 7   **h.** neither a spade nor a 7

4. A student is selected at random from a group in which 30 percent are freshmen, 25 percent are sophomores, 35 percent are juniors, and 10 percent are seniors. Find the probability of selecting each of the following.
  a. a senior
  b. not a senior
  c. a freshman or a sophomore
  d. neither a freshman nor a junior

5. If a woman's chance of winning a prize is $\frac{2}{19}$, then what is her chance of not winning?

6. Which of the following numbers can *not* be a probability?

$$0, 1, -1, \tfrac{4}{3}, 30\%, 300\%, 0.01, 1.01, \tfrac{31}{365}$$

7. Two dice are rolled. What is the probability that the sum of the outcomes on the dice is 5?

8. On two rolls of a pair of dice, what is the probability that at least once the sum of the outcomes on the dice is 5.

9. A boy and a girl are playing a game in which both simultaneously call out a number from 1 through 3. Find the probability for each of the following.
  a. Both call an odd number.
  b. Both call the same number.
  c. One of them calls 2, while the other does not.

10. What is the probability that on 4 flips of a coin the result will be 4 tails?

11. Four pills and three capsules are in a box. One of the pills causes drowsiness, while two of the capsules remedy this side effect. What is the probability that a person who swallows a pill and a capsule will be drowsy?

12. From the letters a, b, c, and d, three letters are selected at random without replacement. What is the probability that a person will draw the word *bad* in the correct order of spelling?

13. From 10 tickets, 3 different winners are to be chosen. If Chris, Amy, and Pat each holds a ticket, then what is the probability they are the 3 winners?

14. If 5 nominees with different last names are introduced randomly, then what is the probability they are introduced in alphabetical order?

15. What is the probability that a family with 3 children contains 2 boys and 1 girl? (Assume boy and girl are equally likely.)

16. From a committee of 6 men and 4 women, a 3-person subcommittee is randomly chosen. What is the probability that the committee contains 2 women and 1 man?

17. If your pocket contains 7 nickels and 5 quarters and you randomly select three coins, then what is the probability of obtaining exactly 35 cents to pay for a newspaper?

18. On a multiple-choice test with 4 choices, what is the probability you guess randomly at the answers on 3 questions and get 2 out of 3 right?

19. On two independent events, A and B, the probability that A will occur is $\frac{1}{4}$ while the probability that B will occur is $\frac{1}{3}$. What is the probability that neither event occurs?

20. If 40 percent of the population has type A blood, then find the following probabilities for two blood donors chosen at random.
   a. Both are type A.
   b. Neither is type A.
   c. 1 out of 2 is type A.

CHAPTER OVERVIEW

Section	Key Concepts to Review
10.1	• Definitions of sequence, terms (of the sequence), arithmetic progression, common difference, geometric progression, common ratio and Fibonacci sequence
	• Formulas for the nth term: arithmetic progression: $a_n = a_1 + (n - 1)d$ geometric progression: $a_n = a_1 r^{n-1}$ Fibonacci sequence: $a_1 = 1$, $a_2 = 1$, $a_n = a_{n-1} + a_{n-2}$ for $n \geq 3$
10.2	• Definitions of series, arithmetic series, and geometric series
	• Formulas for the sum of the first n terms: arithmetic series: $S_n = \dfrac{n}{2}(a_1 + a_n)$ or $S_n = \dfrac{n}{2}[2a_1 + (n-1)d]$ geometric series: $S_n = \dfrac{a_1 - a_1 r^n}{1 - r}$ or $S_n = \dfrac{a_1 - a_n r}{1 - r}$
	• The Greek letter Σ (read "sigma") is used to simplify the notation involved with series. We define sigma notation as follows: $\displaystyle\sum_{i=1}^{n} a_i = a_1 + a_2 + \cdots + a_n.$
10.3	• An infinite geometric series with $|r| < 1$ converges to the value or sum S given by $S = \dfrac{a_1}{1 - r}$.
10.4	• Principle of mathematical induction
	• We prove by mathematical induction that a statement is true for all positive integers by doing the following.
	1. Show by direct substitution that the statement (or formula) is true for $n = 1$.
	2. Show that if the statement is true for any positive integer k, then it is also true for the next highest integer $k + 1$.
	• Two methods for establishing the "$k + 1$" statement from the assumption of the "k" statement

Section	Key Concepts to Review
10.5	• Binomial theorem: For any positive integer n, $$(a + b)^n = a^n + \frac{n}{1!}a^{n-1}b + \frac{n(n-1)}{2!}a^{n-2}b^2 + \cdots$$ $$+ \frac{n(n-1)(n-2)\cdots(n-r+1)}{r!}a^{n-r}b^r + \cdots + b^n,$$ or an alternate form is $$(a + b)^n = \sum_{r=0}^{n} \binom{n}{r}a^{n-r}b^r.$$ • The binomial coefficient $\binom{n}{r}$ is defined by $$\binom{n}{r} = \frac{n!}{r!(n-r)!}.$$ • The rth term of the binomial expansion of $(a + b)^n$ is $$\binom{n}{r-1}a^{n-(r-1)}b^{r-1}.$$ • The symbol $n!$ (read "n factorial") means the product $n \cdot (n-1) \cdot \cdots \cdot 3 \cdot 2 \cdot 1$. • Pascal's triangle

10.6	• Definitions of permutation and combination
	 • Method to construct a tree diagram
	 • Fundamental counting principle
	 • Permutation formulas:
	    n objects, n at a time: $_nP_n = n!$
	    n objects, r at a time: $_nP_r = \underbrace{n(n-1)\cdots(n-r+1)}_{r \text{ factors}} = \dfrac{n!}{(n-r)!}$
	 • Formula for distinguishable permutations (if certain objects are all alike)
	 • Combination formulas:
	    n objects, r at a time: $_nC_r = \dfrac{_nP_r}{r!} = \dfrac{n!}{(n-r)!r!}$
	 • The order of the objects is important in a permutation but does not matter in a combination.

Section	Key Concepts to Review
10.7	• Definitions of probability of an event, independent events, and mutually exclusive events
	• The probability of an event is a number between 0 and 1, inclusive, with impossible events assigned a probability of 0 and events that must take place assigned a probability of 1.
	• Probability formulas
	1. $P(\text{not } E) = 1 - P(E)$
	2. $P(A \text{ and } B) = P(A) \cdot P(B)$ (if A and B are independent events)
	3. a. $P(A \text{ or } B) = P(A) + P(B) - P(A \text{ and } B)$ (for any events A and B)
	b. $P(A \text{ or } B) = P(A) + P(B)$ (if A and B are mutually exclusive)

CHAPTER REVIEW EXERCISES

1. Write the first four terms of the sequence given by $a_n = (-1)^n 2^{n-1}$; also find a_6.
2. Determine a_2, a_4, and a_n so that the following sequence is (a) an arithmetic progression and (b) a geometric progression.

   $$5, a_2, 45, a_4, \ldots, a_n, \ldots$$

3. Find the sum of the first seventy positive integers.
4. Write the series $\sum\limits_{i=2}^{7} 2^{i-1}$ in expanded form and determine the sum.
5. Write the series $1 + (\frac{1}{2})^2 + (\frac{1}{3})^2 + (\frac{1}{4})^2 + (\frac{1}{5})^2$ in sigma notation.
6. Find the sum of the infinite geometric series

   $$1 - \tfrac{1}{4} + \tfrac{1}{16} - \tfrac{1}{64} + \cdots.$$

7. By mathematical induction, prove that the sum of the first n terms of an arithmetic series is given by

   $$a + (a + d) + (a + 2d) + \cdots$$
   $$+ [a + (n - 1)d] = \frac{n}{2}[2a + (n - 1)d].$$

8. Express the repeating decimal $3.\overline{4}$ as the ratio of two integers. (Use infinite geometric series.)
9. **a.** If a, b, and c are the first three terms in an arithmetic progression, express b in terms of a and c.
   b. If a, b, and c are the first three terms in a geometric progression, express b in terms of a and c.

10. \$2,000 is split up into 8 prizes by a lottery system in which each award is \$10 less than the preceding award. How much money is awarded as first prize?
11. Evaluate $\sum\limits_{k=1}^{\infty} \dfrac{1}{3^k}$.
12. Find the 20th term in the arithmetic progression 3, 7, 11, ...
13. Evaluate $\sum\limits_{k=1}^{5} k^2$.
14. What is the sum of the first n odd positive integers?
15. In how many different orders may 7 people stand in a line? How about a circle? In how many different orders can 4 men and 3 women stand in a line so each woman is between two men?
16. How many line segments are needed to connect in all possible ways the 9 points on the circle below?

17. The first term in an arithmetic progression is a_1, the nth term is a_n, the common difference is d, and the number of terms is n. Express d in terms of a_1, a_n, and n.

18. Find the sum of the infinite geometric series
$3^{-1} + 3^{-2} + 3^{-3} + \cdots$.

19. Is $\displaystyle\sum_{i=0}^{n-1} (2i + 1)$ equal to $\displaystyle\sum_{k=1}^{n} (2k - 1)$?

20. Simplify ${}_nC_{n-1}$.

21. If ${}_nC_{12} = {}_nC_3$, find n.

22. Expand $(x + h)^6$ by the binomial theorem.

23. A jack, a queen, and a king are face down on a table. If two of the cards are turned over at random, what is the probability that one of them is a king?

24. If the probability of winning a game is a/b, what is the probability of not winning the game?

25. How many permutations are there of 7 things taken 3 at a time?

26. How many combinations are there of 7 things taken 3 at a time?

27. If ${}_nP_r = 720$ and ${}_nC_r = 120$, find n and r.

28. Write the 10th term in the expansion of $(2x - y)^{11}$.

29. Write the 4th term in the expansion of $(\sqrt{x} + 1)^7$.

30. Prove by mathematical induction that the following formula is true for all positive integer values of n.

$$1 + 5 + 9 + \cdots + (4n - 3) = n(2n - 1)$$

31. If 10 percent of the population has type B blood, then what is the probability that two blood donors chosen at random are type B?

32. If 1 card is selected from a regular deck of 52 cards, then what is the probability this card is a club or a picture card?

33. Write the first four terms in the expansion of $(1 - x)^{15}$.

34. A slot machine contains 3 independent wheels. On each wheel a star, a lemon, a grape, or a cherry may appear. How many outcomes are possible on the slot machine?

In Exercises 35–44 select the choice that completes the statement or answers the question.

35. The sequence $\sqrt{2}, \sqrt{3}, \sqrt{4}, \sqrt{5}$ is
   a. an arithmetic progression
   b. a geometric progression
   c. neither an arithmetic nor geometric progression

36. If the first three terms in a geometric progression are b^2, b^x, and b^8, then x equals
   a. 4   b. $4\sqrt{2}$ c. $2\sqrt{2}$ d. 5

37. The next term in the geometric progression $\sqrt[4]{2}, \sqrt{2}, \sqrt[4]{8}, \ldots$ is
   a. 2   b. $2\sqrt{2}$ c. 4   d. 8

38. Which statement is true?
   a. ${}_{50}C_{10} < {}_{50}C_{40}$ b. ${}_{50}C_{10} = {}_{50}C_{40}$ c. ${}_{50}C_{10} > {}_{50}C_{40}$

39. Which number cannot be a probability?
   a. 1   b. 0   c. 2   d. $\frac{1}{2}$

40. How many telegraphic characters can be made using 3 dots and 2 dashes in each character?
   a. 10   b. 8   c. 32   d. 25

41. The coefficient of the x^9y^2 term in the binomial expansion of $(x + y)^{11}$ is
   a. 55   b. 9   c. 11   d. 165

42. The expression $\displaystyle\sum_{n=1}^{5} c$ (where c is a constant) simplifies to
   a. c b. $5c$ c. cn d. $c + 5$

43. The sum of the infinite geometric series
$5 + \dfrac{5}{3} + \dfrac{5}{3^2} + \cdots$ is
   a. $\frac{25}{3}$ b. $\frac{45}{8}$ c. $\frac{15}{2}$ d. $\frac{125}{9}$

44. The 5th term in the binomial expansion of $(a - b)^7$ is
   a. $35a^3b^4$ b. $-35a^3b^4$
   c. $-21a^2b^5$ d. $21a^2b^5$

Appendix

A.1 SCIENTIFIC NOTATION

Many numbers that appear in scientific work are either very large or very small. For example, the average distance from the earth to the sun is approximately 93,000,000 mi, and the mass of an atom of hydrogen is approximately

0.0000000000000000000000017 g.

To work conveniently with these numbers, we often write them in a form called **scientific notation.**

A positive number is expressed in scientific notation when it is written in the form

$$N = m \times 10^k, \quad \text{where} \quad 1 \leq m < 10 \text{ and } k \text{ is some integer.}$$

For example,

$$93,000,000 = 9.3(10,000,000) = 9.3 \times 10^7$$

$$0.00103 = 1.03\frac{1}{1,000} = 1.03 \times 10^{-3}.$$

To convert a number from regular notation to scientific notation, use the following procedure.

To Convert to Scientific Notation

1. Immediately after the first nonzero digit of the number, place an apostrophe (').
2. Starting at the apostrophe, count the number of places to the decimal point. If you move to the right, your count is expressed as a positive number; if you move to the left, the count is negative.
3. The apostrophe indicates the position of the decimal in the factor between 1 and 10; the count represents the exponent to be used in the factor, which is a power of 10.

The following examples illustrate how this procedure is used. (*Note:* The arrow indicates the direction of the counting.)

Number	=	Number from 1 to 10	×	Power of 10
9'3000000. $\longrightarrow$	=	9.3	×	10^7
1'36 $\rightarrow$	=	1.36	×	10^2
0.0001'36 $\leftarrow$	=	1.36	×	10^{-4}
0.6'2 $\leftarrow$	=	6.2	×	10^{-1}
6'.2	=	6.2	×	10^0

Note from the above examples that a number can be changed from scientific notation to regular notation by adding zeros and moving the decimal point the appropriate number of spaces to the right if the exponent is positive, to the left if the exponent is negative.

Scientific calculators are programmed to work with scientific notation. To enter a number in scientific notation, first enter the significant digits of the number from 1 to 10, press $\boxed{\text{EE}}$ or $\boxed{\text{EXP}}$, and finally enter the exponent of the power of 10. For example, to enter 1.36×10^{-4} press

$$1.36 \ \boxed{\text{EE}} \ 4 \ \boxed{+/-} \ .$$

The display looks like

$$\boxed{1.36 - 04}$$

with the exponent appearing on the right in the display. Read the owner's manual to your calculator to learn its scientific notation capabilities. Here are a few features to check for.

1. If a calculation results in an answer with too many digits for display in ordinary notation, the calculator automatically displays the answer in scientific notation.

    Example: 47^5 $\quad 47 \ \boxed{y^x} \ 5 \ \boxed{=} \qquad \boxed{2.2935 \ 08}$

    Note here that a calculator uses two display places in scientific notation to indicate the exponent and one display place to separate the significant digits from the power of 10. Thus, the display is accurate in this case to only five significant digits. However, $47^5 = 229{,}345{,}007$ and the calculator internally carries this answer. Check your owner's manual to determine the accuracy you can expect in displayed and internal values.

2. Entries in ordinary and scientific notation may be mixed in the same problem.

Example: $(7.6 \times 10^3) + 127$

7.6 $\boxed{\text{EE}}$ 3 $\boxed{+}$ 127 $\boxed{=}$ $\boxed{7.727\ 03}$

3. On some calculators (notably Texas Instruments models) pressing $\boxed{\text{INV}}$ $\boxed{\text{EE}}$ converts back to ordinary notation.

Example: $(7.6 \times 10^3) + 127$

7.6 $\boxed{\text{EE}}$ 3 $\boxed{+}$ 127 $\boxed{=}$ $\boxed{\text{INV}}$ $\boxed{\text{EE}}$ 7,727

EXERCISES A.1

In Exercises 1–10 write the number in scientific notation by replacing the question marks in the following table with the appropriate number.

Number (N)	=	m	$\times$	10^k
1. 740,000	=	?	$\times$	10^5
2. 0.0005	=	?	$\times$	10^{-4}
3. 0.024	=	2.4	$\times$	$10^?$
4. 240	=	2.4	$\times$	$10^?$
5. 12,300,000	=	?	$\times$	?
6. 0.00000123	=	?	$\times$	?
7. ?	=	4.5	$\times$	10^1
8. ?	=	4.5	$\times$	10^{-1}
9. ?	=	9.08	$\times$	10^{-7}
10. ?	=	9.08	$\times$	10^7

In Exercises 11–20 express each number in scientific notation.

11. 42   **12.** 0.6   **13.** 34,251   **14.** 7.21

15. A light year (that is, the distance light travels in 1 year) is about 5,900,000,000,000 mi.

16. A single red cell of human blood contains about 270,000,000 hemoglobin molecules.

17. A certain radio station broadcasts at a frequency of about 1,260,000 hertz (cycles per second).

18. A certain computer can perform an addition in about 0.000014 second.

19. The weight of an oxygen molecule is approximately 0.000000000000000000000053 g.

20. The wavelength of yellow light is about 0.000023 in.

In Exercises 21–30 express each number in regular notation.

21. 9.2×10^4 **22.** 3×10^{-1}

23. 4.21×10^1 **24.** 6.3×10^0

25. The earth travels about 5.8×10^8 mi in its trip around the sun each year.

26. The number of atoms in 1 oz of gold is approximately 8.65×10^{21}.

27. One coulomb is equal to about 6.28×10^{18} electrons.

28. An atom is about 5×10^{-9} in. in diameter.

29. The mass of a molecule of water is about 3×10^{-23} g.

30. The wavelength of red light is approximately 6.6×10^{-5} cm.

31. If light travels about 186,000 mi/second, about how far will it travel in 5 minutes?

32. If one red blood cell contains about 270,000,000 hemoglobin molecules, about how many molecules of hemoglobin are there in 2 million red cells?

33. If the mass of one electron is about 0.0000000000000000000000000009 g, what is, approximately, the mass of 400 electrons?

34. If a certain computer can perform an addition in about 1.4×10^{-5} second, about how long will it take the computer to perform 50 additions?

A.2 APPROXIMATE NUMBERS

Measurements of one kind or another are essential to both scientific and nontechnical work. Weight, distance, time, volume, and temperature are only a few of the quantities for which measurements are required. We measure through the medium of numbers and arbitrary units, such as the foot (British system) or the meter (metric system). These numbers are approximations and can only be as accurate as the measuring instruments allow. For example, if we measure the height of a man to be 68 in., we are saying that his height to the nearest inch is 68 in. This means that his exact height (h) is somewhere in the interval 67.5 in. $\leq h < 68.5$ in. Using a different measuring device, we might record the height to be 68.2 in., so that 68.15 in. $\leq h < 68.25$ in.

The exact height is contained in these intervals because the measurements are estimated by rounding off to some decimal place. We round off by considering the digit in the next place to the right of the desired decimal place. If this digit is less than 5, the digit in the desired decimal place remains the same; if the digit is 5 or greater, the digit in the desired decimal place is increased by one. The digits to the right of the desired decimal place are then dropped and replaced by zeros if the zeros are needed to maintain the position of the decimal point. Rounding off is best understood by considering the following table, which indicates how we round off a measurement of 265.307 ft to various places.

Precision Desired	Measurement of 265.307 Ft Is Given as
Nearest 100 ft	300 ft
Nearest 10 ft	270 ft
Nearest foot	265 ft
Nearest tenth of a foot	265.3 ft
Nearest hundredth of a foot	265.31 ft

The **precision** of a number refers to the place at which we are rounding off. A number rounded off to the nearest hundredth is more precise than a number rounded off to the nearest tenth, and so on.

Another important consideration is the number of **significant digits** in our approximation. All digits, except the zeros that are required to indicate the position of the decimal point, are significant. The following table illustrates this concept.

Approximate Number	Number of Significant Digits
15.13	Four
0.44	Two
0.003	One; the number is *three* thousandths and only the 3 is significant. The zeros are needed to indicate the position of the decimal point.
0.0196	Three
307	Three; the zero is *not* used to indicate the position of the decimal point. Zeros between nonzero digits are always significant.
50.007	Five
8.0	Two; the zero is significant because it shows that the number is rounded off to the nearest tenth. The zero is not required to write the number 8 and should not be written unless it is significant.
0.2600	Four
230	Two; we do not have sufficient information to determine if the number is rounded off in the tens place or to the nearest unit. Unless stated otherwise, we assume the number to be rounded off in the tens place so the zero is needed to indicate the position of the decimal point.
2,000	One
2,00$\bar{0}$	Four; the bar is used to avoid ambiguity by indicating that the number is rounded off at the digit below the bar.
2,$\bar{0}$00	Two

The **accuracy** of a number is determined by the number of significant digits in the number. Thus, 21 (two significant digits) is a more accurate number than 0.07 (one significant digit). It is important to distinguish between the accuracy and the precision of a number. For example, consider the following pairs of numbers.

21 and 0.07   { 21 is more accurate since it contains two significant digits.

0.07 is more precise since it is written to the nearest hundredth. }

123 and 12.3   { Same accuracy since both contain three significant digits.

12.3 is more precise since it is written to the nearest tenth. }

171 and 59 $\Bigg\{$
- 171 is more accurate since it contains three significant digits.
- Same precision since both are written to the nearest integer.

517 and 1.703 $\Bigg\{$
- 1.703 is more accurate since it contains four significant digits.
- 1.703 is more precise since it is written to the nearest thousandth.

Operations with Approximate Numbers

When computing with approximate numbers, we should express our results to a precision or accuracy that is appropriate to the data. We establish guidelines for the result by considering the error involved in computing with such numbers. For example, suppose that $x \approx 123$ and $y \approx 4.27$. Adding these numbers, we have $123 + 4.27 = 127.27$. However, because of rounding off, we know that the exact values of x and y are in the intervals

$$122.5 \leq x < 123.5$$
$$4.265 \leq y < 4.275.$$

Adding the minimum values to find the minimum possible sum, and adding the maximum values to find the maximum possible sum, we have

$$122.5 + 4.265 \leq x + y < 123.5 + 4.275$$
$$126.765 \leq x + y < 127.775.$$

To the nearest integer, the minimum sum is 127 and the maximum sum is 128. Thus, the precision in the result may not be given to more than the nearest integer, and there is even a chance of error if we are that precise with our answer. This example motivates us to the following guideline.

Rounding Off in Addition or Subtraction	When **adding** or **subtracting** approximate numbers, perform the operation and then round off the result to the **precision** of the least precise number.

Thus, if $x \approx 123$ and $y \approx 4.27$, then $x + y \approx 123 + 4.27 = 127.27 \approx 127$. We round off to the nearest integer because 123, the less precise number, is rounded off to the nearest integer.

Before adding or subtracting, it is permissible to round off the more precise numbers since the extra digits are not considered when we round off our result. If this is done, the numbers should be rounded off to one place beyond that of the least precise number.

EXAMPLE 1   Add the approximate numbers: 12.79, 4.3131, 46.2, 9.618.

Solution   46.2 is the least precise number, so before adding we may round off the other numbers to the nearest hundredth.

$$
\begin{array}{r}
12.79 \\
4.31 \\
46.2 \\
\underline{9.62} \\
72.92
\end{array}
$$

Rounding off to the nearest tenth, the result is 72.9.   □

EXAMPLE 2   Perform the indicated operation for the approximate numbers: 0.396 − 17.6 + 150.

Solution   150 is the least precise number. We assume that 150 is rounded off in the tens place, so before performing the operations, we round off the other numbers to the nearest integer. Thus, we have 0 − 18 + 150 = 132. Rounding off in the tens place, the result is 130.   □

To determine the possible error when multiplying or dividing approximate numbers, consider the product of $x \approx 12$ and $y \approx 4.27$.

$$x \cdot y \approx 12(4.27) = 51.24$$

Because of rounding off, we know that the exact values of x and y are in the intervals

$$
\begin{aligned}
11.5 &\le x < 12.5 \\
4.265 &\le y < 4.275.
\end{aligned}
$$

Multiplying the minimum values to find the minimum possible product and the maximum values to find the maximum possible product, we have

$$
\begin{aligned}
(11.5)(4.265) &\le xy < (12.5)(4.275) \\
49.0475 &\le xy < 53.4375.
\end{aligned}
$$

To two significant digits, the minimum product is 49 and the maximum product is 53. Thus, the accuracy of the result may not be given to more than two significant digits, and even that accuracy may be wrong. This example motivates us to the following guideline.

Rounding Off in Multiplication or Division

When **multiplying** or **dividing** approximate numbers, perform the operation and then round off the result to the **accuracy** of the least accurate number.

Thus, if $x \approx 12$ and $y \approx 4.27$, then $x \cdot y \approx 12(4.27) = 51.24 \approx 51$. We round off to two significant digits because 12, the less accurate number, has two significant digits.

Before multiplying or dividing, it is permissible to round off the more accurate numbers to one more significant digit than the least accurate number. If you are using a calculator, you may find it easier to key in the number without rounding off. Either way, you will usually obtain the same final result.

Since finding a power or root of a number involves multiplication, we are concerned with accuracy in such problems. The result should contain the same number of significant digits as the given approximate number.

EXAMPLE 3   Multiply the approximate numbers: $(11.5)(0.042649)$.

Solution   11.5 is the less accurate number, so before multiplying we may round off the other number to four significant digits:

$$(11.5)(0.04265) = 0.490475.$$

Rounding off to three significant digits, the result is 0.490.   □

EXAMPLE 4   Perform the indicated operations for the approximate numbers.

$$\frac{(215.1)(7.63)}{0.0002}$$

Solution   0.0002 is the least accurate number. We round off the other numbers to two significant digits and then perform the operation.

$$\frac{220(7.6)}{0.0002} = 8,360,000$$

Rounding off to one significant digit, the result is 8,000,000.   □

EXAMPLE 5   Find the approximate value of $\sqrt{79.1}$.

Solution   Using a calculator, the initial result is $\sqrt{79.1} \approx 8.8938181$. Since 79.1 contains three significant digits, the final result is 8.89.   □

Exact Numbers

Although most of the numbers in scientific work are approximate, there are some numbers that are exact. Exact numbers are often the result of a counting process or a definition and are commonly seen in formulas. The following are examples of situations in which the numbers are exact.

1. There are 25 students in the class (counting process).
2. Of 100 light bulbs tested, 4 were defective (counting process).
3. There are 60 minutes in 1 degree (definition).
4. A bank pays 6 percent interest (definition).
5. $P = 4s$ (formula for the perimeter of a square).
6. $A = \frac{1}{2}ab$ (formula for the area of a triangle).

Since these numbers are exact, no error is involved when using them in computations. Only approximate numbers must be considered in determining the accuracy or precision of a result.

EXERCISES A.2

In Exercises 1–20 determine the number of significant digits in the approximate number.

1. 123.14
2. 0.59
3. 0.02
4. 0.000491
5. 904
6. 30.03
7. 22.0
8. 0.78000
9. 670
10. 80,000
11. 5.40
12. 540
13. 0.054
14. 54.0
15. 504
16. 540.0
17. $5,40\overline{0}$
18. $5,\overline{4}00$
19. 0.010
20. 0.100

In Exercises 21–32 round off the numbers (a) to four significant digits and (b) to two significant digits.

21. 12.3456
22. 46.054
23. 59,372
24. 72,488
25. 29,697
26. 59,999
27. 84.096
28. 10.998
29. 0.068547
30. 0.0034581
31. 0.0020006
32. 0.39999

In Exercises 33–44 determine the approximate number in each pair that is (a) more accurate and (b) more precise.

33. 423, 0.004
34. 18, 0.700
35. 402, 40.2
36. 312, 98
37. 6.430, 0.304
38. 0.9, 12.0
39. 2,000, 0.0002
40. 1,230, 3,200
41. 9,080, 48.2
42. $4,\overline{0}00, 4\overline{0}$
43. $3\overline{0}0, 30\overline{0}$
44. $8,88\overline{0}, 8.008$

In Exercises 45–64 perform the indicated operations for the approximate numbers.

45. $16.27 + 2.1515 - 4.3$
46. $15 - 3.043 + 103.1$
47. $0.002 - 11.0 + 3.59$
48. $0.200 + 4.6 + 120$
49. $0.3047 + 0.8 + 0.092 + 0.69$
50. $214.32 + 1,000 + 0.75 + 3.2$
51. $(2.1)(0.57892)$
52. $(0.08)(489.0)$
53. $(16.4)(2,001)$
54. $(7.00)(0.094265)$
55. $408 \div 0.002$
56. $602 \div 0.200$
57. $\dfrac{(799.4)(8.6)}{0.04}$
58. $\dfrac{(0.460)(126.1)}{2,200}$
59. $9.714 + (2.3)(0.812)$
60. $(4.50)(0.1234) - 5.8$
61. $\dfrac{4.07}{0.06} + (0.8715)(20)$
62. $5(0.0861) - \dfrac{0.90}{123}$
63. $\sqrt{95.0}$
64. $\sqrt{0.040}$

A.3 LOGARITHMIC COMPUTATIONS AND TABLES

To determine by table the common logarithm of a positive number N (or, equivalently, the exponent above 10 that produces N), we first express N in the scientific notation form

$$N = m \times 10^k, \quad \text{where} \quad 1 \leq m < 10 \text{ and } k \text{ is some integer.}$$

Finding the logarithm to the base 10 of N, we get

$$\begin{aligned}
\log_{10} N = \log_{10}(m \times 10^k) &= \log_{10} m + \log_{10} 10^k \\
&= \log_{10} m + k \log_{10} 10.
\end{aligned}$$

Since $\log_{10} 10 = 1$, we have

$$\log_{10} N = (\log_{10} m) + k.$$

Thus, $\log_{10} N$ is the sum of two numbers: an integer, k, called the **characteristic,** and $\log_{10} m$, called the **mantissa.** Since $1 \leq m < 10$, we find the mantissa by consulting Table 2 at the end of the book, which gives approximations of the logarithms (to the base 10) of numbers between 1 and 10. Consider the excerpt from this table shown in Figure A.1. The first column on the left contains the first two significant digits of m, and the top row contains the third significant digit of m. For example, to find $\log_{10} 5.68$ we look at the intersection of the row opposite 5.6 and the column headed by 8. Thus,

$$\log_{10} 5.68 \approx 0.7543.$$

m	0	1	2	3	4	5	6	7	8	9
5.5	.7404	.7412	.7419	.7427	.7435	.7443	.7451	.7459	.7466	.7474
5.6	.7482	.7490	.7497	.7505	.7513	.7520	.7528	.7536	.7543	.7551
5.7	.7559	.7566	.7574	.7582	.7589	.7597	.7604	.7612	.7619	.7627
5.8	.7634	.7642	.7649	.7657	.7664	.7672	.7679	.7686	.7694	.7701
5.9	.7709	.7716	.7723	.7731	.7738	.7745	.7752	.7760	.7767	.7774

Figure A.1

The following examples should clarify the procedure for finding the logarithm to the base 10 of any positive number. In most cases the equal sign will be used to relate logarithms, with the understanding that these logarithms are only approximations.

EXAMPLE 1   Find $\log_{10} 427$.

Solution

$$\log_{10} 427 = \log_{10}(4.27 \times 10^2)$$
$$= \log_{10} 4.27 + \log_{10} 10^2$$

Since $\log_{10} 10^2 = 2$ and $\log_{10} 4.27 \approx 0.6304$ (Table 2), we have

$$\log_{10} 427 = 0.6304 + 2 = 2.6304.$$ □

EXAMPLE 2   Find $\log_{10} 80$.

Solution

$$\log_{10} 80 = \log_{10}(8.00 \times 10^1)$$
$$= \log_{10} 8.00 + \log_{10} 10^1$$

Since $\log_{10} 10^1 = 1$ and $\log_{10} 8.00 \approx 0.9031$ (Table 2), we have

$$\log_{10} 80 = 0.9031 + 1 = 1.9031.$$ □

EXAMPLE 3   Find $\log_{10} 0.485$.

Solution $\log_{10} 0.485 = \log_{10}(4.85 \times 10^{-1}) = \log_{10} 4.85 + \log_{10} 10^{-1}$
Since $\log_{10} 10^{-1} = -1$ and $\log_{10} 4.85 \approx 0.6857$ (Table 2), we have

$$\log_{10} 0.485 = 0.6857 + (-1) \text{ or } -0.3143.$$ □

EXAMPLE 4   Find $\log_{10} 0.00123$.

Solution

$$\log_{10} 0.00123 = \log_{10}(1.23 \times 10^{-3})$$
$$= \log_{10} 1.23 + \log_{10} 10^{-3}$$

Since $\log_{10} 10^{-3} = -3$ and $\log_{10} 1.23 \approx 0.0899$ (Table 2), we have

$$\log_{10} 0.00123 = 0.0899 + (-3) \text{ or } -2.9101.$$ □

To perform computations, it is necessary to be able to reverse the above procedure and find N if we know $\log_{10} N$. The following examples illustrate this procedure.

EXAMPLE 5   If $\log_{10} N = 2.5250$, find N.

Solution   We know $N = m \times 10^k$, where $1 \le m < 10$ and k is the characteristic.

$$\log_{10} N = 2.5250 = 0.5250 + 2$$

Since $\log_{10} 3.35 \approx 0.5250$ (Table 2), we have $m = 3.35$. Since 2 is the characteristic, we have $k = 2$. Thus,

$$N = 3.35 \times 10^2 = 335. \qquad \square$$

EXAMPLE 6   If $\log_{10} N = 0.9990 + (-3)$, find N.

Solution   We know that $N = m \times 10^k$, where $1 \le m < 10$ and k is the characteristic.

$$\log_{10} N = 0.9990 + (-3)$$

Since $\log_{10} 9.98 \approx 0.9990$ (Table 2), we have $m = 9.98$. Since -3 is the characteristic, we have $k = -3$. Thus,

$$N = 9.98 \times 10^{-3} = 0.00998. \qquad \square$$

EXAMPLE 7   If $\log_{10} N = -1.4567$, find N.

Solution   We must be careful because -1.4567 *does not equal* $0.4567 + (-1)$. We must have a positive mantissa to use Table 2, so we change the form of -1.4567 as follows:

$$\begin{aligned} -1.4567 &= (-1.4567 + 2) + (-2) \\ &= 0.5433 + (-2). \end{aligned}$$

We know that $N = m \times 10^k$, where $1 \le m < 10$ and k is the characteristic. Since $\log_{10} 3.49 \approx 0.5433$ (Table 2), $m = 3.49$. Since -2 is the characteristic, $k = -2$. Thus,

$$N = 3.49 \times 10^{-2} = 0.0349. \qquad \square$$

The following examples illustrate the use of logarithms in difficult computations. The properties of logarithms discussed in Section 5.3 are essential to this work.

EXAMPLE 8   Compute $[(343)(0.678)]^{10}$.

Solution

$$N = [(343)(0.678)]^{10}$$
$$\log_{10} N = \log_{10}[(343)(0.678)]^{10}$$
$$= 10(\log_{10} 343 + \log_{10} 0.678)$$
$$= 10\{2.5353 + [0.8312 + (-1)]\}$$
$$= 10(2.3665)$$
$$= 23.665$$

We know that $N = m \times 10^k$, where $1 \leq m < 10$ and k is the characteristic. Since $\log_{10} 4.62 \approx 0.6650$ (Table 2), $m = 4.62$. Since 23 is the characteristic, $k = 23$. Thus,

$$N = 4.62 \times 10^{23}$$
$$= 462,000,000,000,000,000,000,000.$$ □

EXAMPLE 9   Find an approximate value for $2^{\sqrt{2}}$. (*Note:* $\sqrt{2} \approx 1.41$.)

Solution

$$N = 2^{\sqrt{2}}$$
$$\log_{10} N = \log_{10} 2^{\sqrt{2}} = \sqrt{2} \log_{10} 2$$

Referring to Table 2, we have $\log_{10} 2 \approx 0.3010$.

$$\log_{10} N = \sqrt{2}(0.3010) = (1.41)(0.3010)$$
$$= 0.4244$$

We know that $N = m \times 10^k$, where $1 \leq m < 10$ and k is the characteristic. Since $\log_{10} 2.66 \approx 0.4244$ (Table 2), $m = 2.66$. Since 0 is the characteristic, $k = 0$. Thus,

$$N = 2.66 \times 10^0 = 2.66.$$ □

EXAMPLE 10   Compute $\sqrt[5]{7.13/22.1}$.

Solution

$$N = \sqrt[5]{\frac{7.13}{22.1}}$$
$$\log_{10} N = \log_{10} \left(\frac{7.13}{22.1}\right)^{1/5}$$
$$= \tfrac{1}{5}(\log_{10} 7.13 - \log_{10} 22.1)$$
$$= \tfrac{1}{5}(0.8531 - 1.3444)$$
$$= \tfrac{1}{5}(-0.4913)$$
$$= -0.0983$$

To find N, we must write the logarithm in a form in which the mantissa is positive.

$$\log_{10} N = (-0.0983 + 1) + (-1)$$
$$= 0.9017 + (-1)$$

We know that $N = m \times 10^k$, where $1 \leq m < 10$ and k is the characteristic. Since $\log_{10} 7.97 \approx 0.9017$ (Table 2), $m = 7.97$. Since -1 is the characteristic, $k = -1$. Thus,

$$N = 7.97 \times 10^{-1} = 0.797. \qquad \square$$

One method of determining the natural log of a number is to change the expression to base 10 logarithms.

EXAMPLE 11   Use logarithms to the base 10 to determine ln 170.

Solution

$$\ln 170 = \frac{\log_{10} 170}{\log_{10} e} = \frac{\log_{10} 170}{\log_{10} 2.72} = \frac{2.2304}{0.4346}$$
$$\ln 170 \approx 5.132 \qquad \square$$

A second method, illustrated in the following example, is more accurate and uses a table of natural logarithms (Table 3 at the end of the book).

EXAMPLE 12   Find ln 170.

Solution

$$\ln 170 = \ln(1.70 \times 10^2)$$
$$= \ln 1.7 + \ln 10^2$$
$$= \ln 1.7 + (2) \ln 10$$

Referring to Table 3, we have $\ln 1.7 \approx 0.5306$ and $\ln 10 \approx 2.3026$.

$$\ln 170 = 0.5306 + 2(2.3026)$$
$$\ln 170 \approx 5.136 \qquad \square$$

EXAMPLE 13   Find ln 0.45.

Solution

$$\ln 0.45 = \ln(4.5 \times 10^{-1})$$
$$= \ln 4.5 + \ln 10^{-1}$$
$$= \ln 4.5 + (-1)\ln 10$$

Referring to Table 3, we have $\ln 4.5 \approx 1.5041$ and $\ln 10 \approx 2.3026$. Thus,

$$\ln 0.45 = 1.5041 + (-1)(2.3026)$$
$$= -0.7985. \qquad \square$$

EXAMPLE 14   If $\ln x = 0.95$, find x.

Solution $\ln x = 0.95$. Writing this statement in exponential form, we have

$$e^{0.95} = x$$
$$2.5857 = x. \quad \text{(Table 4)} \qquad \square$$

EXAMPLE 15   If $e^{0.07t} = 2$, find t by using natural logarithms.

Solution   If $e^{0.07t} = 2$, then $\ln 2 = 0.07t$. By Table 3, $\ln 2 = 0.6931$. Thus, $t = 0.6931/0.07 \approx 9.9$. $\qquad \square$

EXERCISES A.3

In Exercises 1–10 find the logarithm to the base 10 of the number. Use Table 2.

1. 2.34
2. 23.4
3. 0.234
4. 0.0234
5. 2,340
6. 187,000
7. 90
8. 0.9
9. 0.111
10. 0.000717

In Exercises 11–20 find N if $\log_{10} N$ equals the number. Use Table 2.

11. 0.8189
12. 2.8189
13. $0.8189 + (-1)$
14. $0.8189 + (-3)$
15. 1.5559
16. 3.9803
17. -1.4123
18. -2.6712
19. -0.5173
20. -4.2990

In Exercises 21–40 compute by means of logarithms.

21. $(413)(2.74)$
22. $(0.0123)(12.7)$
23. $\dfrac{908}{14.3}$
24. $\dfrac{19.2}{604}$
25. $(3.2)^5$
26. $(0.000491)^4$
27. $\sqrt{97}$
28. $\sqrt[5]{0.4}$
29. $\dfrac{(514)(29.6)}{0.02}$
30. $\dfrac{17.9}{(49.6)(0.0821)}$
31. $\sqrt{(16.1)(2.5)}$
32. $\sqrt[3]{(123)(0.22)^2}$
33. $\sqrt[6]{\dfrac{195}{7.6}}$
34. $\sqrt{\dfrac{22.2}{459}}$
35. $\dfrac{(1.8)^2(5.6)^3}{\sqrt{104}}$
36. $\dfrac{\sqrt[3]{17}}{(671)(0.08)^5}$
37. $3^{\sqrt{3}}$ (*Note:* $\sqrt{3} \approx 1.73$.)
38. 2^{π} (*Note:* $\pi \approx 3.14$.)
39. $(12.1)^{-5}$
40. $(1 + 0.01)^{-24}$

In Exercises 41–50 use Table 3 to determine the natural logarithm.

41. ln 7   **42.** ln 3.8   **43.** ln 42

44. ln 123   **45.** ln 6,500   **46.** ln 100

47. ln 0.56   **48.** ln 0.009   **49.** ln 0.00044

50. ln 0.292

In Exercises 51–58 find x if ln x equals the number.

51. 0.19   **52.** 4.6   **53.** 2.3   **54.** 0.55

55. −0.24   **56.** −7.5   **57.** −3.7   **58.** −0.09

A.4 GRAPHS ON LOGARITHMIC PAPER

Sometimes we need to analyze data that range over a large collection of values. In addition, the analysis often requires us to find a formula that fits a set of data. Here we study a scheme that is useful in such situations.

On logarithmic graph paper the axes are laid off at distances proportional to the logarithms of numbers. Consider Figure A.2, in which we construct an axis by marking on the paper the following numbers:

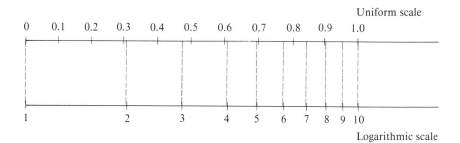

Figure A.2

$$\log_{10} 1 = 0 \qquad \log_{10} 6 = 0.778$$
$$\log_{10} 2 = 0.301 \qquad \log_{10} 7 = 0.845$$
$$\log_{10} 3 = 0.477 \qquad \log_{10} 8 = 0.903$$
$$\log_{10} 4 = 0.602 \qquad \log_{10} 9 = 0.954$$
$$\log_{10} 5 = 0.699 \qquad \log_{10} 10 = 1 \qquad \text{etc.}$$

On the logarithmic scale we use 1 as shorthand for $\log_{10} 1$. Since $\log_{10} x$ is undefined for $x \leq 0$, only positive numbers may be marked off on logarithmic paper. Using this paper, we may obtain accuracy in a graph in which the variables range over a large collection of values.

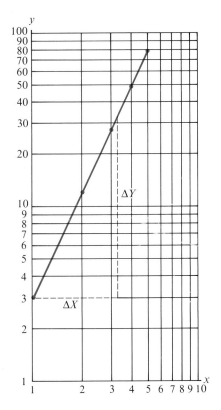

Figure A.3 $y = 3x^2$

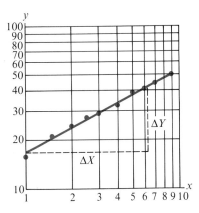

Figure A.4 $y = 17\sqrt{x}$

Logarithmic paper is especially useful when graphing functions of the form $y = ax^m$, which are called **power functions.** By taking the logarithm to the base 10 of a power function, we have

$$\log y = \log ax^m$$
$$= \log a + m \log x.$$

If we let $\log y = Y$, $\log x = X$, and $\log a = b$, the equation becomes

$$Y = b + mX$$

Thus, on logarithmic paper the graph of a power function is a straight line with slope m and Y-intercept b. It also follows that if the graph of experimental data is a straight line on logarithmic paper, the variables are related by an equation of the form $y = ax^m$. The slope m is measured with a ruler and the coefficient a is the intercept on the vertical axis at $x = 1$.

EXAMPLE 1   Use logarithmic paper to graph $y = 3x^2$.

Solution   First, construct a table of corresponding values of x and y.

x	1	2	3	4	5
y	3	12	27	48	75

As shown in Figure A.3, the graph is a straight line. Using a ruler you may verify that $m = \Delta Y/\Delta X = 2$. The intercept at $x = 1$ is 3. These readings agree with the constants in the equation $y = 3x^2$. In this example the y-axis is scaled in two cycles from $\log 1 = 0$ to $\log 10 = 1$ to $\log 100 = 2$. Logarithmic paper may contain any number of cycles. A cycle ranges between any two numbers whose logarithms are consecutive integers.   □

EXAMPLE 2   Determine the equation relating the variables in the following table.

x	1.0	1.5	2.0	2.5	3.0	4.0	5.0	6.0	7.0	9.0
y	16	21	24	27	29	33	38	41	44	50

Solution   Consider Figure A.4. The vertical axis requires one cycle that ranges from $\log 10 = 1$ to $\log 100 = 2$. Since the graph is approximately a straight line, the equation relating the variables is a power function. Using a ruler, $m = \Delta Y/\Delta X \approx 0.5$. The intercept at $x = 1$ is about 17. Thus, the power function is $y \approx 17x^{0.5}$ or $y \approx 17\sqrt{x}$.   □

On semilogarithmic graph paper one axis is scaled as on logarithmic paper and the other axis is scaled as on Cartesian coordinate paper. This paper is used when only one variable ranges over a large collection of values. Semilogarithmic paper is especially useful when graphing exponential functions since the graph is a straight line.

EXAMPLE 3   Use semilogarithmic paper to graph $y = 2^x$.

Solution   First, construct a table of corresponding values of x and y.

x	-2	-1	0	1	2	3
y	0.25	0.5	1	2	4	8

The vertical axis requires two logarithmic cycles. The horizontal axis is equally spaced and is scaled with positive and negative numbers. As shown in Figure A.5, the graph is a straight line.

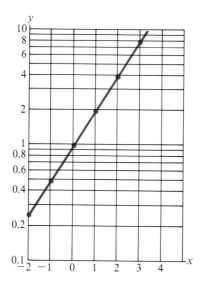

Figure A.5 $y = 2^x$

Although the graphs in the examples are all straight lines, other curves may result. We obtain a straight line only when graphing a power function on logarithmic paper and an exponential function on semilogarithmic paper.

EXERCISES A.4

In Exercises 1–6 graph the functions on logarithmic paper. Also, find the slope and the intercept at $x = 1$.

1. $y = x^2$ **2.** $y = 2x^3$ **3.** $y = 4\sqrt{x}$

4. $y = 5x^{2/3}$ **5.** $xy = 1$ **6.** $x^2y = 10$

In Exercises 7–10 use logarithmic paper to determine the equation relating the variables in the given tables.

7.

x	1	2	3	4	5	6	7
y	7.0	5.5	4.8	4.3	4.0	3.7	3.5

8.

x	1.0	2.0	3.0	4.0	5.0	6.0	7.0
y	5.0	1.4	.66	.39	.26	.19	.14

9.

x	1	3	5	7	9	25	30	70
y	3.0	6.0	8.6	11	13	25	28	49

10.

x	1.5	4.5	7.5	11	16	20	30
y	2.1	8.4	16	25	40	53	88

In Exercises 11–14 graph the functions on semilogarithmic paper.

11. $y = 2^{-x}$ **12.** $y = 5^x$

13. $y = 0.5(3)^x$ **14.** $y = 2(3^{-x})$

Tables

Table 1   Squares, Square Roots, and Prime Factors

No.	Sq.	Sq. rt.	Factors	No.	Sq.	Sq. rt.	Factors
1	1	1.000		51	2,601	7.141	$3 \cdot 17$
2	4	1.414	2	52	2,704	7.211	$2^2 \cdot 13$
3	9	1.732	3	53	2,809	7.280	53
4	16	2.000	2^2	54	2,916	7.348	$2 \cdot 3^3$
5	25	2.236	5	55	3,025	7.416	$5 \cdot 11$
6	36	2.449	$2 \cdot 3$	56	3,136	7.483	$2^3 \cdot 7$
7	49	2.646	7	57	3,249	7.550	$3 \cdot 19$
8	64	2.828	2^3	58	3,364	7.616	$2 \cdot 29$
9	81	3.000	3^2	59	3,481	7.681	59
10	100	3.162	$2 \cdot 5$	60	3,600	7.746	$2^2 \cdot 3 \cdot 5$
11	121	3.317	11	61	3,721	7.810	61
12	144	3.464	$2^2 \cdot 3$	62	3,844	7.874	$2 \cdot 31$
13	169	3.606	13	63	3,969	7.937	$3^2 \cdot 7$
14	196	3.742	$2 \cdot 7$	64	4,096	8.000	2^6
15	225	3.873	$3 \cdot 5$	65	4,225	8.062	$5 \cdot 13$
16	256	4.000	2^4	66	4,356	8.124	$2 \cdot 3 \cdot 11$
17	289	4.123	17	67	4,489	8.185	67
18	324	4.243	$2 \cdot 3^2$	68	4,624	8.246	$2^2 \cdot 17$
19	361	4.359	19	69	4,761	8.307	$3 \cdot 23$
20	400	4.472	$2^2 \cdot 5$	70	4,900	8.367	$2 \cdot 5 \cdot 7$
21	441	4.583	$3 \cdot 7$	71	5,041	8.426	71
22	484	4.690	$2 \cdot 11$	72	5,184	8.485	$2^3 \cdot 3^2$
23	529	4.796	23	73	5,329	8.544	73
24	576	4.899	$2^3 \cdot 3$	74	5,476	8.602	$2 \cdot 37$
25	625	5.000	5^2	75	5,625	8.660	$3 \cdot 5^2$
26	676	5.099	$2 \cdot 13$	76	5,776	8.718	$2^2 \cdot 19$
27	729	5.196	3^3	77	5,929	8.775	$7 \cdot 11$
28	784	5.292	$2^2 \cdot 7$	78	6,084	8.832	$2 \cdot 3 \cdot 13$
29	841	5.385	29	79	6,241	8.888	79
30	900	5.477	$2 \cdot 3 \cdot 5$	80	6,400	8.944	$2^4 \cdot 5$
31	961	5.568	31	81	6,561	9.000	3^4
32	1,024	5.657	2^5	82	6,724	9.055	$2 \cdot 41$
33	1,089	5.745	$3 \cdot 11$	83	6,889	9.110	83
34	1,156	5.831	$2 \cdot 17$	84	7,056	9.165	$2^2 \cdot 3 \cdot 7$
35	1,225	5.916	$5 \cdot 7$	85	7,225	9.220	$5 \cdot 17$
36	1,296	6.000	$2^2 \cdot 3^2$	86	7,396	9.274	$2 \cdot 43$
37	1,369	6.083	37	87	7,569	9.327	$3 \cdot 29$
38	1,444	6.164	$2 \cdot 19$	88	7,744	9.381	$2^3 \cdot 11$
39	1,521	6.245	$3 \cdot 13$	89	7,921	9.434	89
40	1,600	6.325	$2^3 \cdot 5$	90	8,100	9.487	$2 \cdot 3^2 \cdot 5$
41	1,681	6.403	41	91	8,281	9.539	$7 \cdot 13$
42	1,764	6.481	$2 \cdot 3 \cdot 7$	92	8,464	9.592	$2^2 \cdot 23$
43	1,849	6.557	43	93	8,649	9.644	$3 \cdot 31$
44	1,936	6.633	$2^2 \cdot 11$	94	8,836	9.695	$2 \cdot 47$
45	2,025	6.708	$3^2 \cdot 5$	95	9,025	9.747	$5 \cdot 19$
46	2,116	6.782	$2 \cdot 23$	96	9,216	9.798	$2^5 \cdot 3$
47	2,209	6.856	47	97	9,409	9.849	97
48	2,304	6.928	$2^4 \cdot 3$	98	9,604	9.899	$2 \cdot 7^2$
49	2,401	7.000	7^2	99	9,801	9.950	$3^2 \cdot 11$
50	2,500	7.071	$2 \cdot 5^2$	100	10,000	10.000	$2^2 \cdot 5^2$

Table 2   Common Logarithms

m	0	1	2	3	4	5	6	7	8	9
1.0	.0000	.0043	.0086	.0128	.0170	.0212	.0253	.0294	.0334	.0374
1.1	.0414	.0453	.0492	.0531	.0569	.0607	.0645	.0682	.0719	.0755
1.2	.0792	.0828	.0864	.0899	.0934	.0969	.1004	.1038	.1072	.1106
1.3	.1139	.1173	.1206	.1239	.1271	.1303	.1335	.1367	.1399	.1430
1.4	.1461	.1492	.1523	.1553	.1584	.1614	.1644	.1673	.1703	.1732
1.5	.1761	.1790	.1818	.1847	.1875	.1903	.1931	.1959	.1987	.2014
1.6	.2041	.2068	.2095	.2122	.2148	.2175	.2201	.2227	.2253	.2279
1.7	.2304	.2330	.2355	.2380	.2405	.2430	.2455	.2480	.2504	.2529
1.8	.2553	.2577	.2601	.2625	.2648	.2672	.2695	.2718	.2742	.2765
1.9	.2788	.2810	.2833	.2856	.2878	.2900	.2923	.2945	.2967	.2989
2.0	.3010	.3032	.3054	.3075	.3096	.3118	.3139	.3160	.3181	.3201
2.1	.3222	.3243	.3263	.3284	.3304	.3324	.3345	.3365	.3385	.3404
2.2	.3424	.3444	.3464	.3483	.3502	.3522	.3541	.3560	.3579	.3598
2.3	.3617	.3636	.3655	.3674	.3692	.3711	.3729	.3747	.3766	.3784
2.4	.3802	.3820	.3838	.3856	.3874	.3892	.3909	.3927	.3945	.3962
2.5	.3979	.3997	.4014	.4031	.4048	.4065	.4082	.4099	.4116	.4133
2.6	.4150	.4166	.4183	.4200	.4216	.4232	.4249	.4265	.4281	.4298
2.7	.4314	.4330	.4346	.4362	.4378	.4393	.4409	.4425	.4440	.4456
2.8	.4472	.4487	.4502	.4518	.4533	.4548	.4564	.4579	.4594	.4609
2.9	.4624	.4639	.4654	.4669	.4683	.4698	.4713	.4728	.4742	.4757
3.0	.4771	.4786	.4800	.4814	.4829	.4843	.4857	.4871	.4886	.4900
3.1	.4914	.4928	.4942	.4955	.4969	.4983	.4997	.5011	.5024	.5038
3.2	.5051	.5065	.5079	.5092	.5105	.5119	.5132	.5145	.5159	.5172
3.3	.5185	.5198	.5211	.5224	.5237	.5250	.5263	.5276	.5289	.5302
3.4	.5315	.5328	.5340	.5353	.5366	.5378	.5391	.5403	.5416	.5428
3.5	.5441	.5453	.5465	.5478	.5490	.5502	.5514	.5527	.5539	.5551
3.6	.5563	.5575	.5587	.5599	.5611	.5623	.5635	.5647	.5658	.5670
3.7	.5682	.5694	.5705	.5717	.5729	.5740	.5752	.5763	.5775	.5786
3.8	.5798	.5809	.5821	.5832	.5843	.5855	.5866	.5877	.5888	.5899
3.9	.5911	.5922	.5933	.5944	.5955	.5966	.5977	.5988	.5999	.6010
4.0	.6021	.6031	.6042	.6053	.6064	.6075	.6085	.6096	.6107	.6117
4.1	.6128	.6138	.6149	.6160	.6170	.6180	.6191	.6201	.6212	.6222
4.2	.6232	.6243	.6253	.6263	.6274	.6284	.6294	.6304	.6314	.6325
4.3	.6335	.6345	.6355	.6365	.6375	.6385	.6395	.6405	.6415	.6425
4.4	.6435	.6444	.6454	.6464	.6474	.6484	.6493	.6503	.6513	.6522
4.5	.6532	.6542	.6551	.6561	.6571	.6580	.6590	.6599	.6609	.6618
4.6	.6628	.6637	.6646	.6656	.6665	.6675	.6684	.6693	.6702	.6712
4.7	.6721	.6730	.6739	.6749	.6758	.6767	.6776	.6785	.6794	.6803
4.8	.6812	.6821	.6830	.6839	.6848	.6857	.6866	.6875	.6884	.6893
4.9	.6902	.6911	.6920	.6928	.6937	.6946	.6955	.6964	.6972	.6981
5.0	.6990	.6998	.7007	.7016	.7024	.7033	.7042	.7050	.7059	.7067
5.1	.7076	.7084	.7093	.7101	.7110	.7118	.7126	.7135	.7143	.7152
5.2	.7160	.7168	.7177	.7185	.7193	.7202	.7210	.7218	.7226	.7235
5.3	.7243	.7251	.7259	.7267	.7275	.7284	.7292	.7300	.7308	.7316
5.4	.7324	.7332	.7340	.7348	.7356	.7364	.7372	.7380	.7388	.7396

Table 2 *(Continued)*

m	0	1	2	3	4	5	6	7	8	9
5.5	.7404	.7412	.7419	.7427	.7435	.7443	.7451	.7459	.7466	.7474
5.6	.7482	.7490	.7497	.7505	.7513	.7520	.7528	.7536	.7543	.7551
5.7	.7559	.7566	.7574	.7582	.7589	.7597	.7604	.7612	.7619	.7627
5.8	.7634	.7642	.7649	.7657	.7664	.7672	.7679	.7686	.7694	.7701
5.9	.7709	.7716	.7723	.7731	.7738	.7745	.7752	.7760	.7767	.7774
6.0	.7782	.7789	.7796	.7803	.7810	.7818	.7825	.7832	.7839	.7846
6.1	.7853	.7860	.7868	.7875	.7882	.7889	.7896	.7903	.7910	.7917
6.2	.7924	.7931	.7938	.7945	.7952	.7959	.7966	.7973	.7980	.7987
6.3	.7993	.8000	.8007	.8014	.8021	.8028	.8035	.8041	.8048	.8055
6.4	.8062	.8069	.8075	.8082	.8089	.8096	.8102	.8109	.8116	.8122
6.5	.8129	.8136	.8142	.8149	.8156	.8162	.8169	.8176	.8182	.8189
6.6	.8195	.8202	.8209	.8215	.8222	.8228	.8235	.8241	.8248	.8254
6.7	.8261	.8267	.8274	.8280	.8287	.8293	.8299	.8306	.8312	.8319
6.8	.8325	.8331	.8338	.8344	.8351	.8357	.8363	.8370	.8376	.8382
6.9	.8388	.8395	.8401	.8407	.8414	.8420	.8426	.8432	.8439	.8445
7.0	.8451	.8457	.8463	.8470	.8476	.8482	.8488	.8494	.8500	.8506
7.1	.8513	.8519	.8525	.8531	.8537	.8543	.8549	.8555	.8561	.8567
7.2	.8573	.8579	.8585	.8591	.8597	.8603	.8609	.8615	.8621	.8627
7.3	.8633	.8639	.8645	.8651	.8657	.8663	.8669	.8675	.8681	.8686
7.4	.8692	.8698	.8704	.8710	.8716	.8722	.8727	.8733	.8739	.8745
7.5	.8751	.8756	.8762	.8768	.8774	.8779	.8785	.8791	.8797	.8802
7.6	.8808	.8814	.8820	.8825	.8831	.8837	.8842	.8848	.8854	.8859
7.7	.8865	.8871	.8876	.8882	.8887	.8893	.8899	.8904	.8910	.8915
7.8	.8921	.8927	.8932	.8938	.8943	.8949	.8954	.8960	.8965	.8971
7.9	.8976	.8982	.8987	.8993	.8998	.9004	.9009	.9015	.9020	.9025
8.0	.9031	.9036	.9042	.9047	.9053	.9058	.9063	.9069	.9074	.9079
8.1	.9085	.9090	.9096	.9101	.9106	.9112	.9117	.9122	.9128	.9133
8.2	.9138	.9143	.9149	.9154	.9159	.9165	.9170	.9175	.9180	.9186
8.3	.9191	.9196	.9201	.9206	.9212	.9217	.9222	.9227	.9232	.9238
8.4	.9243	.9249	.9253	.9258	.9263	.9269	.9274	.9279	.9284	.9289
8.5	.9294	.9299	.9304	.9309	.9315	.9320	.9325	.9330	.9335	.9340
8.6	.9345	.9350	.9355	.9360	.9365	.9370	.9375	.9380	.9385	.9390
8.7	.9395	.9400	.9405	.9410	.9415	.9420	.9425	.9430	.9435	.9440
8.8	.9445	.9450	.9455	.9460	.9465	.9469	.9474	.9479	.9484	.9489
8.9	.9494	.9499	.9504	.9509	.9513	.9518	.9523	.9528	.9533	.9538
9.0	.9542	.9547	.9552	.9557	.9562	.9566	.9571	.9576	.9581	.9586
9.1	.9590	.9595	.9600	.9605	.9609	.9614	.9619	.9624	.9628	.9633
9.2	.9638	.9643	.9647	.9652	.9657	.9661	.9666	.9671	.9675	.9680
9.3	.9685	.9689	.9694	.9699	.9703	.9708	.9713	.9717	.9722	.9727
9.4	.9731	.9736	.9741	.9745	.9750	.9754	.9759	.9763	.9768	.9773
9.5	.9777	.9782	.9786	.9791	.9795	.9800	.9805	.9809	.9814	.9818
9.6	.9823	.9827	.9832	.9836	.9841	.9845	.9850	.9854	.9859	.9863
9.7	.9868	.9872	.9877	.9881	.9886	.9890	.9894	.9899	.9903	.9908
9.8	.9912	.9917	.9921	.9926	.9930	.9934	.9939	.9943	.9948	.9952
9.9	.9956	.9961	.9965	.9969	.9974	.9978	.9983	.9987	.9991	.9996

Table 3   Natural Logarithms (Base e)

	0	1	2	3	4	5	6	7	8	9
1.0	0.0000	0.0100	0.0198	0.0296	0.0392	0.0488	0.0583	0.0677	0.0770	0.0862
1.1	0.0953	0.1044	0.1133	0.1222	0.1310	0.1398	0.1484	0.1570	0.1655	0.1740
1.2	0.1823	0.1906	0.1989	0.2070	0.2151	0.2231	0.2311	0.2390	0.2469	0.2546
1.3	0.2624	0.2700	0.2776	0.2852	0.2927	0.3001	0.3075	0.3148	0.3221	0.3293
1.4	0.3365	0.3436	0.3507	0.3577	0.3646	0.3716	0.3784	0.3853	0.3920	0.3988
1.5	0.4055	0.4121	0.4187	0.4253	0.4318	0.4383	0.4447	0.4511	0.4574	0.4637
1.6	0.4700	0.4762	0.4824	0.4886	0.4947	0.5008	0.5068	0.5128	0.5188	0.5247
1.7	0.5306	0.5365	0.5423	0.5481	0.5539	0.5596	0.5653	0.5710	0.5766	0.5822
1.8	0.5878	0.5933	0.5988	0.6043	0.6098	0.6152	0.6206	0.6259	0.6313	0.6366
1.9	0.6419	0.6471	0.6523	0.6575	0.6627	0.6678	0.6729	0.6780	0.6831	0.6881
2.0	0.6931	0.6981	0.7031	0.7080	0.7129	0.7178	0.7227	0.7275	0.7324	0.7372
2.1	0.7419	0.7467	0.7514	0.7561	0.7608	0.7655	0.7701	0.7747	0.7793	0.7839
2.2	0.7885	0.7930	0.7975	0.8020	0.8065	0.8109	0.8154	0.8198	0.8242	0.8286
2.3	0.8329	0.8373	0.8416	0.8459	0.8502	0.8544	0.8587	0.8629	0.8671	0.8713
2.4	0.8755	0.8796	0.8838	0.8879	0.8920	0.8961	0.9002	0.9042	0.9083	0.9123
2.5	0.9163	0.9203	0.9243	0.9282	0.9322	0.9361	0.9400	0.9439	0.9478	0.9517
2.6	0.9555	0.9594	0.9632	0.9670	0.9708	0.9746	0.9783	0.9821	0.9858	0.9895
2.7	0.9933	0.9969	1.0006	1.0043	1.0080	1.0116	1.0152	1.0188	1.0225	1.0260
2.8	1.0296	1.0332	1.0367	1.0403	1.0438	1.0473	1.0508	1.0543	1.0578	1.0613
2.9	1.0647	1.0682	1.0716	1.0750	1.0784	1.0818	1.0852	1.0886	1.0919	1.0953
3.0	1.0986	1.1019	1.1053	1.1086	1.1119	1.1151	1.1184	1.1217	1.1249	1.1282
3.1	1.1314	1.1346	1.1378	1.1410	1.1442	1.1474	1.1506	1.1537	1.1569	1.1600
3.2	1.1632	1.1663	1.1694	1.1725	1.1756	1.1787	1.1817	1.1848	1.1878	1.1909
3.3	1.1939	1.1969	1.2000	1.2030	1.2060	1.2090	1.2119	1.2149	1.2179	1.2208
3.4	1.2238	1.2267	1.2296	1.2326	1.2355	1.2384	1.2413	1.2442	1.2470	1.2499
3.5	1.2528	1.2556	1.2585	1.2613	1.2641	1.2669	1.2698	1.2726	1.2754	1.2782
3.6	1.2809	1.2837	1.2865	1.2892	1.2920	1.2947	1.2975	1.3002	1.3029	1.3056
3.7	1.3083	1.3110	1.3137	1.3164	1.3191	1.3218	1.3244	1.3271	1.3297	1.3324
3.8	1.3350	1.3376	1.3403	1.3429	1.3455	1.3481	1.3507	1.3533	1.3558	1.3584
3.9	1.3610	1.3635	1.3661	1.3686	1.3712	1.3737	1.3762	1.3788	1.3813	1.3838
4.0	1.3863	1.3888	1.3913	1.3938	1.3962	1.3987	1.4012	1.4036	1.4061	1.4085
4.1	1.4110	1.4134	1.4159	1.4183	1.4207	1.4231	1.4255	1.4279	1.4303	1.4327
4.2	1.4351	1.4375	1.4398	1.4422	1.4446	1.4469	1.4493	1.4516	1.4540	1.4563
4.3	1.4586	1.4609	1.4633	1.4656	1.4679	1.4702	1.4725	1.4748	1.4771	1.4793
4.4	1.4816	1.4839	1.4861	1.4884	1.4907	1.4929	1.4951	1.4974	1.4996	1.5019
4.5	1.5041	1.5063	1.5085	1.5107	1.5129	1.5151	1.5173	1.5195	1.5217	1.5239
4.6	1.5261	1.5282	1.5304	1.5326	1.5347	1.5369	1.5390	1.5412	1.5433	1.5454
4.7	1.5476	1.5497	1.5518	1.5539	1.5560	1.5581	1.5602	1.5623	1.5644	1.5665
4.8	1.5686	1.5707	1.5728	1.5748	1.5769	1.5790	1.5810	1.5831	1.5851	1.5872
4.9	1.5892	1.5913	1.5933	1.5953	1.5974	1.5994	1.6014	1.6034	1.6054	1.6074
5.0	1.6094	1.6114	1.6134	1.6154	1.6174	1.6194	1.6214	1.6233	1.6253	1.6273
5.1	1.6292	1.6312	1.6332	1.6351	1.6371	1.6390	1.6409	1.6429	1.6448	1.6467
5.2	1.6487	1.6506	1.6525	1.6544	1.6563	1.6582	1.6601	1.6620	1.6639	1.6658
5.3	1.6677	1.6696	1.6715	1.6734	1.6752	1.6771	1.6790	1.6808	1.6827	1.6845
5.4	1.6864	1.6882	1.6901	1.6919	1.6938	1.6956	1.6974	1.6993	1.7011	1.7029

Table 3 (*Continued*)

	0	1	2	3	4	5	6	7	8	9
5.5	1.7047	1.7066	1.7084	1.7102	1.7120	1.7138	1.7156	1.7174	1.7192	1.7210
5.6	1.7228	1.7246	1.7263	1.7281	1.7299	1.7317	1.7334	1.7352	1.7370	1.7387
5.7	1.7405	1.7422	1.7440	1.7457	1.7475	1.7492	1.7509	1.7527	1.7544	1.7561
5.8	1.7579	1.7596	1.7613	1.7630	1.7647	1.7664	1.7681	1.7699	1.7716	1.7733
5.9	1.7750	1.7766	1.7783	1.7800	1.7817	1.7834	1.7851	1.7868	1.7884	1.7901
6.0	1.7918	1.7934	1.7951	1.7967	1.7984	1.8001	1.8017	1.8034	1.8050	1.8066
6.1	1.8083	1.8099	1.8116	1.8132	1.8148	1.8165	1.8181	1.8197	1.8213	1.8229
6.2	1.8245	1.8262	1.8278	1.8294	1.8310	1.8326	1.8342	1.8358	1.8374	1.8390
6.3	1.8405	1.8421	1.8437	1.8453	1.8469	1.8485	1.8500	1.8516	1.8532	1.8547
6.4	1.8563	1.8579	1.8594	1.8610	1.8625	1.8641	1.8656	1.8672	1.8687	1.8703
6.5	1.8718	1.8733	1.8749	1.8764	1.8779	1.8795	1.8810	1.8825	1.8840	1.8856
6.6	1.8871	1.8886	1.8901	1.8916	1.8931	1.8946	1.8961	1.8976	1.8991	1.9006
6.7	1.9021	1.9036	1.9051	1.9066	1.9081	1.9095	1.9110	1.9125	1.9140	1.9155
6.8	1.9169	1.9184	1.9199	1.9213	1.9228	1.9242	1.9257	1.9272	1.9286	1.9301
6.9	1.9315	1.9330	1.9344	1.9359	1.9373	1.9387	1.9402	1.9416	1.9430	1.9445
7.0	1.9459	1.9473	1.9488	1.9502	1.9516	1.9530	1.9544	1.9559	1.9573	1.9587
7.1	1.9601	1.9615	1.9629	1.9643	1.9657	1.9671	1.9685	1.9699	1.9713	1.9727
7.2	1.9741	1.9755	1.9769	1.9782	1.9796	1.9810	1.9824	1.9838	1.9851	1.9865
7.3	1.9879	1.9892	1.9906	1.9920	1.9933	1.9947	1.9961	1.9974	1.9988	2.0001
7.4	2.0015	2.0028	2.0042	2.0055	2.0069	2.0082	2.0096	2.0109	2.0122	2.0136
7.5	2.0149	2.0162	2.0176	2.0189	2.0202	2.0215	2.0229	2.0242	2.0255	2.0268
7.6	2.0281	2.0295	2.0308	2.0321	2.0334	2.0347	2.0360	2.0373	2.0386	2.0399
7.7	2.0412	2.0425	2.0438	2.0451	2.0464	2.0477	2.0490	2.0503	2.0516	2.0528
7.8	2.0541	2.0554	2.0567	2.0580	2.0592	2.0605	2.0618	2.0631	2.0643	2.0656
7.9	2.0669	2.0681	2.0694	2.0707	2.0719	2.0732	2.0744	2.0757	2.0769	2.0782
8.0	2.0794	2.0807	2.0819	2.0832	2.0844	2.0857	2.0869	2.0882	2.0894	2.0906
8.1	2.0919	2.0931	2.0943	2.0956	2.0968	2.0980	2.0992	2.1005	2.1017	2.1029
8.2	2.1041	2.1054	2.1066	2.1078	2.1090	2.1102	2.1114	2.1126	2.1138	2.1150
8.3	2.1163	2.1175	2.1187	2.1199	2.1211	2.1223	2.1235	2.1247	2.1259	2.1270
8.4	2.1282	2.1294	2.1306	2.1318	2.1330	2.1342	2.1353	2.1365	2.1377	2.1389
8.5	2.1401	2.1412	2.1424	2.1436	2.1448	2.1459	2.1471	2.1483	2.1494	2.1506
8.6	2.1518	2.1529	2.1541	2.1552	2.1564	2.1576	2.1587	2.1599	2.1610	2.1622
8.7	2.1633	2.1645	2.1656	2.1668	2.1679	2.1691	2.1702	2.1713	2.1725	2.1736
8.8	2.1748	2.1759	2.1770	2.1782	2.1793	2.1804	2.1815	2.1827	2.1838	2.1849
8.9	2.1861	2.1872	2.1883	2.1894	2.1905	2.1917	2.1928	2.1939	2.1950	2.1961
9.0	2.1972	2.1983	2.1994	2.2006	2.2017	2.2028	2.2039	2.2050	2.2061	2.2072
9.1	2.2083	2.2094	2.2105	2.2116	2.2127	2.2138	2.2148	2.2159	2.2170	2.2181
9.2	2.2192	2.2203	2.2214	2.2225	2.2235	2.2246	2.2257	2.2268	2.2279	2.2289
9.3	2.2300	2.2311	2.2322	2.2332	2.2343	2.2354	2.2364	2.2375	2.2386	2.2396
9.4	2.2407	2.2418	2.2428	2.2439	2.2450	2.2460	2.2471	2.2481	2.2492	2.2502
9.5	2.2513	2.2523	2.2534	2.2544	2.2555	2.2565	2.2576	2.2586	2.2597	2.2607
9.6	2.2618	2.2628	2.2638	2.2649	2.2659	2.2670	2.2680	2.2690	2.2701	2.2711
9.7	2.2721	2.2732	2.2742	2.2752	2.2762	2.2773	2.2783	2.2793	2.2803	2.2814
9.8	2.2824	2.2834	2.2844	2.2854	2.2865	2.2875	2.2885	2.2895	2.2905	2.2915
9.9	2.2925	2.2935	2.2946	2.2956	2.2966	2.2976	2.2986	2.2996	2.3006	2.3016

$\ln 10 \approx 2.3026$

Table 4   Exponential Functions

x	e^x	e^{-x}	x	e^x	e^{-x}
0.00	1.0000	1.0000	1.5	4.4817	0.2231
0.01	1.0101	0.9901	1.6	4.9530	0.2019
0.02	1.0202	0.9802	1.7	5.4739	0.1827
0.03	1.0305	0.9705	1.8	6.0496	0.1653
0.04	1.0408	0.9608	1.9	6.6859	0.1496
0.05	1.0513	0.9512	2.0	7.3891	0.1353
0.06	1.0618	0.9418	2.1	8.1662	0.1225
0.07	1.0725	0.9324	2.2	9.0250	0.1108
0.08	1.0833	0.9231	2.3	9.9742	0.1003
0.09	1.0942	0.9139	2.4	11.023	0.0907
0.10	1.1052	0.9048	2.5	12.182	0.0821
0.11	1.1163	0.8958	2.6	13.464	0.0743
0.12	1.1275	0.8869	2.7	14.880	0.0672
0.13	1.1388	0.8781	2.8	16.445	0.0608
0.14	1.1503	0.8694	2.9	18.174	0.0550
0.15	1.1618	0.8607	3.0	20.086	0.0498
0.16	1.1735	0.8521	3.1	22.198	0.0450
0.17	1.1853	0.8437	3.2	24.533	0.0408
0.18	1.1972	0.8353	3.3	27.113	0.0369
0.19	1.2092	0.8270	3.4	29.964	0.0334
0.20	1.2214	0.8187	3.5	33.115	0.0302
0.21	1.2337	0.8106	3.6	36.598	0.0273
0.22	1.2461	0.8025	3.7	40.447	0.0247
0.23	1.2586	0.7945	3.8	44.701	0.0224
0.24	1.2712	0.7866	3.9	49.402	0.0202
0.25	1.2840	0.7788	4.0	54.598	0.0183
0.30	1.3499	0.7408	4.1	60.340	0.0166
0.35	1.4191	0.7047	4.2	66.686	0.0150
0.40	1.4918	0.6703	4.3	73.700	0.0136
0.45	1.5683	0.6376	4.4	81.451	0.0123
0.50	1.6487	0.6065	4.5	90.017	0.0111
0.55	1.7333	0.5769	4.6	99.484	0.0101
0.60	1.8221	0.5488	4.7	109.95	0.0091
0.65	1.9155	0.5220	4.8	121.51	0.0082
0.70	2.0138	0.4966	4.9	134.29	0.0074
0.75	2.1170	0.4724	5.0	148.41	0.0067
0.80	2.2255	0.4493	5.5	244.69	0.0041
0.85	2.3396	0.4274	6.0	403.43	0.0025
0.90	2.4596	0.4066	6.5	665.14	0.0015
0.95	2.5857	0.3867	7.0	1,096.6	0.0009
1.0	2.7183	0.3679	7.5	1,808.0	0.0006
1.1	3.0042	0.3329	8.0	2,981.0	0.0003
1.2	3.3201	0.3012	8.5	4,914.8	0.0002
1.3	3.6693	0.2725	9.0	8,103.1	0.0001
1.4	4.0552	0.2466	10.0	22,026	0.00005

Table 5   Trigonometric Functions of Angles

Angle	sin	cos	tan	cot	sec	csc	
0°00′	.0000	1.0000	.0000	—	1.000	—	90°00′
10	.0029	1.0000	.0029	343.8	1.000	343.8	50
20	.0058	1.0000	.0058	171.9	1.000	171.9	40
30	.0087	1.0000	.0087	114.6	1.000	114.6	30
40	.0116	.9999	.0116	85.94	1.000	85.95	20
50	.0145	.9999	.0145	68.75	1.000	68.76	10
1°00′	.0175	.9998	.0175	57.29	1.000	57.30	89°00′
10	.0204	.9998	.0204	49.10	1.000	49.11	50
20	.0233	.9997	.0233	42.96	1.000	42.98	40
30	.0262	.9997	.0262	38.19	1.000	38.20	30
40	.0291	.9996	.0291	34.37	1.000	34.38	20
50	.0320	.9995	.0320	31.24	1.001	31.26	10
2°00′	.0349	.9994	.0349	28.64	1.001	28.65	88°00′
10	.0378	.9993	.0378	26.43	1.001	26.45	50
20	.0407	.9992	.0407	24.54	1.001	24.56	40
30	.0436	.9990	.0437	22.90	1.001	22.93	30
40	.0465	.9989	.0466	21.47	1.001	21.49	20
50	.0494	.9988	.0495	20.21	1.001	20.23	10
3°00′	.0523	.9986	.0524	19.08	1.001	19.11	87°00′
10	.0552	.9985	.0553	18.07	1.002	18.10	50
20	.0581	.9983	.0582	17.17	1.002	17.20	40
30	.0610	.9981	.0612	16.35	1.002	16.38	30
40	.0640	.9980	.0641	15.60	1.002	15.64	20
50	.0669	.9978	.0670	14.92	1.002	14.96	10
4°00′	.0698	.9976	.0699	14.30	1.002	14.34	86°00′
10	.0727	.9974	.0729	13.73	1.003	13.76	50
20	.0756	.9971	.0758	13.20	1.003	13.23	40
30	.0785	.9969	.0787	12.71	1.003	12.75	30
40	.0814	.9967	.0816	12.25	1.003	12.29	20
50	.0843	.9964	.0846	11.83	1.004	11.87	10
5°00′	.0872	.9962	.0875	11.43	1.004	11.47	85°00′
10	.0901	.9959	.0904	11.06	1.004	11.10	50
20	.0929	.9957	.0934	10.71	1.004	10.76	40
30	.0958	.9954	.0963	10.39	1.005	10.43	30
40	.0987	.9951	.0992	10.08	1.005	10.13	20
50	.1016	.9948	.1022	9.788	1.005	9.839	10
6°00′	.1045	.9945	.1051	9.514	1.006	9.567	84°00′
10	.1074	.9942	.1080	9.255	1.006	9.309	50
20	.1103	.9939	.1110	9.010	1.006	9.065	40
30	.1132	.9936	.1139	8.777	1.006	8.834	30
40	.1161	.9932	.1169	8.556	1.007	8.614	20
50	.1190	.9929	.1198	8.345	1.007	8.405	10
7°00′	.1219	.9925	.1228	8.144	1.008	8.206	83°00′
10	.1248	.9922	.1257	7.953	1.008	8.016	50
20	.1276	.9918	.1287	7.770	1.008	7.834	40
30	.1305	.9914	.1317	7.596	1.009	7.661	30
40	.1334	.9911	.1346	7.429	1.009	7.496	20
50	.1363	.9907	.1376	7.269	1.009	7.337	10
8°00′	.1392	.9903	.1405	7.115	1.010	7.185	82°00′
10	.1421	.9899	.1435	6.968	1.010	7.040	50
20	.1449	.9894	.1465	6.827	1.011	6.900	40
30	.1478	.9890	.1495	6.691	1.011	6.765	30
40	.1507	.9886	.1524	6.561	1.012	6.636	20
50	.1536	.9881	.1554	6.435	1.012	6.512	10
9°00′	.1564	.9877	.1584	6.314	1.012	6.392	81°00′
	cos	sin	cot	tan	csc	sec	Angle

Table 5 *(Continued)*

Angle	sin	cos	tan	cot	sec	csc	
9°00′	.1564	.9877	.1584	6.314	1.012	6.392	81°00′
10	.1593	.9872	.1614	6.197	1.013	6.277	50
20	.1622	.9868	.1644	6.084	1.013	6.166	40
30	.1650	.9863	.1673	5.976	1.014	6.059	30
40	.1679	.9858	.1703	5.871	1.014	5.955	20
50	.1708	.9853	.1733	5.769	1.015	5.855	10
10°00′	.1736	.9848	.1763	5.671	1.015	5.759	80°00′
10	.1765	.9843	.1793	5.576	1.016	5.665	50
20	.1794	.9838	.1823	5.485	1.016	5.575	40
30	.1822	.9833	.1853	5.396	1.017	5.487	30
40	.1851	.9827	.1883	5.309	1.018	5.403	20
50	.1880	.9822	.1914	5.226	1.018	5.320	10
11°00′	.1908	.9816	.1944	5.145	1.019	5.241	79°00′
10	.1937	.9811	.1974	5.066	1.019	5.164	50
20	.1965	.9805	.2004	4.989	1.020	5.089	40
30	.1994	.9799	.2035	4.915	1.020	5.016	30
40	.2022	.9793	.2065	4.843	1.021	4.945	20
50	.2051	.9787	.2095	4.773	1.022	4.876	10
12°00′	.2079	.9781	.2126	4.705	1.022	4.810	78°00′
10	.2108	.9775	.2156	4.638	1.023	4.745	50
20	.2136	.9769	.2186	4.574	1.024	4.682	40
30	.2164	.9763	.2217	4.511	1.024	4.620	30
40	.2193	.9757	.2247	4.449	1.025	4.560	20
50	.2221	.9750	.2278	4.390	1.026	4.502	10
13°00′	.2250	.9744	.2309	4.331	1.026	4.445	77°00′
10	.2278	.9737	.2339	4.275	1.027	4.390	50
20	.2306	.9730	.2370	4.219	1.028	4.336	40
30	.2334	.9724	.2401	4.165	1.028	4.284	30
40	.2363	.9717	.2432	4.113	1.029	4.232	20
50	.2391	.9710	.2462	4.061	1.030	4.182	10
14°00′	.2419	.9703	.2493	4.011	1.031	4.134	76°00′
10	.2447	.9696	.2524	3.962	1.031	4.086	50
20	.2476	.9689	.2555	3.914	1.032	4.039	40
30	.2504	.9681	.2586	3.867	1.033	3.994	30
40	.2532	.9674	.2617	3.821	1.034	3.950	20
50	.2560	.9667	.2648	3.776	1.034	3.906	10
15°00′	.2588	.9659	.2679	3.732	1.035	3.864	75°00′
10	.2616	.9652	.2711	3.689	1.036	3.822	50
20	.2644	.9644	.2742	3.647	1.037	3.782	40
30	.2672	.9636	.2773	3.606	1.038	3.742	30
40	.2700	.9628	.2805	3.566	1.039	3.703	20
50	.2728	.9621	.2836	3.526	1.039	3.665	10
16°00′	.2756	.9613	.2867	3.487	1.040	3.628	74°00′
10	.2784	.9605	.2899	3.450	1.041	3.592	50
20	.2812	.9596	.2931	3.412	1.042	3.556	40
30	.2840	.9588	.2962	3.376	1.043	3.521	30
40	.2868	.9580	.2994	3.340	1.044	3.487	20
50	.2896	.9572	.3026	3.305	1.045	3.453	10
17°00′	.2924	.9563	.3057	3.271	1.046	3.420	73°00′
10	.2952	.9555	.3089	3.237	1.047	3.388	50
20	.2979	.9546	.3121	3.204	1.048	3.356	40
30	.3007	.9537	.3153	3.172	1.049	3.326	30
40	.3035	.9528	.3185	3.140	1.049	3.295	20
50	.3062	.9520	.3217	3.108	1.050	3.265	10
18°00′	.3090	.9511	.3249	3.078	1.051	3.236	72°00′
	cos	sin	cot	tan	csc	sec	Angle

Table 5 (*Continued*)

Angle	sin	cos	tan	cot	sec	csc	
18°00′	.3090	.9511	.3249	3.078	1.051	3.236	72°00′
10	.3118	.9502	.3281	3.047	1.052	3.207	50
20	.3145	.9492	.3314	3.018	1.053	3.179	40
30	.3173	.9483	.3346	2.989	1.054	3.152	30
40	.3201	.9474	.3378	2.960	1.056	3.124	20
50	.3228	.9465	.3411	2.932	1.057	3.098	10
19°00′	.3256	.9455	.3443	2.904	1.058	3.072	71°00′
10	.3283	.9446	.3476	2.877	1.059	3.046	50
20	.3311	.9436	.3508	2.850	1.060	3.021	40
30	.3338	.9426	.3541	2.824	1.061	2.996	30
40	.3365	.9417	.3574	2.798	1.062	2.971	20
50	.3393	.9407	.3607	2.773	1.063	2.947	10
20°00′	.3420	.9397	.3640	2.747	1.064	2.924	70°00′
10	.3448	.9387	.3673	2.723	1.065	2.901	50
20	.3475	.9377	.3706	2.699	1.066	2.878	40
30	.3502	.9367	.3739	2.675	1.068	2.855	30
40	.3529	.9356	.3772	2.651	1.069	2.833	20
50	.3557	.9346	.3805	2.628	1.070	2.812	10
21°00′	.3584	.9336	.3839	2.605	1.071	2.790	69°00′
10	.3611	.9325	.3872	2.583	1.072	2.769	50
20	.3638	.9315	.3906	2.560	1.074	2.749	40
30	.3665	.9304	.3939	2.539	1.075	2.729	30
40	.3692	.9293	.3973	2.517	1.076	2.709	20
50	.3719	.9283	.4006	2.496	1.077	2.689	10
22°00′	.3746	.9272	.4040	2.475	1.079	2.669	68°00′
10	.3773	.9261	.4074	2.455	1.080	2.650	50
20	.3800	.9250	.4108	2.434	1.081	2.632	40
30	.3827	.9239	.4142	2.414	1.082	2.613	30
40	.3854	.9228	.4176	2.394	1.084	2.595	20
50	.3881	.9216	.4210	2.375	1.085	2.577	10
23°00′	.3907	.9205	.4245	2.356	1.086	2.559	67°00′
10	.3934	.9194	.4279	2.337	1.088	2.542	50
20	.3961	.9182	.4314	2.318	1.089	2.525	40
30	.3987	.9171	.4348	2.300	1.090	2.508	30
40	.4014	.9159	.4383	2.282	1.092	2.491	20
50	.4041	.9147	.4417	2.264	1.093	2.475	10
24°00′	.4067	.9135	.4452	2.246	1.095	2.459	66°00′
10	.4094	.9124	.4487	2.229	1.096	2.443	50
20	.4120	.9112	.4522	2.211	1.097	2.427	40
30	.4147	.9100	.4557	2.194	1.099	2.411	30
40	.4173	.9088	.4592	2.177	1.100	2.396	20
50	.4200	.9075	.4628	2.161	1.102	2.381	10
25°00′	.4226	.9063	.4663	2.145	1.103	2.366	65°00′
10	.4253	.9051	.4699	2.128	1.105	2.352	50
20	.4279	.9038	.4734	2.112	1.106	2.337	40
30	.4305	.9026	.4770	2.097	1.108	2.323	30
40	.4331	.9013	.4806	2.081	1.109	2.309	20
50	.4358	.9001	.4841	2.066	1.111	2.295	10
26°00′	.4384	.8988	.4877	2.050	1.113	2.281	64°00′
10	.4410	.8975	.4913	2.035	1.114	2.268	50
20	.4436	.8962	.4950	2.020	1.116	2.254	40
30	.4462	.8949	.4986	2.006	1.117	2.241	30
40	.4488	.8936	.5022	1.991	1.119	2.228	20
50	.4514	.8923	.5059	1.977	1.121	2.215	10
27°00′	.4540	.8910	.5095	1.963	1.122	2.203	63°00′
	cos	sin	cot	tan	csc	sec	Angle

Table 5 *(Continued)*

Angle	sin	cos	tan	cot	sec	csc	
27°00′	.4540	.8910	.5095	1.963	1.122	2.203	63°00′
10	.4566	.8897	.5132	1.949	1.124	2.190	50
20	.4592	.8884	.5169	1.935	1.126	2.178	40
30	.4617	.8870	.5206	1.921	1.127	2.166	30
40	.4643	.8857	.5243	1.907	1.129	2.154	20
50	.4669	.8843	.5280	1.894	1.131	2.142	10
28°00′	.4695	.8829	.5317	1.881	1.133	2.130	62°00′
10	.4720	.8816	.5354	1.868	1.134	2.118	50
20	.4746	.8802	.5392	1.855	1.136	2.107	40
30	.4772	.8788	.5430	1.842	1.138	2.096	30
40	.4797	.8774	.5467	1.829	1.140	2.085	20
50	.4823	.8760	.5505	1.816	1.142	2.074	10
29°00′	.4848	.8746	.5543	1.804	1.143	2.063	61°00′
10	.4874	.8732	.5581	1.792	1.145	2.052	50
20	.4899	.8718	.5619	1.780	1.147	2.041	40
30	.4924	.8704	.5658	1.767	1.149	2.031	30
40	.4950	.8689	.5696	1.756	1.151	2.020	20
50	.4975	.8675	.5735	1.744	1.153	2.010	10
30°00′	.5000	.8660	.5774	1.732	1.155	2.000	60°00′
10	.5025	.8646	.5812	1.720	1.157	1.990	50
20	.5050	.8631	.5851	1.709	1.159	1.980	40
30	.5075	.8616	.5890	1.698	1.161	1.970	30
40	.5100	.8601	.5930	1.686	1.163	1.961	20
50	.5125	.8587	.5969	1.675	1.165	1.951	10
31°00′	.5150	.8572	.6009	1.664	1.167	1.942	59°00′
10	.5175	.8557	.6048	1.653	1.169	1.932	50
20	.5200	.8542	.6088	1.643	1.171	1.923	40
30	.5225	.8526	.6128	1.632	1.173	1.914	30
40	.5250	.8511	.6168	1.621	1.175	1.905	20
50	.5275	.8496	.6208	1.611	1.177	1.896	10
32°00′	.5299	.8480	.6249	1.600	1.179	1.887	58°00′
10	.5324	.8465	.6289	1.590	1.181	1.878	50
20	.5348	.8450	.6330	1.580	1.184	1.870	40
30	.5373	.8434	.6371	1.570	1.186	1.861	30
40	.5398	.8418	.6412	1.560	1.188	1.853	20
50	.5422	.8403	.6453	1.550	1.190	1.844	10
33°00′	.5446	.8387	.6494	1.540	1.192	1.836	57°00′
10	.5471	.8371	.6536	1.530	1.195	1.828	50
20	.5495	.8355	.6577	1.520	1.197	1.820	40
30	.5519	.8339	.6619	1.511	1.199	1.812	30
40	.5544	.8323	.6661	1.501	1.202	1.804	20
50	.5568	.8307	.6703	1.492	1.204	1.796	10
34°00′	.5592	.8290	.6745	1.483	1.206	1.788	56°00′
10	.5616	.8274	.6787	1.473	1.209	1.781	50
20	.5640	.8258	.6830	1.464	1.211	1.773	40
30	.5664	.8241	.6873	1.455	1.213	1.766	30
40	.5688	.8225	.6916	1.446	1.216	1.758	20
50	.5712	.8208	.6959	1.437	1.218	1.751	10
35°00′	.5736	.8192	.7002	1.428	1.221	1.743	55°00′
10	.5760	.8175	.7046	1.419	1.223	1.736	50
20	.5783	.8158	.7089	1.411	1.226	1.729	40
30	.5807	.8141	.7133	1.402	1.228	1.722	30
40	.5831	.8124	.7177	1.393	1.231	1.715	20
50	.5854	.8107	.7221	1.385	1.233	1.708	10
36°00′	.5878	.8090	.7265	1.376	1.236	1.701	54°00′
	cos	sin	cot	tan	csc	sec	Angle

Table 5 (*Continued*)

Angle	sin	cos	tan	cot	sec	csc	
36°00′	.5878	.8090	.7265	1.376	1.236	1.701	54°00′
10	.5901	.8073	.7310	1.368	1.239	1.695	50
20	.5925	.8056	.7355	1.360	1.241	1.688	40
30	.5948	.8039	.7400	1.351	1.244	1.681	30
40	.5972	.8021	.7445	1.343	1.247	1.675	20
50	.5995	.8004	.7490	1.335	1.249	1.668	10
37°00′	.6018	.7986	.7536	1.327	1.252	1.662	53°00′
10	.6041	.7969	.7581	1.319	1.255	1.655	50
20	.6065	.7951	.7627	1.311	1.258	1.649	40
30	.6088	.7934	.7673	1.303	1.260	1.643	30
40	.6111	.7916	.7720	1.295	1.263	1.636	20
50	.6134	.7898	.7766	1.288	1.266	1.630	10
38°00′	.6157	.7880	.7813	1.280	1.269	1.624	52°00′
10	.6180	.7862	.7860	1.272	1.272	1.618	50
20	.6202	.7844	.7907	1.265	1.275	1.612	40
30	.6225	.7826	.7954	1.257	1.278	1.606	30
40	.6248	.7808	.8002	1.250	1.281	1.601	20
50	.6271	.7790	.8050	1.242	1.284	1.595	10
39°00′	.6293	.7771	.8098	1.235	1.287	1.589	51°00′
10	.6316	.7753	.8146	1.228	1.290	1.583	50
20	.6338	.7735	.8195	1.220	1.293	1.578	40
30	.6361	.7716	.8243	1.213	1.296	1.572	30
40	.6383	.7698	.8292	1.206	1.299	1.567	20
50	.6406	.7679	.8342	1.199	1.302	1.561	10
40°00′	.6428	.7660	.8391	1.192	1.305	1.556	50°00′
10	.6450	.7642	.8441	1.185	1.309	1.550	50
20	.6472	.7623	.8491	1.178	1.312	1.545	40
30	.6494	.7604	.8541	1.171	1.315	1.540	30
40	.6517	.7585	.8591	1.164	1.318	1.535	20
50	.6539	.7566	.8642	1.157	1.322	1.529	10
41°00′	.6561	.7547	.8693	1.150	1.325	1.524	49°00′
10	.6583	.7528	.8744	1.144	1.328	1.519	50
20	.6604	.7509	.8796	1.137	1.332	1.514	40
30	.6626	.7490	.8847	1.130	1.335	1.509	30
40	.6648	.7470	.8899	1.124	1.339	1.504	20
50	.6670	.7451	.8952	1.117	1.342	1.499	10
42°00′	.6691	.7431	.9004	1.111	1.346	1.494	48°00′
10	.6713	.7412	.9057	1.104	1.349	1.490	50
20	.6734	.7392	.9110	1.098	1.353	1.485	40
30	.6756	.7373	.9163	1.091	1.356	1.480	30
40	.6777	.7353	.9217	1.085	1.360	1.476	20
50	.6799	.7333	.9271	1.079	1.364	1.471	10
43°00′	.6820	.7314	.9325	1.072	1.367	1.466	47°00′
10	.6841	.7294	.9380	1.066	1.371	1.462	50
20	.6862	.7274	.9435	1.060	1.375	1.457	40
30	.6884	.7254	.9490	1.054	1.379	1.453	30
40	.6905	.7234	.9545	1.048	1.382	1.448	20
50	.6926	.7214	.9601	1.042	1.386	1.444	10
44°00′	.6947	.7193	.9657	1.036	1.390	1.440	46°00′
10	.6967	.7173	.9713	1.030	1.394	1.435	50
20	.6988	.7153	.9770	1.024	1.398	1.431	40
30	.7009	.7133	.9827	1.018	1.402	1.427	30
40	.7030	.7112	.9884	1.012	1.406	1.423	20
50	.7050	.7092	.9942	1.006	1.410	1.418	10
45°00′	.7071	.7071	1.000	1.000	1.414	1.414	45°00′
	cos	sin	cot	tan	csc	sec	Angle

Table 6   Trigonometric Functions of Real Numbers

Real Number x or θ radians	θ degrees	sin x or sin θ	csc x or csc θ	tan x or tan θ	cot x or cot θ	sec x or sec θ	cos x or cos θ
0.00	0°00′	0.0000	No value	0.0000	No value	1.000	1.000
.01	0°34′	.0100	100.0	.0100	100.0	1.000	1.000
.02	1°09′	.0200	50.00	.0200	49.99	1.000	0.9998
.03	1°43′	.0300	33.34	.0300	33.32	1.000	0.9996
.04	2°18′	.0400	25.01	.0400	24.99	1.001	0.9992
0.05	2°52′	0.0500	20.01	0.0500	19.98	1.001	0.9988
.06	3°26′	.0600	16.68	.0601	16.65	1.002	.9982
.07	4°01′	.0699	14.30	.0701	14.26	1.002	.9976
.08	4°35′	.0799	12.51	.0802	12.47	1.003	.9968
.09	5°09′	.0899	11.13	.0902	11.08	1.004	.9960
0.10	5°44′	0.0998	10.02	0.1003	9.967	1.005	0.9950
.11	6°18′	.1098	9.109	.1104	9.054	1.006	.9940
.12	6°53′	.1197	8.353	.1206	8.293	1.007	.9928
.13	7°27′	.1296	7.714	.1307	7.649	1.009	.9916
.14	8°01′	.1395	7.166	.1409	7.096	1.010	.9902
0.15	8°36′	0.1494	6.692	0.1511	6.617	1.011	0.9888
.16	9°10′	.1593	6.277	.1614	6.197	1.013	.9872
.17	9°44′	.1692	5.911	.1717	5.826	1.015	.9856
.18	10°19′	.1790	5.586	.1820	5.495	1.016	.9838
.19	10°53′	.1889	5.295	.1923	5.200	1.018	.9820
0.20	11°28′	0.1987	5.033	0.2027	4.933	1.020	0.9801
.21	12°02′	.2085	4.797	.2131	4.692	1.022	.9780
.22	12°36′	.2182	4.582	.2236	4.472	1.025	.9759
.23	13°11′	.2280	4.386	.2341	4.271	1.027	.9737
.24	13°45′	.2377	4.207	.2447	4.086	1.030	.9713
0.25	14°19′	0.2474	4.042	0.2553	3.916	1.032	0.9689
.26	14°54′	.2571	3.890	.2660	3.759	1.035	.9664
.27	15°28′	.2667	3.749	.2768	3.613	1.038	.9638
.28	16°03′	.2764	3.619	.2876	3.478	1.041	.9611
.29	16°37′	.2860	3.497	.2984	3.351	1.044	.9582
0.30	17°11′	0.2955	3.384	0.3093	3.233	1.047	0.9553
.31	17°46′	.3051	3.278	.3203	3.122	1.050	.9523
.32	18°20′	.3146	3.179	.3314	3.018	1.053	.9492
.33	18°54′	.3240	3.086	.3425	2.920	1.057	.9460
.34	19°29′	.3335	2.999	.3537	2.827	1.061	.9428
0.35	20°03′	0.3429	2.916	0.3650	2.740	1.065	0.9394
.36	20°38′	.3523	2.839	.3764	2.657	1.068	.9359
.37	21°12′	.3616	2.765	.3879	2.578	1.073	.9323
.38	21°46′	.3709	2.696	.3994	2.504	1.077	.9287
.39	22°21′	.3802	2.630	.4111	2.433	1.081	.9249
0.40	22°55′	0.3894	2.568	0.4228	2.365	1.086	0.9211
.41	23°29′	.3986	2.509	.4346	2.301	1.090	.9171
.42	24°04′	.4078	2.452	.4466	2.239	1.095	.9131
.43	24°38′	.4169	2.399	.4586	2.180	1.100	.9090
.44	25°13′	.4259	2.348	.4708	2.124	1.105	.9048
0.45	25°47′	0.4350	2.299	0.4831	2.070	1.111	0.9004
.46	26°21′	.4439	2.253	.4954	2.018	1.116	.8961
.47	26°56′	.4529	2.208	.5080	1.969	1.122	.8916
.48	27°30′	.4618	2.166	.5206	1.921	1.127	.8870
.49	28°04′	.4706	2.125	.5334	1.875	1.133	.8823
0.50	28°39′	0.4794	2.086	0.5463	1.830	1.139	.8776

Table 6 (Continued)

Real Number x or θ radians	θ degrees	sin x or sin θ	csc x or csc θ	tan x or tan θ	cot x or cot θ	sec x or sec θ	cos x or cos θ
0.50	28°39′	0.4794	2.086	0.5463	1.830	1.139	0.8776
.51	29°13′	.4882	2.048	.5594	1.788	1.146	.8727
.52	29°48′	.4969	2.013	.5726	1.747	1.152	.8678
.53	30°22′	.5055	1.978	.5859	1.707	1.159	.8628
.54	30°56′	.5141	1.945	.5994	1.668	1.166	.8577
0.55	31°31′	0.5227	1.913	0.6131	1.631	1.173	0.8525
.56	32°05′	.5312	1.883	.6269	1.595	1.180	.8473
.57	32°40′	.5396	1.853	.6410	1.560	1.188	.8419
.58	33°14′	.5480	1.825	.6552	1.526	1.196	.8365
.59	33°48′	.5564	1.797	.6696	1.494	1.203	.8309
0.60	34°23′	0.5646	1.771	0.6841	1.462	1.212	0.8253
.61	34°57′	.5729	1.746	.6989	1.431	1.220	.8196
.62	35°31′	.5810	1.721	.7139	1.401	1.229	.8139
.63	36°06′	.5891	1.697	.7291	1.372	1.238	.8080
.64	36°40′	.5972	1.674	.7445	1.343	1.247	.8021
0.65	37°15′	0.6052	1.652	0.7602	1.315	1.256	0.7961
.66	37°49′	.6131	1.631	.7761	1.288	1.266	.7900
.67	38°23′	.6210	1.610	.7923	1.262	1.276	.7838
.68	38°58′	.6288	1.590	.8087	1.237	1.286	.7776
.69	39°32′	.6365	1.571	.8253	1.212	1.297	.7712
0.70	40°06′	0.6442	1.552	0.8423	1.187	1.307	0.7648
.71	40°41′	.6518	1.534	.8595	1.163	1.319	.7584
.72	41°15′	.6594	1.517	.8771	1.140	1.330	.7518
.73	41°50′	.6669	1.500	.8949	1.117	1.342	.7452
.74	42°24′	.6743	1.483	.9131	1.095	1.354	.7385
0.75	42°58′	0.6816	1.467	0.9316	1.073	1.367	0.7317
.76	43°33′	.6889	1.452	.9505	1.052	1.380	.7248
.77	44°07′	.6961	1.436	.9697	1.031	1.393	.7179
.78	44°41′	.7033	1.422	.9893	1.011	1.407	.7109
.79	45°16′	.7104	1.408	1.009	.9908	1.421	.7038
0.80	45°50′	0.7174	1.394	1.030	0.9712	1.435	0.6967
.81	46°25′	.7243	1.381	1.050	.9520	1.450	.6895
.82	46°59′	.7311	1.368	1.072	.9331	1.466	.6822
.83	47°33′	.7379	1.355	1.093	.9146	1.482	.6749
.84	48°08′	.7446	1.343	1.116	.8964	1.498	.6675
0.85	48°42′	0.7513	1.331	1.138	0.8785	1.515	0.6600
.86	49°16′	.7578	1.320	1.162	.8609	1.533	.6524
.87	49°51′	.7643	1.308	1.185	.8437	1.551	.6448
.88	50°25′	.7707	1.297	1.210	.8267	1.569	.6372
.89	51°00′	.7771	1.287	1.235	.8100	1.589	.6294
0.90	51°34′	0.7833	1.277	1.260	0.7936	1.609	0.6216
.91	52°08′	.7895	1.267	1.286	.7774	1.629	.6137
.92	52°43′	.7956	1.257	1.313	.7615	1.651	.6058
.93	53°17′	.8016	1.247	1.341	.7458	1.673	.5978
.94	53°51′	.8076	1.238	1.369	.7303	1.696	.5898
0.95	54°26′	0.8134	1.229	1.398	0.7151	1.719	0.5817
.96	55°00′	.8192	1.221	1.428	.7001	1.744	.5735
.97	55°35′	.8249	1.212	1.459	.6853	1.769	.5653
.98	56°09′	.8305	1.204	1.491	.6707	1.795	.5570
.99	56°43′	.8360	1.196	1.524	.6563	1.823	.5487
1.00	57°18′	0.8415	1.188	1.557	0.6421	1.851	0.5403
1.01	57°52′	.8468	1.181	1.592	.6281	1.880	.5319
1.02	58°27′	.8521	1.174	1.628	.6142	1.911	.5234
1.03	59°01′	.8573	1.166	1.665	.6005	1.942	.5148
1.04	59°35′	.8624	1.160	1.704	.5870	1.975	.5062
1.05	60°10′	0.8674	1.153	1.743	0.5736	2.010	0.4976

Table 6 *(Continued)*

Real Number x or θ radians	θ degrees	sin x or sin θ	csc x or csc θ	tan x or tan θ	cot x or cot θ	sec x or sec θ	cos x or cos θ
1.05	60°10′	0.8674	1.153	1.743	0.5736	2.010	0.4976
1.06	60°44′	.8724	1.146	1.784	.5604	2.046	.4889
1.07	61°18′	.8772	1.140	1.827	.5473	2.083	.4801
1.08	61°53′	.8820	1.134	1.871	.5344	2.122	.4713
1.09	62°27′	.8866	1.128	1.917	.5216	2.162	.4625
1.10	63°02′	0.8912	1.122	1.965	0.5090	2.205	0.4536
1.11	63°36′	.8957	1.116	2.014	.4964	2.249	.4447
1.12	64°10′	.9001	1.111	2.066	.4840	2.295	.4357
1.13	64°45′	.9044	1.106	2.120	.4718	2.344	.4267
1.14	65°19′	.9086	1.101	2.176	.4596	2.395	.4176
1.15	65°53′	0.9128	1.096	2.234	0.4475	2.448	0.4085
1.16	66°28′	.9168	1.091	2.296	.4356	2.504	.3993
1.17	67°02′	.9208	1.086	2.360	.4237	2.563	.3902
1.18	67°37′	.9246	1.082	2.427	.4120	2.625	.3809
1.19	68°11′	.9284	1.077	2.498	.4003	2.691	.3717
1.20	68°45′	0.9320	1.073	2.572	0.3888	2.760	0.3624
1.21	69°20′	.9356	1.069	2.650	.3773	2.833	.3530
1.22	69°54′	.9391	1.065	2.733	.3659	2.910	.3436
1.23	70°28′	.9425	1.061	2.820	.3546	2.992	.3342
1.24	71°03′	.9458	1.057	2.912	.3434	3.079	.3248
1.25	71°37′	0.9490	1.054	3.010	0.3323	3.171	0.3153
1.26	72°12′	.9521	1.050	3.113	.3212	3.270	.3058
1.27	72°46′	.9551	1.047	3.224	.3102	3.375	.2963
1.28	73°20′	.9580	1.044	3.341	.2993	3.488	.2867
1.29	73°55′	.9608	1.041	3.467	.2884	3.609	.2771
1.30	74°29′	0.9636	1.038	3.602	0.2776	3.738	0.2675
1.31	75°03′	.9662	1.035	3.747	.2669	3.878	.2579
1.32	75°38′	.9687	1.032	3.903	.2562	4.029	.2482
1.33	76°12′	.9711	1.030	4.072	.2456	4.193	.2385
1.34	76°47′	.9735	1.027	4.256	.2350	4.372	.2288
1.35	77°21′	0.9757	1.025	4.455	0.2245	4.566	0.2190
1.36	77°55′	.9779	1.023	4.673	.2140	4.779	.2092
1.37	78°30′	.9799	1.021	4.913	.2035	5.014	.1994
1.38	79°04′	.9819	1.018	5.177	.1931	5.273	.1896
1.39	79°38′	.9837	1.017	5.471	.1828	5.561	.1798
1.40	80°13′	0.9854	1.015	5.798	0.1725	5.883	0.1700
1.41	80°47′	.9871	1.013	6.165	.1622	6.246	.1601
1.42	81°22′	.9887	1.011	6.581	.1519	6.657	.1502
1.43	81°56′	.9901	1.010	7.055	.1417	7.126	.1403
1.44	82°30′	.9915	1.009	7.602	.1315	7.667	.1304
1.45	83°05′	0.9927	1.007	8.238	0.1214	8.299	0.1205
1.46	83°39′	.9939	1.006	8.989	.1113	9.044	.1106
1.47	84°13′	.9949	1.005	9.887	.1011	9.938	.1006
1.48	84°48′	.9959	1.004	10.98	.0910	11.03	.0907
1.49	85°22′	.9967	1.003	12.35	.0810	12.39	.0807
1.50	85°57′	0.9975	1.003	14.10	0.0709	14.14	0.0707
1.51	86°31′	.9982	1.002	16.43	.0609	16.46	.0608
1.52	87°05′	.9987	1.001	19.67	.0508	19.69	.0508
1.53	87°40′	.9992	1.001	24.50	.0408	24.52	.0408
1.54	88°14′	.9995	1.000	32.46	.0308	32.48	.0308
1.55	88°49′	0.9998	1.000	48.08	0.0208	48.09	0.0208
1.56	89°23′	.9999	1.000	92.62	.0108	92.63	.0108
1.57	89°57′	1.000	1.000	1256	.0008	1256	.0008

Answers to
Odd-Numbered Problems

Exercises 1.1

1. $0.8\bar{0}$ **3.** $0.\overline{45}$ **5.** $6.1\bar{6}$ **7.** $1.\overline{428571}$ **9.** $\frac{2}{9}$ **11.** $\frac{321}{999} = \frac{107}{333}$ **13.** $\frac{3}{10}$ **15.** 6 (or $\frac{6}{1}$)   **17.** $\frac{1,929}{900} = \frac{643}{300}$

19. Real number, rational number, integer   **21.** None of these   **23.** Real number, rational number, integer

25. Real number, rational number, integer   **27.** Real number, irrational number   **29.** Real number, rational number

31. Commutative property of addition   **33.** Associative property of multiplication   **35.** Multiplication identity property

37. Commutative property of multiplication   **39.** Distributive property   **41.** Associative property of multiplication

43. Commutative property of addition   **45.** Commutative property of multiplication   **47.** Multiplication inverse property

49. Commutative property of addition   **51.** $<$ **53.** $<$ **55.** $<$ **57.** $<$ **59.** $<$

61. $=$ **63.** $<$ **65.** $>$ **67.** $>$ **69.** $>$ **71.** -7 **73.** 5 **75.** -2 **77.** $\frac{14}{5}$ **79.** -0.54

81. -4.1 **83.** -66 **85.** 2 **87.** $-\frac{1}{32}$ **89.** $\frac{20}{9}$ **91.** 0 **93.** -0.04 **95.** 10.8 **97.** -2 **99.** -14

101. 6 **103.** -27 **105.** -16 **107.** $-\frac{1}{16}$ **109.** -126 **111.** -40 **113.** -70

115. $a(b + c)$ [total area] $= ab$ [area 1] $+ ac$ [area 2]   **117.** $y = 1$, x any real number; $x = 0$, y any real number

119. False, $\frac{1}{2}$ **121.** False, 1   **123.** False, $\frac{1}{2}$ **125.** True   **127.** False, 0

Exercises 1.2

1. -3 **3.** -32 **5.** $\frac{13}{16}$ **7.** 10 **9.** 1 **11.** -52 **13.** -30 **15.** -12 **17.** -15 **19.** 14 **21.** $\frac{1}{2}$

23. 0 **25.** $-\frac{1}{2}$ **27.** $19a$ **29.** $14a - 8b$ **31.** $3xy + 5cd$ **33.** $-3x^3 + 8y^2$ **35.** $2x - 5y$

37. $-16a + 11b$ **39.** $10x - 7y$ **41.** $2x^3 + 13x^2y^2 + 9xy^3 - 3x^3y$ **43.** $7y^3 - 3y^2 + 4y + 6$

45. $b^3 - 3b^2 + b + 5$ **47.** $12a + 14$ **49.** $2a + 5b$ **51.** $3^5 = 243$ **53.** 1 **55.** $\frac{1}{9}$ **57.** $\frac{3}{4}$

59. $4^3 = 64$ **61.** $(-3)^5 = -243$ **63.** $1/2^3 = 1/8$ **65.** $2^3 = 8$ **67.** $1/3^2 = 1/9$ **69.** $\frac{3}{4}$ **71.** x^5

73. $4p^2$ **75.** x^2 **77.** $81x^{12}y^8$ **79.** $-27x^4$ **81.** 1 **83.** $-8/x^3$ **85.** $9/(x + h)^2$ **87.** $a^2x^2/4$

89. $2/a^2$ **91.** $-x^4/(4y^5z^8)$ **93.** $x^4/27$ **95.** xz^2/y^2 **97.** $a^4/(8x^{10})$ **99.** $2^{x + y}$ **101.** $2^{1 - n}$ or $1/2^{n - 1}$

103. 5^{2x^2} **105.** y^{7a} **107.** $(a - b)^{x + y}$ **109.** 1 **111.** $x^{1 + x}$ **113.** y^{6x^2} **115.** $1/y$ **117.** $x^{ap - 1}$

119. x^a/y **121.** $(1 - x)^{a + 2}$ **123.** $-akx^{-k - 1}$ **125.** $5 + \frac{4}{3}$; $5\boxed{+}4\boxed{=}\boxed{\div}3\boxed{=}$ **127.** Correct

129. $\frac{2}{9} \cdot 15$; $2\boxed{\div}\boxed{(}9\boxed{\times}15\boxed{)}\boxed{=}$ **131.** Correct   **133.** $5^2 + \sqrt{6^2}$; $5\boxed{x^2}\boxed{+}6\boxed{x^2}\boxed{=}\boxed{\sqrt{}}$

135. $\dfrac{\pi r^2 h}{\frac{4}{3}\pi r^3} = \dfrac{\pi r^2(2r)}{\frac{4}{3}\pi r^3} = \dfrac{2}{\frac{4}{3}} = \dfrac{3}{2}$

Exercises 1.3

1. $2x - 2y$ **3.** $-5x^4 + 5x^3 + 5x^2$ **5.** $-8x^2yz + 2xy^2z - 14xyz^2$ **7.** $p^4q^2 + p^2q^4$ **9.** $4n + 3x$

11. $2x + h$ **13.** $6x^3 + 8x^2 - 10x$ **15.** $a^2 + 7a + 12$ **17.** $6x^2 - 11x + 4$ **19.** $k^2 - 4k + 4$

21. $y^3 + y^2 - 21y + 4$ **23.** $x^3 - y^3$ **25.** $-6y^3 + 5y^2 + 3y - 2$ **27.** $x^{-2} - 1$ **29.** $z^{2a} + 6z^a + 9$

31. $x^{2a} + 2x^ay^b + y^{2b}$ **33.** $x^6 - y^6$ **35.** $x^2 - 2xy + y^2 - 2x + 2y + 1$ **37.** $x^4 - 2x^2y^2 + y^4$

39. $x^{3n} + y^{3n}$ **41.** $y(x + z)$ **43.** $3ax(3ax + 1)$ **45.** $x(2xz^2 + 4y - 5xy^2)$ **47.** $(a - c)(b + d)$

49. $(6x + 1)(6x - 1)$ **51.** $(6r^3 + k^2)(6r^3 - k^2)$ **53.** $(a + 4)(a + 1)$ **55.** $(x + 4)(x - 3)$

57. $(2 + x)(3 - x)$ **59.** $(7z - 6)(z - 1)$ **61.** $(9b + 2)(b - 3)$ **63.** $(x + 4y)(x + y)$ **65.** $(4y - 3)^2$

67. $(3a - 2b)^2$ **69.** $x(1 + y)(1 - y)$ **71.** $3(k - 4)(k + 2)$ **73.** $3x(x - 3)(x - 1)$

75. $(1 - x)(1 + x + x^2)$ **77.** $(ab + c)(a^2b^2 - abc + c^2)$ **79.** $2(y - 2x^2)(y^2 + 2x^2y + 4x^4)$

81. $(x - 1)(x^2 + x + 1)(x + 1)(x^2 - x + 1)$ **83.** $(x - y)(x^2 + xy + y^2)(x + y)(x^2 - xy + y^2)$

85. $(x^a + y^b)(x^a - y^b)$ **87.** $(y + 2)(3x + 1)$ **89.** $(2x - 3)(y + z)$ **91.** $(x + y + z)(x - y - z)$

93. $(x^2 + 1)(x + 1)(x^2 - x + 1)$ **95.** $(a + b)(a^2 - ab + b^2) = a^3 - a^2b + ab^2 + a^2b - ab^2 + b^3 = a^3 + b^3$

97. 1, factorable   **99.** 985, not factorable (with integer coefficients)

101. $\dfrac{x^2 + 2xh + h^2 + 1 - x^2 - 1}{h} = \dfrac{2xh + h^2}{h} = 2x + h$ **103.** $6 - h$ **105.** $2x + h + 2$

107. (a) $a^2 + 2ab + b^2$ (b) $a^2 + 2ab + b^2$ (c) No

109. $(2k_1 + 1)(2k_2 + 1) = 4k_1k_2 + 2k_1 + 2k_2 + 1 = 2(2k_1k_2 + k_1 + k_2) + 1 = 2k + 1$

Exercises 1.4

1. $\frac{4}{11}$ **3.** $\dfrac{1}{3b + 1}$ **5.** -1 **7.** $\dfrac{-1}{ax}$ **9.** $\dfrac{-1}{x + 1}$ **11.** $\dfrac{a + 1}{a - 4}$ **13.** $\dfrac{7 - y}{7 + y}$ **15.** $\dfrac{3x(2x - 5)}{2(x - 4)}$

17. $\dfrac{1}{y - x}$ **19.** $\dfrac{-1}{2h}$ **21.** $x + a - 1$ **23.** $x^2 + ax + a^2$ **25.** $\dfrac{3 - 2x}{x^4}$ **27.** $\dfrac{1 - x}{(x + 1)^3}$ **29.** $\dfrac{y}{4x^2}$

31. $\dfrac{2}{a(a + 3)}$ **33.** $\dfrac{(y - 1)(y - 2)}{(y + 2)(y + 1)}$ **35.** $\dfrac{(x - 1)^2}{(x + 1)^2}$ **37.** $\dfrac{1}{x}$ **39.** $\dfrac{n - 1}{n - 3}$ **41.** $\dfrac{(x + 1)(x - 1)^2}{7x}$

43. $\dfrac{1 - x}{x(a + 1)}$ **45.** 1   **47.** $\dfrac{n + 1}{3n}$ **49.** $\dfrac{x - y}{-4a^2}$ **51.** $-y$ **53.** $\dfrac{-4}{2 + x}$ **55.** 10   **57.** $\dfrac{3a}{xy}$ **59.** $\dfrac{x - 1}{x + 2}$

61. $\dfrac{3n - 2}{n - 6}$ **63.** $\dfrac{9a}{5}$ **65.** $\dfrac{7}{12x}$ **67.** $\dfrac{5k^2 - 3k + 2}{k^3}$ **69.** $\dfrac{2x + 47}{10x}$ **71.** $\dfrac{y - 2}{6(3y + 2)}$ **73.** $\dfrac{3a^2 - 3a + 9}{a(a - 3)}$

75. $\dfrac{6x + 11}{(2x - 3)(2x + 3)}$ **77.** $\dfrac{2n^2 - 2n + 1}{n(n - 1)}$ **79.** $\dfrac{7y - 4}{y^2 - 1}$ **81.** 0   **83.** $\dfrac{a^x - x - 1}{x(a^x - 1)}$ **85.** $\dfrac{x + 2}{x^{n+1}}$

87. $\dfrac{1}{x - y}$ **89.** $\dfrac{-1}{x(x + h)}$ **91.** 1   **93.** $\dfrac{2y^2 + 3y - 4}{y^2}$ **95.** $\dfrac{8x + 5}{(3x + 1)^2}$ **97.** $\frac{65}{72}$ **99.** $\dfrac{x + 1}{x - 1}$

101. $\dfrac{n^2 + n + 2}{n^2 - n - 2}$ **103.** $\dfrac{y + x}{x}$ **105.** $\dfrac{3P}{12 + 2\pi}$ **107.** $\dfrac{x^2 + 3x - 4}{4}$ **109.** $\dfrac{1}{(x + 1)(x + h + 1)}$

111. $\dfrac{x^2 + xh - 1}{x(x + h)}$ **113.** $y + x$ **115.** $\dfrac{xy}{y + x}$ **117.** 24 mi/hour

Exercises 1.5

1. 5   **3.** $\frac{1}{2}$ **5.** 1   **7.** 2   **9.** -3 **11.** 2   **13.** $\frac{1}{2}$ **15.** $\frac{5}{3}$ **17.** $-\frac{1}{2}$ **19.** 5   **21.** -2 **23.** 27

25. 32   **27.** -32 **29.** 9   **31.** $\frac{1}{16}$ **33.** $\frac{27}{8}$ **35.** $\frac{28}{27}$ **37.** $\frac{9}{2}$ **39.** 16   **41.** 4   **43.** -27 **45.** 4

47. $\frac{1}{64}$ **49.** $\frac{1}{5}$ **51.** $x^{1/3}$; $\sqrt[3]{x}$ **53.** $\dfrac{1}{x^{7/6}}$; $\dfrac{1}{\sqrt[6]{x^7}}$ **55.** $2xy^2$ **57.** $\dfrac{a}{2b^3}$ **59.** $\dfrac{x^{1/2}y^2}{z^{2/3}}$; $\dfrac{y^2\sqrt{x}}{\sqrt[3]{z^2}}$ **61.** $x + y$

63. $x^{1/2}y^{1/5}$; $\sqrt{x} \cdot \sqrt[5]{y}$ **65.** $k^{(1-y)/x}$; $\sqrt[x]{k^{1-y}}$ **67.** $x^{(p/q)-1}$ **69.** $6u + 3u^{9/2}$; $6u + 3\sqrt{u^9}$

71. $\dfrac{2x^{1/2}-1}{4}$; $\dfrac{2\sqrt{x}-1}{4}$ **73.** $\dfrac{y^{5/2}}{x}$; $\dfrac{\sqrt{y^5}}{x}$ **75.** $\dfrac{125x^3}{343y^{24}}$ **77.** $\frac{4}{9}x^{5/2}y^{7/12}$; $\frac{4}{9}\sqrt{x^5}\sqrt[12]{y^7}$ **79.** $2^{7/12}$; $\sqrt[12]{2^7}$

81. $16^{2/3}$; $\sqrt[3]{16^2}$ **83.** $3^{3/4}$; $\sqrt[4]{3^3}$ **85.** $a^{1/(mn)}$; $\sqrt[mn]{a}$ **87.** $(2^2 \cdot 3^3)^{1/6}$; $\sqrt[6]{2^2 \cdot 3^3}$

89. $x + x^{-1} - 2 = (x^2 - 2x + 1)/x$ **91.** $x + 2x^{1/2}y^{1/2} + y$; $x + 2\sqrt{x}\sqrt{y} + y$ **93.** $\dfrac{2x-1}{2x^{3/2}}$; $\dfrac{2x-1}{2\sqrt{x^3}}$

95. $\dfrac{x+1}{x-1}$ **97.** $\dfrac{x-2}{2x^{3/2}}$; $\dfrac{x-2}{2\sqrt{x^3}}$ **99.** $\dfrac{x-2}{2x^{3/2}}$; $\dfrac{x-2}{2\sqrt{x^3}}$ **101.** $\dfrac{1+x^3}{x^5}$ **103.** $\dfrac{3(x-5)}{x^{2/3}}$ **105.** $\dfrac{4x^{3/2}+1}{2x^{1/2}}$

107. $\dfrac{x(x^2-5x-1)}{(x-5)^{1/2}}$

Exercises 1.6

1. $6\sqrt{3}$ **3.** $10\sqrt{2}$ **5.** $-2\sqrt[3]{4}$ **7.** $2x^3y^2\sqrt{3x}$ **9.** $xy\sqrt[4]{9x^2y^3}$ **11.** $6\sqrt{3}$ **13.** $\sqrt[4]{10}/2$ **15.** $3\sqrt{x}/x$

17. $\sqrt[3]{xy^2}/y$ **19.** $x\sqrt{2}/2$ **21.** $11\sqrt{2}$ **23.** $5\sqrt{3}$ **25.** $-7\sqrt{2}$ **27.** $-14\sqrt{11}$ **29.** $13\sqrt{3}$ **31.** $7\sqrt{2}$

33. $2\sqrt{6}/3$ **35.** $37\sqrt{6}/6$ **37.** $9\sqrt[3]{3}$ **39.** $-29\sqrt[3]{3}/3$ **41.** $8\sqrt{2x}$ **43.** $(2y - 6x)\sqrt{5x}$ **45.** $5x\sqrt{2xy}/2$

47. $\left(\dfrac{2}{y} - \dfrac{2}{x^2} + \dfrac{5x}{4}\right)\sqrt{2xy}$ **49.** $\left(\dfrac{1}{y} - 1\right)\sqrt[4]{xy^3}$ **51.** $2\sqrt{3}$ **53.** $2\sqrt{2}$ **55.** -4 **57.** $3\sqrt{5}$ **59.** -1

61. $29 - 4\sqrt{30}$ **63.** $3x\sqrt{10}$ **65.** $2xy\sqrt[3]{3xy}$ **67.** $\sqrt{6x}/x$ **69.** $\sqrt[4]{56x}/2$ **71.** $2\sqrt{3}/3$ **73.** $\dfrac{(\sqrt{3}-3)}{3}$

75. $3 - 2\sqrt{2}$ **77.** $\dfrac{x-\sqrt{xy}}{x-y}$ **79.** $\dfrac{\sqrt{x^2-1}}{x+1}$ **81.** $\dfrac{1}{\sqrt{x}+2}$ **83.** $\dfrac{1}{\sqrt{x+h}+\sqrt{x}}$

85. $\dfrac{2}{\sqrt{2x+2h+1}+\sqrt{2x+1}}$ **87.** $1{,}200\sqrt{5}$ **89.** $2x + 6$ **91.** $\dfrac{2\sqrt{x+1}-1}{2\sqrt{x+1}} = \dfrac{2x+2-\sqrt{x+1}}{2x+2}$

93. $\sqrt{a^2+b^2}$ **95.** $\dfrac{\pi(4R^2-2h^2)}{\sqrt{4R^2-h^2}} = \dfrac{\pi(4R^2-2h^2)\sqrt{4R^2-h^2}}{4R^2-h^2}$ **97.** $\dfrac{-1}{x^2\sqrt{x^2+1}} = \dfrac{-\sqrt{x^2+1}}{x^2(x^2+1)}$

99. $|ab| = \sqrt{(ab)^2} = \sqrt{a^2} \cdot \sqrt{b^2} = |a| \cdot |b|$

Chapter 1 Review Exercises

1. 7   **3.** $(1-x)^{2b}$ **5.** $\dfrac{2x}{3y^3}$ **7.** $\dfrac{x^2+1}{4x}$ **9.** $\dfrac{72y^2z^3}{x^4}$ **11.** $16 + 12x - 2x^2 - 2x^3$ **13.** $4\sqrt{3}$ **15.** $2\pi\sqrt{3}$

17. $5\sqrt{3}$ **19.** $2R\sqrt{3}$ **21.** $\frac{3}{10}\sqrt{10}$ **23.** $x^{1/6}$ **25.** $(3 + \sqrt{6})/3$ **27.** $2/h$ **29.** $\dfrac{s-1}{s+1}$

31. $\dfrac{-2}{(x+1)(x+h+1)}$ **33.** -1 **35.** $\dfrac{x+1}{x^n+1}$ **37.** $\dfrac{x-y}{x+y}$ **39.** $\dfrac{x+y}{xy}$ **41.** $x+a+1$ **43.** $\dfrac{2x^{2/3}+1}{2x^{1/3}}$

45. $\dfrac{3x-3}{2x^{1/2}}$ **47.** $\dfrac{2x^2}{\sqrt{1-x^2}} = \dfrac{2x^2\sqrt{1-x^2}}{1-x^2}$ **49.** $\dfrac{-4}{u^2\sqrt{u^2+4}} = \dfrac{-4\sqrt{u^2+4}}{u^2(u^2+4)}$ **51.** $\dfrac{-1}{\sqrt{x}\sqrt{x+h}(\sqrt{x}+\sqrt{x+h})}$

53. -55 **55.** $\frac{1}{10}$ **57.** (a) Commutative property of addition (b) Distributive property **59.** Real, rational

61. $t^3(1+x)(1-x)$ **63.** $(x+1)(x^2+1)$ **65.** $72x^2y^2z^3$ **67.** False, $\sqrt{2}$ **69.** False, 0

71. Perimeter: $16x^3$; Area: $16x^6$ **73.** $\dfrac{1}{(x+2)\sqrt{x-2}}$ **75.** $\dfrac{4x^2-2x+2x-1-2x^2+2x}{(2x+1)^2} = \dfrac{2x^2+2x-1}{(2x+1)^2}$

77. $\left(\dfrac{1}{1+(1/n)}\right)^p = \left(\dfrac{n}{n+1}\right)^p = \dfrac{n^p}{(n+1)^p}$ **79.** $\dfrac{16n^2-4}{n^2}$ **81.** $-\dfrac{1}{ax}$ **83.** $\dfrac{xy(1+x-2y)}{(x-2y)^3}$

85. $x_1+x_2 = -b/a$; $x_1 \cdot x_2 = c/a$ **87.** (a) **89.** (d) **91.** (a) **93.** (c) **95.** (c) **97.** (d) **99.** (a)

Exercises 2.1

1. $\{3\}$ **3.** $\{6\}$ **5.** $\{-1\}$ **7.** $\{3\}$ **9.** $\{2\}$ **11.** $\{-6\}$ **13.** Set of all real numbers **15.** $\emptyset$ **17.** $\{0\}$

19. $\emptyset$ **21.** $\{0\}$ **23.** $\{5\}$ **25.** $\{-5\}$ **27.** $\{-1\}$ **29.** $\{\frac{11}{12}\}$ **31.** $\{-\frac{8}{55}\}$ **33.** $\{4\}$ **35.** $\{\frac{5}{3}\}$

37. Set of all real numbers except 2 **39.** $\emptyset$ **41.** $\{7\}$ **43.** 7, 28 **45.** $40°, 60°, 80°$ **47.** $1.05, 0.05

49. $60,000 **51.** $\frac{18}{5}$ hr **53.** 271 gallons **55.** 93.3 mi **57.** 38 lb of 35 percent tin, 57 lb of 10 percent tin

59. 20 quarts **61.** 24,000 miles **63.** 13, including Sammy **65.** $y = \dfrac{4-x}{2}$ **67.** $x = \dfrac{4-c}{3}$

69. $f = \dfrac{mg-T}{m}$ **71.** $b = \dfrac{2A-hc}{h}$ **73.** $p = \dfrac{a}{1+rt}$ **75.** $d^2 = \dfrac{kL}{R}$ **77.** $R = \dfrac{E-CS}{C}$ **79.** $I_2 = \dfrac{I_1 r_2^2}{r_1^2}$

81. $r = \dfrac{S-a}{S-L}$ **83.** $Z_2 = \dfrac{ZZ_1}{Z_1-Z}$ **85.** $a = \dfrac{fb}{b-f}$ **87.** $x = \dfrac{3}{y-1}$ **89.** $y' = \dfrac{2x+y}{2y-x}$

91. $y' = \dfrac{-x^2-2xy}{x^2+y^2}$ **93.** One solution: $a \neq 0$; no solution: $a = 0, b \neq 0$; infinitely many solutions: $a = 0, b = 0$

Exercises 2.2

1. $5i$ **3.** $8i$ **5.** $i\sqrt{22}$ **7.** $i\sqrt{3}$ **9.** $2-8i\sqrt{2}$ **11.** $2i$ **13.** $2i\sqrt{2}$ **15.** $(1/4)-(\sqrt{23}/4)i$

17. $-2+i$ **19.** $(-1/5)-(\sqrt{14}/5)i$ **21.** $0+9i$ **23.** $0+0i$ **25.** $-3-i$ **27.** $5-6i$

29. $-6+0i$ **31.** $-4+0i$ **33.** $10-10i$ **35.** $13+0i$ **37.** $-3-4i$ **39.** $\frac{80}{9}-2i$

41. $(5i)^2+25 = -25+25 = 0$ **43.** $(-2i\sqrt{2})^2+8 = -8+8 = 0$

45. $(1-i)^2-2(1-i)+2 = 1-2i+i^2-2+2i+2 = 0$

47. $2\left(\dfrac{1}{4}+\dfrac{i\sqrt{23}}{4}\right)^2 - \left(\dfrac{1}{4}+\dfrac{i\sqrt{23}}{4}\right)+3 = \dfrac{1}{8}+\dfrac{i\sqrt{23}}{4}-\dfrac{23}{8}-\dfrac{1}{4}-\dfrac{i\sqrt{23}}{4}+3 = 0$

49. $(2i)^4-16 = 16i^4-16 = 16(1)-16 = 0$ **51.** Solution **53.** Solution **55.** Not a solution **57.** Solution

59. Not a solution **61.** $3-4i$ **63.** $-i$ **65.** -7 **67.** $\frac{1}{5}-\frac{2}{5}i$ **69.** $0+\frac{2}{5}i$ **71.** $-\frac{4}{17}-\frac{1}{17}i$ **73.** $\frac{7}{10}+\frac{1}{10}i$

75. $(1/3)-(2\sqrt{2}/3)i$ **77.** $0+i$ **79.** $\frac{1}{2}-\frac{1}{2}i$

81. (a) $\overline{(1 + i) + (2 - 3i)} = \overline{3 - 2i} = 3 + 2i; \overline{1 + i} + \overline{2 - 3i} = (1 - i) + (2 + 3i) = 3 + 2i$

(b) $\overline{(1 + i)(2 - 3i)} = \overline{5 - i} = 5 + i; \overline{1 + i} \cdot \overline{2 - 3i} = (1 - i)(2 + 3i) = 5 + i$

(c) $\overline{(1 + i)^2} = \overline{2i} = -2i; (\overline{1 + i})^2 = (1 - i)^2 = -2i$

83. $\overline{z + w} = \overline{(a + c) + (b + d)i} = (a + c) - (b + d)i = (a - bi) + (c - di) = \bar{z} + \bar{w}$

85. If z is a real number, $z = a + 0i$, so $\bar{z} = a - 0i$ and $\bar{z} = z$.

87. $\overline{z^2} = \overline{(a + bi)(a + bi)} = \overline{(a^2 - b^2) + 2abi} = (a^2 - b^2) - 2abi = (a - bi)^2 = (\bar{z})^2$

89. (a) $8 - j$ volts **(b)** $\frac{2}{17} - \frac{8}{17}j$ ohms **(c)** $\frac{6}{37} - \frac{1}{37}j$ amps

Exercises 2.3

1. $\{0, -5\}$ **3.** $\{0,5\}$ **5.** $\{2, -2\}$ **7.** $\{4, -2\}$ **9.** $\{4, -1\}$ **11.** $\{5, \frac{1}{3}\}$ **13.** $\{1, \frac{2}{3}\}$ **15.** $\{2, -\frac{1}{2}\}$

17. $\{\pm 3\}$ **19.** $\{\pm 5i\}$ **21.** $\{\pm 2i\sqrt{2}\}$ **23.** $\{3,7\}$ **25.** $\{-1 \pm i\}$ **27.** $\{7 \pm i\sqrt{7}\}$ **29.** 9 **31.** $\frac{9}{16}$

33. $\{1 \pm i\}$ **35.** $\{(-1 \pm \sqrt{13})/2\}$ **37.** $\{(2 \pm \sqrt{10})/3\}$ **39.** $\{(-1 \pm i\sqrt{14})/5\}$ **41.** $\{-4, -1\}$

43. $\{3\}$ **45.** $\{2, -2\}$ **47.** $\{0, \frac{2}{3}\}$ **49.** $\{(1 \pm i\sqrt{2})/3\}$ **51.** $\{-1, \frac{3}{4}\}$ **53.** -15; conjugate complex numbers

55. 0; real, rational, equal **57.** 136; real, irrational, unequal **59.** 84; real, irrational, unequal

61. -3; conjugate complex numbers **63.** 7 **65.** 6 ft, 4 ft **67.** 3.8 in., 7.8 in. **69. (a)** 6 seconds

(b) 5 seconds, 1 second **(c)** The projectile attains the given height going up and coming down. **71.** 2.3 seconds

73. \$5, \$3; The low unit price produces high sales volume, and the higher unit price produces lower sales volume.

75. 11.1 in. **77.** $x_1 + x_2 = \dfrac{-b + \sqrt{b^2 - 4ac}}{2a} + \dfrac{-b - \sqrt{b^2 - 4ac}}{2a} = -\dfrac{2b}{2a} = -\dfrac{b}{a}$

$x_1 \cdot x_2 = \dfrac{-b + \sqrt{b^2 - 4ac}}{2a} \cdot \dfrac{-b - \sqrt{b^2 - 4ac}}{2a} = \dfrac{b^2 - (b^2 - 4ac)}{4a^2} = \dfrac{4ac}{4a^2} = \dfrac{c}{a}$ **79.** $x = \pm R\sqrt{2}/2$

Exercises 2.4

1. $\{-1\}$ **3.** $\emptyset$ **5.** $\{\frac{25}{3}\}$ **7.** $\{6\}$ **9.** $\{-9\}$ **11.** $\{4\}$ **13.** $\{-2\}$ **15.** $\{3\}$ **17.** $\{0\}$ **19.** $\{-3,1\}$

21. $\emptyset$ **23.** $\{0,4\}$ **25.** $\{\frac{1}{2}\}$ **27.** Set of all nonnegative real numbers **29.** $\emptyset$ **31.** $\{1, -9\}$ **33.** $\{1\}$

35. $\{27 \pm 10\sqrt{2}\}$ **37.** $\{-1\}$ **39.** $\{1,16\}$ **41.** $\{0, -2, -7\}$ **43.** $\{0,1, -1\}$ **45.** $\{0, \frac{1}{2}, -\frac{1}{3}\}$

47. $\{2, -2, 1, -1\}$ **49.** $\{1, -1, \frac{11}{3}\}$ **51.** $\{0\}$ **53.** $y = x^3 - 2$ **55.** $\frac{21}{5}$ cm

Exercises 2.5

1. $(-\infty, -\frac{9}{2})$ **3.** $(-\infty, -1]$ **5.** $(-\infty, -6)$

7. $(-\infty, \frac{12}{7})$ **9.** $(-2, \infty)$ **11.** $(-\infty, -3]$

13. $(-8,\infty)$

15. Set of all real numbers

17. $(0,\infty)$

19. $\emptyset$

21. $(-\infty,0]$

23. $\emptyset$

25. $\{x: 1 \le x < 4\}$

27. $(-\infty,3]$

29. $(-3,0)$; $\{x: -3 < x < 0\}$

31. $\{x: x \ge 0\}$

33. $(-\infty,-2) \cup (2,\infty)$; $\{x: x < -2 \text{ or } x > 2\}$ **35.** $(2,6)$

37. $[-5,7]$ **39.** $(\frac{7}{4},4]$ **41.** $(-\infty,-1) \cup (2,\infty)$ **43.** $(-\infty,0)$ **45.** $(-2,3]$ **47.** $[0,1] \cup [2,\infty)$
49. $(1,2)$ **51.** $(-\infty,-1] \cup (0,1]$ **53.** $(-\infty,\infty)$ **55.** $(-3,2)$ **57.** $(-\infty,0) \cup (5,\infty)$ **59.** $[-2,1]$
61. $(-\infty,-\frac{1}{2}] \cup [4,\infty)$ **63.** $(-2,0) \cup (2,\infty)$ **65.** $[-2,-1] \cup [1,2]$ **67.** $[-3,-1] \cup [0,1] \cup [3,\infty)$
69. $(-\infty,-2] \cup [2,3)$ **71.** $(-\infty,-1) \cup (1,\frac{5}{2})$ **73.** $(-\infty,0) \cup (\frac{1}{2},\infty)$ **75.** $(-\infty,-21] \cup (7,\infty)$
77. $(-\infty,4) \cup [12,\infty)$ **79.** $(-3,2)$ **81.** [24,500 mi, 25,500 mi]; [24,855 mi, 24,865 mi]   **83.** $(0°,40°)$
85. $[71,91)$ **87.** $(-7,-3) \cup (2,6)$ **89.** (1 second, 5 seconds)   **91.** $(-\infty,-2] \cup [2,\infty)$
93. $(\sqrt{b} - \sqrt{a})^2 \ge 0$ implies $b - 2\sqrt{ab} + a \ge 0$. Then $-2\sqrt{ab} \ge -(a + b)$, so $\sqrt{ab} \le (a + b)/2$.
95. $a < b$ implies $b - a$ is positive. Then $b - a = (b + c) - (a + c)$, so $(b + c) - (a + c)$ is positive,
which implies $a + c < b + c$.   **97.** If $a < b$ and $b < c$, then $b - a$ and $c - b$ are both positive. Therefore,
the sum $(b - a) + (c - b)$, which simplifies to $c - a$, is positive, which implies $a < c$.

Exercises 2.6

1. $<$ **3.** $=$ **5.** $=$ **7.** $=$ **9.** $>$ **11.** $=$ **13.** $=$ **15.** $<$ **17.** $>$ **19.** $=$ **21.** $\{3,-1\}$
23. $\{34,2\}$ **25.** $\{\frac{3}{2},-\frac{3}{2}\}$ **27.** $\emptyset$ **29.** $[0,\infty)$ **31.** $[0,\infty)$ **33.** $(-\infty,5]$ **35.** $\{1,\frac{1}{3}\}$ **37.** $\{1\}$
39. $\{2,-\frac{4}{3}\}$ **41.** $(-4,4)$ **43.** $[2,8]$ **45.** $(a - 1, a + 1)$ **47.** $(-\infty,-2) \cup (2,\infty)$ **49.** $(-\infty,2] \cup [8,\infty)$
51. $(-\frac{13}{3},1)$ **53.** $[-6,7]$ **55.** $(-\infty,-1) \cup (2,\infty)$ **57.** $\emptyset$ **59.** $(\frac{7}{4},\frac{7}{2})$ **61.** $\{x: |x| < 2\}$
63. $\{x: |x - 5| \le 6\}$ **65.** $\{x: |x - 1.15| < 0.05\}$ **67.** $\{x: |x| > 2\}$ **69.** $\{x: |x - 3| > 3\}$
71. $\{x: |x - 6.5| \ge 0.5\}$ **73.** $\{x: |x| < 5\}$ **75.** $\{x: |x + 1| > 3\}$ **77.** $\{x: |x - 3| < 1\}$ **79.** $\{x: |x - a| < d\}$
81. (1.5 seconds, 6.5 seconds)

Chapter 2 Review Exercises

1. $\{2\}$ **3.** $[0,\infty)$ **5.** $\{\frac{25}{2}\}$ **7.** $(-5,5)$ **9.** $\{-7\}$ **11.** $(-\infty,1] \cup [2,\infty)$ **13.** $\{200,-200\}$ **15.** $\{-1\}$
17. $\{4\}$ **19.** $\{5\sqrt{3}/3\}$ **21.** $\{0,-2\}$ **23.** $(-\infty,-3) \cup (0,3)$ **25.** $\{\frac{3}{2},-4\}$ **27.** $\{1,-1,5\}$ **29.** $(-\infty,1)$

31. $\{6,-1\}$ **33.** $(-3,1) \cup [5,\infty)$ **35.** False, $2 < 3$ but $2(-1) > 3(-1)$ **37.** False, $|-1 + 3| < |-1| + |3|$

39. True   **41.** $y = x - z - 10P$ **43. (a)** $-4 + 2i$ **(b)** $\frac{3}{10} - \frac{1}{10}i$ **(c)** 17 **(d)** $10i$ **45.** 7   **47.** $6 - 8i$

49. (a) $(-\infty,0] \cup (6,\infty)$, $\{x: x \le 0 \text{ or } x > 6\}$ **(b)** $(-4,-1)$, $\{x: -4 < x < -1\}$ **51.** $\{20\}$

53. $y' = \dfrac{12x - 5y + 3}{5x + 1}$ **55.** $x = \dfrac{1 \pm \sqrt{1 - 4s}}{2}$ **57.** (d)   **59.** (c)   **61.** (a)   **63.** (a)   **65.** $522

67. $R = E/I$ **69.** 9 ft   **71.** $85,106   **73.** 2 seconds; [0 seconds, 1.5 seconds]   **75.** If $x - 2 = 1$, then $x = 3$. Thus, we cannot divide both sides of the equation by $x - 3$ since we would be dividing by zero.   **77.** $(1 + \sqrt{5})/2$

Exercises 3.1

1. Not a function   **3.** Function   **5.** Function   **7.** Not a function   **9.** Function   **11.** Not a function

13. *D:* $(0,\infty)$; *R:* $(0,\infty)$ **15.** *D:* $(-\infty,\infty)$; *R:* $[3,\infty)$ **17.** *D:* $(-\infty,\infty)$; *R:* $[0,\infty)$ **19.** *D:* $(-\infty,\infty)$; *R:* $(-\infty,1]$

21. *D:* Set of all real numbers except -1; *R:* Set of all real numbers except 0   **23.** *D:* $(-\infty,\infty)$; *R:* $\{0\}$

25. *D:* $[1,\infty)$; *R:* $[0,\infty)$ **27.** *D:* $[-2,2]$; *R:* $[-2,0]$ **29.** *D:* $(-\infty,4]$; *R:* $(0,4]$ **31.** $(-2,-5)$, $(1,4)$

33. $(1,1)$, $(-1,1)$ **35.** $(0,7)$, $(\frac{7}{3},0)$, $(-5,22)$, $(\frac{2}{3},5)$ **37.** 0   **39.** 2   **41.** Function   **43.** Function

45. Not a function   **47.** *D:* $\{-2,-1,1,2\}$; *R:* $\{1,4\}$ **49.** 11, 5, -3 **51.** 19, 4, 7   **53.** $\frac{2}{5}$, -2, undefined

55. (a) 3 **(b)** 1 **(c)** -1 **57. (a)** 2 **(b)** -10 **(c)** 10 **(d)** -4 **(e)** 3 **(f)** $\frac{1}{2}$ **(g)** 4 **(h)** 4 **(i)** 2 **(j)** $x^2 + 1$

59. (a) 3 **(b)** -1 **(c)** No   **61. (a)** -2 **(b)** -1 **(c)** Undefined   **63. (a)** 3 **(b)** 0 **(c)** 3

65. (a) $2(x + h)^2 - 1$ **(b)** $4xh + 2h^2$ **(c)** $4x + 2h$ **67.** 7   **69.** $2x + h - 2$ **71.** $A = s^2$; $(0,\infty)$

73. $L = A/5$; $[25,\infty)$ **75.** $s = P/4$; $(0,\infty)$ **77.** $d = 40t$; $[0,\infty)$ **79.** $e = 28n$; $[0,\infty)$ **81.** $e = 0.06p$; $[0,\infty)$

83. $c = 5x + 400$; $\{x: x \ge 0$, where x is an integer$\}$ **85.** $a = 10,000 - 50n$; $[0,200]$

87. $t = 310 + 0.19(i - 2,000)$; $[2,000, 4,000]$ **89.** $c = \begin{cases} 1.75 & \text{if } 0 \le n \le 12 \\ 1.75 + 0.0382(n - 12) & \text{if } 12 < n \le 48; \end{cases}$ $[0,48]$

91. $A = (12 - 2x)x$; $(0,6)$ **93.** $A = (100 - 2x)x$; $(0,50)$ **95.** $A = \left(\dfrac{100}{w} + 10\right)(w + 6)$; $(0,10)$

Exercises 3.2

1.

3. (a) 2 **(b)** 3 **(c)** 4 **(d)** 5 **(e)** 13 **(f)** $\sqrt{17}$ **(g)** $\sqrt{40}$ or $2\sqrt{10}$ **(h)** $\sqrt{41}$ **(i)** $\sqrt{2}$ **(j)** $\sqrt{17}$ **(k)** 10 **(l)** 13   **5.** $(4,-4)$; 30

7. $d_1 = \sqrt{40}$, $d_2 = \sqrt{90}$, $d_3 = \sqrt{130}$; since $(d_3)^2 = (d_1)^2 + (d_2)^2$, the Pythagorean theorem ensures a right triangle.   **9.** $\sqrt{13}$

11.

13.

15.

17.

19.

21.

23.

25.

27.

29.

31.

33.

35. (a)

(b)

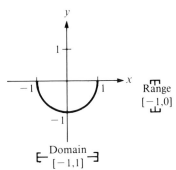

Range
$[-1,0]$

Domain
$[-1,1]$

37. (a) 0 **(b)** 1 **(c)** -2 **(d)** -4
(e)

$f(x)$

39.

41.

43.

$g(x)$

Range
$\{-1,1\}$

Domain
$(-\infty,\infty)$

45.

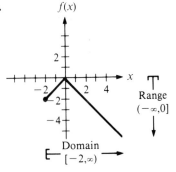

$f(x)$

Range
$(-\infty,0]$

Domain
$[-2,\infty)$

47. Each graph is the reflection of the other about the x-axis.

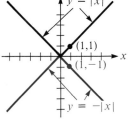

$y = |x|$

$(1,1)$
$(1,-1)$

$y = -|x|$

49. (a) $(-\infty,\infty)$ **(b)** $(-\infty,4]$ **(c)** $f(0) = -5; f(5) = 0$ **(d)** $\{3\}$ **(e)** $\{1,5\}$ **(f)** $\{x: 1 < x < 5\}$
(g) $\{x: x < 1 \text{ or } x > 5\}$ **(h)** $\{x: 0 < x < 6\}$ **51.** Function **53.** Not a function **55.** Not a function
57. Function **59.** Function

Exercises 3.3

1. Even   **3.** Odd

5. Neither   **7.** Even

9. (a)

(b)

11. (a)

(b)

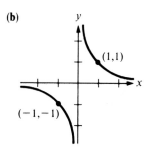

13. Odd   **15.** Even   **17.** Neither

19. Even   **21.** Odd

23. Even

25.

27.

29.

31.

33.

35.

37.

39.

41.

43.

45.

Exercises 3.4

1. $(f + g)(x) = 3x - 1$, $(f - g)(x) = x + 1$, $(f \cdot g)(x) = 2x^2 - 2x$, $(f/g)(x) = 2x/(x - 1)$, $(f \circ g)(x) = 2x - 2$,
$(g \circ f)(x) = 2x - 1$; $D_{f/g}$ is the set of all real numbers except 1, otherwise the domain is $(-\infty,\infty)$.
3. $(f + g)(x) = 4x^2 - x - 2$, $(f - g)(x) = 4x^2 + x - 8$, $(f \cdot g)(x) = -4x^3 + 12x^2 + 5x - 15$,
$(f/g)(x) = (4x^2 - 5)/(-x + 3)$, $(f \circ g)(x) = 4x^2 - 24x + 31$, $(g \circ f)(x) = -4x^2 + 8$; $D_{f/g}$ is the set of all real
numbers except 3, otherwise the domain is $(-\infty,\infty)$.   **5.** $(f + g)(x) = x^2 + 1$, $(f - g)(x) = x^2 - 1$, $(f \cdot g)(x) = x^2$,
$(f/g)(x) = x^2$, $(f \circ g)(x) = 1$, $(g \circ f)(x) = 1$; Domain in all cases is $(-\infty,\infty)$.
7. $(f + g)(x) = (x^2 + x + 1)/(x^2 + x)$, $(f - g)(x) = (1 + x - x^2)/(x^2 + x)$, $(f \cdot g)(x) = 1/(1 + x)$,
$(f/g)(x) = (1 + x)/x^2$, $(f \circ g)(x) = (1 + x)/x$, $(g \circ f)(x) = 1/(x + 1)$; Domain in all cases is the set of all real
numbers except 0 and -1.   **9.** $(f + g)(x) = x^2 - 1 + \sqrt{x + 1}$, $(f - g)(x) = x^2 - 1 - \sqrt{x + 1}$,
$(f \cdot g)(x) = (x^2 - 1)\sqrt{x + 1}$, $(f/g)(x) = (x - 1)\sqrt{x + 1}$, $(f \circ g)(x) = x$, $(g \circ f)(x) = |x|$;
$D_{f + g} = D_{f - g} = D_{f \cdot g} = D_{f \circ g} = [-1,\infty)$, $D_{f/g} = (-1,\infty)$, $D_{g \circ f} = (-\infty,\infty)$ **11.** $f + g = \{(2,3), (3,5)\}$,
$f - g = \{(2,5), (3,5)\}$, $f \cdot g = \{(2,-4), (3,0)\}$, $f/g = \{(2,-4)\}$, $f \circ g = \{(3,2), (4,3), (5,4)\}$,
$g \circ f = \{(0,-1), (1,0), (2,1), (3,2)\}$; $D_{f + g} = D_{f - g} = D_{f \cdot g} = \{2,3\}$, $D_{f/g} = \{2\}$, $D_{f \circ g} = \{3,4,5\}$, $D_{g \circ f} = \{0,1,2,3\}$
13. (a) -3 **(b)** 20   **(c)** Undefined   **(d)** 4   **15.** $g(x) = 4x - 1$; $f(x) = x^3$ **17.** $g(x) = 3 - x$; $f(x) = 2x^4$
19. $g(x) = 2x + 1$; $f(x) = \sqrt[3]{x}$ **21.** $g(x) = 2x^2 + 2$ **23.** -12 **25.** $(f \circ g \circ h)(x) = 2x^2 - 2$
27. $(f \circ g \circ h)(x) = 1/(1 - \sqrt{x}) = (1 + \sqrt{x})/(1 - x)$ **29.** Odd function; $(f \circ g)(-x) = f[g(-x)] = f[-g(x)]$
$= -f[g(x)] = -(f \circ g)(x)$ **31.** $(f \circ g)(x) = \sqrt{-x^2 - 5}$; No value for x produces a range element that is a real
number.   **33. (a)** $d = s\sqrt{2}$ **(b)** $s = P/4$ **(c)** $d = P\sqrt{2}/4$

Exercises 3.5

1. $f^{-1} = \{(7,1),(8,2),(9,3)\}$; $D_{f^{-1}} = \{7,8,9\}$; $R_{f^{-1}} = \{1,2,3\}$; $D_f = \{1,2,3\}$; $R_f = \{7,8,9\}$

3. $f^{-1} = \{(10,-2),(0,0),(-1,5)\}$; $D_{f^{-1}} = \{10,0,-1\}$; $R_{f^{-1}} = \{-2,0,5\}$; $D_f = \{-2,0,5\}$; $R_f = \{10,0,-1\}$

5. $f^{-1} = \{(4,4),(5,5),(6,6)\}$; $D_{f^{-1}} = R_{f^{-1}} = D_f = R_f = \{4,5,6\}$ **7.** No   **9.** Yes   **11.** No   **13.** No

15. Yes   **17.** Yes

19.

21.

23.

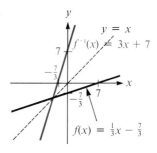

25. $y = x^2$ is not a one-to-one function, so no inverse function exists.

27. $f(x) = 2$ is not a one-to-one function, so no inverse function exists.

29.

31.

33.

35.

37. $f(x) = \sqrt{1 - x^2}$ is not a one-to-one function, so no inverse function exists.

39. $\frac{2}{7}$ **41.** $f^{-1}(x) = x^2 + 1$, $x \ge 0$; $D_f = R_{f^{-1}} = [1,\infty)$, $R_f = D_{f^{-1}} = [0,\infty)$

43. $f[g(x)] = 3(\frac{1}{3}x + \frac{2}{3}) - 2 = x$; $g[f(x)] = \frac{1}{3}(3x - 2) + \frac{2}{3} = x$

45. $f[g(x)] = \dfrac{1}{[(1 - 4x)/x] + 4} = \dfrac{x}{1 - 4x + 4x} = x$; $g[f(x)] = \dfrac{1 - 4[1/(x + 4)]}{1/(x + 4)} = \dfrac{(x + 4) - 4}{1} = x$

47. $f[g(x)] = \sqrt{(x^2 - 1) + 1} = \sqrt{x^2} = x$ (since $x \ge 0$); $g[f(x)] = (\sqrt{x + 1})^2 - 1 = (x + 1) - 1 = x$

49. $f[f(x)] = \dfrac{2[(2x+1)/(3x-2)]+1}{3[(2x+1)/(3x-2)]-2} = \dfrac{2(2x+1)+(3x-2)}{3(2x+1)-2(3x-2)} = \dfrac{7x}{7} = x$

51. Not inverses since $f[g(x)] = 3(\frac{1}{3}x+1)-1 = x+2$ **53.** $y = \sqrt{x}, x > 0$; This is the formula for the side length (y) of a square in terms of the area (x).

Exercises 3.6

1. 1   **3.** $\frac{4}{7}$ **5.** $-\frac{11}{2}$ **7.** 0   **9.** Undefined   **17.**

11. -1 **13.** $\frac{1}{2}$ **15.** (a) -1 (b) $\frac{1}{2}$ (c) 0
(d) Undefined   (e) 3

19. $\frac{1}{6}$; For each 6-lb increase of the earth weight of an object, the moon weight increases 1 lb.
21. -32; For each second (up to 10 seconds) the velocity decreases by 32 ft/second.   **23.** 0.12; For each brochure the cost increases \$0.12.   **25.** 7; For each mile the cost increases \$7.   **27.** 4   **29.** 2   **31.** $2x+h$ **33.** $-2x-h$
35. 2   **37.** 0   **39.** $-1/[x(x+h)]$ **41.** $2\pi r + \pi h$

Exercises 3.7

1. $y = \frac{7}{3}x; \frac{70}{3}$ **3.** $y = \frac{3}{4}x^2; 12$ **5.** $y = 72/x; 3$ **7.** $y = 24/x^3; \frac{8}{9}$ **9.** 60   **11.** 49   **13.** 4 in.
15. 2,000,000 tons   **17.** 3 in.   **19.** 600 rpm   **21.** $500\pi/3$ cubic units   **23.** 181 lb   **25.** Exposure time is multiplied by 4.   **27.** $\frac{5}{3}$ atm   **29.** Force is multiplied by 24.

Chapter 3 Review Exercises

1. $(0,4), (\frac{5}{2},-1), (-1,6), (2,0), (3,-2)$ **3.** Yes   **5.** Yes; no   **7.** $-\frac{1}{2}$ **9.** $y = \frac{5}{12}x; \frac{25}{3}$ **11.** Set of all real numbers except 0 and -1 **13.** $A = d^2/2$ **15.** Yes   **17.** No   **19.** Yes   **21.** Odd function
23.   **25.**   **27.** $\frac{9}{4}$ **29.** $A = \pi d^2/4$

31.

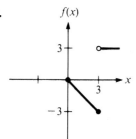

33. $\{y: y = 3 \text{ or } -3 \le y \le 0\}$

35.

37.

39.

41.

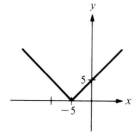

43. Even function   **45. (a)** $g^{-1}(x) = -2x + 5$ **(b)** 7   **47.** $[-\pi/2, \pi/2]$ **49.** 0   **51.** (0,1]

53.

55. $6x + 3h$ **57.** -1 **59.** $A = (\sqrt{3}/4)x^2$

61. $(f \cdot g)(x) = x;$ set of all real numbers except 0

63. $e = \begin{cases} 500 & \text{if } 0 \le a \le 2,000 \\ 500 + 0.09(a - 2,000) & \text{if } a > 2,000 \end{cases}$ **65.** $[-c,c]$

67. Even function   **69.** $\{-c,0,c\}$

71.

73. (a) $D: (-\infty,\infty); R: (0,\infty)$ **(b)** $D: (0,\infty); R: (-\infty,\infty)$

(c) $D: (-\infty,\infty); R: [-1,1]$ **(d)** $D: [-3,3]; R: [-3,0]$

75. $-\frac{19}{12}$ **77. (b)**   **79. (d)**   **81. (b)**

83. (a)   **85. (a)**

Exercises 4.1

1. Yes; 3   **3.** Yes; 1   **5.** No   **7.** No   **9.** Yes; 0   **11.** $y = 5x - 17$ **13.** $y = \frac{1}{2}x + 1$ **15.** $y = -\frac{1}{2}x$

17. $y = -x + 5$ **19.** $y = -x - 6$ **21.** 1; (0,7)   **23.** 5; (0,0)   **25.** 0; (0,−2)   **27.** $-\frac{2}{3}$; $(0,-\frac{2}{3})$

29. 6; (0,7)

31.

33.

35.
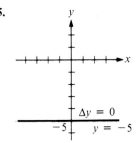

37. $f(x) = \frac{2}{3}x - 2$ **39.** $f(x) = -\frac{4}{3}x - \frac{13}{3}$ **41.** $f(x) = 1$ **43.** −1   **45.** $-\frac{11}{3}$

47. (a) $y = x + 1$ (b) $y = -x + 3$ **49.** (a) $y = -2x + 6$ (b) $y = \frac{1}{2}x - \frac{3}{2}$

51. (a) $y = -\frac{1}{7}x$ (b) $y = 7x$ **53.** (a) $y = \frac{5}{3}x - \frac{14}{3}$ (b) $y = -\frac{3}{5}x - \frac{12}{5}$

55. (a) $y = -\frac{1}{4}x + \frac{13}{4}$ (b) $y = 4x - 18$ **57.** (a) $y = \frac{3}{2}x - \frac{13}{12}$ (b) $y = -\frac{2}{3}x$

59. (a) $y = 0.6x + 40$ (b) \$55   (c) \$40   (d) \$0.60   **61.** (a) $v = -32t - 220$ (b) −348 ft/second

(c) −220 ft/second   (d) + ball rising, − ball falling   **63.** $m_{AB} = 4$, $m_{BC} = -\frac{1}{4}$; Since $m_{AB} \cdot m_{BC} = -1$,

the segments are perpendicular, ensuring a right triangle.   **65.** 5   **67.** (a) B (b) C (c) E (d) D (e) A

69. $m_1 = -A_1/B_1$, $m_2 = -A_2/B_2$; Since the lines are parallel, $-A_1/B_1 = -A_2/B_2$, so $A_1B_2 = A_2B_1$.

71. m; The rate of change is a constant.

Exercises 4.2

1.

3.

5.

7.

9.

11.

13.

15.

17.

19.

21. $(-2,-3)$; $x = -2$ **23.** $(0,-4)$; $x = 0$ **25.** $(1,1)$; $x = 1$

27. $(\frac{3}{2}, -\frac{23}{4})$; $x = \frac{3}{2}$ **29.** $(\frac{1}{2}, \frac{9}{4})$; $x = \frac{1}{2}$

31. Set of all real numbers   **33.** 144 ft; 6 seconds   **35.** 360 watts

37. 10, 10   **39.** 10, 10   **41.** 200 yd by 100 yd   **43.** 3 in.

45. \$15   **47.** $(-\infty,-1) \cup (5,\infty)$ **49.** $(-1,5)$

51. $(-\infty,-1] \cup [2,\infty)$ **53.** $(-\infty,-2] \cup [2,\infty)$ **55.** $(0,3)$

57. $[-2,\frac{3}{2}]$ **59.** $(-1,\frac{5}{3})$ **61.** Set of all real numbers except -1

63. $\emptyset$ **65.** $[-1,1]$

Exercises 4.3

1. $x^2 + 7x - 2 = (x + 5)(x + 2) - 12$ **3.** $6x^3 - 3x^2 + 14x - 7 = (2x - 1)(3x^2 + 7) + 0$

5. $3x^4 + x - 2 = (x^2 - 1)(3x^2 + 3) + x + 1$ **7.** $x^3 - 5x^2 + 2x - 3 = (x - 1)(x^2 - 4x - 2) - 5$

9. $2x^3 + x - 5 = (x + 1)(2x^2 - 2x + 3) - 8$ **11.** $x^4 - 16 = (x - 2)(x^3 + 2x^2 + 4x + 8) + 0$ **13.** 0, 0

15. $-9, 87$ **17.** 0, 0   **19.** $-\frac{176}{27}, -\frac{10}{3}$

Exercises 4.4

1. degree 2; -3 is a zero of multiplicity 2.   **3.** degree 3; -4, $\sqrt{2}$, and $-\sqrt{2}$ are zeros of multiplicity 1.

5. $P(x) = (x - 1)(x - 2)(x - 3)$ **7.** $P(x) = x(x + 4)(x - i)(x + i)$ **9.** $P(x) = (x - 1)^2(x - 5)^2$

11.

13.

15.

17.

19.

21. $P(x) = (x - 2)(x - \sqrt{3})(x + \sqrt{3})$

23. $P(x) = x(x - (4 - \sqrt{3}))(x - (4 + \sqrt{3}))(x - (2 + 3i))(x - (2 - 3i))$

25. $P(x) = (x - 3)(x + 3)(x - 2i)(x + 2i)$ **27.** $2 - \sqrt{7}$ **29.** $-i\sqrt{2}$ **31.** $2 - 2i$

33. $-\frac{1}{2}$ and -1 (multiplicity 2)   **35.** $\pm\sqrt{2}$ **37.** $\pm i\sqrt{2}, -\sqrt{2}$ **39.** $1, -1, -i$

41.

43. True   **45.** False   **47.** Since $x - b$ is a factor of $P(x)$, we know $P(x) = (x - b)Q(x)$. Then $P(b) = (b - b)Q(b) = 0$, so b is a zero of $y = P(x)$.

Exercises 4.5

1. $\pm 20, \pm 10, \pm 5, \pm 4, \pm 2, \pm 1$ **3.** $\pm 6, \pm 3, \pm 2, \pm \frac{3}{2}, \pm 1, \pm \frac{3}{4}, \pm \frac{1}{2}, \pm \frac{1}{4}$ **5.** positive: 3; negative: 0

7. positive: 1; negative: 3   **9.** $4, -1, -2$ **11.** $-\frac{1}{3}, \pm i\sqrt{5}$ **13.** $\frac{1}{2}$ (multiplicity 3)   **15.** $-1, 2, (1 \pm i\sqrt{79})/8$

17. $-2, -\frac{1}{2}, (-1 \pm i\sqrt{35})/2$ **19.** $P(0) = -8, P(-1) = 5$; Since $P(0)$ and $P(-1)$ have opposite signs, the location theorem guarantees at least one real zero between 0 and -1.   **21.** $P(1) = -4, P(2) = 2$; Since $P(1)$ and $P(2)$ have opposite signs, the location theorem guarantees at least one real zero between 1 and 2. Similarly, since $P(-1) = -16$ and $P(-2) = 50$, there is at least one real zero between -1 and -2.   **23.** 0.7   **25.** 0.3   **27.** -1.3

29. By the rational zero theorem, the only possible rational zeros of the function $P(x) = x^2 - 2$ are $1, -1, 2,$ and -2. Since $P(\sqrt{2}) = (\sqrt{2})^2 - 2 = 0, \sqrt{2}$ is a zero of the function. But $\sqrt{2}$ is not one of the possible rational zeros. Thus, $\sqrt{2}$ is not a rational number.

Exercises 4.6

1. $x = \frac{3}{2}$ **3.** $x = 6, x = -1$

5.

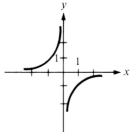

Vertical asymptote: $x = 0$
Horizontal asymptote: $y = 0$

7.

Vertical asymptote: $x = -1$
Horizontal asymptote: $y = 0$

9.

Vertical asymptote: $x = -2$
Horizontal asymptote: $y = 1$

11.

Vertical asymptote: $x = -\frac{5}{3}$
Horizontal asymptote: $y = \frac{2}{3}$

13.

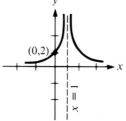

Vertical asymptote: $x = 1$
Horizontal asymptote: $y = 0$

15.

Vertical asymptotes: $x = -1,$
                        $x = 4$
Horizontal asymptote: $y = 0$

17.

Vertical asymptote: $x = 0$
Horizontal asymptote: $y = 0$

19.

Vertical asymptote: $x = 0$
Horizontal asymptote: $y = 1$

21.

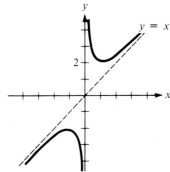

Vertical asymptote: $x = 0$
Horizontal asymptote: none

23.

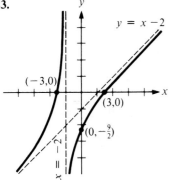

Vertical asymptote: $x = -2$
Horizontal asymptote: none

25.

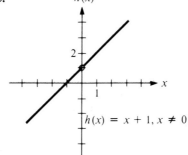

$h(x) = x + 1, x \neq 0$

27.

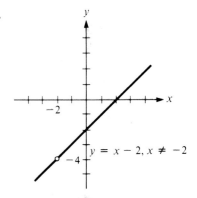

$y = x - 2, x \neq -2$

29.

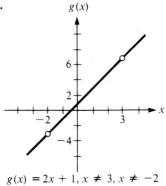

$g(x) = 2x + 1, x \neq 3, x \neq -2$

31.

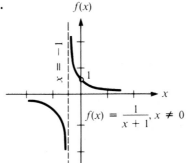

$f(x) = \dfrac{1}{x + 1}, x \neq 0$

Vertical asymptote: $x = -1$
Horizontal asymptote: $y = 0$

33.

$y = \dfrac{x - 1}{(x - 1)^2} = \dfrac{1}{x - 1}$

Vertical asymptote: $x = 1$
Horizontal asymptote: $y = 0$

Chapter 4 Review Exercises

1. Yes   **3.** No   **5.** No   **7.** $0, \frac{3}{2}$ **9.** $2, i, -i$ **11.** $(5 \pm \sqrt{17})/4$

13.

15.

17.

19.

21. $(-\infty, -3] \cup [2, \infty)$ **23.** Set of all real numbers except $-\frac{1}{2}$

25. $(-5, -1)$; $x = -5$ **27.** $y = \frac{2}{3}$; $(\frac{13}{2}, \frac{2}{3})$ **29.** $y = \frac{2}{3}x - 2$

31. **(a)** -16 **(b)** $\frac{7}{5}$ **33.** $P(x) = x(x - 5)(x + 2)(x - 3)$ **35.** 16

37. $x - 3$ **39.** **(a)** 4 **(b)** 0   **41.** Yes; there can be no real zeros greater than 2 if and only if the numbers in the bottom row of the synthetic division all have the same sign.   **43.** $x^2 - x - 6 = 0$ **45.** 1   **47.** 2.3

49. 108 ft (front) by 81 ft   **51.** **(a)** $v = -32t + 320$ **(b)** 32 ft/second

(c) 320 ft/second **(d)** 10 seconds; Projectile reaches its highest point when $t = 10$ seconds.   **53.** (b)   **55.** (a)   **57.** (b)   **59.** (c)   **61.** (a)

Exercises 5.1

1.

$(-3, \frac{1}{8})$, $(0,1)$, $(-1, \frac{1}{2})$, $(2,4)$

3.

$(2, 0.01)$, $(0,1)$,
$(-1, 10)$, $(-2, 100)$

5.

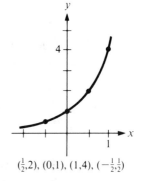

$(\frac{1}{2}, 2)$, $(0,1)$, $(1,4)$, $(-\frac{1}{2}, \frac{1}{2})$

7.

$(0,1), (-2,\frac{1}{2}), (1,\sqrt{2}), (2,2)$

9.

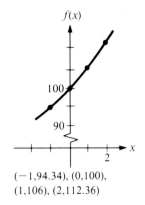

$(-1,94.34), (0,100),$
$(1,106), (2,112.36)$

11.

13.

15.

17.

19.

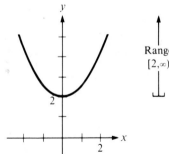

21. (a) 4   (b) 3   (c) $\frac{1}{3}$ (d) 5   (e) 16   (f) 100   (g) $\frac{1}{8}$ (h) 9   (i) $\frac{1}{8}$ (j) 27

23. If $f(x) = b^x$, then $f(x_1 + x_2) = b^{x_1 + x_2} = b^{x_1} \cdot b^{x_2} = f(x_1)f(x_2)$.

25. (a) $y = 100(1.05)^t$ (b) \$115.76   **27.** (a) $f(t) = 10,000(0.8)^t$ (b) \$5,120   **29.** (a) $y = 16(\frac{1}{2})^{2t}$ (b) 4 g

Exercises 5.2

1. $\log_3 9 = 2$ **3.** $\log_{1/2}(\frac{1}{4}) = 2$ **5.** $\log_4(\frac{1}{16}) = -2$ **7.** $\log_{25} 5 = \frac{1}{2}$ **9.** $\log_7 1 = 0$ **11.** $5^1 = 5$

13. $(\frac{1}{3})^2 = \frac{1}{9}$ **15.** $2^{-2} = \frac{1}{4}$ **17.** $49^{1/2} = 7$ **19.** $100^0 = 1$ **21.** 2   **23.** 1   **25.** -1 **27.** $\frac{1}{2}$ **29.** 0

31. 3   **33.** $\frac{1}{5}$ **35.** 5   **37.** 3   **39.** $b > 0, b \neq 1$ **41.** $3; -1; 0$ **43.** $4; 0; -7$

45.   **47.**   **49.**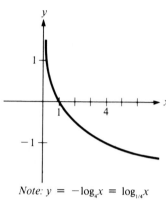

Note: $y = -\log_4 x = \log_{1/4} x$
(see Exercise 47)

51.   **53.**   **55.**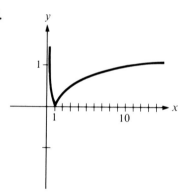

$x = -1$

57. $(-\infty, -3) \cup (3, \infty)$ **59.** Set of all real numbers except 1   **61.** 7   **63.** 3.6   **65.** 1   **67.** 7.9×10^{-3}

69. (a) 0   (b) 10   (c) 20   (d) 30; If the power ratio is multiplied by 10, add 10 dB to loudness.   **71.** 25 dB

Exercises 5.3

1. $\log_{10} 7 + \log_{10} 5$ **3.** $\log_6 3 - \log_6 5$ **5.** $16 \log_5 11$ **7.** $\frac{1}{2} \log_2 3$ **9.** $\frac{1}{5} \log_{10} 16$ **11.** $2 + 3 \log_4 3$

13. $\frac{1}{2}(\log_b x + \log_b y)$ **15.** $\frac{5}{3} \log_b x$ **17.** $\frac{1}{4}(\log_b x + 2 \log_b y - 3 \log_b z)$ **19.** $\log_2 12$ **21.** $\log_4 4$ or 1

23. $\log_7 9$ **25.** $\log_{10} \frac{3}{8}$ **27.** $\log_b \sqrt[3]{xy^2}$ **29.** $\log_b(x - 1)$ **31.** $\log_b \sqrt{\dfrac{xz^3}{y^5}}$ **33.** 5   **35.** $k + \log_{10} m$

37. 3   **39.** $\dfrac{n}{x} x^n = nx^{n-1}$ **41.** (a) 0.7781   (b) 0.6020   (c) 0.2386   (d) -0.1761 (e) -0.3010 (f) 0.6990

43. $\log_b(1/a) = \log_b a^{-1} = -\log_b a$ **45.** Let $x_1 = b^y$ and $x_2 = b^z$, then $y = \log_b x_1$ and $z = \log_b x_2$.

(a) $\dfrac{x_1}{x_2} = \dfrac{b^y}{b^z} = b^{y-z}$; Thus, $\log_b \dfrac{x_1}{x_2} = y - z = \log_b x_1 - \log_b x_2$.   **(b)** $(x_1)^k = (b^y)^k = b^{ky}$;

Thus, $\log_b(x_1)^k = ky = k \log_b x_1$.   **47.** $\log_{10} 10^x = x \log_{10} 10 = x$, where x is any real number

Exercises 5.4

1. $\{3.322\}$ **3.** $\{1.044\}$ **5.** $\{-3.170\}$ **7.** $\{3.376\}$ **9.** $\{-0.7304\}$ **11.** $\{5.825\}$ **13.** $\{0.0159\}$
15. $\{4.351\}$ **17.** $\{0.0830\}$ **19.** $\{18.58\}$ **21.** $\{100\}$ **23.** $\{10\}$ **25.** $\{\frac{9}{10}\}$ **27.** $\{5\}$ **29.** $\{2\sqrt{2}\}$ **31.** $\{\frac{1}{2}\}$
33. $\{2\}$ **35.** $\{\pm 3\}$ **37.** $\{2\}$ **39.** $(0,\infty)$ **41.** 3.170   **43.** 1.465   **45.** -4.248 **47.** -3.190 **49.** -3
51. 2   **53.** 18.9 percent

Exercises 5.5

1. 1.1735   **3.** $\{3.9318\}$ **5.** $\{0.00552\}$ **7.** $\{5\}$ **9.** 2   **11.** $\{0.4343\}$ **13.** $\{9\}$ **15.** $\{0.3466\}$ **17.** $\{7.259\}$
19. $-0.6931/k$ **21. (a)** \$1,194.05   **(b)** \$6,333.85   **(c)** \$3,166.07   **23.** 4.1 yr   **25. (a)** \$1,197.22   **(b)** \$57,466.23
(c) \$245,325.30   **(d)** \$216,224.53   **27.** 9.9 percent   **29. (a)** 1,078   **(b)** 24.6 hr   **31.** 111,430   **33.** 7.6 g
35. (a) -0.000124 **(b)** 78 percent   **37.** 7:10 A.M.   **39.** $-\ln x$ **41.** $\ln x + \ln(x - 1) - \ln(x + 2)$

Chapter 5 Review Exercises

1. $\{7\}$ **3.** $\{32\}$ **5.** $\{3.43 \times 10^7\}$ **7.** $\{0.00384\}$
9. $\{3\}$ **11.** $\ln(x^2 \cdot \sqrt[3]{y})$ **13.** $\frac{1}{2}[3 \log_b x - \log_b y]$
15. $(\frac{7}{3},\infty)$ **17.** $2^3 = 8$ **19.**

$f(x) = (\frac{1}{2})^x$
$D_f : (-\infty,\infty)$
$R_f : (0,\infty)$

$g(x) = \log_{1/2} x$
$D_g : (0,\infty)$
$R_g : (-\infty,\infty)$

21. $(3, \frac{1}{64})$, $(-1, 4)$, $(0, 1)$, $(\frac{3}{2}, \frac{1}{8})$, $(-2, 16)$ **23.** \$3,664.21
25. $f^{-1} = \{(-1,3), (-2,4)\}$; $D_f = R_{f^{-1}} = \{3,4\}$; $D_{f^{-1}} = R_f = \{-1,-2\}$ **27.** M^2 **29.** $\frac{1}{3}(x + y)$ **31.** 0
33. 8.85   **35.** $y = r + ce^{-kt}$; $\dfrac{y - r}{c} = e^{-kt}$; $-kt = \ln \dfrac{y - r}{c}$; $t = \dfrac{-1}{k} \ln \dfrac{y - r}{c} = \dfrac{1}{k} \ln \left(\dfrac{y - r}{c}\right)^{-1} = \dfrac{1}{k} \ln \dfrac{c}{y - r}$

37.

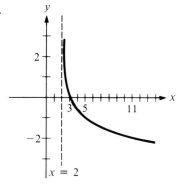

39. (pH of X) = (pH of Y) $-$ 1   **41.** 1   **43.** $\frac{13}{2}$ **45.** -0.098 **47.** (a)
49. (d)   **51.** (b)   **53.** (d)

Exercises 6.1

1. $\sin A = \cos B = \frac{5}{13}$, $\cos A = \sin B = \frac{12}{13}$, $\tan A = \cot B = \frac{5}{12}$, $\cot A = \tan B = \frac{12}{5}$, $\sec A = \csc B = \frac{13}{12}$, $\csc A = \sec B = \frac{13}{5}$ **3.** $\sin \theta = \cos \beta = \frac{4}{5}$, $\cos \theta = \sin \beta = \frac{3}{5}$, $\tan \theta = \cot \beta = \frac{4}{3}$, $\cot \theta = \tan \beta = \frac{3}{4}$,

$\sec \theta = \csc \beta = \frac{5}{3}$, $\csc \theta = \sec \beta = \frac{5}{4}$ **5.** $\dfrac{x}{y}$ **7.** $\dfrac{40}{9}$ **9.** $\dfrac{y}{x}$ **11.** $\sec X$ or $\csc Z$ **13.** 20   **15.** 36

17. $\cos A = \dfrac{12}{13}$ **19.** $\csc \theta = 2$, $\cos \theta = \dfrac{\sqrt{3}}{2}$, $\sec \theta = \dfrac{2}{\sqrt{3}}$, $\tan \theta = \dfrac{1}{\sqrt{3}}$, $\cot \theta = \sqrt{3}$ **21.** $\tan \theta = \dfrac{5}{3}$,

$\sin \theta = \dfrac{5}{\sqrt{34}}$, $\csc \theta = \dfrac{\sqrt{34}}{5}$, $\cos \theta = \dfrac{3}{\sqrt{34}}$, $\sec \theta = \dfrac{\sqrt{34}}{3}$ **23.** $\cot \theta = \dfrac{1}{2}$, $\sin \theta = \dfrac{2}{\sqrt{5}}$, $\csc \theta = \dfrac{\sqrt{5}}{2}$, $\cos \theta = \dfrac{1}{\sqrt{5}}$,

$\sec \theta = \sqrt{5}$ **25.** $\cos 73°$ **27.** $\cot 8°30'$ **29.** $\sec 21.9°$ **31.** No   **33.** No   **35.** Yes   **37.** $2\sqrt{3}$,
4 (hypotenuse)   **39.** 0.9925   **41.** 0.1228   **43.** 5.396   **45.** 1.287   **47.** 0.2130   **49.** $45°00'$ or $45.0°$

51. $38°20'$ or $38.3°$ **53.** $55°50'$ or $55.8°$ **55.** $52°30'$ or $52.6°$ **57.** $3°40'$ or $3.6°$ **59.** $\sin \theta = \dfrac{\text{opp}}{\text{hyp}}$ and

$0 <$ opp $<$ hyp   **61.** $b = 87$ ft, $c = 1\overline{0}0$ ft, $B = 60°$ **63.** $a = 13$ ft, $b = 7.5$ ft, $B = 30°$ **65.** $a = 8.1$ ft,
$b = 24$ ft, $A = 19°$ **67.** $a = 2\overline{0}$ ft, $b = 14$ ft, $B = 35°$ **69.** $b = 85.1$ ft, $c = 121$ ft, $B = 44.5°$
71. $a = 379$ ft, $b = 497$ ft, $A = 37°20'$ **73.** $b = 975$ ft, $c = 975$ ft, $A = 1°50'$ **75.** $b = 14$ ft, $A = 24°$,
$B = 66°$ **77.** $c = 1.4$ ft, $A = 45°$, $B = 45°$ **79.** $a = 23.1$ ft, $A = 62°30'$, $B = 27°30'$ **81.** 25 ft
83. 52 yd   **85.** 89 ft   **87.** $4\overline{0}0$ ft   **89.** $10°$ **91.** 22 ft   **93.** 14.1 in.   **95.** $60°$ **97.** 530 ft
99. 9.8 ft

Exercises 6.2

	$\sin \theta$	$\csc \theta$	$\cos \theta$	$\sec \theta$	$\tan \theta$	$\cot \theta$
1.	$-4/5$	$-5/4$	$3/5$	$5/3$	$-4/3$	$-3/4$
3.	$5/13$	$13/5$	$-12/13$	$-13/12$	$-5/12$	$-12/5$
5.	$-4/5$	$-5/4$	$3/5$	$5/3$	$-4/3$	$-3/4$
7.	$1/\sqrt{2}$	$\sqrt{2}$	$1/\sqrt{2}$	$\sqrt{2}$	1	1
9.	$2/\sqrt{5}$	$\sqrt{5}/2$	$-1/\sqrt{5}$	$-\sqrt{5}$	-2	$-1/2$

11. Q_4 **13.** Q_3 **15.** Q_2

	$\sin \theta$	$\csc \theta$	$\cos \theta$	$\sec \theta$	$\tan \theta$	$\cot \theta$
17.	$-3/5$	$-5/3$	$-4/5$	$-5/4$	$3/4$	$4/3$
19.	$-\sqrt{3}/2$	$-2/\sqrt{3}$	$1/2$	2	$-\sqrt{3}$	$-1/\sqrt{3}$
21.	$-3/4$	$-4/3$	$-\sqrt{7}/4$	$-4/\sqrt{7}$	$3/\sqrt{7}$	$\sqrt{7}/3$
23.	$1/3$	3	$\sqrt{8}/3$	$3/\sqrt{8}$	$1/\sqrt{8}$	$\sqrt{8}$

25. $25°$ **27.** $60°$ **29.** $83°$ **31.** $25°30'$ **33.** $-\frac{1}{2}$ **35.** $2/\sqrt{3}$ **37.** -1 **39.** $\frac{1}{2}$ **41.** $-1/\sqrt{2}$
43. 0 **45.** $2/\sqrt{3}$ **47.** $-\sqrt{3}$ **49.** $-1/\sqrt{2}$ **51.** $-1/\sqrt{2}$ **53.** $-\sqrt{3}/2$ **55.** $\sqrt{3}$ **57.** 1
59. $2/\sqrt{3}$ **61.** 1 **63.** -1 **65.** -0.5299 **67.** 3.487 **69.** 1.923 **71.** -0.8557 **73.** -0.5354
75. 0.0494 **77.** -1.360 **79.** -0.9983 **81.** -0.6111 **83.** 0.3346 **85.** 0.1564 **87.** 1.150
89. 0.4848 **91.** -0.6787 **93.** 1.881

Exercises 6.3

1. 5 **3.** 8 yd   **5.** 9.6 meters   **7.** 5 **9.** $\dfrac{11}{2}, \dfrac{11}{4\pi}$ **11.** $\dfrac{25\pi}{2}$ in.^{2} **13.** 8 ft^2 **15.** $\dfrac{\pi}{6}$ **17.** $\dfrac{\pi}{3}$

19. $\dfrac{2\pi}{3}$ **21.** $\dfrac{5\pi}{6}$ **23.** $\dfrac{7\pi}{6}$ **25.** $\dfrac{4\pi}{3}$ **27.** $\dfrac{5\pi}{3}$ **29.** $\dfrac{11\pi}{6}$ **31.** $\dfrac{10\pi}{9}$ **33.** $\dfrac{\pi}{9}$ **35.** $\dfrac{61\pi}{36}$ **37.** $60°$

39. $40°$ **41.** $120°$ **43.** $140°$ **45.** $432°$ **47.** $70°$ **49.** $115°$ **51.** $229°$ **53.** $332°$

55. 6 rad/second   **57.** (a) $(200\pi/3)$ rad/minute   (b) 400π in./minute   **59.** 300π in./second   **61.** π in.

63. $860,000$ mi

Exercises 6.4

1. 1　　**3.** 0　　**5.** 0　　**7.** 0　　**9.** Undefined　　**11.** 1　　**13.** -1　　**15.** $\frac{1}{2}$　　**17.** 1　　**19.** $\frac{1}{2}$　　**21.** $\dfrac{1}{\sqrt{3}}$

23. $\dfrac{\sqrt{3}}{2}$　　**25.** $-\dfrac{2}{\sqrt{3}}$　　**27.** $\dfrac{1}{\sqrt{3}}$　　**29.** $-\dfrac{\sqrt{3}}{2}$　　**31.** $-\dfrac{\sqrt{3}}{2}$　　**33.** $\sqrt{3}$　　**35.** $-\dfrac{1}{\sqrt{2}}$　　**37.** 2　　**39.** $\frac{1}{2}$

41. $\frac{1}{2}$　　**43.** -2　　**45.** $-\dfrac{1}{\sqrt{2}}$　　**47.** $-\sqrt{3}$　　**49.** $-\dfrac{\sqrt{3}}{2}$　　**51.** -0.4176 (C: -0.4161)　　**53.** 1.162 (C: 1.158)

55. -0.7643 (C: -0.7664)　　**57.** 1.560 (C: 1.566)　　**59.** 0.2085 (C: 0.2116)　　**61.** 0.6749 (C: 0.6702)

63. -1.041 (C: -1.042)　　**65.** 0.9521 (C: 0.9511)　　**67.** 0.2190 (C: 0.2225)　　**69.** -1.081 (C: -1.082)

71. (a) 0.8417, 0.8415　　**(b)** 0.5417, 0.5403　　**(c)** 1,1.000　　**(d)** 0.4794, 0.4794

73.

$$d_1 = d_2$$
$$\sqrt{x^2 + (y-1)^2} = \sqrt{(2y)^2}$$
$$x^2 + y^2 - 2y + 1 = 4y^2$$
since $x^2 + y^2 = 1$ we have
$$1 - 2y + 1 = 4y^2$$
$$0 = 4y^2 + 2y - 2$$
$$0 = 2(2y - 1)(y + 1)$$
$$y = \tfrac{1}{2}$$
$$x^2 + (\tfrac{1}{2})^2 = 1$$
$$x = \dfrac{\sqrt{3}}{2}$$

Exercises 6.5

1.

Amplitude: 2
Period: 2π

3.

Amplitude: 3
Period: 2π

5.

Amplitude: 2
Period: $2\pi/3$

7.

Amplitude: 1
Period: $\pi/9$

9.

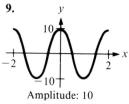

Amplitude: 10
Period: 2

11.

Amplitude: 3
Period: $\pi/2$

13.

Amplitude: 1
Period: 4π

15.

Amplitude: $\frac{1}{2}$
Period: $2\pi/3$

17.

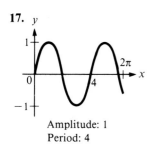

Amplitude: 1
Period: 4

19.

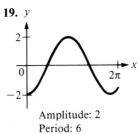

Amplitude: 2
Period: 6

21.

Amplitude: 1
Period: 2π
Phase shift: $-\pi/2$

23.

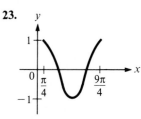

Amplitude: 1
Period: 2π
Phase shift: $\pi/4$

25.

Amplitude: $\frac{1}{2}$
Period: π
Phase shift: $-\pi/8$

27.

Amplitude: 1
Period: 8π
Phase shift: -2π

29.

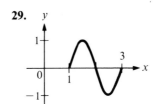

Amplitude: 1
Period: 2
Phase shift: 1

31. $y = 3 \sin 2x$ **33.** $y = -4 \sin 3x$ **35.** $y = 1.5 \cos \frac{1}{2}x$ **37.** $y = -2 \cos 10x$ **39.** $y = 10 \sin \pi x$

41. (d)   **43.** (d)   **45.** (c)   **47.** (d)   **49.** (d)   **51.** (d)   **53.** (b)   **55.** (d)   **57.** (d)   **59.** (a)

Exercises 6.6

x	0	$\dfrac{\pi}{6}$	$\dfrac{\pi}{3}$	$\dfrac{\pi}{2}$	$\dfrac{2\pi}{3}$	$\dfrac{5\pi}{6}$	π	$\dfrac{7\pi}{6}$	$\dfrac{4\pi}{3}$	$\dfrac{3\pi}{2}$	$\dfrac{5\pi}{3}$	$\dfrac{11\pi}{6}$	2π
1. $\cot x$	und.	1.7	0.6	0	-0.6	-1.7	und.	1.7	0.6	0	-0.6	-1.7	und.
3. $\csc x$	und.	2	1.2	1	1.2	2	und.	-2	-1.2	-1	-1.2	-2	und.

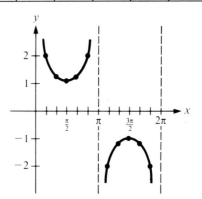

5. *D:* set of all real numbers except $x = \dfrac{\pi}{2} + k\pi$ (k any integer); *R:* $(-\infty, -1] \cup [1, \infty)$; period: 2π **7.** Inc., dec., dec.,

inc.   **9.** Always increasing   **11.** Inc., inc., dec., dec.   **13.** The other trigonometric functions do not attain a maximum
value.

15.   **17.**   **19.**

21.

Exercises 6.7

(*Note:* Answers give suggested intermediate steps.)

1. $\sin x \dfrac{1}{\cos x} = \dfrac{\sin x}{\cos x} = \tan x$ **3.** $\sin^2 x \dfrac{1}{\sin x} = \sin x$ **5.** $\cos x \dfrac{\sin x}{\cos x} = \sin x = \dfrac{1}{\csc x}$ **7.** $\cos x \dfrac{\sin x}{\cos x} \dfrac{1}{\sin x} = 1$

9. $\sin^2 x \dfrac{\cos^2 x}{\sin^2 x} = \cos^2 x$ **11.** $\sin^2 x \dfrac{1}{\cos^2 x} = \tan^2 x$ **13.** $\dfrac{\cos^2 x}{\cos^2 x/\sin^2 x} = \sin^2 x$

15. $\dfrac{(\cos x/\sin x)(\sin x/\cos x)}{1/\cos x} = \cos x$ **17.** $1 - \cos^2 x = \sin^2 x = \dfrac{1}{\csc^2 x}$ **19.** $\dfrac{(1 + \sin x)(1 - \sin x)}{1 + \sin x} = 1 - \sin x$

21. $\dfrac{1 - \sin^2 x}{\sin^2 x} = \dfrac{\cos^2 x}{\sin^2 x} = \cot^2 x$ **23.** $\dfrac{\sec^2 x}{\csc^2 x} = \dfrac{1/\cos^2 x}{1/\sin^2 x} = \dfrac{\sin^2 x}{\cos^2 x} = \tan^2 x$

25. $\sin x \dfrac{\sin x}{\cos x} + \cos x = \dfrac{\sin^2 x + \cos^2 x}{\cos x} = \dfrac{1}{\cos x} = \sec x$ **27.** $1 - \dfrac{\sin^2 x}{\tan^2 x} = 1 - \dfrac{\sin^2 x}{\sin^2 x/\cos^2 x} = 1 - \cos^2 x = \sin^2 x$

29. $\dfrac{1/\sin x}{(\sin x/\cos x) + (\cos x/\sin x)} = \dfrac{\cos x}{\sin^2 x + \cos^2 x} = \cos x$

31. $(1 - \sin x)\left(\dfrac{1}{\cos x} + \dfrac{\sin x}{\cos x}\right) = \dfrac{1 - \sin^2 x}{\cos x} = \dfrac{\cos^2 x}{\cos x} = \cos x$ **33.** $\sin x \cos x(\cos^2 x + \sin^2 x) = \sin x \cos x$

35. $\sin^2 x + 2 \sin x \cos x + \cos^2 x + \sin^2 x - 2 \sin x \cos x + \cos^2 x = 2(\sin^2 x + \cos^2 x) = 2$

37. $\dfrac{1}{\cos x} - \cos x = \dfrac{1 - \cos^2 x}{\cos x} = \dfrac{\sin^2 x}{\cos x} = \sin x\left(\dfrac{\sin x}{\cos x}\right) = \sin x \tan x$ **39.** $\dfrac{\sin x}{1/\sin x} + \dfrac{\cos x}{1/\cos x} = \sin^2 x + \cos^2 x = 1$

Exercises 6.8

1. $\cos\left(\dfrac{\pi}{2} + x\right) = \cos\dfrac{\pi}{2}\cos x - \sin\dfrac{\pi}{2}\sin x$
$$= (0)\cos x - (1)\sin x$$
$$= -\sin x$$

3. $\cos(\pi - x) = \cos\pi\cos x + \sin\pi\sin x$
$$= (-1)\cos x + (0)\sin x$$
$$= -\cos x$$

5. $\cos\left(\dfrac{3\pi}{2} + x\right) = \cos\dfrac{3\pi}{2}\cos x - \sin\dfrac{3\pi}{2}\sin x$
$$= (0)\cos x - (-1)\sin x$$
$$= \sin x$$

7. $\sin(x_1 - x_2) = \sin[x_1 + (-x_2)]$
$$= \sin x_1\cos(-x_2) + \cos x_1\sin(-x_2)$$
Since $\cos(-x_2) = \cos x_2$ and $\sin(-x_2) = -\sin x_2$, we have
$$\sin(x_1 - x_2) = \sin x_1\cos x_2 - \cos x_1\sin x_2$$

9. $\sin\left(\dfrac{\pi}{2} - x\right) = \sin\dfrac{\pi}{2}\cos x - \cos\dfrac{\pi}{2}\sin x$
$$= (1)\cos x - (0)\sin x$$
$$= \cos x$$

11. $\sin(\pi + x) = \sin\pi\cos x + \cos\pi\sin x$
$$= (0)\cos x + (-1)\sin x$$
$$= -\sin x$$

13. $\sin\left(\dfrac{3\pi}{2} - x\right) = \sin\dfrac{3\pi}{2}\cos x - \cos\dfrac{3\pi}{2}\sin x$
$$= (-1)\cos x - (0)\sin x$$
$$= -\cos x$$

15. $\cos 2x = \cos^2 x - \sin^2 x$ (Example 4)
Since $\cos^2 x = 1 - \sin^2 x$, we have
$$\cos 2x = (1 - \sin^2 x) - \sin^2 x$$
$$= 1 - 2\sin^2 x$$

17. $\tan(x_1 + x_2) = \dfrac{\sin(x_1 + x_2)}{\cos(x_1 + x_2)}$
$$= \dfrac{\sin x_1\cos x_2 + \cos x_1\sin x_2}{\cos x_1\cos x_2 - \sin x_1\sin x_2}$$
$$= \dfrac{[(\sin x_1\cos x_2)/(\cos x_1\cos x_2)] + [(\cos x_1\sin x_2)/(\cos x_1\cos x_2)]}{[(\cos x_1\cos x_2)/(\cos x_1\cos x_2)] - [(\sin x_1\sin x_2)/(\cos x_1\cos x_2)]}$$
$$= \dfrac{\tan x_1 + \tan x_2}{1 - \tan x_1\tan x_2}$$

19. $\tan(x_1 - x_2) = \tan[x_1 + (-x_2)]$
$$= \dfrac{\tan x_1 + \tan(-x_2)}{1 - \tan x_1\tan(-x_2)}$$
Since $\tan(-x_2) = -\tan x_2$, we have
$$\tan(x_1 - x_2) = \dfrac{\tan x_1 - \tan x_2}{1 + \tan x_1\tan x_2}$$

21. (a) $\dfrac{5}{13}$ (b) $\dfrac{5}{12}$ (c) $\dfrac{13}{12}$ (d) $-\dfrac{119}{169}$ (e) $\dfrac{120}{169}$ (f) $\dfrac{3\sqrt{13}}{13}$ (g) $\dfrac{2\sqrt{13}}{13}$ **23.** (a) $\dfrac{\sqrt{3}}{2}$ (b) $\dfrac{1}{2}$ (c) -1 (d) $\dfrac{-\sqrt{3}}{2}$

25. $\sin 3x = \sin(2x + x) = \sin 2x\cos x + \cos 2x\sin x$
$$= (2\sin x\cos x)\cos x + (1 - 2\sin^2 x)\sin x$$
$$= 2\sin x\cos^2 x + \sin x - 2\sin^3 x$$
$$= 2\sin x(1 - \sin^2 x) + \sin x - 2\sin^3 x$$
$$= 2\sin x - 2\sin^3 x + \sin x - 2\sin^3 x$$
$$= 3\sin x - 4\sin^3 x$$

27. $\dfrac{1}{\sin x}\,2\sin x\cos x = 2\cos x$ **29.** $\dfrac{2\cos^2 x - 1 + 1}{2} = \cos^2 x$ **31.** $(1 - 2\sin^2 x) + 2\sin^2 x = 1$

33. $\dfrac{1 + 2\cos^2 x - 1}{2\sin x \cos x} = \dfrac{\cos x}{\sin x} = \cot x$

35. $\sin(x + y) + \sin(x - y) = (\sin x \cos y + \cos x \sin y) + (\sin x \cos y - \cos x \sin y)$
$= 2 \sin x \cos y$

37. $\dfrac{1}{3}$

Exercises 6.9

1. $\left\{\dfrac{\pi}{3}, \dfrac{2\pi}{3}\right\}$ **3.** $\left\{\dfrac{2\pi}{3}, \dfrac{4\pi}{3}\right\}$ **5.** $\left\{\dfrac{3\pi}{4}, \dfrac{7\pi}{4}\right\}$ **7.** $\{0.12, 3.02\}$ **9.** $\{1.88, 5.02\}$ **11.** $\left\{\dfrac{\pi}{4}, \dfrac{7\pi}{4}\right\}$ **13.** $\emptyset$

15. $\{0.32, 3.46\}$ **17.** $\left\{x: x = \dfrac{\pi}{6} + k2\pi \text{ or } x = \dfrac{5\pi}{6} + k2\pi, k \text{ any integer}\right\}$ **19.** $\{x: x = 3.60 + k2\pi \text{ or }$

$x = 5.82 + k2\pi, k \text{ any integer}\}$ **21.** $\left\{\dfrac{\pi}{4}, \dfrac{3\pi}{4}, \dfrac{5\pi}{4}, \dfrac{7\pi}{4}\right\}$ **23.** $\left\{0, \dfrac{\pi}{2}, \dfrac{3\pi}{2}\right\}$ **25.** $\left\{\dfrac{\pi}{6}, \dfrac{\pi}{2}, \dfrac{5\pi}{6}, \dfrac{3\pi}{2}\right\}$

27. $\{1.25, 1.71, 4.39, 4.85\}$ **29.** $\left\{\dfrac{\pi}{6}, \dfrac{\pi}{2}, \dfrac{5\pi}{6}, \dfrac{7\pi}{6}, \dfrac{3\pi}{2}, \dfrac{11\pi}{6}\right\}$ **31.** $\left\{\dfrac{\pi}{12}, \dfrac{5\pi}{12}, \dfrac{13\pi}{12}, \dfrac{17\pi}{12}\right\}$ **33.** $\left\{\dfrac{3\pi}{4}\right\}$

35. $\left\{0, \dfrac{\pi}{4}, \dfrac{\pi}{2}, \pi, \dfrac{5\pi}{4}, \dfrac{3\pi}{2}\right\}$ **37.** $\{0.90, 5.38\}$ **39.** $\emptyset$ **41.** $\left\{\dfrac{\pi}{4}, \dfrac{5\pi}{4}\right\}$ **43.** $\{0, \pi\}$ **45.** $\left\{\dfrac{\pi}{4}, \dfrac{5\pi}{4}\right\}$ **47.** $\left\{\dfrac{\pi}{6}, \dfrac{5\pi}{6}\right\}$

49. $\left\{\dfrac{\pi}{2}, \dfrac{7\pi}{6}, \dfrac{11\pi}{6}\right\}$ **51.** $\left\{\dfrac{7\pi}{6}, \dfrac{3\pi}{2}, \dfrac{11\pi}{6}\right\}$ **53.** $\{1.84, 2.20, 4.98, 5.34\}$

Exercises 6.10

1. $\dfrac{\pi}{3}$ **3.** $-\dfrac{\pi}{3}$ **5.** $\dfrac{\pi}{4}$ **7.** $-\dfrac{\pi}{2}$ **9.** $\dfrac{\pi}{4}$ **11.** $-\dfrac{\pi}{6}$ **13.** 0.32 **15.** 2.15 **17.** 1.05 **19.** 2.50

21. 1 **23.** $\dfrac{1}{2}$ **25.** 0 **27.** 1 **29.** -0.6552 (C: -0.6616)   **31.** $\dfrac{3}{5}$ **33.** $\dfrac{12}{5}$ **35.** $\dfrac{117}{125}$ **37.** $\dfrac{\sqrt{1 - x^2}}{x}$

39. $\dfrac{x - 2}{\sqrt{9 - (x - 2)^2}}$ **41.** $\text{arcsec}\left(\dfrac{x - 2}{3}\right)$ **43.** $\ln\left|\dfrac{1 - \sqrt{1 - x^2}}{x}\right|$ **45.** $-\dfrac{\sqrt{x^2 + 4}}{4x}$

47. $\dfrac{13}{4}\arctan\left(\dfrac{x - 5}{4}\right) - 2\ln\left|\dfrac{4}{\sqrt{(x - 5)^2 + 16}}\right|$ **49.** $\theta = \arcsin\left(\dfrac{1}{m}\right)$ **51.** $\theta = \arcsin\left(\dfrac{Tg}{2v_0}\right)$

53. $t = \pi + \arccos b$ **55.** $\theta = \arctan\left(\dfrac{a + b}{c}\right) - \arctan\left(\dfrac{b}{c}\right)$ **57.**

59. $[-\pi/2, \pi/2]$ **61.** $[0, \pi]$ **63.** $(-\infty, \infty)$

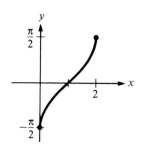

Chapter 6 Review Exercises

1. 0.8415   **3.** No solution   **5.** 0   **7.** 1.353   **9.** 0   **11.**

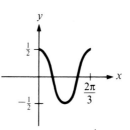

Amplitude: $\frac{1}{2}$
Period: $2\pi/3$
Phase shift: 0

13.

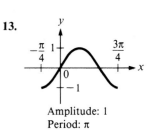

Amplitude: 1
Period: π
Phase shift: $-\pi/4$

15. $\{7\pi/6,11\pi/6\}$ **17.** $\{0,\pi/3,\pi,5\pi/3\}$

19. $\dfrac{\sin x + (\sin x/\cos x)}{1 + (1/\cos x)} = \dfrac{\sin x \cos x + \sin x}{\cos x + 1} = \dfrac{\sin x(\cos x + 1)}{\cos x + 1} = \sin x$

21. $(\sin x - \cos x)^2 = \sin^2 x - 2 \sin x \cos x + \cos^2 x$
$= 1 - 2 \sin x \cos x$
$= 1 - \sin 2x$

23. $50\pi/3$ in.²   **25.** $\sin(3\pi - x) = \sin 3\pi \cos x - \cos 3\pi \sin x = (0) \cos x - (-1) \sin x = \sin x$

27. $y = 4 \sin \frac{3}{2}x$ **29.** (a) $\frac{7}{6}\pi$ (b) $72°$ **31.** $\frac{8}{17}$ **33.** $38°40'$ **35.** $2a^2 - 1$ **37.** $\sin 68°$ **39.** 2 radians

41. (a) $\{x\colon x = \pi/2 + k2\pi\}$ (b) $\{x\colon x = \pi/2 + k2\pi \text{ radians}\}$ (c) $\{x\colon x = 90° + k360°\}$ (In each case, k represents integers.)   **43.** (b)   **45.** (c)   **47.** (b)   **49.** (a)   **51.** (a)   **53.** (b)   **55.** (c)   **57.** (d)

59. 78 ft   **61.** 49 mi

Exercises 7.1

1. $b = 35$ ft, $c = 48$ ft, $C = 105°$ **3.** $b = 140$ ft, $c = 1\overline{0}0$ ft, $A = 20°$ **5.** $b = 34.6$ ft, $c = 23.5$ ft, $C = 41°20'$

7. $c = 110$ ft, $B = 26°$, $C = 109°$ **9.** $c = 57$ ft, $A = 42°$, $C = 108°$ or $c = 12$ ft, $A = 138°$, $C = 12°$

11. $a = 26$ ft, $A = 8°$, $B = 22°$ **13.** 433 ft   **15.** 95 mi   **17.** $\theta = 15°$, $\alpha = 57°$, $y = 23$, $\beta = 12°$, $x = 110$

(*Note:* $\theta = 81°$, $\alpha = 123°$, $y = 86$, $\beta = 49°$, $x = 16$ also satisfy the data but do not correspond to the figure and the condition that α is an acute angle.)

Exercises 7.2

1. $c = 14$ ft, $A = 49°$, $B = 71°$ **3.** $b = 27.4$ ft, $A = 162°30'$, $C = 7°10'$ **5.** $c = 23.1$ ft, $A = 55°50'$, $B = 28°30'$ **7.** $A = 128°$, $B = 20°$, $C = 32°$ **9.** $A = 37°40'$, $B = 93°30'$, $C = 48°50'$ **11.** 31 m

13. 1,300 ft   **15.** $a = 11$ ft, $c = 43$ ft, $B = 78°$ **17.** $c = 78$ ft, $A = 26°$, $B = 34°$ **19.** $b = 243$ ft, $c = 152$ ft, $A = 33°00'$ **21.** $A = 104°$, $B = 29°$, $C = 47°$ **23.** $a = 165$ ft, $A = 164°00'$, $C = 8°50'$ or $a = 17.4$ ft, $A = 1°40'$, $C = 171°10'$ **25.** $A = 46°30'$, $B = 58°00'$, $C = 75°30'$ **27.** 99 mi   **29.** 34 m   **31.** 401 ft

Exercises 7.3

1. (a) 13 lb **(b)** 23°   **3. (a)** 18 lb **(b)** 34°   **5. (a)** 16 mi/hour **(b)** 18°   **7. (a)** 510 mi **(b)** 81°

9. (a) 13° **(b)** 310 mi/hour   **11.** v, 5$\bar{0}$ lb; h, 87 lb   **13.** v, 8.2 lb; h, 16 lb   **15.** $R = 7.1$ lb, $\theta = 47°$

17. $R = 22$ lb, $\theta = 149°$ **19.** $R = 24$ lb, $\theta = 196°$ **21. (a)** 340 mi/hour **(b)** 91 mi/hour   **23.** 34 lb

25. 26 lb   **27. (a)** 4$\bar{0}$ lb **(b)** 12°   **29. (a)** 321 lb **(b)** 5°50′   **31.** 23 lb   **33.** 30°, 56°, 86°

Exercises 7.4

1. $3(\cos 0° + i \sin 0°)$ **3.** $2(\cos 270° + i \sin 270°)$ **5.** $\sqrt{2}(\cos 315° + i \sin 315°)$

7. $5(\cos 36°50′ + i \sin 36°50′)$ **9.** $\sqrt{13}(\cos 213°40′ + i \sin 213°40′)$ **11.** $2(\cos 120° + i \sin 120°)$

13. $2(\cos 315° + i \sin 315°)$ **15.** $0 + 3i$ **17.** $-4 + 0i$ **19.** $-\dfrac{\sqrt{3}}{2} + \dfrac{3}{2}i$ **21.** $-\sqrt{2} - \sqrt{2}i$

23. $0.6157 + 0.7880i$ **25. (a)** $8(\cos 63° + i \sin 63°)$ **(b)** $\frac{1}{2}(\cos 41° + i \sin 41°)$ **(c)** $2[\cos(-41°) + i \sin(-41°)]$

27. (a) $\cos 270° + i \sin 270°$ **(b)** $\cos(-90°) + i \sin(-90°)$ **(c)** $\cos 90° + i \sin 90°$

29. (a) $36(\cos 336° + i \sin 336°)$ **(b)** $\frac{1}{4}[\cos(-74°) + i \sin(-74°)]$ **(c)** $4(\cos 74° + i \sin 74°)$

31. (a) $2\sqrt{2}(\cos 345° + i \sin 345°)$ **(b)** $\sqrt{2}(\cos 255° + i \sin 255°)$ **(c)** $\left(\dfrac{\sqrt{2}}{2}\right)[\cos(-255°) + i \sin(-255°)]$

33. (a) $\cos(-65°) + i \sin(-65°)$ **(b)** $\cos 25° + i \sin 25°$ **(c)** $\cos(-25°) + i \sin(-25°)$ **35.** $16 + 0i$

37. $16 + 16\sqrt{3}i$ **39.** $1 + 0i$

41. $1 + 0i$

$-\dfrac{1}{2} + \dfrac{\sqrt{3}}{2}i$

$-\dfrac{1}{2} - \dfrac{\sqrt{3}}{2}i$

43. $2(\cos 67°30′ + i \sin 67°30′)$
$2(\cos 157°30′ + i \sin 157°30′)$
$2(\cos 247°30′ + i \sin 247°30′)$
$2(\cos 337°30′ + i \sin 337°30′)$

45. $\sqrt[5]{2}(\cos 12° + i \sin 12°)$
$\sqrt[5]{2}(\cos 84° + i \sin 84°)$
$\sqrt[5]{2}(\cos 156° + i \sin 156°)$
$\sqrt[5]{2}(\cos 228° + i \sin 228°)$
$\sqrt[5]{2}(\cos 300° + i \sin 300°)$

47. $1 + \sqrt{3}i$
$-2 + 0i$
$1 - \sqrt{3}i$

49. $1 + 0i$
$0 + i$
$-1 + 0i$
$0 - i$

Exercises 7.5

1.

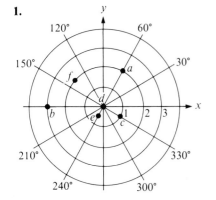

3. (a) $(1,330°)$ **(b)** $(2,11\pi/6)$ **(c)** $(1,\pi/3)$ **(d)** $(3,225°)$

(e) $(1,5\pi/4)$ **(f)** $(2,180°)$

5.

Circle

7.

Circle

9.

Circle

11.

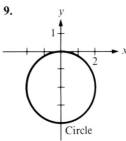

$r = 1 + \sin \theta$

Cardioid

13.

$r = \sin \theta - 1$

Cardioid

15.

Cardioid

17.

Rose

19.

Rose

21.

23.

25.

27.

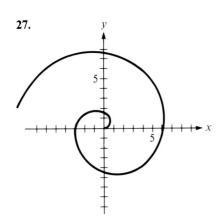

29. $(-1,0)$ **31.** $(3,0)$ **33.** $(-1,-1)$ **35.** $(1,180°)$ **37.** $(\sqrt{2},315°)$
39. $(2,150°)$ **41.** $r = 2 \sec \theta$ **43.** $r = 3$ **45.** $r = \cos \theta$
47. $r = \tan \theta \sec \theta$ **49.** $r^2 = 4 \sec 2\theta$ **51.** $x^2 + y^2 = 9$ **53.** $x = 2$
55. $x^2 + y^2 = -3y$ **57.** $y = 2$ **59.** $3y + 2x = 1$

Chapter 7 Review Exercises

1. $109°$ **3.** $48°50'$ **5.** $\sqrt{37} \approx 6.1$ **7.** $8(\cos 90° + i \sin 90°); 0 + 8i$ **9.** $\cos 450° + i \sin 450°; 0 + i$

11.

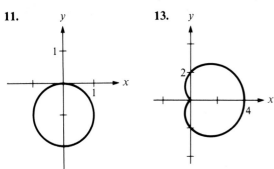

13.

15. $(\sqrt{2}, 225°)$ **17.** 21 lb

19. Since $\cos 90° = 0$, $c^2 = a^2 + b^2 - 2ab \cos 90°$ becomes $c^2 = a^2 + b^2$ **21.** $2x - y = 3$ **23.** $3(\cos 60° + i \sin 60°)$, $3(\cos 180° + i \sin 180°)$, $3(\cos 300° + i \sin 300°)$

25. LF: 193 ft; CF: 276 ft; RF: 266 ft   **27.** $30°, 56°, 94°$

29. $116°$ **31.** (a)   **33.** (b)   **35.** (b)   **37.** (c)

39. (c)

Exercises 8.1

1. $x - 3y + 11 = 0$ **3.** $x^2 + y^2 + 4x + 6y - 3 = 0$ **5.** $y^2 + 4y + 8x - 12 = 0$ **7.** $3x^2 + 4y^2 - 12 = 0$

9. $4x^2 - 12y^2 + 27 = 0$ **11.** $(2,1)$ **13.** $(0, -\frac{7}{2})$ **15.** $(0, \frac{3}{2})$ **17.** $y = -\frac{3}{2}x + 10$ **19.** $y = -x - 9$

21. $y = \frac{5}{7}x + \frac{6}{7}$ **23.** $y = \frac{7}{3}x - \frac{16}{3}$ **25.** $\overline{OB} = \sqrt{a^2 + b^2}$; $\overline{AC} = \sqrt{a^2 + b^2}$; thus, $\overline{OB} = \overline{AC}$ **27.** $m_{OB} = b/a$

and $m_{AC} = -b/a$. If the diagonals are perpendicular, $(b/a)(-b/a) = -1$, so $a^2 = b^2$, which implies $a = b$. Thus, the

rectangle is a square.

29. Midpoint of OB is $\left(\dfrac{a + b}{2}, \dfrac{c}{2}\right)$; midpoint of AC is $\left(\dfrac{a + b}{2}, \dfrac{c}{2}\right)$; thus, OB and AC bisect each other.

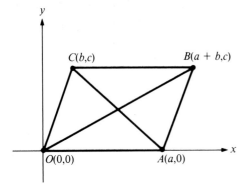

31. If C and D are the midpoints of OB and AB,

then $C = \left(\dfrac{b}{2}, \dfrac{c}{2}\right)$ and $D = \left(\dfrac{a + b}{2}, \dfrac{c}{2}\right)$. Since OA

and CD both have slope 0, CD is parallel to OA.

Also, $\overline{OA} = a$ and $\overline{CD} = \left|\dfrac{a + b}{2} - \dfrac{b}{2}\right| = \dfrac{a}{2}$,

so $\overline{CD}$ is one-half of $\overline{OA}$.

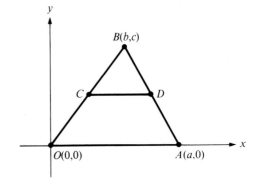

Exercises 8.2

1. $(x + 3)^2 + (y - 4)^2 = 4$ **3.** $x^2 + y^2 = 1$ **5.** $(x + 4)^2 + (y + 1)^2 = 9$ **7.** $(x - 3)^2 + (y + 3)^2 = 18$

9. $(x + 1)^2 + (y + 1)^2 = 58$ **11.** $(2,5); 4$ **13.** $(3,0); \sqrt{20}$ **15.** $(0,0); 3$ **17.** $(5,3); 7$ **19.** $(-4,1); \sqrt{18}$

21. $(-\frac{7}{2}, -\frac{3}{2}); \dfrac{\sqrt{42}}{2}$ **23.** $(-1,2); \frac{7}{2}$

Exercises 8.3

1.

Foci: $(3,0), (-3,0)$

3.

Foci: $(0,\sqrt{3}), (0,-\sqrt{3})$

5.

Foci: $(\sqrt{32},0), (-\sqrt{32},0)$

7.

Foci: $(0,\sqrt{3}/2), (0,-\sqrt{3}/2)$

9.

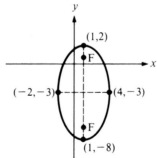

Foci: $(1,1), (1,-7)$
Center: $(1,-3)$

11.

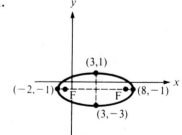

Foci: $(3 - \sqrt{21},-1), (3 + \sqrt{21},-1)$
Center: $(3,-1)$

13.

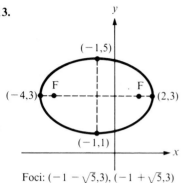

Foci: $(-1 - \sqrt{5}, 3)$, $(-1 + \sqrt{5}, 3)$
Center: $(-1, 3)$

15. $\dfrac{x^2}{16} + \dfrac{y^2}{25} = 1$ **17.** $\dfrac{x^2}{36} + \dfrac{y^2}{20} = 1$ **19.** $\dfrac{x^2}{9} + \dfrac{y^2}{4} = 1$

21. $\dfrac{(x-2)^2}{16} + \dfrac{(y-2)^2}{25} = 1$ **23.** $\dfrac{(x-2)^2}{25} + \dfrac{(y-3)^2}{16} = 1$

25. **(a)** 186,000,000 mi   **(b)** 3,000,000 mi

Exercises 8.4

1.

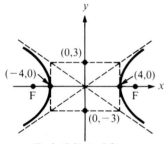

Foci: $(5, 0)$, $(-5, 0)$
Asymptotes: $y = \pm\frac{3}{4}x$

3.

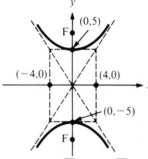

Foci: $(0, \sqrt{41})$, $(0, -\sqrt{41})$
Asymptotes: $y = \pm\frac{5}{4}x$

5.

Foci: $(\sqrt{40}, 0)$, $(-\sqrt{40}, 0)$
Asymptotes: $y = \pm\frac{1}{3}x$

7.

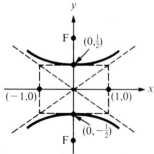

Foci: $(0, \sqrt{5}/2)$, $(0, -\sqrt{5}/2)$
Asymptotes: $y = \pm\frac{1}{2}x$

9.

Foci: $(-2 + \sqrt{34}, 3)$, $(-2 - \sqrt{34}, 3)$
Center: $(-2, 3)$
Asymptotes: $y - 3 = \pm\frac{5}{3}(x + 2)$

11.

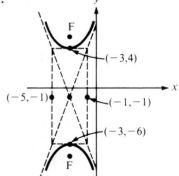

Foci: $(-3, -1 + \sqrt{29})$, $(-3, -1 - \sqrt{29})$
Center: $(-3, -1)$
Asymptotes: $y + 1 = \pm\frac{5}{2}(x + 3)$

13.

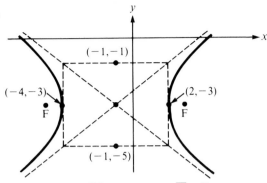

Foci: $(-1 + \sqrt{13}, -3), (-1 - \sqrt{13}, -3)$
Center: $(-1, -3)$
Asymptotes: $y + 3 = \pm\frac{2}{3}(x + 1)$

15. $\dfrac{y^2}{16} - \dfrac{x^2}{9} = 1$ **17.** $\dfrac{x^2}{9} - \dfrac{y^2}{7} = 1$

19. $\dfrac{x^2}{9} - \dfrac{y^2}{25} = 1$ **21.** $\dfrac{(x + 3)^2}{9} - \dfrac{(y + 4)^2}{16} = 1$

23. $\dfrac{(y - 1)^2}{4} - \dfrac{(x - 3)^2}{5} = 1$

Exercises 8.5

1.

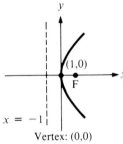

Vertex: $(0,0)$
Focus: $(1,0)$
Directrix: $x = -1$

3.

Vertex: $(0,0)$
Focus: $(0, -\frac{1}{2})$
Directrix: $y = \frac{1}{2}$

5.

Vertex: $(0,0)$
Focus: $(-3,0)$
Directrix: $x = 3$

7.

Vertex: $(0,0)$
Focus: $(0,\frac{3}{16})$
Directrix: $y = -\frac{3}{16}$

9.

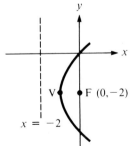

Vertex: $(-1,-2)$
Focus: $(0,-2)$
Directrix: $x = -2$

11.

Vertex: $(2,-4)$
Focus $(2, -\frac{15}{4})$
Directrix: $y = -\frac{17}{4}$

13.

Vertex: $(1,-2)$
Focus: $(1,-1)$
Directrix: $y = -3$

15. $x^2 = 12y$ **17.** $y^2 = 2x$ **19.** $x^2 = 16y$ **21.** $(y-1)^2 = 12(x+1)$
23. $(x-3)^2 = 12y$ **25. (a)** $x^2 = 12y$ if the vertex is at $(0,0)$ **(b)** $(0,3)$
27. 21 ft

Exercises 8.6

1. Parabola   **3.** Ellipse   **5.** Hyperbola   **7.** Parabola   **9.** Circle   **11.** Two intersecting straight lines:
$y + 2 = \pm x$ **13.** Circle   **15.** Parabola   **17.** Point: $(-1,1)$ **19.** No graph   **21. (1)** c   **(2)** d   **(3)** e
(4) a   **(5)** b

Chapter 8 Review Exercises

1. $(x+1)^2 = -16(y-1)$
3. $(-2,3)$; 5   **5.** $(0,3), (0,-3)$
7. $y = \pm\frac{4}{5}x$ **9.** $(0,-\frac{7}{4})$
11. $(x-1)^2 + (y-2)^2 = 9$

13.

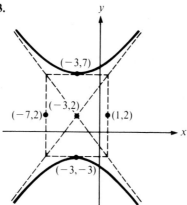

15. $(-1,-1), (7,-1)$
17. (c)   **19.** (d)

Exercises 9.1

1. $(-2,-4)$ **3.** $(-\frac{3}{2},-\frac{1}{2})$ **5.** $(0,4)$ **7.** $(-6,-6)$ **9.** Dependent   **11.** $(4,21)$ **13.** $(-1,1)$
15. No solution, inconsistent   **17.** $(5,1)$ **19.** $(\frac{67}{41},-\frac{155}{41})$ **21.** 46, 24   **23.** $140°, 40°$ **25.** 200 g
27. $v_0 = 100$ ft/second; $a = -32$ ft/second^2 **29.** $F_1 = 12$ lb; $F_2 = 6$ lb   **31.** 3 mi/hour; 9 mi/hour
33. $8,000 at 6 percent, $2,000 at 11 percent   **35.** More than 1,000   **37.** $p = (d-b)/(a-c)$; positive: a, d;
negative: b, c **39.** Tax $\approx$ $28,037, bonus $\approx$ $6,542

Exercises 9.2

1. -10 **3.** 2 **5.** 26 **7.** -3 **9.** 97 **11.** 0 **13.** -52 **15.** 17 **17.** 0 **19.** 254 **21.** $(4,21)$

23. $(-1,1)$ **25.** No solution, inconsistent   **27.** $(5,1)$ **29.** $(\frac{67}{41}, -\frac{155}{41})$ **31.** $x = 1, y = 2, z = 0$

33. $x = -3, y = 1, z = 4$ **35.** $x = \frac{1}{2}, y = -\frac{3}{2}, z = 1$ **37.** $x = 2, y = 2, z = -2$ **39.** No unique solution

41. $I_1 = -\frac{8}{73}$ amp; $I_2 = \frac{55}{73}$ amp; $I_3 = -\frac{47}{73}$ amp   **43.** $I_1 = \frac{73}{101}$ amp; $I_2 = -\frac{35}{101}$ amp; $I_3 = -\frac{38}{101}$ amp

45. $y = -3x + 3$

Exercises 9.3

1. $(\frac{5}{2}, \frac{9}{2})$ **3.** $a = \frac{1}{4}, b = \frac{3}{4}, c = 1$ **5.** $(-27,59)$ **7.** No unique solution (dependent)   **9.** $x = 7, y = -3,$

$z = -5$ **11.** $x = 3, y = -1, z = -1$ **13.** $x_1 = 2, x_2 = -2, x_3 = -2$ **15.** $A = \frac{1}{2}, B = -1, C = \frac{1}{2}$

17. No unique solution (inconsistent)   **19.** $a = 2, b = -1, c = -1, d = 0$ **21.** $a = -1, b = 5, c = -2$

23. $a = -1, b = 0, c = 7, d = -5$ **25.** $r_1 = 11$ m, $r_2 = 3$ m, $r_3 = 5$ m   **27.** 700

Exercises 9.4

1. 2×4 (read 2 by 4)   **3.** 2×1 (read 2 by 1)   **5.** $a = 10, b = -3, c = 5, d = 4$ **7.** $a = 2, b = 0, c = -3$

9. $\begin{bmatrix} -1 & -3 \\ 4 & 10 \end{bmatrix}$ **11.** $\begin{bmatrix} 0 & 2 & -1 \\ 4 & -3 & 1 \end{bmatrix}$ **13.** $\begin{bmatrix} 10 & 2 \\ -1 & -5 \end{bmatrix}$ **15.** Both AB and BA are 2×2.

17. AB is 2×2, BA is 3×3.   **19.** Both AB and BA are undefined.   **21.** $\begin{bmatrix} 11 & 0 \\ -2 & 3 \end{bmatrix}$ **23.** $\begin{bmatrix} 1 & 3 \\ -1 & 0 \end{bmatrix}$

25. $\begin{bmatrix} -10 \\ 7 \\ 0 \end{bmatrix}$ **27.** $\begin{bmatrix} 1 & -4 & -3 \\ -2 & -4 & -9 \end{bmatrix}$ **29.** Undefined   **31.** B **33.** $\begin{bmatrix} 0 & 4 \\ -6 & -10 \end{bmatrix}$ **35.** I

37. $\begin{bmatrix} -3 & -2 & -9 \\ -2 & -6 & -5 \end{bmatrix}$ **39.** $\begin{aligned} x + y &= 0 \\ 5x - 2y &= 3 \end{aligned}$ **41.** $\begin{bmatrix} 3 & 1 \\ -7 & -2 \end{bmatrix}$ **43.** $\frac{1}{13}\begin{bmatrix} 2 & 3 \\ 1 & -5 \end{bmatrix}$ **45.** A^{-1} does not exist.

47. $\begin{bmatrix} 1 & 0 & -1 \\ 1 & 1 & -1 \\ 1 & 1 & 0 \end{bmatrix}$ **49.** A^{-1} does not exist.   **51.** $(4,21)$ **53.** $(-27,59)$ **55.** $(\frac{67}{41}, -\frac{155}{41})$

57. $x = 2, y = 2, z = -2$ **59.** (a) $(29,17)$ (b) $(-25,-15)$

Exercises 9.5

1. $A = -\frac{1}{4}, B = \frac{1}{4}$ **3.** $A = 3, B = -3, C = 3$ **5.** $A = 2, B = -4$ **7.** $A = 1, B = -2, C = -5, D = 5$

9. $A = -2, B = 1, C = 2, D = 1$ **11.** $\dfrac{1/2}{1+x} + \dfrac{1/2}{1-x}$ **13.** $\dfrac{1}{x+1} + \dfrac{-1}{(x+1)^2}$ **15.** $x + \dfrac{2}{x+2} + \dfrac{2}{x-2}$

17. $\dfrac{1}{x} + \dfrac{-3}{x-2} + \dfrac{2}{x+3}$ **19.** $\dfrac{1}{x} + \dfrac{-x}{x^2+1}$ **21.** $\dfrac{1}{x+1} + \dfrac{1}{x-1} + \dfrac{-2x}{x^2+1}$ **23.** $\dfrac{x+4}{x^2+1} + \dfrac{-x}{x^2+2}$

Exercises 9.6

1. $(0,0)$, $(1,1)$ **3.** $(3,1)$, $(-3,1)$ **5.** $(2,0)$, $(-2,0)$ **7.** $(0,10)$, $(6,-8)$ **9.** $(2,2)$ $(2,-2)$, $(-2,2)$, $(-2,-2)$

11. $(2\sqrt{2},4)$, $(2\sqrt{2},-4)$, $(-2\sqrt{2},4)$, $(-2\sqrt{2},-4)$ **13.** $(5,3)$, $(3,5)$ **15.** $(1,2)$, $(-1,-2)$, $(2\sqrt{2},\sqrt{2}/2)$,

$(-2\sqrt{2},-\sqrt{2}/2)$ **17.** $(1,1)$, $(1,-1)$ **19.** $(0,-2)$, $(5,3)$ **21.** $(1,4)$, $(2,1)$, $(-\frac{2}{3},9)$

23. $\left(\sqrt{\dfrac{-1+\sqrt{21}}{2}},\dfrac{-1+\sqrt{21}}{2}\right)$, $\left(-\sqrt{\dfrac{-1+\sqrt{21}}{2}},\dfrac{-1+\sqrt{21}}{2}\right)$ **25.** $(2,0)$, $(-2,0)$ **27.** $(\frac{1}{7},-\frac{1}{3})$

29. $(\frac{1}{2},-\frac{1}{3},-\frac{1}{2})$ **31.** $(1,1)$, $(1,-1)$, $(-1,1)$, $(-1,-1)$ **33.** $(2.4,1.7)$, $(2.4,-1.7)$, $(-2.4,1.7)$, $(-2.4,-1.7)$

35. $\dfrac{x^2}{a^2}-\left(\dfrac{b^2x^2}{a^2}\right)\dfrac{1}{b^2}=0$, which never equals 1   **37.** $4\sqrt{6}$ ft by $3\sqrt{6}$ ft   **39.** $b=9$ m, $h=8$ m

41. *A:* 4 hours, *B:* 12 hours, *C:* 6 hours

Exercises 9.7

1.

3.

5.

7.

9.

11.

13.

15.
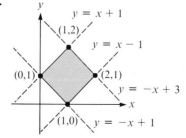

17. Max: 8; Min: -12 **19.** $(2,1)$ **21.** 38,000 **23.** $C = 70$ at $x = 20, y = 30$

25. $0 \le 90x + 50y + 30z \le 20,000$; $x, y, z \ge 0$ **27.** $0 \le Bx + Cy \le A$; $x, y \ge 0$ **29.** 80 type A, 400 type B,

max profit: \$37,200 **31.** Food A: none; Food B: $\frac{9}{2}$ ounces

Chapter 9 Review Exercises

1. $(-1,-5)$ **3.** $(6,0)$ **5.** $(13,-3)$ **7.** $x = 5, y = -1, z = -1$ **9.** $x = 15, y = 13, z = -5$

11. $(1,2)$ **13.** $(\frac{1}{2}, \frac{1}{3})$ **15.** $(3 + \sqrt{5}, 3 - \sqrt{5}), (3 - \sqrt{5}, 3 + \sqrt{5})$ **17.** $(4,2)$ **19.** $x = 1, y = -1, z = 0$

21. $x = 0, y = 0, z = 0$ **23.** -11 **25.** 0 **27.** $-\begin{vmatrix} c & d \\ a & b \end{vmatrix} = -(cb - da) = ad - bc = \begin{vmatrix} a & b \\ c & d \end{vmatrix}$

29. $\begin{bmatrix} -7 & -5 \\ -3 & -7 \end{bmatrix}$ **31.** Undefined **33.** $\begin{bmatrix} 12 & 15 \\ 9 & 12 \end{bmatrix}$ **35.** $\begin{bmatrix} 13 & 17 \\ 8 & 25 \end{bmatrix}$

37.

39. $P = 13,750$ at $x = 50, y = 375$ **41.** $a = 2, b = -3, c = 4$

43. For $r \ge \sqrt{5}$ the curves intersect at the four points given by

$(\pm \sqrt{(r^2 + 5)/2}, \pm \sqrt{(r^2 - 5)/2})$. **45.** 8 **47.** (a) **49.** False

51. $A = -4, B = 7$ **53.** $1 + \dfrac{-3/2}{x + 3} + \dfrac{3/2}{x - 3}$

55. $\frac{10}{9}$ quarts, $\frac{10}{3}$ quarts, $\frac{20}{3}$ quarts

Exercises 10.1

1. 1, 3, 5, 7; 99 **3.** 1, 4, 9, 16; 49 **5.** 1, $-\frac{1}{2}, \frac{1}{3}, -\frac{1}{4}$; $-\frac{1}{10}$ **7.** $\frac{1}{2}, \frac{1}{4}, \frac{1}{8}, \frac{1}{16}$; $\frac{1}{32}$ **9.** 1, 3, 6, 10; 36

11. AP; $d = 5$; 27, 32 **13.** Neither **15.** GP; $r = 1.02$; $(1.02)^5, (1.02)^6$ **17.** GP; $r = 0.1$; 0.00003, 0.000003

19. Neither **21.** $a_n = 4n - 1$; 159 **23.** $a_n = -5n + 14$; -11 **25.** $a_n = 0.3n + 0.8$; 8.3

27. $a_n = \frac{4}{3}n - 1$; 39 **29.** $a_n = \frac{3}{8}n + \frac{29}{8}$; $\frac{53}{8}$ **31.** $a_n = 2(3)^{n-1}$; 1,458 **33.** $a_n = (-\frac{1}{2})^{n-1}$; $-\frac{1}{32}$

35. $a_n = \sqrt{2}(\sqrt{2})^{n-1}$; 16 **37.** $a_n = 6(\frac{2}{3})^{n-1}$; $\frac{128}{243}$ **39.** $a_n = (-\frac{1}{3})^{n-1}$; $-\frac{1}{243}$ **41.** 4, $\frac{9}{2}$, 5, $\frac{11}{2}$ **43.** 6 **45.** 7

47. $\frac{5}{9}$ or $-\frac{5}{9}$ **49.** $a_1 = \frac{15}{2}, d = \frac{3}{2}$ **51.** 1, 1, 2, 3, 5, 8, 13, 21, 34, 55, 89, 144, 233, 377, 610

Exercises 10.2

1. 210 **3.** 2,186 **5.** About 5.42 **7.** 3,367/1,024 **9.** 31.9 **11.** $1 + 2 + 3 + 4 + \cdots + 9 + 10 = 55$

13. $-1 + 1 - 1 + 1 - 1 + 1 = 0$ **15.** $1 + 3 + 5 + 7 + \cdots + 21 + 23 = 144$ **17.** $4 + \frac{9}{2} + 5 + \frac{11}{2} = 19$

19. $1 + \frac{1}{3} + \frac{1}{9} + \frac{1}{27} + \frac{1}{81} = 121/81$ **21.** $\sum_{i=3}^{10} i$ **23.** $\sum_{i=1}^{4} \frac{1}{2^i}$ **25.** $\sum_{i=1}^{5} \frac{i}{i+1}$ **27.** $\sum_{i=1}^{6} x^i$ **29.** $\sum_{i=1}^{n} i^3$

31. $n(n + 1)$ **33.** 1,600 ft **35.** 254 **37.** 3,389/81 ft **39.** \$8,318.13

Exercises 10.3

1. 2 **3.** $\frac{1}{4}$ **5.** $\frac{10}{3}$ **7.** $\frac{9}{10}$ **9.** $\frac{50}{9}$ **11.** $\frac{2}{9}$ **13.** $\frac{7}{99}$ **15.** $\frac{107}{333}$ **17.** 6 **19.** $\frac{643}{300}$ **21.** 45 ft
23. $-\frac{3}{2} < x < -\frac{1}{2}$; $(x + 1)/(-2x - 1)$ **25.** No number can be assigned as the sum of this series. The series does not converge to a sum because as n gets larger and larger, the terms in the series do not approach zero. Method 3 is incorrect because the formula applies only when $|r| < 1$.

Exercises 10.4

1. *Step 1:* $1 \overset{?}{=} \dfrac{1(1 + 1)}{2}$; $1 = 1$

    Step 2: $\quad \underbrace{1 + 2 + 3 + \cdots + k}_{\text{induction assumption}} + (k + 1)$

$$= \quad \frac{k(k + 1)}{2} \quad + (k + 1)$$

$$= (k + 1)\left(\frac{k}{2} + 1\right)$$

$$= \frac{(k + 1)(k + 2)}{2}$$

$$= \frac{(k + 1)[(k + 1) + 1]}{2}$$

3. *Step 1:* $2(1) \overset{?}{=} 1(1 + 1)$; $2 = 2$

    Step 2: $\quad \underbrace{2 + 4 + \cdots + 2k}_{\text{induction assumption}} + 2(k + 1)$

$$= \quad k(k + 1) \quad + 2(k + 1)$$

$$= (k + 1)(k + 2)$$

$$= (k + 1)[(k + 1) + 1]$$

5. *Step 1:* $4(1) \overset{?}{=} 2(1)(1 + 1)$; $4 = 4$

    Step 2: $\quad \underbrace{4 + 8 + \cdots + 4k}_{\text{induction assumption}} + 4(k + 1)$

$$= \quad 2k(k + 1) \quad + 4(k + 1)$$

$$= 2(k + 1)(k + 2)$$

$$= 2(k + 1)[(k + 1) + 1]$$

7. *Step 1:* $2^1 \overset{?}{=} 2(2^1 - 1)$; $2 = 2$

    Step 2: $\quad \underbrace{2 + 2^2 + \cdots + 2^k}_{\text{induction assumption}} + 2^{k + 1}$

$$= \quad 2(2^k - 1) \quad + 2^{k + 1}$$

$$= 2^{k + 1} - 2 + 2^{k + 1}$$

$$= 2(2^{k + 1}) - 2$$

$$= 2(2^{k + 1} - 1)$$

9. *Step 1:* $1^3 \overset{?}{=} \dfrac{1^2(1 + 1)^2}{4}$; $1 = 1$

   Step 2: $\underbrace{1^3 + 2^3 + \cdots + k^3}_{\text{induction assumption}} + (k + 1)^3$

   $= \dfrac{k^2(k + 1)^2}{4} + (k + 1)^3$

   $= (k + 1)^2 \left[\dfrac{k^2}{4} + (k + 1) \right]$

   $= (k + 1)^2 \left[\dfrac{k^2 + 4k + 4}{4} \right]$

   $= \dfrac{(k + 1)^2(k + 2)^2}{4}$

   $= \dfrac{(k + 1)^2[(k + 1) + 1]^2}{4}$

11. *Step 1:* If $n = 1$, $2^1 > 1$, which is true.

    Step 2: $2^k > k$ (induction assumption)

    $2 \cdot 2^k > 2 \cdot k$

    $2^{k + 1} > 2k$

    $2^{k + 1} > k + 1$ (since $2k \geq k + 1$, for $k \geq 1$)

13. *Step 1:* If $n = 5$, $5^2 < 2^5$, which is true.

    Step 2: $k^2 < 2^k$ (induction assumption)

    $k^2 + (2k + 1) < 2^k + (2k + 1)$

    $(k + 1)^2 < 2^k + 2k + k$ (since $k > 1$)

    $= 2^k + 3k$

    $< 2^k + k^2$ (since $k > 3$)

    $< 2^k + 2^k$ (induction assumption)

    $= 2(2^k) = 2^{k + 1}$

15. *Step 1:* If $n = 2$, $\log(x_1 \cdot x_2) = \log x_1 + \log x_2$, which is true.

    Step 2: $\log (x_1 x_2 \cdots x_n x_{n + 1})$

    $= \log[(x_1 x_2 \cdots x_n)x_{n + 1}]$

    $= \underbrace{\log(x_1 x_2 \cdots x_n)}_{\text{induction assumption}} + \log x_{n + 1}$

    $= (\log x_1 + \log x_2 + \cdots + \log x_n) + \log x_{n + 1}$

    $= \log x_1 + \log x_2 + \cdots + \log x_n + \log x_{n + 1}$

17. *Step 1:* If $n = 1$, $0 < a < 1$, which is given.

    Step 2: $0 < a^k < 1$ (induction assumption)

        $a \cdot 0 < a \cdot a^k < a \cdot 1$

        $0 < a^{k+1} < a$

        $0 < a^{k+1} < 1$ (since $0 < a < 1$)

19. *Step 1:* True for $n = 1$, since $x - a$ is divisible by $x - a$.

    Step 2: $x^{k+1} - a^{k+1}$

        $= x^{k+1} - xa^k + xa^k - a^{k+1}$

        $= x(x^k - a^k) + a^k(x - a)$

Then $a^k(x - a)$ is obviously divisible by $x - a$, while $x(x^k - a^k)$ is divisible by $x - a$ since $x - a$ divides $x^k - a^k$ by the induction assumption. Thus, $x^{k+1} - a^{k+1}$ is divisible by $x - a$.

Exercises 10.5

1. $(x + y)^6 = x^6 + 6x^5y + 15x^4y^2 + 20x^3y^3 + 15x^2y^4 + 6xy^5 + y^6$

3. $(x - y)^5 = x^5 - 5x^4y + 10x^3y^2 - 10x^2y^3 + 5xy^4 - y^5$ **5.** $(x + h)^4 = x^4 + 4x^3h + 6x^2h^2 + 4xh^3 + h^4$

7. $(x - 1)^7 = x^7 - 7x^6 + 21x^5 - 35x^4 + 35x^3 - 21x^2 + 7x - 1$ **9.** $(2x + y)^3 = 8x^3 + 12x^2y + 6xy^2 + y^3$

11. $(3c - 4d)^4 = (3c)^4 + 4(3c)^3(-4d) + 6(3c)^2(-4d)^2 + 4(3c)(-4d)^3 + (-4d)^4$

        $= 81c^4 - 432c^3d + 864c^2d^2 - 768cd^3 + 256d^4$

13. $(x + y)^{15} = x^{15} + 15x^{14}y + 105x^{13}y^2 + 455x^{12}y^3 + \cdots$

15. $(x - 3y)^{12} = x^{12} + 12x^{11}(-3y) + 66x^{10}(-3y)^2 + 220x^9(-3y)^3 + \cdots$

        $= x^{12} - 36x^{11}y + 594x^{10}y^2 - 5,940x^9y^3 + \cdots$

17. 1, 3, 3, 1; The binomial coefficients 1, 3, 3, 1 match the entries in row 3 of Pascal's triangle.

19. $\binom{6}{0}, \binom{6}{1}, \binom{6}{2}, \binom{6}{3}, \binom{6}{4}, \binom{6}{5}, \binom{6}{6}$ **21.** 10   **23.** 126   **25.** 1   **27.** 190   **29.** 43,758

31. $18x^{17}y$ **33.** $-702x^2y^{11}$ **35.** $126x^8$ **37.** $80x$ **39.** $4,320x^6y^9$ **41.** 330   **43.** -960

45. $(1.2)^4 = 1 + 4(0.2) + 6(0.2)^2 + 4(0.2)^3 + (0.2)^4 = 2.0736$

47. $\binom{n}{n-1} = \dfrac{n!}{(n-1)![n-(n-1)]!} = \dfrac{n!}{(n-1)!} = \dfrac{n(n-1)!}{(n-1)!} = n$

49. (a) $2^n = (1 + 1)^n = \binom{n}{0}1^n + \binom{n}{1}1^{n-1} \cdot 1 + \binom{n}{2}1^{n-2} \cdot 1^2 + \cdots + \binom{n}{n}1^n;$

thus, $\binom{n}{0} + \binom{n}{1} + \binom{n}{2} + \cdots + \binom{n}{n} = 2^n$ **(b)** 4,096

51. (a) $\binom{4}{3} = 4, \binom{4}{2} = 6, \binom{5}{3} = 10$; thus, $\binom{4}{3} + \binom{4}{2} = \binom{5}{3}$

(b) Except for the 1's, each entry in Pascal's triangle is the sum of the two numbers on either side of it in the preceding row.

(c) $\binom{k}{r} + \binom{k}{r-1} = \dfrac{k!}{r!(k-r)!} + \dfrac{k!}{(r-1)![k-(r-1)]!} = \dfrac{k![k-(r-1)] + k!r}{r![k-(r-1)]!}$

        $= \dfrac{k!(k+1)}{r![k-(r-1)]!} = \dfrac{(k+1)!}{r![(k+1)-r]!} = \binom{k+1}{r}$

Exercises 10.6

1. 720   **3.** 126   **5.** 120   **7.** 6,720   **9.** 10   **11. (a)** $5!/0! = 120$ **(b)** $5!/(0!5!) = 1$ **(c)** $5!/(5!0!) = 1$

13. 8   **15.** 12   **17.** 34,650   **19.** 720   **21.** 56   **23.** 495

HHH HHT HTH HTT   THH THT   TTH TTT

25. 18,564   **27.** 15   **29.** 45   **31.** 10   **33.** 80   **35.** 113,400   **37.** 36   **39.** 120   **41.** 5

43. $_nC_{n-r} = \dfrac{n!}{[n-(n-r)]!(n-r)!} = \dfrac{n!}{r!(n-r)!} = {}_nC_r$ **45. (a)**

47. $(a+b)^4 = \displaystyle\sum_{r=0}^{4} {}_4C_r a^{4-r} b^r = {}_4C_0 a^4 b^0 + {}_4C_1 a^3 b^1 + {}_4C_2 a^2 b^2 + {}_4C_3 a^1 b^3 + {}_4C_4 a^0 b^4$

$$= a^4 + 4a^3b + 6a^2b^2 + 4ab^3 + b^4$$

49. $15x^4y^2$

Exercises 10.7

1. $\frac{5}{12}$ **3. (a)** $\frac{1}{52}$ **(b)** $\frac{1}{4}$ **(c)** $\frac{3}{13}$ **(d)** $\frac{25}{26}$ **(e)** $\frac{2}{13}$ **(f)** $\frac{11}{13}$ **(g)** $\frac{4}{13}$ **(h)** $\frac{9}{13}$ **5.** $\frac{17}{19}$ **7.** $\frac{1}{9}$ **9. (a)** $\frac{4}{9}$ **(b)** $\frac{1}{3}$ **(c)** $\frac{4}{9}$

11. $\frac{1}{12}$ **13.** $\frac{1}{120}$ **15.** $\frac{3}{8}$ **17.** $\frac{21}{44}$ **19.** $\frac{1}{2}$

Chapter 10 Review Exercises

1. $-1, 2, -4, 8; 32$ **3.** 2,485   **5.** $\displaystyle\sum_{i=1}^{5} \frac{1}{i^2}$

7.   *Step 1:* If $n = 1$, $a = \frac{1}{2}[2a + (1-1)d]$, which is true.

   Step 2: $a + (a+d) + \cdots + [a+(k-1)d] + (a+kd)$

$$\overbrace{}^{\text{induction assumption}}$$

$$= \frac{k}{2}[2a + (k-1)d] \quad + (a+kd)$$

$$= ka + \frac{k^2d}{2} - \frac{kd}{2} + a + kd$$

$$= \frac{k^2d + kd + 2ka + 2a}{2}$$

$$= \frac{kd(k+1) + 2a(k+1)}{2}$$

$$= \frac{k+1}{2}(2a + kd)$$

9. (a) $b = (a + c)/2$ (b) $b = \pm\sqrt{ac}$ **11.** $\frac{1}{2}$ **13.** 55 **15.** 5,040; 720; 144 **17.** $d = \dfrac{a_n - a_1}{n - 1}$ **19.** Yes
21. 15 **23.** $\frac{2}{3}$ **25.** 210 **27.** $n = 10, r = 3$ **29.** $35x^2$ **31.** 0.01 **33.** $1 - 15x + 105x^2 - 455x^3$
35. (c) **37.** (a) **39.** (c) **41.** (a) **43.** (c)

Exercises A.1

1. 7.4 **3.** -2 **5.** 1.23×10^7 **7.** 45 **9.** 0.000000908 **11.** 4.2×10^1 **13.** 3.4251×10^4
15. 5.9×10^{12} mi **17.** 1.26×10^6 **19.** 5.3×10^{-23} g **21.** 92,000 **23.** 42.1 **25.** 580,000,000 mi
27. 6,280,000,000,000,000,000 **29.** 0.0000000000000000000000003 g **31.** 5.58×10^7 mi **33.** 3.6×10^{-25}

Exercises A.2

1. Five **3.** One **5.** Three **7.** Three **9.** Two **11.** Three **13.** Two **15.** Three **17.** Four
19. Two **21.** (a) 12.35 (b) 12 **23.** (a) 59,370 (b) 59,000 **25.** (a) $29,7\bar{0}0$ (b) $3\bar{0},000$
27. (a) 84.10 (b) 84 **29.** (a) 0.06855 (b) 0.069 **31.** (a) 0.002001 (b) 0.0020 **33.** (a) 423 (b) 0.004
35. (a) Same accuracy (b) 40.2 **37.** (a) 6.430 (b) Same precision **39.** (a) Same accuracy (b) 0.0002
41. (a) Same accuracy (b) 48.2 **43.** (a) $30\bar{0}$ (b) $30\bar{0}$ **45.** 14.1 **47.** -7.4 **49.** 1.9 **51.** 1.2
53. 32,800 **55.** 200,000 **57.** 200,000 **59.** 11.6 **61.** 90 **63.** 9.75

Exercises A.3

1. 0.3692 **3.** -0.6308 **5.** 3.3692 **7.** 1.9542 **9.** -0.9547 **11.** 6.59 **13.** 0.659 **15.** 36.0
17. 0.0387 **19.** 0.304 **21.** 1,130 **23.** 63.5 **25.** 335 **27.** 9.85 **29.** 761,000 **31.** 6.34 **33.** 1.72
35. 55.8 **37.** 6.69 **39.** 3.85×10^{-6} **41.** 1.9459 **43.** 3.7377 **45.** 8.7796 **47.** -0.5798
49. -7.7288 **51.** 1.2092 **53.** 9.9742 **55.** 0.7866 **57.** 0.0247

Exercises A.4

1. $y = x^2$ is a power function with $m = 2$ and $a = 1$. The graph is a straight line with slope 2. The intercept at $x = 1$ is 1.
3. $y = 4\sqrt{x}$ is a power function with $m = \frac{1}{2}$ and $a = 4$. The graph is a straight line with slope $\frac{1}{2}$. The intercept at $x = 1$
is 4. **5.** $xy = 1$, so $y = 1/x = x^{-1}$. The function is a power function with $m = -1$ and $a = 1$. The graph is a straight
line with slope -1. The intercept at $x = 1$ is 1. **7.** $y = 7.0x^{-0.35}$ **9.** $y = 3.0x^{0.65}$ **11.** The function $y = 2^{-x}$ is an
exponential function and its graph on semilogarithmic paper is a straight line with negative slope. **13.** The function
$y = 0.5(3^x)$ is an exponential function and its graph on semilogarithmic paper is a straight line with positive slope.

Index

Trigonometry

General Angle Definitions
If θ is an angle in standard position and if (x,y) is any point on the terminal ray of θ [except $(0,0)$], then

$\sin \theta = \dfrac{y}{r}$ $\leftarrow$ reciprocals $\rightarrow$ $\csc \theta = \dfrac{r}{y}$

$\cos \theta = \dfrac{x}{r}$ $\leftarrow$ reciprocals $\rightarrow$ $\sec \theta = \dfrac{r}{x}$

$\tan \theta = \dfrac{y}{x}$ $\leftarrow$ reciprocals $\rightarrow$ $\cot \theta = \dfrac{x}{y}$.

Note: $r = \sqrt{x^2 + y^2}$

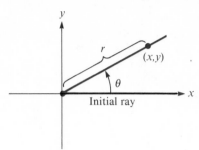

Unit Circle Definitions
Consider a point (x,y) on the unit circle $x^2 + y^2 = 1$ at arc length s from $(1,0)$. Then

$\sin s = y$ $\leftarrow$ reciprocals $\rightarrow$ $\csc s = \dfrac{1}{y}$

$\cos s = x$ $\leftarrow$ reciprocals $\rightarrow$ $\sec s = \dfrac{1}{x}$

$\tan s = \dfrac{y}{x}$ $\leftarrow$ reciprocals $\rightarrow$ $\cot s = \dfrac{x}{y}$.

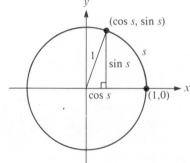

Right Triangle Definitions
If θ is an acute angle in a right triangle as shown below, then

$\sin \theta = \dfrac{\text{opposite}}{\text{hypotenuse}}$ $\leftarrow$ reciprocals $\rightarrow$ $\csc \theta = \dfrac{\text{hypotenuse}}{\text{opposite}}$

$\cos \theta = \dfrac{\text{adjacent}}{\text{hypotenuse}}$ $\leftarrow$ reciprocals $\rightarrow$ $\sec \theta = \dfrac{\text{hypotenuse}}{\text{adjacent}}$

$\tan \theta = \dfrac{\text{opposite}}{\text{adjacent}}$ $\leftarrow$ reciprocals $\rightarrow$ $\cot \theta = \dfrac{\text{adjacent}}{\text{opposite}}$.

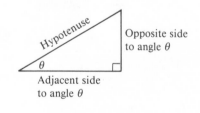

Signs of the Trigonometric Ratios

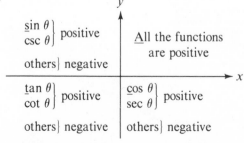